Springer Optimization and Its Applications

VOLUME 115

Aims and Scope
Optimization has been expanding in all directions at an astonishing rate during the last few decades. New algorithmic and theoretical techniques have been developed, the diffusion into other disciplines has proceeded at a rapid pace, and our knowledge of all aspects of the field has grown even more profound. At the same time, one of the most striking trends in optimization is the constantly increasing emphasis on the interdisciplinary nature of the field. Optimization has been a basic tool in all areas of applied mathematics, engineering, medicine, economics, and other sciences.

The series *Springer Optimization and Its Applications* publishes undergraduate and graduate textbooks, monographs and state-of-the-art expository work that focus on algorithms for solving optimization problems and also study applications involving such problems. Some of the topics covered include nonlinear optimization (convex and nonconvex), network flow problems, stochastic optimization, optimal control, discrete optimization, multi-objective programming, description of software packages, approximation techniques and heuristic approaches.

More information about this series at http://www.springer.com/series/7393

Boris Goldengorin
Editor

Optimization and Its Applications in Control and Data Sciences

In Honor of Boris T. Polyak's 80th Birthday

Editor
Boris Goldengorin
Department of Industrial
and Systems Engineering
Ohio University
Athens, OH, USA

ISSN 1931-6828 ISSN 1931-6836 (electronic)
Springer Optimization and Its Applications
ISBN 978-3-319-82490-1 ISBN 978-3-319-42056-1 (eBook)
DOI 10.1007/978-3-319-42056-1

Printed on acid-free paper

This Springer imprint is published by Springer Nature
The registered company is Springer International Publishing AG Switzerland

This book is dedicated to Professor Boris T. Polyak on the occasion of his 80th birthday.

Preface

This book is a collection of papers related to the International Conference "Optimization and Its Applications in Control and Data Sciences" dedicated to Professor Boris T. Polyak on the occasion of his 80th birthday, which was held in Moscow, Russia, May 13–15, 2015.

Boris Polyak obtained his Ph.D. in mathematics from Moscow State University, USSR, in 1963 and the Dr.Sci. degree from Moscow Institute of Control Sciences, USSR, in1986. Between 1963 and 1971 he worked at Lomonosov Moscow State University, and in 1971 he moved to the V.A. Trapeznikov Institute of Control Sciences, Russian Academy of Sciences. Professor Polyak was the Head of Tsypkin Laboratory and currently he is a Chief Researcher at the Institute. Professor Polyak has held visiting positions at universities in the USA, France, Italy, Israel, Finland, and Taiwan; he is currently a professor at Moscow Institute for Physics and Technology. His research interests in optimization and control have an emphasis in stochastic optimization and robust control. Professor Polyak is IFAC Fellow, and a recipient of Gold Medal EURO-2012 of European Operational Research Society. Currently, Boris Polyak's h-index is 45 with 11807 citations including 4390 citations since 2011.

This volume contains papers reflecting developments in theory and applications rooted by Professor Polyak's fundamental contributions to constrained and unconstrained optimization, differentiable and nonsmooth functions including stochastic optimization and approximation, optimal and robust algorithms to solve many problems of estimation, identification, and adaptation in control theory and its applications to nonparametric statistics and ill-posed problems.

This book focus is on the recent research in modern optimization and its implications in control and data analysis. Researchers, students, and engineers will benefit from the original contributions and overviews included in this book. The book is of great interest to researchers in large-scale constraint and unconstrained, convex and non-linear, continuous and discrete optimization. Since it presents open problems in optimization, game and control theories, designers of efficient algorithms and software for solving optimization problems in market and data analysis will benefit from new unified approaches in applications from managing portfolios of financial instruments to finding market equilibria. The book is also beneficial to theoreticians in operations research, applied mathematics, algorithm design, artificial intelligence, machine learning, and software engineering. Graduate students will be updated with the state-of-the-art in modern optimization, control theory, and data analysis.

Athens, OH, USA
March 2016

Boris Goldengorin

Acknowledgements

This volume collects contributions presented within the International Conference "Optimization and Its Applications in Control and Data Sciences" held in Moscow, Russia, May 13–15, 2015 or submitted by an open call for papers to the book "Optimization and Its Applications in Control Sciences and Data Analysis" announced at the same conference.

I would like to express my gratitude to Professors Alexander S. Belenky (National Research University Higher School of Economics and MIT) and Panos M. Pardalos (University of Florida) for their support in organizing the publication of this book including many efforts with invitations of top researches in contributing and reviewing the submitted papers.

I am thankful to the reviewers for their comprehensive feedback on every submitted paper and their timely replies. They greatly improved the quality of submitted contributions and hence of this volume. Here is the list of all reviewers:

1. **Anatoly Antipin**, Federal Research Center "Computer Science and Control" of Russian Academy of Sciences, Moscow, Russia
2. **Saman Babaie-Kafaki**, Faculty of Mathematics, Statistics, and Computer Science Semnan University, Semnan, Iran
3. **Amit Bhaya**, Graduate School of Engineering (COPPE), Federal University of Rio de Janeiro (UFRJ), Rio de Janeiro, Brazil
4. **Lev Bregman**, Department of Mathematics, Ben Gurion University, Beer Sheva, Israel
5. **Arkadii A. Chikrii**, Optimization Department of Controlled Processes, Cybernetics Institute, National Academy of Sciences, Kiev, Ukraine
6. **Giacomo Como**, The Department of Automatic Control, Lund University, Lund, Sweden
7. **Xiao Liang Dong**, School of Mathematics and Statistics, Xidian University, Xi'an, People's Republic of China
8. **Trevor Fenner**, School of Computer Science and Information Systems, Birkbeck College, University of London, London, UK

9. **Sjur Didrik Flåm**, Institute of Economics, University of Bergen, Bergen, Norway
10. **Sergey Frenkel**, The Institute of Informatics Problems, Russian Academy of Science, Moscow, Russia
11. **Piyush Grover**, Mitsubishi Electric Research Laboratories, Cambridge, MA, USA
12. **Jacek Gondzio**, School of Mathematics The University of Edinburgh, Edinburgh, Scotland, UK
13. **Rita Giuliano**, Dipartimento di Matematica Università di Pisa, Pisa, Italy
14. **Grogori Kolesnik**, Department of Mathematics, California State University, Los Angeles, CA, USA
15. **Pavlo S. Knopov**, Department of Applied Statistics, Faculty of Cybernetics, Taras Shevchenko National University, Kiev, Ukraine
16. **Arthur Krener**, Mathematics Department, University of California, Davis, CA, USA
17. **Bernard C. Levy**, Department of Electrical and Computer Engineering, University of California, Davis, CA, USA
18. **Vyacheslav I. Maksimov**, Institute of Mathematics and Mechanics, Ural Branch of the Russian Academy of Sciences, Ekaterinburg, Russia
19. **Yuri Merkuryev**, Department of Modelling and Simulation, Riga Technical University, Riga, Latvia
20. **Arkadi Nemorovski**, School of Industrial and Systems Engineering, Atlanta, GA, USA
21. **José Valente de Oliveira**, Faculty of Science and Technology, University of Algarve Campus de Gambelas, Faro, Portugal
22. **Alex Poznyak**, Dept. Control Automatico CINVESTAV-IPN, Mexico D.F., Mexico
23. **Vladimir Yu. Protasov**, Faculty of Mechanics and Mathematics, Lomonosov Moscow State University, and Faculty of Computer Science of National Research University Higher School of Economics, Moscow, Russia
24. **Simeon Reich**, Department of Mathematics, Technion-Israel Institute of Technology, Haifa, Israel
25. **Alessandro Rizzo**, Computer Engineering, Politecnico di Torino, Torino, Italy
26. **Carsten W. Scherer**, Institute of Mathematical Methods in Engineering, University of Stuttgart, Stuttgart, Germany
27. **Alexander Shapiro**, School of Industrial and Systems Engineering, Atlanta, GA, USA
28. **Lieven Vandenberghe**, UCLA Electrical Engineering Department, Los Angeles, CA, USA
29. **Yuri Yatsenko**, School of Business, Houston Baptist University, Houston, TX, USA

I would like to acknowledge the superb assistance that the staff of Springer has provided (thank you Razia Amzad). Also I would like to acknowledge help in preparation of this book from Silembarasanh Panneerselvam.

Technical assistance with reformatting some papers and compilation of this book's many versions by Ehsan Ahmadi (PhD student, Industrial and Systems Engineering Department, Ohio University, Athens, OH, USA) is greatly appreciated.

Finally, I would like to thank all my colleagues from the Department of Industrial and Systems Engineering, The Russ College of Engineering and Technology, Ohio University, Athens, OH, USA for providing me with a pleasant atmosphere to work within C. Paul Stocker Visiting Professor position.

Contents

Contributors

Neculai Andrei Center for Advanced Modeling and Optimization, Research Institute for Informatics, Bucharest, Romania

Academy of Romanian Scientists, Bucharest, Romania

Anatoly Antipin Federal Research Center "Computer Science and Control" of Russian Academy of Sciences, Dorodnicyn Computing Centre, Moscow, Russia

Alexander S. Belenky National Research University Higher School of Economics, Moscow, Russia

Center for Engineering Systems Fundamentals, Massachusetts Institute of Technology, Cambridge, MA, USA

Daniel Berend Departments of Mathematics and Computer Science, Ben-Gurion University, Beer Sheva, Israel

Shankar P. Bhattacharyya Department of Electrical and Computer Engineering, Texas A&M University, College Station, TX, USA

Paolo Bolzern Politecnico di Milano, DEIB, Milano, Italy

Stephen Boyd Department of Electrical Engineering, Stanford University, Stanford, CA, USA

Giuseppe C. Calafiore Politecnico di Torino, Torino, Italy

Luca Carlone Massachusetts Institute of Technology, Cambridge, MA, USA

Patrizio Colaneri Politecnico di Milano, DEIB, IEIIT-CNR, Milano, Italy

Grace S. Deaecto School of Mechanical Engineering, UNICAMP, Campinas, Brazil

Frank Dellaert Georgia Institute of Technology, Atlanta, GA, USA

Steven Diamond Department of Computer Science, Stanford University, Stanford, CA, USA

Lyudmila G. Egorova National Research University Higher School of Economics, Moscow, Russia

José C. Geromel School of Electrical and Computer Engineering, UNICAMP, Campinas, Brazil

Zoltán Horváth Department of Mathematics and Computational Sciences, Széchenyi István University, Győr, Hungary

Lee H. Keel Department of Electrical and Computer Engineering, Tennessee State University, Nashville, USA

Elena Khoroshilova Faculty of Computational Mathematics and Cybernetics, Lomonosov Moscow State University, Moscow, Russia

Dmitry A. Klyushin Kiev National Taras Shevchenko University, Kiev, Ukraine

Anders Lindquist Shanghai Jiao Tong University, Shanghai, China

Royal Institute of Technology, Stockholm, Sweden

Sergey I. Lyashko Department of Computational Mathematics, Kiev National Taras Shevchenko University, Kiev, Ukraine

Daniel N. Mohsenizadeh Department of Electrical and Computer Engineering, Texas A&M University, College Station, TX, USA

Susana Nascimento Department of Computer Science and NOVA Laboratory for Computer Science and Informatics (NOVA LINCS), Faculdade de Ciências e Tecnologia, Universidade Nova de Lisboa, Caparica, Portugal

Yurii Nesterov Center for Operations Research and Econometrics (CORE), Catholic University of Louvain (UCL), Louvain-la-Neuve, Belgium

Vilma A. Oliveira Department of Electrical and Computer Engineering, University of Sao Paulo at Sao Carlos, Sao Carlos, SP, Brazil

Giorgio Picci University of Padova, Padova, Italy

Roman A. Polyak Department of Mathematics, The Technion – Israel Institute of Technology, Haifa, Israel

Maryna V. Prysiazhna Kiev National Taras Shevchenko University, Kiev, Ukraine

Luba Sapir Department of Mathematics, Ben-Gurion University and Deutsche Telekom Laboratories at Ben-Gurion University, Beer Sheva, Israel

Vladimir V. Semenov Department of Computational Mathematics, Kiev National Taras Shevchenko University, Kiev, Ukraine

Vladimir Shikhman Center for Operations Research and Econometrics (CORE), Catholic University of Louvain (UCL), Louvain-la-Neuve, Belgium

Maksym P. Shlykov Kiev National Taras Shevchenko University, Kiev, Ukraine

Yunfei Song Department of Industrial and Systems Engineering, Lehigh University, Bethlehem, PA, USA

Tamás Terlaky Department of Industrial and Systems Engineering, Lehigh University, Bethlehem, PA, USA

A New Adaptive Conjugate Gradient Algorithm for Large-Scale Unconstrained Optimization

Neculai Andrei

This paper is dedicated to Prof. Boris T. Polyak on the occasion of his 80th birthday. Prof. Polyak's contributions to linear and nonlinear optimization methods, linear algebra, numerical mathematics, linear and nonlinear control systems are well-known. His articles and books give careful attention to both mathematical rigor and practical relevance. In all his publications he proves to be a refined expert in understanding the nature, purpose and limitations of nonlinear optimization algorithms and applied mathematics in general. It is my great pleasure and honour to dedicate this paper to Prof. Polyak, a pioneer and a great contributor in his area of interests.

Abstract An adaptive conjugate gradient algorithm is presented. The search direction is computed as the sum of the negative gradient and a vector determined by minimizing the quadratic approximation of objective function at the current point. Using a special approximation of the inverse Hessian of the objective function, which depends by a positive parameter, we get the search direction which satisfies both the sufficient descent condition and the Dai-Liao's conjugacy condition. The parameter in the search direction is determined in an adaptive manner by clustering the eigenvalues of the matrix defining it. The global convergence of the algorithm is proved for uniformly convex functions. Using a set of 800 unconstrained optimization test problems we prove that our algorithm is significantly more efficient and more robust than CG-DESCENT algorithm. By solving five applications from the MINPACK-2 test problem collection, with 10^6 variables, we show that the suggested adaptive conjugate gradient algorithm is top performer versus CG-DESCENT.

Keywords Unconstrained optimization • Adaptive conjugate gradient method • Sufficient descent condition • Conjugacy condition • Eigenvalues clustering • Numerical comparisons

---

N. Andrei (✉)
Center for Advanced Modeling and Optimization, Research Institute for Informatics, 8-10, Averescu Avenue, Bucharest, Romania

Academy of Romanian Scientists, Splaiul Independentei Nr. 54, Sector 5, Bucharest, Romania
e-mail: nandrei@ici.ro

B. Goldengorin (ed.), *Optimization and Its Applications in Control and Data Sciences*, Springer Optimization and Its Applications 115,
DOI 10.1007/978-3-319-42056-1_1

1 Introduction

For solving the large-scale unconstrained optimization problem

$$\min\{f(x) : x \in R^n\}, \tag{1}$$

where $f : R^n \to R$ is a continuously differentiable function, we consider the following algorithm

$$x_{k+1} = x_k + \alpha_k d_k, \tag{2}$$

where the step size α_k is positive and the directions d_k are computed using the updating formula:

$$d_{k+1} = -g_{k+1} + u_{k+1}. \tag{3}$$

Here, $g_k = \nabla f(x_k)$, and $u_{k+1} \in R^n$ is a vector to be determined. Usually, in (2), the steplength α_k is computed using the Wolfe line search conditions [34, 35]:

$$f(x_k + \alpha_k d_k) \leq f(x_k) + \rho \alpha_k g_k^T d_k, \tag{4}$$

$$g_{k+1}^T d_k \geq \sigma g_k^T d_k, \tag{5}$$

where $0 < \rho \leq \sigma < 1$. Also, the strong Wolfe line search conditions consisting of (4) and the following strengthened version of (5):

$$\left|g_{k+1}^T d_k\right| \leq -\sigma g_k^T d_k \tag{6}$$

can be used.

Observe that (3) is a general updating formula for the search direction computation. The following particularizations of (3) can be presented. If $u_{k+1} = 0$, then we get the steepest descent algorithm. If $u_{k+1} = (I - \nabla^2 f(x_{k+1})^{-1})g_{k+1}$, then the Newton method is obtained. Besides, if $u_{k+1} = (I - B_{k+1}^{-1})g_{k+1}$, where B_{k+1} is an approximation of the Hessian $\nabla^2 f(x_{k+1})$ then we find the quasi-Newton methods. On the other hand, if $u_{k+1} = \beta_k d_k$, where β_k is a scalar and $d_0 = -g_0$, the family of conjugate gradient algorithms is generated.

In this paper we focus on conjugate gradient method. This method was introduced by Hestenes and Stiefel [21] and Stiefel [31], ($\beta_k^{HS} = g_{k+1}^T y_k / y_k^T d_k$), to minimize positive definite quadratic objective functions (Here$y_k = g_{k+1} - g_k$.) This algorithm for solving positive definite linear algebraic systems of equations is known as *linear conjugate gradient*. Later, the algorithm was generalized to *nonlinear conjugate gradient* in order to minimize arbitrary differentiable nonlinear functions, by Fletcher and Reeves [14], ($\beta_k^{FR} = \|g_{k+1}\|^2 / \|g_k\|^2$), Polak and Ribière [27] and Polyak [28], ($\beta_k^{PRP} = g_{k+1}^T y_k / \|g_k\|^2$), Dai and Yuan [10],

($\beta_k^{DY} = \|g_{k+1}\|^2/y_k^T d_k$), and many others. An impressive number of nonlinear conjugate gradient algorithms have been established, and a lot of papers have been published on this subject insisting both on theoretical and computational aspects. An excellent survey of the development of different versions of nonlinear conjugate gradient methods, with special attention to global convergence properties, is presented by Hager and Zhang [20].

In this paper we consider another approach to generate an efficient and robust conjugate gradient algorithm. We suggest a procedure for u_{k+1} computation by minimizing the quadratic approximation of the function f in x_{k+1} and using a special representation of the inverse Hessian which depends on a positive parameter. The parameter in the matrix representing the search direction is determined in an adaptive manner by minimizing the largest eigenvalue of it. The idea, taken from the linear conjugate gradient, is to cluster the eigenvalues of the matrix representing the search direction.

The algorithm and its properties are presented in Sect. 2. We prove that the search direction used by this algorithm satisfies both the sufficient descent condition and the Dai and Liao conjugacy condition [11]. Using standard assumptions, Sect. 3 presents the global convergence of the algorithm for uniformly convex functions. In Sect. 4 the numerical comparisons of our algorithm versus the CG-DESCENT conjugate gradient algorithm [18] are presented. The computational results, for a set of 800 unconstrained optimization test problems, show that this new algorithm substantially outperform CG-DESCENT, being more efficient and more robust. Considering five applications from the MINPACK-2 test problem collection [4], with 10^6 variables, we show that our algorithm is way more efficient and more robust than CG-DESCENT.

2 The Algorithm

In this section we describe the algorithm and its properties. Let us consider that at the kth iteration of the algorithm an inexact Wolfe line search is executed, that is the step-length α_k satisfying (4) and (5) is computed. With these the following elements $s_k = x_{k+1} - x_k$ and $y_k = g_{k+1} - g_k$ are computed. Now, let us take the quadratic approximate of function f in x_{k+1} as

$$\Phi_{k+1}(d) = f_{k+1} + g_{k+1}^T d + \frac{1}{2} d^T B_{k+1} d, \tag{7}$$

where B_{k+1} is an approximation of the Hessian $\nabla^2 f(x_{k+1})$ of function f and d is the direction to be determined. The search direction d_{k+1} is computed as in (3), where u_{k+1} is computed as solution of the following minimizing problem

$$\min_{u_{k+1} \in R^n} \Phi_{k+1}(d_{k+1}). \tag{8}$$

Introducing d_{k+1} from (3) in the minimizing problem (8), then u_{k+1} is obtained as

$$u_{k+1} = (I - B_{k+1}^{-1})g_{k+1}. \tag{9}$$

Clearly, using different approximations B_{k+1} of the Hessian $\nabla^2 f(x_{k+1})$ different search directions d_{k+1} can be obtained. In this paper we consider the following expression of B_{k+1}^{-1}:

$$B_{k+1}^{-1} = I - \frac{s_k y_k^T - y_k s_k^T}{y_k^T s_k} + \omega_k \frac{s_k s_k^T}{y_k^T s_k}, \tag{10}$$

where ω_k is a positive parameter which follows to be determined. Observe that B_{k+1}^{-1} is the sum of a skew symmetric matrix with zero diagonal elements $(s_k y_k^T - y_k s_k^T)/y_k^T s_k$, and a pure symmetric and positive definite one $I + \omega_k s_k s_k^T / y_k^T s_k$. The expression of B_{k+1}^{-1} in (10) is a small modification of the BFGS quasi-Newton updating formula without memory. This is considered here in order to get the sufficient descent and the conjugacy conditions of the corresponding search direction. Now, from (9) we get:

$$u_{k+1} = \left[\frac{s_k y_k^T - y_k s_k^T}{y_k^T s_k} - \omega_k \frac{s_k s_k^T}{y_k^T s_k} \right] g_{k+1}. \tag{11}$$

Denote $H_{k+1} = B_{k+1}^{-1}$. Therefore, using (11) in (3) the search direction can be expressed as

$$d_{k+1} = -H_{k+1} g_{k+1}, \tag{12}$$

where

$$H_{k+1} = I - \frac{s_k y_k^T - y_k s_k^T}{y_k^T s_k} + \omega_k \frac{s_k s_k^T}{y_k^T s_k}. \tag{13}$$

Observe that the search direction (12), where H_{k+1} is given by (13), obtained by using the expression (10) of the inverse Hessian B_{k+1}^{-1}, is given by:

$$d_{k+1} = -g_{k+1} + \left(\frac{y_k^T g_{k+1}}{y_k^T s_k} - \omega_k \frac{s_k^T g_{k+1}}{y_k^T s_k} \right) s_k - \frac{s_k^T g_{k+1}}{y_k^T s_k} y_k. \tag{14}$$

Proposition 2.1. *Consider $\omega_k > 0$ and the step length α_k in (2) is determined by the Wolfe line search conditions (4) and (5). Then the search direction (14) satisfies the descent condition $g_{k+1}^T d_{k+1} \le 0$.*

Proof. By direct computation, since $\omega_k > 0$, we get:

$$g_{k+1}^T d_{k+1} = -\|g_{k+1}\|^2 - \omega_k \frac{(g_{k+1}^T s_k)^2}{y_k^T s_k} \leq 0. \blacksquare$$

Proposition 2.2. *Consider $\omega_k > 0$ and the step length α_k in (2) is determined by the Wolfe line search conditions (4) and (5). Then the search direction (14) satisfies the Dai and Liao conjugacy condition $y_k^T d_{k+1} = -v_k(s_k^T g_{k+1})$, where $v_k \geq 0$.*

Proof. By direct computation we have

$$y_k^T d_{k+1} = -\left[\omega_k + \frac{\|y_k\|^2}{y_k^T s_k}\right](s_k^T g_{k+1}) \equiv -v_k(s_k^T g_{k+1}),$$

where $v_k \equiv \omega_k + \frac{\|y_k\|^2}{y_k^T s_k}$. By Wolfe line search conditions (4) and (5) it follows that $y_k^T s_k > 0$, therefore $v_k > 0$. $\blacksquare$

Observe that, although we have considered the expression of the inverse Hessian as that given by (10), which is a non-symmetric matrix, the search direction (14), obtained in this manner, satisfies both the descent condition and the Dai and Liao conjugacy condition. Therefore, the search direction (14) leads us to a genuine conjugate gradient algorithm. The expression (10) of the inverse Hessian is only a technical argument to get the search direction (14). It is remarkable to say that from (12) our method can be considered as a quasi-Newton method in which the inverse Hessian, at each iteration, is expressed by the non-symmetric matrix H_{k+1}. More than this, the algorithm based on the search direction given by (14) can be considered as a three-term conjugate gradient algorithm.

In this point, to define the algorithm the only problem we face is to specify a suitable value for the positive parameter ω_k. As we know, the convergence rate of the nonlinear conjugate gradient algorithms depend on the structure of the eigenvalues of the Hessian and the condition number of this matrix. The standard approach is based on a singular value study on the matrix H_{k+1} (see for example [6, 7]), i.e. the numerical performances and the efficiency of the quasi-Newton methods are based on the condition number of the successive approximations of the inverse Hessian. A matrix with a large condition number is called an ill-conditioned matrix. Ill-conditioned matrices may produce instability in numerical computation with them. Unfortunately, many difficulties occur when applying this approach to general nonlinear optimization problems. Mainly, these difficulties are associated to the condition number computation of a matrix. This is based on the singular values of the matrix, which is a difficult and laborious task. However, if the matrix H_{k+1} is a normal matrix, then the analysis is simplified because the condition number of a normal matrix is based on its eigenvalues, which are easier to be computed.

As we know, generally, in a small neighborhood of the current point, the nonlinear objective function in the unconstrained optimization problem (1) behaves like a quadratic one for which the results from linear conjugate gradient can apply. But, for faster convergence of linear conjugate gradient algorithms some approaches can be considered like: the presence of isolated smallest and/or largest eigenvalues of the matrix H_{k+1}, as well as gaps inside the eigenvalues spectrum [5], clustering of the eigenvalues about one point [33] or about several points [23], or preconditioning [22]. If the matrix has a number of certain distinct eigenvalues contained in m disjoint intervals of very small length, then the linear conjugate gradient method will produce a very small residual after m iterations [24]. This is an important property of linear conjugate gradient method and we try to use it in nonlinear case in order to get efficient and robust conjugate gradient algorithms. Therefore, we consider the extension of the method of clustering the eigenvalues of the matrix defining the search direction from linear conjugate gradient algorithms to nonlinear case.

The idea is to determine ω_k by clustering the eigenvalues of H_{k+1}, given by (13), by minimizing the largest eigenvalue of the matrix H_{k+1} from the spectrum of this matrix. The structure of the eigenvalues of the matrix H_{k+1} is given by the following theorem.

Theorem 2.1. *Let H_{k+1} be defined by* (13). *Then H_{k+1} is a nonsingular matrix and its eigenvalues consist of 1 ($n-2$ multiplicity), λ_{k+1}^{+}, and λ_{k+1}^{-}, where*

$$\lambda_{k+1}^{+} = \frac{1}{2}\left[(2+\omega_k b_k) + \sqrt{\omega_k^2 b_k^2 - 4a_k + 4}\right], \tag{15}$$

$$\lambda_{k+1}^{-} = \frac{1}{2}\left[(2+\omega_k b_k) - \sqrt{\omega_k^2 b_k^2 - 4a_k + 4}\right], \tag{16}$$

and

$$a_k = \frac{\|y_k\|^2\|s_k\|^2}{(y_k^T s_k)^2} > 1, \quad b_k = \frac{\|s_k\|^2}{y_k^T s_k} \geq 0. \tag{17}$$

Proof. By the Wolfe line search conditions (4) and (5) we have that $y_k^T s_k > 0$. Therefore, the vectors y_k and s_k are nonzero vectors. Let V be the vector space spanned by $\{s_k, y_k\}$. Clearly, $\dim(V) \leq 2$ and $\dim(V^{\perp}) \geq n-2$. Thus, there exist a set of mutually unit orthogonal vectors $\{u_k^i\}_{i=1}^{n-2} \subset V^{\perp}$ such that

$$s_k^T u_k^i = y_k^T u_k^i = 0, \; i = 1, \ldots, n-2,$$

which from (13) leads to

$$H_{k+1} u_k^i = u_k^i, \; i = 1, \ldots, n-2.$$

Therefore, the matrix H_{k+1} has $n-2$ eigenvalues equal to 1, which corresponds to $\{u_k^i\}_{i=1}^{n-2}$ as eigenvectors.

Now, we are interested to find the rest of the two remaining eigenvalues, denoted as λ_{k+1}^{+} and λ_{k+1}^{-}, respectively. From the formula of algebra (see for example [32])

$$\det(I + pq^T + uv^T) = (1 + q^T p)(1 + v^T u) - (p^T v)(q^T u),$$

where $p = \frac{y_k + \omega_k s_k}{y_k^T s_k}$, $q = s_k$, $u = -\frac{s_k}{y_k^T s_k}$ and $v = y_k$, it follows that

$$\det(H_{k+1}) = \frac{\|s_k\|^2 \|y_k\|^2}{(y_k^T s_k)^2} + \omega_k \frac{\|s_k\|^2}{y_k^T s_k} \equiv a_k + \omega_k b_k. \tag{18}$$

But, $a_k > 1$ and $b_k \geq 0$, therefore, H_{k+1} is a nonsingular matrix.
On the other hand, by direct computation

$$tr(H_{k+1}) = n + \omega_k \frac{\|s_k\|^2}{y_k^T s_k} \equiv n + \omega_k b_k. \tag{19}$$

By the relationships between the determinant and the trace of a matrix and its eigenvalues, it follows that the other eigenvalues of H_{k+1} are the roots of the following quadratic polynomial

$$\lambda^2 - (2 + \omega_k b_k)\lambda + (a_k + \omega_k b_k) = 0. \tag{20}$$

Clearly, the other two eigenvalues of the matrix H_{k+1} are determined from (20) as (15) and (16), respectively. Observe that $a_k > 1$ follows from Wolfe conditions and the inequality

$$\frac{y_k^T s_k}{\|s_k\|^2} \leq \frac{\|y_k\|^2}{y_k^T s_k}. \blacksquare$$

In order to have both λ_{k+1}^{+} and λ_{k+1}^{-} as real eigenvalues, from (15) and (16) the following condition must be fulfilled $\omega_k^2 b_k^2 - 4a_k + 4 \geq 0$, out of which the following estimation of the parameter ω_k can be determined:

$$\omega_k \geq \frac{2\sqrt{a_k - 1}}{b_k}. \tag{21}$$

Since $a_k > 1$, if $\|s_k\| > 0$, it follows that the estimation of ω_k given in (21) is well defined. From (20) we have

$$\lambda_{k+1}^{+} + \lambda_{k+1}^{-} = 2 + \omega_k b_k > 0, \tag{22}$$

$$\lambda_{k+1}^{+} \lambda_{k+1}^{-} = a_k + \omega_k b_k > 0. \tag{23}$$

Therefore, from (22) and (23) we have that both λ_{k+1}^{+} and λ_{k+1}^{-} are positive eigenvalues. Since $\omega_k^2 b_k^2 - 4a_k + 4 \geq 0$, from (15) and (16) we have that $\lambda_{k+1}^{+} \geq \lambda_{k+1}^{-}$. By direct computation, from (15), using (21) we get

$$\lambda_{k+1}^{+} \geq 1 + \sqrt{a_k - 1} > 1. \tag{24}$$

A simple analysis of Eq. (20) shows that $1 \leq \lambda_{k+1}^{-} \leq \lambda_{k+1}^{+}$. Therefore H_{k+1} is a positive definite matrix. The maximum eigenvalue of H_{k+1} is λ_{k+1}^{+} and its minimum eigenvalue is 1.

Proposition 2.3. *The largest eigenvalue*

$$\lambda_{k+1}^{+} = \frac{1}{2}\left[(2 + \omega_k b_k) + \sqrt{\omega_k^2 b_k^2 - 4a_k + 4}\right] \tag{25}$$

gets its minimum $1 + \sqrt{a_k - 1}$, *when* $\omega_k = \frac{2\sqrt{a_k - 1}}{b_k}$.

Proof. Observe that $a_k > 1$. By direct computation the minimum of (25) is obtained for $\omega_k = (2\sqrt{a_k - 1})/b_k$, for which its minimum value is $1 + \sqrt{a_k - 1}$. ■

We see that according to proposition 2.3 when $\omega_k = (2\sqrt{a_k - 1})/b_k$ the largest eigenvalue of H_{k+1} arrives at the minimum value, i.e. the spectrum of H_{k+1} is clustered. In fact for $\omega_k = (2\sqrt{a_k - 1})/b_k$, $\lambda_{k+1}^{+} = \lambda_{k+1}^{-} = 1 + \sqrt{a_k - 1}$. Therefore, from (17) the following estimation of ω_k can be obtained:

$$\omega_k = 2\frac{y_k^T s_k}{\|s_k\|^2}\sqrt{a_k - 1} \leq 2\frac{\|y_k\|}{\|s_k\|}\sqrt{a_k - 1}. \tag{26}$$

From (17) $a_k > 1$, hence if $\|s_k\| > 0$ it follows that the estimation of ω_k given by (26) is well defined. However, we see that the minimum of λ_{k+1}^{+} obtained for $\omega_k = 2\sqrt{a_k - 1}/b_k$, is given by $1 + \sqrt{a_k - 1}$. Therefore, if a_k is large, then the largest eigenvalue of the matrix H_{k+1} will be large. This motivates the parameter ω_k to be computed as:

$$\omega_k = \begin{cases} 2\sqrt{\tau - 1}\frac{\|y_k\|}{\|s_k\|}, & \text{if } a_k \geq \tau, \\ 2\sqrt{a_k - 1}\frac{\|y_k\|}{\|s_k\|}, & \text{otherwise}, \end{cases} \tag{27}$$

where $\tau > 1$ is a positive constant. Therefore, our algorithm is an adaptive conjugate gradient algorithm in which the value of the parameter ω_k in the search direction (14) is computed as in (27) trying to cluster all the eigenvalues of H_{k+1} defining the search direction of the algorithm.

Now, as we know, Powell [30] constructed a three dimensional nonlinear unconstrained optimization problem showing that the PRP and HS methods could cycle infinitely without converging to a solution. Based on the insight gained by his example, Powell [30] proposed a simple modification of PRP method where the conjugate gradient parameterβ_k^{PRP} is modified as $\beta_k^{PRP+} = \max\{\beta_k^{PRP}, 0\}$. Later on, for general nonlinear objective functions Gilbert and Nocedal [15] studied the theoretical convergence and the efficiency of PRP+ method. In the following, to attain a good computational performance of the algorithm we apply the Powell's idea and consider the following modification of the search direction given by (14) as:

$$d_{k+1} = -g_{k+1} + \max\left(\frac{y_k^T g_{k+1} - \omega_k s_k^T g_{k+1}}{y_k^T s_k}, 0\right) s_k - \frac{s_k^T g_{k+1}}{y_k^T s_k} y_k. \tag{28}$$

where ω_k is computed as in (27).

Using the procedure of acceleration of conjugate gradient algorithms presented in [1], and taking into consideration the above developments, the following algorithm can be presented.

NADCG Algorithm (New Adaptive Conjugate Gradient Algorithm)

Step 1. Select a starting point $x_0 \in R^n$ and compute: $f(x_0)$, $g_0 = \nabla f(x_0)$. Select some positive values for ρ and σ used in Wolfe line search conditions. Consider a positive value for the parameter τ. ($\tau > 1$) Set $d_0 = -g_0$ and $k = 0$.
Step 2. Test a criterion for stopping the iterations. If this test is satisfied, then stop; otherwise continue with step 3.
Step 3. Determine the steplength α_k by using the Wolfe line search (4) and (5).
Step 4. Compute $z = x_k + \alpha_k d_k$, $g_z = \nabla f(z)$ and $y_k = g_k - g_z$.
Step 5. Compute: $\bar{a}_k = \alpha_k g_z^T d_k$ and $\bar{b}_k = -\alpha_k y_k^T d_k$.
Step 6. Acceleration scheme. If $\bar{b}_k > 0$, then compute $\xi_k = -\bar{a}_k / \bar{b}_k$ and update the variables as $x_{k+1} = x_k + \xi_k \alpha_k d_k$, otherwise update the variables as $x_{k+1} = x_k + \alpha_k d_k$.
Step 7. Compute ω_k as in (27).
Step 8. Compute the search direction as in (28).
Step 9. Powell restart criterion. If $\left|g_{k+1}^T g_k\right| > 0.2\|g_{k+1}\|^2$, then set $d_{k+1} = -g_{k+1}$.
Step 10. Consider $k = k + 1$ and go to step 2. ■

If function f is bounded along the direction d_k, then there exists a stepsize α_k satisfying the Wolfe line search (see for example [13] or [29]). In our algorithm when the Beale-Powell restart condition is satisfied, then we restart the algorithm with the negative gradient $-g_{k+1}$. More sophisticated reasons for restarting the algorithms have been proposed in the literature [12], but we are interested in the performance of a conjugate gradient algorithm that uses this restart criterion associated to a direction satisfying both the descent and the conjugacy conditions. Under reasonable assumptions, the Wolfe conditions and the Powell restart criterion are sufficient to prove the global convergence of the algorithm. The first trial of the step length crucially affects the practical behavior of the algorithm. At every iteration $k \geq 1$ the starting guess for the step α_k in the line search is computed as $\alpha_{k-1} \|d_{k-1}\| / \|d_k\|$. For uniformly convex functions, we can prove the linear convergence of the acceleration scheme used in the algorithm [1].

3 Global Convergence Analysis

Assume that:

i. *The level set* $S = \{x \in R^n : f(x) \leq f(x_0)\}$ *is bounded.*
ii. *In a neighborhood* N *of* S *the function* f *is continuously differentiable and its gradient is Lipschitz continuous, i.e. there exists a constant* $L > 0$ *such that* $\|\nabla f(x) - \nabla f(y)\| \leq L\|x - y\|$, *for all* $x, y \in N$.

Under these assumptions on f there exists a constant $\Gamma \geq 0$ such that $\|\nabla f(x)\| \leq \Gamma$ for all $x \in S$. For any conjugate gradient method with strong Wolfe line search the following general result holds [26].

Proposition 3.1. *Suppose that the above assumptions hold. Consider a conjugate gradient algorithm in which, for all* $k \geq 0$, *the search direction* d_k *is a descent direction and the steplength* α_k *is determined by the Wolfe line search conditions. If*

$$\sum_{k \geq 0} \frac{1}{\|d_k\|^2} = \infty, \tag{29}$$

then the algorithm converges in the sense that

$$\liminf_{k \to \infty} \|g_k\| = 0. \tag{30}$$

For *uniformly convex functions* we can prove that the norm of the direction d_{k+1} computed as in (28) with (27) is bounded above. Therefore, by proposition 3.1 we can prove the following result.

Theorem 3.1. *Suppose that the assumptions (i) and (ii) hold. Consider the algorithm NADCG where the search direction* d_k *is given by* (28) *and* ω_k *is computed as in* (27). *Suppose that* d_k *is a descent direction and* α_k *is computed by the strong Wolfe line search. Suppose that* f *is a uniformly convex function on* S *i.e. there exists a constant* $\mu > 0$ *such that*

$$(\nabla f(x) - \nabla f(y))^T (x - y) \geq \mu \|x - y\|^2 \tag{31}$$

for all $x, y \in N$. *Then*

$$\lim_{k \to \infty} \|g_k\| = 0. \tag{32}$$

Proof. From Lipschitz continuity we have $\|y_k\| \leq L\|s_k\|$. On the other hand, from uniform convexity it follows that $y_k^T s_k \geq \mu \|s_k\|^2$. Now, from (27)

$$\omega_k = 2\sqrt{\tau - 1}\frac{\|y_k\|}{\|s_k\|} \leq 2\sqrt{\tau - 1}\frac{L\|s_k\|}{\|s_k\|} = 2L\sqrt{\tau - 1}.$$

On the other hand, from (28) we have

$$\|d_{k+1}\| \leq \|g_{k+1}\| + \frac{\left|y_k^T g_{k+1}\right|}{y_k^T s_k} \|s_k\| + \omega_k \frac{\left|s_k^T g_{k+1}\right|}{y_k^T s_k} \|s_k\| + \frac{\left|s_k^T g_{k+1}\right|}{y_k^T s_k} \|y_k\|$$

$$\leq \Gamma + \frac{\|y_k\| \, \Gamma \, \|s_k\|}{\mu \|s_k\|^2} + 2L\sqrt{\tau - 1}\frac{\|s_k\| \, \Gamma \, \|s_k\|}{\mu \|s_k\|^2} + \frac{\|s_k\| \, \Gamma \, \|y_k\|}{\mu \|s_k\|^2}$$

$$\leq \Gamma + 2\frac{L\Gamma}{\mu} + 2L\sqrt{\tau - 1}\frac{\Gamma}{\mu},$$

showing that (29) is true. By proposition 3.1 it follows that (30) is true, which for uniformly convex functions is equivalent to (32). ■

4 Numerical Results and Comparisons

The NADCG algorithm was implemented in double precision Fortran using loop unrolling of depth 5 and compiled with f77 (default compiler settings) and run on a Workstation Intel Pentium 4 with 1.8 GHz. We selected a number of 80 large-scale unconstrained optimization test functions in generalized or extended form presented in [2]. For each test function we have considered 10 numerical experiments with the number of variables increasing as $n = 1000, 2000, \ldots, 10000$. The algorithm uses the Wolfe line search conditions with cubic interpolation, $\rho = 0.0001$, $\sigma = 0.8$ and the same stopping criterion $\|g_k\|_\infty \leq 10^{-6}$,where $\|.\|_\infty$is the maximum absolute component of a vector.

Since, CG-DESCENT [19] is among the best nonlinear conjugate gradient algorithms proposed in the literature, but not necessarily the best, in the following we compare our algorithm NADCG versus CG-DESCENT. The algorithms we compare in these numerical experiments find local solutions. Therefore, the comparisons of algorithms are given in the following context. Let $f_i^{ALG1}$and f_i^{ALG2} be the optimal value found by ALG1 and ALG2, for problem $i = 1, \ldots, 800$, respectively. We say that, in the particular problem i, the performance of ALG1 was better than the performance of ALG2 if:

$$\left\|f_i^{ALG1} - f_i^{ALG2}\right\| < 10^{-3} \tag{33}$$

and the number of iterations (#iter), or the number of function-gradient evaluations (#fg), or the CPU time of ALG1 was less than the number of iterations, or the number of function-gradient evaluations, or the CPU time corresponding to ALG2, respectively.

Figure 1 shows the Dolan-Moré's performance profiles subject to CPU time metric for different values of parameter τ. Form Fig. 1, for example for$\tau = 2$,

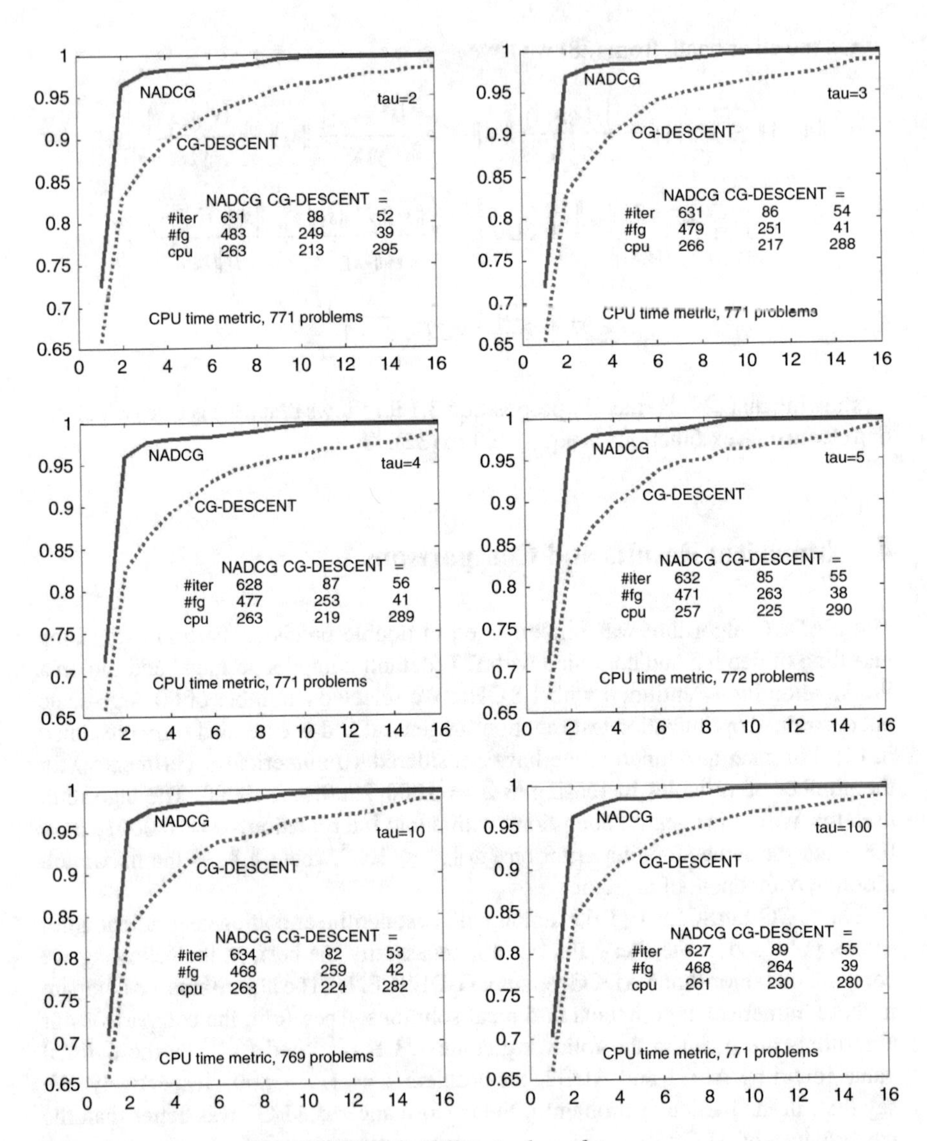

Fig. 1 NADCG versus CG-DESCENT for different values of τ

comparing NADCG versus CG-DESCENT with Wolfe line search (version 1.4), subject to the number of iterations, we see that NADCG was better in 631 problems (i.e. it achieved the minimum number of iterations for solving 631 problems), CG-DESCENT was better in 88 problems and they achieved the same number of iterations in 52 problems, etc. Out of 800 problems, we considered in this numerical study, only for 771 problems does the criterion (33) hold. From Fig. 1 we see that for different values of the parameterτ NADCG algorithm has similar performances versus CG-DESCENT. Therefore, in comparison with CG-DESCENT, on average, NADCG appears to generate the best search direction and the best step-length. We see that this very simple adaptive scheme lead us to a conjugate gradient algorithm which substantially outperform the CG-DESCENT, being way more efficient and more robust.

From Fig. 1 we see that NADCG algorithm is very little sensitive to the values of the parameter τ. In fact, for $a_k \geq \tau$, from (28) we get:

$$\frac{\partial d_{k+1}}{\partial \tau} = -\frac{1}{\sqrt{\tau - 1}} \frac{\|y_k\|}{\|s_k\|} \frac{s_k^T g_{k+1}}{y_k^T s_k} s_k, \tag{34}$$

where $\tau > 1$. Therefore, since the gradient of the function f is Lipschitz continuous and the quantity $s_k^T g_{k+1}$ is going to zero it follows that along the iterations $\partial d_{k+1}/\partial \tau$ tends to zero, showing that along the iterations the search direction is less and less sensitive subject to the value of the parameter τ. For uniformly convex functions, using the assumptions from Sect. 3 we get:

$$\left\| \frac{\partial d_{k+1}}{\partial \tau} \right\| \leq \frac{1}{\sqrt{\tau - 1}} \frac{L\Gamma}{\mu}. \tag{35}$$

Therefore, for example, for larger values of τ the variation of d_{k+1} subject to τ decreases showing that the NADCG algorithm is very little sensitive to the values of the parameter τ. This is illustrated in Fig. 1 where the performance profiles have the same allure for different values of τ.

In the following, in the second set of numerical experiments, we present comparisons between NADCG and CG-DESCENT conjugate gradient algorithms for solving some applications from the MINPACK-2 test problem collection [4]. In Table 1 we present these applications, as well as the values of their parameters.

The infinite-dimensional version of these problems is transformed into a finite element approximation by triangulation. Thus a finite-dimensional minimization problem is obtained whose variables are the values of the piecewise linear function at the vertices of the triangulation. The discretization steps are $nx = 1,000$ and $ny = 1,000$, thus obtaining minimization problems with 1,000,000 variables. A comparison between NADCG (Powell restart criterion, $\|\nabla f(x_k)\|_\infty \leq 10^{-6}$, $\rho = 0.0001, \sigma = 0.8, \tau = 2$) and CG-DESCENT (version 1.4, Wolfe line search, default settings, $\|\nabla f(x_k)\|_\infty \leq 10^{-6}$) for solving these applications is given in Table 2.

Table 1 Applications from the MINPACK-2 collection

A1	Elastic–plastic torsion [16, pp. 41–55], $c = 5$
A2	Pressure distribution in a journal bearing [9], $b = 10,\ \varepsilon = 0.1$
A3	Optimal design with composite materials [17], $\lambda = 0.008$
A4	Steady-state combustion [3, pp. 292–299], [8], $\lambda = 5$
A5	Minimal surfaces with Enneper conditions [25, pp. 80–85]

Table 2 Performance of NADCG versus CG-DESCENT

	NADCG			CG-DESCENT		
	#iter	#fg	cpu	#iter	#fg	cpu
A1	1111	2253	352.14	1145	2291	474.64
A2	2845	5718	1136.67	3370	6741	1835.51
A3	4270	8573	2497.35	4814	9630	3949.71
A4	1413	2864	2098.74	1802	3605	3786.25
A5	1548	3116	695.59	1225	2451	753.75
TOTAL	11187	22524	6780.49	12356	24718	10799.86

1,000,000 variables. CPU seconds

From Table 2, we see that, subject to the CPU time metric, the NADCG algorithm is top performer and the difference is significant, about **4019.37 s** for solving all these five applications.

The NADCG and CG-DESCENT algorithms (and codes) are different in many respects. Since both of them use the Wolfe line search (however, implemented in different manners), these algorithms mainly differ in their choice of the search direction. The search direction d_{k+1} given by (27) and (28) used in NADCG is more elaborate: it is adaptive in the sense to cluster the eigenvalues of the matrix defining it and it satisfies both the descent condition and the conjugacy condition in a restart environment.

5 Conclusions

An adaptive conjugate gradient algorithm has been presented. The idea of this paper is to compute the search direction as the sum of the negative gradient and an arbitrary vector which was determined by minimizing the quadratic approximation of objective function at the current point. The solution of this quadratic minimization problem is a function of the inverse Hessian. In this paper we introduce a special expression of the inverse Hessian of the objective function which depends by a positive parameter ω_k. For any positive values of this parameter the search direction satisfies both the sufficient descent condition and the Dai-Liao's conjugacy condition. Thus, the algorithm is a conjugate gradient one. The parameter in the search direction is determined in an adaptive manner, by clustering the spectrum of the matrix defining the search direction. This idea is taken from the linear conjugate gradient, where clustering the eigenvalues of the matrix is very benefic subject to the convergence. Mainly, in our nonlinear case, clustering the eigenvalues reduces to determine the value of the parameter ω_k to minimize the largest eigenvalue of the matrix. The adaptive computation of the parameter ω_k in the search direction is subject to a positive constant which has a very little impact on the performances of our algorithm. The steplength is computed using the classical Wolfe line search conditions with a special initialization. In order to improve the reducing the values of the objective function to be minimized an acceleration scheme is used. For uniformly convex functions, under classical assumptions, the algorithm is globally convergent. Thus, we get an accelerated adaptive conjugate gradient algorithm. Numerical experiments and intensive comparisons using 800 unconstrained optimization problems of different dimensions and complexity proved that this adaptive conjugate gradient algorithm is way more efficient and more robust than CG-DESCENT algorithm. In an effort to see the performances of this adaptive conjugate gradient we solved five large-scale nonlinear optimization applications from MINPACK-2 collection, up to 10^6 variables, showing that NADCG is obvious more efficient and more robust than CG-DESCENT.

References

1. Andrei, N.: Acceleration of conjugate gradient algorithms for unconstrained optimization. Appl. Math. Comput. **213**, 361–369 (2009)
2. Andrei, N.: Another collection of large-scale unconstrained optimization test functions. ICI Technical Report, January 30 (2013)
3. Aris, R.: The Mathematical Theory of Diffusion and Reaction in Permeable Catalysts. Oxford University Press, New York (1975)
4. Averick, B.M., Carter, R.G., Moré, J.J., Xue, G.L.: The MINPACK-2 test problem collection. Mathematics and Computer Science Division, Argonne National Laboratory. Preprint MCS-P153-0692, June (1992)
5. Axelsson, O., Lindskog, G.: On the rate of convergence of the preconditioned conjugate gradient methods. Numer. Math. **48**, 499–523, (1986)
6. Babaie-Kafaki, S.: A eigenvalue study on the sufficient descent property of a modified Polak-Ribière-Polyak conjugate gradient method. Bull. Iran. Math. Soc. **40**(1) 235–242 (2014)
7. Babaie-Kafaki, S., Ghanbari, R.: A modified scaled conjugate gradient method with global convergence for nonconvex functions. Bull. Belgian Math. Soc. Simon Stevin **21**(3), 465–47 (2014)
8. Bebernes, J., Eberly, D.: Mathematical problems from combustion theory. In: Applied Mathematical Sciences, vol. 83. Springer, New York (1989)
9. Cimatti, G.: On a problem of the theory of lubrication governed by a variational inequality. Appl. Math. Optim. **3**, 227–242 (1977)
10. Dai, Y.H., Yuan, Y.: A nonlinear conjugate gradient method with a strong global convergence property. SIAM J. Optim. **10**, 177–182 (1999)
11. Dai, Y.H., Liao, L.Z.: New conjugacy conditions and related nonlinear conjugate gradient methods. Appl. Math. Optim. **43**, 87–101, (2001)
12. Dai, Y.H., Liao, L.Z., Duan, L.: On restart procedures for the conjugate gradient method. Numer. Algorithms **35**, 249–260 (2004)
13. Dennis, J.E., Schnabel, R.B.: Numerical Methods for Unconstrained Optimization and Nonlinear Equations. Prentice-Hall, Englewood Cliffs (1983)
14. Fletcher, R., Reeves, C.M.: Function minimization by conjugate gradients. Comput. J. **7**, 149–154 (1964)
15. Gilbert, J.C., Nocedal, J.: Global convergence properties of conjugate gradient methods for optimization. SIAM J. Optim. **2**(1), 21–42 (1992)
16. Glowinski, R.: Numerical Methods for Nonlinear Variational Problems. Springer, Berlin (1984)
17. Goodman, J., Kohn, R., Reyna, L.: Numerical study of a relaxed variational problem from optimal design. Comput. Methods Appl. Mech. Eng. **57**, 107–127 (1986)

18. Hager, W.W., Zhang, H.: A new conjugate gradient method with guaranteed descent and an efficient line search. SIAM J. Optim. **16**, 170–192 (2005)
19. Hager, W.W., Zhang, H.: Algorithm 851: CG-DESCENT, a conjugate gradient method with guaranteed descent. ACM Trans. Math. Softw. **32**, 113–137 (2006)
20. Hager, W.W., Zhang, H.: A survey of nonlinear conjugate gradient methods. Pac. J. Optim. **2**(1), 35–58 (2006)
21. Hestenes, M.R., Steifel, E.: Metods of conjugate gradients for solving linear systems. J. Res. Natl. Bur. Stand. Sec. B. **48**, 409–436 (1952)
22. Kaporin, I.E.: New convergence results and preconditioning strategies for the conjugate gradient methods. Numer. Linear Algebra Appl. **1**(2), 179–210 (1994)
23. Kratzer, D., Parter, S.V., Steuerwalt, M.: Bolck splittings for the conjugate gradient method. Comp. Fluid **11**, 255–279 (1983)
24. Luenberger, D.G., Ye, Y.: Linear and Nonlinear Programming. International Series in Operations Research & Management Science, 3rd edn. Springer Science+Business Media, New York (2008)
25. Nitsche, J.C.C.: Lectures On Minimal Surfaces, vol. 1. Cambridge University Press, Cambridge (1989)
26. Nocedal, J.: Conjugate gradient methods and nonlinear optimization. In: Adams, L., Nazareth, J.L. (eds.) Linear and Nonlinear Conjugate Gradient Related Methods, pp. 9–23. SIAM, Philadelphia (1996)
27. Polak, E., Ribière, G.: Note sur la convergence de directions conjuguée. Rev. Fr. Informat Recherche Oper. 3e Année **16**, 35–43 (1969)
28. Polyak, B.T.: The conjugate gradient method in extreme problems. USSR Comp. Math. Math. Phys. **9**, 94–112 (1969)
29. Polyak, B.T.: Introduction to Optimization. Optimization Software, Publications Division, New York (1987)
30. Powell, M.J.D.: Nonconvex minimization calculations and the conjugate gradient method. In: Griffiths, D.F. (ed.) Numerical Analysis (Dundee, 1983). Lecture Notes in Mathematics, vol. 1066, pp. 122–141. Springer, Berlin (1984)
31. Stiefel, E.: Über einige methoden der relaxationsrechnung. Z. Angew. Math. Phys. **3**, 1–33 (1952)
32. Sun, W., Yuan, Y.X.: Optimization Theory and Methods: Nonlinear Programming. Springer Science + Business Media, New York (2006)
33. Winther, R.: Some superlinear convergence results for the conjugate gradient method. SIAM J. Numer. Anal. **17**, 14–17 (1980)
34. Wolfe, P.: Convergence conditions for ascent methods. SIAM Rev. **11**, 226–235 (1969)
35. Wolfe, P.: Convergence conditions for ascent methods. II: some corrections. SIAM Rev. **13**, 185–188 (1971)

On Methods of Terminal Control with Boundary-Value Problems: Lagrange Approach

Anatoly Antipin and Elena Khoroshilova

Abstract A dynamic model of terminal control with boundary value problems in the form of convex programming is considered. The solutions to these finite-dimensional problems define implicitly initial and terminal conditions at the ends of time interval at which the controlled dynamics develops. The model describes a real situation when an object needs to be transferred from one state to another. Based on the Lagrange formalism, the model is considered as a saddle-point controlled dynamical problem formulated in a Hilbert space. Iterative saddle-point method has been proposed for solving it. We prove the convergence of the method to saddle-point solution in all its components: weak convergence—in controls, strong convergence—in phase and conjugate trajectories, and terminal variables.

Keywords Terminal control • Boundary values problems • Controllability • Lagrange function • Saddle-point method • Convergence

1 Introduction

Terminal control problem is considered in this article. The problem consists of two main components in the form of linear controlled dynamics and two finite-dimensional convex boundary value problems. The problem consists in choosing such a control that the corresponding phase trajectory (the solution of differential equation) is to connect the solutions of two boundary value problems, which are tied to the ends of the time interval. The terminal control problem can be viewed as a generalization of one of the main problems in the controllability theory for the

A. Antipin (✉)
Federal Research Center "Computer Science and Control" of Russian Academy of Sciences, Dorodnicyn Computing Centre, Moscow, Russia
e-mail: asantip@yandex.ru

E. Khoroshilova
Faculty of Computational Mathematics and Cybernetics, Lomonosov Moscow State University, Moscow, Russia
e-mail: khorelena@gmail.com

B. Goldengorin (ed.), *Optimization and Its Applications in Control and Data Sciences*, Springer Optimization and Its Applications 115,
DOI 10.1007/978-3-319-42056-1_2

case where the boundary conditions are defined implicitly as solutions of convex programming problems. Such models have countless varieties of applications.

To solve this problem, we propose an iterative process of the saddle-point type, and its convergence to the solution of the problem is proved. This solution includes the following components: optimal control, optimal phase trajectory, conjugate trajectory, and solutions of terminal boundary value problems. The method of solving as an iterative process builds sequences of controls, trajectories, conjugate trajectories, and similar sequences in terminal spaces. Here, the subtlety of the situation is that trajectories are expected to tie the solutions of boundary value problems. To achieve this, we organize special (additional) finite-dimensional iterative process at the ends of time interval. These iterative processes in finite-dimensional spaces ensure the convergence to terminal solutions.

The proposed approach [2–12, 17, 18] is considered in the framework of the Lagrange formalism in contrast to the Hamilton formalism, the top of which is the Pontryagin maximum principle. Although the Lagrange approach assumes the convexity of problems, this assumption is not dominant fact, since the class of problems to be solved remains quite extensive. This class includes problems with linear controlled dynamics and convex integral and terminal objective functions. Furthermore, the idea of linearization significantly reduce the pressure of convexity. The class of possible models is greatly enriched by the use of different kinds of boundary value problems. The proposed method is based on a saddle-point structure of the problem, and converges to the solution of the problem as to a saddle point of the Lagrange function. The convergence of iterative process to the solution is proved. Namely, the convergence in controls is weak, but the convergence in other components of the solution is strong. Other approaches are shown in [22, 23].

2 Problem Statement

Consider a boundary value problem of optimal control on a fixed time interval $[t_0, t_1]$ with a movable right end. Dynamics of controllable trajectories $x(\cdot)$ is described by a linear system of ordinary differential equations

$$\frac{d}{dt}x(t) = D(t)x(t) + B(t)u(t), \quad t_0 \le t \le t_1,$$

where $D(t), B(t)$ are $n \times n, n \times r$ continuous matrices ($r < n$). Controls $u(\cdot) \in \mathrm{U}$ are assumed to be bounded in the norm L_2^r

$$\mathrm{U} = \left\{ u(\cdot) \in \mathrm{L}_2^r[t_0, t_1] \mid \frac{1}{2}\|u(\cdot)\|_{\mathrm{L}_2^r}^2 \le \mathrm{C}^2 \right\}.$$

While controls are taking all admissible values from U, the ODE system for a given $x_0 = x(t_0)$ generates a set of trajectories $x(\cdot)$, the right ends $x_1 = x(t_1)$ of which describe the attainability set $X(t_1) \subset \mathrm{R}^n$.

Any function $x(\cdot) \in L_2^n[t_0, t_1]$ satisfying this system for almost all $t \in [t_0, t_1]$ can be considered as a solution. In particular, it may occur that the Cantor staircase function (see [19, p. 361]), which is not an absolutely continuous function, is a solution. This function is differentiable almost everywhere, but it cannot be recovered from its derivative. Therefore, instead of examining differential system on the entire space of trajectories $x(\cdot) \in L_2^n[t_0, t_1]$, we restrict ourselves to its subset of absolutely continuous functions [19]. Every absolutely continuous function satisfies the identity

$$x(t) = x(t_0) + \int_{t_0}^{t} (D(\tau)x(\tau) + B(\tau)u(\tau))d\tau, \quad t_0 \leq t \leq t_1.$$

It is shown in [26, Book 2, p. 443] that a unique trajectory $x(\cdot)$ is associated with any control $u(\cdot) \in \mathrm{U} \subset L_2^r[t_0, t_1]$ in the above system of equations. The class of absolutely continuous functions forms a linear variety that is everywhere dense in $L_2^n[t_0, t_1]$. We denote this linear variety by $\mathrm{AC}^n[t_0, t_1]$. Its closure is $\overline{\mathrm{AC}}^n[t_0, t_1] = L_2^n[t_0, t_1]$. The Newton-Leibniz formula and the integration-by-parts formula hold for every pair of functions $x(\cdot), u(\cdot) \in \mathrm{AC}^n[t_0, t_1] \times \mathrm{U}$.[1]

In applications, a control $u(\cdot)$ is often a piecewise continuous function. The presence of jump points in control $u(\cdot)$ has no effect on trajectory $x(\cdot)$. Moreover, this trajectory will not change even if we change the values of $u(\cdot)$ on a set of measure zero.

Now we are ready to formulate the problem. Namely, we need to find the initial value x_0^* and control function $u^*(\cdot) \in \mathrm{U}$ such that the corresponding trajectory $x^*(\cdot)$, being the solution of the differential system, starts from the point x_0^* at the left end of the time interval and comes to the point $x^*(t_1)$ at the right end:

$$x_0^* \in \mathrm{Argmin}\{\varphi_0(x_0) \mid A_0 x_0 \leq a_0, \; x_0 \in \mathbb{R}^n\},$$

$$(x^*(t_1), x^*(\cdot), u^*(\cdot)) \in \mathrm{Argmin}\{\varphi_1(x(t_1)) \mid A_1 x(t_1) \leq a_1, \; x(t_1) \in X(t_1) \subset \mathbb{R}^n,$$

$$\frac{d}{dt}x(t) = D(t)x(t) + B(t)u(t), \quad x(t_0) = x_0^*, \; x(t_1) = x^*(t_1),$$

$$x(\cdot) \in \mathrm{AC}^n[t_0, t_1], \; u(\cdot) \in \mathrm{U}\}.$$

---

[1] Scalar products and norms are defined, respectively, as

$$\langle x(\cdot), y(\cdot)\rangle = \int_{t_0}^{t_1} \langle x(t), y(t)\rangle dt, \; \|x(\cdot)\|^2 = \int_{t_0}^{t_1} |x(t)|^2 dt,$$

where $\langle x(t), y(t)\rangle = \sum_1^n x_i(t)y_i(t), \; |x(t)|^2 = \sum_1^n x_i^2(t), \quad t_0 \leq t \leq t_1,$

$$x(t) = (x_1(t), \ldots, x_n(t))^{\mathrm{T}}, \; y(t) = (y_1(t), \ldots, y_n(t))^{\mathrm{T}}.$$

Here A_0, A_1 are constant $m \times n$-matrices ($m < n$); a_0, a_1 are given m-vectors; scalar functions $\varphi_0(x_0)$, $\varphi_1(x_1)$ are convex and differentiable with gradients satisfying the Lipschitz condition.

In the convex case, optimization problems for $\varphi_0(x_0)$ and $\varphi_1(x_1)$ are equivalent to variational inequalities $\langle \nabla\varphi_0(x_0^*), x_0^* - x_0 \rangle \le 0$ and $\langle \nabla\varphi_1(x_1^*), x_1^* - x_1 \rangle \le 0$. As a result of linearization, the original problem is reduced to the following equivalent problem:

$$x_0^* \in \text{Argmin}\{\langle \nabla\varphi_0(x_0^*), x_0\rangle \mid A_0 x_0 \le a_0,\ x_0 \in \mathrm{R}^n\}, \tag{1}$$

$$(x^*(t_1), x^*(\cdot), u^*(\cdot)) \in \text{Argmin}\{\langle \nabla\varphi_1(x^*(t_1)), x(t_1)\rangle \mid A_1 x(t_1) \le a_1,$$
$$x(t_1) \in X(t_1) \subset \mathrm{R}^n, \tag{2}$$

$$\frac{d}{dt}x(t) = D(t)x(t) + B(t)u(t), \quad x(t_0) = x_0^*,\ x(t_1) = x^*(t_1), \tag{3}$$

$$x(\cdot) \in \mathrm{AC}^n[t_0, t_1],\ u(\cdot) \in \mathrm{U}\}. \tag{4}$$

It is proved that the solution $(x_0^*, x^*(t_1), x^*(\cdot), u^*(\cdot)) \in \mathrm{R}^n \times \mathrm{R}^n \times \mathrm{L}_2^n[t_0, t_1] \times \mathrm{U}$ of the problem exists [26]. We focus once again that the symbol $x(t_1)$ denotes the right end of the phase trajectory $x(t)$, i.e., the element of reachability set. Classical linear controlled systems for dynamics were studied in [24].

3 Classic Lagrangian for Original Problem

The considered problem is a terminal control problem formulated in the Hilbert space. As we know, in convex programming theory for finite-dimensional spaces, there is always a dual problem in the dual (conjugate) space, corresponding to the primal problem. Through appropriate analogy, we will try to get explicit dual problem for (1)–(4) in the functional spaces. To this end, we scalarize systems and introduce a linear convolution known as the Lagrangian:

$$\begin{aligned}\mathcal{L}(x_0, x(t_1), x(\cdot), u(\cdot); p_0, p_1, \psi(\cdot)) = {} & \langle \nabla\varphi_0(x_0^*), x_0\rangle + \langle \nabla\varphi_1(x^*(t_1)), x(t_1)\rangle \\ & + \langle p_0, A_0 x_0 - a_0\rangle + \langle p_1, A_1 x(t_1) - a_1\rangle \\ & + \int_{t_0}^{t_1} \langle \psi(t), D(t)x(t) + B(t)u(t) - \frac{d}{dt}x(t)\rangle dt\end{aligned} \tag{5}$$

for all $(x_0, x(t_1), x(\cdot), u(\cdot)) \in \mathrm{R}^n \times \mathrm{R}^n \times \mathrm{AC}^n[t_0, t_1] \times \mathrm{U}$, $(p_0, p_1, \psi(\cdot)) \in \mathrm{R}_+^m \times \mathrm{R}_+^m \times \Psi_2^n[t_0, t_1]$.[2] Here $\Psi_2^n[t_0, t_1]$ is a linear variety of absolutely continuous functions from the conjugate space. This set is everywhere dense in $\mathrm{L}_2^n[t_0, t_1]$, i.e., $\overline{\Psi_2^n}[t_0, t_1] = \mathrm{L}_2^n[t_0, t_1]$.

[2]For simplicity, the positive orthant R_+^m hereinafter will also be referred to $p \ge 0$.

Saddle point $(x^*(t_0), x^*(t_1), x^*(\cdot), u^*(\cdot); p_0^*, p_1^*, \psi^*(\cdot))$ of the Lagrange function is formed by primal $(x^*(t_0), x^*(t_1), x^*(\cdot), u^*(\cdot))$ and dual $(p_0^*, p_1^*, \psi^*(\cdot))$ variables, the first of which is a solution of (1)–(4). By definition, the saddle point satisfies the system of inequalities

$$\begin{aligned}
&\langle \nabla\varphi_0(x_0^*), x_0^*\rangle + \langle \nabla\varphi_1(x^*(t_1)), x^*(t_1)\rangle + \langle p_0, A_0x_0^* - a_0\rangle + \langle p_1, A_1x^*(t_1) - a_1\rangle \\
&\quad + \int_{t_0}^{t_1} \langle \psi(t), D(t)x^*(t) + B(t)u^*(t) - \frac{d}{dt}x^*(t)\rangle dt \\
&\quad \le \langle \nabla\varphi_0(x_0^*), x_0^*\rangle + \langle \nabla\varphi_1(x^*(t_1)), x^*(t_1)\rangle + \langle p_0^*, A_0x_0^* - a_0\rangle \\
&\quad + \langle p_1^*, A_1x^*(t_1) - a_1\rangle + \int_{t_0}^{t_1} \langle \psi^*(t), D(t)x^*(t) + B(t)u^*(t) \\
&\quad - \frac{d}{dt}x^*(t)\rangle dt \le \langle \nabla\varphi_0(x_0^*), x_0\rangle + \langle \nabla\varphi_1(x^*(t_1)), x(t_1)\rangle \\
&\quad + \langle p_0^*, A_0x_0 - a_0\rangle + \langle p_1^*, A_1x(t_1) - a_1\rangle + \int_{t_0}^{t_1} \langle \psi^*(t), D(t)x(t) \\
&\quad + B(t)u(t) - \frac{d}{dt}x(t)\rangle dt
\end{aligned} \tag{6}$$

for all $(x_0, x(t_1), x(\cdot), u(\cdot)) \in \mathrm{R}^n \times \mathrm{R}^n \times \mathrm{AC}^n[t_0, t_1] \times \mathrm{U}$, $(p_0, p_1, \psi(\cdot)) \in \mathrm{R}_+^m \times \mathrm{R}_+^m \times \Psi_2^n[t_0, t_1]$.

According to the Kuhn-Tucker theorem, but for functional spaces, we can say that if the original problem (1)–(4) has primal and dual solutions, they form a saddle point for the Lagrangian. It was shown that the converse is also true: a saddle point of the Lagrange function (5) contains primal and dual solutions of the original problem (1)–(4).

In fact, the left-hand inequality of (6) is a problem of maximizing the linear function in variables $(p_0, p_1, \psi(\cdot))$ on the whole space $\mathrm{R}_+^m \times \mathrm{R}_+^m \times \Psi_2^n[t_0, t_1]$:

$$\begin{aligned}
&\langle p_0 - p_0^*, A_0x_0^* - a_0\rangle + \langle p_1 - p_1^*, A_1x^*(t_1) - a_1\rangle \\
&\quad + \int_{t_0}^{t_1} \langle \psi(t) - \psi^*(t), D(t)x^*(t) + B(t)u^*(t) \\
&\quad - \frac{d}{dt}x^*(t)\rangle dt \le 0,
\end{aligned} \tag{7}$$

where $(p_0, p_1, \psi(\cdot)) \in \mathrm{R}_+^m \times \mathrm{R}_+^m \times \Psi_2^n[t_0, t_1]$. From (7), we have

$$\langle p_0 - p_0^*, A_0x_0^* - a_0\rangle \le 0, \quad \langle p_1 - p_1^*, A_1x^*(t_1) - a_1\rangle \le 0, \tag{8}$$

$$\int_{t_0}^{t_1} \langle \psi(t) - \psi^*(t), D(t)x^*(t) + B(t)u^*(t) - \frac{d}{dt}x^*(t)\rangle dt \le 0 \tag{9}$$

for all $(p_0, p_1, \psi(\cdot)) \in \mathrm{R}^m_+ \times \mathrm{R}^m_+ \times \Psi^n_2[t_0, t_1]$. Putting $p_0 = 0$ and $p_0 = 2p_0^*$ in the first inequality of (8); $p_1 = 0$ and $p_1 = 2p_1^*$ in the second inequality of (8), and $\psi(t) \equiv 0$ and $\psi(t) = 2\psi^*(t)$ in (9), we obtain the system

$$\begin{gathered}\langle p_0^*, A_0x_0^* - a_0\rangle = 0, \quad A_0x_0^* - a_0 \le 0,\\ \langle p_1^*, A_1x^*(t_1) - a_1\rangle = 0, \quad A_1x^*(t_1) - a_1 \le 0,\\ D(t)x^*(t) + B(t)u^*(t) - \frac{d}{dt}x^*(t) = 0, \;\; x^*(t_0) = x_0^*. \end{gathered} \tag{10}$$

The right-hand inequality of (6) is a problem of minimizing the Lagrangian in variables $(x_0, x_1, x(\cdot), u(\cdot))$ with fixed values of $p_0 = p_0^*, p_1 = p_1^*, \psi(\cdot) = \psi^*(\cdot)$. Show below that the set of primal variables from $(x_0^*, x^*(t_1), x^*(\cdot), u^*(\cdot); p_0^*, p_1^*, \psi^*(\cdot))$ is the solution of (1)–(4). Indeed, in view of (10), from the right-hand inequality of (6), we have

$$\begin{aligned}&\langle \nabla\varphi_0(x_0^*), x_0^*\rangle + \langle \nabla\varphi_1(x^*(t_1)), x^*(t_1)\rangle\\ &\qquad \le \langle \nabla\varphi_0(x_0^*), x_0\rangle + \langle \nabla\varphi_1(x^*(t_1)), x(t_1)\rangle\\ &\qquad + \langle p_0^*, A_0x_0 - a_0\rangle + \langle p_1^*, A_1x(t_1) - a_1\rangle\\ &\qquad + \int_{t_0}^{t_1} \langle \psi^*(t), D(t)x(t) + B(t)u(t) - \frac{d}{dt}x(t)\rangle dt \end{aligned} \tag{11}$$

for all $(x_0, x(t_1), x(\cdot), u(\cdot)) \in \mathrm{R}^n \times \mathrm{R}^n \times \mathrm{AC}^n[t_0, t_1] \times \mathrm{U}$.

Considering the inequality (11) under additional scalar constraints

$$\begin{gathered}\langle p_0^*, A_0x_0 - a_0\rangle \le 0, \;\; \langle p_1^*, A_1x(t_1) - a_1\rangle \le 0,\\ \int_{t_0}^{t_1} \langle \psi^*(t), D(t)x(t) + B(t)u(t) - \tfrac{d}{dt}x(t)\rangle dt = 0, \end{gathered} \tag{12}$$

we obtain the optimization problem

$$\langle \nabla\varphi_0(x_0^*), x_0^*\rangle + \langle \nabla\varphi_1(x^*(t_1)), x^*(t_1)\rangle \le \langle \nabla\varphi_0(x_0^*), x_0\rangle + \langle \nabla\varphi_1(x^*(t_1)), x(t_1)\rangle.$$

But from (10) we see that the solution $(x^*(\cdot), u^*(\cdot))$ belongs to a narrower set than (12) . Therefore, this point is also a minimum on a subset of solutions of the system (12), i.e.,

$$\begin{aligned}&\langle \nabla\varphi_0(x_0^*), x_0^*\rangle + \langle \nabla\varphi_1(x^*(t_1)), x^*(t_1)\rangle \le \langle \nabla\varphi_0(x_0^*), x_0\rangle + \langle \nabla\varphi_1(x^*(t_1)), x(t_1)\rangle,\\ &A_0x_0 - a_0 \le 0, \;\; A_1x(t_1) - a_1 \le 0, \end{aligned} \tag{13}$$

$$\frac{d}{dt}x(t) = D(t)x(t) + B(t)u(t) \tag{14}$$

for all $(x_0, x(t_1), x(\cdot), u(\cdot)) \in \mathrm{R}^n \times \mathrm{R}^n \times \mathrm{AC}^n[t_0, t_1] \times \mathrm{U}$.

Thus, if the Lagrangian (5) has a saddle point then primal components of this point form the solution of (1)–(4), and therefore of the original problem of convex programming in infinite-dimensional space.

4 Dual Lagrangian for Dual Problem

Show how the Lagrangian in linear dynamic problems provides a dual problem in dual (conjugate) space. Using formulas for the transition to conjugate linear operators

$$\langle p_0, A_0 x_0\rangle = \langle A_0^{\mathrm{T}} p_0, x_0\rangle,\ \langle p_1, A_1 x_1\rangle = \langle A_1^{\mathrm{T}} p_1, x_1\rangle,$$
$$\langle \psi(t), D(t)x(t)\rangle = \langle D^{\mathrm{T}}(t)\psi(t), x(t)\rangle,$$
$$\langle \psi(t), B(t)u(t)\rangle = \langle B^{\mathrm{T}}(t)\psi(t), u(t)\rangle,$$

and the integration-by-parts formula on the interval $[t_0, t_1]$

$$\langle \psi(t_1), x(t_1)\rangle - \langle \psi(t_0), x(t_0)\rangle$$
$$= \int_{t_0}^{t_1} \langle \tfrac{d}{dt}\psi(t), x(t)\rangle dt + \int_{t_0}^{t_1} \langle \psi(t), \tfrac{d}{dt}x(t)\rangle dt,$$

we write out the dual Lagrangian to (5):

$$\begin{aligned}\mathscr{L}^{\mathrm{T}}(p_0, p_1, \psi(\cdot); x_0, x(t_1), x(\cdot), u(\cdot)) &= \langle \nabla\varphi_0(x_0^*) + A_0^{\mathrm{T}} p_0 + \psi_0, x_0\rangle \\ &+ \langle \nabla\varphi_1(x^*(t_1)) + A_1^{\mathrm{T}} p_1 - \psi_1, x(t_1)\rangle \\ &+ \langle -p_0, a_0\rangle + \langle -p_1, a_1\rangle + \int_{t_0}^{t_1} \langle D^{\mathrm{T}}(t)\psi(t) \\ &+ \frac{d}{dt}\psi(t), x(t)\rangle dt + \int_{t_0}^{t_1} \langle B^{\mathrm{T}}(t)\psi(t), u(t)\rangle dt,\end{aligned} \tag{15}$$

for all $(p_0, p_1, \psi(\cdot)) \in \mathrm{R}_+^m \times \mathrm{R}_+^m \times \Psi_2^n[t_0, t_1]$, $(x_0, x(t_1), x(\cdot), u(\cdot)) \in \mathrm{R}^n \times \mathrm{R}^n \times \mathrm{AC}^n[t_0, t_1] \times \mathrm{U}$, where $\psi_0 = \psi(t_0)$, $\psi_1 = \psi(t_1)$.

Primal and dual Lagrangians (5) and (15) have the same saddle points $(x_0^*, x^*(t_1), x^*(\cdot), u^*(\cdot); p_0^*, p_1^*, \psi^*(\cdot))$. These saddle points satisfy the saddle-point system (6), the dual version of which has the form

$$\langle \nabla\varphi_0(x_0^*) + A_0^{\mathrm{T}} p_0 + \psi_0, x_0^*\rangle + \langle \nabla\varphi_1(x^*(t_1)) + A_1^{\mathrm{T}} p_1 - \psi_1, x^*(t_1)\rangle$$
$$+\langle -p_0, a_0\rangle + \langle -p_1, a_1\rangle$$

$$+\int_{t_0}^{t_1}\langle D^{\mathrm{T}}(t)\psi(t)+\frac{d}{dt}\psi(t),x^*(t)\rangle dt+\int_{t_0}^{t_1}\langle B^{\mathrm{T}}(t)\psi(t),u^*(t)\rangle dt$$
$$\leq\langle\nabla\varphi_0(x_0^*)+A_0^{\mathrm{T}}p_0^*+\psi_0^*,x_0^*\rangle+\langle\nabla\varphi_1(x^*(t_1))+A_1^{\mathrm{T}}p_1^*-\psi_1^*,x^*(t_1)\rangle$$
$$+\langle -p_0^*,a_0\rangle+\langle -p_1^*,a_1\rangle$$
$$+\int_{t_0}^{t_1}\langle D^{\mathrm{T}}(t)\psi^*(t)+\frac{d}{dt}\psi^*(t),x^*(t)\rangle dt+\int_{t_0}^{t_1}\langle B^{\mathrm{T}}(t)\psi^*(t),u^*(t)\rangle dt$$
$$\leq\langle\nabla\varphi_0(x_0^*)+A_0^{\mathrm{T}}p_0^*+\psi_0^*,x_0\rangle+\langle\nabla\varphi_1(x^*(t_1))+A_1^{\mathrm{T}}p_1^*-\psi_1^*,x(t_1)\rangle$$
$$+\langle -p_0^*,a_0\rangle+\langle -p_1^*,a_1\rangle$$
$$+\int_{t_0}^{t_1}\langle D^{\mathrm{T}}(t)\psi^*(t)+\frac{d}{dt}\psi^*(t),x(t)\rangle dt+\int_{t_0}^{t_1}\langle B^{\mathrm{T}}(t)\psi^*(t),u(t)\rangle dt \tag{16}$$

for all $(x_0,x(t_1),x(\cdot),u(\cdot))\in\mathrm{R}^n\times\mathrm{R}^n\times\mathrm{AC}^n[t_0,t_1]\times\mathrm{U}$, $(p_0,p_1,\psi(\cdot))\in\mathrm{R}^m_+\times\mathrm{R}^m_+\times\Psi_2^n[t_0,t_1]$.

Repeat now the same transformations that were carried out in the previous paragraph. It was shown there that the saddle-point system leads to the original problem. In a similar way, we will get the dual problem. From the right-hand inequality of (16), we have

$$\langle\nabla\varphi_0(x_0^*)+A_0^{\mathrm{T}}p_0^*+\psi_0^*,x_0^*-x_0\rangle+\langle\nabla\varphi_1(x^*(t_1))+A_1^{\mathrm{T}}p_1^*-\psi_1^*,x^*(t_1)-x(t_1)\rangle$$
$$+\int_{t_0}^{t_1}\langle D^{\mathrm{T}}(t)\psi^*(t)+\frac{d}{dt}\psi^*(t),x^*(t)-x(t)\rangle dt$$
$$+\int_{t_0}^{t_1}\langle B^{\mathrm{T}}(t)\psi^*(t),u^*(t)-u(t)\rangle dt\leq 0$$

for all $(x_0,x(t_1),x(\cdot),u(\cdot))\in\mathrm{R}^n\times\mathrm{R}^n\times\mathrm{AC}^n[t_0,t_1]\times\mathrm{U}$.

Since the variables $(x_0,x(t_1),x(\cdot),u(\cdot))$ independently vary (each within its admissible subspace or set), the last inequality is decomposed into four independent inequalities

$$\langle\nabla\varphi_0(x_0^*)+A_0^{\mathrm{T}}p_0^*+\psi_0^*,x_0^*-x_0\rangle\leq 0,\quad x_0\in\mathrm{R}^n,$$
$$\langle\nabla\varphi_1(x^*(t_1))+A_1^{\mathrm{T}}p_1^*-\psi_1^*,x^*(t_1)-x(t_1)\rangle\leq 0,\quad x_1\in\mathrm{R}^n,$$
$$\int_{t_0}^{t_1}\langle D^{\mathrm{T}}(t)\psi^*(t)+\frac{d}{dt}\psi^*(t),x^*(t)-x(t)\rangle dt\leq 0,\quad x(\cdot)\in\mathrm{AC}^n[t_0,t_1],$$
$$\int_{t_0}^{t_1}\langle B^{\mathrm{T}}(t)\psi^*(t),u^*(t)-u(t)\rangle dt\leq 0,\quad u(\cdot)\in\mathrm{U}.$$

It is well known that a linear functional reaches a finite extremum on the whole subspace only when its gradient vanishes. So, we come to the system of problems

$$\nabla\varphi_0(x_0^*) + A_0^{\mathrm{T}} p_0^* + \psi_0^* = 0, \tag{17}$$

$$D^{\mathrm{T}}(t)\psi^*(t) + \frac{d}{dt}\psi^*(t) = 0,\ \nabla\varphi_1(x^*(t_1)) + A_1^{\mathrm{T}} p_1^* - \psi_1^* = 0, \tag{18}$$

$$\int_{t_0}^{t_1} \langle B^{\mathrm{T}}(t)\psi^*(t), u^*(t) - u(t)\rangle dt \le 0,\ \ \forall u(\cdot) \in \mathrm{U}. \tag{19}$$

Given (17) and (18), we rewrite the left-hand inequality (16) in the form

$$\begin{aligned}\langle \nabla\varphi_0(x_0^*) + A_0^{\mathrm{T}} p_0 + \psi_0, x_0^*\rangle + \langle \nabla\varphi_1(x^*(t_1)) + A_1^{\mathrm{T}} p_1 - \psi_1, x^*(t_1)\rangle \\ +\langle -p_0, a_0\rangle + \langle -p_1, a_1\rangle \\ + \int_{t_0}^{t_1} \langle D^{\mathrm{T}}(t)\psi(t) + \frac{d}{dt}\psi(t), x^*(t)\rangle dt + \int_{t_0}^{t_1} \langle B^{\mathrm{T}}(t)\psi(t), u^*(t)\rangle dt \\ \le \langle -p_0^*, a_0\rangle + \langle -p_1^*, a_1\rangle + \int_{t_0}^{t_1} \langle B^{\mathrm{T}}(t)\psi^*(t), u^*(t)\rangle dt.\end{aligned}$$

Considering this inequality under performing scalar constraints

$$\langle \nabla\varphi_0(x_0^*) + A_0^{\mathrm{T}} p_0 + \psi_0, x_0^*\rangle = 0, \quad \langle \nabla\varphi_1(x^*(t_1)) + A_1^{\mathrm{T}} p_1 - \psi_1, x^*(t_1)\rangle = 0,$$
$$\int_{t_0}^{t_1} \langle D^{\mathrm{T}}(t)\psi(t) + \frac{d}{dt}\psi(t), x^*(t)\rangle dt = 0,$$

we arrive at the problem of maximization for scalar function

$$\begin{aligned}\langle -p_0, a_0\rangle + \langle -p_1, a_1\rangle + \int_{t_0}^{t_1} \langle B^{\mathrm{T}}(t)\psi(t), u^*(t)\rangle dt \\ \le \langle -p_0^*, a_0\rangle + \langle -p_1^*, a_1\rangle + \int_{t_0}^{t_1} \langle B^{\mathrm{T}}(t)\psi^*(t), u^*(t)\rangle dt,\end{aligned}$$

where $(p_0, p_1, \psi(\cdot)) \in \mathrm{R}_+^m \times \mathrm{R}_+^m \times \Psi_2^n[t_0, t_1]$.

Combining with (17)–(19), we get the dual with respect to (1)–(4) problem:

$$\begin{aligned}(p_0^*, p_1^*, \psi^*(\cdot)) \in \mathrm{Argmax}\,\{\langle -p_0, a_0\rangle + \langle -p_1, a_1\rangle \\ + \int_{t_0}^{t_1} \langle B^{\mathrm{T}}(t)\psi(t), u^*(t)\rangle dt \ |\end{aligned} \tag{20}$$

$$\nabla\varphi_0(x_0^*) + A_0^{\mathrm{T}} p_0 + \psi_0 = 0,\ \ (p_0, p_1, \psi(\cdot)) \in \mathrm{R}_+^m \times \mathrm{R}_+^m \times \Psi_2^n[t_0, t_1], \tag{21}$$

$$\left. D^{\mathrm{T}}(t)\psi(t) + \frac{d}{dt}\psi(t) = 0,\ \ \psi_1 = \nabla\varphi_1(x^*(t_1)) + A_1^{\mathrm{T}} p_1 \right\},\tag{22}$$

$$\int_{t_0}^{t_1} \langle B^{\mathrm{T}}(t)\psi^*(t), u^*(t) - u(t)\rangle dt \le 0,\ \ u(\cdot) \in \mathrm{U}. \tag{23}$$

5 Mutually Dual Problems

Write out together a pair of mutually dual problems.

The primal problem:

$$(x_0^*, x^*(t_1), x^*(\cdot), u^*(\cdot)) \in \text{Argmin}\{\langle \nabla\varphi_0(x_0^*), x_0\rangle + \langle \nabla\varphi_1(x^*(t_1)), x(t_1)\rangle \mid$$
$$A_0x_0 \le a_0,\ A_1x(t_1) \le a_1,\ (x_0, x(t_1)) \in \mathrm{R}^n \times \mathrm{R}^n,$$
$$\frac{d}{dt}x(t) = D(t)x(t) + B(t)u(t),\ x(t_0) = x_0^*,\ x(t_1) = x^*(t_1),$$
$$x(\cdot) \in \mathrm{AC}^n[t_0, t_1],\ u(\cdot) \in \mathrm{U}\}.$$

The dual problem:

$$(p_0^*, p_1^*, \psi^*(\cdot)) \in \text{Argmax}\{\langle -p_0, a_0\rangle + \langle -p_1, a_1\rangle$$
$$+ \int_{t_0}^{t_1} \langle B^{\mathrm{T}}(t)\psi(t), u^*(t)\rangle dt \mid$$
$$\nabla\varphi_0(x_0^*) + A_0^{\mathrm{T}}p_0 + \psi_0 = 0,\ \ (p_0, p_1) \in \mathrm{R}_+^m \times \mathrm{R}_+^m,\ \ \psi(\cdot) \in \Psi_2^n[t_0, t_1],$$
$$D^{\mathrm{T}}(t)\psi(t) + \frac{d}{dt}\psi(t) = 0,\ \ \psi_1 = \nabla\varphi_1(x^*(t_1)) + A_1^{\mathrm{T}}p_1\},$$
$$\int_{t_0}^{t_1} \langle B^{\mathrm{T}}(t)\psi^*(t), u^*(t) - u(t)\rangle dt \le 0,\ \ u(\cdot) \in \mathrm{U}.$$

If this system is not dynamic, it takes the form of primal and dual problems known in the finite-dimensional optimization:

$$\begin{aligned}(x_0^*, x_1^*) &\in \text{Argmin}\{\langle \nabla\varphi_0(x_0^*), x_0\rangle + \langle \nabla\varphi_1(x^*(t_1)), x(t_1)\rangle \mid \\ &\qquad A_0x_0 \le a_0,\quad A_1x(t_1) \le a_1,\quad (x_0, x(t_1)) \in \mathrm{R}^n \times \mathrm{R}^n\},\\ (p_0^*, p_1^*) &\in \text{Argmax}\{\langle -p_0, a_0\rangle + \langle -p_1, a_1\rangle \mid \nabla\varphi_0(x_0^*) + A_0^{\mathrm{T}}p_0 = 0,\\ &\qquad \nabla\varphi_1(x^*(t_1)) + A_1^{\mathrm{T}}p_1 = 0,\quad (p_0, p_1) \in \mathrm{R}_+^m \times \mathrm{R}_+^m\}.\end{aligned}$$

Primal and dual problems (separately or together) can serve as a basis for the development of a whole family of methods for computing the saddle points of the primal or dual Lagrange functions [2–8, 27]. It is possible to construct saddle-point methods, which will converge monotonically in norm to the saddle points of Lagrangians. With regard to initial boundary value problems of terminal control it will mean weak convergence in controls, and strong convergence in trajectories, dual trajectories and terminal variables.

In this paper, we consider an iterative process for solving the boundary-value differential system. On the one hand, this process will be obtained from the saddle-point inequalities. On the other hand, it will be close to a differential system, derived from the integral form of the Pontryagin maximum principle.

6 Boundary-Value Dynamic Controlled System

Now we unite together the left-hand inequality of the saddle-point system (6) (for the classical Lagrangian) and the right-hand inequality of the saddle-point system (16) (for the dual Lagrangian). Subsystems (10), (12), (18), (19) and (17) were obtained as consequences of these systems. Writing them here, we arrive at the following boundary-value dynamic controlled system:

$$\frac{d}{dt}x^*(t) = D(t)x^*(t) + B(t)u^*(t), \quad x^*(t_0) = x_0^*, \tag{24}$$

$$\langle p_0 - p_0^*, A_0x_0^* - a_0\rangle \le 0, \quad p_0 \ge 0, \tag{25}$$

$$\langle p_1 - p_1^*, A_1x^*(t_1) - a_1\rangle \le 0, \quad p_1 \ge 0, \tag{26}$$

$$\frac{d}{dt}\psi^*(t) + D^{\mathrm{T}}(t)\psi^*(t) = 0, \quad \psi_1^* = \nabla\varphi_1(x^*(t_1)) + A_1^{\mathrm{T}}p_1^*, \tag{27}$$

$$\int_{t_0}^{t_1} \langle B^{\mathrm{T}}(t)\psi^*(t), u^*(t) - u(t)\rangle dt \le 0, \quad u(\cdot) \in \mathrm{U}, \tag{28}$$

$$\nabla\varphi_0(x_0^*) + A_0^{\mathrm{T}}p_0^* + \psi_0^* = 0. \tag{29}$$

Variational inequalities of the system can be rewritten in equivalent form of operator equations with operators of projection onto the corresponding convex closed sets. Then we obtain the following system of differential and operator equations:

$$\frac{d}{dt}x^*(t) = D(t)x^*(t) + B(t)u^*(t), \quad x^*(t_0) = x_0^*, \tag{30}$$

$$p_0^* = \pi_+(p_0^* + \alpha(A_0x_0^* - a_0)), \tag{31}$$

$$p_1^* = \pi_+(p_1^* + \alpha(A_1x^*(t_1) - a_1)), \tag{32}$$

$$\frac{d}{dt}\psi^*(t) + D^{\mathrm{T}}(t)\psi^*(t) = 0, \quad \psi_1^* = \nabla\varphi_1(x^*(t_1)) + A_1^{\mathrm{T}}p_1^*, \tag{33}$$

$$u^*(t) = \pi_U(u^*(t) - \alpha B^{\mathrm{T}}(t)\psi^*(t)), \tag{34}$$

$$x_0^* = x_0^* - \alpha(\nabla\varphi_0(x_0^*) + A_0^{\mathrm{T}}p_0^* + \psi_0^*), \tag{35}$$

where $\pi_+(\cdot), \pi_U(\cdot)$ are projection operators onto the positive orthant R_+^m and the set of controls U, $\alpha > 0$.

7 Saddle-Point Method for Solving Boundary Value Controlled System

Construct now an iterative process based on the system (30)–(35). Suppose that the values of dual variables p_0^k, p_1^k, the initial value x_0^k of trajectory, and the control $u^k(\cdot) \in \mathrm{U}$ are known on the k-th iteration. Describe below how to find the values of these variables on the next iteration. Formally, the process is as follows (parameter $\alpha > 0$ characterizes the value of iteration step):

$$\frac{d}{dt}x^k(t) = D(t)x^k(t) + B(t)u^k(t), \quad x^k(t_0) = x_0^k, \tag{36}$$

$$p_0^{k+1} = \pi_+(p_0^k + \alpha(A_0 x_0^k - a_0)), \tag{37}$$

$$p_1^{k+1} = \pi_+(p_1^k + \alpha(A_1 x^k(t_1) - a_1)), \tag{38}$$

$$\frac{d}{dt}\psi^k(t) + D^{\mathrm{T}}(t)\psi^k(t) = 0, \quad \psi_1^k = \nabla\varphi_1(x^k(t_1)) + A_1^{\mathrm{T}} p_1^k, \tag{39}$$

$$u^{k+1}(t) = \pi_U(u^k(t) - \alpha B^{\mathrm{T}}(t)\psi^k(t)), \tag{40}$$

$$x_0^{k+1} = x_0^k - \alpha(\nabla\varphi_0(x_0^k) + A_0^{\mathrm{T}} p_0^k + \psi_0^k), \quad k = 0, 1, 2 \ldots \tag{41}$$

Here, using the initial value x_0^k and control $u^k(\cdot)$, we solve the differential equation (36) and find the trajectory $x^k(\cdot)$. Then, using p_0^k and x_0^k, we calculate p_0^{k+1} from (37). Finding $x_1^k = x^k(t)|_{t=t_1}$ and using p_1^k, we can then determine p_1^{k+1} in (38). With p_1^k and x_1^k we calculate the terminal value ψ_1^k, and find the conjugate trajectory $\psi^k(t)$ on the whole interval from the differential system (39). Further, using $u^k(\cdot)$ together with $\psi^k(\cdot)$ we define $u^{k+1}(\cdot)$ from (40). Finally, calculating $\psi_0^k = \psi^k(t_0)$ and taking into account x_0^k, p_0^k, we define x_0^{k+1} from (41).

The process (36)–(41) refers to methods of simple iteration and is the simplest of the known computational processes. For strictly contraction mappings this process converges at a geometric rate. However, in the case of saddle-point object the simple iteration method does not converge to the saddle point (only their analogues in optimization—the gradient projection methods—are converging). Therefore, to solve the saddle-point problem we use the saddle-point extragradient approach developed in [1, 21]. Other gradient-type approaches have been investigated by many authors [13, 14, 16, 25]. As for variational inequalities, we can point to [15, 20].

The proposed method for solving the problem (1)–(2) is a controlled process (36)–(41), each iteration of which breaks down into two half-steps, providing convergence. Formulas of this iterative method are as follows:

1. *the predictive half-step*

$$\frac{d}{dt}x^k(t) = D(t)x^k(t) + B(t)u^k(t), \quad x^k(t_0) = x_0^k, \tag{42}$$

$$\bar{p}_0^k = \pi_+(p_0^k + \alpha(A_0x_0^k - a_0)), \tag{43}$$

$$\bar{p}_1^k = \pi_+(p_1^k + \alpha(A_1x^k(t_1) - a_1)), \tag{44}$$

$$\frac{d}{dt}\psi^k(t) + D^{\mathrm{T}}(t)\psi^k(t) = 0, \quad \psi_1^k = \nabla\varphi_1(x^k(t_1)) + A_1^{\mathrm{T}}p_1^k, \tag{45}$$

$$\bar{u}^k(t) = \pi_U(u^k(t) - \alpha B^{\mathrm{T}}(t)\psi^k(t)), \tag{46}$$

$$\bar{x}_0^k = x_0^k - \alpha(\nabla\varphi_0(x_0^k) + A_0^{\mathrm{T}}p_0^k + \psi_0^k); \tag{47}$$

2. *the basic half-step*

$$\frac{d}{dt}\bar{x}^k(t) = D(t)\bar{x}^k(t) + B(t)\bar{u}^k(t), \quad \bar{x}^k(t_0) = \bar{x}_0^k, \tag{48}$$

$$p_0^{k+1} = \pi_+(p_0^k + \alpha(A_0\bar{x}_0^k - a_0)), \tag{49}$$

$$p_1^{k+1} = \pi_+(p_1^k + \alpha(A_1\bar{x}^k(t_1) - a_1)), \tag{50}$$

$$\frac{d}{dt}\bar{\psi}^k(t) + D^{\mathrm{T}}(t)\bar{\psi}^k(t) = 0, \quad \bar{\psi}_1^k = \nabla\varphi_1(\bar{x}^k(t_1)) + A_1^{\mathrm{T}}\bar{p}_1^k, \tag{51}$$

$$u^{k+1}(t) = \pi_U(u^k(t) - \alpha B^{\mathrm{T}}(t)\bar{\psi}^k(t)), \tag{52}$$

$$x_0^{k+1} = x_0^k - \alpha(\nabla\varphi_0(\bar{x}_0^k) + A_0^{\mathrm{T}}\bar{p}_0^k + \bar{\psi}_0^k), \quad k = 0, 1, 2\ldots \tag{53}$$

At each half-step, two differential equations are solved, and an iterative step in controls, trajectories, conjugate trajectories, initial values and finite-dimensional dual variables is implemented.

From formulas of this process we can see that differential equations (42), (48) and (45), (51) are only used to calculate functions $x^k(t)$, $\bar{x}^k(t)$ and $\psi^k(t)$, $\bar{\psi}^k(t)$, so the process can be rewritten in a more compact form

$$\bar{p}_0^k = \pi_+(p_0^k + \alpha(A_0x_0^k - a_0)), \tag{54}$$

$$p_0^{k+1} = \pi_+(p_0^k + \alpha(A_0\bar{x}_0^k - a_0)), \tag{55}$$

$$\bar{p}_1^k = \pi_+(p_1^k + \alpha(A_1x^k(t_1) - a_1)), \tag{56}$$

$$p_1^{k+1} = \pi_+(p_1^k + \alpha(A_1\bar{x}^k(t_1) - a_1)), \tag{57}$$

$$\bar{u}^k(t) = \pi_U(u^k(t) - \alpha B^{\mathrm{T}}(t)\psi^k(t)), \tag{58}$$

$$u^{k+1}(t) = \pi_U(u^k(t) - \alpha B^{\mathrm{T}}(t)\bar{\psi}^k(t)), \tag{59}$$

$$\bar{x}_0^k = x_0^k - \alpha(\nabla\varphi_0(x_0^k) + A_0^{\mathrm{T}}p_0^k + \psi_0^k), \tag{60}$$

$$x_0^{k+1} = x_0^k - \alpha(\nabla\varphi_0(\bar{x}_0^k) + A_0^{\mathrm{T}}\bar{p}_0^k + \bar{\psi}_0^k), \tag{61}$$

where $x^k(\cdot)$, $\bar{x}^k(\cdot)$, $\psi^k(\cdot)$ and $\bar{\psi}^k(\cdot)$ are calculated in (42), (48), (45) and (51).

For auxiliary estimates required further to prove the convergence of the method, we present operator equations (54)–(61) in the form of variational inequalities

$$\langle \bar{p}_0^k - p_0^k - \alpha(A_0x_0^k - a_0), p_0 - \bar{p}_0^k\rangle \geq 0, \tag{62}$$

$$\langle p_0^{k+1} - p_0^k - \alpha(A_0\bar{x}_0^k - a_0), p_0 - p_0^{k+1}\rangle \geq 0, \tag{63}$$

$$\langle \bar{p}_1^k - p_1^k - \alpha(A_1x^k(t_1) - a_1), p_1 - \bar{p}_1^k\rangle \geq 0, \tag{64}$$

$$\langle p_1^{k+1} - p_1^k - \alpha(A_1\bar{x}^k(t_1) - a_1), p_1 - p_1^{k+1}\rangle \geq 0, \tag{65}$$

$$\int_{t_0}^{t_1} \langle \bar{u}^k(t) - u^k(t) + \alpha B^{\mathrm{T}}(t)\psi^k(t), u(t) - \bar{u}^k(t)\rangle dt \geq 0, \tag{66}$$

$$\int_{t_0}^{t_1} \langle u^{k+1}(t) - u^k(t) + \alpha B^{\mathrm{T}}(t)\bar{\psi}^k(t), u(t) - u^{k+1}(t)\rangle dt \geq 0, \tag{67}$$

$$\langle \bar{x}_0^k - x_0^k + \alpha(\nabla\varphi_0(x_0^k) + A_0^{\mathrm{T}}p_0^k + \psi_0^k), x_0 - \bar{x}_0^k\rangle \geq 0, \tag{68}$$

$$\langle x_0^{k+1} - x_0^k + \alpha(\nabla\varphi_0(\bar{x}_0^k) + A_0^{\mathrm{T}}\bar{p}_0^k + \bar{\psi}_0^k), x_0 - x_0^{k+1}\rangle \geq 0 \tag{69}$$

for all $p_0 \in \mathrm{R}_+^m$, $p_1 \in \mathrm{R}_+^m$, $u(\cdot) \in \mathrm{U}$, $x_0 \in \mathrm{R}^n$. The inequalities (62)–(69) leads to estimates

$$|\bar{p}_0^k - p_0^{k+1}| \leq \alpha\|A_0\||x_0^k - \bar{x}_0^k|, \tag{70}$$

$$|\bar{p}_1^k - p_1^{k+1}| \leq \alpha\|A_1\||x^k(t_1) - \bar{x}^k(t_1)|, \tag{71}$$

$$\begin{aligned} |\bar{x}_0^k - x_0^{k+1}| &\leq \alpha(|\nabla\varphi_0(x_0^k) - \nabla\varphi_0(\bar{x}_0^k)| + |A_0^{\mathrm{T}}(p_0^k - \bar{p}_0^k)| + |\psi_0^k - \bar{\psi}_0^k|) \\ &\leq \alpha(L_0|x_0^k - \bar{x}_0^k| + \|A_0^{\mathrm{T}}\||p_0^k - \bar{p}_0^k| + |\psi_0^k - \bar{\psi}_0^k|), \end{aligned} \tag{72}$$

$$\|\bar{u}^k(\cdot) - u^{k+1}(\cdot)\| \leq \alpha\|B^{\mathrm{T}}(t)(\psi^k(\cdot) - \bar{\psi}^k(\cdot))\| \leq \alpha B_{\max}\|\psi^k(\cdot) - \bar{\psi}^k(\cdot)\|, \tag{73}$$

where $B_{\max} = \max\|B(t)\|$ for all $t \in [t_0, t_1]$; L_0 is a Lipschitz constant for $\nabla\varphi_0(x_0)$.

1. To prove the theorem on convergence of the method we need two more estimates. We mean estimates of deviations $|x^k(t) - \bar{x}^k(t)|$, $|\psi^k(t) - \bar{\psi}^k(t)|$, $t \in [t_0, t_1]$, and accordingly $|x^k(t_1) - \bar{x}^k(t_1)|$, $|\psi_1^k - \bar{\psi}_1^k|$. By the linearity of the Eqs. (42) and (48), we have

$$\frac{d}{dt}\left(x^k(t) - \bar{x}^k(t)\right) = D(t)(x^k(t) - \bar{x}^k(t)) + B(t)(u^k(t) - \bar{u}^k(t)),$$
$$x^k(t_0) - \bar{x}^k(t_0) = x_0^k - \bar{x}_0^k.$$

Integrate the resulting equation from t_0 to t:

$$(x^k(t) - \bar{x}^k(t)) - (x^k(t_0) - \bar{x}^k(t_0)) =$$
$$= \int_{t_0}^{t} D(\tau)(x^k(\tau) - \bar{x}^k(\tau))d\tau + \int_{t_0}^{t} B(\tau)(u^k(\tau) - \bar{u}^k(\tau))d\tau.$$

From last equation, we obtain the estimate

$$|x^k(t) - \bar{x}^k(t)| \leq D_{\max} \int_{t_0}^{t} |x^k(\tau) - \bar{x}^k(\tau)|d\tau$$
$$+ B_{\max} \int_{t_0}^{t_1} |u^k(\tau) - \bar{u}^k(\tau)|d\tau + |x_0^k - \bar{x}_0^k|, \tag{74}$$

where $D_{\max} = \max \|D(t)\|$, $t \in [t_0, t_1]$. Now we can apply the Gronwall lemma [26, Book 1, p. 472] in the form: if $0 \leq \varphi(t) \leq a \int_{t_0}^{t} \varphi(\tau)d\tau + b$ then $\varphi(t) \leq be^{a(t_1 - t_0)}$, $t_0 \leq t \leq t_1$, where $\varphi(t)$ is continuous function, $a, b \geq 0$ are constants. Using this lemma, we obtain from (74)

$$|x^k(t) - \bar{x}^k(t)| \leq e^{D_{\max}(t_1 - t_0)} \left(B_{\max} \int_{t_0}^{t_1} |u^k(t) - \bar{u}^k(t)|dt + |x_0^k - \bar{x}_0^k| \right).$$

Squaring and using the Cauchy-Bunyakovskii inequality, we find for every t

$$|x^k(t) - \bar{x}^k(t)|^2 \leq 2e^{2D_{\max}(t_1 - t_0)}$$
$$\times \left(B_{\max}^2 (t_1 - t_0)\|u^k(\cdot) - \bar{u}^k(\cdot)\|^2 + |x_0^k - \bar{x}_0^k|^2\right). \tag{75}$$

Putting $t = t_1$ in (75), we obtain an estimate of terminal value deviations for trajectories

$$|x^k(t_1) - \bar{x}^k(t_1)|^2 \leq 2e^{2D_{\max}(t_1 - t_0)}$$
$$\times \left(B_{\max}^2 (t_1 - t_0)\|u^k(\cdot) - \bar{u}^k(\cdot)\|^2 + |x_0^k - \bar{x}_0^k|^2\right). \tag{76}$$

We have already mentioned that the differential system (3) produces a linear single-valued mapping which assigns to each control $u(\cdot)$ the single trajectory $x(\cdot)$.

Using the above-mentioned Gronwall lemma, one can show that a linear mapping is bounded on any bounded set, i.e., continuous. Indeed, we write down the difference between two linear equations (42) and (24):

$$\frac{d}{dt}\left(x^k(t) - x^*(t)\right) = D(t)(x^k(t) - x^*(t)) + B(t)(u^k(t) - u^*(t)),$$
$$x^k(t_0) - x^*(t_0) = x_0^k - x_0^*.$$

Passing from this difference to analogue of (74), we have

$$|x^k(t) - x^*(t)| \leq D_{\max} \int_{t_0}^{t} |x^k(\tau) - x^*(\tau)| d\tau$$
$$+ B_{\max} \int_{t_0}^{t_1} |u^k(\tau) - u^*(\tau)| d\tau + |x_0^k - x_0^*|.$$

Concluding these considerations, we obtain an estimate similar to (75):

$$\begin{aligned} &|x^k(t) - x^*(t)|^2 \leq 2e^{2D_{\max}(t_1 - t_0)} \\ &\times \left(B_{\max}^2 (t_1 - t_0)\|u^k(\cdot) - u^*(\cdot)\|^2 + |x_0^k - x_0^*|^2\right). \end{aligned} \tag{77}$$

This means that the considered above linear operator transforms a bounded set of controls and initial conditions in a bounded set of trajectories.

2. Finally, from (45), (51) we get similar estimates for conjugate trajectories $|\psi^k(t) - \bar{\psi}^k(t)|$:

$$\frac{d}{dt}\left(\psi^k(t) - \bar{\psi}^k(t)\right) + D^{\mathrm{T}}(t)(\psi^k(t) - \bar{\psi}^k(t)) = 0, \tag{78}$$

where $\psi_1^k - \bar{\psi}_1^k = \nabla\varphi_1(x^k(t_1)) - \nabla\varphi_1(\bar{x}^k(t_1)) + A_1^{\mathrm{T}}(p_1^k - \bar{p}_1^k)$.

Integrate (78) from t to t_1:

$$\psi^k(t) - \bar{\psi}^k(t) = \int_t^{t_1} D^{\mathrm{T}}(t)(\psi^k(t) - \bar{\psi}^k(t))dt + \psi_1^k - \bar{\psi}_1^k.$$

We have the following estimate

$$\begin{aligned} |\psi^k(t) - \bar{\psi}^k(t)| &\leq \textstyle\int_t^{t_1} |D^{\mathrm{T}}(t)(\psi^k(t) - \bar{\psi}^k(t))|dt + |\psi_1^k - \bar{\psi}_1^k| \\ &\leq D_{\max} \textstyle\int_t^{t_1} |\psi^k(t) - \bar{\psi}^k(t)|dt + b, \end{aligned} \tag{79}$$

where $t \in [t_0, t_1]$, $b = |\psi_1^k - \bar{\psi}_1^k|$. Here we again use the Gronwall lemma [26, Book 1, p. 472]: if $0 \leq \varphi(t) \leq a \int_t^{t_1} \varphi(\tau)d(\tau) + b$ then $\varphi(t) \leq be^{a(t_1 - t)}$, $t_0 \leq t \leq t_1$, where $\varphi(t)$ is a continuous function, $a, b \geq 0$ are constants. Based on this statement, we obtain from (79)

$$|\psi^k(t) - \bar{\psi}^k(t)|^2 \le e^{2D_{\max}(t_1-t)}|\psi_1^k - \bar{\psi}_1^k|^2. \tag{80}$$

Since the estimate (80) is true for all $t \in [t_0, t_1]$, it holds, in particular, at $t = t_0$:

$$|\psi_0^k - \bar{\psi}_0^k|^2 \le e^{2D_{\max}(t_1-t_0)}|\psi_1^k - \bar{\psi}_1^k|^2. \tag{81}$$

For terminal values, we obtain from (45) and (51)

$$\begin{aligned}|\psi_1^k - \bar{\psi}_1^k| &\le |\nabla\varphi_1(x^k(t_1)) - \nabla\varphi_1(\bar{x}^k(t_1))| + \|A_1^{\mathrm{T}}\||p_1^k - \bar{p}_1^k| \\ &\le L_1|x^k(t_1) - \bar{x}^k(t_1)| + \|A_1^{\mathrm{T}}\||p_1^k - \bar{p}_1^k|,\end{aligned}$$

or squaring,

$$|\psi_1^k - \bar{\psi}_1^k|^2 \le (L_1|x^k(t_1) - \bar{x}^k(t_1)| + \|A_1^{\mathrm{T}}\||p_1^k - \bar{p}_1^k|)^2, \tag{82}$$

where L_1 is a Lipschitz constant for $\nabla\varphi_1(x(t_1))$. Comparing (81) and (82), we obtain

$$\begin{aligned}|\psi_0^k - \bar{\psi}_0^k|^2 &\le e^{2D_{\max}(t_1-t_0)}(L_1|x^k(t_1) - \bar{x}^k(t_1)| + \|A_1^{\mathrm{T}}\||p_1^k - \bar{p}_1^k|)^2 \\ &\le 2e^{2D_{\max}(t_1-t_0)}(L_1^2|x^k(t_1) - \bar{x}^k(t_1)|^2 + \|A_1^{\mathrm{T}}\|^2|p_1^k - \bar{p}_1^k|^2).\end{aligned} \tag{83}$$

Substitute (82) in (80)

$$|\psi^k(t) - \bar{\psi}^k(t)|^2 \le e^{2D_{\max}(t_1-t)}(L_1|x^k(t_1) - \bar{x}^k(t_1)| + \|A_1^{\mathrm{T}}\||p_1^k - \bar{p}_1^k|)^2,$$

and integrate this inequality from t_0 to t_1:

$$\begin{aligned}\|\psi^k(\cdot) - \bar{\psi}^k(\cdot)\|^2 \le &\left(e^{2D_{\max}(t_1-t_0)} - 1\right)/(2D_{\max}) \\ &\times \left(L_1|x^k(t_1) - \bar{x}^k(t_1)| + \|A_1^{\mathrm{T}}\||p_1^k - \bar{p}_1^k|\right)^2.\end{aligned} \tag{84}$$

Finally, it remains to prove the boundedness of conjugate trajectories. From (45) and (33), we have

$$\frac{d}{dt}\left(\psi^k(t) - \psi^*(t)\right) + D^{\mathrm{T}}(t)(\psi^k(t) - \psi^*(t)) = 0.$$

Proceeding in a similar way as was done with (78), we obtain an analogue of estimation (84):

$$\begin{aligned}\|\psi^k(\cdot) - \psi^*(\cdot)\|^2 \le &\left(e^{2D_{\max}(t_1-t_0)} - 1\right)/(2D_{\max}) \\ &\times \left(L_1|x^k(t_1) - x^*(t_1)| + \|A_1^{\mathrm{T}}\||p_1^k - p_1^*|\right)^2.\end{aligned} \tag{85}$$

8 Proof of Method Convergence

We show below that process (42)–(53) converges in all the variables. Moreover, it converges monotonically in the norm of the Cartesian product for some variables.

Theorem 1. *Suppose the set of solutions* $(x_0^*, x^*(t_1), x^*(\cdot), u^*(\cdot); p_0^*, p_1^*, \psi^*(\cdot))$ *to the problem* (30)–(35) *is not empty and belongs to* $\mathrm{R}^n \times \mathrm{R}^n \times \mathrm{AC}^n[t_0,t_1] \times \mathrm{U} \times \mathrm{R}^m_+ \times \mathrm{R}^m_+ \times \Psi^n_2[t_0,t_1]$, *the functions* $\varphi_0(x_0)$, $\varphi_1(x_1)$ *are differentiable with gradients satisfying the Lipschitz condition, the step length* α *is chosen from* $0 < \alpha < \min\left(\frac{1}{\gamma_1}, \frac{1}{\gamma_2}, \frac{1}{\gamma_3}, \frac{1}{\gamma_4}, \frac{1}{\gamma_5}\right)$, *where* γ_i *are determined in* (104).[3]

Then the sequence $\{(x_0^k, x^k(t_1), x^k(\cdot), u^k(\cdot); p_0^k, p_1^k, \psi^k(\cdot))\}$ *generated by* (42)–(53) *contains a subsequence which converges to a solution of the problem in controls, phase and conjugate trajectories, as well as to solutions of terminal problems at both ends of the time interval. In particular, the sequence*

$$\{|x_0^k - x_0^*|^2 + \|u^k(\cdot) - u^*(\cdot)\|^2 + |p_0^k - p_0^*|^2 + |p_1^k - p_1^*|^2\}$$

decreases monotonically on the Cartesian product $\mathrm{R}^n \times \mathrm{L}^r_2[t_0,t_1] \times \mathrm{R}^m_+ \times \mathrm{R}^m_+$.

Proof. The main efforts in the proof are focused on obtaining estimates $|u^k(t) - u^*(t)|^2$, $|x_0^k - x_0^*|^2$, $|p_0^k - p_0^*|^2$ and $|p_1^k - p_1^*|^2$. In our method, some of formulas are written in the form of variational inequalities, while others are written in the form of differential equations. So, for uniformity of reasoning we will also write the differential equations in the form of variational inequalities.

1. Rewrite Eq. (51) as the variational inequality

$$\langle \nabla\varphi_1(\bar{x}^k(t_1)) + A_1^{\mathrm{T}}\bar{p}_1^k - \bar{\psi}_1^k, x^*(t_1) - \bar{x}^k(t_1)\rangle$$
$$+ \int_{t_0}^{t_1} \langle D^{\mathrm{T}}(t)\bar{\psi}^k(t) + \frac{d}{dt}\bar{\psi}^k(t), x^*(t) - \bar{x}^k(t)\rangle dt \geq 0.$$

Similarly we proceed with (33):

$$-\langle \nabla\varphi_1(x^*(t_1)) + A_1^{\mathrm{T}}p_1^* - \psi_1^*, x^*(t_1) - \bar{x}^k(t_1)\rangle$$
$$- \int_{t_0}^{t_1} \langle D^{\mathrm{T}}(t)\psi^*(t) + \frac{d}{dt}\psi^*(t), x^*(t) - \bar{x}^k(t)\rangle dt \geq 0.$$

Sum together these inequalities

$$\langle \nabla\varphi_1(\bar{x}^k(t_1)) - \nabla\varphi_1(x^*(t_1)) + A_1^{\mathrm{T}}(\bar{p}_1^k - p_1^*) - (\bar{\psi}_1^k - \psi_1^*), x^*(t_1) - \bar{x}^k(t_1)\rangle$$
$$+ \int_{t_0}^{t_1} \langle D^{\mathrm{T}}(t)(\bar{\psi}^k(t) - \psi^*(t)) + \frac{d}{dt}(\bar{\psi}^k(t) - \psi^*(t)), x^*(t) - \bar{x}^k(t)\rangle dt \geq 0. \tag{86}$$

[3] See below the proof of the theorem.

Using the integration-by-parts formula

$$\int_{t_0}^{t_1} \langle \frac{d}{dt}(\bar{\psi}^k(t) - \psi^*(t)), x^*(t) - \bar{x}^k(t)\rangle dt$$
$$= - \int_{t_0}^{t_1} \langle \bar{\psi}^k(t) - \psi^*(t), \frac{d}{dt}(x^*(t) - \bar{x}^k(t))\rangle dt$$
$$+\langle \bar{\psi}_1^k - \psi_1^*, x^*(t_1) - \bar{x}^k(t_1)\rangle - \langle \bar{\psi}_0^k - \psi_0^*, x_0^* - \bar{x}_0^k\rangle,$$

we transform the differential term in the left-hand part of (86) (this transformation means the transition to the conjugate differential operator):

$$\langle \nabla\varphi_1(\bar{x}^k(t_1)) - \nabla\varphi_1(x^*(t_1)), x^*(t_1) - \bar{x}^k(t_1)\rangle$$
$$+\langle A_1^{\mathrm{T}}(\bar{p}_1^k - p_1^*), x^*(t_1) - \bar{x}^k(t_1)\rangle - \langle \bar{\psi}_1^k - \psi_1^*, x^*(t_1) - \bar{x}^k(t_1)\rangle$$
$$+ \int_{t_0}^{t_1} \langle \bar{\psi}^k(t) - \psi^*(t), D(t)(x^*(t) - \bar{x}^k(t)) - \frac{d}{dt}(x^*(t) - \bar{x}^k(t))\rangle dt$$
$$+\langle \bar{\psi}_1^k - \psi_1^*, x^*(t_1) - \bar{x}^k(t_1)\rangle - \langle \bar{\psi}_0^k - \psi_0^*, x_0^* - \bar{x}_0^k\rangle \geq 0.$$

Reducing similar terms and taking into account that the gradient of the convex function $\varphi_1(x)$ is a monotone operator, i.e., $\langle \nabla\varphi_1(y) - \nabla\varphi_1(x), y - x\rangle \geq 0$, $\forall x, y \in \mathrm{R}^n$, we obtain

$$\langle A_1^{\mathrm{T}}(\bar{p}_1^k - p_1^*), x^*(t_1) - \bar{x}^k(t_1)\rangle + \int_{t_0}^{t_1} \langle \bar{\psi}^k(t) - \psi^*(t), D(t)(x^*(t) - \bar{x}^k(t))$$
$$- \frac{d}{dt}(x^*(t) - \bar{x}^k(t))\rangle dt - \langle \bar{\psi}_0^k - \psi_0^*, x_0^* - \bar{x}_0^k\rangle \geq 0. \tag{87}$$

2. Now, we get the inequality for the variable p_1. To do this, we put $p_1 = p_1^{k+1}$ in (64):

$$\langle \bar{p}_1^k - p_1^k - \alpha(A_1 x^k(t_1) - a_1), p_1^{k+1} - \bar{p}_1^k\rangle \geq 0.$$

Add and subtract $\alpha\langle A_1\bar{x}^k(t_1) - a_1, p_1^{k+1} - \bar{p}_1^k\rangle$:

$$\langle \bar{p}_1^k - p_1^k, p_1^{k+1} - \bar{p}_1^k\rangle + \alpha\langle (A_1\bar{x}^k(t_1) - a_1) - (A_1 x^k(t_1) - a_1), p_1^{k+1} - \bar{p}_1^k\rangle$$
$$-\alpha\langle A_1\bar{x}^k(t_1) - a_1, p^{k+1}(t_1) - \bar{p}_1^k\rangle \geq 0.$$

Using (71), we estimate the second term

$$\langle \bar{p}_1^k - p_1^k, p_1^{k+1} - \bar{p}_1^k\rangle + \alpha^2\|A_1\|^2|\bar{x}^k(t_1) - x^k(t_1)|^2 - \alpha\langle A_1\bar{x}^k(t_1) - a_1, p_1^{k+1} - \bar{p}_1^k\rangle \geq 0.$$

Put $p_1 = p_1^*$ in (65):

$$\langle p_1^{k+1} - p_1^k, p_1^* - p_1^{k+1}\rangle - \alpha\langle A_1\bar{x}^k(t_1) - a_1, p_1^* - p_1^{k+1}\rangle \geq 0.$$

Add up these inequalities

$$\langle \bar{p}_1^k - p_1^k, p_1^{k+1} - \bar{p}_1^k\rangle + \langle p_1^{k+1} - p_1^k, p_1^* - p_1^{k+1}\rangle$$
$$+\alpha^2\|A_1\|^2|\bar{x}^k(t_1) - x^k(t_1)|^2 - \alpha\langle A_1\bar{x}^k(t_1) - a_1, p_1^* - \bar{p}_1^k\rangle \geq 0.$$

Assuming $p_1 = \bar{p}_1^k$ in (26), we have

$$\alpha\langle p_1^* - \bar{p}_1^k, A_1x^*(t_1) - a_1\rangle \geq 0.$$

Summarize the last two inequalities

$$\langle \bar{p}_1^k - p_1^k, p_1^{k+1} - \bar{p}_1^k\rangle + \langle p_1^{k+1} - p_1^k, p_1^* - p_1^{k+1}\rangle$$
$$+\alpha^2\|A_1\|^2|\bar{x}^k(t_1) - x^k(t_1)|^2 - \alpha\langle A_1(\bar{x}^k(t_1) - x^*(t_1)), p_1^* - \bar{p}_1^k\rangle \geq 0. \quad (88)$$

3. We obtain a similar inequality for the variable p_0. To this end, we put $p_0 = p_0^{k+1}$ in (62):

$$\langle \bar{p}_0^k - p_0^k - \alpha(A_0x_0^k - a_0), p_0^{k+1} - \bar{p}_0^k\rangle \geq 0.$$

Add and subtract $\alpha\langle A_0\bar{x}_0^k - a_0, p_0^{k+1} - \bar{p}_0^k\rangle$:

$$\langle \bar{p}_0^k - p_0^k, p_0^{k+1} - \bar{p}_0^k\rangle + \alpha\langle (A_0\bar{x}_0^k - a_0) - (A_0x_0^k - a_0), p_0^{k+1} - \bar{p}_0^k\rangle$$
$$-\alpha\langle A_0\bar{x}_0^k - a_0, p_0^{k+1} - \bar{p}_0^k\rangle \geq 0.$$

Using (70), we estimate the second term

$$\langle \bar{p}_0^k - p_0^k, p_0^{k+1} - \bar{p}_0^k\rangle + \alpha^2\|A_0\|^2|\bar{x}_0^k - x_0^k|^2 - \alpha\langle A_0\bar{x}_0^k - a_0, p_0^{k+1} - \bar{p}_0^k\rangle \geq 0.$$

Put $p_0 = p_0^*$ in (63):

$$\langle p_0^{k+1} - p_0^k, p_0^* - p_0^{k+1}\rangle - \alpha\langle A_0\bar{x}_0^k - a_0, p_0^* - p_0^{k+1}\rangle \geq 0.$$

Add up these inequalities

$$\langle \bar{p}_0^k - p_0^k, p_0^{k+1} - \bar{p}_0^k\rangle + \langle p_0^{k+1} - p_0^k, p_0^* - p_0^{k+1}\rangle$$
$$+\alpha^2\|A_0\|^2|\bar{x}_0^k - x_0^k|^2 - \alpha\langle A_0\bar{x}_0^k - a_0, p_0^* - \bar{p}_0^k\rangle \geq 0.$$

Putting $p_0 = \bar{p}_0^k$ in (25), we get

$$\alpha \langle p_0^* - \bar{p}_0^k, A_0 x_0^* - a_0 \rangle \geq 0.$$

Add up the last two inequalities

$$\begin{aligned} &\langle \bar{p}_0^k - p_0^k, p_0^{k+1} - \bar{p}_0^k \rangle + \langle p_0^{k+1} - p_0^k, p_0^* - p_0^{k+1} \rangle \\ &+\alpha^2 \|A_0\|^2 |\bar{x}_0^k - x_0^k|^2 - \alpha \langle A_0(\bar{x}_0^k - x_0^*), p_0^* - \bar{p}_0^k \rangle \geq 0. \end{aligned} \tag{89}$$

4. We summarize (88), (89) and (87), pre-multiplying the latter by α:

$$\begin{aligned} &\langle \bar{p}_1^k - p_1^k, p_1^{k+1} - \bar{p}_1^k \rangle + \langle p_1^{k+1} - p_1^k, p_1^* - p_1^{k+1} \rangle \\ &+\langle \bar{p}_0^k - p_0^k, p_0^{k+1} - \bar{p}_0^k \rangle + \langle p_0^{k+1} - p_0^k, p_0^* - p_0^{k+1} \rangle \\ &+\alpha^2 \|A_1\|^2 |\bar{x}^k(t_1) - x^k(t_1)|^2 + \alpha^2 \|A_0\|^2 |\bar{x}_0^k - x_0^k|^2 \\ &-\alpha \langle A_0(\bar{x}_0^k - x_0^*), p_0^* - \bar{p}_0^k \rangle - \alpha \langle \bar{\psi}_0^k - \psi_0^*, x_0^* - \bar{x}_0^k \rangle \\ &+\alpha \int_{t_0}^{t_1} \langle \bar{\psi}^k(t) - \psi^*(t), D(t)(x^*(t) - \bar{x}^k(t)) - \frac{d}{dt}(x^*(t) - \bar{x}^k(t)) \rangle dt \geq 0. \end{aligned} \tag{90}$$

5. Consider now inequalities with respect to controls. Put $u(\cdot) = u^{k+1}(\cdot)$ in (66)

$$\int_{t_0}^{t_1} \langle \bar{u}^k(t) - u^k(t) + \alpha B^{\mathrm{T}}(t) \psi^k(t), u^{k+1}(t) - \bar{u}^k(t) \rangle dt \geq 0.$$

Add and subtract the term $\bar{\psi}^k(t)$ under the sign of scalar product

$$\begin{aligned} &\int_{t_0}^{t_1} \langle \bar{u}^k(t) - u^k(t), u^{k+1}(t) - \bar{u}^k(t) \rangle dt \\ &-\alpha \int_{t_0}^{t_1} \langle B^{\mathrm{T}}(t)(\bar{\psi}^k(t) - \psi^k(t)), u^{k+1}(t) - \bar{u}^k(t) \rangle dt \\ &+\alpha \int_{t_0}^{t_1} \langle B^{\mathrm{T}}(t) \bar{\psi}^k(t), u^{k+1}(t) - \bar{u}^k(t) \rangle dt \geq 0. \end{aligned} \tag{91}$$

Put $u = u^*(\cdot)$ in (67)

$$\int_{t_0}^{t_1} \langle u^{k+1}(t) - u^k(t) + \alpha B^{\mathrm{T}}(t) \bar{\psi}^k(t), u^*(t) - u^{k+1}(t) \rangle dt \geq 0. \tag{92}$$

Add up (91) and (92)

$$\int_{t_0}^{t_1} \langle \bar{u}^k(t) - u^k(t), u^{k+1}(t) - \bar{u}^k(t) \rangle dt$$
$$+ \int_{t_0}^{t_1} \langle u^{k+1}(t) - u^k(t), u^*(t) - u^{k+1}(t) \rangle dt$$
$$-\alpha \int_{t_0}^{t_1} \langle B^{\mathrm{T}}(t)(\bar{\psi}^k(t) - \psi^k(t)), u^{k+1}(t) - \bar{u}^k(t) \rangle dt$$
$$+\alpha \int_{t_0}^{t_1} \langle B^{\mathrm{T}}(t)\bar{\psi}^k(t), u^*(t) - \bar{u}^k(t) \rangle dt \geq 0. \tag{93}$$

Substituting $u(t) = \bar{u}^k(t)$ in (28), we have

$$\int_{t_0}^{t_1} \langle B^{\mathrm{T}}(t)\psi^*(t), \bar{u}^k(t) - u^*(t) \rangle dt \geq 0. \tag{94}$$

Summarize (93) and (94)

$$\int_{t_0}^{t_1} \langle \bar{u}^k(t) - u^k(t), u^{k+1}(t) - \bar{u}^k(t) \rangle dt$$
$$+ \int_{t_0}^{t_1} \langle u^{k+1}(t) - u^k(t), u^*(t) - u^{k+1}(t) \rangle dt$$
$$-\alpha \int_{t_0}^{t_1} \langle B^{\mathrm{T}}(t)(\bar{\psi}^k(t) - \psi^k(t)), u^{k+1}(t) - \bar{u}^k(t) \rangle dt$$
$$+\alpha \int_{t_0}^{t_1} \langle \bar{\psi}^k(t) - \psi^*(t), B(t)(u^*(t) - \bar{u}^k(t)) \rangle dt \geq 0. \tag{95}$$

Add up (90) with (95)

$$\langle \bar{p}_1^k - p_1^k, p_1^{k+1} - \bar{p}_1^k \rangle + \langle p_1^{k+1} - p_1^k, p_1^* - p_1^{k+1} \rangle$$
$$+\langle \bar{p}_0^k - p_0^k, p_0^{k+1} - \bar{p}_0^k \rangle + \langle p_0^{k+1} - p_0^k, p_0^* - p_0^{k+1} \rangle$$
$$+\alpha^2 \|A_1\|^2 |\bar{x}^k(t_1) - x^k(t_1)|^2 + \alpha^2 \|A_0\|^2 |\bar{x}_0^k - x_0^k|^2$$
$$-\alpha \langle A_0(\bar{x}_0^k - x_0^*), p_0^* - \bar{p}_0^k \rangle - \alpha \langle \bar{\psi}_0^k - \psi_0^*, x_0^* - \bar{x}_0^k \rangle$$
$$+\alpha \int_{t_0}^{t_1} \langle \bar{\psi}^k(t) - \psi^*(t), D(t)(x^*(t) - \bar{x}^k(t))$$
$$+B(t)(u^*(t) - \bar{u}^k(t)) - \frac{d}{dt}(x^*(t) - \bar{x}^k(t)) \rangle dt$$

$$+\int_{t_0}^{t_1}\langle\bar{u}^k(t)-u^k(t),u^{k+1}(t)-\bar{u}^k(t)\rangle dt$$
$$+\int_{t_0}^{t_1}\langle u^{k+1}(t)-u^k(t),u^*(t)-u^{k+1}(t)\rangle dt$$
$$-\alpha\int_{t_0}^{t_1}\langle B^{\mathrm{T}}(t)(\bar{\psi}^k(t)-\psi^k(t)),u^{k+1}(t)-\bar{u}^k(t)\rangle dt\geq 0. \quad (96)$$

6. Subtract (48) from (30):

$$D(t)(x^*(t)-\bar{x}^k(t))+B(t)(u^*(t)-\bar{u}^k(t))-\frac{d}{dt}(x^*(t)-\bar{x}^k(t))=0.$$

Given the resulting equation, the first of integrals in (96) is zeroed, and as a result, we get

$$\langle\bar{p}_1^k-p_1^k,p_1^{k+1}-\bar{p}_1^k\rangle+\langle p_1^{k+1}-p_1^k,p_1^*-p_1^{k+1}\rangle$$
$$+\langle\bar{p}_0^k-p_0^k,p_0^{k+1}-\bar{p}_0^k\rangle+\langle p_0^{k+1}-p_0^k,p_0^*-p_0^{k+1}\rangle$$
$$+\alpha^2\|A_1\|^2|\bar{x}^k(t_1)-x^k(t_1)|^2+\alpha^2\|A_0\|^2|\bar{x}_0^k-x_0^k|^2$$
$$-\alpha\langle A_0(\bar{x}_0^k-x_0^*),p_0^*-\bar{p}_0^k\rangle-\alpha\langle\bar{\psi}_0^k-\psi_0^*,x_0^*-\bar{x}_0^k\rangle$$
$$+\int_{t_0}^{t_1}\langle\bar{u}^k(t)-u^k(t),u^{k+1}(t)-\bar{u}^k(t)\rangle dt$$
$$+\int_{t_0}^{t_1}\langle u^{k+1}(t)-u^k(t),u^*(t)-u^{k+1}(t)\rangle dt$$
$$-\alpha\int_{t_0}^{t_1}\langle B^{\mathrm{T}}(t)(\bar{\psi}^k(t)-\psi^k(t)),u^{k+1}(t)-\bar{u}^k(t)\rangle dt\geq 0. \quad (97)$$

7. Proceed as done above, but in relation to the variable x_0. To do this, we put $x_0=x_0^{k+1}$ in (68):

$$\langle\bar{x}_0^k-x_0^k+\alpha(\nabla\varphi_0(x_0^k)+A_0^{\mathrm{T}}p_0^k+\psi_0^k),x_0^{k+1}-\bar{x}_0^k\rangle\geq 0.$$

Add and subtract $\alpha(\nabla\varphi_0(\bar{x}_0^k)+A_0^{\mathrm{T}}\bar{p}_0^k+\bar{\psi}_0^k)$ under the sign of scalar product:

$$\langle\bar{x}_0^k-x_0^k+\alpha((\nabla\varphi_0(x_0^k)+A_0^{\mathrm{T}}p_0^k+\psi_0^k)-(\nabla\varphi_0(\bar{x}_0^k)+A_0^{\mathrm{T}}\bar{p}_0^k+\bar{\psi}_0^k),x_0^{k+1}-\bar{x}_0^k\rangle$$
$$+\alpha\langle\nabla\varphi_0(\bar{x}_0^k)+A_0^{\mathrm{T}}\bar{p}_0^k+\bar{\psi}_0^k,x_0^{k+1}-\bar{x}_0^k\rangle\geq 0.$$

Put $x_0=x_0^*$ in (69)

$$\langle x_0^{k+1}-x_0^k+\alpha(\nabla\varphi_0(\bar{x}_0^k)+A_0^{\mathrm{T}}\bar{p}_0^k+\bar{\psi}_0^k),x_0^*-x_0^{k+1}\rangle\geq 0.$$

Add up the inequalities obtained

$$\langle \bar{x}_0^k - x_0^k, x_0^{k+1} - \bar{x}_0^k \rangle + \langle x_0^{k+1} - x_0^k, x_0^* - x_0^{k+1} \rangle$$
$$+\alpha \langle (\nabla \varphi_0(x_0^k) + A_0^{\mathrm{T}} p_0^k + \psi_0^k) - (\nabla \varphi_0(\bar{x}_0^k) + A_0^{\mathrm{T}} \bar{p}_0^k + \bar{\psi}_0^k), x_0^{k+1} - \bar{x}_0^k \rangle$$
$$+\alpha \langle \nabla \varphi_0(\bar{x}_0^k) + A_0^{\mathrm{T}} \bar{p}_0^k + \bar{\psi}_0^k, x_0^* - \bar{x}_0^k \rangle \geq 0. \tag{98}$$

We rewrite the Eq. (35) as a variational inequality, putting in it $x_0 = \bar{x}_0^k$:

$$-\alpha \langle \nabla \varphi_0(x_0^*) + A_0^{\mathrm{T}} p_0^* + \psi_0^*, x_0^* - \bar{x}_0^k \rangle \geq 0.$$

Add up this inequality with (98)

$$\langle \bar{x}_0^k - x_0^k, x_0^{k+1} - \bar{x}_0^k \rangle + \langle x_0^{k+1} - x_0^k, x_0^* - x_0^{k+1} \rangle$$
$$+\alpha \langle (\nabla \varphi_0(x_0^k) + A_0^{\mathrm{T}} p_0^k + \psi_0^k) - (\nabla \varphi_0(\bar{x}_0^k) + A_0^{\mathrm{T}} \bar{p}_0^k + \bar{\psi}_0^k), x_0^{k+1} - \bar{x}_0^k \rangle$$
$$+\alpha \langle \nabla \varphi_0(\bar{x}_0^k) - \nabla \varphi_0(x_0^*), x_0^* - \bar{x}_0^k \rangle + \alpha \langle A_0^{\mathrm{T}} (\bar{p}_0^k - p_0^*), x_0^* - \bar{x}_0^k \rangle$$
$$+\alpha \langle \bar{\psi}_0^k - \psi_0^*, x_0^* - \bar{x}_0^k \rangle \geq 0. \tag{99}$$

We add up (97) with (99)

$$\langle \bar{p}_1^k - p_1^k, p_1^{k+1} - \bar{p}_1^k \rangle + \langle p_1^{k+1} - p_1^k, p_1^* - p_1^{k+1} \rangle + \alpha^2 \|A_1\|^2 |\bar{x}^k(t_1) - x^k(t_1)|^2$$
$$+\langle \bar{p}_0^k - p_0^k, p_0^{k+1} - \bar{p}_0^k \rangle + \langle p_0^{k+1} - p_0^k, p_0^* - p_0^{k+1} \rangle + \alpha^2 \|A_0\|^2 |\bar{x}_0^k - x_0^k|^2$$
$$-\alpha \langle A_0(\bar{x}_0^k - x_0^*), p_0^* - \bar{p}_0^k \rangle - \alpha \langle \bar{\psi}_0^k - \psi_0^*, x_0^* - \bar{x}_0^k \rangle$$
$$+\int_{t_0}^{t_1} \langle \bar{u}^k(t) - u^k(t), u^{k+1}(t) - \bar{u}^k(t) \rangle dt$$
$$+\int_{t_0}^{t_1} \langle u^{k+1}(t) - u^k(t), u^*(t) - u^{k+1}(t) \rangle dt$$
$$-\alpha \int_{t_0}^{t_1} \langle B^{\mathrm{T}}(t)(\bar{\psi}^k(t) - \psi^k(t)), u^{k+1}(t) - \bar{u}^k(t) \rangle dt$$
$$+\langle \bar{x}_0^k - x_0^k, x_0^{k+1} - \bar{x}_0^k \rangle + \langle x_0^{k+1} - x_0^k, x_0^* - x_0^{k+1} \rangle$$
$$+\alpha \langle (\nabla \varphi_0(x_0^k) + A_0^{\mathrm{T}} p_0^k + \psi_0^k) - (\nabla \varphi_0(\bar{x}_0^k) + A_0^{\mathrm{T}} \bar{p}_0^k + \bar{\psi}_0^k), x_0^{k+1} - \bar{x}_0^k \rangle$$
$$+\alpha \langle \nabla \varphi_0(\bar{x}_0^k) - \nabla \varphi_0(x_0^*), x_0^* - \bar{x}_0^k \rangle + \alpha \langle A_0^{\mathrm{T}} (\bar{p}_0^k - p_0^*), x_0^* - \bar{x}_0^k \rangle$$
$$+\alpha \langle \bar{\psi}_0^k - \psi_0^*, x_0^* - \bar{x}_0^k \rangle \geq 0.$$

Given the monotony of gradient $\nabla \varphi_0(x_0)$ in penultimate line (the negative term $\alpha \langle \nabla \varphi_0(\bar{x}_0^k) - \nabla \varphi_0(x_0^*), x_0^* - \bar{x}_0^k \rangle$ is discarded) and collecting similar terms (the terms $\pm \alpha \langle \bar{\psi}_0^k - \psi_0^*, x_0^* - \bar{x}_0^k \rangle$ and $\pm \alpha \langle A_0^{\mathrm{T}} (\bar{p}^k - p^*), x_0^* - \bar{x}_0^k \rangle$ cancel each other), we get:

$$\langle \bar{p}_1^k - p_1^k, p_1^{k+1} - \bar{p}_1^k \rangle + \langle p_1^{k+1} - p_1^k, p_1^* - p_1^{k+1} \rangle + \alpha^2 \|A_1\|^2 |\bar{x}^k(t_1) - x^k(t_1)|^2$$
$$+ \langle \bar{p}_0^k - p_0^k, p_0^{k+1} - \bar{p}_0^k \rangle + \langle p_0^{k+1} - p_0^k, p_0^* - p_0^{k+1} \rangle + \alpha^2 \|A_0\|^2 |\bar{x}_0^k - x_0^k|^2$$
$$+ \int_{t_0}^{t_1} \langle \bar{u}^k(t) - u^k(t), u^{k+1}(t) - \bar{u}^k(t) \rangle dt$$
$$+ \int_{t_0}^{t_1} \langle u^{k+1}(t) - u^k(t), u^*(t) - u^{k+1}(t) \rangle dt$$
$$-\alpha \int_{t_0}^{t_1} \langle B^{\mathrm{T}}(t)(\bar{\psi}^k(t) - \psi^k(t)), u^{k+1}(t) - \bar{u}^k(t) \rangle dt$$
$$+ \langle \bar{x}_0^k - x_0^k, x_0^{k+1} - \bar{x}_0^k \rangle + \langle x_0^{k+1} - x_0^k, x_0^* - x_0^{k+1} \rangle$$
$$+\alpha \langle \nabla\varphi_0(x_0^k) - \nabla\varphi_0(\bar{x}_0^k), x_0^{k+1} - \bar{x}_0^k \rangle + \alpha \langle A_0^{\mathrm{T}}(p_0^k - \bar{p}_0^k), x_0^{k+1} - \bar{x}_0^k \rangle$$
$$+\alpha \langle \psi_0^k - \bar{\psi}_0^k, x_0^{k+1} - \bar{x}_0^k \rangle \geq 0. \tag{100}$$

8. Using the identity $|y_1 - y_2|^2 = |y_1 - y_3|^2 + 2\langle y_1 - y_3, y_3 - y_2 \rangle + |y_3 - y_2|^2$, we transform the scalar product in (100) into the sum (difference) of squares

$$|p_1^{k+1} - p_1^k|^2 - |p_1^{k+1} - \bar{p}_1^k|^2 - |\bar{p}_1^k - p_1^k|^2$$
$$+|p_1^k - p_1^*|^2 - |p_1^{k+1} - p_1^*|^2 - |p_1^{k+1} - p_1^k|^2$$
$$+|p_0^{k+1} - p_0^k|^2 - |p_0^{k+1} - \bar{p}_0^k|^2 - |\bar{p}_0^k - p_0^k|^2$$
$$+|p_0^k - p_0^*|^2 - |p_0^{k+1} - p_0^*|^2 - |p_0^{k+1} - p_0^k|^2$$
$$+2\alpha^2 \|A_0\|^2 |\bar{x}_0^k - x_0^k|^2 + 2\alpha^2 \|A_1\|^2 |\bar{x}^k(t_1) - x^k(t_1)|^2$$
$$+\|u^{k+1}(\cdot) - u^k(\cdot)\|^2 - \|\bar{u}^k(\cdot) - u^{k+1}(\cdot)\|^2 - \|u^k(\cdot) - \bar{u}^k(\cdot)\|^2$$
$$+\|u^k(\cdot) - u^*(\cdot)\|^2 - \|u^{k+1}(\cdot) - u^*(\cdot)\|^2 - \|u^{k+1}(\cdot) - u^k(\cdot)\|^2$$
$$-2\alpha \int_{t_0}^{t_1} \langle B^{\mathrm{T}}(t)(\bar{\psi}^k(t) - \psi^k(t)), u^{k+1}(t) - \bar{u}^k(t) \rangle dt$$
$$+|x_0^{k+1} - x_0^k|^2 - |x_0^{k+1} - \bar{x}_0^k|^2 - |\bar{x}_0^k - x_0^k|^2$$
$$+|x_0^k - x_0^*|^2 - |x_0^{k+1} - x_0^*|^2 - |x_0^{k+1} - x_0^k|^2$$
$$+2\alpha \langle \nabla\varphi_0(x_0^k) - \nabla\varphi_0(\bar{x}_0^k), x_0^{k+1} - \bar{x}_0^k \rangle + 2\alpha \langle A_0^{\mathrm{T}}(p_0^k - \bar{p}_0^k), x_0^{k+1} - \bar{x}_0^k \rangle$$
$$+2\alpha \langle \psi_0^k - \bar{\psi}_0^k, x_0^{k+1} - \bar{x}_0^k \rangle \geq 0. \tag{101}$$

Rewrite the inequality as follows

$$\begin{aligned}
&|p_1^{k+1} - p_1^*|^2 + |p_0^{k+1} - p_0^*|^2 + \|u^{k+1}(\cdot) - u^*(\cdot)\|^2 + |x_0^{k+1} - x_0^*|^2 \\
&+|p_1^{k+1} - \bar{p}_1^k|^2 + |\bar{p}_1^k - p_1^k|^2 + |p_0^{k+1} - \bar{p}_0^k|^2 + |\bar{p}_0^k - p_0^k|^2 \\
&-2\alpha^2\|A_0\|^2|\bar{x}_0^k - x_0^k|^2 - 2\alpha^2\|A_1\|^2|\bar{x}^k(t_1) - x^k(t_1)|^2 \\
&+\|\bar{u}^k(\cdot) - u^{k+1}(\cdot)\|^2 + \|u^k(\cdot) - \bar{u}^k(\cdot)\|^2 + |x_0^{k+1} - \bar{x}_0^k|^2 + |\bar{x}_0^k - x_0^k|^2 \\
&+2\alpha\int_{t_0}^{t_1}\langle B^{\mathrm{T}}(t)(\bar{\psi}^k(t) - \psi^k(t)), u^{k+1}(t) - \bar{u}^k(t)\rangle dt \\
&-2\alpha\langle\nabla\varphi_0(x_0^k) - \nabla\varphi_0(\bar{x}_0^k), x_0^{k+1} - \bar{x}_0^k\rangle - 2\alpha\langle A_0^{\mathrm{T}}(p_0^k - \bar{p}_0^k), x_0^{k+1} - \bar{x}_0^k\rangle \\
&-2\alpha\langle\psi_0^k - \bar{\psi}_0^k, x_0^{k+1} - \bar{x}_0^k\rangle \\
&\le |p_1^k - p_1^*|^2 + |p_0^k - p_0^*|^2 + \|u^k(\cdot) - u^*(\cdot)\|^2 + |x_0^k - x_0^*|^2.
\end{aligned}$$

Using the Cauchy-Bunyakovskii inequality, we estimate the remaining terms in the form of scalar products

$$\begin{aligned}
&|p_1^{k+1} - p_1^*|^2 + |p_0^{k+1} - p_0^*|^2 + \|u^{k+1}(\cdot) - u^*(\cdot)\|^2 + |x_0^{k+1} - x_0^*|^2 \\
&+|p_1^{k+1} - \bar{p}_1^k|^2 + |\bar{p}_1^k - p_1^k|^2 + |p_0^{k+1} - \bar{p}_0^k|^2 + |\bar{p}_0^k - p_0^k|^2 \\
&-2\alpha^2\|A_0\|^2|\bar{x}_0^k - x_0^k|^2 - 2\alpha^2\|A_1\|^2|\bar{x}^k(t_1) - x^k(t_1)|^2 \\
&+\|\bar{u}^k(\cdot) - u^{k+1}(\cdot)\|^2 + \|u^k(\cdot) - \bar{u}^k(\cdot)\|^2 + |x_0^{k+1} - \bar{x}_0^k|^2 + |\bar{x}_0^k - x_0^k|^2 \\
&-2\alpha B_{\max}\|\bar{\psi}^k(\cdot) - \psi^k(\cdot)\|\|u^{k+1}(\cdot) - \bar{u}^k(\cdot)\| \\
&-2\alpha|\nabla\varphi_0(x_0^k) - \nabla\varphi_0(\bar{x}_0^k)||x_0^{k+1} - \bar{x}_0^k| - 2\alpha|\|A_0^{\mathrm{T}}\||p_0^k - \bar{p}_0^k||x_0^{k+1} - \bar{x}_0^k| \\
&-2\alpha|\psi_0^k - \bar{\psi}_0^k||x_0^{k+1} - \bar{x}_0^k| \\
&\le |p_1^k - p_1^*|^2 + |p_0^k - p_0^*|^2 + \|u^k(\cdot) - u^*(\cdot)\|^2 + |x_0^k - x_0^*|^2. \qquad (102)
\end{aligned}$$

We continue to estimate the individual terms in the left-hand side of last inequality.

(a) Involving (73) and (84), we have

$$\begin{aligned}
&2\alpha B_{\max}\|\bar{\psi}^k(\cdot) - \psi^k(\cdot)\|\|u^{k+1}(\cdot) - \bar{u}^k(\cdot)\| \le 2(\alpha B_{\max})^2\|\bar{\psi}^k(\cdot) - \psi^k(\cdot)\|^2 \\
&\le (\alpha B_{\max})^2\left(e^{2D_{\max}(t_1-t_0)} - 1\right)/D_{\max}\cdot(L_1|x^k(t_1) - \bar{x}^k(t_1)| + \|A_1^{\mathrm{T}}\||p_1^k - \bar{p}_1^k|)^2 \\
&\le 2(\alpha B_{\max})^2\left(e^{2D_{\max}(t_1-t_0)} - 1\right)/D_{\max}\cdot(L_1^2|x^k(t_1) - \bar{x}^k(t_1)|^2 + \|A_1^{\mathrm{T}}\|^2|p_1^k - \bar{p}_1^k|^2) \\
&= \alpha^2 d_1|x^k(t_1) - \bar{x}^k(t_1)|^2 + \alpha^2 d_2|p_1^k - \bar{p}_1^k|^2,
\end{aligned}$$

where

$$d_1 = 2L_1^2 B_{\max}^2 \left(e^{2D_{\max}(t_1-t_0)} - 1\right)/D_{\max},$$
$$d_2 = 2\|A_1^{\mathrm{T}}\|^2 B_{\max}^2 \left(e^{2D_{\max}(t_1-t_0)} - 1\right)/D_{\max}.$$

(b) Due to Lipschitz condition and obvious inequality $2|a||b| \le a^2+b^2$, we have

$$2\alpha|\nabla\varphi_0(x_0^k) - \nabla\varphi_0(\bar{x}_0^k)||x_0^{k+1} - \bar{x}_0^k| \le 2\alpha L_0|x_0^k - \bar{x}_0^k||x_0^{k+1} - \bar{x}_0^k|$$
$$\le \alpha L_0|x_0^k - \bar{x}_0^k|^2 + \alpha L_0|x_0^{k+1} - \bar{x}_0^k|^2,$$
$$2\alpha||A_0^{\mathrm{T}}|||p_0^k - \bar{p}_0^k||x_0^{k+1} - \bar{x}_0^k| \le \alpha||A_0^{\mathrm{T}}|||p_0^k - \bar{p}_0^k|^2 + \alpha||A_0^{\mathrm{T}}|||x_0^{k+1} - \bar{x}_0^k|^2,$$
$$2\alpha|\psi_0^k - \bar{\psi}_0^k||x_0^{k+1} - \bar{x}_0^k| \le \alpha|\psi_0^k - \bar{\psi}_0^k|^2 + \alpha|x_0^{k+1} - \bar{x}_0^k|^2.$$

By virtue of (83), we receive

$$|\psi_0^k - \bar{\psi}_0^k|^2 \le 2e^{2D_{\max}(t_1-t_0)}(L_1^2|x^k(t_1) - \bar{x}^k(t_1)|^2 + \|A_1^{\mathrm{T}}\|^2|p_1^k - \bar{p}_1^k|^2)$$
$$= d_3|x^k(t_1) - \bar{x}^k(t_1)|^2 + d_4|p_1^k - \bar{p}_1^k|^2,$$

where $d_3 = 2L_1^2 e^{2D_{\max}(t_1-t_0)}$, $d_4 = 2\|A_1^{\mathrm{T}}\|^2 e^{2D_{\max}(t_1-t_0)}$.

Then the inequality (102) takes the form

$$|p_1^{k+1} - p_1^*|^2 + |p_0^{k+1} - p_0^*|^2 + \|u^{k+1}(\cdot) - u^*(\cdot)\|^2 + |x_0^{k+1} - x_0^*|^2$$
$$+|p_1^{k+1} - \bar{p}_1^k|^2 + |\bar{p}_1^k - p_1^k|^2 + |p_0^{k+1} - \bar{p}_0^k|^2 + |\bar{p}_0^k - p_0^k|^2$$
$$-2\alpha^2\|A_0\|^2|\bar{x}_0^k - x_0^k|^2 - 2\alpha^2\|A_1\|^2|\bar{x}^k(t_1) - x^k(t_1)|^2$$
$$+\|\bar{u}^k(\cdot) - u^{k+1}(\cdot)\|^2 + \|u^k(\cdot) - \bar{u}^k(\cdot)\|^2 + |x_0^{k+1} - \bar{x}_0^k|^2 + |\bar{x}_0^k - x_0^k|^2$$
$$-\alpha^2 d_1|x^k(t_1) - \bar{x}^k(t_1)|^2 - \alpha^2 d_2|p_1^k - \bar{p}_1^k|^2 - \alpha L_0|x_0^k - \bar{x}_0^k|^2 - \alpha L_0|x_0^{k+1} - \bar{x}_0^k|^2$$
$$-\alpha||A_0^{\mathrm{T}}|||p_0^k - \bar{p}_0^k|^2 - \alpha||A_0^{\mathrm{T}}|||x_0^{k+1} - \bar{x}_0^k|^2$$
$$-\alpha(d_3|x^k(t_1) - \bar{x}^k(t_1)|^2 + d_4|p_1^k - \bar{p}_1^k|^2) - \alpha|x_0^{k+1} - \bar{x}_0^k|^2$$
$$\le |p_1^k - p_1^*|^2 + |p_0^k - p_0^*|^2 + \|u^k(\cdot) - u^*(\cdot)\|^2 + |x_0^k - x_0^*|^2.$$

Collecting similar terms, we obtain

$$|p_1^{k+1} - p_1^*|^2 + |p_0^{k+1} - p_0^*|^2 + \|u^{k+1}(\cdot) - u^*(\cdot)\|^2 + |x_0^{k+1} - x_0^*|^2 + |p_1^{k+1} - \bar{p}_1^k|^2$$
$$+(1 - \alpha^2 d_2 - \alpha d_4)|\bar{p}_1^k - p_1^k|^2 + |p_0^{k+1} - \bar{p}_0^k|^2$$
$$+(1 - \alpha||A_0^{\mathrm{T}}||)|\bar{p}_0^k - p_0^k|^2 + (1 - 2\alpha^2\|A_0\|^2 - \alpha L_0)|\bar{x}_0^k - x_0^k|^2$$
$$+\|\bar{u}^k(\cdot) - u^{k+1}(\cdot)\|^2 + \|u^k(\cdot) - \bar{u}^k(\cdot)\|^2 + (1 - \alpha(1 + L_0 + ||A_0^{\mathrm{T}}||))|x_0^{k+1} - \bar{x}_0^k|^2$$

$$-(\alpha^2 d_1 + \alpha d_3 + 2\alpha^2\|A_1\|^2)|x^k(t_1) - \bar{x}^k(t_1)|^2$$
$$\le |p_1^k - p_1^*|^2 + |p_0^k - p_0^*|^2 + \|u^k(\cdot) - u^*(\cdot)\|^2 + |x_0^k - x_0^*|^2. \tag{103}$$

Using the estimate (76)

$$|x^k(t_1) - \bar{x}^k(t_1)|^2 \le 2e^{2D_{\max}(t_1-t_0)}\left(B_{\max}^2(t_1-t_0)\|u^k(\cdot) - \bar{u}^k(\cdot)\|^2 + |x_0^k - \bar{x}_0^k|^2\right)$$
$$= d_5\|u^k(\cdot) - \bar{u}^k(\cdot)\|^2 + d_6|x_0^k - \bar{x}_0^k|^2,$$

where $d_5 = 2B_{\max}^2(t_1-t_0)e^{2D_{\max}(t_1-t_0)}$, $d_6 = 2e^{2D_{\max}(t_1-t_0)}$, we get

$$|p_1^{k+1} - p_1^*|^2 + |p_0^{k+1} - p_0^*|^2 + \|u^{k+1}(\cdot) - u^*(\cdot)\|^2 + |x_0^{k+1} - x_0^*|^2$$
$$+|p_1^{k+1} - \bar{p}_1^k|^2 + (1 - \alpha(\alpha d_2 + d_4))|\bar{p}_1^k - p_1^k|^2$$
$$+|p_0^{k+1} - \bar{p}_0^k|^2 + (1 - \alpha\|A_0^{\mathrm{T}}\|)|\bar{p}_0^k - p_0^k|^2$$
$$+(1 - \alpha(2\alpha\|A_0\|^2 + L_0 + (\alpha d_1 + d_3 + 2\alpha\|A_1\|^2)d_6))|\bar{x}_0^k - x_0^k|^2$$
$$+\|\bar{u}^k(\cdot) - u^{k+1}(\cdot)\|^2 + (1 - \alpha(\alpha d_1 + d_3 + 2\alpha\|A_1\|^2)d_5)\|u^k(\cdot) - \bar{u}^k(\cdot)\|^2$$
$$+(1 - \alpha(1 + L_0 + ||A_0^{\mathrm{T}}\|))|x_0^{k+1} - \bar{x}_0^k|^2$$
$$\le |p_1^k - p_1^*|^2 + |p_0^k - p_0^*|^2 + \|u^k(\cdot) - u^*(\cdot)\|^2 + |x_0^k - x_0^*|^2.$$

Introducing notations

$$\gamma_1 = d_4 + \alpha d_2, \quad \gamma_2 = 1 + L_0 + ||A_0^{\mathrm{T}}\|, \quad \gamma_5 = \|A_0^{\mathrm{T}}\|,$$
$$\gamma_3 = 2\alpha\|A_0\|^2 + L_0 + (\alpha d_1 + d_3 + 2\alpha\|A_1\|^2)d_6,$$
$$\gamma_4 = (\alpha d_1 + d_3 + 2\alpha\|A_1\|^2)d_5, \tag{104}$$

we arrive at the inequality of the form

$$|p_1^{k+1} - p_1^*|^2 + |p_0^{k+1} - p_0^*|^2 + \|u^{k+1}(\cdot) - u^*(\cdot)\|^2 + |x_0^{k+1} - x_0^*|^2$$
$$+|p_1^{k+1} - \bar{p}_1^k|^2 + |p_0^{k+1} - \bar{p}_0^k|^2 + \|\bar{u}^k(\cdot) - u^{k+1}(\cdot)\|^2$$
$$+(1 - \alpha\gamma_1)|\bar{p}_1^k - p_1^k|^2 + (1 - \alpha\gamma_5)|\bar{p}_0^k - p_0^k|^2 + (1 - \alpha\gamma_2)|x_0^{k+1} - \bar{x}_0^k|^2$$
$$+(1 - \alpha\gamma_3)|\bar{x}_0^k - x_0^k|^2 + (1 - \alpha\gamma_4)\|u^k(\cdot) - \bar{u}^k(\cdot)\|^2$$
$$\le |p_1^k - p_1^*|^2 + |p_0^k - p_0^*|^2 + \|u^k(\cdot) - u^*(\cdot)\|^2 + |x_0^k - x_0^*|^2. \tag{105}$$

Choosing the value of α from the condition

$$0 < \alpha < \min\left(\frac{1}{\gamma_1}, \frac{1}{\gamma_2}, \frac{1}{\gamma_3}, \frac{1}{\gamma_4}, \frac{1}{\gamma_5}\right), \tag{106}$$

it is possible to provide strict positiveness of all the terms in (105). Discarding in the left-hand side of this inequality all lines except the first and last, we obtain

$$|p_1^{k+1} - p_1^*|^2 + |p_0^{k+1} - p_0^*|^2 + \|u^{k+1}(\cdot) - u^*(\cdot)\|^2 + |x_0^{k+1} - x_0^*|^2$$
$$\leq |p_1^k - p_1^*|^2 + |p_0^k - p_0^*|^2 + \|u^k(\cdot) - u^*(\cdot)\|^2 + |x_0^k - x_0^*|^2, \tag{107}$$

that means a monotonous decrease of the sequence

$$\{|p_1^k - p_1^*|^2 + |p_0^k - p_0^*|^2 + \|u^k(\cdot) - u^*(\cdot)\|^2 + |x_0^k - x_0^*|^2\}$$

on the Cartesian product $\mathrm{R}_+^m \times \mathrm{R}_+^m \times \mathrm{L}_2^r[t_0, t_1] \times \mathrm{R}^n$.

9. We sum up the inequality (105) from $k = 0$ to $k = N$:

$$|p_1^{N+1} - p_1^*|^2 + |p_0^{N+1} - p_0^*|^2 + \|u^{N+1}(\cdot) - u^*(\cdot)\|^2 + |x_0^{N+1} - x_0^*|^2$$
$$+ \sum_{k=0}^{N} |p_1^{k+1} - \bar{p}_1^k|^2 + \sum_{k=0}^{N} |p_0^{k+1} - \bar{p}_0^k|^2 + \sum_{k=0}^{N} \|\bar{u}^k(\cdot) - u^{k+1}(\cdot)\|^2$$
$$+(1 - \alpha\gamma_1) \sum_{k=0}^{N} |\bar{p}_1^k - p_1^k|^2 + (1 - \alpha\gamma_5) \sum_{k=0}^{N} |\bar{p}_0^k - p_0^k|^2$$
$$+(1 - \alpha\gamma_2) \sum_{k=0}^{N} |x_0^{k+1} - \bar{x}_0^k|^2 + (1 - \alpha\gamma_3) \sum_{k=0}^{N} |\bar{x}_0^k - x_0^k|^2$$
$$+(1 - \alpha\gamma_4) \sum_{k=0}^{N} \|u^k(\cdot) - \bar{u}^k(\cdot)\|^2$$
$$\leq |p_1^0 - p_1^*|^2 + |p_0^0 - p_0^*|^2 + \|u^0(\cdot) - u^*(\cdot)\|^2 + |x_0^0 - x_0^*|^2. \tag{108}$$

Provided (106), this inequality implies that the sequence is bounded for any N

$$|p_1^{N+1} - p_1^*|^2 + |p_0^{N+1} - p_0^*|^2 + \|u^{N+1}(\cdot) - u^*(\cdot)\|^2 + |x_0^{N+1} - x_0^*|^2$$
$$\leq |p_1^0 - p_1^*|^2 + |p_0^0 - p_0^*|^2 + \|u^0(\cdot) - u^*(\cdot)\|^2 + |x_0^0 - x_0^*|^2, \tag{109}$$

and the following series converge

$$\sum_{k=0}^{\infty} |p_1^{k+1} - \bar{p}_1^k|^2 < \infty, \quad \sum_{k=0}^{\infty} |\bar{p}_1^k - p_1^k|^2 < \infty,$$

$$\sum_{k=0}^{\infty} |p_0^{k+1} - \bar{p}_0^k|^2 < \infty, \quad \sum_{k=0}^{\infty} |\bar{p}_0^k - p_0^k|^2 < \infty,$$

$$\sum_{k=0}^{\infty}|x_0^{k+1}-\bar{x}_0^k|^2<\infty,\quad \sum_{k=0}^{\infty}|\bar{x}_0^k-x_0^k|^2<\infty,$$

$$\sum_{k=0}^{\infty}\|\bar{u}^k(\cdot)-u^{k+1}(\cdot)\|^2<\infty,\quad \sum_{k=0}^{\infty}\|u^k(\cdot)-\bar{u}^k(\cdot)\|^2<\infty.$$

The convergence of the series implies the vanishing of the terms for these series

$$|p_1^{k+1}-\bar{p}_1^k|\to 0,\ |\bar{p}_1^k-p_1^k|\to 0,\ |p_0^{k+1}-\bar{p}_0^k|\to 0,\ |\bar{p}_0^k-p_0^k|\to 0,$$
$$|x_0^{k+1}-\bar{x}_0^k|\to 0,\ |\bar{x}_0^k-x_0^k|\to 0,$$
$$\|\bar{u}^k(\cdot)-u^{k+1}(\cdot)\|\to 0,\ \|u^k(\cdot)-\bar{u}^k(\cdot)\|\to 0. \tag{110}$$

Hence, by the triangle inequality, we obtain

$$|p_1^{k+1}-p_1^k|\to 0,\ |p_0^{k+1}-p_0^k|\to 0,\ |x_0^{k+1}-x_0^k|\to 0,\ \|u^{k+1}(\cdot)-u^k(\cdot)\|\to 0.$$

From (75), (76) and (84) it follows that

$$|x^k(t)-\bar{x}^k(t)|\to 0,\ |x^k(t_1)-\bar{x}^k(t_1)|\to 0,\ \|\psi^k(\cdot)-\bar{\psi}^k(\cdot)\|\to 0\ \text{ as } k\to\infty. \tag{111}$$

Moreover, every term in the left-hand side of (109) is bounded

$$|p_1^k-p_1^*|\le \text{const},\ |p_0^k-p_0^*|\le \text{const},$$
$$|x_0^k-x_0^*|\le \text{const},\ \|u^k(\cdot)-u^*(\cdot)\|\le \text{const}.$$

Finally, as it follows from (77), (76) and (85), the sequences below also are bounded

$$\|x^k(\cdot)-x^*(\cdot)\|\le \text{const},\ |x^k(t_1)-x^*(t_1)|\le \text{const},\ \|\psi^k(\cdot)-\psi^*(\cdot)\|\le \text{const}.$$

10. Since the sequence $\{(x_0^k,x^k(t_1),x^k(\cdot),u^k(\cdot);p_0^k,p_1^k,\psi^k(\cdot))\}$ is bounded on the Cartesian product $\mathrm{R}^n\times\mathrm{R}^n\times \mathrm{AC}^n[t_0,t_1]\times \mathrm{U}\times \mathrm{R}_+^m\times \mathrm{R}_+^m\times \Psi_2^n[t_0,t_1]$ then this sequence is weakly compact [19]. The latter means that there exists the subsequence $\{(x_0^{k_i},x^{k_i}(t_1),x^{k_i}(\cdot),u^{k_i}(\cdot);\ p_0^{k_i},p_1^{k_i},\psi^{k_i}(\cdot))\}$ and the point $(x_0',x'(t_1),x'(\cdot),u'(\cdot);p_0',p_1',\psi'(\cdot))$, which is the weak limit of this subsequence. Note that in finite-dimensional (Euclidean) spaces of variables p_0,p_1 and $x_0,x(t_1)$ strong and weak convergences coincide.

Note also that all linear differential operators of system (42)–(53) are weakly continuous [19], and therefore the transition to weak limit is possible. Passing to weak limit as $k_i\to\infty$ in the whole system with the exception of Eqs. (46) and (52), we obtain

$$\frac{d}{dt}x'(t) = D(t)x'(t) + B(t)u'(t), \quad x'(t_0) = x'_0,$$

$$p'_0 = \pi_+(p'_0 + \alpha(A_0 x'_0 - a_0)),$$
$$p'_1 = \pi_+(p'_1 + \alpha(A_1 x'(t_1) - a_1)),$$
$$\frac{d}{dt}\psi'(t) + D^{\mathrm{T}}(t)\psi'(t) = 0, \quad \psi'_1 = \nabla\varphi_1(x'(t_1)) + A_1^{\mathrm{T}} p'_1,$$
$$\nabla\varphi_0(x'_0) + A_0^{\mathrm{T}} p'_0 + \psi'_0 = 0. \tag{112}$$

Thus, it was shown that weak limit of the subsequence is the solution of incomplete system (112). This system does not contain a limit expression for the iterative formula (52) (or (46)). To obtain a complete picture, it is necessary to show that the weak limit point satisfies this limit relation, i.e., the variational inequality

$$\int_{t_0}^{t_1} \langle B^{\mathrm{T}}(t)\psi'(t), u'(t) - u(t)\rangle dt \le 0, \quad u(\cdot) \in \mathrm{U}, \tag{113}$$

or, that is the same, the operator equation

$$u'(t) = \pi_U(u'(t) - \alpha B^{\mathrm{T}}(t)\bar{\psi}'(t)). \tag{114}$$

Let us prove this fact. The right-hand side of (52) (or (46)) as an operator is not weakly continuous, therefore the transition to weak limit is, generally speaking, impossible. We will use a different approach. The first three equations of the system (112) coincide with (10). The solution of system (10) is a solution of (1)–(4).

The saddle point $(x_0^*, x_1^*, x^*(\cdot), u^*(\cdot); p_0^*, p_1^*, \psi^*(\cdot))$ of system (6) for Lagrangian (5) is formed by primal $(x_0^*, x_1^*, x^*(\cdot), u^*(\cdot))$ and dual $(p_0^*, p_1^*, \psi^*(\cdot))$ variables for problem (1)–(4). In Sects. 3 and 4 it was shown that the above saddle point is also a saddle point for dual Lagrangian (15). Consequently, this point satisfies the dual system of saddle-point inequalities (16). In turn, this system generates the dual (conjugate) problem (20)–(23). In this case, the saddle point $(x_0^*, x_1^*, x^*(\cdot), u^*(\cdot); p_0^*, p_1^*, \psi^*(\cdot))$ is solution of dual problem. In particular, the variational inequality (23) is true. In considered case, it takes the form (113).

Thus, combining (112) and (113), it can be argued that weak limit point is the solution of primal and dual system (24)–(29), i.e.,

$$(x'_0, x'_1, x'(\cdot), u'(\cdot), p'_0, p'_1, \psi'(\cdot)) = (x_0^*, x_1^*, x^*(\cdot), u^*(\cdot), p_0^*, p_1^*, \psi^*(\cdot)).$$

The system (24)–(29) is necessary and sufficient condition for solving the problem in form of Lagrange principle (the saddle-point principle). It remains to note that the process converges monotonically in variables $(p_0, p_1, x_0, u(\cdot))$. As it is well-known [26], the convergence in $(x(\cdot), \psi(\cdot))$ is strong. The convergence in finite-dimensional variables $(p_0, p_1, x_0, x(t_1))$ is also strong. The theorem is proved. □

9 Conclusions

In this paper, the dynamic model of terminal control with boundary value problems at the ends of time interval is interpreted as a saddle-point problem. The problem is considered in functional Hilbert space. Solution of the problem (as a saddle point) satisfies the saddle-point system of inequalities with respect to primal and dual variables. In the linear-convex case, this system can be considered as a strengthening of the maximum principle. This enhancement allows us to expand the possibilities of dynamic modeling of real situations both due to the large variety of finite-dimensional boundary value problems, and due to the diversity of the new saddle-point methods.

New quality of the proposed technique is that it allows us to prove the strong (in norm) convergence of computing process in all the variables of the problem, except for controls, where the convergence is weak. Features of the new technology are demonstrated with the example of linear-convex dynamic problem with boundary conditions described by convex programming problems. Such problems as n-person games, variational inequalities, extreme mappings, equilibrium economic models and others can successfully play the role of boundary value problems.

Acknowledgements The work was carried out with financial support from RFBR (project No. 15-01-06045-a), and the Ministry of Education and Science of the Russian Federation in the framework of Increase Competitiveness Program of NUST "MISiS" (agreement Ń02.A03.21.0004).

References

1. Antipin, A.S.: On method to find the saddle point of the modified Lagrangian. Ekonomika i Matem. Metody **13**(3), 560–565 (1977) [in Russian]
2. Antipin, A.S.: Equilibrium programming: proximal methods. Comp. Maths. Math. Phys. **37**(11), 1285–1296 (1997)
3. Antipin, A.S.: Saddle-point problem and optimization problem as one system. Trudy Instituta Matematiki i Mekhaniki UrO RAN **14**(2), 5–15 (2008) [in Russian]
4. Antipin, A.S.: Equilibrium programming: models and methods for solving. Izvestiya IGU. Matematika. **2**(1), 8–36 (2009) [in Russian], http://isu.ru/izvestia
5. Antipin, A.S.: Two-person game wish Nash equilibrium in optimal control problems. Optim. Lett. **6**(7), 1349–1378 (2012). doi:10.1007/s11590-011-0440-x
6. Antipin, A.S.: Terminal control of boundary models. Comp. Maths. Math. Phys. **54**(2), 275–302 (2014)
7. Antipin, A.S., Khoroshilova, E.V.: Linear programming and dynamics. Trudy Instituta Matematiki i Mekhaniki UrO RAN **19**(2), 7–25 (2013) [in Russian]
8. Antipin, A.S., Khoroshilova, E.V.: On boundary-value problem for terminal control with quadratic criterion of quality. Izvestiya IGU. Matematika **8**, 7–28 (2014) [in Russian]. http://isu.ru/izvestia
9. Antipin, A.S., Khoroshilova, E.V.: Multicriteria boundary value problem in dynamics. Trudy Instituta Matematiki i Mekhaniki UrO RAN **21**(3), 20–29 (2015) [in Russian]
10. Antipin, A.S., Khoroshilova, E.V.: Optimal control with connected initial and terminal conditions. Proc. Steklov Inst. Math. **289**(1), Suppl. 9–25 (2015)

11. Antipin, A.S., Vasilieva, O.O.: Dynamic method of multipliers in terminal control. Comput. Math. Math. Phys. **55**(5), 766–787 (2015)
12. Antipin, A.S., Vasiliev, F.P., Artemieva, L.A.: Regularization method for searching for an equilibrium point in two-person saddle-point games with approximate input data. Dokl. Math. **1**, 49–53 (2014). Pleiades Publishing
13. Chinchuluun, A., Pardalos, P.M., Migdalas, A., Pitsoulis, L. (eds.): Pareto Optimality, Game Theory and Equilibria. Springer Optimization and Its Applications, vol. 17. Springer, New York (2008)
14. Chinchuluun, A., Pardalos, P.M., Enkhbat, R., Tseveendorj, I. (eds.): Optimization and Optimal Control, Theory and Applications, vol. 39. Springer, New York (2010)
15. Facchinei, F., Pang, J.-S.: Finite-Dimensional Variational Inequalities and Complementarity Problems, vol. 1. Springer, Berlin (2003)
16. Hager, W.W., Pardalos, P.M. Optimal Control: Theory, Algorithms and Applications. Kluwer Academic, Boston (1998)
17. Khoroshilova, E.V.: Extragradient method of optimal control with terminal constraints. Autom. Remote. Control **73**(3), 517–531 (2012). doi:10.1134/S0005117912030101
18. Khoroshilova, E.V.: Extragradient-type method for optimal control problem with linear constraints and convex objective function. Optim. Lett. **7**(6), 1193–1214 (2013). doi:10.1007/s11590-012-0496-2
19. Kolmogorov, A.N., Fomin, S.V.: Elements of the Theory of Functions and Functional Analysis, 7th edn. FIZMATLIT, Moscow (2009)
20. Konnov, I.V.: Nonlinear Optimization and Variational Inequalities. Kazan University, Kazan (2013) [in Russian]
21. Korpelevich, G.M.: Extragradient method for finding saddle points and other applications. Ekonomika i Matem. Metody **12**(6), 747–756 (1976) [in Russian]
22. Krishchenko, A.P., Chetverikov, V.N.: The covering method for the solution of terminal control problems. Dokl. Math. **92**(2), 646–650 (2015)
23. Mcintyre, J.E. Neighboring optimal terminal control with discontinuous forcing functions. AIAA J. **4**(1), 141–148 (1966). doi:10.2514/3.3397
24. Polyak, B.T., Khlebnikov, M.B., Shcherbakov, P.S.: Control by Liner Systems Under External Perturbations. LENAND, Moscow (2014)
25. Stewart, D.E.: Dynamics with inequalities. In: Impacts and Hard Constraints. SIAM, Philadelphia (2013)
26. Vasiliev, F.P.: Optimization Methods: In 2 books. MCCME, Moscow (2011) [in Russian]
27. Vasiliev, F.P., Khoroshilova, E.V., Antipin, A.S.: Regularized extragradient method of searching a saddle point in optimal control problem. Proc. Steklov Inst. Math. **275**(Suppl. 1), 186–196 (2011) DOI: 10.1134/S0081543811090148

Optimization of Portfolio Compositions for Small and Medium Price-Taking Traders

Alexander S. Belenky and Lyudmila G. Egorova

Dedicated to the 80th birthday of Professor Boris Teodorovich Polyak.

Abstract The paper proposes two new approaches to designing efficient mathematical tools for quantitatively analyzing decision-making processes that small and medium price-taking traders undergo in forming and managing their portfolios of financial instruments traded in a stock exchange. Two mathematical models underlying these approaches are considered. If the trader can treat price changes for each financial instrument of her interest as those of a random variable with a known (for instance, a uniform) probability distribution, one of these models allows the trader to formulate the problem of finding an optimal composition of her portfolio as an integer programming problem. The other model is suggested to use when the trader does not possess any particular information on the probability distribution of the above-mentioned random variable for financial instruments of her interest while being capable of estimating the areas to which the prices of groups of financial instruments (being components of finite-dimensional vectors for each group) are likely to belong. When each such area is a convex polyhedron described by a finite set of compatible linear equations and inequalities of a balance kind, the use of this model allows one to view the trader's decision on her portfolio composition as that of a player in an antagonistic game on sets of disjoint player strategies. The payoff function of this game is a sum of a linear and a bilinear function of two vector arguments, and the trader's guaranteed financial result in playing against the stock exchange equals the exact value of the maximin of this function. This value, along with the vectors at which it is attained, can be found by solving a mixed

A.S. Belenky (✉)
National Research University Higher School of Economics, Moscow, Russia
Institute for Data, Systems, and Society, Massachusetts Institute of Technology, Cambridge, MA, USA
e-mail: abelenky@hse.ru

L.G. Egorova
National Research University Higher School of Economics, Moscow, Russia
e-mail: legorova@hse.ru

B. Goldengorin (ed.), *Optimization and Its Applications in Control and Data Sciences*, Springer Optimization and Its Applications 115,
DOI 10.1007/978-3-319-42056-1_3

programming problem. Finding an upper bound for this maximin value (and the vectors at which this upper bound is attained) is reducible to finding saddle points in an auxiliary antagonistic game with the same payoff function on convex polyhedra of disjoint player strategies. These saddle points can be calculated by solving linear programming problems forming a dual pair.

Keywords Convex polyhedron • Equilibrium points • Financial instruments • Integer programming • Linear programming • Mixed programming • Price-taking traders • Random variable probability distribution • Two-person games on sets of disjoint player strategies

JEL Classification: G11, C6

1 Introduction

Stock exchanges as markets of a special structure can be viewed as economic institutions whose functioning affects both the global economy and economic developments in every country. This fact contributes to a great deal of attention to studying the stock exchange behavior, which has been displayed for years by a wide spectrum of experts, especially by financiers, economists, sociologists, psychologists, politicians, and mathematicians. What becomes known as a result of their studies, what these experts can (and wish to) explain and interpret from findings of their studies to both interested individuals and society as a whole to help them understand how the stock exchanges work, and how good (or bad) these explanations are make a difference. Indeed, economic issues and policies, the financial stability and the financial security of every country, and the financial status of millions of individuals in the world who invest their personal money in sets of financial instruments traded in stock exchanges are affected by the stock exchange behavior. The existing dependency of so many "customers" on the above-mentioned ability (or inability) of the experts to provide trustworthy explanations of this behavior makes advances in developing tools for quantitatively analyzing the work of stock exchanges important for both the financial practice and economic science.

These tools seem indispensible, first of all, for specialists in economics and finance, since they let them (a) receive, process, analyze, and interpret available information on the behavior of both stock exchanges and their participants, (b) form, test, and analyze both scientific and experience-based hypotheses on the stock exchange behavior, along with mathematical models for its description, and (c) study, evaluate, and generalize the experience of successful traders. However, since the quality of the above analyses heavily affects financial decisions of so many individuals whose well-being substantially depends on the quality of decisions on forming and managing their portfolios, it is clear that developing the tools presents interest for a sizable number of these individuals as well, especially if the tools are easy to operate, are widely available, and the results of the tools' work are easy to understand.

Developing such tools for quantitatively studying the financial behavior and strategies of price-taking traders and those of any groups of them presents particular interest, since these strategies and this behavior, in fact, (a) determine the behavior of stock exchanges, (b) reflect both the state of the global economy and that of the economy in every country in which particular stock exchanges function, and (c) let one draw and back up conclusions on the current investors' mood. However, the development of such tools requires substantial efforts from researchers to make the tools helpful in studying particular characteristics attributed to stock exchanges, for instance, regularities of the dynamics of financial instrument values depending on the financial behavior of so-called "bulls" and "bears" [27]. The same is true for studying the reaction of particular stock exchanges in particular countries on forming financial "bubbles," on crushes of particular stocks, and on financial and general economic crises, especially taking into account large volumes of the available data, interdependencies of particular ingredients of this data, and the probabilistic nature of the data.

Three questions on tools for quantitatively studying the financial behavior and strategies of price-taking traders are in order: (1) can the above-mentioned tools be developed in principle, and if yes, what can they help analyze, (2) who and how can benefit from their development, and (3) is there any need for developing such tools while so many different tools for studying stock exchanges have already been developed (and have been recognized at least by the scientific community)?

1. Every stock exchange is a complicated system whose behavior is difficult to predict, since this behavior depends on (a) decisions made by its numerous participants, (b) political and economic situations and tendencies both in the world and in particular countries, (c) breakthroughs in science and technology, and (d) environmental issues associated with the natural anomalies and disasters that may affect agriculture, industry, and people's everyday life. However, there are examples of global systems having a similar degree of complexity whose behavior has been studied and even successfully forecast. Weather, agricultural systems, electoral systems, certain kinds of service systems, including those supplying energy (electricity, water, and gas), and particular markets, where certain goods are traded, can serve as such examples. Indeed, for instance, the dynamics of the land productivity with respect to particular agricultural crops in a geographic region, which substantially depends on both local weather and human activities relating to cultivating the crops and which is difficult to study, has been successfully researched. The dynamics of changing priorities of the electorate in a country, which substantially depends on the political climate both there and in the world, as well as on actions undertaken by candidates on the ballot and their teams to convince the voters to favor these candidates, is successfully monitored and even predicted in the course of, for instance, U.S.

presidential election campaigns, despite the obvious complexity of studying it. The dynamics of energy consumption by a resident in a region, which depends on the financial status of this resident and her family, on the climate and local weather in the region, on the time of the day, on her life style and habits, etc., is successfully forecast in calculating parameters of regional electrical grids though its formalized description presents substantial difficulties.

While there are obvious similarities in the dependency of all the above-mentioned global systems on the nature and human behavior, the conventional wisdom suggests that developing any universal decision-support systems applicable to studying and analyzing these systems from any common, uniform positions is hardly possible. However, the authors believe that certain regularities detected in studying these systems [6] can successfully be applied in studying and analyzing stock exchanges and financial strategies of their participants [7].

At the same time, one should bear in mind that according to [22, 44], with all the tools available to world financial analysts, they correctly predict the behavior of financial instruments approximately in 50 % of the cases. This may suggest that either the tools adequately describing the stock exchange regularities have not so far been developed, or not everything in stock exchanges can be predicted with a desirable accuracy in principle though the existing tools seem helpful for understanding regularities underlying the tendencies of the stock exchange behavior. In any case, it seems that the tools allowing one to analyze the "potential" of a price-taking trader and the impact of her decisions on both the composition of her portfolio of financial instruments traded in a stock exchange and on the behavior of this stock exchange as a whole are needed the most.

2. Economists, financial analysts, and psychologists are direct beneficiaries of developing tools for quantitatively studying the financial behavior and decision-making strategies of price-taking traders, whereas these traders themselves are likely to benefit from developing these tools at least indirectly, by using results of the studies that can be undertaken by the above-mentioned direct beneficiaries. A set of mathematical models describing the process of making investment decisions by price-taking traders and software implementing existing or new techniques for solving problems formulated with the use of these models, along with available statistical data reflecting the stock exchange behavior, should constitute the core of the tools for analyzing the psychology of making decisions by price-taking traders possessing abilities to divine the market price dynamics with certain probabilities. One should expect that the use of the tools by price-taking traders for improving their decisions (no matter whether such an improvement can be attained by any of them) is likely to change the behavior of every stock exchange as a whole, and this is likely to affect both the economy of a particular country and the global economy.
3. The financial theory in general and financial mathematics in particular offer models describing the financial behavior of price-taking traders assuming that these traders are rational and make their decisions in an attempt to maximize their utility functions under a reasonable estimate of both the current and the future market status. In all these models, including the Black-Scholes model

for derivative investment instruments and those developed by Markowitz for the stocks [33], their authors assume that the trader possesses information on the probability distribution of the future prices of the financial instruments (at least for those being of interest to the trader). The use of these models leads to solving quite complicated mathematical problems that can in principle be solved only approximately, and even approximate solutions can be obtained only for a limited number of the variables that are to be taken into consideration. Since the parameters of the above-mentioned probability distributions are also known only approximately, the question on how correctly these models may reflect the real financial behavior of the traders in their everyday work with financial instruments, and to what extent these models are applicable as tools for quantitatively analyzing the work of any stock exchange as a whole seems to remain open.

The present paper discusses a new approach to developing a package of mathematical tools for quantitatively analyzing the financial behavior of small and medium price-taking traders (each possessing the ability to divine future price values of certain financial instruments traded in any particular stock exchange) by means of integer, mixed, and linear programming techniques (the latter being the most powerful techniques for solving optimization problems). It is assumed that each such trader forms her portfolio of only those financial instruments of her interest traded in a stock exchange for which the above ability has been confirmed by the preliminary testing that the trader undergoes using the publicly available data on the dynamics of all the financial instruments traded there. Once the initial trader's portfolio has been formed, at each moment, the trader gets involved in making decisions on which financial instruments from the portfolio (and in which volumes) to sell and to hold, as well as on which financial instruments traded in the stock exchange (and in which volumes) to buy to maximize the portfolio value. The paper concerns such decisions that a price-taking trader might make if she had tools for analyzing the dynamics of financial instruments being of her interest (at the time of making these decisions) in the following two situations:

(a) For each financial instrument, the trader believes that its price values will increase and decrease within a segment as two random variables each uniformly distributed on one of the two parts into which its current price value divides the segment, and
(b) no information on the probability distribution of the above-mentioned random variable is available to or can be obtained by the trader though she can estimate the areas to which the price values of groups of financial instruments from her portfolio, considered as components of vectors in finite-dimensional spaces, are likely to belong.

It is shown that in the first situation, the deployment of one of the two mathematical models, proposed in the paper, allows the trader to reduce the problem of finding an optimal composition of her portfolio to solving an integer programming problem. In the second situation, the use of the other model allows the trader to find her optimal investment strategy as that of a player in a two-person game on sets of disjoint player strategies, analogous to the game with the nature, in which the payoff function is a sum of a linear and a bilinear function of two vector variables. It is proven that in the second situation, finding an optimal investment strategy of a trader is equivalent to solving a mixed programming problem, whereas finding an upper bound for the trader's guaranteed result in the game is reducible to calculating an equilibrium point in an auxiliary antagonistic game, which can be done by solving linear programming problems forming a dual pair.

The structure of the rest of the paper is as follows: Sect. 2 presents a brief survey of existing approaches to studying the financial behavior of small and medium price-taking traders. Section 3 briefly discusses ideas underlying the development of the tools for detecting a set of financial instruments for which a price-taking trader is able to divine their future price values. Section 4 addresses the problem of forming an optimal portfolio of financial securities by a price-taking trader assuming that the trader knows only the area within which the price value of each particular financial security of her interest (considered as the value of a uniformly distributed random variable) changes. Section 5 presents a game model for finding strategies of a price-taking trader with respect to managing her portfolio when the trader cannot consider price values of financial securities of her interest as random variables with known probability distributions. In this case, finding a global maximum of the minimum function describing the guaranteed financial result for a trader, associated with her decision to buy, hold, and sell financial securities is reducible to solving a mixed programming problem, whereas finding an upper bound for this guaranteed result is reducible to finding Nash equilibria in an auxiliary antagonistic game. Finding these equilibria is, in turn, reducible to solving linear programming problems forming a dual pair. Section 6 provides two numerical programming problems forming a dual pair. Section 6 provides two numerical examples illustrative of using the game model, presented in Sect. 5, in calculating optimal investment strategies of a trader in both forming a new and in managing an existing portfolio of financial securities traded in a stock exchange. Section 7 presents concluding remarks on the problems under consideration in the paper.

2 A Brief Review of Publications on Modeling the Financial Behavior of Small and Medium Price-Taking Traders in a Stock Exchange

2.1 *Mathematical Models for Developing an Optimal Portfolio of a Price-Taking Trader*

There are three basic approaches to forming an optimal portfolio of a price-taking trader with the use of mathematical models, proposed by economists H. Markowitz, W. Sharpe, and S. Ross.

Markowitz considered the problem of forming a trader's optimal portfolio under the budget limitations in which two parameters—the growth of the portfolio value (to be maximized) and the risks associated with the instability of the stock exchange (to be minimized)—are optimized. He assumed that (a) mathematically, the first parameter is modeled by the expectation of the portfolio's return, whereas the second one is modeled by the portfolio's return variance, (b) the value of each security considered for the inclusion in the portfolio is a random variable with a normal probability distribution, and (c) the shares of securities in the portfolio are non-negative, real numbers. The underlying idea of the Markowitz approach to finding an optimal portfolio composition consists of determining a Pareto-optimal border (efficient frontier) of attainable values of the above two parameters on a plane with the "expectation-variance" coordinates and finding on this border an optimal trader's portfolio by solving an auxiliary problem of calculating a tangent to the above border. Markowitz proved [33] that (a) components of the solutions to the auxiliary problem are piecewise functions, (b) there are so-called "corner" portfolios, corresponding to discontinuity points of the functions being derivatives of the components of the auxiliary problem solutions, and (c) these "corner" portfolios are sufficient for describing all the portfolios from the Pareto-optimal border, since all the portfolios from the Pareto-optimal border are linear combinations of these "corner" portfolios. Here, the choice of a particular portfolio by a particular trader depends on the preferences that the trader has on both the return and the risk associated with the inclusion of a particular security in the portfolio, and the pair of these two parameters determines a point on the above Pareto-optimal border.

Though the Markowitz model presents obvious theoretical interest, it does not seem to have received attention in the practice of analyzing the stock exchange work, apparently, due to (a) the assumptions under which this model has been developed (particularly, normal probability distributions for the above-mentioned random variables), (b) difficulties in obtaining data for calculating both the expectation of the return and the variance of the return under unstable conditions of the stock market, (c) the need to use approximate methods in calculating the Pareto-optimal border when the number of securities is large, etc. Nor has it received enough attention from the practice of using stock exchange work models though models that employ the Markowitz idea in practical calculations but use other parameters instead of the expectation and the variance of the return have been implemented in particular practical calculations. For instance, mean-semivariance [34], mean-absolute deviation [25], mean-VaR (Value-at-Risk) [21], mean-CVaR (Conditional Value-at-Risk) [45], and chance-variance [28] are among such models, despite the presence of the above-mentioned Markowitz model deficiencies in them.

In 1964 Sharpe developed another model for choosing a composition of the investment portfolio, called Capital Asset Pricing Model (CAPM) [49], which incorporated some ideas of the Markowitz model. That is, all the traders are assumed to be rational and developing their strategies proceeding from the expected revenue and the values of standard deviations of the return rate for securities from the portfolio. Sharpe's model assumes that all the traders (a) have access to the same information on the stock exchange, (b) have the same horizon of planning and estimate the expected rates of return and standard deviations of these rates in the same manner, and (c) proceed from the same risk-free interest rate at which every trader can borrow money (for instance, from the interest rate of the Treasury Bills). He suggested that under the above assumptions, the traders, who estimate the risk-free interest rate, co-variations, dispersions, and the expected rates of return for every security uniformly, would choose the same portfolio, called the market portfolio, in which the share of each security corresponds to its market value. Since each trader has her own utility function, she would allocate her capital among the risk-free securities and the market portfolio in a manner that guarantees her a desirable level of the yield while not exceeding a certain level of risk. Thus, due to the same rate of return and to the same market portfolio for all the traders, CAPM-effective portfolios on the plane with the risk and the rate of return as coordinates will occupy a line, called the market line, and all the portfolios below this line will not be effective. Sharpe proposed a mathematical expression for the above line whose coefficients are determined by the risk-free interest rate and by the expectation and the dispersion of the rate of the market portfolio return. For each security from the trader's portfolio, the expected premium for the risk of investing in this security can be described by a function of its rate of return and of the risk-free rate so that the expected premium for the risk associated with investing in a particular security is proportional to the expectation of the premium for investing in securities with the above risk-free interest rate. Moreover, Sharpe proposed to measure "the sensitivity" of a particular security to changes in the market of securities by a coefficient "beta" reflecting the level of coincidence of the direction in which the security changes and the direction in which the market as a whole does. When "beta" equals 1, the value of the security increases or decreases with the same rate as does the whole market so that the inclusion of securities with the coefficient values exceeding 1 increases the risk of investment in securities from the portfolio while increasing the return of the portfolio.

Though the Sharpe model for all the traders (a) assumes their rationality, the uniformity of information on the stock exchange that they possess, and the uniformity of their evaluation of the return and the risk associated with each security from the portfolio, (b) does not take into consideration a number of factors affecting the return of securities from the portfolio besides the market portfolio, and (c) shows a substantial deviation between the actual and the calculated data and the instability of the above coefficient "beta," which can be viewed as a "sensitivity coefficient," this model is widely used in practical calculations. In these calculations, the "sensitivity coefficient" is determined based upon the available statistical data on the monthly yield of securities and on the values of some stock exchange indices.

The Arbitrage Pricing Theory, APT, proposed by Ross in 1976 [47], uses a multifactor mathematical model for describing the dependence of the return of a security on a number of factors in forecasting values of the expected return. For instance, the trader may consider the values of the stock indices, the level of interest rates, the level of inflation, the GDP growth rates, etc. as the factors in the APT model with coefficients reflecting the sensitivity of the return of a particular security to the changes of a particular factor in the model. Ross considered a trader's problem of composing an investment portfolio that is insensitive to any factor changes, does not need any additional trader's resources, and has a positive return. Ross proved that such a portfolio, called an arbitrage portfolio (where the arbitrage is understood in the sense of receiving a risk-free profit as a result of dealing with the same securities, but with different prices) allows one to get a positive profit with zero initial capital and in the absence of risk. He proved that the arbitrage is impossible if the dependence among a substantial number of securities considered to be included in the portfolio is "almost linear" [47]. He also proposed to calculate the expected return of a particular security as a linear combination of summands each of which depends on the expected return of the portfolio having the unit sensitivity to a particular factor and zero sensitivity to the other factors.

The recommendations of the Ross model present mostly theoretical interest, and they may be considered realistic only for a large number of different securities in the trader's portfolio, i.e., for extremely large markets. In any case, the use of these recommendations implies the determination of the composition of factors affecting the return of a security.

2.2 *Modeling the Dynamics of Security Price Values*

Traders who do not use mathematical models for optimizing their investment portfolios usually make decisions on forming and managing these portfolios based upon forecasts of the future price values of the financial instruments that interest them, which can be done with the use of several statistical methods.

Regression Analysis Linear regression models, which are used for estimating parameters in the above-mentioned CAPM and APT models, are widely used for describing the dependences of the security yield on (a) its dividend yield ratio (the ratio of the annual dividend per security share to the market value of the share price of this security at the end of the fiscal year) [24], (b) P/E coefficient (price-earning ratio, equaling the ratio of the current market capitalization of the company to its carrying value) [31], (c) the banking interest rate and the spread value (the difference between the best price values of market orders for buying and selling a particular security at one and the same moment) [19], etc. Though the regression models are widely used, their use requires processing a substantial volume of data to adequately estimate parameters of the model. The absence of the adequacy of the model (for models presented in the form of any numerical dependency between explanatory variables and those to be explained), a relatively low accuracy of the forecasts, and the impossibility to analyze the interrelation among the data available for processing in the form of finite time series are shortcomings/limitations of the regression models that are in use for forecasting future price values of securities traded in stock exchanges. To provide a correct interpretation of the forecast with the use of these models, one should determine the full set of explanatory variables to avoid (a) the bias of the coefficient estimates, (b) the incorrectness of t-statistics and that of other quality indices in the absence of the significant variables, and (c) the risk of multicollinearity (i.e., a linear dependency among the factors). Finally, as shown in [38, 39], for almost all the data describing U.S. macroeconomic indices, all the regressions obtained with the use of the least square methods have turned out to be spurious, i.e., while the dependence between the above-mentioned variables was present (since the regression coefficients were significant, and all the assumptions under which the use of linear regressions is justified were present), there were no dependences between the explanatory variables and those to be explained.

Time Series Analysis Time series of various special structures are widely used for describing macroeconomic indices. However, due to the non-stationarity of, particularly, financial time series, and the presence of both the autoregression and the moving average, time series models cannot describe these indices adequately. Nevertheless, one can transform the initial time series into those whose values are the differrences of a certain degree for the initial time series, and these time series of the differences can be used in forecasting security price values [8, 20, 42]. Autoregression heteroskedasticity models (ARCH), which describe changes of the dispersion of the time series, have been proposed to take into consideration the feature of stock exchange time series associated with the sequencing of the periods of high and small volatility of the parameters described by these time series [15]. Other econometric models for studying time series that describe financial indices, along with tests for verifying and estimating coefficients in these models, the quality of the forecast, and the assumptions underlying the models, are discussed in [56]. The practical use of the time series techniques requires (a) processing a substantial volume of data, (b) an adequate choice of the structure of a particular time series for forecasting a particular parameter, and (c) establishing the applicability of the models under the feature of non-stationarity of the parameters to be forecast, which usually presents considerable difficulties.

Stochastic Differential Equations Stochastic differential equations are widely used in the description of the dynamics of both the price of and the yield for a security, and the most general description is usually offered in the form

$$dX_t = \mu(X_t, t)dt + \sigma(X_t, t)dW_t,$$

where W_t is a Wiener process, being a continuous-time analog of a random walk. Here, the random process X_t, describing the dynamics of the share price or that of the yield, can be modeled under different particular variants of the functions μ and σ. The most known models are those used for describing the dynamics of the interest rate r_t (in which the equality $X_t = r_t$ holds) in determining the value of the share price of a bond, and they include

- the Merton model, describing the capital assets in which $dr_t = \alpha dt + \gamma dW_t$ [35],

- the Vasicek model $dr_t = \alpha(\beta - r_t)dt + \gamma dW_t$ [57], reflecting the tendency of the interest rate to return to its average level β with the velocity described by the parameter α,
- Cox-Ingersoll-Ross model $dr_t = (\alpha - \beta r_t)dt + \gamma(r_t)^{1/2}dW_t$ [11], which takes into consideration the dependence of the volatility ($\sigma(r_t, t) = \gamma(r_t)^{1/2}$) on the interest rate r_t,
- the Chen model [10] of the interest rate dynamics $dr_t = (\alpha_t - r_t)dt + (\gamma_t r_t)^{1/2}dW_t$ in which the coefficients are also stochastic processes described by analogous stochastic differential equations $d\alpha_t = (\alpha - \alpha_t)dt + (\alpha_t)^{1/2}dW_t^1$, $d\gamma_t = (\gamma - \gamma_t)dt + (\gamma_t)^{1/2}dW_t^2$.

The Levy processes deliver another type of stochastic differential equations [18, 23, 59] that are used for describing the dynamics of the share price (or that of the yield) of a security. These processes can be represented by a sum of three summands, where the first and the second ones describe the trend of the dynamics and fluctuations around this trend, whereas the third one models the jumps of the value of the security share price as a reaction on any events being external to the stock exchange

$$dX_t = \mu dt + \sigma dW_t + dJ_t,$$

where J_t is a compound Poisson process with a known arrival rate of the external events. The complexity of (a) obtaining both analytical and approximate solutions to the equations adequately describing real stock exchange processes, (b) understanding mathematical models underlying the description of the processes under study and the solution techniques of the problems formulated on the basis of these models, and (c) implementing the solutions, are among major disadvantages of modeling the dynamics of the share price of a security with the use of stochastic differential equations.

Other Methods Non-parametric methods [56], particularly, neural networks [36, 40], data mining [16], and machine learning [54] should be mentioned among other methods used for describing the dynamics of the share price of a security. However, besides disadvantages attributed to the parametric methods, mentioned earlier, there are some additional ones associated with the specifics of non-parametric methods. For instance, the choice of the topology (the number of layers and neurons and their connections) of the neural networks presents considerable difficulties, and the training of a neural network requires processing large volumes of data while even the correspondence of the model to the historical data does not necessarily guarantee a high level of quality of the forecasts on the basis of the neural network models.

2.3 *Some Empirical Facts on the Financial Behavior of Price-Taking Traders and Financial Analysts*

Some Statistical Facts Illustrative of the Ability (or of the Inability) of Price-taking Traders To Make Successful Investment Decisions According to the information on transactions of ten thousand investors in a seven year period of time, the average, securities sold by the traders had a higher interest rate than those bought by them later on [41]. Also, an analysis of transactions related to more than sixty six thousand broker accounts showed that the average yield of a trader was equal 16,4 %, and the average yield of active traders reached only 11,4 % while the yield of the market portfolio was 17,9 % [1]. Based on this data, the authors of the above-mentioned publications concluded that the traders adopted wrong financial decisions in quite a number of cases. The authors of [13] believe that adopting wrong financial decisions by the traders is associated with traders' emotions, and several tests have been conducted to analyze how the physiology status of a trader (measured by, for instance, the frequency of her pulse and the electrical conductivity of her skin) affected these decisions. That is, the values of these trader's medical parameters were measured in the course of such market events as changes in the trend and an increase of the volatility or that of the volume of transactions with respect to particular financial instruments [29], and they were compared with the values in the periods of market stability. It has turned out that, generally, experienced traders have been less sensitive emotionally to the stock exchange events than inexperienced ones. Similar to the authors of [1, 37], the authors of [48] believe that the tendency of both professional private investors and traders to sell profitable securities more often than unprofitable ones (the disposition effect) should be interpreted as the inability of the traders to adopt right financial decisions in a substantial number of cases. The authors of [26] assert that private investors are overconfident and that they are sure that they interpret all the available financial information correctly, which leads to their excessive participation in stock exchange bids and causes them to sustain big losses. Neither a correlation between the success of a trader and her personal qualities (such as age, experience, the size of her capital, the level of education, etc.) [30], nor a correlation between her success and the gender have been detected [44]. Moreover, conclusions of some researchers on the performance of the investment funds managed by Ph. D. holders, as well as by those having scientific publications in the top journals on economics and finance, suggest that the viewpoint that these funds achieve better financial results should be considered controversial [9]. The fate of the Long-Term Capital Management—the hedge fund which crashed in 1999, despite the fact that more than a half of its participants and partners had scientific degrees, and Nobel economists R. Merton and M. Scholes were among its partners—seems to be illustrative of this statement.

As is known, traders who do not have any specific knowledge or time for making decisions based upon analyzing available information are inclined to follow the actions of other traders, called gurus (the so-called herding effect) [46, 55, 58]. Usually, lucky and experienced traders or the insiders, who have access to information inaccessible to others, are chosen as such gurus, and the herding effect is present even among experienced traders (though to a much lower degree) [58]. In [55], the authors suggest that in the market of currency futures contracts, small speculators follow the actions of large ones if this looks profitable to them. This conclusion on the herding effect is also suggested in [14], where the authors explain such a behavior of the traders by (a) expectations of substantial profits to come, (b) personal motives of particular market participants, and (c) the so-called cascade model, when an agent possessing negative information on a particular financial instrument, but contemplating the actions of other market players may change her mind and ignore the available negative information. An excessive attention to the actions of the neighbors (even if a trader follows and takes into account the actions of only successful ones) leads to forming financial "bubbles" and stock exchange crashes [51]. Some authors believe that people are inclined to trust today's information on the market status more than to evaluate the current situation there taking into account the prehistory of this information [12]. The authors of [43] believe that even if some financial information is public, but is not accessible to all the market players, the market will not react to this information (in particular, will not react to the actions undertaken by those possessing this information) by changing either the volumes of transactions with respect to particular financial instruments or the market prices. According to [4], private investors often buy securities discussed in the media while do not have a chance to estimate all the available financial instruments and to comprehend all the available financial information. Results of quite a number of empirical studies on the herding effect are surveyed in [52].

Some Facts Illustrative of the Ability (or of the Inability) of Financial Analysts to Form Correct Recommendations and Forecasts As mentioned earlier, the analysis of forecasts offered by financial analysts [22] showed that in a majority of all the forecasts analyzed, the share of correct forecasts is approximately 50 %. In [44] the authors estimated the effectiveness of the forecasts of analysts and experts on the Russian stock market, and they concluded that only 56.8 % of the experts offered correct forecasts. An analysis of the forecasts offered by economists working in the government, banks, entrepreneurship, and in the field of science, conducted by the Federal Bank of Philadelphia since 1946 (Livingston Survey), caused the authors of the survey to conclude that professional economists apparently unable to correctly predict the behavior of financial markets in principle [50].

Some Facts on Financial Results Achieved by Investment Funds An analysis of financial activities of hedge funds in the period of time from 1994 through 2003 [32] showed that more than a half of them could not achieve the yield of their portfolios higher than the market one. Particularly, this means that even a passive investment strategy (buy and hold) could have brought to the clients of the fund a larger yield than that they received by investing a comparable amount of money in hedge funds. Also, the authors of [32] estimated the number of funds that displayed the results exceeding the average in the two consequent years and concluded that the share of these funds had been about only 50 %, whereas only 11 out of 18 funds that beat the market in 1995 repeated this result in 1996.

2.4 *A Few Remarks Regarding Directions of Studies on Modeling the Financial Behavior of Small and Medium Price-Taking Traders in a Stock Exchange*

The presented brief survey of publications suggests that studying the financial behavior of small and medium price-taking traders interacting with a stock exchange with the use of mathematical models should be conducted at least in two major directions (a) modeling short term strategies of the transaction participants, and (b) modeling long-term investments in financial instruments (mostly, in securities).

It seems obvious that strategies of a trader with respect to her short-term transactions substantially depend on both the trader's knowledge of models and methods for analyzing the market of financial instruments and her willingness to use these analytical tools. For traders who do not have necessary mathematical knowledge, the latter mostly depends on their ability to quickly understand only the substance of the above-mentioned models and methods and on the ability of these tools to solve large-scale problems in an acceptable time. However, the surveyed publications bear evidence that comprehending the existing models and methods implies the involvement of consultants possessing the necessary knowledge. Thus, the need in developing new mathematical tools for studying the financial behavior of traders and for helping them achieve their investment goals seems obvious, and, since the use of these tools will likely affect the behavior of stock exchanges, the development of these tools presents interest for both society and financial science.

A game-theoretic approach to studying trader's financial behavior, which has successfully been applied in studying phenomena in the nature and society in general, and particular results in studying classes of games with nonlinear payoff functions on convex polyhedral sets of player strategies [6] seem promising in solving large-scale problems that the traders face in strategizing their everyday transactions, especially taking into account the simplicity of interpreting the strategies (obtained as a result of solving these games) in the stock exchange terms [7].

At the same time, the above-presented survey suggests that even if particular effective tools for developing financial strategies of a trader existed, not all the traders would use them. For quite a number of traders, especially for those who have only recently started interacting with a stock exchange and do not have enough experience, their own intuition is likely to remain the most reliable tool in making financial decisions. For these traders, as well as for all the others who would like to detect the dynamics of values of which financial instruments they can forecast most accurately, the tools allowing any interested trader to estimate her chances to succeed in including a particular financial instrument in her portfolio seem indispensable. The availability of such tools may allow many of those who think about trading in a stock exchange to avoid unjustified risks of investing in general or those with respect to particular financial instruments both in the short term and in the long run.

3 Detecting the Ability of a Price-Taking Trader to Divine Future Price Values of a Financial Instrument and to Succeed in Using this Ability in a Standard and in a Margin Trading

The ability of a trader to divine the price value dynamics of a set of particular financial instruments traded in a stock exchange matters a great deal in forming her optimal portfolio. However, even for a person gifted in divining either future values of any time series in general or only those of time series describing the dynamics of particular financial instruments with a probability exceeding 50 %, it is clear that this ability as such may turn out to be insufficient for successfully trading securities either in a long run or even in a short period of time. Thus, tools for both detecting the ability of a potential trader to divine the values of the share prices of, for instance, securities from a particular set of securities with a probability exceeding 50 % and testing this ability (from the viewpoint of a final financial result that the trader may expect to achieve by trading corresponding financial securities within any particular period of time) are needed. These tools should help the potential trader develop confidence in her ability to succeed by trading particular financial securities and evaluate whether this ability is safe to use in trading with risks associated with the margin trading at least under certain leverage rates, offered by brokers working at a stock exchange. It seems obvious that, apparently, no tools can guarantee in principle that the results that they produce are directly applicable in a real trading. However, they may (a) give the interested person (i.e., a potential trader) the impression on what she should expect by embarking the gamble of trading in stock exchanges, and (b) advise those who do not display the ability to succeed in trading financial securities (either in general or in a margin trading with a particular leverage) to abstain from participating in these activities.

The above-mentioned tools for testing the ability of a trader to divine the upward and downward directions of changing the value of the share price of a financial security and those for evaluating possible financial results of trading this security in a particular stock exchange consist of two separate parts. The first part (for testing the trader's ability to divine) is a software complex in which (a) the trader is offered a time series describing the dynamics of the price value of the security share for a particular financial security that interests her, and (b) her prediction made at the endpoint of a chosen segment of the time series is compared with the real value of the share price of this financial security at the point next to that endpoint. It is clear that to estimate the probability of the random event consisting of correctly predicting this value of the share price, one should first find the frequency of correct answers offered by the trader (provided the trials are held under the same conditions) and make sure that the outcome of each trial does not depend on the outcomes of the other trials (i.e., that the so-called Bernoulli scheme [17] of conducting the trails takes place). If these conditions are met, one can calculate the frequency of this event as a ratio of the correct predictions to the total number of trials, and this ratio can be considered as an estimate of the probability under consideration [17]. A possible approach to making the trials independent of each other and to securing the same conditions for the trials may look as follows: one may prepare a set of randomly chosen segments of the time series having the same length and let the trader make her prediction at the endpoint of each segment proceeding from the observation of the time series values within the segment.

The second part of the testing tools (for estimating possible final financial results of trading a particular security with a detected probability to divine the directions of changing the value of the share price of this security) is also a software complex in which trading experiments can be conducted. For instance, the trader can be offered a chance to predict the direction of changing the value of the share price of a security at any consequent set of moments (at which real values of the share price of the security constitute a time series) and to choose the number of shares that she wishes to trade (to buy or to sell) at each moment from the set. By comparing the results of the trader's experiments with real values of the share price of a security from the sets (time series segments) of various lengths at the trader's choice, she concludes about her potential to succeed or to fail in working with the security under consideration.

Finally, the complex allows the trader to make sure that at each testing step (i.e., at each moment t) of evaluating financial perspectives of working with each particular financial security of her interest, the probability with which the trader divines the value of the share price of this financial security at the moment $t+1$ does coincide with the one detected earlier (or at least is sufficiently close to it). This part of the software is needed to avoid unjustified recommendations on including a particular financial security in the trader's portfolio if for whatever reasons, the above coincidence (or closeness) does not take place.

4 Finding Optimal Trader's Strategies of Investing in Standard Financial Securities. Model 1: The Values of Financial Securities Are Random Variables with Uniform Probability Distributions

In considering the financial behavior of a price-taking trader who at the moment t wants to trade financial instruments that are traded in a particular stock exchange, two situations should be analyzed.

Situation 1

The trader does not possess any financial instruments at the moment t while possessing a certain amount of cash that can be used both for buying financial instruments and for borrowing them from a broker (to sell the borrowed financial instruments short).

Situation 2

The trader has a portfolio of financial instruments, along with a certain amount of cash, and she tries to increase the value of her portfolio by selling and buying financial instruments of her interest, as well as by borrowing them from the brokers (to sell the borrowed financial instruments short).

To simplify the material presentation and to avoid the repetition of parts of the reasoning to follow, in Model 1, which is studied in Sect. 4, Situation 2 is considered first. Moreover, it is assumed that the trader's portfolio consists of financial securities only; cases in which derivative financial instruments are parts of the trader's portfolio are not considered in this paper. Remark 1 at the end of Sect. 4 explains how the model developed for finding the best investment strategy of the trader in Situation 2 (Model 1) can be used in finding such a strategy in Situation 1.

4.1 Notation

Let

$N = \{1, 2, \ldots, n\}$ be a set of (the names of) financial securities comprising the portfolio of a trader that are traded in a particular stock exchange and interest the trader;

$t_0 < \ldots < t < t+1 < t+2 < \ldots$ be a set of the time moments at which the trader adopts decisions on changing the structure of her portfolio;

m_t be the amount of cash that the trader possesses at the moment t;

W_t be the total value of the trader's portfolio at the moment t (in the form of cash and financial securities), i.e., the trader's welfare at the moment t;

$s_{i,t}$ be the spot value of the share price of financial security i at the moment t, i.e., the price at which a share of financial security i is traded at the moment t at the stock exchange under the conditions of an immediate financial operation;

$v_{i,t}$ be the (non-negative, integer) number of shares of financial security i that the trader possesses at the moment t.

The following four assumptions on how the trader makes decisions on changing her portfolio at the moment t seem natural:

1. The trader possesses a confirmed (tested) ability of estimating the probability p_i with which the future value of the share price of financial security i may change at the moment $t+1$ in a particular direction for each $i \in \overline{1,n}$, i.e., the ability to predict whether this value will increase or will not increase. (See some of the details further in Sect. 4.)
2. At each moment t (from the above set of moments), the trader can divide the set of financial securities N into three subsets I_t^+, I_t^-, I_t^0 for which $N = I_t^+ \cup I_t^- \cup I_t^0$, and $I_t^+ \cap I_t^- = \emptyset$, $I_t^- \cap I_t^0 = \emptyset$, $I_t^+ \cap I_t^0 = \emptyset$, where

   I_t^+ is the set of financial securities on which the trader is confident that the values of their share prices will increase at the moment $t+1$ (so she intends to buy securities from this set at the moment t),

I_t^- is the set of financial securities on which the trader is confident that the values of their share prices will decrease at the moment $t + 1$ (so she intends to sell securities from this set at the moment t),

I_t^0 is the set of financial securities on which the trader is confident that the values of their share prices will not change at the moment $t + 1$ or will change insignificantly (so she does not intent to either sell or buy securities from this set at the moment t).

3. For buying financial securities from the set I_t^+ , the trader can spend both the available cash and the money to be received as a result of selling financial securities from the set I_t^- at the moment t, as well as finances that the trader can borrow from any lenders (if such finances are available to the trader). Analogously, for selling financial securities from the set $i \in I_t^-$, the trader may use her own reserves of this security (of the size $v_{i,t}$), as well as to borrow a certain number of shares of this security from a broker to open a short position (if this service is available to the trader from the broker);
4. The trader does not undertake any actions with financial securities from the set I_t^0.

To simplify the mathematical formulation of problems to be considered in this section of the paper in the framework of Model 1, in the reasoning to follow, it is assumed that the trader (a) works only with shares and bonds as financial securities (called standard securities further in this paper), and (b) puts only market orders, i.e., those that can be implemented immediately, at the spot market prices.

4.2 The Statement and Mathematical Formulation of the Problem of Finding a Trader's Optimal Investment Strategy

Let at the moment t, the trader have a portfolio of standard securities $v_{i,t}, i \in \overline{1,n}$ and a certain amount of cash m_t so that her welfare at the moment t equals $W_t = m_t + \sum_{i=1}^{n} v_{i,t}s_{i,t}$. The problem of finding optimal investment strategies of the trader consists of choosing (a) the numbers of shares of securities $x_{i,t}^+$ (integers) from the set I_t^+ to buy (about which the trader expects the increase of the values of their share prices at the moment $t + 1$), (b) the numbers of shares of securities $x_{i,t}^-$ (integers) from the set I_t^- in her current portfolio to sell (about which the trader expects the decrease of the values of their share prices at the moment $t + 1$), and (c) the numbers of shares of securities $z_{i,t}^-$ (integers) from the set I_t^- to sell, which are to be borrowed from a broker at the value of the share price equaling $s_{i,t}$ to open a short position at the moment t with the return of these securities to the broker at the moment $t + 1$ at the share price value $s_{i,t+1}$ (for which the trader expects the inequality $s_{i,t} > s_{i,t+1}$ to hold), to maximize the increment of her welfare at the moment $t + 1$.

The welfare that the trader expects to have at the moment $t + 1$ thus equals

$$W_{t+1} = \sum_{i\in I_t^0} v_{i,t}s_{i,t+1} + \sum_{i\in I_t^+} (v_{i,t} + x_{i,t}^+)s_{i,t+1} + \sum_{i\in I_t^-} (v_{i,t} - x_{i,t}^-)s_{i,t+1} +$$
$$+ \left(m_t - \sum_{i\in I_t^+} x_{i,t}^+ s_{i,t} + \sum_{i\in I_t^-} x_{i,t}^- s_{i,t} + \sum_{i\in I_t^-} z_{i,t}^-(s_{i,t} - s_{i,t+1})\right),$$

where the first three summands determine a part of the trader's welfare formed by the value of the securities from her portfolio at the moment $t + 1$, and the last summand determines the amount of cash remaining after the finalization of all the deals on buying and selling securities by the moment $t + 1$, including the return of the borrowed securities to the broker.

The (positive or negative) increment of the trader's welfare that she expects to attain at the moment $t + 1$ compared with that at the moment t after the completion of all the transactions equals

$$\Delta W_{t+1} = \sum_{i\in I_t^0} v_{i,t}(s_{i,t+1} - s_{i,t}) + \sum_{i\in I_t^+} (v_{i,t} + x_{i,t}^+)(s_{i,t+1} - s_{i,t}) +$$
$$+ \sum_{i\in I_t^-} (v_{i,t} - x_{i,t}^-)(s_{i,t+1} - s_{i,t}) + \sum_{i\in I_t^-} z_{i,t}^-(s_{i,t} - s_{i,t+1}).$$

Here, $v_{i,t}$, $s_{i,t}$, m_t, $i \in I_t^+, I_t^-$ are known real numbers (the numbers $v_{i,t}$ are integers), and the numbers $s_{i,t+1}$, $i \in I_t^+, I_t^-$ are the values of random variables. Further, it is assumed that the values of the share prices of securities $i, j \in N$ at the moment $t+1$ are independent random variables.

The trader conducts her transactions taking into consideration the following constraints:

1. The numbers of shares of financial securities bought, sold, and borrowed are integers,
2. $x_{i,t}^-$, the number of shares of security i sold from the trader's portfolio, cannot exceed the available number of shares $v_{i,t}$ of this security that the trader possesses at the moment t, i.e., the inequalities

$$x_{i,t}^- \leq v_{i,t}, i \in I_t^-,$$

hold (one should notice that if the trader plans to sell any number of shares of security i that exceeds $v_{i,t}$, then she borrows the number of shares of this security $z_{i,t}^-$ from a broker to open a short position to sell security i in the number of shares additional to the number $v_{i,t}$),

3. the limitations on the accessibility to the capital to borrow while using a credit with a particular (credit) leverage cannot be exceeded; these limitations may be described, for instance, by the inequality

$$\sum_{i \in I_t^+} x_{i,t}^+ s_{i,t} + \sum_{i \in I_t^-} z_{i,t}^- s_{i,t} - \left(m_t + \sum_{i \in I_t^-} x_{i,t}^- s_{i,t} \right) \leq k_t \left(m_t + \sum_{i=1}^{n} v_{i,t} s_{i,t} \right).$$

Here, k_t is the size of the credit leverage, the first two summands on the left hand side of this inequality represent the sum of the total expenses bore by the trader at the moment t (that are associated with buying securities in the market and with the trader's debt to the brokers who lent her securities from the set I_t^- to open a short position). The third summand (on the left hand side of the above inequality) reflects the amount of cash that the trader will possess as a result of selling securities from the set I_t^- that the trader has as part of her own portfolio at the moment t. The right hand side of the inequality reflects the maximal amount of money (that is assumed to be) available to the trader for borrowing with the credit leverage of the size k_t, and this amount depends on the total amount of the capital that the trader possesses at the moment t before she makes any of the above-mentioned transactions. One should bear in mind that particular mathematical relations reflecting the limitations on the accessibility of a particular trader to the capital to borrow may vary, and such relations much depend on the situation in the stock exchange at the moment t and on the ability of the trader to convince particular brokers to lend her securities and on the ability to convince particular lenders to lend her cash (or on both).

It is also assumed that in making investment decisions at the moment t, the trader proceeds from the value α of a threshold, determining whether to make transactions in the stock exchange in principle. That is, she makes the transactions if the inequality $W_{t+1} \geq \alpha W_t$ holds, meaning that the trader tends to keep the level of the ratio of her welfare at every moment compared with that at the previous moment not lower than a particular value α of the threshold, $\alpha > 0$.

4.3 Transforming the Problem of Finding an Optimal Investment Strategy of a Trader into an Integer Programming Problem

Let at the moment t, the trader be able to estimate $s_{i,t+1}^{max}$ and $s_{i,t+1}^{min}$, the boarders of a segment to which the values of the share price of security $i \in I_t^+ \cup I_t^- \cup I_t^0$ will belong at the moment $t+1$ (based upon either the previous data or any fundamental assumptions on the dynamics of the value of the share price that this security may have). If the trader can make no assumptions on a particular probability distribution of the values of the share price that security i may have within these boarders, it is natural to consider that these values change upwards and downwards (with respect to the value $s_{i,t}$) as continuous random variables u and v uniformly distributed on the segments $[s_{i,t}, s^{max}_{i,t+1}]$ and $[s^{min}_{i,t+1}, s_{i,t}]$, respectively, with the probability distribution densities

$$f_1(u) = \begin{cases} \dfrac{1}{s^{max}_{i,t+1} - s_{i,t}}, & \text{if } u \in [s_{i,t}, s^{max}_{i,t+1}], \\ 0, & \text{if } u \notin [s_{i,t}, s^{max}_{i,t+1}], \end{cases}$$

$$f_2(v) = \begin{cases} \dfrac{1}{s_{i,t} - s^{min}_{i,t+1}}, & \text{if } v \in [s^{min}_{i,t+1}, s_{i,t}], \\ 0, & \text{if } v \notin [s^{min}_{i,t+1}, s_{i,t}]. \end{cases}$$

Thus, if the trader assumes that the value of the share price of security i will increase at the moment $t+1$ compared with its current value, i.e., that the inequality $s_{i,t+1} > s_{i,t}$ will hold, then the expectation of the value of the share price that this security will have at the moment $t+1$ equals $Ms_{i,t+1} = \frac{s_{i,t}+s^{max}_{i,t+1}}{2}$. On the contrary, if the trader assumes that this value of the share price will decrease at the moment $t+1$, i.e., that the inequality $s_{i,t+1} < s_{i,t}$ will hold, then the expectation of the value of the share price that security i will have at the moment $t+1$ equals $Ms_{i,t+1} = \frac{s^{min}_{i,t+1}+s_{i,t}}{2}$. Finally, if the trader cannot make either assumption about the value of the share price that security i will have at the moment $t+1$, it is natural to consider that the value of the share price of this security will not change, i.e., that the equality $Ms_{i,t+1} = s_{i,t+1} = s_{i,t}$ will hold.

If at the moment t, the trader expects with the probability p_i that the value of the share price of security i will increase at the moment $t+1$, i.e., that the inclusion $i \in I_t^+$ will hold (event A_1), then the expectation of the value of the share price that this security will have at the moment $t+1$ assumes the value $\frac{s_{i,t}+s^{max}_{i,t+1}}{2}$, and the probability of the event A_1 equals p_i. Otherwise, two events are possible at the moment $t+1$: (a) the value of the share price of security i at the moment $t+1$ will decrease (event A_2), and (b) the value of the share price of security i at the moment $t+1$ will remain equal to the one at the moment t (event A_3), and it is natural to assume that these two events are equally possible with the probability $\frac{1-p_i}{2}$.

Thus, the expectation of the value of the share price that security $i \in I_t^+$ will have at the moment $t+1$ can be calculated proceeding from the probabilities of the three incompatible events A_1, A_2, A_3, reflected in Table 1.

If at the moment t, the trader expects with the probability p_i that the value of the share price of security i will decrease at the moment $t+1$, i.e., that the inclusion $i \in I_t^-$ will hold, then the reasoning similar to the previous one allows one to conclude

Table 1 The values of the conditional expectation $M(s_{i,t+1}/A_k), i \in I_t^+, k \in \overline{1,3}$

$M(s_{i,t+1}/A_k), i \in I_t^+$	$\frac{s_{i,t}+s^{max}_{i,t+1}}{2}$	$\frac{s^{min}_{i,t+1}+s_{i,t}}{2}$	$s_{i,t}$
$P(A_k)$	p_i	$\frac{1-p_i}{2}$	$\frac{1-p_i}{2}$

Table 2 The values of the conditional expectation $M(s_{i,t+1}/B_k), i \in I_t^-, k \in \overline{1,3}$

$M(s_{i,t+1}/B_k), i \in I_t^-$	$\frac{s_{i,t+1}^{min}+s_{i,t}}{2}$	$\frac{s_{i,t}+s_{i,t+1}^{max}}{2}$	$s_{i,t}$
$P(B_k)$	p_i	$\frac{1-p_i}{2}$	$\frac{1-p_i}{2}$

Table 3 The values of the conditional expectation $M(s_{i,t+1}/C_k), i \in I_t^0, k \in \overline{1,3}$

$M(s_{i,t+1}/C_k), i \in I_t^0$	$s_{i,t}$	$\frac{s_{i,t+1}^{min}+s_{i,t}}{2}$	$\frac{s_{i,t}+s_{i,t+1}^{max}}{2}$
$P(C_k)$	p_i	$\frac{1-p_i}{2}$	$\frac{1-p_i}{2}$

that the expectation of the value of the share price that security $i \in I_t^-$ will have at the moment $t+1$ can be calculated proceeding from the probabilities of the three incompatible events B_1, B_2, B_3, reflected in Table 2.

Finally, if the trader expects with the probability p_i that for security i the inclusion $i \in I_t^0$ will hold at the moment $t+1$, the expectation of the value of the share price that security $i \in I_t^0$ will have at the moment $t+1$ can be calculated proceeding from the probabilities of the three incompatible events C_1, C_2, C_3, reflected in Table 3.

Thus, in the above three cases for security i to belong to one of the three subsets of the set N, the expectations $Ms_{i,t+1}$ are calculated as follows [17]:

$$Ms_{i,t+1} = p_i \frac{s_{i,t} + s_{i,t+1}^{max}}{2} + \frac{1-p_i}{2} \frac{s_{i,t+1}^{min} + s_{i,t}}{2} + \frac{1-p_i}{2} s_{i,t}, i \in I_t^+,$$

$$Ms_{i,t+1} = p_i \frac{s_{i,t+1}^{min} + s_{i,t}}{2} + \frac{1-p_i}{2} \frac{s_{i,t} + s_{i,t+1}^{max}}{2} + \frac{1-p_i}{2} s_{i,t}, i \in I_t^-,$$

$$Ms_{i,t+1} = p_i s_{i,t} + \frac{1-p_i}{2} \frac{s_{i,t+1}^{min} + s_{i,t}}{2} + \frac{1-p_i}{2} \frac{s_{i,t} + s_{i,t+1}^{max}}{2}, i \in I_t^0,$$

Certainly, generally, the trader can make any particular assumptions on the regularities that probability distributions of the future values of the share prices may have for securities from the sets I_t^+, I_t^-, I_t^0 at the moment $t+1$ (for instance, that these distributions will be normal). Such assumptions may let her more accurately calculate the expectations of the values of these share prices using the same logic that was employed under the assumption on the uniform distributions of these values.

To calculate an optimal trader's strategy of changing her portfolio at the moment t, one should choose the value of the threshold α and formulate the problem of finding such a strategy as, for instance, that of maximizing the expectation of the portfolio value increment, provided all the constraints associated with this choice hold. In the simplest case of such a formulation, one can assume that (a) the trader deals with and is interested in only those securities that are present in her portfolio at the moment t, (b) she may buy securities only from the set I_t^+, and she may sell securities only from the set I_t^-, and c) the trader does not make any transactions with securities from the set I_t^0 (see assumption 4 at the end of Sect. 4.1). Then, this maximization problem can be formulated, for instance, as follows [7]:

$$M[\triangle W_{t+1}] = \sum_{i \in I_t^0} v_{i,t}(Ms_{i,t+1} - s_{i,t}) + \sum_{i \in I_t^+} (v_{i,t} + x_{i,t}^+)(Ms_{i,t+1} - s_{i,t}) + \quad (1)$$

$$+ \sum_{i \in I_t^-} (v_{i,t} - x_{i,t}^-)(Ms_{i,t+1} - s_{i,t}) + \sum_{i \in I_t^-} z_{i,t}^-(s_{i,t} - Ms_{i,t+1}) \rightarrow max.$$

$$\sum_{i \in I_t^0} v_{i,t}Ms_{i,t+1} + \sum_{i \in I_t^+} (v_{i,t} + x_{i,t}^+)Ms_{i,t+1} + \sum_{i \in I_t^-} (v_{i,t} - x_{i,t}^-)Ms_{i,t+1} +$$

$$+ \left(m_t - \sum_{i \in I_t^+} x_{i,t}^+ s_{i,t} + \sum_{i \in I_t^-} x_{i,t}^- s_{i,t} + \sum_{i \in I_t^-} z_{i,t}^-(s_{i,t} - Ms_{i,t+1}) \right) \geq \alpha \left(m_t + \sum_{i=1}^{n} v_{i,t}s_{i,t} \right),$$

$$\sum_{i \in I_t^+} x_{i,t}^+ s_{i,t} + \sum_{i \in I_t^-} z_{i,t}^- s_{i,t} - \left(m_t + \sum_{i \in I_t^-} x_{i,t}^- s_{i,t} \right) \leq k_t \left(m_t + \sum_{i=1}^{n} v_{i,t}s_{i,t} \right),$$

$$x_{i,t}^- \leq v_{i,t}, i \in I_t^-,$$

where $x_{i,t}^+$, $i \in I_t^+$, $x_{i,t}^-$, $i \in I_t^-$, $z_{i,t}^-$, $i \in I_t^-$ are integers. This problem is an integer programming one in which $x_{i,t}^+$, $i \in I_t^+$, $x_{i,t}^-$, $i \in I_t^-$, and $z_{i,t}^-$, $i \in I_t^-$ are the variables.

Generally, (a) the set of standard securities $\tilde{N}$ (which contains N as a subset) that are of the trader's interest may include those that are not necessarily present in her portfolio at the moment t, and (b) the trader may proceed from the estimates of $Ms_{i,t+1}$ for all the securities from the set $\tilde{N}$ and make decisions of changing the composition of her portfolio based upon the values of the differences $Ms_{i,t+1} - s_{i,t}$ for all of these securities (so that assumption at the end of Sect. 4.1 does not hold).

Let the trader divide the whole set $\tilde{N}$ of standard securities that interest her at the moment t into the subsets $\tilde{I}_t^+$, $\tilde{I}_t^-$, and $\tilde{I}_t^0$, where $\tilde{I}_t^+$ is a set of standard securities for which the trader believes with the probability $p_i > 0.5$ that the share price values that these securities will have at the moment $t+1$ will increase, $\tilde{I}_t^-$ is a set of standard securities for which the trader believes with the probability $p_i > 0.5$ that the share price values that these securities will have at the moment $t + 1$ will decrease, and $\tilde{I}_t^0$ is a set of standard securities for which the trader believes with the probability $p_i > 0.5$ that the share price values that these securities will have at the moment $t + 1$ will not change.

Let the trader know the boarders of the segment $[s_{i,t+1}^{min}, s_{i,t+1}^{max}]$ within which the value of $s_{i,t+1}, i \in I_t^+ \cup I_t^- \cup I_t^0$ will change at the moment $t + 1$ while the trader can make no assumptions on a particular probability distribution that the value of $s_{i,t+1}$, considered as that of a random variable, may have (within these borders). Then, as before, it seems natural to assume that this value changes upwards as a continuous random variable u uniformly distributed on the segment $[s_{i,t}, s_{i,t+1}^{max}]$ and changes downwards as a continuous random variable v distributed uniformly on the segment $[s_{i,t+1}^{min}, s_{it}]$. The latter assumption allows one to calculate the expectations $Ms_{i,t+1}$ in just the same manner this was done earlier for standard securities from the set N.

First, consider standard securities that the trader may be interested in buying, including securities with particular names that some of the standard securities in her portfolio have. Let $\hat{I}_t^+ \subset \tilde{I}_t^+, \hat{I}_t^- \subset \tilde{I}_t^-$, and $\hat{I}_t^0 \subset \tilde{I}_t^0$ be the sets of standard securities for which the differences $Ms_{i,t+1} - s_{i,t}$ are strictly positive. If at least one of the three sets $\hat{I}_t^+, \hat{I}_t^-$, and $\hat{I}_t^0$ is not empty, the trader may consider buying new standard securities from the set $\hat{I}_t^+ \cup \hat{I}_t^- \cup \hat{I}_t^0$ at the moment t.

Second, consider standard securities that are already in the trader's portfolio at the moment t. Let $\tilde{I}_t^+(av) \subset \tilde{I}_t^+$, $\tilde{I}_t^-(av) \subset \tilde{I}_t^-$, $\tilde{I}_t^0(av) \subset \tilde{I}_t^0$ be the sets of names of the standard securities that the trader possesses at the moment t, and let $v_{i,t}$ be the number of shares of standard security i, $i \in \tilde{I}_t^+(av) \cup \tilde{I}_t^-(av) \cup \tilde{I}_t^0(av)$. Let $\hat{I}_t^+(av) \subset \tilde{I}_t^+(av)$, $\hat{I}_t^-(av) \subset \tilde{I}_t^-(av)$, and $\hat{I}_t^0(av) \subset \tilde{I}_t^0(av)$ be the sets of i for which the differences $Ms_{i,t+1} - s_{i,t}$ are strictly positive.

It is clear that the trader may consider a) holding the standard securities from the sets $\hat{I}_t^+(av)$, $\hat{I}_t^-(av)$, and $\hat{I}_t^0(av)$, and b) selling all the standard securities from the sets $\tilde{I}_t^+(av) \setminus \hat{I}_t^+(av)$, $\tilde{I}_t^-(av) \setminus \hat{I}_t^-(av)$, and $\tilde{I}_t^0(av) \setminus \hat{I}_t^0(av)$ and borrowing standard securities from these sets from brokers. Since the trader believes that selling standard securities, in particular, from the sets $\tilde{I}_t^+(av) \setminus \hat{I}_t^+(av)$, $\tilde{I}_t^-(av) \setminus \hat{I}_t^-(av)$, and $\tilde{I}_t^0(av) \setminus \hat{I}_t^0(av)$ short leads to receiving the money that can be spent, particularly, for buying new standard securities from the sets $\hat{I}_t^+$, $\hat{I}_t^-$, and $\hat{I}_t^0$ (provided these sets are not empty), the trader needs to find an optimal investment strategy of changing her portfolio. This problem can be formulated as follows:

$$M[\Delta W_{t+1}] = \sum_{i \in \hat{I}_t^+} x_{i,t}^+ (Ms_{i,t+1} - s_{i,t}) + \sum_{i \in \hat{I}_t^-} x_{i,t}^+ (Ms_{i,t+1} - s_{i,t}) + \tag{2}$$

$$\sum_{i \in \hat{I}_t^0} x_{i,t}^+ (Ms_{i,t+1} - s_{i,t}) + \sum_{i \in \hat{I}_t^+(av)} v_{it}(Ms_{i,t+1} - s_{i,t}) + \sum_{i \in \hat{I}_t^-(av)} v_{it}(Ms_{i,t+1} - s_{i,t}) +$$

$$\sum_{i \in \hat{I}_t^0(av)} v_{it}(Ms_{i,t+1} - s_{i,t}) + \sum_{i \in (\tilde{I}_t^+ \setminus \hat{I}_t^+) \cup (\tilde{I}_t^- \setminus \hat{I}_t^-) \cup (\tilde{I}_t^0 \setminus \hat{I}_t^0)} z_{i,t}^- (s_{i,t} - Ms_{i,t+1}) \rightarrow max.$$

$$\sum_{i \in \hat{I}_t^+} x_{i,t}^+ Ms_{i,t+1} + \sum_{i \in \hat{I}_t^-} x_{i,t}^+ Ms_{i,t+1} + \sum_{i \in \hat{I}_t^0} x_{i,t}^+ Ms_{i,t+1} +$$

$$+ \sum_{i \in \hat{I}_t^+(av)} v_{i,t} Ms_{i,t+1} + \sum_{i \in \hat{I}_t^-(av)} v_{i,t} Ms_{i,t+1} + \sum_{i \in \hat{I}_t^0(av)} v_{i,t} Ms_{i,t+1}) +$$

$$+ \left(m_t - \sum_{i \in \hat{I}_t^+} x_{i,t}^+ s_{i,t} - \sum_{i \in \hat{I}_t^-} x_{i,t}^+ s_{i,t} - \sum_{i \in \hat{I}_t^0} x_{i,t}^+ s_{i,t} \right) +$$

$$+ \sum_{i \in \tilde{I}_t^+(av) \setminus \hat{I}_t^+(av)} v_{i,t} s_{i,t} + \sum_{i \in \tilde{I}_t^-(av) \setminus \hat{I}_t^-(av)} v_{i,t} s_{i,t} + \sum_{i \in \tilde{I}_t^0(av) \setminus \hat{I}_t^0(av)} v_{i,t} s_{i,t} +$$

$$+ \sum_{i \in (\tilde{I}_t^+ \setminus \hat{I}_t^+) \cup (\tilde{I}_t^- \setminus \hat{I}_t^-) \cup (\tilde{I}_t^0 \setminus \hat{I}_t^0)} z_{i,t}^-(s_{i,t} - M s_{i,t+1}) \geq \alpha \left(m_t + \sum_{i \in \tilde{I}_t^+(av) \cup \tilde{I}_t^-(av) \cup \tilde{I}_t^0(av)} v_{i,t} s_{i,t} \right),$$

$$\sum_{i \in \hat{I}_t^+} x_{i,t}^+ s_{i,t} + \sum_{i \in \hat{I}_t^-} x_{i,t}^+ s_{i,t} + \sum_{i \in \hat{I}_t^0} x_{i,t}^+ s_{i,t} + \sum_{i \in (\tilde{I}_t^+ \setminus \hat{I}_t^+) \cup (\tilde{I}_t^- \setminus \hat{I}_t^-) \cup (\tilde{I}_t^0 \setminus \hat{I}_t^0)} z_{i,t}^- s_{i,t} -$$

$$- \left(m_t + \sum_{i \in \tilde{I}_t^+(av) \setminus \hat{I}_t^+(av)} v_{i,t} s_{i,t} + \sum_{i \in \tilde{I}_t^-(av) \setminus \hat{I}_t^-(av)} v_{i,t} s_{i,t} + \sum_{i \in \tilde{I}_t^0(av) \setminus \hat{I}_t^0(av)} v_{i,t} s_{i,t} \right) \leq$$

$$\leq k_t \left(m_t + \sum_{i \in \tilde{I}_t^+(av) \cup \tilde{I}_t^-(av) \cup \tilde{I}_t^0(av)} v_{i,t} s_{i,t} \right),$$

where $x_{i,t}^+$, $i \in \hat{I}_t^+ \cup \hat{I}_t^- \cup \hat{I}_t^0 \subset \tilde{I}_t^+ \cup \tilde{I}_t^- \cup \tilde{I}_t^0$, are the numbers of shares of securities from the set $\tilde{I}_t^+ \cup \tilde{I}_t^- \cup \tilde{I}_t^0$ that are bought at the moment t.

As before, the (expected) increment of the trader's welfare is calculated as the difference between the expected trader's welfare at the moment $t + 1$ as a result of buying and selling securities in the stock exchange and her welfare at the moment t (with respect to the activities related to the interaction with the stock exchange). That is, at the moment $t+1$, the expected trader's welfare is a sum of (a) the expected value of new securities bought at the moment t, (b) the expected value of securities from her portfolio that have been held since the moment t, (c) the amount of cash remaining at the moment $t + 1$ after spending a part of cash that is available at the moment t for buying new securities and receiving cash as a result of selling securities from the set $i \in (\tilde{I}_t^+(av) \setminus \hat{I}_t^+(av)) \cup (\tilde{I}_t^-(av) \setminus \hat{I}_t^-(av)) \cup (\tilde{I}_t^0(av) \setminus \hat{I}_t^0(av))$, and (d) the amount of cash expected to be received as a result of selling short securities borrowed from brokers. This problem is also an integer programming one in which $x_{i,t}^+$, $i \in \hat{I}_t^+ \cup \hat{I}_t^- \cup \hat{I}_t^0$, and $z_{i,t}^-$, $i \in (\tilde{I}_t^+ \setminus \hat{I}_t^+) \cup (\tilde{I}_t^- \setminus \hat{I}_t^-) \cup (\tilde{I}_t^0 \setminus \hat{I}_t^0)$ are the variables.

Both problem (1) and problem (2) can be solved exactly, with the use of software for solving integer programming problems, if the number of the variables allows one to solve this problem in an acceptable time.

As is known, in solving applied integer programming problems, integer variables are often considered as continuous ones, i.e., a relaxation of the problem is solved instead of the initial integer programming problem, and all the non-integer components of the solution are rounded-off [61] in line with any methodology. Such a transformation is mostly used when the constraints in the integer programming problem have the form of inequalities (which is the case in the problem under consideration). One should notice that the problem of rounding-off non-integer solutions in relaxed linear programming problems (with respect to the initial integer programming ones) and an approach to estimating the accuracy of this rounding-off are discussed in scientific publications, in particular, in [2].

Thus, the conditions for the variables to be integer are to be replaced in problem (1) with those of non-negativity for the volumes of securities to be bought, sold, and borrowed

$$x_{i,t}^{+} \geq 0, i \in I_t^{+},\ x_{i,t}^{-} \geq 0, i \in I_t^{-},\ z_{i,t}^{-} \geq 0, i \in I_t^{-},$$

which transforms this integer programming problem into a linear programming one. Analogously, for problem (2), conditions for the variables to be integer are to be replaced with those of non-negativity

$$x_{i,t}^{+} \geq 0, i \in \hat{I}_t^{+} \cup \hat{I}_t^{-} \cup \hat{I}_t^{0},\ z_{i,t}^{-} \geq 0, i \in (\tilde{I}_t^{+} \setminus \hat{I}_t^{+}) \cup (\tilde{I}_t^{-} \setminus \hat{I}_t^{-}) \cup (\tilde{I}_t^{0} \setminus \hat{I}_t^{0}).$$

Adding these conditions to the system of constraints of problem (2) transforms this problem into a linear programming one.

Both the system of constraints and the goal function of problem (1) and those of problem (2) are substantially different. Particularly, there are no inequalities of the kind $x_{i,t}^{-} \leq v_{i,t}, i \in I_t^{-}$ in the system of constraints of problem (2). Under the assumptions made in formulating problem (1), the trader may or may not sell all the standard securities from the set I_t^{-}. On the contrary, in problem (2), the suggested division of the set $\tilde{N}$ into the subsets implies that the trader will sell all the standard securities from the set $(\tilde{I}_t^{+}(av) \setminus \hat{I}_t^{+}(av)) \cup (\tilde{I}_t^{-}(av) \setminus \hat{I}_t^{-}(av)) \cup (\tilde{I}_t^{0}(av) \setminus \hat{I}_t^{0}(av))$.

Also, in comparing the mathematical formulations of problems (1) and (2), one should bear in mind that the trader's division of the set N into the three subsets I_t^{+}, I_t^{-}, I_t^{0}, for which $N = I_t^{+} \cup I_t^{-} \cup I_t^{0}$ and $I_t^{+} \cap I_t^{-} = \emptyset$, $I_t^{-} \cap I_t^{0} = \emptyset$, $I_t^{+} \cap I_t^{0} = \emptyset$, is, generally, purely intuitive and may turn out to be wrong. This division is not based on any mathematical analysis of either directions of potential changes in which the share price values of particular securities may move or on any numerical relations among the probabilities with which these moves may take place and the limits within which the changes are possible. In contrast, the division of the set $\tilde{N}$ of standard securities that interest the trader at the moment t into the subsets $\tilde{I}_t^{+}$, $\tilde{I}_t^{-}$, and $\tilde{I}_t^{0}$ and dealing only with those securities i from this set for which the differences $Ms_{i,t+1} - s_{i,t}$ are strictly positive are a result of such an analysis. In the framework of this analysis, solving problem (2) may, in fact, be viewed as a means for testing the intuition of an interested trader with respect to her ability to properly choose the set of securities to consider for potential transactions.

Example 1. Consider security A from the set I_t^{+} whose current share price value (at the moment t) equals 10.00 US dollars. Let the trader expect that at the moment $t+1$, the share price value of security A a) will be between 10.00 US dollars and 12.00 US dollars with the probability 0.6, b) will be between 2.00 US dollars and 10.00 US dollars with the probability 0.2, and c) will remain equal to 10.00 US dollars (i.e., will remain unchanged) with the probability 0.2.

Then using the above formulae for calculating the expectation of the share price value for a security from the set I_t^{+}, one can easily be certain that the expectation of the share price value of security A at the moment $t + 1$ equals 9.80 US dollars, i.e., contrary to the trader initial analysis, the expectation of the share price value of this security will decrease.

Example 2. Consider security B from the set I_t^-, whose current share price value (at the moment t) equals 100.00 US dollars. Let the trader expect that at the moment $t + 1$, the share price value of security B a) will be between 90.00 US dollars and 100.00 US dollars with the probability 0.6, b) will be between 100.00 US dollars and 160.00 US dollars with the probability 0.2, and c) will remain equal to 100.00 US dollars (i.e., will remain unchanged) with the probability 0.2.

Then using the above formulae for calculating the expectation of the share price value for a security from the set I_t^-, one can easily be certain that the expectation of the share price value of security B at the moment $t + 1$ equals 103.00 US dollars, i.e., contrary to the trader initial analysis, the expectation of the share price value of this security will increase.

Remark 1. It is clear that finding an optimal investment strategy of the trader in Situation 1, one should add the equalities $v_{i,t} = 0, i \in \overline{1,n}$ and $x_{i,t}^- = 0,\ i \in I_t^-$ to the system of constraints of problem (1) and set $I_t^+(av) = \emptyset,\ I_t^-(av) = \emptyset,\ I_t^0(av) = \emptyset$ in the system of constraints of problem (2). Also, one should bear in mind that in the formulation of problems (1) and (2), it is assumed that the value of the money at which standard securities are sold at the moment t remains unchanged at the moment $t + 1$. However, if this is not the case, it is easy to reformulate problem (2) taking into consideration the difference in this value.

5 Finding Optimal Trader's Strategies of Investing in Standard Financial Securities. Model 2: The Trader Can Numerically Estimate Only the Areas in Which the Values of the Share Prices of all the Securities That Interest Her May Change

Let N be a set of (names of) standard securities that interest a trader at the moment t. Further, let us assume that at the moment t, a trader can choose (1) a set of securities $I_t^+ \subseteq N$ whose share price values (as she believes) will increase at the moment $t+1$ compared with their share price values at the moment t, and (2) a set of securities $I_t^- \subseteq N$ whose share price values (as she believes) will decrease at the moment $t+1$ compared with those at the moment t. Finally, let the trader correctly forecast that the share price values of securities from the set I_t^+ will increase with the probability $p^+ > 0.5$ (so that the share price values of securities from the set I_t^+ will not increase with the probability $1-p^+$). Analogously, let the trader correctly forecast that the share price values of securities from the set I_t^- will decrease with the probability $p^- > 0.5$ (so that the share price values of securities from the set I_t^- will not decrease with the probability $1-p^-$).

If (a) the set of securities N also contains standard securities forming the subset $I_t^0 = N \setminus (I_t^+ \cup I_t^-)$, (b) she believes that the share price values of securities forming this set may change at the moment $t+1$, and (c) she does not have any assumptions on the direction in which the share price values of these securities will change at the moment $t+1$, it seems natural to assume that both the increasing and not increasing of the share price values of these securities are equally possible with the probability 0.5. It is natural to assume that $I_t^+ \cap I_t^- = \emptyset$, $I_t^- \cap I_t^0 = \emptyset$, and $I_t^+ \cap I_t^0 = \emptyset$.

Let

1. $x_t = (x_t^+, x_t^-, x_t^0)$ be the vector of volumes (numbers of shares) of securities from the set N that the trader intends to buy and to sell at the moment t (based on her beliefs), where $x_t^+ \in X_t^+ \subset R_+^{|I_t^+|}$ is the vector of volumes (numbers of shares) of such securities from the set I_t^+, $x_t^- \in X_t^- \subset R_+^{|I_t^-|}$ is the vector of volumes (numbers of shares) of such securities from the set I_t^-, and $x_t^0 \in X_t^0 \subset R_+^{|I_t^0|}$ is the vector of volumes (numbers of shares) of such securities from the set I_t^0;
2. $y_{t+1} = (y_{t+1}^+, y_{t+1}^-, y_{t+1}^0) \in Y_{t+1}^+ \times Y_{t+1}^- \times Y_{t+1}^0 \subset R_+^{|I_t^+|+|I_t^-|+|I_t^0|}$ be the vector whose components are the values of the share prices of securities from the set N at the moment $t+1$ if the trader correctly determines directions in which the values of the share prices of these securities may change, where $y_{t+1}^+ \in Y_{t+1}^+ \subset R_+^{|I_t^+|}$ is the vector whose components are the values of the share prices of securities from the set I_t^+ at the moment $t+1$ if the trader correctly determines directions in which these values of the share prices may change (with the probability $p^+ > 0.5$), $y_{t+1}^- \in Y_{t+1}^- \subset R_+^{|I_t^-|}$ is the vector whose components are the values of the share prices of securities from the set I_t^- at the moment $t+1$ if the trader correctly determines directions in which these values of the share prices may change (with the probability $p^- > 0.5$), and $y_{t+1}^0 \in Y_{t+1}^0 \subset R_+^{|I_t^0|}$ is the vector whose components are the values of the share prices of securities from the set I_t^0 at which they will be available in the stock exchange at the moment $t+1$, if the trader correctly determines the areas in which these values of the share prices may change (with the probability $p^0 = 0.5$);
3. $z_{t+1} = (z_{t+1}^+, z_{t+1}^-, z_{t+1}^0) \in Z_{t+1}^+ \times Z_{t+1}^- \times Z_{t+1}^0 \subset R_+^{|I_t^+|+|I_t^-|+|I_t^0|}$ be the vector whose components are the values of the share prices of securities from the set N at the moment $t+1$ if the trader incorrectly determines directions in which the values of the share prices of these securities may change, where $z_{t+1}^+ \in Z_{t+1}^+ \subset R_+^{|I_t^+|}$ is the vector whose components are the values of the share prices of securities from the set I_t^+ at the moment $t+1$ if the trader incorrectly determines directions in which these values of the share prices may change (with the probability $1-p^+$), $z_{t+1}^- \in Z_{t+1}^- \subset R_+^{|I_t^-|}$ is the vector whose components are the values of the share prices of securities from the set I_t^- at the moment $t+1$ if the trader incorrectly determines directions in which these values of the share prices may change (with the probability $1-p^-$), and $z_{t+1}^0 \in Z_{t+1}^0 \subset R_+^{|I_t^0|}$ is the vector whose components are the values of the share prices of securities from

the set I_t^0 at which they will be available in the stock exchange at the moment $t+1$ if the trader incorrectly determines the areas in which these values of the share prices may change (with the probability $1-p^0=0.5$).

Throughout Sect. 5, the optimality of the strategy to be exercised at the moment t is understood in the sense of maximizing the value of the trader' s portfolio at the moment $t+1$.

As mentioned earlier (see Sect. 4), the trader may consider finding an optimal investment strategy in two situations: (a) in forming a new portfolio (Situation 1), and (b) in changing a composition of the existing portfolio (Situation 2). Unlike in Sect. 5, for Model 2, Situation 1 is considered first, and based upon the analysis of the results obtained for Situation 1, Situation 2 is considered.

Situation 1

Let the trader possess no securities at the moment t, and let N be a set of (the names of) standard securities that interest the trader at the moment t. As before, let $N = I_t^+ \cup I_t^- \cup I_t^0$, $I_t^+ \cap I_t^- = \emptyset$, $I_t^+ \cap I_t^0 = \emptyset$, $I_t^- \cap I_t^0 = \emptyset$, where all the three sets have the same meaning as described at the beginning of Sect. 4, and let $|N| = |I_t^+| + |I_t^-| + |I_t^0| = n$.

It is obvious that if the trader does not possess any standard securities at the moment t, she can only either buy these securities (by investing cash that she possesses at the moment t) or borrow money or securities or both (under certain conditions offered by potential lenders or/and brokers that the trader views to be acceptable) and use the borrowed money (or/and the money to be received as a result of selling the borrowed securities short) to invest it in securities from the set N. With respect to Situation 1, the vectors x_t^- and x_t^0 should be understood as volumes of those securities (from the set $I_t^- \cup I_t^0$) that are the only securities that the trader may eventually consider to borrow from brokers to sell these securities short to receive the above-mentioned cash. However, at the moment t, the trader also has a certain amount of cash (see the description of the underlying conditions of Situation 1 at the beginning of Sect. 4). So her major problem is to find the best variant of investing all the cash that she can afford to invest in securities (i.e., in buying securities) at the moment t in such a manner that the value of her portfolio of securities, which is formed as a result of this investment, will be maximal at the moment $t+1$.

Thus, in Situation 1, all the three sets I_t^+, I_t^-, and I_t^0 are those from which the trader may buy securities at the moment t, and the trader forms all these three sets at any moment t at her own discretion, proceeding from her financial abilities at the moment t. One should also emphasize that if the trader decides to borrow securities from a broker to sell them short (provided the broker offers such a transaction to the trader), and the trader can choose which particular securities to borrow within financial limitations agreed upon by both parties, she will do this in the course of forming the above three sets.

It is clear that if at the moment t, the trader were absolutely sure that the share price values of all the securities from the set I_t^+ would only increase, the share price values of all securities from the set I_t^- would only decrease, and if the set I_t^0 were empty, then she would invest all the cash available at the moment t in securities from the set I_t^+. The trader would certainly not buy any securities from the set I_t^- though she would borrow securities from this set to sell them short at the moment t (provided such a transaction is offered to her by any broker or brokers) and to invest the money received (from this selling) in securities from the set I_t^+ by adding the money received to all the cash available to the trader at the moment t (for the purpose of investing in standard securities). As one will have a chance to be certain, mathematically, the corresponding problem is a particular case of the problem under consideration in this section of the paper.

However, in the rest of this section, it is assumed that the trader believes that (a) the share price values of each of securities from the set I_t^+ may increase only with a certain probability p^+, whereas these values may decrease with the probability $1 - p^+$, and (b) there is a non-empty set I_t^0 of securities for each of which its share price value may increase or decrease with the same probability $p^0 = 0.5$. Analogously, the trader believes that the share price values of securities from the set I_t^- may decrease also only with a certain probability p^-, whereas they may increase with the probability $1 - p^-$.

Examples at the end of Sect. 4 are illustrative of such relations between the values of the probabilities $p^+ (p^+ > 0.5)$, $p^- (p^- > 0.5)$, and the values of coordinates of the vectors from the set $X_t^+ \cup X_t^-$ that the expectations of the share price values of some securities from the set I_t^+ at the moment $t + 1$ are lower than their current values (i.e., those at the moment t), whereas the expectations of the share price values of some securities from the set I_t^- at the moment $t + 1$ exceed their current values. The same reasons are applicable to the set I_t^0 as well, which explains the trader's interest to securities from this set.

While it seems quite clear how the trader may form the sets X_t^+ and X_t^0, one may raise a natural question: what should be considered as the set X_t^- in Situation 1? The trader does not possess any securities at the moment t at all, and she assumes that if she possessed securities from the set I_t^-, she would certainly have sold at least some of them trying to protect her welfare. When the optimality of the strategy to be exercised at the moment t is understood in the sense of maximizing the value of the trader's portfolio at the moment $t + 1$ (which is the case under consideration in this paper), at least three approaches to what the trader may view as the set X_t^- are possible.

Approach 1. Taking into consideration the above-mentioned examples at the end of Sect. 4, the trader considers spending a part of the available (her own) cash for buying securities from the set I_t^- at the moment t at the share price values of these securities that exist at the moment t (while she has no access to credits in any form), and possible (feasible) variants of allocating this money among these securities determine the set X_t^-. The determination of the set X_t^- should be followed by making a decision on which securities from the set I_t^- (or from its subset) and in which volumes to buy to maximize the trader portfolio's value at the moment $t + 1$. (In contrast, choosing particular volumes of securities (to buy and to sell) proceeding from a particular vector of them in the already determined set X_t^- corresponds to considering Situation 2 in which the trader already has securities from the set X_t^- in her portfolio.)

Approach 2. The trader can get a cash credit at the moment t on certain acceptable conditions, and she is to return this credit at the moment $t + 1$ or later (possibly, with some interest in both cases). Once again, taking into consideration the above-mentioned examples at the end of Sect. 4, the trader may consider spending a part of this credit for buying some securities from the set I_t^- at the moment t in an attempt to maximize the trader portfolio's value at the moment $t + 1$ in just the same way this is done under Approach 1.

Approach 3. At the moment t, the trader borrows securities from a subset of the set I_t^- from a broker to sell the borrowed securities short at the moment t; however, she is to return the borrowed securities to the broker (possibly, along with a certain percentage of the cost of the borrowed securities to be paid to the broker for using them as a form of a credit from this broker) later than at the moment $t + 1$. This move is based on the hope that at the time of returning the borrowed securities, their share price values will be lower than those at which these securities were borrowed. The trader uses the money received as a result of selling the borrowed securities short for buying securities from the set N. Here, as under Approaches 1 and 2, the trader's aim is to maximize the value of her portfolio at the moment $t + 1$, and securities to borrow are chosen from among those from the set I_t^- that are offered by the broker. The trader is interested in borrowing such securities from the broker whose share price values at the moment t would allow her to sell these securities at the maximal possible amount of money to be added to the trader's cash (that she can afford to spend for buying securities at the moment t). This borrowing is done with the aim of spending all the cash (that the trader can spend for buying securities from the whole set N at the moment t) to maximize the trader portfolio's value at the moment $t + 1$.

Thus, under any of these three approaches, one may consider that at the moment t, the trader has a certain amount of cash that she can spend in forming her portfolio in such a manner that this portfolio would have the maximal market value at the moment $t + 1$. (Here, some strategies of allocating a portion of the cash available at the moment t for buying some securities to be returned to the broker (or to the brokers) later than at the moment $t + 1$ can be exercised.) It is the allocation of this cash either among securities from the set I_t^+ only (and taking the risk shown in examples at the end of Sect. 4) or among securities from the set $I_t^+ \cup I_t^- \cup I_t^0$ that determines the set X_t^-.

Let us first consider Situation 1 assuming that one of the above three approaches to determining the set X_t^- is used, which means that no matter what particular approach is employed, taking into consideration examples at the end of Sect. 4, the trader chooses which securities to buy from all the three sets I_t^+, I_t^-, and I_t^0 to form her portfolio.

Further, let at each moment t the trader proceed from the existence of linear constraints of the balance kind imposed on the components of the vector x_t, including bilateral constraints-inequalities imposed on each component of each of the three vectors forming the vector x_t. It is natural to assume that these constraints, which, in particular, reflect the trader's financial status at the moment t, are compatible. The presence of such constraints allows one to consider, for instance, that the sets X_t^+ (the set of feasible values of the vector x_t^+), X_t^- (the set of feasible values of the vector x_t^-) , and X_t^0 (the set of feasible values of the vector x_t^0) are formed by the vectors from subsets of convex polyhedra $M_t^+ \subset R_+^{|I_t^+|}$, $M_t^- \subset R_+^{|I_t^-|}$, and $M_t^0 \subset R_+^{|I_t^0|}$, respectively. In this case, each of these three polyhedra is described by a system of compatible linear constraints binding variables forming vectors from the corresponding space only, and the above-mentioned subset of the polyhedron is determined by the requirement for all the coordinates of the vectors from this subset to be non-negative integers so that (a) the above mentioned subsets take the form $M_t^+ = \{x_t^+ \in R_+^{|I_t^+|} : B_t^+ x_t^+ \geq d_t^+, x_t^+ \in Q_+^{|I_t^+|}\}$, $M_t^- = \{x_t^- \in R_+^{|I_t^-|} : B_t^- x_t^- \geq d_t^-, x_t^- \in Q_+^{|I_t^-|}\}$, and $M_t^0 = \{x_t^0 \in R_+^{|I_t^0|} : B_t^0 x_t^0 \geq d_t^0, x_t^0 \in Q_+^{|I_t^0|}\}$, where B_t^+, B_t^-, B_t^0 are matrices, d_t^+, d_t^-, d_t^0 are vectors of corresponding sizes, and Q_+^k is a direct product of k sets of the set of all non-negative integers Q_+, and (b) X_t, a set of feasible values of the vectors $x_t = (x_t^+, x_t^-, x_t^0)$, has the form $X_t = M_t^+ \times M_t^- \times M_t^0$. According to the assumptions on the bilateral constraints-inequalities, the sets M_t^+, M_t^-, and M_t^0 are either subsets of the corresponding parallelepipeds or coincide with them.

However, generally, the sets of feasible values X_t^+ , X_t^- , and X_t^0 may be determined by a set of linear equations and inequalities binding together the variables being coordinates of all the vectors x_t^+, x_t^-, and x_t^0 so that X_t (the set of feasible values of the vectors $x_t = (x_t^+, x_t^-, x_t^0)$), may have the form $X_t = M_t = \{x_t \in R_+^n : B_t x_t \geq d_t, x_t \in Q_+^n\} \subset M_t^+ \times M_t^- \times M_t^0$, where each of these three sets is non-empty and contains the zero vector. Analogously, it is natural to assume that each of the sets Y_{t+1}^+, Y_{t+1}^-,Y_{t+1}^0 and Z_{t+1}^+, Z_{t+1}^-, Z_{t+1}^0 is a (non-empty) convex polyhedron, since the values of the share prices of securities from the set N are non-negative, real numbers bounded from above. Finally, let the trader believe that at each moment t, the directions in which the values of the share prices of securities from the set N may change are "connected" within each of the three sets I_t^+, I_t^-, and I_t^0. Here, this "connection" is understood in the sense that the values of the share prices of all the securities from the set I_t^+ will change in one and the same direction at the moment $t + 1$, and the same is true for the values of the share prices of all the securities from each of the two sets I_t^- and I_t^0. Also, let the trader believe that the share price values within each of the sets Y_{t+1}^+, Y_{t+1}^-,Y_{t+1}^0 and Z_{t+1}^+, Z_{t+1}^-, Z_{t+1}^0 change independently of those in the other five sets.

At each moment, one may view the interaction between the trader and the stock exchange in Situation 1 as an antagonistic game between them. In this game, a strategy of the trader is to choose a) how many shares of securities from the sets I_t^+, I_t^-, and I_t^0 should be bought at the moment t, and b) how many shares of securities from the set N to borrow from a broker to sell them short at the moment t (see Remark 3 at the end of Sect. 5) with the intent of both types of the transactions to form her portfolio with the maximum possible value at the moment $t + 1$. The stock exchange's strategy in this game is "to choose" the values of the share prices of securities from the set N the most unfavorably to the trader. This game can be viewed to be analogous to the game with the nature in which "the nature" (the stock exchange in the game under consideration) may offer the trader the most unfavorable combinations of the values of the share prices that securities from the sets I_t^+, I_t^-, and I_t^0 may assume at the moment $t + 1$ (while the trader chooses the volumes of security shares to buy from each of these three sets at the moment t). These combinations (of the share price values) are chosen in the form of vectors from the (non-empty) convex polyhedra Y_{t+1}^+, Y_{t+1}^-,Y_{t+1}^0 and Z_{t+1}^+, Z_{t+1}^-, Z_{t+1}^0, and (as mentioned earlier) vectors from these convex polyhedra are chosen independently of each other.

The structure of this game allows one to find an optimal trader's strategy by solving a mixed programming problem. Finding an upper bound of the trader's guaranteed result in this game can be done by solving linear programming problems forming a dual pair [6].

Theorem. *There exists an antagonistic game describing the interaction between the trader and the stock exchange at each moment t, and this game is the one on (non-empty) sets of disjoint player strategies one of which is* $X_t = M_t = \{x_t \in R_+^n : B_t x_t \geq d_t, x_t \in Q_+^n\} \subseteq M_t^+ \times M_t^- \times M_t^0$, *and the other is* $\theta_{t+1} = \{w_{t+1} \in R_+^{2n} : A_t w_{t+1} \geq b_t\}$ *with the bilinear payoff function* $\langle x_t, D_t w_{t+1}\rangle$, *where*

$$D_t = \begin{pmatrix} D^{|I_t^+|}(p^+) & D^{|I_t^+|}(1-p^+) & 0 & 0 & 0 & 0 \\ 0 & 0 & D^{|I_t^-|}(p^-) & D^{|I_t^-|}(1-p^-) & 0 & 0 \\ 0 & 0 & 0 & 0 & D^{|I_t^0|}(\frac{1}{2}) & D^{|I_t^0|}(\frac{1}{2}) \end{pmatrix},$$

$x_t = (x_t^+, x_t^-, x_t^0) \in X_t$, $w_{t+1} = (w_{t+1}^+, w_{t+1}^-, w_{t+1}^0) \in \theta_{t+1} = \theta_{t+1}^+ \times \theta_{t+1}^- \times \theta_{t+1}^0$, *$D_t$ is a* $(|I_t^+| + |I_t^-| + |I_t^0|) \times 2(|I_t^+| + |I_t^-| + |I_t^0|)$ *matrix,* $D^{|I|}$ *is a* $(|I| \times |I|)$ *diagonal matrix all whose elements on the main diagonal equal x, X_t is a set of the trader's strategies,* θ_{t+1} *is a set of the stock exchange strategies,* $\theta_{t+1}^+ = Y_{t+1}^+ \times Z_{t+1}^+$, $\theta_{t+1}^- = Y_{t+1}^- \times Z_{t+1}^-$, $\theta_{t+1}^0 = Y_{t+1}^0 \times Z_{t+1}^0$ *are (non-empty) convex polyhedra,* $w_{t+1}^+ = (y_{t+1}^+, z_{t+1}^+) \in \theta_{t+1}^+$, $w_{t+1}^- = (y_{t+1}^-, z_{t+1}^-) \in \theta_{t+1}^-$, $w_{t+1}^0 = (y_{t+1}^0, z_{t+1}^0) \in \theta_{t+1}^0$ *are vectors,* Q_+ *is the set of all non-negative, integer numbers,* Q_+^n *is a direct product of n sets* Q_+, *and the payoff function is maximized with respect to the vector x and is minimized with respect to the vector* w_{t+1}. *In this game, an optimal trader's strategy is the one at which the maximin of the payoff function of the game is attained, and finding the exact value of this maximin is reducible to solving a mixed programming problem. Finding an upper bound of this maximin is reducible to solving linear programming problems forming a dual pair [7].*

Proof. Let us first assume that the set of trader's strategies X_t is a direct product of the three subsets of vectors with all integer components from disjoint polyhedra M_t^+, M_t^-, and M_t^0, i.e., $X_t = X_t^+ \times X_t^- \times X_t^0 = M_t^+ \times M_t^- \times M_t^0$ in the spaces $R_+^{|I_t^+|}$,

$R_+^{|I_t^-|}$, and $R_+^{|I_t^0|}$, respectively, where $M_t^+ = \{x_t \in R_+^{|I_t^+|} : B_t^+ x_t^+ \geq d_t^+, x_t \in Q_+^{|I_t^+|}\}$, $M_t^- = \{x_t^- \in R_+^{|I_t^-|} : B_t^- x_t^- \geq d_t^-, x_t^- \in Q_+^{|I_t^-|}\}$, and $M_t^+ = \{x_t^0 \in R_+^{|I_t^0|} : B_t^0 x_t^0 \geq d_t^0, x_t^0 \in Q_+^{|I_t^0|}\}$.

1. Let us consider securities forming the set I_t^+ at the moment t. If the trader correctly forecast directions in which the values of the share prices of securities from this set may change, then a) by buying securities from the set I_t^+ in the volumes (numbers of shares) being components of the vector x_t^+, and b) by expecting the values of the share prices of these securities at the moment $t+1$ to be components of the vector y_{t+1}^+, the trader would hope to invest the money available to her at the moment t in such a manner that would maximize the value of the part of her portfolio (associated with securities from the set I_t^+) at the moment $t+1$. Here, the trader's best investment strategy in the game with the stock exchange (with "the nature") with respect to securities from the set X_t^+ consists of choosing such volumes of securities from the set I_t^+ to buy that can be found by solving the problem

$$\min_{y_{t+1}^+ \in Y_{t+1}^+} \langle x_t^+, y_{t+1}^+ \rangle \to \max_{x_t^+ \in X_t^+}.$$

If the trader did not correctly forecast the directions in which the values of the share prices of securities from the set I_t^+ may change, i.e., if the values of the share prices of securities from the set I_t^+ did not increase at the moment $t+1$, the best investment strategy of the trader in her game with the stock exchange with respect to securities from the set X_t^+ would be determined by solutions to the problem

$$\min_{z_{t+1}^+ \in Z_{t+1}^+} \langle x_t^+, z_{t+1}^+ \rangle \to \max_{x_t^+ \in X_t^+}.$$

Since the trader correctly forecasts the directions in which the values of the share prices of securities from the set I_t^+ may change only with the probability p^+, the worst financial result of the trader's choice of the volumes of securities from the set I_t^+ to be bought at the moment t, i.e., the worst financial result of choosing the vector $x_t^+ \in X_t^+$ at the moment t by the trader, can be viewed as a discrete random variable taking the values $\min_{y_{t+1}^+ \in Y_{t+1}^+} \langle x_t^+, y_{t+1}^+ \rangle$ and $\min_{z_{t+1}^+ \in Z_{t+1}^+} \langle x_t^+, z_{t+1}^+ \rangle$ with the probabilities p^+ and $1 - p^+$, respectively. It is clear that an optimal trader's strategy in the case under consideration may consist of choosing a vector $x_t^+ \in X_t^+$ that maximizes the expectation of this discrete random variable. If this is the case, the optimal trader's strategy is found by solving the problem

$$p^+ \min_{y_{t+1}^+ \in Y_{t+1}^+} \langle x_t^+, y_{t+1}^+ \rangle + (1 - p^+) \min_{z_{t+1}^+ \in Z_{t+1}^+} \langle x_t^+, z_{t+1}^+ \rangle \to \max_{x_t^+ \in X_t^+}.$$

One can easily be certain that the equality

$$\max_{x_t^+ \in X_t^+} \left[p^+ \min_{y_{t+1}^+ \in Y_{t+1}^+} \langle x_t^+, y_{t+1}^+ \rangle + (1-p^+) \min_{z_{t+1}^+ \in Z_{t+1}^+} \langle x_t^+, z_{t+1}^+ \rangle \right] =$$

$$\max_{x_t^+ \in X_t^+} \left[\min_{y_{t+1}^+ \in Y_{t+1}^+} \langle x_t^+, D^{|I_t^+|}(p^+) y_{t+1}^+ \rangle + \min_{z_{t+1}^+ \in Z_{t+1}^+} \langle x_t^+, D^{|I_t^+|}(1-p^+) z_{t+1}^+ \rangle \right]$$

holds, and since the vectors y_{t+1}^+ and z_{t+1}^+ from the sets Y_{t+1}^+ and Z_{t+1}^+ are chosen independently of each other, the following equalities also hold:

$$\max_{x_t^+ \in X_t^+} \left[\min_{y_{t+1}^+ \in Y_{t+1}^+} \langle x_t^+, D^{|I_t^+|}(p^+) y_{t+1}^+ \rangle + \min_{z_{t+1}^+ \in Z_{t+1}^+} \langle x_t^+, D^{|I_t^+|}(1-p^+) z_{t+1}^+ \rangle \right] =$$

$$\max_{x_t^+ \in X_t^+} \left[\min_{(y_{t+1}^+, z_{t+1}^+) \in Y_{t+1}^+ \times Z_{t+1}^+} \langle x_t^+, D^{|I_t^+|}(p^+) D^{|I_t^+|}(1-p^+)(y_{t+1}^+, z_{t+1}^+) \rangle \right] =$$

$$\max_{x_t^+ \in X_t^+} \left[\min_{w_{t+1}^+ \in \theta_{t+1}^+} \langle x_t^+, D^{2|I_t^+|}(p^+, 1-p^+) w_{t+1}^+ \rangle \right],$$

where $w_{t+1}^+ = (y_{t+1}^+, z_{t+1}^+)$, $\theta_{t+1}^+ = Y_{t+1}^+ \times Z_{t+1}^+$, $D^{2|I_t^+|}(p^+, 1-p^+) = D^{|I_t^+|}(p^+) D^{|I_t^+|}(1-p^+)$, $D^{|I_t^+|}(p^+)$ is a $|I_t^+| \times |I_t^+|$ diagonal matrix all whose elements on the main diagonal equal p^+, $D^{|I_t^+|}(1-p^+)$ is a $|I_t^+| \times |I_t^+|$ diagonal matrix all whose elements on the main diagonal equal $1-p^+$, and $D^{2|I_t^+|}(p^+, 1-p^+)$ is a $|I_t^+| \times 2|I_t^+|$ matrix formed by accessing the matrix $D^{|I_t^+|}(1-p^+)$ to the matrix $D^{|I_t^+|}(p^+)$ from the right.

2. Let us consider securities forming the set I_t^- at the moment t. If the trader correctly forecast directions in which the values of the share prices of securities from this set may change, then a) by buying securities from the set I_t^- in the volumes (numbers of shares) being components of the vector x_t^-, and b) by expecting the values of the share prices of these securities at the moment $t+1$ to be components of the vector y_{t+1}^-, the trader would hope to invest the money available to her at the moment t in such a manner that would maximize the value of the part of her portfolio (associated with securities from the set I_t^-) at the moment $t+1$. Here, the trader's best investment strategy in the game with the stock exchange with respect to securities from the set X_t^- consists of choosing such volumes (numbers of shares) of securities to buy that can be found by solving the problem

$$\min_{y_{t+1}^- \in Y_{t+1}^-} \langle x_t^-, y_{t+1}^- \rangle \to \max_{x_t^- \in X_t^-}.$$

If the trader did not correctly forecast directions in which the values of the share prices of securities from the set I_t^- may change, i.e., if the values of the share prices of securities from the set I_t^- did not decrease at the moment $t+1$, the best investment strategy of the trader in her game with the stock exchange with respect to securities from the set X_t^- would be determined by solutions to the problem

$$\min_{z_{t+1}\in Z_{t+1}^-} \langle x_t^-, z_{t+1}^- \rangle \to \max_{x_t^-\in X_t^-} .$$

The reasoning analogous to the one presented in part 1 of this Proof lets one write the expression for the expectation of the worst financial result of the trader's decision to buy securities from the set I_t^- in the volumes (numbers of shares) determined by the vector x_t^- in the form

$$\min_{w_{t+1}^-\in\theta_{t+1}^-} \langle x_t^-, D^{2|I_t^-|}(p^-, 1-p^-)w_{t+1}^- \rangle.$$

Under the assumption on the optimality of the trader's strategy that was made with respect to securities from the set I_t^+, one can be certain that the trader tries to maximize her expected financial result associated with choosing the vector $x_t^- \in X_t^-$ by solving the problem

$$\max_{x_t^-\in X_t^-} \left[\min_{w_{t+1}^-\in\theta_{t+1}^-} \langle x_t^-, D^{2|I_t^-|}(p^-, 1-p^-)w_{t+1}^- \rangle \right],$$

where $w_{t+1}^- = (y_{t+1}^-, z_{t+1}^-)$, $\theta_{t+1}^- = Y_{t+1}^- \times Z_{t+1}^-$, $D^{2|I_t^-|}(p^-, 1-p^-) = D^{|I_t^-|}(p^-)D^{|I_t^-|}(1-p^-)$, $D^{|I_t^-|}(p^-)$ is a $|I_t^-| \times |I_t^-|$ diagonal matrix all whose elements on the main diagonal equal p^-, $D^{|I_t^-|}(1-p^-)$ is a $|I_t^-| \times |I_t^-|$ diagonal matrix all whose elements on the main diagonal equal $1-p^-$, and $D^{2|I_t^-|}(p^-, 1-p^-)$ is a $|I_t^-| \times 2|I_t^-|$ matrix formed by accessing the matrix $D^{|I_t^-|}(1-p^-)$ to the matrix $D^{|I_t^-|}(p^-)$ from the right.

3. Let us consider securities forming the set I_t^0 at the moment t for which the trader determines the direction in which the values of their share prices at the moment $t+1$ may change with the probability $p^0 = 1/2$. The best investment strategy of the trader in her game with the stock exchange with respect to securities from the set I_t^0 would be to choose the volumes (numbers of shares) of securities from this set that are determined by solutions to the problems

$$\min_{y_{t+1}^0\in Y_{t+1}^0} \langle x_t^0, y_{t+1}^0 \rangle \to \max_{x_t^0\in X_t^0} .$$

and

$$\min_{z_{t+1}^0\in Z_{t+1}^0} \langle x_t^0, z_{t+1}^0 \rangle \to \max_{x_t^0\in X_t^0} .$$

A reasoning similar to that presented in parts 1 and 2 of this Proof allows one to write the expression for the expectation of the financial result associated with choosing (buying) by the trader the volumes (numbers of shares) of securities from the set I_t^0 (being components of the vector x_t^0) in the form

$$\min_{w_{t+1}^0 \in \theta_{t+1}^0} \langle x_t^0, D^{2|I_t^0|}(p^0, 1-p^0) w_{t+1}^0 \rangle,$$

Under the same assumption on the optimality of the trader's strategy that was made with respect to securities from the set I_t^+ and I_t^-, the trader tries to maximize this minimum by choosing the vector $x_t^0 \in X_t^0$ as a vector component of a solution to the problem

$$\max_{x_t^0 \in X_t^0} \left[\min_{w_{t+1}^0 \in \theta_{t+1}^0} \langle x_t^0, D^{2|I_t^0|}(p^0, 1-p^0) w_{t+1}^0 \rangle \right],$$

where $w_{t+1}^0 = (y_{t+1}^0, z_{t+1}^0)$, $\theta_{t+1}^0 = Y_{t+1}^0 \times Z_{t+1}^0$, $D^{2|I_t^0|}(p^0, 1-p^0) = D^{|I_t^0|}(p^0) D^{|I_t^0|}(1-p^0)$, $D^{|I_t^0|}(p^0)$ is a $|I_t^0| \times |I_t^0|$ diagonal matrix all whose elements on the main diagonal equal p^0, $D^{|I_t^0|}(1-p^0)$ is a $|I_t^0| \times |I_t^0|$ diagonal matrix all whose elements on the main diagonal equal $1-p^0$, and $D^{2|I_t^0|}(p^0, 1-p^0)$ is a $|I_t^0| \times 2|I_t^0|$ matrix formed by accessing the matrix $D^{|I_t^0|}(1-p^0)$ to the matrix $D^{|I_t^0|}(p^0)$ from the right.

4. Since the financial results of choosing the volumes (numbers of shares) of securities from the sets I_t^+, I_t^-, and I_t^0 are random variables (since the trader forecasts the directions in which the values of their share prices at the moment $t+1$ will change within the polyhedra Y_{t+1}^+, Y_{t+1}^-,Y_{t+1}^0 and Z_{t+1}^+, Z_{t+1}^-, Z_{t+1}^0 only with certain probabilities), the expectations of the worst compound financial result is a sum of the above three expectations [17].

   Let the matrix D_t have the form

$$D_t = \begin{pmatrix} D^{2|I_t^+|}(p^+, 1-p^+) & 0 & 0 \\ 0 & D^{2|I_t^-|}(p^-, 1-p^-) & 0 \\ 0 & 0 & D^{2|I_t^0|}(p^0, 1-p^0) \end{pmatrix} =$$

$$= \begin{pmatrix} D^{|I_t^+|}(p^+)\ D^{|I_t^+|}(1-p^+) & 0 & 0 & 0 & 0 \\ 0 & 0 & D^{|I_t^-|}(p^-)\ D^{|I_t^-|}(1-p^-) & 0 & 0 \\ 0 & 0 & 0 & 0 & D^{|I_t^0|}(\frac{1}{2})\ D^{|I_t^0|}(\frac{1}{2}) \end{pmatrix},$$

while $x_t = (x_t^+, x_t^-, x_t^0)$ belongs to the set X_t, and $w_{t+1} = (w_{t+1}^+, w_{t+1}^-, w_{t+1}^0)$ belongs to the convex polyhedron $\theta_{t+1} = \theta_{t+1}^+ \times \theta_{t+1}^- \times \theta_{t+1}^0$. Further, let linear inequalities describing the convex polyhedron $\theta_{t+1} = \{w_{t+1} \in R_+^{2n} : A_t w_{t+1} \geq b_t\}$ be compatible so that A_t, B_t are matrices, and b_t, d_t are vectors of corresponding dimensions, whose elements are formed by the coefficients of the above two compatible systems of linear equations and inequalities. Then, when the trader chooses a particular vector x_t from the set X_t, the expectation of the compound worst financial result determined by this choice can be calculated as

$$\min_{w_{t+1}\in\theta_{t+1}} \langle x_t, D_t w_{t+1}\rangle.$$

5. Let now $X_t = M_t = \{x_t \in R^n_+ : B_t x_t \geq d_t, x_t \in Q^n_+\} \subset M_t^+ \times M_t^- \times M_t^0$, where B_t is a matrix of a general structure, not necessarily corresponding to the structure of the set $M_t = M_t^+ \times M_t^- \times M_t^0$ as a direct product of subsets of the three polyhedra from the spaces $R_+^{|I_t^+|}$, $R_+^{|I_t^-|}$, and $R_+^{|I_t^0|}$, respectively. This means that the system of linear equations and inequalities in the description of the set M_t contains at least one that binds together components of all the three vectors x_t^+, x_t^-, and x_t^0.

Let the trader choose the vector $x_t = (x_t^+, x_t^-, x_t^0) \in X_t$. Depending on in which direction the share price values of securities from the sets I_t^+, I_t^-, and I_t^0 may change, the trader may obtain the following worst financial results:

1. $\min_{y_{t+1}^+\in Y_{t+1}^+} \langle x_t^+, y_{t+1}^+\rangle$ or $\min_{z_{t+1}^+\in Z_{t+1}^+} \langle x_t^+, z_{t+1}^+\rangle$ for securities from the set I_t^+,
2. $\min_{y_{t+1}^-\in Y_{t+1}^-} \langle x_t^-, y_{t+1}^-\rangle$ or $\min_{z_{t+1}^-\in Z_{t+1}^-} \langle x_t^-, z_{t+1}^-\rangle$ for securities from the set I_t^-,
3. $\min_{y_{t+1}^0\in Y_{t+1}^0} \langle x_t^0, y_{t+1}^0\rangle$ or $\min_{z_{t+1}^0\in Z_{t+1}^0} \langle x_t^0, z_{t+1}^0\rangle$ for securities from the set I_t^0.

According to the (earlier made) assumptions on the sets Y_{t+1}^+, Y_{t+1}^-,Y_{t+1}^0 and Z_{t+1}^+, Z_{t+1}^-,Z_{t+1}^0,

(a) non-empty convex polyhedra in each of which all the components of the vectors belonging to the sets I_t^+, I_t^-, and I_t^0, respectively, change in one and the same direction, and
(b) the direction of changing the values for all the components of the vectors y_{t+1}^+, y_{t+1}^-,y_{t+1}^0 and z_{t+1}^+, z_{t+1}^-, z_{t+1}^0 are chosen (by the stock exchange) randomly, with the probabilities p^+, p^-, p^0 and $(1-p^+)$, $(1-p^-)$, $(1-p^0)$, respectively, independently of each other for all the components of these six vectors,

the above six worst financial results can be viewed as the values of three random variables ξ^+, ξ^-, ξ^0.

Each of these three random variables is, in turn, a discrete random variable with two possible values for each variable. That is, the discrete random variable ξ^+ assumes the values $\min_{y_{t+1}^+\in Y_{t+1}^+} \langle x_t^+, y_{t+1}^+\rangle$ and $\min_{z_{t+1}^+\in Z_{t+1}^+} \langle x_t^+, z_{t+1}^+\rangle$ with the probabilities p^+ and $1-p^+$, respectively (since, in line with assumption a), the probability with which all the components of those vectors whose components belong to the set I_t^+ hit the sets Y_{t+1}^+ and Z_{t+1}^+ with the probabilities p^+ and $1-p^+$, respectively). Analogously, the discrete random variable ξ^- assumes two values $\min_{y_{t+1}^-\in Y_{t+1}^-} \langle x_t^-, y_{t+1}^-\rangle$ and $\min_{z_{t+1}^-\in Z_{t+1}^-} \langle x_t^-, z_{t+1}^-\rangle$ with the probabilities p^- and $1-p^-$, respectively, whereas the discrete random variable ξ^0 assumes two values $\min_{y_{t+1}^0\in Y_{t+1}^0} \langle x_t^0, y_{t+1}^0\rangle$ and $\min_{z_{t+1}^0\in Z_{t+1}^0} \langle x_t^0, z_{t+1}^0\rangle$ with the probabilities p^o and $1-p^0$, respectively.

Since the expectation of the sum of the random variables ξ^+, ξ^-, ξ^0 equals the sum of their expectations [17], the equality

$$M[\xi^+ + \xi^- + \xi^0] = p^+(\min_{y_{t+1}^+ \in Y_{t+1}^+} \langle x_t^+, y_{t+1}^+ \rangle) + (1-p^+)(\min_{z_{t+1}^+ \in Z_{t+1}^+} \langle x_t^+, z_{t+1}^+ \rangle) +$$
$$p^-(\min_{y_{t+1}^- \in Y_{t+1}^-} \langle x_t^-, y_{t+1}^- \rangle) + (1-p^-)(\min_{z_{t+1}^- \in Z_{t+1}^-} \langle x_t^-, z_{t+1}^- \rangle) +$$
$$p^0(\min_{y_{t+1}^0 \in Y_{t+1}^0} \langle x_t^0, y_{t+1}^0 \rangle) + (1-p^0)(\min_{z_{t+1}^0 \in Z_{t+1}^0} \langle x_t^+, z_{t+1}^0 \rangle)$$

holds, which, in line with the notation from the formulation of the Theorem, takes the form

$$M[\xi^+ + \xi^- + \xi^0] = \min_{w_{t+1} \in \theta_{t+1}} \langle x_t, D_t w_{t+1} \rangle$$

for any $x_t \in X_t$.

6. It seems natural to consider that the best trader's choice of the vector x_t is the vector at which the maximin

$$\max_{x_t \in X_t} \left[\min_{w_{t+1} \in \theta_{t+1}} \langle x_t, D_t w_{t+1} \rangle \right]$$

is attained. Though all the components of the vector x_t are integers, the same logic that was applied in [6] in finding the maximum of the minimum function similar to the above one (but with all the components of the vector variable under the maximum sign assuming non-negative, real values) allows one to be certain that the equality

$$\max_{x_t \in X_t} \left[\min_{w_{t+1} \in \theta_{t+1}} \langle x_t, D_t w_{t+1} \rangle \right] = \max_{x_t \in X_t} \left[\max_{z_{t+1} \in \{z_{t+1} \geq 0 \ : z_{t+1} A_t \leq x_t D_t\}} \langle b_t, z_{t+1} \rangle \right]$$

holds. Indeed, since the set $\theta_{t+1} = \{w_{t+1} \in R_+^{2n} : A_t w_{t+1} \geq b_t\}$ is a (non-empty) convex polyhedron so that the linear function $\langle x_t, D_t w_{t+1} \rangle$ attains its minimum on this convex polyhedron for any $x_t \in X_t$, the set $\{z_{t+1} \geq 0 : z_{t+1} A_t \leq x_t D_t\}$, which is a set of feasible solutions to the linear programming problem that is dual to the problem $\min_{w_{t+1} \in \theta_{t+1}} \langle x_t, D_t w_{t+1} \rangle$, is nonempty for any $x_t \in X_t$ [60]. Thus, in both problems

$$\langle x_t, D_t w_{t+1} \rangle \to \min_{w_{t+1} \in \theta_{t+1}}$$

and

$$\langle b_t, z_{t+1} \rangle \to \max_{z_{t+1} \in \{z_{t+1} \geq 0 : z_{t+1} A_t \leq x_t D_t\}},$$

which form a dual pair of linear programming problems for any $x_t \in X_t$, the goal functions attain their extreme values at certain points of the sets $\theta_{t+1} = \{w_{t+1} \in R_+^{2n} : A_t w_{t+1} \geq b_t\}$ and $\{z_{t+1} \geq 0 : z_{t+1} A_t \leq x_t D_t\}$, respectively, for every $x_t \in X_t$ due to the duality theorem of linear programming [60]. Thus, the equality

$$\min_{w_{t+1} \in \theta_{t+1}} \langle x_t, D_t w_{t+1} \rangle = \max_{z_{t+1} \in \{z_{t+1} \geq 0 : z_{t+1} A_t \leq x_t D_t\}} \langle b_t, z_{t+1} \rangle$$

holds for every $x_t \in X_t$, and since the set X_t is finite, the equality

$$\max_{x_t \in X_t} \left[\min_{w_{t+1} \in \theta_{t+1}} \langle x_t, D_t w_{t+1} \rangle \right] = \max_{x_t \in X_t} \left[\max_{z_{t+1} \in \{z_{t+1} \geq 0 : z_{t+1} A_t \leq x_t D_t\}} \langle b_t, z_{t+1} \rangle \right]$$

also holds, which means that the equality

$$\max_{x_t \in X_t} \left[\min_{w_{t+1} \in \theta_{t+1}} \langle x_t, D_t w_{t+1} \rangle \right] = \max_{\{x_t \in R_+^n : B_t x_t \geq d_t, x_t \in Q_+^n\}} \left[\max_{z_{t+1} \in \{z_{t+1} \geq 0 : z_{t+1} A_t \leq x_t D_t\}} \langle b_t, z_{t+1} \rangle \right],$$

where Q_+ is a set of all non-negative, integer numbers, and Q_+^n is a direct product of n sets Q_+, holds. This means that the value

$$\max_{x_t \in M_t} \left[\min_{w_{t+1} \in \theta_{t+1}} \langle x_t, D_t w_{t+1} \rangle \right]$$

can be found by solving the problem

$$\langle b_t, z_{t+1} \rangle \to \max_{\{(x_t, z_{t+1}) \in R_+^n \times R_+^m : B_t x_t \geq d_t, z_{t+1} A_t \leq x_t D_t, x_t \in Q_+^n\}},$$

where m is the number of rows in the matrix A_t, which is a mixed programming problem.

7. It is clear that if the numbers of securities in the sets I_t^+, I_t^-, and I_t^0 are large, solving this problem may present considerable difficulties. At the same time, since the values of components of the vector x_t usually substantially exceed 1, one can consider these numbers as non-negative, real ones, solve the problem of finding

$$\max_{x_t \in \tilde{M}_t} \left[\min_{w_{t+1} \in \theta_{t+1}} \langle x_t, D_t w_{t+1} \rangle \right]$$

with the vector variables x_t belonging to the above-mentioned convex polyhedron $\tilde{M}_t = \{x_t \in R_+^n : B_t x_t \geq d_t\}$, which contains the set M_t (and is described by a compatible system of linear equations and inequalities), and round off all the non-integer components of the vector x_t in the solution in just the same way it was mentioned in Sect. 4 in considering problems (1) and (2). Thus (if the number of shares in the trader's portfolio is large), the trader may decide to calculate the above maximum of the minimum function, which is an upper bound for the number $\max_{x_t \in M_t} \left[\min_{w_{t+1} \in \theta_{t+1}} \langle x_t, D_t w_{t+1} \rangle\right]$. The value of this upper bound is attained at a saddle point of an antagonistic game on the convex polyhedra $\tilde{M}_t$ and θ_{t+1} with the payoff function

$$\langle x_t, D_t w_{t+1} \rangle. \tag{3}$$

Let

$$Q_t = \{(x_t, h_t) \geq 0 : h_t A_t \leq x_t D_t, B_t x_t \geq d_t\},$$

$$P_{t,t+1} = \{(w_{t+1}, \pi_{t+1}) \geq 0 : \pi_{t+1} B_t \leq -D_t w_{t+1}, A_t w_{t+1} \geq b_t\}.$$

Then the optimal values of the vectors $(x_t)^*$ and $(w_{t+1})^*$, forming a saddle point of function (3) on the set $\tilde{M}_t \times \theta_{t+1}$, are found as components of the solution vectors to linear programming problems

$$\langle b_t, h_t \rangle \to \max_{(x_t, h_t) \in Q_t},$$

$$\langle -d_t, \pi_{t+1} \rangle \to \min_{(w_{t+1}, \pi_{t+1}) \in P_{t+1}},$$

forming a dual pair.

If $((x_t)^*, (h_t)^*, (w_{t+1})^*, (\pi_{t+1})^*)$ is a solution of the above pair of linear programming problems, then the values of the vectors $(x_t^+)^*$, $(x_t^-)^*$ and $(x_t^0)^*$, where $(x_t)^* = ((x_t^+)^*, (x_t^-)^*, (x_t^0)^*)$, are completely determined by the values of the vector $(x_t)^*$ [6]. The Theorem is proved. □

Remark 2. As mentioned in the course of proving the Theorem, all the variables x_t are integers so that the value of the maximin of the function (3) when $x_t \in \tilde{M}_t$—which is attained at a saddle point of the game on the sets $\tilde{M}_t$ and θ_{t+1} with the payoff function $\langle x_t, D_t w_{t+1} \rangle$ that is maximized with respect to $x \in \tilde{M}_t$ and is minimized with respect to $w_{t+1} \in \theta_{t+1}$—is only an upper bound of the maximin of this function when $x_t \in M_t$. Also, as shown there, finding the exact value of this maximin is reducible to solving a mathematical programming problem with mixed variables and a linear goal function. However, it is clear that solving this mixed programming problem within an acceptable period of time may present considerable difficulties for the problems with sizes being of interest for both theoretical studies and practical calculations while solving linear programming problems in finding a saddle point of the game on $\tilde{M}_t \times \theta_{t+1}$ with the payoff function described by (3) does not present any computational difficulties in such calculations. Moreover, quickly finding an upper bound of the maximin of the function (3) may interest small and medium price-taking traders for their practical calculations the most. Also, in theoretical studies of the interaction between a trader and a stock exchange (to which the present paper belongs), traditionally (see, for instance, the seminal publication of Markowitz [33]), volumes of shares to be bought and sold by a trader are assumed to be non-negative, real numbers (variables). Finally, generally, the coefficients in the systems of linear equations and inequalities describing the convex polyhedra that participate in the mathematical formulation of the mixed programming problem under consideration are known only approximately. With all this in mind, the replacement of the problem of finding the exact value of the maximin of the function (3) when $x_t \in M_t$ with finding an upper bound of this value seems justifiable in practical calculations.

Situation 2

There are two cases to be considered in Situation 2. In the first case, the trader does not have any intent to keep particular securities that she possesses at the moment t (either based on her own beliefs or at someone's advice), whereas in the second case, the trader has this intent with respect to particular securities. It is clear that in the first case, to estimate what portfolio would have the maximum value at the moment $t+1$, the trader should first estimate the total cash that she would have if she sold all the securities from her portfolio at the moment t proceeding from the share price values that these securities have at the moment t. Then the trader should solve the same problem that she would solve in Situation 1 in forming a portfolio (a) proceeding from the total amount of cash available to her at the moment t, and (b) taking into account that she can borrow cash and/or securities from a broker to be returned later. If the borrowed cash or securities should be returned later than at the moment $t+1$, then in the first case of Situation 2, finding the trader's best investment strategies (in the sense of maximizing the value of her portfolio at the moment $t+1$) is either reducible to solving a mixed programming problem (for finding the exact value of the maximin of the function (3) when $x_t \in M_t$) or to finding saddle points in an antagonistic game (for finding an upper bound of the above-mentioned maximin) that are similar to those considered in finding such strategies earlier, in Situation 1.

In the second case of Situation 2, one can easily show that the considered game of changing the portfolio of securities is formulated as the game on the sets $M_t^+ \times M_t^- \times M_t^0$ or M_t and $\theta_{t+1} = \{w_{t+1} \in R_+^{2n} : A_t w_{t+1} \geq b_t\}$ of the player strategies with the payoff function $\langle x_t, D_t w_{t+1}\rangle + \langle q, w_{t+1}\rangle$. Here, $q \in R_+^{2n}$ is a particular vector, A_t, B_t are matrices, and d_t, b_t are vectors of corresponding dimensions. Their elements are formed by coefficients of compatible systems of linear equations and inequalities of the balance kind that describe sets of feasible values of the variables forming the vectors x_t and w_{t+1}. Two subcases should then be considered.

In the first subcase, the trader does not borrow any securities (from a broker) from the set I_t^-.

Let $v_t = (v_t^+, v_t^-, v_t^0) \in R_+^{|I_t^+|+|I_t^-|+|I_t^0|}$ be the vector of volumes (numbers of shares) of securities from the set N that the trader has in her portfolio at the moment t and would like to keep at the moment $t+1$ for whatever reasons. As in the Proof of the Theorem, let us first consider the case in which $X_t = X_t^+ \times X_t^- \times X_t^0 = M_t^+ \times M_t^- \times M_t^0$ in the spaces $R_+^{|I_t^+|}$, $R_+^{|I_t^-|}$, and $R_+^{|I_t^0|}$, respectively, where $M_t^+ = \{x_t \in R_+^{|I_t^+|} : B_t^+ x_t^+ \geq d_t^+, x_t \in Q_+^{|I_t^+|}\}$, $M_t^- = \{x_t^- \in R_+^{|I_t^-|} : B_t^- x_t^- \geq d_t^-, x_t^- \in Q_+^{|I_t^-|}\}$, and $M_t^0 = \{x_t^0 \in R_+^{|I_t^0|} : B_t^0 x_t^0 \geq d_t^0, x_t^0 \in Q_+^{|I_t^0|}\}$. Then the optimal trader's strategy of choosing the volumes of securities from the set X_t^+ to buy is found by maximizing the expectation of the discrete random variable

$$p^+ \min_{y_{t+1}^+ \in Y_{t+1}^+} [\langle x_t^+, y_{t+1}^+ \rangle + \langle v_t^+, y_{t+1}^+ \rangle] + (1-p^+) \min_{z_{t+1}^+ \in Z_{t+1}^+} [\langle x_t^+, z_{t+1}^+ \rangle + \langle v_t^+, z_{t+1}^+ \rangle],$$

which describes the expectation of the financial result associated with buying securities from the set I_t^+.

Since the equality

$$\max_{x_t^+ \in X_t^+} \left[p^+ \min_{y_{t+1}^+ \in Y_{t+1}^+} \left(\langle x_t^+, y_{t+1}^+ \rangle + \langle v_t^+, y_{t+1}^+ \rangle \right) + (1-p^+) \min_{z_{t+1}^+ \in Z_{t+1}^+} \left(\langle x_t^+, z_{t+1}^+ \rangle + \langle v_t^+, z_{t+1}^+ \rangle \right) \right] =$$

$$= \max_{x_t^+ \in X_t^+} \left[\min_{y_{t+1}^+ \in Y_{t+1}^+} \left(\langle x_t^+, D^{|I_t^+|}(p^+) y_{t+1}^+ \rangle + \langle p^+ v_t^+, y_{t+1}^+ \rangle \right) + \right.$$

$$\left. \min_{z_{t+1}^+ \in Z_{t+1}^+} \left(\langle x_t^+, D^{|I_t^+|}(1-p^+) z_{t+1}^+ \rangle + \langle (1-p^+) v_t^+, z_{t+1}^+ \rangle \right) \right]$$

holds, and the since the vectors y_{t+1}^+ and z_{t+1}^+ from the sets Y_{t+1}^+ and Z_{t+1}^+ are chosen independently of each other, the equalities

$$\max_{x_t^+ \in X_t^+} \left[\min_{y_{t+1}^+ \in Y_{t+1}^+} \left(\langle x_t^+, D^{|I_t^+|}(p^+) y_{t+1}^+ \rangle + \langle p^+ v_t^+, y_{t+1}^+ \rangle \right) + \right.$$

$$\left. + \min_{z_{t+1}^+ \in Z_{t+1}^+} \left(\langle x_t^+, D^{|I_t^+|}(1-p^+) z_{t+1}^+ \rangle + \langle (1-p^+) v_t^+, z_{t+1}^+ \rangle \right) \right] =$$

$$= \max_{x_t^+ \in X_t^+} \left[\min_{(y_{t+1}^+, z_{t+1}^+) \in Y_{t+1}^+ \times Z_{t+1}^+} \left(\langle x_t^+, D^{|I_t^+|}(p^+) y_{t+1}^+ \rangle + \langle p^+ v_t^+, y_{t+1}^+ \rangle \right) + \right.$$

$$\left. + \langle x_t^+, D^{|I_t^+|}(1-p^+) z_{t+1}^+ \rangle + \langle (1-p^+) v_t^+, z_{t+1}^+ \rangle \right] =$$

$$= \max_{x_t^+ \in X_t^+} \left[\min_{w_{t+1}^+ \in \theta_{t+1}^+} \left(\langle x_t^+, D^{2|I_t^+|}(p^+, 1-p^+) w_{t+1}^+ \rangle + \langle (p^+ v_t^+, (1-p^+) v_t^+), w_{t+1}^+ \rangle \right) \right],$$

hold.

Analogously, the maximum of the expectation of the financial result associated with buying securities from the set I_t^- at the moment t can be written as written as

$$\max_{x_t^- \in X_t^-} \left[\min_{w_{t+1}^- \in \theta_{t+1}^-} \left(\langle x_t^-, D^{2|I_t^-|}(p^-, 1-p^-) w_{t+1}^- \rangle + \langle (p^- v_t^-, (1-p^-) v_t^-), w_{t+1}^- \rangle \right) \right],$$

whereas the maximum of the expectation of the financial result associated with choosing (buying) the volumes of securities from the set I_t^0 can be written as

$$\max_{x_t^0 \in X_t^0} \left[\min_{w_{t+1}^0 \in \theta_{t+1}^0} \left(\langle x_t^0, D^{2|I_t^0|} \left(\frac{1}{2}, \frac{1}{2} \right) w_{t+1}^0 \rangle + \langle \left(\frac{1}{2} v_t^0, \frac{1}{2} v_t^0 \right), w_{t+1}^0 \rangle \right) \right].$$

Thus, if the trader's best strategy of choosing the volumes of financial securities from the set M_t is understood as that maximizing the expectation of the financial result associated with buying securities being components of the vector $x_t \in X_t^+ \times X_t^- \times X_t^0$, this strategy can be found by calculating

$$\max_{x_t \in X_t} \left[\min_{w_{t+1} \in \theta_{t+1}} \left(\langle x_t, D_t w_{t+1} \rangle + \langle q, w_{t+1} \rangle \right) \right],$$

where $q = \left((p^+ v_t^+, (1-p^+) v_t^+), (p^- v_t^-, (1-p^-) v_t^-), \left(\frac{1}{2} v_t^0, \frac{1}{2} v_t^0 \right) \right)$.

In just the same way this was done in the course of proving the Theorem, one can be certain that this strategy remains optimal if $X_t = M_t = \{x_t \in R_+^n : B_t x_t \geq d_t, x_t \in Q_+^n\} \subset M_t^+ \times M_t^- \times M_t^0$, where B_t is a matrix of a general structure, not necessarily corresponding to the structure of the set $M_t = M_t^+ \times M_t^- \times M_t^0$ as a direct product of the above-mentioned subsets of the three polyhedra from the spaces $R_+^{|I_t^+|}$, $R_+^{|I_t^-|}$, and $R_+^{|I_t^0|}$, respectively (see earlier in Sect. 5).

One can easily be certain that the equalities

$$\max_{x_t \in M_t} \left[\min_{w_{t+1} \in \theta_{t+1}} \left(\langle x_t, D_t w_{t+1} \rangle + \langle q, w_{t+1} \rangle \right) \right] =$$

$$= \max_{\{x_t \in R_+^n : B_t x_t \geq d_t, x_t \in Q_+^n\}} \left[\max_{\{z_{t+1} \geq 0 : z_{t+1} A_t \leq x_t D_t + q\}} \langle b_t, z_{t+1} \rangle \right] =$$

$$= \max_{\{(x_t,\, z_{t+1}) \geq 0:\ B_t x_t \geq d_t,\ z_{t+1} A_t \leq x_t D_t + q,\ x_t \in Q_+^n\}} \langle b_t, z_{t+1} \rangle$$

hold for both types of the structure of the set $X_t = M_t$ so that the maximin

$$\max_{x_t \in M_t} \left[\min_{w_{t+1} \in \theta_{t+1}} \left(\langle x_t, D_t w_{t+1} \rangle + \langle q, w_{t+1} \rangle \right) \right]$$

is found by solving a mixed programming problem of finding the maximum of the linear function $\langle b_t, z_{t+1} \rangle$ on the set $\{(x_t, z_{t+1}) \geq 0 : B_t x_t \geq d_t, z_{t+1} A_t \leq x_t D_t + q, x_t \in Q_+^n\}$.

In just the same way it was done in considering Situation 1, if one treats components of the vector x_t as non-negative, real numbers, finding the maximin

$$\max_{\{x_t \in R_+^n : B_t x_t \geq d_t\}} \left[\min_{w_{t+1} \in \theta_{t+1}} \left(\langle x_t, D_t w_{t+1} \rangle + \langle q, w_{t+1} \rangle \right) \right],$$

which is an upper bound of the maximin

$$\max_{x_t \in M_t} \left[\min_{w_{t+1} \in \theta_{t+1}} (\langle x_t, D_t w_{t+1} \rangle + \langle q, w_{t+1} \rangle) \right],$$

is reducible to finding a saddle point in the antagonistic game on the sets of player strategies $\tilde{M}_t$ and θ_{t+1} with the payoff function $\langle x_t, D_t w_{t+1} \rangle + \langle q, w_{t+1} \rangle$.

A saddle point in this game can be found [6] by solving linear programming problems

$$\langle b_t, h_t \rangle \to \max_{(x_t, h_t) \in Q_t(q)},$$

$$\langle -d_t, \pi_{t+1} \rangle + \langle q, w_{t+1} \rangle \to \min_{(w_{t+1}, \pi_{t+1}) \in P_{t,t+1}},$$

forming a dual pair, where $Q_t(q) = \{(x_t, h_t) \geq 0 : h_t A_t \leq q + x_t D_t, B_t x_t \geq d_t\}$ and $P_{t,t+1} = \{(w_{t+1}, \pi_{t+1}) \geq 0 : \pi_{t+1} B_t \leq -D_t w_{t+1}, A_t w_{t+1} \geq b_t\}$.

In the second subcase, the trader borrows securities from the broker to sell them at the moment t to have additional cash for buying those securities at the moment t whose share price values she expects to decrease at the moment later than $t + 1$ (and the trader should return the borrowed securities later than at the moment $t + 1$). The only difference between this subcase and the first subcase is in the amount of cash available for buying securities that interest the trader at the moment t, i.e., in the parameters determining the set M_t.

Remark 3. One should bear in mind that both the trader's guaranteed result and its upper estimate in her game with the stock exchange determine only the trader's investment strategies at the moment t, and they do not determine the total financial result of applying these strategies. This is the case, since neither the goal function in the maximin problem nor the payoff function, for instance, in game (3) (when $x_t \in \tilde{M}_t$) take into consideration such components of the trader's welfare at the moment $t + 1$ as, for instance, the amount of cash remaining after finalizing all the transactions associated with buying securities from the sets I_t^+, I_t^- and I_t^0. However, the above-mentioned financial result can easily be calculated based upon the solutions to the mixed programming problems and games considered for both Situation 1 and Situation 2.

One should also bear in mind that if the trader borrows securities from a broker, and she needs to return them to the broker at the moment $t + 1$, other approaches to what should be chosen as the set X_t^- are to be considered. The deployment of such approaches leads to a different structure of the payoff functions in the games describing the interaction of the trader with the stock exchange, including the structure of the matrix D_t. One can, however, show that in the framework of this interaction, finding corresponding maximin values or saddle points in corresponding games can be done based on the same theoretical foundation developed in [6]. Certainly, in some cases, the interaction between the trader and the stock exchange is formalized in the form of maximin problems and games of more complicated structures than those studied in Sect. 5; however, their consideration lies beyond the scope of the present publication.

Finally, one should notice that by solving either the above-mentioned problem (i.e., the problem of finding the trader's guaranteed result or that of finding an upper estimate of this result), the trader determines which share price values she should expect to deal with at the moment t with respect to all the standard securities from the set N. This information can be used, in particular, in making decisions on borrowing standard securities to be returned to brokers at the moment $t + 1$.

6 Illustrative Examples

The aim of this section is to illustrate how a price-taking trader may make decisions on forming her portfolio out of standard securities when at the moment t, she can make no assumptions on probability distributions of the values of the share prices that (standard) securities of her interest may have at the moment $t + 1$. As shown in Sect. 5, if, nevertheless, the trader can estimate the areas in which the values of the share prices of these securities may change at the moment $t + 1$, game models of a special kind may help the trader calculate her optimal investment strategies at the moment t aimed at increasing the value of her portfolio at the moment $t + 1$. Particularly, the present section illustrates how the games described in Sect. 5 are formed, and what linear programming problems are solved to find an upper estimate of this value by calculating saddle points in one of these games with the use of standard software packages for solving linear programming problems. To this end, two numerical examples are considered in both situations mentioned in the text of Sect. 5, and in the description of these examples, the notation from Sect. 5 is used.

As mentioned in Remark 2 (see Sect. 5), solving the above-mentioned linear programming problems lets the trader determine only an upper bound of the expected increment value of her portfolio by considering volumes (numbers of shares) of securities to be bought and sold as non-negative, real numbers. Such a consideration is, however, in line with traditional approaches exercised in theoretical studies of stock exchanges [33]. Moreover, even from a practical viewpoint— when the number of different securities that interest the trader is large—solving mixed programming problems to calculate the exact (integer) numbers of shares for each (standard) security to buy and to sell to secure the exact value of the expected increment of the trader portfolio's value may present substantial difficulties. If this is the case, finding the exact numbers of shares of the above-mentioned standard securities will hardly interest the traders in making decisions on forming and managing their investment portfolios.

In just the same way as in Sect. 5, in the illustrative examples to follow, the optimality of the trader's investment strategy is considered in the sense of the value of her portfolio at the moment $t + 1$.

6.1 Illustrative Example 1

Consider a trader who plans to interact with a stock exchange by forming a portfolio of financial instruments. Let us assume that at the moment t, the trader (a) is interested in only two particular standard securities that are traded in the stock exchange (so that $N = \{1, 2\}$ for this trader), (b) does not have a portfolio of financial instruments traded in this stock exchange (so that $v_1 = v_2 = 0$ for this trader), (c) has the amount of cash equaling $m_t = 10,000.00$ financial units, for instance, US dollars, and (d) has a broker who is ready to provide her with a credit. It is assumed that (a) the credit leverage equals $k_t = 0.5$ for borrowing standard securities from the broker to let the trader open short positions there, and (b) the broker is ready to offer the trader securities from the set N (which are the only securities that interest the trader at the moment t) to borrow.

Let at the moment t, the values of the share prices equal $s_{1,t} = 100$ US dollars for security 1 and $s_{2,t} = 50$ US dollars for security 2. Further, let the trader believe that the value of the share price of security 1 will increase at the moment $t + 1$, whereas the value of the share price of security 2 will decrease at the moment $t + 1$ so that $I_t^+ = \{1\}, I_t^- = \{2\}$ and $I_t^0 = \emptyset$. Moreover, let the trader be confident that the price values of the above two securities will change the way she believes they will with the probabilities $p^+ = 0.6$ and $p^- = 0.7$, respectively. Finally, let the trader adhere to Approach 3 to the understanding of what should be viewed as the set X_t^- (see Sect. 5).

The first step in finding the trader's best investment strategy is to find out how much of additional cash she can have as a result of borrowing securities from the broker and selling them short at the moment t. Further, since security 2 is the only one that the trader should be interested in borrowing from the broker (hoping that the share price value of this security will decrease in the future), the trader should determine how many shares of security 2 she should borrow to sell them at the moment t. It is obvious that since the total cost of the shares of security 2 that the trader can borrow from the broker at the moment t cannot exceed 5,000,00, and the share price value of one share of security 2 equals 50.00 US dollars at the moment t, the maximum number of shares of security 2 that the trader can borrow equals 100.

Let x_1^+ and x_2^- be the numbers of shares of security 1 and security 2, respectively, that the trader plans to have in her portfolio at the moment $t + 1$, which means that the trader plans to buy x_1^+ shares of security 1 and x_2^- shares of security 2 at the moment t. According to the description of the trader's actions in forming her portfolio at the moment t, presented in Sect. 5, the trader should estimate how many shares and of which securities from the set N she should have at the moment $t+1$ that would maximize the value of her portfolio at the moment $t+1$. It is clear that in this particular example, one should expect the trader not to buy any shares of security 2. However, one should bear in mind that, generally, despite the fact that at the moment t, the trader borrows (from the broker) at least some securities from the set X_t^- to receive additional cash, it may happen that the portfolio with the maximum value at the moment $t + 1$ may include at least some of the securities that were borrowed at the moment t (security 2 in the example under consideration). Thus, for the purpose of illustrating the trader's actions in the general case, buying both shares of security 1 and shares of security 2 are considered.

As mentioned in Sect. 5, the trader determines the description of the sets X_t^+ and X_t^- at her own discretion, so let the trader describe them with the following system of linear inequalities (proceeding from her financial abilities at the moment t):

$$x_1^+ \geq 0;$$

$$x_2^- \geq 0;$$

$$s_{1,t}x_1^+ + s_{2,t}x_2^- \leq m_t + 5000.$$

Here, the first two of the above three inequalities reflect the condition of non-negativity of the transaction volumes, whereas the third one puts the limit on the volume of securities 1 and 2 that the trader can buy with her own money and with the money to be received from selling at the moment t shares of security 2 (borrowed from the broker).

Thus, $M_t = \{x_t \in R_+^2 : B_t x_t \geq d_t\}$, the set of the volumes of securities 1 and 2 that the trader can buy at the moment t, where $x_t = (x_{1,t}^+, x_{2,t}^-) = (x_1^+, x_2^-)$, is such that

$$B_t = \begin{pmatrix} 1 & 0 \\ 0 & 1 \\ -s_{1,t} & -s_{2,t} \end{pmatrix} = \begin{pmatrix} 1 & 0 \\ 0 & 1 \\ -100 & -50 \end{pmatrix}, d_t = \begin{pmatrix} 0 \\ 0 \\ -m_t - 5000 \end{pmatrix} = \begin{pmatrix} 0 \\ 0 \\ -15000 \end{pmatrix},$$

and the inequality

$$\begin{pmatrix} 1 & 0 \\ 0 & 1 \\ -100 & -50 \end{pmatrix} (x_1^+, x_2^-) \geq \begin{pmatrix} 0 \\ 0 \\ -15000 \end{pmatrix}$$

holds (see Sect. 5). To simplify the notation in the description of the illustrative example to follow, let also

$$y_{1,t+1}^+ = y_1^+,\ y_{2,t+1}^- = y_2^-,\ z_{1,t+1}^+ = z_1^+,\ z_{2,t+1}^- = z_2^-.$$

While $x_t = (x_1^+, x_2^-)$ is the vector of the trader's strategies in her game with the stock exchange (see Sect. 5), the strategies of the stock exchange can be represented by the vector $w_{t+1} = (y_1^+, z_1^+, y_2^-, z_2^-)$ whose components are the (expected) values of the share prices of securities 1 and 2 at the moment $t+1$. Here, y_1^+, y_2^- are the (expected) values of the share prices of securities 1 and 2 at the moment $t+1$, respectively, if the trader has correctly predicted directions in which the values of these two securities will change, and z_1^+, z_2^- are the (expected) values of the share prices of securities 1 and 2 at the moment $t+1$, respectively, if the trader has failed to predict these directions correctly.

Let the trader believe that the maximum and the minimum values of the share prices of securities 1 and 2 at the moment $t+1$ will be $s^{max}_{1,t+1} = 115$, $s^{max}_{2,t+1} = 65$, $s^{min}_{1,t+1} = 75$, $s^{min}_{2,t+1} = 35$ US dollars, respectively. Further, let the trader put stop orders on the above maximum and minimum price values of securities 2 and 1 at the moment $t+1$ to avoid unexpected financial losses associated with increasing the value of the share price of security 2 beyond $s^{max}_{2,t+1}$ and with decreasing the value of the share price of security 1 below $s^{min}_{1,t+1}$, respectively. Then, $\theta_{t+1} = \{w_{t+1} \in R^4_+ : A_t w_{t+1} \geq b_t\}$, the set of possible strategies of the stock exchange in the game, can be described by the system of inequalities

$$
\begin{aligned}
s_{1,t} &\leq y_1^+ \leq s^{max}_{1,t+1},\\
s^{min}_{1,t+1} &\leq z_1^+ \leq s_{1,t},\\
s^{min}_{2,t+1} &\leq y_2^- \leq s_{2,t},\\
s_{2,t} &\leq z_2^- \leq s^{max}_{2,t+1},
\end{aligned}
$$

which takes the following vector-matrix form:

$$
A_t = \begin{pmatrix} 1 & 0 & 0 & 0\\ -1 & 0 & 0 & 0\\ 0 & 1 & 0 & 0\\ 0 & -1 & 0 & 0\\ 0 & 0 & 1 & 0\\ 0 & 0 & -1 & 0\\ 0 & 0 & 0 & 1\\ 0 & 0 & 0 & -1 \end{pmatrix}, \quad b_t = \begin{pmatrix} s_{1,t}\\ -s^{max}_{1,t+1}\\ s^{min}_{1,t+1}\\ -s_{1,t}\\ s^{min}_{2,t+1}\\ -s_{2,t}\\ s_{2,t}\\ -s^{max}_{2,t+1} \end{pmatrix} = \begin{pmatrix} 100\\ -115\\ 75\\ -100\\ 35\\ -50\\ 50\\ -65 \end{pmatrix},
$$

$$
\begin{pmatrix} 1 & 0 & 0 & 0\\ -1 & 0 & 0 & 0\\ 0 & 1 & 0 & 0\\ 0 & -1 & 0 & 0\\ 0 & 0 & 1 & 0\\ 0 & 0 & -1 & 0\\ 0 & 0 & 0 & 1\\ 0 & 0 & 0 & -1 \end{pmatrix} (y_1^+, z_1^+, y_2^-, z_2^-) \geq \begin{pmatrix} 100\\ -115\\ 75\\ -100\\ 35\\ -50\\ 50\\ -65 \end{pmatrix}.
$$

According to the Theorem (see Sect. 5), the payoff function of the game between the trader and the stock exchange takes the form $\langle x_t, D_t w_{t+1} \rangle$, where

$$
D_t = \begin{pmatrix} p^+ & 1-p^+ & 0 & 0\\ 0 & 0 & p^- & 1-p^- \end{pmatrix} = \begin{pmatrix} 0.6 & 0.4 & 0 & 0\\ 0 & 0 & 0.7 & 0.3 \end{pmatrix}.
$$

To simplify the notation further, let

$$h_{1,t} = h_1,\ h_{2,t} = h_2,\ h_{3,t} = h_3,\ h_{4,t} = h_4,\ h_{5,t} = h_5,\ h_{6,t} = h_6,\ h_{7,t} = h_7,\ h_{8,t} = h_8,$$

and let

$$\pi_{1,t+1} = u_1,\ \pi_{2,t+1} = u_2,\ \pi_{3,t+1} = u_3.$$

As shown in Sect. 5, saddle points in the game under consideration can be found by solving linear programming problems

$$100h_1 - 115h_2 + 75h_3 - 100h_4 + 35h_5 - 50h_6 + 50h_7 - 65h_8 \to \\ \to \max_{(h_1,h_2,h_3,h_4,h_5,h_6,h_7,h_8;x_1^+,x_2^-)}, \tag{4}$$

$$\begin{aligned} h_1 - h_2 &\leq 0.6x_1^+, \\ h_3 - h_4 &\leq 0.4x_1^+, \\ h_5 - h_6 &\leq 0.7x_2^-, \\ h_7 - h_8 &\leq 0.3x_2^-, \\ -100x_1^+ - 50x_2^- &\geq -15000, \\ h_i &\geq 0, i = \overline{1,8}, \\ x_1^+ &\geq 0, x_2^- \geq 0, \end{aligned}$$

and

$$15000u_3 \to \min_{(u_1,u_2,u_3;y_1^+,z_1^+,y_2^-,z_2^-)}, \tag{5}$$

$$\begin{aligned} u_1 - 100u_3 &\leq -0.6y_1^+ - 0.4z_1^+, \\ u_2 - 50u_3 &\leq -0.7y_2^- - 0.3z_2^-, \\ 100 &\leq y_1^+ \leq 115, \\ 75 &\leq z_1^+ \leq 100, \\ 35 &\leq y_2^- \leq 50, \\ 50 &\leq z_2^- \leq 65, \\ u_i &\geq 0, i = \overline{1,3}, \end{aligned}$$

forming a dual pair.

Solutions to problems (4) and (5) were found with the use of a computer program implemented on the Maple 7 computing platform, which includes software for solving linear programming problems. These solutions are

$$x_1^+ = 150, x_2^- = 0,$$

$$h_1 = 90, h_2 = 0, h_3 = 60, h_4 = 0, h_5 = 0, h_6 = 0, h_7 = 0, h_8 = 0,$$

for problem (4), and

$$u_1 = 0, u_2 = 0, u_3 = 0.9,$$

$$y_1^+ = 100, z_1^+ = 75, y_2^- = 35, z_2^- = 50,$$

for problem (5).

Thus, the trader's optimal strategy consists of (a) borrowing from a broker 100 units of security 2 and selling them at the moment t, and (b) buying 150 units of security 1 at the moment t. As a result of the deployment of this optimal strategy, the expectation of the value of the trader's portfolio at the moment $t+1$ equals 13500.

6.2 *Illustrative Example 2*

Consider now a trader who interacts with a stock exchange by maintaining a portfolio of financial instruments and who at the moment t (a) is interested in six particular securities that are traded in the stock exchange, (b) has a portfolio of financial instruments that consists of only these six securities so that the numbers of shares of these securities in her portfolio at the moment t equal $v_1 = 10$, $v_2 = 30$, $v_3 = 50$, $v_4 = 0$, $v_5 = 4$, $v_6 = 12$, respectively, (c) has her own cash at the amount l of $m_t = \$1000$, and (d) has a broker who is ready to provide her with securities traded in the stock exchange for opening short positions there with the leverage $k_t = 2$.

Let the share price values of these six securities at the moment t be 50, 90, 10, 22, 49, and \$50, respectively, and let the trader believe that the share price values of securities 1 and 2 will increase at the moment $t+1$ (with the probability $p^+ > 0.5$), whereas the share price values of securities 3 and 4 will decrease at the moment $t+1$ (with the probability $p^- > 0.5$). Further, let the trader believe that share price values of securities 5 and 6 may increase or decrease (with the same probability $p^0 = 0.5$) so that $I_t^+ = \{1, 2\}$, $I_t^- = \{3, 4\}$ and $I_t^0 = \{5, 6\}$. Moreover, let the trader be confident that the share price values of securities from the groups $I_t^+ = \{1, 2\}$, $I_t^- = \{3, 4\}$ will change the way she believes they will with the probabilities $p^+ = 0.56$, $p^- = 0.6$, respectively. Finally, let Table 4 reflect the trader's estimates of the maximum and the minimum share price values of all the six securities at the moment $t+1$.

Table 4 Parameters for security prices

	1	2	3	4	5	6
Current price value	50	90	10	22	49	50
Minimal expected price value of the security at the moment $t+1$	40	75	8	18	42	30
Maximal expected price value of the security at the moment $t+1$	60	120	13	25	53	70

As in Illustrative Example 1, throughout Illustrative Example 2, it is assumed that (a) the trader adheres to Approach 3 to the understanding of what should be viewed as the set X_t^-, and (b) the trader's goal is to maximize the value of her portfolio at the moment $t+1$.

First, consider case 1 in Situation 2 (see Sect. 5) in which the trader does not plan to keep at the moment $t+1$ particular securities that she possesses at the moment t in her portfolio. In this case, the trader should estimate the total amount of cash that she may have by selling all the securities that she possesses at the moment t at their current price values (that exist on the market at the moment t), which equals

$$10 \times 50 + 30 \times 90 + 50 \times 10 + 0 \times 22 + 4 \times 49 + 12 \times 50 + 1000 = 5496.$$

Consider the first subcase of case 1 in which the trader does not plan to borrow any securities from the broker.

Should the trader use systems of inequalities analogous to those from Illustrative Example 1 for describing the set of her feasible strategies $M_t = \{x_t \in R_+^2 : B_t x_t \geq d_t\}$ and the set of those for the stock exchange $\theta_{t+1} = \{w_{t+1} \in R_+^4 : A_t w_{t+1} \geq b_t\}$, one can easily be certain that the matrices and the vectors in the description of the sets M_t and θ_{t+1} and in that of the payoff function $\langle x_t, D_t w_{t+1}\rangle$ are as follows:

$$B_t = \begin{pmatrix} 1 & 0 & 0 & 0 & 0 & 0 \\ 0 & 1 & 0 & 0 & 0 & 0 \\ 0 & 0 & 1 & 0 & 0 & 0 \\ 0 & 0 & 0 & 1 & 0 & 0 \\ 0 & 0 & 0 & 0 & 1 & 0 \\ 0 & 0 & 0 & 0 & 0 & 1 \\ -s_{1,t} & -s_{2,t} & -s_{3,t} & -s_{4,t} & -s_{5,t} & -s_{6,t} \end{pmatrix} = \begin{pmatrix} 1 & 0 & 0 & 0 & 0 & 0 \\ 0 & 1 & 0 & 0 & 0 & 0 \\ 0 & 0 & 1 & 0 & 0 & 0 \\ 0 & 0 & 0 & 1 & 0 & 0 \\ 0 & 0 & 0 & 0 & 1 & 0 \\ 0 & 0 & 0 & 0 & 0 & 1 \\ -50 & -90 & -10 & -22 & -49 & -50 \end{pmatrix},$$

$$d_t = \left(0, 0, 0, 0, 0, 0, -m_t - \sum_{i \in N} v_{i,t} s_{i,t}\right) = (0, 0, 0, 0, 0, 0, -5496),$$

$$A_t = \begin{pmatrix}
1 & 0 & 0 & 0 & 0 & 0 & 0 & 0 & 0 & 0 & 0 & 0 \\
-1 & 0 & 0 & 0 & 0 & 0 & 0 & 0 & 0 & 0 & 0 & 0 \\
0 & 1 & 0 & 0 & 0 & 0 & 0 & 0 & 0 & 0 & 0 & 0 \\
0 & -1 & 0 & 0 & 0 & 0 & 0 & 0 & 0 & 0 & 0 & 0 \\
0 & 0 & 1 & 0 & 0 & 0 & 0 & 0 & 0 & 0 & 0 & 0 \\
0 & 0 & -1 & 0 & 0 & 0 & 0 & 0 & 0 & 0 & 0 & 0 \\
0 & 0 & 0 & 1 & 0 & 0 & 0 & 0 & 0 & 0 & 0 & 0 \\
0 & 0 & 0 & -1 & 0 & 0 & 0 & 0 & 0 & 0 & 0 & 0 \\
0 & 0 & 0 & 0 & 1 & 0 & 0 & 0 & 0 & 0 & 0 & 0 \\
0 & 0 & 0 & 0 & -1 & 0 & 0 & 0 & 0 & 0 & 0 & 0 \\
0 & 0 & 0 & 0 & 0 & 1 & 0 & 0 & 0 & 0 & 0 & 0 \\
0 & 0 & 0 & 0 & 0 & -1 & 0 & 0 & 0 & 0 & 0 & 0 \\
0 & 0 & 0 & 0 & 0 & 0 & 1 & 0 & 0 & 0 & 0 & 0 \\
0 & 0 & 0 & 0 & 0 & 0 & -1 & 0 & 0 & 0 & 0 & 0 \\
0 & 0 & 0 & 0 & 0 & 0 & 0 & 1 & 0 & 0 & 0 & 0 \\
0 & 0 & 0 & 0 & 0 & 0 & 0 & -1 & 0 & 0 & 0 & 0 \\
0 & 0 & 0 & 0 & 0 & 0 & 0 & 0 & 1 & 0 & 0 & 0 \\
0 & 0 & 0 & 0 & 0 & 0 & 0 & 0 & -1 & 0 & 0 & 0 \\
0 & 0 & 0 & 0 & 0 & 0 & 0 & 0 & 0 & 1 & 0 & 0 \\
0 & 0 & 0 & 0 & 0 & 0 & 0 & 0 & 0 & -1 & 0 & 0 \\
0 & 0 & 0 & 0 & 0 & 0 & 0 & 0 & 0 & 0 & 1 & 0 \\
0 & 0 & 0 & 0 & 0 & 0 & 0 & 0 & 0 & 0 & -1 & 0 \\
0 & 0 & 0 & 0 & 0 & 0 & 0 & 0 & 0 & 0 & 0 & 1 \\
0 & 0 & 0 & 0 & 0 & 0 & 0 & 0 & 0 & 0 & 0 & -1
\end{pmatrix},$$

$$b_t = (s_{1,t}, -s^{max}_{1,t+1}, s_{2,t}, -s^{max}_{2,t+1}, s^{min}_{1,t+1}, -s_{1,t}, s^{min}_{2,t+1}, -s_{2,t}, s^{min}_{3,t+1}, -s_{3,t}, s^{min}_{4,t+1}, -s_{4,t},$$

$$s_{3,t}, -s^{max}_{3,t+1}, s_{4,t}, -s^{max}_{4,t+1}, s_{5,t}, -s^{max}_{5,t+1}, s_{6,t}, -s^{max}_{6,t+1}, s^{min}_{5,t+1}, -s_{5,t}, s^{min}_{6,t+1}, -s_{6,t}) =$$

$$(50, -60, 90, -120, 40, -50, 75, -90, 8, -10, 18, -22, 10, -13,$$

$$22, -25, 49, -53, 50, -70, 42, -49, 30, -50),$$

$$x_t = (x_1^+, x_2^+, x_3^-, x_4^-, x_5^0, x_6^0),$$

$$w_{t+1} = (y_1^+, y_2^+, z_1^+, z_2^+, y_3^-, y_4^-, z_3^-, z_4^-, y_5^0, y_6^0, z_5^0, z_6^0),$$

$$D_t = \begin{pmatrix}
0.56 & 0 & 0.44 & 0 & 0 & 0 & 0 & 0 & 0 & 0 & 0 & 0 \\
0 & 0.56 & 0 & 0.44 & 0 & 0 & 0 & 0 & 0 & 0 & 0 & 0 \\
0 & 0 & 0 & 0 & 0.6 & 0 & 0.4 & 0 & 0 & 0 & 0 & 0 \\
0 & 0 & 0 & 0 & 0 & 0.6 & 0 & 0.4 & 0 & 0 & 0 & 0 \\
0 & 0 & 0 & 0 & 0 & 0 & 0 & 0 & 0.5 & 0 & 0.5 & 0 \\
0 & 0 & 0 & 0 & 0 & 0 & 0 & 0 & 0 & 0.5 & 0 & 0.5
\end{pmatrix},$$

Here, to simplify the notation in the description of Illustrative Example 2, it is implied that

$$x_{1,t} = x_1^+,\ x_{2,t} = x_2^+,\ x_{3,t} = x_3^-,\ x_{4,t} = x_4^-,\ x_{5,t} = x_5^0,\ x_{6,t} = x_6^0$$

The first (prime) linear programming problem in the above pair, which is analogous to problem (4), contains 30 variables and 19 constraints-inequalities (besides the inequalities reflecting the non-negativity of a part of the prime variables for which their non-negativity is not reflected in the above-mentioned 19 constraints-inequalities). The second (dual) problem contains 19 variables and 30 constraints-inequalities (besides the inequalities reflecting the non-negativity of a part of the dual variables for which their non-negativity does not follow from the corresponding part of the above-mentioned 30 constraints-inequalities).

The first (prime) linear programming problem is formulated as follows:

$$50h_1 - 60h_2 + 90h_3 - 120h_4 + 40h_5 - 50h_6 + 75h_7 - 90h_8 +$$
$$+ 8h_9 - 10h_{10} + 18h_{11} - 22h_{12} + 10h_{13} - 13h_{14} + 22h_{15} - 25h_{16} +$$
$$+ 49h_{17} - 53h_{18} + 50h_{19} - 70h_{20} + 42h_{21} - 49h_{22} + 30h_{23} - 50h_{24} \rightarrow$$
$$\rightarrow \max_{(h_1,h_2,h_3,h_4,h_5,h_6,h_7,h_8,h_9,h_{10},h_{11},h_{12},h_{13},h_{14},h_{15},h_{16},h_{17},h_{18},h_{19},h_{20},h_{21},h_{22},h_{23},h_{24};x_1^+,x_2^+,x_3^-,x_4^-,x_5^0,x_6^0)},$$

$$h_1 - h_2 \le 0.56x_1^+,$$
$$h_3 - h_4 \le 0.56x_2^+,$$
$$h_5 - h_6 \le 0.44x_1^+,$$
$$h_7 - h_8 \le 0.44x_2^+,$$
$$h_9 - h_{10} \le 0.6x_3^-,$$
$$h_{11} - h_{12} \le 0.6x_4^-,$$
$$h_{13} - h_{14} \le 0.4x_3^-,$$
$$h_{15} - h_{16} \le 0.4x_4^-,$$
$$h_{17} - h_{18} \le 0.5x_5^0,$$
$$h_{19} - h_{20} \le 0.5x_6^0,$$
$$h_{21} - h_{22} \le 0.5x_5^0,$$
$$h_{23} - h_{24} \le 0.5x_6^0,$$
$$h_i \ge 0, i = \overline{1,24},$$
$$x_1^+ \ge 0, x_2^+ \ge 0, x_3^- \ge 0, x_4^- \ge 0, x_5^0 \ge 0, x_6^0 \ge 0,$$
$$50x_1^+ + 90x_2^+ + 10x_3^- + 22x_4^- + 49x_5^0 + 50x_6^0 \le 5496.$$

Here, the last inequality puts the limit on the volumes of securities 1, 2, 3, 4, 5, and 6 that the trader can buy with the money that she may have at the moment t by selling all the securities from her portfolio that she possesses at the moment t.

The second (dual) linear programming problem is formulated as follows:

$$5496u_7 \to \min_{(u_1,u_2,u_3,u_4,u_5,u_6,u_7;y_1^+,y_2^+,z_1^+,z_2^+,y_3^-,y_4^-,z_3^-,z_4^-,y_5^0,y_6^0,z_5^0,z_6^0)},$$

$$u_1 - 50u_7 \leq -0.56y_1^+ - 0.44z_1^+,$$

$$u_2 - 90u_7 \leq -0.56y_2^+ - 0.44z_2^+,$$

$$u_3 - 10u_7 \leq -0.6y_3^- - 0.4z_3^-$$

$$u_4 - 22u_7 \leq -0.6y_4^- - 0.z_4^-$$

$$u_5 - 49u_7 \leq -0.5y_5^0 - 0.5z_5^0$$

$$u_6 - 50u_7 \leq -0.5y_6^0 - 0.5z_6^0$$

$$50 \leq y_1^+ \leq 60,$$

$$90 \leq y_2^+ \leq 120,$$

$$40 \leq z_1^+ \leq 50,$$

$$75 \leq z_2^+ \leq 90,$$

$$8 \leq y_3^- \leq 10,$$

$$18 \leq y_4^- \leq 22,$$

$$10 \leq z_3^- \leq 13,$$

$$22 \leq z_4^- \leq 25,$$

$$49 \leq y_5^0 \leq 53,$$

$$50 \leq y_6^0 \leq 70,$$

$$42 \leq z_5^0 \leq 49,$$

$$30 \leq z_6^0 \leq 50,$$

$$u_i \geq 0, i = \overline{1,7}.$$

The solution to the first problem is

$$x_1^+ = 0, x_2^+ = 0, x_3^- = 0, x_4^- = 0, x_5^0 = 112.16, x_6^0 = 0,$$

$$h_1 = 0, h_2 = 0, h_3 = 0, h_4 = 0, h_5 = 0, h_6 = 0, h_7 = 0, h_8 = 0,$$

$$h_9 = 0, h_{10} = 0, h_{11} = 0, h_{12} = 0, h_{13} = 0, h_{14} = 0, h_{15} = 0, h_{16} = 0,$$

$$h_{17} = 56.08, h_{18} = 0, h_{19} = 0, h_{20} = 0, h_{21} = 56.08, h_{22} = 0, h_{23} = 0, h_{24} = 0,$$

and the solution to the second problem, which is analogous to problem (5), is

$$u_1 = 0, u_2 = 0, u_3 = 0, u_4 = 0, u_5 = 0, u_6 = 0, u_7 = 0.93,$$

$$y_1^+ = 50, y_2^+ = 90, z_1^+ = 40, z_2^+ = 75.39, y_3^- = 8, y_4^- = 18, z_3^- = 10, z_4^- = 22,$$

$$y_5^0 = 49, y_6^0 = 50, z_5^0 = 42, z_6^0 = 30.$$

It seems interesting to compare the composition of the trader's portfolio at the moment t with the optimal composition of this portfolio at the moment $t+1$ proceeding from the expectations of the share price values of all the securities that are of interest to the trader at the moment t. As one can see, the optimal portfolio composition at the moment $t+1$ consists of the following numbers of shares of securities 1–6:

$$v_1 = 0, v_2 = 0, v_3 = 0, v_4 = 0, v_5 = 112.16, v_6 = 0,$$

whereas the composition of the trader's portfolio at the moment t is

$$v_1 = 10, v_2 = 30, v_3 = 50, v_4 = 0, v_5 = 4, v_6 = 12.$$

Consider now the second subcase of case 1 in which the trader plans to borrow securities from the set $I_t^- = \{3, 4\}$ from the broker to open short positions and to sell the borrowed securities at the moment t, along with all the securities that she possesses in her portfolio at the moment t. The only difference with the first subcase (of case 1) consists of the amount of cash that the trader may have at the moment t in addition to the amount 5496.00 US dollars, which the trader may have by selling securities from her portfolio that she has at the moment t. That is, the amount of cash that the trader may have as a result of selling securities from the set I_t^- that she borrows from the broker equals $k_t\left(m_t + \sum_{i=1}^{n} v_{i,t}s_{i,t}\right) = 2 \times 5496 = 10992$. Thus, the total amount of cash that trader may have at the moment t equals $3 \times 5496 = 16488$, and the vector $(0, 0, 0, 0, 0, 0, -16488)$ is the vector d_t in subcase 2 of case 1 in the description of the set $M_t = \{x_t \in R_+^2 : B_t x_t \geq d_t\}$.

The formulation of the prime linear programming problem in subcase 2 of case 1 differs from that in subcase 1 of case 1 only by the last inequality in the system of its constraints, and this inequality takes the form

$$50x_1^+ + 90x_2^+ + 10x_3^- + 22x_4^- + 49x_5^0 + 50x_6^0 \leq 16488.$$

The solution to this problem is

$$x_1^+ = 0, x_2^+ = 0, x_3^- = 0, x_4^- = 0, x_5^0 = 336.49, x_6^0 = 0,$$

$$h_1 = 0, h_2 = 0, h_3 = 0, h_4 = 0, h_5 = 0, h_6 = 0, h_7 = 0, h_8 = 0,$$

$$h_9 = 0, h_{10} = 0, h_{11} = 0, h_{12} = 0, h_{13} = 0, h_{14} = 0, h_{15} = 0, h_{16} = 0,$$

$$h_{17} = 168.24, h_{18} = 0, h_{19} = 0, h_{20} = 0, h_{21} = 168.24, h_{22} = 0, h_{23} = 0, h_{24} = 0.$$

The value of the prime problem is equal to 15310.28.

The formulation of the dual linear programming problem in subcase 2 of case 1 differs from that of subcase 1 of case 1 only by the coefficient for the variable u_7, which now equals 16488. The solution to the dual problem is

$$u_1 = 0, u_2 = 0, u_3 = 0, u_4 = 0, u_5 = 0, u_6 = 0, u_7 = 0.93,$$

$$y_1^+ = 50, y_2^+ = 90, z_1^+ = 40, z_2^+ = 75.39, y_3^- = 8, y_4^- = 18, z_3^- = 10, z_4^- = 22,$$

$$y_5^0 = 49, y_6^0 = 50, z_5^0 = 42, z_6^0 = 30.$$

Consider now case 2 of Situation 2 in which the trader plans to keep at the moment $t+1$ securities that she possesses at the moment t while buying additional securities 1–6 and using the amount of her own cash that she has at the moment t. As before, two subcases are possible. That is, the trader does not borrow any securities from the broker (subcase 1), and the trader borrows securities from the set I_t^- from the broker (subcase 2).

The only difference between the subcase 1 of case 2 and subcase 1 of case 1 consists of the presence of a linear function $\langle q, w_{t+1}\rangle$ in the payoff function of the game between the trader and the stock exchange, where

$$q = \Big(p^+v_1, p^+v_2, (1-p^+)v_1, (1-p^+)v_2, p^-v_3, p^-v_4, (1-p^-)v_3, (1-p^-)v_4,$$

$$\frac{1}{2}v_5, \frac{1}{2}v_6, \frac{1}{2}v_5, \frac{1}{2}v_6\Big) = (5.6, 16.8, 4.4, 13.2, 30, 0, 20, 0, 2, 6, 2, 6).$$

This difference, however, leads to a slightly different system of constraints in the first (prime) linear programming problem to be solved to determine saddle points of the game so that this linear programming problem takes the form

$$50h_1 - 60h_2 + 90h_3 - 120h_4 + 40h_5 - 50h_6 + 75h_7 - 90h_8 +$$
$$+ 8h_9 - 10h_{10} + 18h_{11} - 22h_{12} + 10h_{13} - 13h_{14} + 22h_{15} - 25h_{16} +$$
$$+ 49h_{17} - 53h_{18} + 50h_{19} - 70h_{20} + 42h_{21} - 49h_{22} + 30h_{23} - 50h_{24} \rightarrow$$
$$\rightarrow \max_{(h_1,h_2,h_3,h_4,h_5,h_6,h_7,h_8,h_9,h_{10},h_{11},h_{12},h_{13},h_{14},h_{15},h_{16},h_{17},h_{18},h_{19},h_{20},h_{21},h_{22},h_{23},h_{24};x_1^+,x_2^+,x_3^-,x_4^-,x_5^0,x_6^0)},$$

$$h_1 - h_2 \le 0.56x_1^+ + 5.6,$$
$$h_3 - h_4 \le 0.56x_2^+ + 16.8,$$
$$h_5 - h_6 \le 0.44x_1^+ + 4.4,$$
$$h_7 - h_8 \le 0.44x_2^+ + 13.2,$$
$$h_9 - h_{10} \le 0.6x_3^- + 30,$$

$$h_{11} - h_{12} \leq 0.6x_4^-,$$
$$h_{13} - h_{14} \leq 0.4x_3^- + 20,$$
$$h_{15} - h_{16} \leq 0.4x_4^-,$$
$$h_{17} - h_{18} \leq 0.5x_5^0 + 2,$$
$$h_{19} - h_{20} \leq 0.5x_6^0 + 6,$$
$$h_{21} - h_{22} \leq 0.5x_5^0 + 2,$$
$$h_{23} - h_{24} \leq 0.5x_6^0 + 6,$$
$$h_i \geq 0, i = \overline{1, 24},$$
$$x_1^+ \geq 0, x_2^+ \geq 0, x_3^- \geq 0, x_4^- \geq 0, x_5^0 \geq 0, x_6^0 \geq 0,$$
$$50x_1^+ + 90x_2^+ + 10x_3^- + 22x_4^- + 49x_5^0 + 50x_6^0 \leq 1000,$$

and in the goal function in the second (dual) linear programming problem, which takes the form (see Sect. 5)

$$1000u_7 + 5.6y_1^+ + 16.8y_2^+ + 4.4z_1^+ + 13.2z_2^+ + 30y_3^- + 20z_3^- + 2y_5^0 + 6y_6^0 + 2z_5^0 + 6z_6^0 \to$$
$$\to \min_{(u_1,u_2,u_3,u_4,u_5,u_6,u_7;y_1^+,y_2^+,z_1^+,z_2^+,y_3^-,y_4^-,z_3^-,z_4^-,y_5^0,y_6^0,z_5^0,z_6^0)},$$
$$u_1 - 50u_7 \leq -0.56y_1^+ - 0.44z_1^+,$$
$$u_2 - 90u_7 \leq -0.56y_2^+ - 0.44z_2^+,$$
$$u_3 - 10u_7 \leq -0.6y_3^- - 0.4z_3^-$$
$$u_4 - 22u_7 \leq -0.6y_4^- - 0.z_4^-$$
$$u_5 - 49u_7 \leq -0.5y_5^0 - 0.5z_5^0$$
$$u_6 - 50u_7 \leq -0.5y_6^0 - 0.5z_6^0$$
$$50 \leq y_1^+ \leq 60,$$
$$90 \leq y_2^+ \leq 120,$$
$$40 \leq z_1^+ \leq 50,$$
$$75 \leq z_2^+ \leq 90,$$
$$8 \leq y_3^- \leq 10,$$
$$18 \leq y_4^- \leq 22,$$
$$10 \leq z_3^- \leq 13,$$

$22 \le z_4^- \le 25,$

$49 \le y_5^0 \le 53,$

$50 \le y_6^0 \le 70,$

$42 \le z_5^0 \le 49,$

$30 \le z_6^0 \le 50,$

$u_i \ge 0, i = \overline{1,7}.$

The solution to the first problem is

$$x_1^+ = 0, x_2^+ = 0, x_3^- = 0, x_4^- = 0, x_5^0 = 20.41, x_6^0 = 0,$$
$$h_1 = 5.6, h_2 = 0, h_3 = 16.8, h_4 = 0, h_5 = 4.4, h_6 = 0, h_7 = 13.2, h_8 = 0,$$
$$h_9 = 30, h_{10} = 0, h_{11} = 0, h_{12} = 0, h_{13} = 20, h_{14} = 0, h_{15} = 0, h_{16} = 0,$$
$$h_{17} = 12.2, h_{18} = 0, h_{19} = 6, h_{20} = 0, h_{21} = 12.2, h_{22} = 0, h_{23} = 6, h_{24} = 0,$$

and the solution to the second problem, which is analogous to problem (5), is

$$u_1 = 0, u_2 = 0, u_3 = 0, u_4 = 0, u_5 = 0, u_6 = 0, u_7 = 0.93,$$
$$y_1^+ = 50, y_2^+ = 90, z_1^+ = 40, z_2^+ = 75.39, y_3^- = 8, y_4^- = 18, z_3^- = 10, z_4^- = 22,$$
$$y_5^0 = 49, y_6^0 = 50, z_5^0 = 42, z_6^0 = 30.$$

Thus, the optimal trader's strategy consists of buying 20.41 units of security 5 while not buying and not selling other securities from her portfolio at the moment t. The expected value of the trader's portfolio as a result of the deployment of this optimal strategy equals 4988.57.

Finally, consider subcase 2 of case 2 in which the trader (a) plans to borrow securities from the broker, and (b) plans to keep at the moment $t+1$ securities that she possesses at the moment t while buying additional securities 1–6 and using the total amount of cash being at her disposal at the moment t. As in case 1, the only difference between the subcase 1 and subcase 2 consists of the total amount of cash that the trader can use for buying new securities from the set N, which equals $k_t \left(m_t + \sum_{i=1}^{n} v_{i,t} s_{i,t}\right) + 1000 = 2 \times 5496 + 1000 = 11992.$

The formulation of the prime linear programming problem in subcase 2 of case 2 differs from that in subcase 1 only by the last inequality in the system of problem's constraints, and this inequality takes the form

$$50x_1^+ + 90x_2^+ + 10x_3^- + 22x_4^- + 49x_5^0 + 50x_6^0 \le 11992.$$

The solution to this problem is

$$x_1^+ = 0, x_2^+ = 0, x_3^- = 0, x_4^- = 0, x_5^0 = 244.73, x_6^0 = 0,$$
$$h_1 = 5.6, h_2 = 0, h_3 = 16.8, h_4 = 0, h_5 = 4.4, h_6 = 0, h_7 = 13.2, h_8 = 0,$$
$$h_9 = 30, h_{10} = 0, h_{11} = 0, h_{12} = 0, h_{13} = 20, h_{14} = 0, h_{15} = 0, h_{16} = 0,$$
$$h_{17} = 124.37, h_{18} = 0, h_{19} = 0, h_{20} = 0, h_{21} = 124.37, h_{22} = 0, h_{23} = 6, h_{24} = 0.$$

The formulation of the dual linear programming problem in subcase 2 of case 1 differs from that of subcase 1 of case 1 only by the coefficient for the variable u_7, which now equals 11992. The solution to the dual problem is

$$u_1 = 0, u_2 = 0, u_3 = 0, u_4 = 0, u_5 = 0, u_6 = 0, u_7 = 0.93,$$
$$y_1^+ = 50, y_2^+ = 90, z_1^+ = 40, z_2^+ = 75.39, y_3^- = 8, y_4^- = 18, z_3^- = 10, z_4^- = 22,$$
$$y_5^0 = 49, y_6^0 = 50, z_5^0 = 42, z_6^0 = 30.$$

In just the same way it was done in case 1, it is interesting to compare the composition of the trader's portfolio at the moment t with the optimal composition of this portfolio at the moment $t + 1$ proceeding from the expectations of the share price values of all the securities that are of interest to the trader at the moment t. As one can see, the optimal portfolio composition at the moment $t + 1$ consists of the following numbers of shares of securities 1–6:

$$v_1 = 10, v_2 = 30, v_3 = 50, v_4 = 0, v_5 = 248.73, v_6 = 12,$$

whereas the composition of the trader's portfolio at the moment t is

$$v_1 = 10, v_2 = 30, v_3 = 50, v_4 = 0, v_5 = 4, v_6 = 12.$$

Finding solutions to the dual pair of linear programming problems (to solving which the finding of the optimal trader's strategy is reducible) was done with the use of another specially developed computer program, implemented on Maple 7 computing platform. (As mentioned earlier, in Illustrative Examples 1 and 2, the number of shares to be bought by the trader can be rounded-off to make them integers.)

7 Concluding Remarks

1. Studying the financial behavior of small and medium price-taking traders in their interaction with a stock exchange presents both scientific and practical interest. As a result of these studies, (a) viewpoints of both researchers of stock markets

and successful stock market players on how the stock exchange functions, and (b) their explanations of why the market players act as they do become known. In addition to that, recommendations on (a) how the market players should act to succeed, and (b) what decision-making models can be viewed as those adequately describing the interaction of individual market players with the stock exchange become available.

The authors believe that currently, two competing viewpoints on what models should be considered adequate prevail in both scientific and mass media publications.

Fundamental scientific approaches to mathematically modeling the interaction of a trader and a particular stock exchange, briefly surveyed, for instance, in [7], underlie the first one. This viewpoint is based on the belief that an adequate model is the one of the so-called representative agent, who is rational in adopting decisions on forming and managing her portfolio of securities and derivative financial instruments and tries to maximize her welfare. This belief is accompanied by the assumption that this "rational" agent (a) knows the probability distribution of the values of future prices for every financial instrument that is of her interest and is traded in the stock exchange (with which this trader interacts), and (b) makes her decisions based upon this knowledge. However, the real life does not seem to support either the above assumption or the above belief underlying this viewpoint. As mentioned earlier, deviations of the trader's financial behavior from a rational one [5, 22, 38], as well as the inability of even financial analysts to make rational investment decisions and forecast directions in which the values of the share prices of particular securities (considered as random variables) will change (under any assumptions on the probability distributions of the values of these share prices), have widely been reported in scientific publications [3, 32, 41, 44, 50].

The other viewpoint on the decision-making models adequately describing the interaction of a trader with a stock exchange is "pushed" by particular "lucky traders" who have managed to make money on adopting non-standard financial decisions. Some of them, particularly, N.Taleb [53], even deny the effectiveness of any economic and mathematical theories describing the functioning of the stock market for forming a trader's decision on managing her portfolio. Instead of adhering to such theories in managing the portfolio of a trader, N. Taleb suggests the trader to focus her attention exceptionally on the crises that happen in a stock exchange and in the world. He believes that only at the time of these crises can a trader expect to attain significant financial results. However, as shown in [1], at least under quite natural assumptions, a price-taking trader who is capable of recognizing regular events with a probability even slightly exceeding 50 % is almost guaranteed to receive a positive average gain. It is clear that attaining such a result may or may not be the case if all the trader's activities consist of waiting for "black swan" events to occur .

The authors believe that both viewpoints on the adequacy of the decision-making models are extreme, and neither reflects the practice of the interaction of a trader with a stock exchange. This state of affairs raises the following two groups of questions:

(a) Can any alternative to the above extreme views on the adequacy of the decision-making models be proposed? Can mathematical models capable of facilitating the decision-making process that small and medium price-taking traders undergo in estimating the expected financial results be proposed? Can such models work successfully in the absence of knowledge on any probability distribution of future price values of financial instruments traded in a particular stock exchange?
(b) Can one propose mathematical models the use of which would allow a trader (with a confirmed ability to correctly estimate directions of changing the price values of financial instruments of her interest) to make rational decisions on the structure of her portfolio at a particular moment t in principle? Can such models be proposed if the trader can indicate a segment within which the future values of the price of a particular financial instrument will change being uniformly distributed? Can one propose such models if the trader can estimate only the expected areas in which the values of the prices for the groups of financial instruments forming together the whole set of the financial instruments of her interest (into which this set is divided by the trader) may change? Can one develop these models with the use of only the simplest linear equations and inequalities of a balance type?

The present paper offers positive answers to all the above questions. However, the authors believe that the proposed mathematical models and approaches to finding trader's optimal investment strategies need to be tested and researched by both economists and other analysts studying financial aspects of the interaction between a trader and a stock exchange. The authors consider the tools proposed in this paper mostly as a powerful instrument allowing interested researchers to study particular aspects of the stock exchange behavior in the framework of a large-scale decision-support system. These tools allow one to use the models with millions of variables and constraints, which distinguishes the authors' approach to modeling stock exchanges from those already proposed.

2. As is well known, global optimization problems are difficult to solve, and there are no uniform approaches allowing one to find global extrema in problems mathematically formalizing many of theoretical and practical optimization problems. Thus, detecting classes of problems in which not only global extrema can be found in principle, but those in which these extrema can be found with the use of the most powerful computational techniques, linear programming being one of them, undoubtedly presents both scientific and applied interest. As shown in the paper, finding a point of the global maximum of a particular nonlinear function (the minimum function on a convex polyhedron described by a compatible system of linear equations and inequalities) on a subset of another convex polyhedron formed by vectors with all the coordinates being non-negative integers is

reducible to solving a mixed programming problem. It was also shown that finding the global maximum of the above function on this another convex polyhedron (described by another compatible system of linear equations and inequalities) is reducible to solving linear programming problems forming a dual pair.

3. While there are numerous schemes for and approaches to forecasting time series, the need in tools helping a potential or an acting small or medium price-taking trader reliably estimate the ability to divine future values of the share prices of securities remains high. Such tools can save a lot of money to private investors and even prevent personal financial tragedies. It is clear that (a) a detected ability to divine future values of the share prices of particular securities by processing results of the trials according to the Bernoulli scheme, and (b) the ability to divine the actual values of the share prices of particular securities in dealing with these prices in real life may not be the same. So the availability of the tool that allows one to compare both abilities seems critical at least from a practical viewpoint.
4. In two mathematical models proposed in this paper, the authors assumed that for all the securities being of interest to a trader, the trader either (a) can indicate a segment within which the values of the prices of a particular financial instrument will change being uniformly distributed, or (b) can only estimate the areas in which the expected values of the prices for the whole set of financial instruments that interest her may change. However, it is possible that there are two groups of securities that interest the trader, and for one group, her ability to divine future values of the share prices of particular securities corresponds to case (a) from point 3 of this section, whereas for the other group, the ability to divine directions in which the price values of securities from this group will change corresponds to case (b) from the same point of this section. If the trader is firm in dividing financial resources available to her between these two groups (in dealing with securities from these groups), then both models can be used separately. If this is the case, the trader's optimal investment strategies can be determined by solving corresponding mathematical programming problems considered in Sects. 4 and 5 of this paper. Otherwise, the trader faces a complicated problem of dividing financial resources available to her at the moment t between the two groups, which leads to considering models whose structure and features are completely different from those considered in the present paper.

The authors would like to emphasize that in the models formalizing the interaction of a trader with the stock exchange in the form of mathematical programming problems with Boolean variables, presented in Sects. 4 and 5 of the paper, they did not consider some particular risks that the trader may be interested in taking into consideration in making her decision on developing or changing her portfolio of securities. Though such risks are traditionally considered in publications on modeling the behavior of traders trading securities in a stock exchange, the inclusion of the risks considered, for instance, in [33], in the models proposed in this paper would lead to solving large-scale nonlinear programming problems with integer or mixed variables (formulated on the basis of these models). Such problems are difficult to solve even for relatively small problem sizes, and from the authors' viewpoint, this inclusion would hardly make corresponding models and problems an effective tool of studying stock exchanges and traders' behavior in interacting with them. At the same time, the authors would like to make it clear that their search for the models that could be considered an effective tools for studying the stock exchange behavior continues, and models of the mentioned kind presented in this paper should be viewed as no more than only the first step towards this direction.

5. Finally, only the modeling of the decision-making process that individual price-taking traders undergo in the course of their interaction with a stock exchange was the subject of this paper. However, one should bear in mind that both small and medium price-taking traders may form coalitions and act either as one legal entity or as groups in the framework of which the interests of all the group members within each group are to be observed. Moreover, such groups are implicitly formed when some (and, possibly, quite a substantial number of) small price-taking traders exercise the strategy of following someone's decisions (for instance, those of large traders or "lucky" ones) independently of their (groups') sizes. Studying aspects of the financial behavior of these groups presents obvious interest in an attempt to understand the mechanisms of the interacting between individual traders and a stock exchange. However, such studies require both a particular use of known and the development of new mathematical tools, and the discussion of these issues, which invokes that of a set of fundamental modeling problems, lies beyond the scope of the present paper.

Acknowledgements The financial support from the Russian Federation Government within the framework of the implementation of the 5–100 Programme Roadmap of the National Research University Higher School of Economics is acknowledged. The authors are grateful for financial support of their work to DeCAn Laboratory at the National Research University Higher School of Economics, headed by Prof. Fuad Aleskerov, with whom the authors have had fruitful discussions on problems of developing tools for quantitatively analyzing large-scale systems in general and those for stock exchanges in particular. L. Egorova expresses her gratitude to LATNA Laboratory, NRU HSE, headed by Prof. Panos Pardalos, RF government grant, ag.11.G34.31.0057 and A. Belenky expresses his gratitude to the MIT Institute for Data, Systems, and Society and to Dr. Richard Larson, MIT's Mitsui Professor of Data, Systems, and Society.

References

1. Aleskerov F.T., Egorova, L.G.: Is it so bad that we cannot recognize black swans? Econ. Lett. **117**(3), 563–565 (2012)
2. Asratyan, A.S., Kuzyurin, N.N.: Analysis of randomized rounding for integer programs. Discret. Math. Appl. **14**(6), 543–554 (2004) (in Russian)
3. Barber, B., Odean, T.: Trading is hazardous to your wealth: The common stock investment performance of individual investors J. Financ. **55**(2), 773–806 (2000)

4. Barber, B.M., Odcan, T.: All that glitters: The effect of attention and news on the buying behavior of individual and institutional investors. Rev. Financ. Stud. **21**(2), 785–818 (2008)
5. Barberis, N., Thaler, R.: A survey of behavioral finance, NBER Working Paper Series (2002). http://www.nber.org/papers/w9222
6. Belenky, A.S.: Minimax planning problems with linear constraints and methods of their solution. Autom. Remote Control **42**(10), 1409–1419 (1981)
7. Belenky, A.S., Egorova, L.G.: An approach to forming and managing a portfolio of financial securities by small and medium price-taking traders in a stock exchange, advances in intelligent systems and computing. In: Proceedings of the 3rd International Conference on Modelling, Computation and Optimization in Information Systems and Management Sciences MCO (2015), pp. 257–268
8. Bhardwaj, G., Swanson, N.R.: An empirical investigation of the usefulness of ARFIMA models for predicting macroeconomic and financial time series. J. Econ. **131**, 539–578 (2006)
9. Chaudhuri, R., Ivković, Z., Pollet, J., Trzcinka, C.: What a difference a Ph.D. makes: more than three little letters (2013). http://ssrn.com/abstract=2344938
10. Chen, L.: Stochastic mean and stochastic volatility- a three-factor model of the term structure of interest rates and its application to the pricing of interest rate derivatives. Financ. Mark. Inst. Instrum. **5**, 1–88 (1996)
11. Cox, J.C., Ingersoll, J.E., Ross, S.A.: The relation between forward prices and futures prices. J. Financ. Econ. **9**(4), 321–346 (1981)
12. De Bondt, W., Thaler, R.: Does the stock market overreact? J. Financ. **XL**(3), 793–805 (1985)
13. De Bondt, W., Mayoral, R.M., Vallelado, E.: Behavioral decision-making in finance: an overview and assessment of selected research. Rev. Espanola de Financiacion y Contabilidad **XLII**(157), 99–118 (2013)
14. Devenow, A, Welch, I.: Rational herding in financial economics. Eur. Econ. Rev. **40**(3–5), 603–615 (1996)
15. Engle, R.F.: Autoregressive conditional heteroskedastisity with estimates of the variance of U.K. inflation. Econometrica **50**, 987–1007 (1982)
16. Enke, D., Thawornwong, S.: The use of data mining and neural networks for forecasting stock market returns. Expert Syst. Appl. **29**, 927–940 (2005)
17. Feller, W.: An Introduction to Probability Theory and Its Applications, vol. 1 and 2. Wiley, New York (1991)
18. Geman, H.: Pure jump Levy processes for asset price modelling. J. Bank. Finance. **26**, 1297–1316 (2002)
19. Hjalmarsson, E.: Predicting global stock returns. J. Financ. Quant. Anal.**45**(1), 49–80 (2010)
20. Jarrett, J.: Daily variation and predicting stock market returns for the frankfurter borse (stock market). J. Bus. Econ. Manag. **9**(3), 189–198 (2008)
21. Jorion, P.: Value at Risk: A New Benchmark for Managing Derivatives Risk. Irwin Professional Publishers, Chicago (2000)
22. Kahneman, D.: Thinking, Fast and Slow. Penguin, New York (2011)
23. Kim, Y.S., Rachev, S.T., Bianchi, M.L., Fabozzi, F.J.: Financial market models with Levy processes and time-varying volatility. J. Bank. Financ. **32**(7), 1363–1378 (2008)
24. Koijen, R., van Nieuwerburgh, S.: Predictability of returns and cash flows. Ann. Rev. Financ. Econ. **3**, 467–491 (2011)
25. Konno, H., Yamakazi, H.: Mean-absolute deviation portfolio optimization model and its applications to tokyo stock market. Manag. Sci. **37**(5), 519–531
26. Kuo, W.Y., Lin, T.C.: Overconfident individual day traders: evidence from the Taiwan futures market. J. Bank. Financ. **37**(9), 3548–3561 (2013)
27. Lin, W., Engle, R., Ito, T.: Do bulls and bears move across borders? International transmission of stock returns and volatility. Rev. Financ. Stud. **7**(3), 507–538 (1994)
28. Liu, Y.K., Wyu, X.L, Hao, F.F.: A new chance-variance optimization criterion for portfolio selection in uncertain decision systems. Expert Syst. Appl. **39**(7), 6514–6526 (2012)
29. Lo, A.W., Repin, D.V.: The psychophysiology of real-time financial risk processing. J. Cogn. Neurosci. **14**(3), 323–339 (2002)

30. Lo, A.W., Repin, D.V., Steenbarger, B.N.: Fear and greed in financial markets: a clinical study of day-traders. Am. Econ. Rev. **95**(2), 352–359 (2005)
31. Lewellen, J.: Predicting returns with financial ratios. J. Financ. Econ. **74**, 209–235 (2004)
32. Malkiel, B.G., Saha, A.: Hedge funds: risk and return. Financ. Anal. J. **61**(6), 80–88 (2005)
33. Markowitz, H.: Portfolio selection. J. Financ. **VII**(1), 77–91 (1952)
34. Mao, J.: Models of capital budgeting, E-V vs. E-S. J. Financ. Quant. Anal. **4**(05), 657–675 (1970)
35. Merton, R.: An intertemporal capital asset pricing model. Econometrica **41**(5), 867–887 (1973)
36. Morelli, M., Montagna, G., Nicrosini, G., Treccani, M., Farina, D., Amato, P.: Pricing financial derivatives with neural networks. Phys. A Stat. Mech. Appl. **338**(1–2), 160–165 (2004)
37. Mullainathan, S., Thaler, R.H.: Behavioral Economics, NBER Working paper (2000). http://www.nber.org/papers/w7948
38. Nelson, C.R., Kang, H.: Pitfalls in the use of time as an explanatory variable in regression. J. Bus. Econ. Stat. **2**, 73–82 (1984)
39. Nelson, C.R., Plosser, C.I.: Trends and random walks in macroeconomic time series: some evidence and implication. J. Monet. Econ. **10**, 139–62 (1982)
40. O'Connor, N., Madden, M.: A neural network approach to predicting stock exchange movements using external factors. Knowl.-Based Syst. **19**(5), 371–378 (2006)
41. Odean, T.: Do investors trade too much? Am. Econ. Rev. **89**(5), 1279–1298 (1999)
42. Ostermark, R.: Predictability of Finnish and Swedish stock returns. Omega **17**(3), 223–236 (1989)
43. Palomino, F., Renneboog, L., Zhang, C.: Information salience, investor sentiment, and stock returns: the case of British soccer betting. J. Corp. Financ. **15**(3), 368–387 (2009)
44. Penikas, H., Proskurin, S.: How well do analysts predict stock prices? Evidence from Russia, Working papers by NRU Higher School of Economics. Series FE "Financial Economics", WP BRP 18/FE/2013 (2013)
45. Rockafellar, R.T., Uryasev, S.: Optimization of conditional value-at-risk. J. Risk **2**, 21–41 (2000)
46. Rothig, A., Chiarella, C.: Small traders in currency futures markets. J. Futur. Mark. **31**(9), 898–914 (2011)
47. Ross, S.: The arbitrage theory of capital asset pricing. J. Econ. Theory **13**(3), 341–360 (1976)
48. Shapira, Z., Venezia, I.: Patterns of behavior of professionally managed and independent investors. J. Bank. Financ. **25**(8), 1573–1587 (2001)
49. Sharpe, W.F.: Capital asset prices: a theory of market equilibrium under conditions of risk. J. Financ. **19**(3), 425–442 (1964)
50. Soderlind, P.: Predicting stock price movements: regressions versus economists. Appl. Econ. Lett. **17**, 869–874 (2010)
51. Sornette, D.: Dragon-kings, black swans and the prediction of crises. Int. J. Terraspace Sci. Eng. **2**(1), 1–18 (2009)
52. Stracca, L.: Behavioral finance and asset prices: where do we stand? J. Econ. Psychol. **25**, 373–405 (2004)
53. Taleb, N.N.: The Black Swan: The Impact of The Highly Improbable. Penguin Books, London (2008)
54. Tay, F.E.H., Lijuan, C.: Application of support vector machines in financial time series forecasting. Omega Int. J. Manag. Sci. **29**(4), 309–317 (2001)
55. Tedeschi, G., Iori, G., Gallegati, M.: Herding effects in order driven markets: the rise and fall of gurus. J. Econ. Behav. Organ. **81**, 82–96 (2012)
56. Tsay, R.S.: Analysis of Financial Time Series. Wiley, New York (2002)
57. Vasicek, O.: An equilibrium characterization of the term structure. J. Financ. Econ. **5**(2), 177–188 (1977)
58. Venezia, I., Nashikkar, A., Shapira, Z.: Firm specific and macro herding by professional and amateur investors and their effects on market volatility. J. Bank. Financ. **35**, 1599–1609 (2011)

59. Yu, C.L., Li, H., Wells, M.T.: MCMC estimation of levy jump models using stock and option prices. Math. Financ. **21**(3), 383–422 (2011)
60. Yudin, D., Golshtein, E.: Linear Programming. Israel Program of Scientific Translations, Jerusalem (1965)
61. Yudin, D.B., Yudin, A.D.: Extreme Models in Economics. LIBROKOM, Moscow (2009) (in Russian)

Indirect Maximum Likelihood Estimation

Daniel Berend and Luba Sapir

Abstract We study maximum likelihood estimators (henceforth MLE) in experiments consisting of two stages, where the first-stage sample is unknown to us, but the second-stage samples are known and depend on the first-stage sample. The setup is similar to that in parametric empirical Bayes models, and arises naturally in numerous applications. However, problems arise when the number of second-level observations is not the same for all first-stage observations. As far as we know, this situation has been discussed in very few cases (see Brandel, Empirical Bayes methods for missing data analysis. Technical Report 2004:11, Department of Mathematics, Uppsala University, Sweden, 2004 and Carlin and Louis, Bayes and Empirical Bayes Methods for Data Analysis, 2nd edn. Chapman & Hall, Boca Raton, 2000) and no analytic expression for the indirect maximum likelihood estimator was derived there. The novelty of our paper is that it details and exemplifies this point. Specifically, we study in detail two situations:

1. Both levels correspond to normal distributions; here we are able to find an explicit formula for the MLE and show that it forms uniformly minimum-variance unbiased estimator (henceforth UMVUE).
2. Exponential first-level and Poissonian second-level; here the MLE can usually be expressed only implicitly as a solution of a certain polynomial equation. It seems that the MLE is usually not a UMVUE.

In both cases we discuss the intuitive meaning of our estimator, its properties, and show its advantages vis-à-vis other natural estimators.

D. Berend (✉)
Departments of Mathematics and Computer Science, Ben-Gurion University, Beer Sheva, Israel
e-mail: berend@cs.bgu.ac.il

L. Sapir
Department of Mathematics, Ben-Gurion University and Deutsche Telekom Laboratories at Ben-Gurion University, Beer Sheva, Israel
e-mail: lsapir@bgu.ac.il

B. Goldengorin (ed.), *Optimization and Its Applications in Control and Data Sciences*, Springer Optimization and Its Applications 115,
DOI 10.1007/978-3-319-42056-1_4

Keywords Unobserved observations • Indirect observations • Indirect maximum likelihood estimator • Empirical Bayes estimation • Fisher information • Two-level setup • Normal-Normal distributions • Exponential-Poissonian distributions

1 Introduction

In parametric statistics, the probability distribution generating the experimental data is completely known, except for the values of the parameters. The classical setup in point estimation problems is loosely defined as follows. Assume that a population can be represented by a random variable X, whose density is $f(\cdot\,;\theta)$, where the form of the density is known, except for the unknown parameter θ. Let $x_1, x_2, \ldots, x_k$ be the values of a random sample $X_1, X_2, \ldots, X_k$ from $f(\cdot\,;\theta)$. On the basis of the observations $x_1, x_2, \ldots, x_k$, it is required to estimate the value of the unknown parameter θ, or of some function thereof $\tau(\theta)$.

There are several methods of finding point estimators of a parameter: the Bayes method, the method of moments, the method of least squares, etc. One of these methods, and probably the most popular one, is the method of *maximal likelihood.* This method, introduced by Fisher in 1912, can be applied in most problems, has a strong intuitive appeal, and usually yields a reasonable estimator of θ. Furthermore, if the sample is large, the method typically yields an excellent estimator of the parameter. For these reasons, the maximal likelihood method is probably the most widely used method of estimation in statistics [1, 4, 11, 13, 16, 20, 23, 24].

However, the methods of obtaining point estimators of θ were developed mostly for the classical setup, where the value of the parameter θ of the distribution function is unknown, but the observations (almost by definition) are. A more complicated situation occurs when inferences about the unknown parameter should be done based on noisy observations. In this situation, before estimating the parameter, various noise-subtraction techniques are employed [3, 5, 12, 21].

This paper is motivated by even more problematic situations. Namely, the observations $x_1, x_2, \ldots, x_k$ are unknown to us, but we do have some related information. (One is almost tempted to refer to the x_i's as "unobserved observations".) Consider the following, somewhat simplified, example. There is a machine in a certain factory which produces sensory devices for harsh environments. The quality of the devices produced by the machine is distributed according to some known distribution function with unknown (perhaps multi-dimensional) parameter, say $P_i \sim \text{Beta}(\alpha, \beta), \quad 1 \le i \le k$, with α, β unknown. We need to estimate the parameter (α, β) of the machine as best we can. For example, each device has to signal in case it encounters radiation exceeding a certain threshold, and it performs correctly with a probability P_i. If a device can be tested infinitely many times, the value p_i of P_i for this particular device will be exactly known. However, due to the harsh environment, a device may supply us with only few indications until it wears out. Thus the exact values of the observations, namely the correctness probabilities p_i, are unknown, and we have to make our inferences based on the observations from the Bernoulli distribution representing the correctness of the indications. We would like to estimate the parameter (α, β) of the original machine, based on the few indications at hand.

Clearly, we may try to estimate each p_i based on the small sample available for the i-th sensor, and then conclude about the parameter (α, β) of the machine. To realize that this approach is non-optimal, consider the situation where, say, device number 1 supplies us with 20 indications until it is out of use, whereas device number 2 supplies only 5 of them. The estimate of p_1 is more reliable than that of p_2, which should be taken into account when the estimation of the machine's parameter takes place. However, the method above fails to give each estimate a weight, corresponding to its reliability.

In this paper we consider the problem of estimating the unknown parameter of the distribution of the first level, based on the data of the second level. We note that the classical setup of the parameter estimation problem is a special instance of our setup, namely when the second distribution is constant. We propose in the two-levels case an approach for calculating (precisely or approximately) the maximum likelihood estimate. Our method may be termed *indirect maximum likelihood estimation*. In particular, we provide a detailed discussion of the indirect maximum likelihood estimate calculation for several specific pairs of distributions.

The same setup is encountered not only in various practical applications, but also in empirical Bayes inference. Namely, it appears in the empirical Bayes method for data analysis as a particular step dealing with marginal maximum likelihood estimation of the parameter (see Sect. 5). Our approach details this issue for the situation where the number of second-stage (observed) observations is not necessarily the same for each first-stage (unobserved) observation. As far as we know, the case of distinct number of second-level observations has been discussed in very few cases (see [8, 9]). However, no analytic expression for the indirect maximum likelihood estimator was derived there in this situation.

In this paper, we illustrate our method in two situations. In one of these, MLE may be obtained in closed-form; in the other, it may be obtained only implicitly as a solution of a certain equation. In the first case we are able to explain how and why our formula takes into account the number of observations in the second stage, related to each observation of the first stage. In the second case, where we are unable to obtain an analytic formula for the MLE, we show the MLE value is confined to a certain interval and various iterative approximation methods can be used to find the value. In both cases, we discuss the properties of our estimator and show its advantages versus other natural estimators.

There does exist quite a bit of research on estimation in situations where one does not have direct observations from the distribution whose parameter is the object for estimation. We refer, for example, to [6, 10, 14, 25]. However, those models, which usually originate in economics, seem to be very different from our model, and we shall not dwell on their setup here.

The setup is formally defined in Sect. 2. Section 3 contains the main results dealing with several specific pairs of first- and second-level distributions. Section 4 provides the proofs. In Sect. 5 we discuss some potential applications of our approach. In particular, our approach may improve one of the steps of empirical Bayes inference. In Sect. 6 we summarize and raise some questions for future research.

2 Setup of the Problem and Approach for Solution

The mathematical formulation of the problem is as follows:

Problem 1. *A random sample $X_1, X_2, \ldots, X_k$ is taken from a distribution $f(x;\theta)$, where the parameter $\theta \in \Theta$ is unknown. The corresponding observations $x_1, x_2, \ldots, x_k$ are* ***unknown****. We are given a second distribution $g(t;x)$ and samples $T_{i1}, T_{i2}, \ldots, T_{in_i}$ from $g(t;x_i)$ for $1 \le i \le k$, with corresponding observed values, which may be organized in the following table:*

$$\mathbf{t} = \begin{pmatrix} t_{11}\ t_{12}\ \ldots\ldots\ t_{1n_1} \\ t_{21}\ t_{22}\ \ldots\ t_{2n_2} \\ \ldots\ldots\ldots \\ t_{k1}\ t_{k2}\ \ldots\ldots\ldots\ t_{kn_k} \end{pmatrix}.$$

Goal: *Find the MLE for θ or for some function $\tau(\theta)$.*

The required estimator $\hat{\theta}$ is a function of the statistics T_{ij}, $1 \le i \le k$, $1 \le j \le n_i$. The table $\mathbf{t}$ of observations may be regarded as the value observed for the multi-dimensional statistic $\mathbf{T} = (T_{ij})_{1 \le i \le k, 1 \le j \le n_i}$. Note that the rows of the tables $\mathbf{t}$ and $\mathbf{T}$ are in general of varying lengths. The statistic $\hat{\theta}$ may be explicitly defined by

$$\hat{\theta} = \arg\max_{\theta \in \Theta} P(\mathbf{T};\theta). \tag{1}$$

(Here $P(\mathbf{T};\theta)$ signifies either a probability or a density.) Note that the parameter θ may be multi-dimensional.

Remark 1. In the theoretical level, one may consider various generalizations of the problem. For example, the values t_{ij} may also be unknown, and information regarding their values can be derived only from samples taken from other populations in which the t_{ij}'s serve as parameters. Another generalization is where there are several distributions $g_i(t;x)$ and the T_{ij}'s in each row of $\mathbf{T}$ are distributed $g_i(t;x_i)$. We shall not deal with these extensions here.

The problem of finding the value of θ maximizing $P(\mathbf{T} = \mathbf{t};\theta)$ on the right-hand side of (1) can be made more transparent by replacing $P(\mathbf{T} = \mathbf{t};\theta)$ with an explicit expression (adapted to the case of continuous variables). Denote:

$$\mathbf{T}_i = (T_{i1}, T_{i2}, \ldots, T_{in_i}), \qquad 1 \le i \le k.$$

The likclihood function $L(\theta)$, for any given observations $\mathbf{t} = (t_{ij})_{1\le i\le k, 1\le j\le n_i}$, is given by

$$L(\theta) = \underbrace{\int \dots \int}_{\mathscr{D}_X^k} \left(\prod_{j=1}^{n_1} g(t_{1j}; x_1) \cdot \ldots \cdot \prod_{j=1}^{n_k} g(t_{kj}; x_k) \right) \cdot \prod_{i=1}^{k} f(x_i; \theta) dx_i$$

$$= \prod_{i=1}^{k} \int_{\mathscr{D}_X} \prod_{j=1}^{n_i} g(t_{ij}; x_i) f(x_i; \theta) dx_i,$$

where $\mathscr{D}_X$ is the support of the first-level random variables $X_i, \quad i = 1, \ldots, k$. Thus, for a given table $\mathbf{t}$, the MLE for θ is defined by:

$$\hat{\theta} = \arg\max_{\theta\in\Theta} \prod_{i=1}^{k} \int_{\mathscr{D}_X} f(x_i; \theta) \prod_{j=1}^{n_i} g(t_{ij}; x_i) dx_i.$$

3 The Main Results

In this section we illustrate our method for dealing with Problem 1 for several pairs of distributions $f(x; \theta)$ and $g(t; x)$. In one of these, we are able to give a closed-form formula for the MLE. However, even in the classical setup, there is usually no closed-form formula for the MLE. In our, more complicated, setup there seem to be very few situations where such a formula may be obtained. In most cases, the MLE is defined implicitly as the solution of a certain equation. Our second instance of study falls into this category, and we are able to present, under some assumptions, a procedure for finding an approximate solution.

While one may study any pair of distributions $f(x; \theta)$ and $g(t; x)$; it is natural to take the first as a conjugate prior of the second ([9, 11]). The first two cases studied here deal with such pairs (or special cases thereof).

3.1 MLE of the Mean in the Normal-Normal Case

We start with a case where the MLE is unique and can be explicitly found.

Theorem 1. *Consider the setup in Problem 1, where the distribution $f(x; \mu)$ is $N(\mu, \sigma_1^2)$ with unknown μ and known σ_1^2, and the distribution $g(t; x)$ is $N(x, \sigma_2^2)$ with known σ_2^2. Then the MLE for μ is given by the statistic*

$$M = \sum_{i=1}^{k} w_i \overline{T}_i, \tag{2}$$

where $w_i = \dfrac{\frac{1}{\sigma_1^2+\sigma_2^2/n_i}}{\sum_{r=1}^{k} \frac{1}{\sigma_1^2+\sigma_2^2/n_r}}$ *and* $\overline{T}_i = \dfrac{1}{n_i} \sum_{j=1}^{n_i} T_{ij}, \quad 1 \le i \le k.$

Corollary 1. *If the number of observations at the second stage is the same for each observation of the first stage, i.e.,* $n_i = n$ *for* $1 \le i \le k$*, then (2) reduces to*

$$M = \frac{1}{kn} \cdot \sum_{i=1}^{k} \sum_{j=1}^{n} T_{ij}.$$

Let us explain the intuitive meaning of the theorem. The most natural estimator for each x_i is the sample mean $\overline{T}_i = \frac{1}{n} \sum_{j=1}^{n_i} T_{ij}$ of the observations related to this specific x_i. At first glance, the natural estimator for μ is the mean $\overline{T} = \frac{1}{k} \sum_{i=1}^{k} \overline{T}_i$ of these means. The estimator M in the theorem is also an average of the $\overline{T}_i$'s, but a weighted average rather than a simple average. The reason for the weighting is that, the larger n_i is, the more reliable is $\overline{T}_i$ as an estimator for x_i. The weight assigned to $\overline{T}_i$ is inversely proportional to the variance of $\overline{T}_i$, which is $\sigma_1^2 + \sigma_2^2/n_i$ (see (20) below), and thus increases with n_i. Observe also the relation between our weights and the variances σ_1^2, σ_2^2. If σ_2^2 is much smaller than σ_1^2, then the t_{ij}'s are good estimates for x_i, and our estimator is close to $\overline{T}$. As σ_2^2 grows relative to σ_1^2, the effect of the noise in the measurements t_{ij} becomes more significant, and M deviates more from $\overline{T}$. If σ_2^2 is much larger than σ_1^2, then the w_i's are almost proportional to the n_i's. Finally, we note also that both M and $\overline{T}$ are functions of the $\overline{T}_i$'s only; this is natural, as each $\overline{T}_i$ is a sufficient statistic for the parameter x_i of $g(t; x_i)$.

Corollary 1 deals with the case where the samples taken to estimate the x_i's are of the same size, and thus these estimates are equally reliable. Unsurprisingly, in this case M coincides with $\overline{T}$.

Clearly, the estimator M in the theorem is an unbiased estimator of μ. Of course, there are many other unbiased estimators of μ, for example $\overline{T} = \frac{1}{k} \sum_{i=1}^{k} \overline{T}_i$, and, more generally, any weighted average

$$M^* = \sum_{i=1}^{k} \alpha_i \overline{T}_i, \qquad \left(\sum_{i=1}^{k} \alpha_i = 1 \right),$$

of the $\overline{T}_i$'s. However, the following proposition proves that M is the most efficient among all unbiased estimators.

Proposition 1. *The estimator M is a uniformly minimum-variance unbiased estimator. Its variance is given by:*

$$V(M) = \frac{1}{\sum_{i=1}^{k} \frac{1}{\sigma_1^2 + \sigma_2^2/n_i}}. \tag{3}$$

Remark 2. It is amusing to observe what Proposition 1 yields in the special case of the estimator $\overline{T} = \frac{1}{k}\sum_{i=1}^{k} T_i$. Since $V(\overline{T}) = \sum_{i=1}^{k} \frac{1}{k^2} V(\overline{T}_i)$, and (20) and (21) imply $V(M) = \frac{1}{\sum_{i=1}^{k} 1/V(\overline{T}_i)}$, the inequality $V(M) \le V(\overline{T})$ is equivalent to

$$\frac{k}{\sum_{i=1}^{k} 1/V(\overline{T}_i)} \le \frac{1}{k}\sum_{i=1}^{k} V(\overline{T}_i). \tag{4}$$

Thus, the variances of M and $\overline{T}$ are proportional to the harmonious and the arithmetic means, respectively, of the numbers $V(\overline{T}_i)$. Consequently, the inequality $V(M) \le V(\overline{T})$, with strict inequality unless all the n_i's are equal, follows from the classical means inequality [7, 15].

As in the classical situation, it seems natural to expect the consistency of the MLE [20, 23]. In our setup, (3) immediately implies that M is a mean-squared-error consistent estimator of μ as the number of rows in the table **t** tends to infinity. (In particular, M is also a weakly consistent estimator.) The following corollary states it more formally.

Corollary 2. *Let* $(M_k)_{k=1}^{\infty}$ *be a sequence of estimators of* μ*, where each* M_k *is as on the right-hand side of (2), and is based on a table* **t** *with k rows. Then* $(M_k)_{k=1}^{\infty}$ *forms a (mean-squared-error) consistent sequence of estimators of* μ*.*

3.2 MLE in the Exponential-Poissonian Case

In this section we present a case where the MLE can in general be only implicitly defined as the zero of a certain polynomial. We then test its performance by Monte-Carlo simulations. The distribution in question is exponential, and we try to estimate its expectation (Of course, since the expectation is a one-to-one function of the parameter, it does not matter if we deal with the MLE of the parameter or with the MLE of the expectation, but later in the section we will also discuss unbiased estimators, where it does matter.).

Theorem 2. *Consider the setup in Problem 1, where the distribution* $f(x;\theta)$ *is* $\mathrm{Exp}(\theta)$ *with an unknown parameter* θ*, and the distribution* $g(t;x)$ *is* $\mathscr{P}(x)$*. Then the value* $\hat{\tau}$ *of the MLE of* $\tau = 1/\theta$ *is the solution of the equation*

$$\sum_{i=1}^{k} \frac{\bar{t}_i + 1/n_i}{\tau + 1/n_i} = k, \tag{5}$$

where $\bar{t}_i = \frac{1}{n_i}\sum_{j=1}^{n_i} t_{ij}, \quad 1 \le i \le k.$

Our next result shows that, usually, the MLE exists and is unique.

Proposition 2. *If at least one of the* $t_{ij}, \quad 1 \le i \le k, \quad 1 \le j \le n_i$, *is non-zero, then (5) has a unique solution* $\hat{\tau}$ *in the positive axis. Moreover, the solution* $\hat{\tau}$ *is confined to the interval* $[c - 1/n_{\min}, c - 1/n_{\max}]$, *where* $n_{\min} = \min_{1\le i\le k} n_i, \quad n_{\max} = \max_{1\le i\le k} n_i$, *and* $c = \frac{1}{k}\sum_{i=1}^{k}(\bar{t}_i + 1/n_i)$.

Remark 3. If $t_{ij} = 0$ for $1 \le i \le k, \quad 1 \le j \le n_i$, then (5) has the unique solution $\hat{\tau} = 0$, which is not in the allowed range.

Obviously, the length of the interval containing the solution is $\frac{1}{n_{\min}} - \frac{1}{n_{\max}} < 1$, and the value $\hat{\tau}$ can be arbitrarily approximated by various iterative schemes (such as the bisection method or Newton's method).

Note that (5) depends on the observations of the table only through the sample averages of each row, that, as in the normal-normal case of the preceding section, form sufficient statistics for the unobserved observations $x_i, \quad 1 \le i \le k$, from the first stage. Thus, $\hat{\tau}$ depends on the data only through the values of the sufficient statistics $\overline{T}_i$. As in the preceding section, there is a "natural" estimator of the parameter, also depending only on the sufficient statistics $\overline{T}_i$, namely

$$\overline{\overline{T}} = \frac{1}{k}\sum_{i=1}^{k} \overline{T}_i. \tag{6}$$

In the sequel, we will compare (by simulation) this estimator with the MLE.

The following corollary of Theorem 2 lists a few cases, in which the MLE can be obtained explicitly.

Corollary 3. *If the table of observations in not identically* 0, *and it consists of*

1. *a single row, i.e.,* $k = 1$, *then* $\hat{\tau} = \overline{T}_1$;
2. *only two rows, i.e.,* $k = 2$, *then*

$$\hat{\tau} = \frac{2\left(\frac{1}{n_2}\overline{T}_1 + \frac{1}{n_1}\overline{T}_2\right)}{\frac{1}{n_1} + \frac{1}{n_2} - \overline{T}_1 - \overline{T}_2 + \sqrt{\left(\frac{1}{n_1} + \frac{1}{n_2} - \overline{T}_1 - \overline{T}_2\right)^2 + 8\left(\frac{1}{n_2}\overline{T}_1 + \frac{1}{n_1}\overline{T}_2\right)}};$$

3. *the same number of observations in all rows, i.e.,* $n_i = n, \quad 1 \le i \le k$, *then*

$$\hat{\tau} = \frac{1}{k}\sum_{i=1}^{k} \overline{T}_i = \frac{1}{kn}\sum_{i=1}^{k}\sum_{j=1}^{n} T_{ij}.$$

The last part follows straightforwardly from Proposition 2, and the first is a special case of the last. The second part follows from a routine calculation, which we omit here. Note that the intuitive reason in part 3 is basically that, since the $\overline{T}_i$'s are based on samples of the same size, they are equally reliable. Hence the natural estimator, giving them equal weight, coincides with the MLE. We mention that, by Proposition 2, if all n_i's tend to ∞, then the difference between the MLE estimator and $\overline{T}$ tends to 0.

The following example illustrates the calculation of the MLE by Theorem 2.

Example 1. Under the setup in Theorem 2 we generated $k = 4$ "unobserved" observations x_1, x_2, x_3, x_4 from $\mathrm{Exp}(\theta_{\mathrm{true}})$ with $\theta_{\mathrm{true}} = 1$. At the second stage, a few observations were taken from each $\mathscr{P}(x_i)$, producing the table:

$$\mathbf{t} = \begin{pmatrix} 0\ 2\ 1\ 0\ 2\ 0\ 1\ 1 \\ 1\ 0\ 1\ 0\ 3\ 1\ 2\ 2\ 1\ 0 \\ 0\ 0\ 3\ 1 \\ 0\ 0\ 0\ 0 \end{pmatrix}.$$

Equation (5), corresponding to this data, reduces to:

$$\frac{4}{8\tau + 1} + \frac{6}{10\tau + 1} + \frac{3}{4\tau + 1} = 2. \tag{7}$$

Figure 1 depicts the expression on the left-hand side of (7) as a function of τ. The maximum value of the likelihood function (calculated by applying Maple's `fsolve`

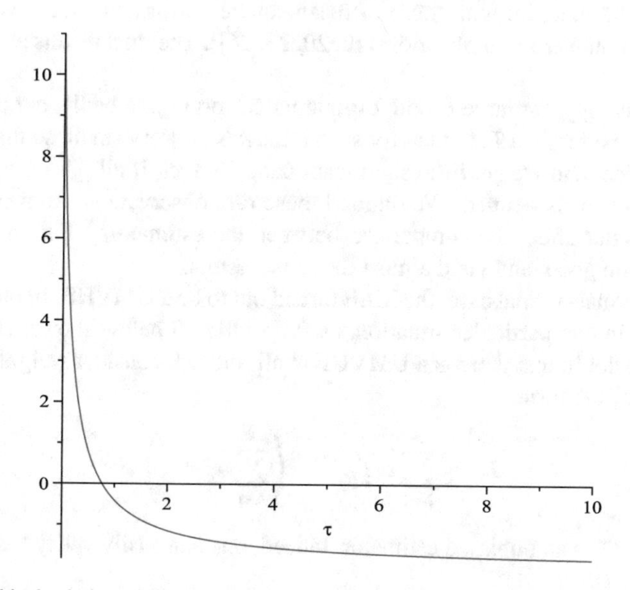

Fig. 1 Graphical solution of (7)

Table 1 Errors of the MLE and of $\overline{T}$

Error	θ 0.01	0.1	1	10	100
$\text{err}(\hat{\tau}, \tau)$	0.1419	0.1437	0.1471	0.1793	0.3437
$\text{err}(\bar{t}, \tau)$	0.1419	0.1437	0.1473	0.1897	0.4232

method to (7)) is attained at the point 0.762, and thus the MLE value is $\hat{\tau} = 0.762$. (Note that the endpoints of the interval in Proposition 2, which contains the solution $\hat{\tau}$, are 0.675 and 0.825.) We mention that the value of the estimator $\overline{T}$ from (6) for our data is $\bar{t} = 0.744$.

It is not easy to compare the MLE with $\overline{T}$ theoretically, since the closed-form of the MLE cannot be obtained in general. However, we can compare the estimators by Monte-Carlo simulations. Table 1 provides such a comparison. We have tried the parameter values $\theta = 0.01,\ 0.1,\ 1,\ 10,\ 100$ (and $\tau = 100,\ 10,\ 1,\ 0.1,\ 0.01$, respectively). For each θ we took a random sample $X_1, X_2, \ldots, X_{50}$ from $\text{Exp}(\theta)$, and for each value x_i of X_i – the values $t_{i1}, t_{i2}, \ldots, t_{i,n_i}$ of a random sample from $\mathscr{P}(x_i)$, where n_i is selected uniformly between 1 and 40. The procedure was repeated 10000 times (with the same n_i values). For each iteration s, $1 \le s \le 10000$, we calculated the MLE and $\overline{T}$ estimates $\hat{\tau}_s = 1/\hat{\theta}_s$ and $\bar{t}_s$, respectively. As a measure of the error for the estimators, we took the quantities

$$\text{err}(\hat{\tau}, \tau) = \sqrt{\frac{1}{10^4}\sum_{s=1}^{10^4}\frac{(\tau - \hat{\tau}_s)^2}{\tau^2}}, \qquad \text{err}(\bar{t}, \tau) = \sqrt{\frac{1}{10^4}\sum_{s=1}^{10^4}\frac{(\tau - \bar{t}_s)^2}{\tau^2}}.$$

(Compare this measure with those of mean-squared-error and root mean-squared-error, used in numerous applications; [2, 20, 23, 26]). The final results are presented in Table 1.

Unsurprisingly, for large θ both estimators perform quite badly, but the MLE is consistently better than $\overline{T}$. In fact, for such θ the x_i's are very small, so that most t_{ij}'s tend to vanish, and we get little significant data. (In fact, if all t_{ij}'s do vanish, then neither estimator is defined. We omitted these rare observations from our sample, which does not affect the comparison between the estimators.) For small θ, both estimators are good, and yield almost the same results.

In the normal-normal case, the MLE turned out to be a UMVUE. In our case, this is still true in one particular situation, as Proposition 3 below shows. However, in general, we doubt that there is a UMVUE at all. Indeed, consider weighted average estimators of the form:

$$T^* = \sum_{i=1}^{k}\alpha_i \overline{T}_i, \qquad \left(\sum_{i=1}^{k}\alpha_i = 1\right). \tag{8}$$

Obviously, T^* is an unbiased estimator. Indeed, one can easily verify that

$$T_{ij} + 1 \sim G\left(\frac{\theta}{\theta+1}\right), \quad 1 \le i \le k,\ 1 \le j \le n_i, \tag{9}$$

and therefore

$$E(T^*) = \sum_{i=1}^{k} \alpha_i E(\overline{T}_i) = \sum_{i=1}^{k} \alpha_i E(T_{i1}) = \sum_{i=1}^{k} \alpha_i \frac{1}{\theta} = \tau.$$

We note that $\overline{T}$ is the special case $\alpha_1 = \alpha_2 = \ldots = \alpha_k = \frac{1}{k}$ of T^*.

Proposition 3. *1. If all n_i's are equal, then $\hat{\tau} = \overline{T}$ is a UMVUE.*
2. If not all n_i's are the same, then T^* is not a UMVUE (for any choice of α_i's).

4 Proofs

Proof of Theorem 1. The likelihood function in our case is

$$\begin{aligned} L(\mu) &= \prod_{i=1}^{k} \int_{-\infty}^{\infty} \left(\prod_{i=1}^{n_i} \frac{1}{\sqrt{2\pi}\sigma_2} \cdot e^{-\frac{(x_i - t_{ij})^2}{2\sigma_2^2}} \right) \cdot \frac{1}{\sqrt{2\pi}\sigma_1} e^{-\frac{(\mu - x_i)^2}{2\sigma_1^2}} dx_i \\ &= \prod_{i=1}^{k} \left(\frac{1}{\sqrt{2\pi}\sigma_2} \right)^{n_i} \cdot \frac{1}{\sqrt{2\pi}\sigma_1} \cdot \int_{-\infty}^{\infty} e^{-\frac{1}{2}\left(\frac{(\mu - x_i)^2}{\sigma_1^2} + \sum_{j=1}^{n_i} \frac{(x_i - t_{ij})^2}{\sigma_2^2} \right)} dx_i. \end{aligned} \tag{10}$$

A routine calculation shows that the expression in the exponent may be written in the form

$$\frac{(\mu - x_i)^2}{\sigma_1^2} + \sum_{j=1}^{n_i} \frac{(x_i - t_{ij})^2}{\sigma_2^2} = (A_i x_i - B_i)^2 + C_i, \tag{11}$$

where

$$\begin{aligned} A_i &= \frac{\sqrt{\sigma_2^2 + n_i \sigma_1^2}}{\sigma_2 \sigma_1}, \\ B_i &= \frac{\mu \sigma_2^2 + \sigma_1^2 \sum_{j=1}^{n_i} t_{ij}}{\sqrt{\sigma_2^2 + n_i \sigma_1^2}} \cdot \frac{1}{\sigma_2 \sigma_1}, \\ C_i &= \frac{n_i}{\sigma_2^2 + n_i \sigma_1^2} \mu^2 - \frac{2 \sum_{j=1}^{n_i} t_{ij}}{\sigma_2^2 + n_i \sigma_1^2} \mu - \frac{1}{\sigma_2^2} \left(\frac{\sigma_1^2 (\sum_{j=1}^{n_i} t_{ij})^2}{\sigma_2^2 + n_i \sigma_1^2} - \sum_{j=1}^{n_i} t_{ij}^2 \right). \end{aligned} \tag{12}$$

Substituting (11) in (10), we obtain

$$L(\mu) = \prod_{i=1}^{k} \left(\frac{1}{\sqrt{2\pi\sigma_2}} \right)^{n_i} \cdot \frac{1}{\sqrt{2\pi\sigma_1}} \cdot \int_{-\infty}^{\infty} e^{-\frac{1}{2}(A_i x_i - B_i)^2 - \frac{1}{2}C_i} dx_i$$

$$= (2\pi\sigma_2)^{-\frac{1}{2}\sum_{i=1}^{k} n_i} (2\pi\sigma_1)^{-\frac{k}{2}} \prod_{i=1}^{k} \frac{\sqrt{2\pi} e^{-\frac{1}{2}C_i}}{A_i} \int_{-\infty}^{\infty} \frac{1}{\sqrt{2\pi}} e^{-\frac{1}{2}(y_i - B_i)^2} dy_i$$

$$= e^{-\frac{1}{2}\sum_{i=1}^{k} C_i} \cdot (2\pi\sigma_2)^{-\frac{1}{2}\sum_{i=1}^{k} n_i} \sigma_1^{-\frac{k}{2}} \cdot \prod_{i=1}^{k} A_i^{-1}. \tag{13}$$

Passing to logarithms we find that

$$\ln L(\mu) = -\frac{1}{2}\sum_{i=1}^{k} C_i - \frac{1}{2}\sum_{i=1}^{k} n_i \ln(2\pi\sigma_2) - \frac{k}{2}\ln\sigma_1 - \sum_{i=1}^{k} \ln A_i. \tag{14}$$

Note that, on the right-hand side of (14), only the C_i's depends on the parameter μ. In fact, $\ln L(\mu)$ is a quadratic function of μ:

$$\ln L(\mu) = -\frac{1}{2}\left(\sum_{i=1}^{k} \frac{n_i}{\sigma_2^2 + n_i\sigma_1^2}\right) \cdot \mu^2 + \sum_{i=1}^{k} \frac{\sum_{j=1}^{n_i} t_{ij}}{\sigma_2^2 + n_i\sigma_1^2} \cdot \mu + D, \tag{15}$$

where

$$D = \frac{1}{2\sigma_2^2}\sum_{i=1}^{k}\left(\frac{\sigma_1^2(\sum_{j=1}^{n_i} t_{ij})^2}{\sigma_2^2 + n_i\sigma_1^2} - \sum_{j=1}^{n_i} t_{ij}^2\right) - \sum_{i=1}^{k}\frac{n_i}{2}\ln(2\pi\sigma_2) - \frac{k}{2}\ln\sigma_1 - \sum_{i=1}^{k}\ln A_i.$$

Thus $\ln L(\mu)$ has a unique maximum, obtained at the point

$$\hat{\mu} = \sum_{i=1}^{k} \frac{\frac{1}{\sigma_2^2/n_i + \sigma_1^2}}{\sum_{r=1}^{k} \frac{1}{\sigma_2^2/n_r + \sigma_1^2}} \cdot \frac{\sum_{j=1}^{n_i} t_{ij}}{n_i}. \tag{16}$$

Hence, the statistic M, which corresponds to (16), is the MLE of μ. This completes the proof.

Proof of Proposition 1. Clearly,

$$V(M) = \sum_{i=1}^{k} w_i^2 V(\overline{T}_i), \tag{17}$$

where

$$w_i = \frac{\frac{1}{\sigma_1^2+\sigma_2^2/n_i}}{\sum_{r=1}^{k} \frac{1}{\sigma_1^2+\sigma_2^2/n_r}}. \tag{18}$$

Now:

$$V(\overline{T}_i) = \frac{1}{n_i^2} \cdot \left(\sum_{j=1}^{n_i} V(T_{ij}) + 2 \sum_{1 \le j < s \le n_i}^{n_i} \mathrm{Cov}(T_{ij}, T_{is}) \right), \qquad 1 \le i \le k. \tag{19}$$

Obviously, $V(T_{ij}) = \sigma_1^2 + \sigma_2^2$ for $1 \le i \le k, \ \ 1 \le j \le n_i$, and $\mathrm{Cov}(T_{ij}, T_{is}) = \sigma_1^2$ for $1 \le i \le k, \ \ 1 \le j < s \le n_i$. Thus,

$$V(\overline{T}_i) = \frac{1}{n_i^2} \cdot \left(n_i(\sigma_2^2 + \sigma_1^2) + 2\binom{n_i}{2}\sigma_1^2 \right) = \sigma_1^2 + \sigma_2^2/n_i, \qquad 1 \le i \le k. \tag{20}$$

Substituting (20) and (18) in (17), we see that

$$V(M) = \sum_{i=1}^{k} \left(\frac{\frac{1}{\sigma_1^2+\sigma_2^2/n_i}}{\sum_{r=1}^{k} \frac{1}{\sigma_1^2+\sigma_2^2/n_r}} \right)^2 \cdot \left(\sigma_1^2 + \sigma_2^2/n_i\right) = \frac{1}{\sum_{r=1}^{k} \frac{1}{\sigma_1^2+\sigma_2^2/n_r}}. \tag{21}$$

To show that M is the UMVUE, one can easily calculate by (15) the Fisher information on μ contained in $\mathbf{T} = (\mathbf{T}_1, \mathbf{T}_2, \ldots, \mathbf{T}_k)$, which is

$$I_{\mathbf{T}}(\mu) = \sum_{r=1}^{k} \frac{1}{\sigma_1^2 + \sigma_2^2/n_r}. \tag{22}$$

Thus the Cramér-Rao lower bound $\frac{1}{I_{\mathbf{T}}(\mu)}$ coincides with (21), which completes the proof.

Proof of Theorem 2. The likelihood function is

$$\begin{aligned} L(\theta) &= \prod_{i=1}^{k} \int_0^{\infty} \left(\prod_{j=1}^{n_i} \frac{x_i^{t_{ij}} e^{-x_i}}{t_{ij}!} \right) \cdot \theta e^{-\theta x_i} dx_i \\ &= \prod_{i=1}^{k} \int_0^{\infty} \frac{x_i^{\sum_{j=1}^{n_i} t_{ij}} e^{-(n_i+\theta)x_i} \cdot \theta}{\prod_{j=1}^{n_i} t_{ij}!} dx_i. \end{aligned} \tag{23}$$

Recall that $\int_0^\infty x^{\alpha-1}e^{-\lambda x}dx = \Gamma(\alpha)\lambda^{-\alpha}$ for $\alpha, \lambda > 0$. Therefore

$$\int_0^\infty x_i^{\sum_{j=1}^{n_i} t_{ij}} e^{-(n_i+\theta)x_i} dx_i = \frac{\Gamma(1+\sum_{j=1}^{n_i} t_{ij})}{(n_i+\theta)^{1+\sum_{j=1}^{n_i} t_{ij}}},$$

which yields

$$\begin{aligned} L(\theta) &= \prod_{i=1}^{k} \left(\frac{\Gamma(1+\sum_{j=1}^{n_i} t_{ij})\theta}{(n_i+\theta)^{1+\sum_{j=1}^{n_i} t_{ij}} \prod_{j=1}^{n_i} t_{ij}!} \right) \\ &= \theta^k \prod_{i=1}^{k} \frac{(\sum_{j=1}^{n_i} t_{ij})!}{(n_i+\theta)^{1+\sum_{j=1}^{n_i} t_{ij}} \prod_{j=1}^{n_i} t_{ij}!}. \end{aligned} \tag{24}$$

Passing to logarithms and differentiating, we obtain:

$$\frac{d\ln L(\theta)}{d\theta} = \frac{k}{\theta} - \sum_{i=1}^{k} \left(1 + \sum_{j=1}^{n_i} t_{ij}\right) \frac{1}{n_i+\theta}. \tag{25}$$

The value of the MLE of θ maximizes $\ln L(\theta)$, and thus satisfies $\frac{d\ln L(\theta)}{d\theta} = 0$ and forms a solution of

$$\sum_{i=1}^{k} \frac{\bar{t}_i + 1/n_i}{1/\theta + 1/n_i} = k.$$

In terms of τ, we obtain (5). This completes the proof.

Proof of Proposition 2. Put $f(\tau) = \sum_{i=1}^{k} \frac{\bar{t}_i+1/n_i}{\tau+1/n_i}$. Clearly,

$$f(0) = \sum_{i=1}^{k} (n_i\bar{t}_i + 1) = k + \sum_{i=1}^{k}\sum_{j=1}^{n_i} t_{ij} > k,$$

and $f(\tau) \underset{\tau\longrightarrow\infty}{\longrightarrow} 0$. Since, $f(\tau)$ is a continuous monotonically decreasing function of τ in $[0, \infty)$, this implies that (5) has a unique solution on the positive axis.

Now, since

$$k = f(\hat{\tau}) \geq \frac{\sum_{i=1}^{k} (\bar{t}_i + 1/n_i)}{\hat{\tau} + 1/n_{\min}},$$

we obtain

$$\hat{\tau} \geq \frac{\sum_{i=1}^{k}(\bar{t}_i + 1/n_i)}{k} - \frac{1}{n_{\min}} = c - \frac{1}{n_{\min}}.$$

Similarly,

$$\hat{\tau} \leq \frac{\sum_{i=1}^{k}(\bar{t}_i + 1/n_i)}{k} - \frac{1}{n_{\max}} = c - \frac{1}{n_{\max}}.$$

This completes the proof.

Before proving Proposition 3, we need

Lemma 1.

1. $V(T^*) = \sum_{i=1}^{k} \alpha_i^2 \cdot \frac{n_i + \theta}{n_i \theta^2}$.
2. *The Fisher information $I_{\mathbf{T}}(\tau)$ on τ contained in $\mathbf{T} = (\mathbf{T}_1, \mathbf{T}_2, \ldots, \mathbf{T}_k)$, is given by*

$$I_{\mathbf{T}}(\tau) = \sum_{r=1}^{k} \frac{n_r \theta^2}{n_r + \theta}.$$

Proof of Lemma 1. 1. We have

$$V(T^*) = \sum_{i=1}^{k} \alpha_i^2 \cdot V(\overline{T}_i), \tag{26}$$

where

$$V(\overline{T}_i) = \frac{1}{n_i^2} \cdot \left(\sum_{j=1}^{n_i} V(T_{ij}) + 2 \sum_{1 \leq j < s \leq n_i} \mathrm{Cov}(T_{ij}, T_{is}) \right), \qquad 1 \leq i \leq k. \tag{27}$$

By (9) we obtain $V(T_{ij}) = \frac{\theta+1}{\theta^2}$ for $1 \leq i \leq k, \;\; 1 \leq j \leq n_i$, and a routine calculation shows that $\mathrm{Cov}(T_{ij}, T_{is}) = \frac{1}{\theta^2}$ for $1 \leq i \leq k, \;\; 1 \leq j < s \leq n_i$. Thus,

$$V(\overline{T}_i) = \frac{1}{n_i^2} \cdot \left(n_i \frac{\theta + 1}{\theta^2} + 2 \binom{n_i}{2} \frac{1}{\theta^2} \right) = \frac{n_i + \theta}{n_i \theta^2}, \qquad 1 \leq i \leq k. \tag{28}$$

Substituting (28) in (26), we find that

$$V(T^*) = \sum_{i=1}^{k} \alpha_i^2 \cdot \frac{n_i + \theta}{n_i \theta^2}. \tag{29}$$

2. We start from (24). A routine calculation yields:

$$\frac{d^2 \ln L(\tau)}{d^2\tau} = \frac{k}{\tau^2} - \sum_{i=1}^{k}\left(\sum_{j=1}^{n_i} t_{ij} + 1\right)\frac{2\tau n_i + 1}{\tau^2(\tau n_i + 1)^2}. \tag{30}$$

Since $E(T_{ij}) = \tau$, the Fisher information on τ contained in $\mathbf{T} = (\mathbf{T}_1, \mathbf{T}_2, \ldots, \mathbf{T}_k)$, is

$$\begin{aligned} I_{\mathbf{T}}(\tau) &= -\frac{k}{\tau^2} + \sum_{i=1}^{k}\left(\sum_{j=1}^{n_i} \tau + 1\right)\frac{2\tau n_i + 1}{\tau^2(\tau n_i + 1)^2} \\ &= \sum_{i=1}^{k} \frac{n_i}{\tau(\tau n_i + 1)} = \sum_{i=1}^{k} \frac{n_i\theta^2}{n_i + \theta}. \end{aligned} \tag{31}$$

This completes the proof.

Proof of Proposition 3. We start with the second part. Suppose that T^* is the UMVUE. Hence, its variance is equal to the Cramér-Rao lower bound $\frac{1}{I_{\mathbf{T}}(\tau(\theta))}$. By Lemma 1:

$$\sum_{i=1}^{k} \alpha_i^2 \cdot \frac{n_i + \theta}{n_i\theta^2} = \frac{1}{\sum_{r=1}^{k} \frac{n_r\theta^2}{n_r+\theta}}. \tag{32}$$

Denoting $\beta_i = \frac{n_i\theta^2}{n_i+\theta}$, $1 \le i \le k$, we rewrite (32) in the form:

$$\sum_{i=1}^{k} \frac{\alpha_i^2}{\beta_i} = \frac{1}{\sum_{r=1}^{k} \beta_r}.$$

Hence, by the Cauchy-Schwarz inequality:

$$\sum_{i=1}^{k} \alpha_i = \sum_{i=1}^{k} \frac{\alpha_i}{\sqrt{\beta_i}} \cdot \sqrt{\beta_i} \le \sqrt{\sum_{i=1}^{k} \frac{\alpha_i^2}{\beta_i}} \cdot \sqrt{\sum_{i=1}^{k} \beta_i} = \sqrt{\frac{1}{\sum_{r=1}^{k} \beta_r}} \cdot \sqrt{\sum_{i=1}^{k} \beta_i} = 1.$$

Since $\sum_{i=1}^{k} \alpha_i = 1$, the weak inequality in the last chain is actually an equality. Thus, the $\frac{\alpha_i}{\sqrt{\beta_i}}$ and the $\sqrt{\beta_i}$ are proportional:

$$\alpha_i = c \cdot \beta_i, \qquad 1 \le i \le k.$$

Clearly, $c = 1/\sum_{i=1}^{k} \beta_i$. Therefore,

$$\alpha_i = \frac{\beta_i}{\sum_{r=1}^{k} \beta_r} = \frac{\frac{n_i\theta^2}{n_i+\theta}}{\sum_{r=1}^{k} \frac{n_r\theta^2}{n_r+\theta}}, \qquad 1 \le i \le k.$$

If not all n_i's are the same, then α_i depends in a non-trivial way on the parameter θ. This contradicts the assumption that T^* is an estimator and completes the proof of the second part.

If all n_i's are equal, we easily see that (32) indeed holds for $\alpha_i = \frac{1}{k}$ $1 \leq i \leq k$, and $T^* = \overline{T} = \hat{\tau}$, and $V(\hat{\tau})$ is equal to the Cramér-Rao lower bound $\frac{1}{I_{\mathbf{T}}(\tau(\theta))} = \frac{n+\theta}{kn\theta^2}$. Hence $\hat{\tau}$ is a UMVUE.

This proves the proposition.

5 Potential Applications and Empirical Bayes

Problems fitting naturally the indirect setup arise in various seemingly unrelated domains, such as e-commerce, survey analysis, data mining, etc. Consider the following example, taken from the area of reputation systems. The indirect setup arises here in the context of the beta and Dirichlet reputation systems (see [17, 18]). Namely, we have a set of objects of the same type (movies, hotels, department stores, web providers), for which we collect users' ratings. The basic idea of reputation systems is to have a mechanism allowing users to submit these ratings and enabling a computation of an aggregated rating of these objects based on the individual ratings. In the beta reputation system, suggested by Jøsang and Ismail [17], for example, each object receives a binary grade ("good"/"bad") from an unknown number of people. (The model is slightly more complex, but the simplified model retains the point we want to elaborate on.) Here, the probability P_i of a random object to obtain a "good" grade is distributed Beta(α, β) for some parameter (α, β). Thus, one may consider the problem of estimating the unknown parameter of the Beta distribution corresponding to indirect observations p_i, $1 \leq i \leq k$, based on the rating data of the second level. Note that, from the point of view of the user of the system, the average reputation score of the population is very important; by comparing the score of a particular object with that average he decides whether to use this object or look for another. As this average is a function of the parameter, it is important to estimate the parameter well. One problem that naive methods fail to address is that, in practice, some popular objects receive many ratings, while others receive only few. This difference should be taken into account when estimating the parameter. A naive approach may try to estimate each unobserved observation on the basis of the second-stage observations attached to it, and then estimate the parameter on the basis of these estimates. As we will see, the indirect maximum likelihood estimator does better in the sense that unobserved observations with many second-stage observations obtain a larger weight than those with few second-stage observations.

Recently, Jøsang and Haller [18] suggested a generalization of the beta reputation system, rating objects by $m \geq 2$ discrete levels, known as the Dirichlet reputation system. This model is based on using Dirichlet's density function (see [18]) to combine feedback and derive reputation scores as a function of m parameters corresponding to the density function. The same kind of problems and a similar indirect setup appears also in this model.

Another real-life situation fitting the two-stage setup arises in measurements with errors. Suppose we have a group of items of some type, and want to know their mean weight μ. Suppose the weight is known to be $N(\mu, 1)$-distributed, and we are allowed to examine k random items from the population. Since our scales are not completely accurate, we weigh each item in the sample several times, say we weigh n_i times item i. Thus, we obtain a table of the second-stage observations t_{ij}, $1 \le i \le k$, $1 \le j \le n_i$, where each row $1 \le i \le k$ of the table contains the measurements t_{ij}, $1 \le j \le n_i$, of the i-th item with unknown actual weight x_i. We need to estimate μ on the basis of the observed measurements.

The two-stage setup appears also in Bayesian analysis of missing data. Unlike in our situation, the purpose there is to estimate the unobserved observations of the first stage. This is done with the help of the posterior distribution of the unobserved observations, given the evidence of the second-stage observations. The posterior distribution is equal to the likelihood times the prior distribution of the unobserved observations, divided by the marginal distribution of the second-stage observations [9]. Thus, to calculate the posterior distribution one should know the value of the parameter. In the Bayesian approach we assume that this value is known. If the parameter is unknown (as in our setup) one can use a known hyperprior distribution of the parameter to compute the posterior distribution or to use the marginal maximum likelihood estimate (henceforth MMLE) of the parameter, and plug it into the calculation of the posterior distribution. The last procedure is called Empirical Bayes, various aspects of which were studied in numerous papers [9, 22, 27]. However, it is often problematic to find the MMLE in closed form, especially if the number of the second-stage observations is not the same for all first-stage observations. Our paper details and exemplifies this issue.

6 Conclusions and Future Research

This paper focuses on the parameter estimation problem under indirect information, formally defined in Problem 1. The setup is motivated by practical contemporary applications from various domains, such as classification, reputation systems in e-commerce, survey analysis, etc. We propose a maximum likelihood approach for estimating the unknown parameter of the distribution of the first level, based on the known observations of the second level, and illustrate it for two pairs of distributions. Our approach raises several questions to be explored in theoretical and applied aspects of the problem, such as:

- Developing our method for real-life applications. For example, suppose one wants to estimate the parameters of beta binomial or Dirichlet reputation systems, suggested by Jøsang and Ismail [17] and Jøsang and Haller [18], based on large-scale databases. Our approach takes into account the reliability of the data provided by each agent (namely, the number of ratings given by this agent) to the parameter estimation, which is essential in the context of data sets.

- An evaluation of the loss of accuracy of the MLE due to our indirect setup, as compared to the "ideal" MLE corresponding to the classical case. For this purpose it seems interesting to compare the Fisher information on θ contained in $\mathbf{T}_1, \mathbf{T}_2, \ldots, \mathbf{T}_k$ with that contained in $X_1, X_2, \ldots, X_k$. For mixture models, this was done by Kagan and Li [19], who obtained a universal upper bound for the loss of the accuracy.

Acknowledgements The authors express their gratitude to E. Gudes and N. Gal-Oz for encouraging them to look into various questions regarding reputation systems, which eventually led to this research, and for their comments on the first draft of the paper. The authors also thank I. Gertsbakh for many discussions related to this topic, and A. Kagan and L. Stefanski for their comments on the first draft of the paper.

The authors acknowledge Deutsche Telekom Laboratories at Ben-Gurion University for support of this research.

References

1. Aït-Sahalia, Y., Kimmela, R.: Maximum likelihood estimation of stochastic volatility models. J. Financ. Econ. **83**(2), 413–452 (2007)
2. Anderson, M.P., Woessner, W.W.: Applied Groundwater Modeling: Simulation of Flow and Advective Transport, 2nd edn. Academc, New York (1992)
3. Assefi, T.: Stochastic Processes and Estimation Theory with Applications. Wiley, New York (1979)
4. Ayebo, A., Kozubowski, T.J.: An asymmetric generalization of Gaussian and Laplace laws. J. Probab. Stat. Sci. **1**(2), 187–210 (2003)
5. Berouti, M., Schwartz, R., Makhoul, J.: Enhancement of speech corrupted by acoustic noise. In: IEEE International Conference on Acoustics, Speech, and Signal Processing (ICASSP), vol. 4, pp.208–211 (1979)
6. Bishop, C.M.: Latent variable models. In: Jordan, M. (ed.) Learning in Graphical Models, pp. 371–403. MIT, London (1999)
7. Brenner, J.L.: A unified treatment and extension of some means of classical analysis–I: comparison theorems. J. Combin. Inform. Syst. Sci. **3**, 175–199 (1978)
8. Brandel, J.: Empirical Bayes methods for missing data analysis. Technical Report 2004:11, Department of Mathematics, Uppsala University, Sweden (2004)
9. Carlin, B.P., Louis, T.A.: Bayes and Empirical Bayes Methods for Data Analysis, 2nd edn. Chapman & Hall, Boca Raton (2000)
10. Chen, X., Hong, H., Tamer, E.: Measurement error models with auxiliary data. Rev. Econ. Stud. **72**(2), 343–366 (2005)
11. DeGroot, M.: Probability and Statistics, 2nd edn. Addison-Wesley, Boston (1986)
12. Douarche, F., Buisson, L., Ciliberto, S., Petrosyan, A.: A simple noise subtraction technique. Rev. Sci. Instrum. **75**(12), 5084–5089 (2004)
13. Gertsbakh, I.: Reliability Theory: With Applications to Preventive Maintenance. Springer, New York (2000)
14. Gourieroux, C., Monfort, A., Renault, E.: Indirect inference. J. Appl. Econ. **8**, S85–S118 (1993)
15. Hardy, G.H., Littlewood, J.E., Pólya, G.: Inequalities, 2nd edn. Cambridge University Press, Cambridge (1952)
16. Ideker, T., Thorsson, V., Siegel, A.F., Hood, L.E.: Testing for differentially-expressed genes by maximum-likelihood analysis of microarray data. J. Comput. Biol. **7**(6), 805–817 (2000)

17. Jøsang, A., Ismail, R.: The beta reputation system. In: Proceedings of the 15-th Bled Conference on Electronic Commerce, Bled, Slovenia, pp.17–19 (2002)
18. Jøsang, A., Haller, J.: Dirichlet reputation systems. In: Proceedings of the Second International Conference on Availability, Reliability and Security (ARES 2007), Vienna, pp.112–119 (2007)
19. Kagan, A., Li, B.: An identity for the Fisher information and Mahalanobis distance. J. Stat. Plann. Inference **138**(12), 3950–3959 (2008)
20. Lehmann, E.L., Casella, G.: Theory of Point Estimation, 2nd edn. Springer, New York (1998)
21. Linhard, K., Haulick, T.: Spectral noise subtraction with recursive gain curves. In: Fifth International Conference on Spoken Language Processing, (ICSLP-1998), paper 0109, Sydney (1998)
22. Maritz, J. S., Lwin, T.: Assessing the performance of empirical Bayes estimators. Ann. Inst. Stat. Math. **44**(4), 641–657 (1992)
23. Mood, A.M., Graybill, F., Boes, D.: Introduction to the Theory of Statistics, 3rd edn. McGraw Hill, Singapore (1974)
24. Shi, G., Nehorai, A.: Maximum likelihood estimation of point scatterers for computational time-reversal imaging. Commun. Inform. Syst. **5**(2), 227–256 (2005)
25. Smith, Jr. A.A.: Indirect inference. In: Steven, N.D., Lawrence, E.B. (eds.) The New Palgrave Dictionary of Economics. Palgrave Macmillan (2008). DOI:10.1057/9780230226203.0778
26. Tuchler, M., Singer, A.C., Koetter, R.: Minimum mean squared error equalization using a priori information. IEEE Trans. Signal Process. **50**(3), 673–683 (2002)
27. Walter, G.G., Hamedani, G.G.: Bayes empirical Bayes estimation for discrete exponential families. Ann. Inst. Stat. Math. **41**(1), 101–119 (1989)

Lagrangian Duality in Complex Pose Graph Optimization

Giuseppe C. Calafiore, Luca Carlone, and Frank Dellaert

Abstract Pose Graph Optimization (PGO) is the problem of estimating a set of poses from pairwise relative measurements. PGO is a nonconvex problem, and currently no known technique can guarantee the efficient computation of a global optimal solution. In this paper, we show that Lagrangian duality allows computing a globally optimal solution, under certain conditions that are satisfied in many practical cases. Our first contribution is to frame the PGO problem in the complex domain. This makes analysis easier and allows drawing connections with the recent literature on *unit gain graphs*. Exploiting this connection we prove nontrival results about the spectrum of the matrix underlying the problem. The second contribution is to formulate and analyze the properties of the Lagrangian dual problem in the complex domain. The dual problem is a semidefinite program (SDP). Our analysis shows that the duality gap is connected to the number of eigenvalues of the *penalized pose graph matrix*, which arises from the solution of the SDP. We prove that if this matrix has a *single eigenvalue in zero*, then (1) the duality gap is zero, (2) the primal PGO problem has a unique solution, and (3) the primal solution can be computed by *scaling* an eigenvector of the penalized pose graph matrix. The third contribution is algorithmic: we exploit the dual problem and propose an algorithm that computes a guaranteed optimal solution for PGO when the penalized pose graph matrix satisfies the Single Zero Eigenvalue Property (SZEP). We also propose a variant that deals with the case in which the SZEP is not satisfied. This variant, while possibly suboptimal, provides a very good estimate for PGO in practice. The fourth contribution is a numerical analysis. Empirical evidence shows that in the vast majority of cases (100 % of the tests under noise regimes of practical robotics

G.C. Calafiore
Politecnico di Torino, Torino, Italy
e-mail: giuseppe.calafiore@polito.it

L. Carlone (✉)
Massachusetts Institute of Technology, Cambridge, MA, USA
e-mail: lcarlone@mit.edu

F. Dellaert
Georgia Institute of Technology, Atlanta, GA, USA
e-mail: dellaert@cc.gatech.edu

B. Goldengorin (ed.), *Optimization and Its Applications in Control and Data Sciences*, Springer Optimization and Its Applications 115,
DOI 10.1007/978-3-319-42056-1_5

applications) the penalized pose graph matrix does satisfy the SZEP, hence our approach allows computing the global optimal solution. Finally, we report simple counterexamples in which the duality gap is nonzero, and discuss open problems.

Keywords Maximum likelihood estimation • Mobile robots • Motion estimation • Position measurement • Rotation measurement • Simultaneous localization and mapping • Duality

1 Introduction

Pose graph optimization (PGO) consists in the estimation of the poses (positions and orientations) of a mobile robot, from relative pose measurements. The problem can be formulated as the minimization of a nonconvex cost, and can be conveniently visualized as a graph, in which a (to-be-estimated) pose is attached to each vertex, and a given relative pose measurement is associated to each edge.

PGO is a key problem in many application endeavours. In robotics, it lies at the core of state-of-the-art algorithms for localization and mapping in both single robot [11, 13, 14, 26, 27, 33, 36, 43, 54, 58] and multi robot [1, 42, 46–48] systems. In computer vision and control, problems that are closely related to PGO need to be solved for structure from motion [2, 32, 34, 35, 37, 56, 69], attitude synchronization [38, 57, 74], camera network calibration [75], sensor network localization [59, 60], and distributed consensus on manifolds [65, 76]. Moreover, similar formulations arise in molecule structure determination from microscopy imaging [3, 71].

A motivating example in robotics is the one pictured in Fig. 1a. A mobile robot is deployed in an unknown environment at time $t = 0$. The robot traverses the environment and at each discrete time step acquires a sensor measurement (e.g., distances from obstacles within the sensing radius). From wheel rotation, the robot is capable of measuring the relative motion between two consecutive poses (say, at time i and j). Moreover, comparing the sensor measurement, acquired at different times, the robot can also extrapolate relative measurements between non consecutive poses (e.g., between i and k in the figure). PGO uses these measurements to estimate robot poses. The graph underlying the problem is shown in Fig. 1b, where we draw in different colors the edges due to relative motion measurements (the *odometric edges*, in black) and the edges connecting non-consecutive poses (the *loop closures*, in red). The importance of estimating the robot poses is two-fold. First, the knowledge of the current robot pose is often needed for performing high-level tasks within the environment. Second, from the knowledge of all past poses, the robot can *register* all sensor footprints in a common frame, and obtain a *map* of the environment, which is needed for model-based navigation and path planning.

Related Work in Robotics Since the seminal paper [54], PGO attracted large attention from the robotics community. Most state-of-the-art techniques currently

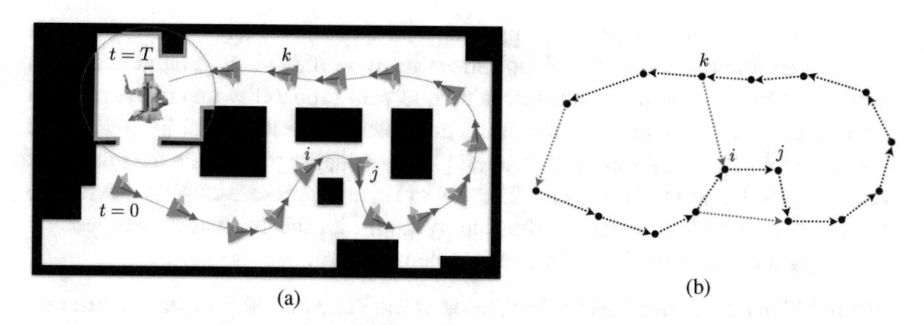

Fig. 1 (**a**) Pose graph optimization in robotics. A mobile robot is deployed in an unknown environment at time $t = 0$. At each time step the robot measures distances from obstacles within the sensing radius (*red circle*). The sensor footprint (i.e., the set of measurements) at time T is visualized in *orange*. By matching sensor footprints acquired at different time steps, the robot establishes relative measurements between poses along its trajectory. PGO consists in the estimation of robot poses from these relative measurements. (**b**) Directed graph underlying the problem

rely on iterative nonlinear optimization, which refines a given initial guess. The Gauss-Newton method is a popular choice [44, 45, 51], as it converges quickly when the initialization is close to a minimum of the cost function. Trust region methods (e.g., the Levenberg-Marquart method, or Powell's Dog-Leg method [52]) have also been applied successfully to PGO [62, 63]; the gradient method has been shown to have a large convergence basin, while suffering from long convergence tails [36, 48, 58]. A large body of literature focuses on speeding up computation. This includes exploiting sparsity [31, 44], using reduction schemes to limit the number of poses [9, 50], faster linear solvers [22, 33], or approximate solutions [14, 24].

PGO is a nonconvex problem and iterative optimization techniques can only guarantee local convergence. State-of-the-art iterative solvers fail to converge to a global minimum of the cost for relatively small noise levels [11, 15]. This fact recently triggered efforts towards the design of more robust techniques, together with a theoretical analysis of PGO. Huang et al.[41] discuss the number of local minima in small PGO problems. Knuth and Barooah [49] investigate the growth of the error in absence of loop closures. Carlone [10] provides conservative approximations of the basin of convergence for the Gauss-Newton method. Huang et al. [40] and Wang et al.[78] discuss the nonlinearities in PGO. In order to improve global convergence, a successful strategy consists in solving for the rotations first, and then using the resulting estimate to bootstrap iterative methods for PGO [11, 13–15]. This is convenient because the *rotation subproblem*[1] can be solved globally, with performance guarantees, in 2D [11], and many heuristic algorithms for rotation estimation also perform well in 3D [15, 32, 34, 56]. Despite the empirical success

[1]We use the term "rotation subproblem" to denote the problem of associating a rotation to each node in the graph, using relative rotation measurements. This corresponds to disregarding the translation measurements in PGO.

of state-of-the-art techniques, no approach can guarantee global convergence. It is not even known if the global optimizer itself is unique in general instances (while it is known that the minimizer is unique with probability one in the rotation subproblem [11]). The lack of guarantees promoted a recent interest in *verification* techniques for PGO. Carlone and Dellaert [12] use duality to evaluate the quality of a candidate solution in planar PGO. The work [12] also provides empirical evidence that in many problem instances the duality gap, i.e., the mismatch between the optimal cost of the primal and the dual problem, is zero.

Related Work in Other Fields Variations of the PGO problem appear in different research fields. In computer vision, a somehow more difficult variant of the problem is known as *bundle adjustment* [2, 32, 34, 35, 37, 56, 69]. Contrarily to PGO, in bundle adjustment the relative measurements between the (camera) poses are only known up to scale. While no closed-form solution is known for bundle adjustment, many authors focused on the solution of the rotation subproblem [2, 32, 34, 35, 37, 56, 69]. The corresponding algorithms have excellent performance in practice, but they come with little guarantees, as they are based on relaxation. Fredriksson and Olsson [32] use duality theory to design a verification technique for quaternion-based rotation estimation.

Related work in multi robot systems and sensor networks also includes contributions on rotation estimation (also known as *attitude synchronization* [17, 38, 57, 74, 79]). Borra et al.[6] propose a distributed algorithm for planar rotation estimation. Tron and Vidal [75] provide convergence results for distributed attitude consensus using gradient descent; distributed consensus on manifold [65] is related to estimation from relative measurements, as discussed in [76]. A problem that is formally equivalent to PGO is discussed in [59, 60] with application to sensor network localization. Piovan et al.[60] provide observability conditions and discuss iterative algorithms that reduce the effect of noise. Peters et al.[59] study pose estimation in graphs with a single loop (related closed-form solutions also appear in other literatures [25, 69]), and provide an estimation algorithm over general graphs, based on the limit of a set of continuous-time differential equations, proving its effectiveness through numerical simulations. We only mention that a large literature in sensor network localization also deals with other types of relative measurements [55], including relative positions (with known rotations) [4, 64], relative distances [5, 8, 19–21, 23, 29, 70] and relative bearing measurements [30, 73, 77].

A less trivial connection can be established with related work in molecular structure determination from *cryo-electron microscopy* [71, 72], which offers very lucid and mature treatment of rotation estimation. Singer and Shkolnisky [71, 72] provide two approaches for rotation estimation, based on relaxation and semidefinite programming (SDP). Another merit of [71] is to draw connections between planar rotation estimation and the "MAX-2-LIN MOD L" problem in combinatorial optimization, and "MAX-K-CUT" problem in graph theory. Bandeira et al. [3] provide a Cheeger-like inequality that establishes performance bounds for the SDP relaxation. Saunderson et al. [66, 67] propose a tighter SDP relaxation, based on a spectrahedral representation of the convex hull of the rotation group.

Contribution This paper shows that the use of Lagrangian duality allows computing a *guaranteed* globally optimal solution for PGO in many practical cases, and proves that in those cases the solution is unique.

Section 2 recalls preliminary concepts, and discusses the properties of a particular set of 2×2 matrices, which are scalar multiples of a planar rotation matrix. These matrices are omnipresent in planar PGO and acknowledging this fact allows reformulating the problem over complex variables.

Section 3 frames PGO as a problem in complex variables. This makes analysis easier and allows drawing connections with the recent literature on *unit gain graphs* [61]. Exploiting this connection we prove nontrival results about the spectrum of the matrix underlying the problem (the *pose graph matrix*), such as the number of zero eigenvalues in particular graphs.

Section 4 formulates the Lagrangian dual problem in the complex domain. Moreover it presents an SDP relaxation of PGO, interpreting the relaxation as the dual of the dual problem. Our SDP relaxation is related to the one of [32, 71], but we deal with 2D poses, rather than rotations; moreover, we only use the SDP relaxation to complement our discussion on duality and to support some of the proofs. Section 4.3 contains keys results that relate the solution of the dual problem to the primal PGO problem. We show that the *duality gap* is connected to the zero eigenvalues of the *penalized pose graph matrix*, which arises from the solution of the dual problem. We prove that if this matrix has a *single eigenvalue in zero*, then (1) the duality gap is zero, (2) the primal PGO problem has a unique solution (up to an arbitrary roto-translation), and (3) the primal solution can be computed by *scaling* the eigenvector of the penalized pose graph matrix corresponding to the zero eigenvalue. To the best of our knowledge, this is the first work to discuss the uniqueness of the PGO solution for general graphs and to provide a provably optimal solution.

Section 5 exploits our analysis of the dual problem to devise computational approaches for PGO. We propose an algorithm that computes a guaranteed optimal solution for PGO when the penalized pose graph matrix satisfies the Single Zero Eigenvalue Property (SZEP). We also propose a variant that deals with the case in which the SZEP is not satisfied. This variant, while possibly suboptimal, is shown to perform well in practice, outperforming related approaches.

Section 6 elucidates on our theoretical results with numerical tests. In practical regimes of operation (rotation noise < 0.3 rad and translation noise < 0.5 m), our Monte Carlo runs always produced a penalized pose graph matrix satisfying the SZEP. Hence, in all tests with reasonable noise our approach enables the computation of the optimal solution. For larger noise levels (e.g., 1 rad standard deviation for rotation measurements), we observed cases in which the penalized pose graph matrix has multiple eigenvalues in zero. To stimulate further investigation towards structural results on duality (e.g., maximum level of noise for which the duality gap is provably zero) we report simple examples in which the duality gap is nonzero.

2 Notation and Preliminary Concepts

Section 2.1 introduces our notation. Section 2.2 recalls standard concepts from graph theory, and can be safely skipped by the expert reader. Section 2.3, instead, discusses the properties of the set of 2×2 matrices that are multiples of a planar rotation matrix. We denote this set with the symbol $\alpha SO(2)$. The set $\alpha SO(2)$ is of interest in this paper since the action of any matrix $Z \in \alpha SO(2)$ can be conveniently represented as a multiplication between complex numbers, as discussed in Sect. 3.3. Table 1 summarizes the main symbols used in this paper.

Table 1 Symbols used in this paper

Graph	
$\mathcal{G} = (\mathcal{V}, \mathcal{E})$	Directed graph
m	Number of edges
n	Number of nodes
$\mathcal{V}$	Vertex set; $\lvert\mathcal{V}\rvert = n$
$\mathcal{E}$	Edge set; $\lvert\mathcal{E}\rvert = m$
$e = (i, j) \in \mathcal{E}$	Edge between nodes i and j
$\mathcal{A} \in \mathbb{R}^{n \times m}$	Incidence matrix of $\mathcal{G}$
$A \in \mathbb{R}^{(n-1) \times m}$	Anchored incidence matrix of $\mathcal{G}$
$\mathcal{L} = \mathcal{A}^\top \mathcal{A}$	Laplacian matrix of $\mathcal{G}$
$L = A^\top A$	Anchored Laplacian matrix of $\mathcal{G}$
Real PGO formulation	
$\bar{\mathcal{A}} = \mathcal{A} \otimes I_2$	Augmented incidence matrix
$\bar{A} = A \otimes I_2$	Augmented anchored incidence matrix
$\bar{\mathcal{L}} = \mathcal{L} \otimes I_2$	Augmented Laplacian matrix
$\mathcal{W} \in \mathbb{R}^{4n \times 4n}$	Real pose graph matrix
$W \in \mathbb{R}^{(4n-2) \times (4n-2)}$	Real anchored pose graph matrix
$p \in \mathbb{R}^{2n}$	Node positions
$\rho \in \mathbb{R}^{2(n-1)}$	Anchored node positions
$r \in \mathbb{R}^{2n}$	Node rotations
Complex PGO formulation	
$\tilde{W} \in \mathbb{C}^{(2n-1) \times (2n-1)}$	Complex anchored pose graph matrix
$\tilde{\rho} \in \mathbb{C}^{n-1}$	Anchored complex node positions
$\tilde{r} \in \mathbb{C}^{n}$	Complex node rotations
Miscellanea	
$SO(2)$	2D rotation matrices
$\alpha SO(2)$	Scalar multiple of a 2D rotation matrix
$\lvert\mathcal{V}\rvert$	Cardinality of the set $\mathcal{V}$
I_n	$n \times n$ identity matrix
0_n (1_n)	Column vector of zeros (ones) of dimension n
$\mathrm{Tr}(X)$	Trace of the matrix X

2.1 Notation

The cardinality of a set $\mathcal{V}$ is written as $|\mathcal{V}|$. The sets of real and complex numbers are denoted with $\mathbb{R}$ and $\mathbb{C}$, respectively. I_n denotes the $n \times n$ identity matrix, 1_n denotes the (column) vector of all ones of dimension n, $0_{n\times m}$ denotes the $n \times m$ matrix of all zeros (we also use the shorthand $0_n \doteq 0_{n\times 1}$). For a matrix $M \in \mathbb{C}^{m\times n}$, M_{ij} denotes the element of M in row i and column j. The Frobenius norm of a matrix $M \in \mathbb{C}^{m\times n}$ is denoted as $\|M\|_F \doteq \sqrt{\sum_{i=1}^{m}\sum_{j=1}^{n}|M_{ij}|^2}$. For matrices with a block structure we use $[M]_{ij}$ to denote the $d \times d$ block of M at the block row i and block column j. In this paper we only deal with matrices that have 2×2 blocks, i.e., $d = 2$, hence the notation $[M]_{ij}$ is unambiguous.

2.2 Graph Terminology

A *directed graph* $\mathcal{G}$ is a pair $(\mathcal{V}, \mathcal{E})$, where the *vertices* or *nodes* $\mathcal{V}$ are a finite set of elements, and $\mathcal{E} \subset \mathcal{V} \times \mathcal{V}$ is the set of *edges*. Each edge is an ordered pair $e = (i,j)$. We say that e is *incident* on nodes i and j, *leaves* node i, called *tail*, and is *directed towards* node j, called *head*. The number of nodes is denoted with $n \doteq |\mathcal{V}|$, while the number of edges is $m \doteq |\mathcal{E}|$.

A directed graph $\mathcal{G}(\mathcal{V}, \mathcal{E})$ is *(weakly) connected* if the underlying undirected graph, obtained by disregarding edge orientations in $\mathcal{G}$, contains a path from i to j for any pairs of nodes $i,j \in \mathcal{V}$. A directed graph is *strongly connected* if it contains a directed path from i to j for any $i,j \in \mathcal{V}$.

The *incidence matrix* $\mathcal{A}$ of a directed graph is a $m \times n$ matrix with elements in $\{-1, 0, +1\}$ that exhaustively describes the graph topology. Each row of $\mathcal{A}$ corresponds to an edge and has exactly two non-zero elements. For the row corresponding to edge $e = (i,j)$, there is a -1 on the i-th column and a $+1$ on the j-th column.

The set of *outgoing neighbors* of node i is $\mathcal{N}_i^{\text{out}} \doteq \{j : (i,j) \in \mathcal{E}\}$. The set of *incoming neighbors* of node i is $\mathcal{N}_i^{\text{in}} \doteq \{j : (j,i) \in \mathcal{E}\}$. The set of *neighbors* of node i is the union of outgoing and incoming neighbors $\mathcal{N}_i \doteq \mathcal{N}_i^{\text{out}} \cup \mathcal{N}_i^{\text{in}}$.

2.3 The Set $\alpha SO(2)$

The set $\alpha SO(2)$ is defined as

$$\alpha SO(2) \doteq \{\alpha R : \alpha \in \mathbb{R},\ R \in SO(2)\},$$

where $SO(2)$ is the set of 2D rotation matrices. Recall that $SO(2)$ can be parametrized by an angle $\theta \in (-\pi, +\pi]$, and any matrix $R \in SO(2)$ is in the form:

$$R = R(\theta) = \begin{bmatrix} \cos(\theta) & -\sin(\theta) \\ \sin(\theta) & \cos(\theta) \end{bmatrix}. \tag{1}$$

Clearly, $SO(2) \subset \alpha SO(2)$. The set $\alpha SO(2)$ is closed under standard matrix *multiplication*, i.e., for any $Z_1, Z_2 \in \alpha SO(2)$, also the product $Z_1 Z_2 \in \alpha SO(2)$. In full analogy with $SO(2)$, it is also trivial to show that the multiplication is commutative, i.e., for any $Z_1, Z_2 \in \alpha SO(2)$ it holds that $Z_1 Z_2 = Z_2 Z_1$. Moreover, for $Z = \alpha R$ with $R \in SO(2)$ it holds that $Z^\top Z = |\alpha|^2 I_2$. The set $\alpha SO(2)$ is also closed under matrix *addition*, since for $R_1, R_2 \in SO(2)$, we have that

$$\alpha_1 R_1 + \alpha_2 R_2 = \alpha_1 \begin{bmatrix} c_1 & -s_1 \\ s_1 & c_1 \end{bmatrix} + \alpha_2 \begin{bmatrix} c_2 & -s_2 \\ s_2 & c_2 \end{bmatrix} = \tag{2}$$

$$= \begin{bmatrix} \alpha_1 c_1 + \alpha_2 c_2 & -(\alpha_1 s_1 + \alpha_2 s_2) \\ \alpha_1 s_1 + \alpha_2 s_2 & \alpha_1 c_1 + \alpha_2 c_2 \end{bmatrix} = \begin{bmatrix} a & -b \\ b & a \end{bmatrix} = \alpha_3 R_3 ,$$

where we used the shorthands c_i and s_i for $\cos(\theta_i)$ and $\sin(\theta_i)$, and we defined $a \doteq \alpha_1 c_1 + \alpha_2 c_2$ and $b \doteq \alpha_1 s_1 + \alpha_2 s_2$. In (2), the scalar $\alpha_3 \doteq \pm\sqrt{a^2 + b^2}$ (if nonzero) normalizes $\begin{bmatrix} a & -b \\ b & a \end{bmatrix}$, such that $R_3 \doteq \begin{bmatrix} a/\alpha_3 & -b/\alpha_3 \\ b/\alpha_3 & a/\alpha_3 \end{bmatrix}$ is a rotation matrix; if $\alpha_3 = 0$, then $\alpha_1 R_1 + \alpha_2 R_2 = 0_{2\times 2}$, which also falls in our definition of $\alpha SO(2)$. From this reasoning, it is clear that an alternative definition of $\alpha SO(2)$ is

$$\alpha SO(2) \doteq \left\{ \begin{bmatrix} a & -b \\ b & a \end{bmatrix} : a, b \in \mathbb{R} \right\}. \tag{3}$$

$\alpha SO(2)$ is tightly coupled with the set of complex numbers $\mathbb{C}$. Indeed, a matrix in the form (3) is also known as a *matrix representation* of a complex number [39]. We explore the implications of this fact for PGO in Sect. 3.3.

3 Pose Graph Optimization in the Complex Domain

3.1 Standard PGO

PGO estimates n poses from m relative pose measurements. We focus on the planar case, in which the i-th pose x_i is described by the pair $x_i \doteq (p_i, R_i)$, where $p_i \in \mathbb{R}^2$ is a position in the plane, and $R_i \in SO(2)$ is a planar rotation. The pose measurement between two nodes, say i and j, is described by the pair (Δ_{ij}, R_{ij}), where $\Delta_{ij} \in \mathbb{R}^2$ and $R_{ij} \in SO(2)$ are the relative position and rotation measurements, respectively.

The problem can be visualized as a directed graph $\mathcal{G}(\mathcal{V}, \mathcal{E})$, where an unknown pose is attached to each node in the set $\mathcal{V}$, and each edge $(i, j) \in \mathcal{E}$ corresponds to a relative pose measurement between nodes i and j (Fig. 2).

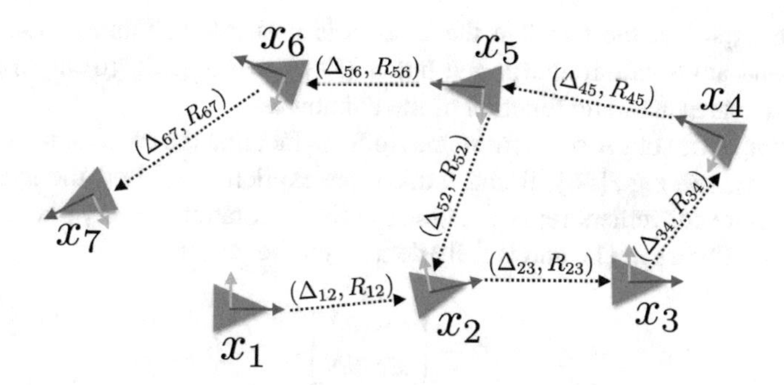

Fig. 2 Schematic representation of Pose Graph Optimization: the objective is to associate a pose x_i to each node of a directed graph, given relative pose measurements (Δ_{ij}, R_{ij}) for each edge (i, j) in the graph

In a noiseless case, the measurements satisfy:

$$\Delta_{ij} = R_i^\top (p_j - p_i), \qquad R_{ij} = R_i^\top R_j, \tag{4}$$

and we can compute the unknown rotations $\{R_1, \ldots, R_n\}$ and positions $\{p_1, \ldots, p_n\}$ by solving a set of linear equations (relations (4) become linear after rearranging the rotation R_i to the left-hand side). In absence of noise, the problem admits a unique solution as long as one fixes the pose of a node (say $p_1 = 0_2$ and $R_1 = I_2$) and the underling graph is connected.

In this work we focus on connected graphs, as these are the ones of practical interest in PGO (a graph with k connected components can be split in k subproblems, which can be solved and analyzed independently).

Assumption 1 (Connected Pose Graph). *The graph $\mathcal{G}$ underlying the pose graph optimization problem is (weakly) connected.*

In presence of noise, the relations (4) cannot be met exactly and pose graph optimization looks for a set of positions $\{p_1, \ldots, p_n\}$ and rotations $\{R_1, \ldots, R_n\}$ that minimize the mismatch with respect to the measurements. This mismatch can be quantified by different cost functions. We adopt the formulation proposed in [12]:

$$\min_{\{p_i\},\{R_i\} \in SO(2)^n} \sum_{(i,j)\in\mathcal{E}} \|\Delta_{ij} - R_i^\top (p_j - p_i)\|_2^2 + \frac{1}{2}\|R_{ij} - R_i^\top R_j\|_F^2, \tag{5}$$

where $\|\cdot\|_2$ is the standard Euclidean distance and $\|\cdot\|_F$ is the Frobenius norm. The Frobenius norm $\|R_a - R_b\|_F$ is a standard measure of distance between two rotations R_a and R_b, and it is commonly referred to as the *chordal* distance, see, e.g., [37]. In (5), we used the short-hand notation $\{p_i\}$ (resp. $\{R_i\}$) to denote the set of unknown positions $\{p_1 \ldots, p_n\}$ (resp. rotations).

Rearranging the terms, problem (5) can be rewritten as:

$$\min_{\{p_i\},\{R_i\} \in SO(2)^n} \sum_{(i,j)\in\mathcal{E}} \|(p_j - p_i) - R_i \Delta_{ij}\|_2^2 + \frac{1}{2}\|R_j - R_i R_{ij}\|_F^2, \tag{6}$$

where we exploited the fact that the 2-norm is invariant to rotation, i.e., for any vector v and any rotation matrix R it holds $\|Rv\|_2 = \|v\|_2$. Eq. (6) highlights that the objective is a quadratic function of the unknowns.

The complexity of the problem stems from the fact that the constraint $R_i \in SO(2)$ is nonconvex, see, e.g., [66]. To make this more explicit, we follow the line of [12], and use a more convenient representation for nodes' rotations. Every planar rotation R_i can be written as in (1), and is fully defined by the vector

$$r_i = \begin{bmatrix} \cos(\theta_i) \\ \sin(\theta_i) \end{bmatrix}. \tag{7}$$

Using this parametrization and with simple matrix manipulation, Eq. (6) becomes (*cf.* with Eq. (11) in [12]):

$$\begin{aligned} \min_{\{p_i\},\{r_i\}} \quad & \textstyle\sum_{(i,j)\in\mathcal{E}} \|(p_j - p_i) - D_{ij} r_i\|_2^2 + \|r_j - R_{ij} r_i\|_2^2 \\ \text{s.t.:} \quad & \|r_i\|_2^2 = 1, \ \ i = 1, \ldots, n \end{aligned} \tag{8}$$

where we defined:

$$D_{ij} = \begin{bmatrix} \Delta_{ij}^x & -\Delta_{ij}^y \\ \Delta_{ij}^y & \Delta_{ij}^x \end{bmatrix}, \quad (\text{with } \Delta_{ij} \doteq [\Delta_{ij}^x \ \Delta_{ij}^y]^\top), \tag{9}$$

and where the constraints $\|r_i\|_2^2 = 1$ specify that we look for vectors r_i that represent admissible rotations (i.e., such that $\cos(\theta_i)^2 + \sin(\theta_i)^2 = 1$).

Problem (8) is a quadratic problem with quadratic equality constraints. The latter are nonconvex, hence computing a local minimum of (8) is hard in general. There are two problem instances, however, for which it is easy to compute a global minimizer, which attains zero optimal cost. These two cases are recalled in Propositions 1 and 2, while procedures to compute the corresponding optimal solutions are given in sections "Proof of Proposition 1: Zero Cost in Trees" and "Proof of Proposition 2: Zero Cost in Balanced Graphs" in Appendix.

Proposition 1 (Zero Cost in Trees). *An optimal solution for a PGO problem in the form* (8) *whose underlying graph is a tree attains zero cost.*

The proof is given in section "Proof of Proposition 1: Zero Cost in Trees" in Appendix. Roughly speaking, in a tree, we can build an optimal solution by concatenating the relative pose measurements, and this solution annihilates the cost function. This comes with no surprises, as the *chords* (i.e., the extra edges, added to a spanning tree) are indeed the elements that create redundancy and improve the pose estimate. However, also for graphs with chords, it is possible to attain the zero cost in problem (8).

Definition 1 (Balanced Pose Graph). A pose graph is *balanced* if the pose measurements compose to the identity along each cycle in the graph.[2,3]

In a balanced pose graph, there exists a configuration that explains exactly the measurements, as formalized in the following proposition.

Proposition 2 (Zero Cost in Balanced Pose Graphs). *An optimal solution for a balanced pose graph optimization problem attains zero cost.*

The proof is given in section "Proof of Proposition 2: Zero Cost in Balanced Graphs" in Appendix. The concept of balanced graph describes a noiseless setup, while in real instances the measurements do not compose to the identity along cycles, because of noise. Note that a tree can be considered a special case of a balanced graph with no cycles.

We note the following fact, which will be useful in Sect. 3.2.

Proposition 3 (Coefficient Matrices in PGO). *Matrices* $D_{ij}, I_2, -I_2, R_{ij}$ *appearing in* (8) *belong to* $\alpha SO(2)$.

This fact is trivial, since $R_{ij}, I_2 \in SO(2) \subset \alpha SO(2)$ (the latter also implies $-I_2 \in \alpha SO(2)$). Moreover, the structure of D_{ij} in (9) clearly falls in the definition of matrices in $\alpha SO(2)$ given in (3).

3.2 Matrix Formulation and Anchoring

In this section we rewrite the cost function (8) in a more convenient matrix form. The original cost is:

$$f(p,r) \doteq \sum_{(i,j)\in\mathcal{E}} \|(p_j - p_i) - D_{ij} r_i\|_2^2 + \|r_j - R_{ij} r_i\|_2^2 \tag{10}$$

where we denote with $p \in \mathbb{R}^{2n}$ and $r \in \mathbb{R}^{2n}$ the vectors stacking all nodes positions and rotations, respectively. Now, let $\mathcal{A} \in \mathbb{R}^{m\times n}$ denote the *incidence matrix* of the graph underlying the problem: if (i,j) is the k-th edge, then $\mathcal{A}_{ki} = -1$, $\mathcal{A}_{kj} = +1$. Let $\bar{\mathcal{A}} = \mathcal{A} \otimes I_2 \in \mathbb{R}^{2m\times 2n}$, and denote with $\bar{\mathcal{A}}_k \in \mathbb{R}^{2\times 2n}$ the k-th block row of $\bar{\mathcal{A}}$. From the structure of $\bar{\mathcal{A}}$, it follows that $\bar{\mathcal{A}}_k p = p_j - p_i$. Also, we define $\bar{D} \in \mathbb{R}^{2m\times 2n}$ as a block matrix where the k-th block row $\bar{D}_k \in \mathbb{R}^{2\times 2n}$ corresponding to the k-th

[2]We use the somehow standard term "composition" to denote the group operation for $SE(2)$. For two poses $T_1 \doteq (p_1, R_1)$ and $T_2 \doteq (p_2, R_2)$, the composition is $T_1 \cdot T_2 = (p_1 + R_1 p_2, R_1 R_2)$ [16]. Similarly, the *identity* element is $(0_2, I_2)$.

[3]When composing measurements along the loop, edge direction is important: for two consecutive edges (i,k) and (k,j) along the loop, the composition is $T_{ij} = T_{ik} \cdot T_{kj}$, while if the second edge is in the form (j,k), the composition becomes $T_{ij} = T_{ik} \cdot T_{jk}^{-1}$.

edge (i,j) is all zeros, except for a 2×2 block $-D_{ij}$ in the i-th block column. Using the matrices $\bar{\mathscr{A}}$ and $\bar{D}$, the first sum in (10) can be written as:

$$\sum_{(i,j)\in\mathscr{E}} \|(p_j - p_i) - D_{ij}r_i\|_2^2 = \sum_{k=1}^{m} \|\bar{\mathscr{A}}_k p + \bar{D}_k r\|_2^2 = \|\bar{\mathscr{A}} p + \bar{D} r\|_2^2 \tag{11}$$

Similarly, we define $\bar{U} \in \mathbb{R}^{2m \times 2n}$ as a block matrix where the k-th block row $\bar{U}_k \in \mathbb{R}^{2 \times 2n}$ corresponding to the k-th edge (i,j) is all zeros, except for 2×2 blocks in the i-th and j-th block columns, which are equal to R_{ij} and I_2, respectively. Using $\bar{U}$, the second sum in (10) becomes:

$$\sum_{(i,j)\in\mathscr{E}} \|r_j - R_{ij}r_i\|_2^2 = \sum_{k=1}^{m} \|\bar{U}_k r\|_2^2 = \|\bar{U} r\|_2^2 \tag{12}$$

Combining (11) and (12), the cost in (10) becomes:

$$f(p,r) = \left\| \begin{bmatrix} \bar{\mathscr{A}} & \bar{D} \\ 0 & \bar{U} \end{bmatrix} \begin{bmatrix} p \\ r \end{bmatrix} \right\|_2^2 = \begin{bmatrix} p \\ r \end{bmatrix}^\top \begin{bmatrix} \bar{\mathscr{A}}^\top \bar{\mathscr{A}} & \bar{\mathscr{A}}^\top \bar{D} \\ \bar{D}^\top \bar{\mathscr{A}} & \bar{D}^\top \bar{D} + \bar{U}^\top \bar{U} \end{bmatrix} \begin{bmatrix} p \\ r \end{bmatrix}$$
$$= \begin{bmatrix} p \\ r \end{bmatrix}^\top \begin{bmatrix} \bar{\mathscr{L}} & \bar{\mathscr{A}}^\top \bar{D} \\ \bar{D}^\top \bar{\mathscr{A}} & \bar{Q} \end{bmatrix} \begin{bmatrix} p \\ r \end{bmatrix}, \tag{13}$$

where we defined $\bar{Q} \doteq \bar{D}^\top \bar{D} + \bar{U}^\top \bar{U}$ and $\bar{\mathscr{L}} \doteq \bar{\mathscr{A}}^\top \bar{\mathscr{A}}$, to simplify notation. Note that, since $\bar{\mathscr{A}} \doteq \mathscr{A} \otimes I_2$, it is easy to show that $\bar{\mathscr{L}} = \mathscr{L} \otimes I_2$, where $\mathscr{L} \doteq \mathscr{A}^\top \mathscr{A}$ is the Laplacian matrix of the graph underlying the problem. A pose graph optimization instance is thus completely defined by the matrix

$$\mathscr{W} \doteq \begin{bmatrix} \bar{\mathscr{L}} & \bar{\mathscr{A}}^\top \bar{D} \\ \bar{D}^\top \bar{\mathscr{A}} & \bar{Q} \end{bmatrix} \in \mathbb{R}^{4n \times 4n} \tag{14}$$

From (13), $\mathscr{W}$ can be easily seen to be symmetric and positive semidefinite. Other useful properties of $\mathscr{W}$ are stated in the next proposition.

Proposition 4 (Properties of $\mathscr{W}$). *Matrix $\mathscr{W}$ in* (14) *is positive semidefinite, and*

1. *has at least two eigenvalues in zero;*
2. *is composed by 2×2 blocks $[\mathscr{W}]_{ij}$, and each block is a multiple of a rotation matrix, i.e., $[\mathscr{W}]_{ij} \in \alpha SO(2)$, $\forall i,j = 1,\ldots,2n$. Moreover, the diagonal blocks of $\mathscr{W}$ are nonnegative multiples of the identity matrix, i.e., $[\mathscr{W}]_{ii} = \alpha_{ii} I_2$, $\alpha_{ii} \geq 0$.*

A formal proof of Proposition 4 is given in section "Proof of Proposition 4: Properties of $\mathscr{W}$" in Appendix. An intuitive explanation of the second claim follows from the fact that (1) $\mathscr{W}$ contains sums and products of the matrices in the original formulation (8) (which are in $\alpha SO(2)$ according to Lemma 3), and (2) the set $\alpha SO(2)$ is closed under matrix sum and product (Sect. 2.3).

The presence of two eigenvalues in zero has a very natural geometric interpretation: the cost function encodes inter-nodal measurements, hence it is invariant to global translations of node positions, i.e., $f(p, r) = f(p + p_a, r)$, where $p_a \doteq (1_n \otimes I_2)a = [a^\top \ \ldots \ a^\top]^\top$ (n copies of a), with $a \in \mathbb{R}^2$. Algebraically, this translates to the fact that the matrix $(1_n \otimes I_2) \in \mathbb{R}^{2n\times 2}$ is in the null space of the augmented incidence matrix $\mathscr{A}$, which also implies a two dimensional null space for $\mathscr{W}$.

Position Anchoring In this paper we show that the duality properties in pose graph optimization are tightly coupled with the spectrum of the matrix $\mathscr{W}$. We are particularly interested in the eigenvalues at zero, and from this perspective it is not convenient to carry on the two null eigenvalues of $\mathscr{W}$ (claim 1 of Proposition 4), which are always present, and are due to an intrinsic observability issue.

We remove the translation ambiguity by fixing the position of an arbitrary node. Without loss of generality, we fix the position p_1 of the first node to the origin, i.e., $p_1 = 0_2$. This process is commonly called *anchoring*. Setting $p_1 = 0$ is equivalent to removing the corresponding columns and rows from $\mathscr{W}$, leading to the following "anchored" PGO problem:

$$f(r, \rho) = \begin{bmatrix} 0_2 \\ \rho \\ r \end{bmatrix}^\top \mathscr{W} \begin{bmatrix} 0_2 \\ \rho \\ r \end{bmatrix} = \begin{bmatrix} \rho \\ r \end{bmatrix}^\top W \begin{bmatrix} \rho \\ r \end{bmatrix} \tag{15}$$

where ρ is the vector p without its first two-elements vector p_1, and W is obtained from $\mathscr{W}$ by removing the rows and the columns corresponding to p_1. The structure of W is as follows:

$$W = \begin{bmatrix} \bar{A}^\top \bar{A} & \bar{A}^\top \bar{D} \\ \bar{D}^\top \bar{A} & \bar{Q} \end{bmatrix} \doteq \begin{bmatrix} \bar{L} & \bar{S} \\ \bar{S}^\top & \bar{Q} \end{bmatrix} \tag{16}$$

where $\bar{A} = A \otimes I_2$, and A is the *anchored* (or *reduced*) incidence matrix, obtained by removing the first column from $\mathscr{A}$, see, e.g., [14]. On the right-hand-side of (16) we defined $\bar{S} \doteq \bar{A}^\top \bar{D}$ and $\bar{L} \doteq \bar{A}^\top \bar{A}$.

We call W the *real (anchored) pose graph matrix*. W is still symmetric and positive semidefinite (it is a principal submatrix of a positive semidefinite matrix). Moreover, since W is obtained by removing a $2 \times 4n$ block row and a $4n \times 2$ block column from $\mathscr{W}$, it is still composed by 2×2 matrices in $\alpha SO(2)$, as specified in the following remark.

Remark 1 (Properties of W). The positive semidefinite matrix W in (16) is composed by 2×2 blocks $[W]_{ij}$, that are such that $[W]_{ij} \in \alpha SO(2)$, $\forall i, j = 1, \ldots, 2n-1$. Moreover, the diagonal blocks of W are nonnegative multiples of the identity matrix, i.e., $[W]_{ii} = \alpha_{ii} I_2$, $\alpha \geq 0$.

After anchoring, our PGO problem becomes:

$$f^\star = \min_{\rho, r} \begin{bmatrix} \rho \\ r \end{bmatrix}^\top W \begin{bmatrix} \rho \\ r \end{bmatrix} \quad \text{s.t.:} \quad \|r_i\|_2^2 = 1, \quad i = 1, \ldots, n \tag{17}$$

3.3 To Complex Domain

In this section we rewrite problem (17), in which the decision variables are real vectors, into a problem in complex variables. The main motivation for this choice is that the real representation (17) is somehow redundant: as we show in Proposition 7, each eigenvalue of W is repeated twice (multiplicity 2), while the complex representation does not have this redundancy, making analysis easier. In the rest of this chapter, quantities marked with a tilde $(\tilde{\cdot})$ live in the complex domain $\mathbb{C}$.

Any real vector $v \in \mathbb{R}^2$ can be represented by a complex number $\tilde{v} = \eta e^{j\varphi}$, where $j^2 = -1$ is the *imaginary unit*, $\eta = \|v\|_2$ and φ is the angle that v forms with the horizontal axis. We use the operator $(\cdot)^\vee$ to map a 2-vector to the corresponding complex number, $\tilde{v} = v^\vee$. When convenient, we adopt the notation $v \sim \tilde{v}$, meaning that v and $\tilde{v}$ are the vector and the complex representation of the same number.

The action of a real 2×2 matrix Z on a vector $v \in \mathbb{R}^2$ cannot be represented, in general, as a scalar multiplication between complex numbers. However, if $Z \in \alpha SO(2)$, this is possible. To show this, assume that $Z = \alpha R(\theta)$, where $R(\theta)$ is a counter-clockwise rotation of angle θ. Then,

$$Z\,v = \alpha R(\theta) v \sim \tilde{z}\,\tilde{v}, \quad \text{where } \tilde{z} = \alpha e^{j\theta}. \tag{18}$$

With slight abuse of notation we extend the operator $(\cdot)^\vee$ to $\alpha SO(2)$, such that, given $Z = \alpha R(\theta) \in \alpha SO(2)$, then $Z^\vee = \alpha e^{j\theta} \in \mathbb{C}$. By inspection, one can also verify the following relations between the sum and product of two matrices $Z_1, Z_2 \in \alpha SO(2)$ and their complex representations $Z_1^\vee, Z_2^\vee \in \mathbb{C}$:

$$(Z_1\,Z_2)^\vee = \tilde{Z}_1^\vee\,Z_2^\vee \qquad (Z_1 + Z_2)^\vee = Z_1^\vee + Z_2^\vee. \tag{19}$$

We next discuss how to apply the machinery introduced so far to reformulate problem (17) in the complex domain. The variables in problem (17) are the vectors $\rho \in \mathbb{R}^{2(n-1)}$ and $r \in \mathbb{R}^{2n}$ that are composed by 2-vectors, i.e., $\rho = [\rho_1^\top, \ldots, \rho_{n-1}^\top]^\top$ and $r = [r_1^\top, \ldots, r_n^\top]^\top$, where $\rho_i, r_i \in \mathbb{R}^2$. Therefore, we define the *complex positions* and the *complex rotations*:

$$\begin{aligned} \tilde{\rho} &= [\tilde{\rho}_1, \ldots, \tilde{\rho}_{n-1}]^* \in \mathbb{C}^{n-1}, \text{ where: } \tilde{\rho}_i = \rho_i^\vee \\ \tilde{r} &= [\tilde{r}_1, \ldots, \tilde{r}_n]^* \quad \in \mathbb{C}^n, \quad \text{where: } \tilde{r}_i = r_i^\vee \end{aligned} \tag{20}$$

Using the complex parametrization (20), the constraints in (17) become:

$$|\tilde{r}_i|^2 = 1, \quad i = 1, \ldots, n. \tag{21}$$

Similarly, we would like to rewrite the objective as a function of $\tilde{\rho}$ and $\tilde{r}$. This re-parametrization is formalized in the following proposition, whose proof is given in section "Proof of Proposition 5: Cost in the Complex Domain" in Appendix.

Proposition 5 (Cost in the Complex Domain). *For any pair* (ρ, r)*, the cost function in* (17) *is such that:*

$$f(\rho, r) = \begin{bmatrix} \rho \\ r \end{bmatrix}^{\top} W \begin{bmatrix} \rho \\ r \end{bmatrix} = \begin{bmatrix} \tilde{\rho} \\ \tilde{r} \end{bmatrix}^{*} \tilde{W} \begin{bmatrix} \tilde{\rho} \\ \tilde{r} \end{bmatrix} \tag{22}$$

where the vectors $\tilde{\rho}$ *and* $\tilde{r}$ *are built from* ρ *and* r *as in* (20)*, and the matrix* $\tilde{W} \in \mathbb{C}^{(2n-1)\times(2n-1)}$ *is such that* $\tilde{W}_{ij} = [W]_{ij}^{\vee}$*, with* $i, j = 1, \ldots, 2n-1$.

Remark 2 (Real Diagonal Entries for $\tilde{W}$*).* According to Remark 1, the diagonal blocks of W are multiples of the identity matrix, i.e., $[W]_{ii} = \alpha_{ii} I_2$. Therefore, the diagonal elements of $\tilde{W}$ are $\tilde{W}_{ii} = [W]_{ii}^{\vee} = \alpha_{ii} \in \mathbb{R}$.

Proposition 5 enables us to rewrite problem (17) as:

$$f^{\star} = \min_{\tilde{\rho}, \tilde{r}} \begin{bmatrix} \tilde{\rho} \\ \tilde{r} \end{bmatrix}^{*} \tilde{W} \begin{bmatrix} \tilde{\rho} \\ \tilde{r} \end{bmatrix} \tag{23}$$

$$\text{s.t.:} \quad |\tilde{r}_i|^2 = 1, \quad i = 1, \ldots, n.$$

We call $\tilde{W}$ the *complex (anchored) pose graph matrix*. Clearly, the matrix $\tilde{W}$ preserves the same block structure of W in (16):

$$\tilde{W} \doteq \begin{bmatrix} L & \tilde{S} \\ \tilde{S}^{*} & \tilde{Q} \end{bmatrix} \tag{24}$$

where $\tilde{S}^{*}$ is the conjugate transpose of $\tilde{S}$, and $L \doteq A^{\top} A$ where A is the anchored incidence matrix. In Sect. 4 we apply Lagrangian duality to the problem (23). Before that, we provide results to characterize the spectrum of the matrices W and $\tilde{W}$, drawing connections with the recent literature on unit gain graphs, [61].

3.4 Analysis of the Real and Complex Pose Graph Matrices

In this section we take a closer look at the structure and the properties of the real and the complex pose graph matrices W and $\tilde{W}$. In analogy with (13) and (16), we write $\tilde{W}$ as

$$\tilde{W} = \begin{bmatrix} A^{\top} A & A^{\top} \tilde{D} \\ (A^{\top} \tilde{D})^{*} & \tilde{U}^{*} \tilde{U} + \tilde{D}^{*} \tilde{D} \end{bmatrix} = \begin{bmatrix} A & \tilde{D} \\ 0 & \tilde{U} \end{bmatrix}^{*} \begin{bmatrix} A & \tilde{D} \\ 0 & \tilde{U} \end{bmatrix} \tag{25}$$

where $\tilde{U} \in \mathbb{C}^{m \times n}$ and $\tilde{D} \in \mathbb{C}^{m \times n}$ are the "complex versions" of $\bar{U}$ and $\bar{D}$ in (13), i.e., they are obtained as $\tilde{U}_{ij} = [\bar{U}]_{ij}^{\vee}$ and $\tilde{D}_{ij} = [\bar{D}]_{ij}^{\vee}$, $\forall i, j$.

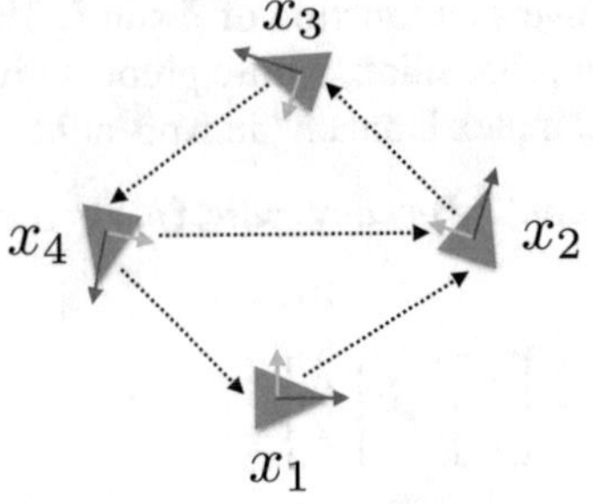

Incidence matrix:

$$\mathscr{A} = \begin{bmatrix} -1 & +1 & 0 & 0 \\ 0 & -1 & +1 & 0 \\ 0 & 0 & -1 & +1 \\ +1 & 0 & 0 & -1 \\ 0 & +1 & 0 & -1 \end{bmatrix} \begin{matrix} (1,2) \\ (2,3) \\ (3,4) \\ (4,1) \\ (4,2) \end{matrix}$$

Anchored Incidence matrix:

$$A = \begin{bmatrix} +1 & 0 & 0 \\ -1 & +1 & 0 \\ 0 & -1 & +1 \\ 0 & 0 & -1 \\ +1 & 0 & -1 \end{bmatrix}$$

Complex Incidence matrix:

$$\tilde{U} = \begin{bmatrix} -e^{j\theta_{12}} & +1 & 0 & 0 \\ 0 & -e^{j\theta_{23}} & +1 & 0 \\ 0 & 0 & -e^{j\theta_{34}} & +1 \\ +1 & 0 & 0 & -e^{j\theta_{41}} \\ 0 & +1 & 0 & -e^{j\theta_{42}} \end{bmatrix}$$

Fig. 3 Example of incidence matrix, anchored incidence matrix, and complex incidence matrix, for the toy PGO problem on the *top left*. If $R_{ij} = R(\theta_{ij})$ is the relative rotation measurement associated to edge (i,j), then the matrix $\tilde{U}$ can be seen as the incidence matrix of a unit gain graph with gain $e^{j\theta_{ij}}$ associated to each edge (i,j)

The factorization (25) is interesting, as it allows to identify two important matrices that compose $\tilde{W}$: the first is A, the anchored incidence matrix that we introduced earlier; the second is $\tilde{U}$ which is a generalization of the incidence matrix, as specified by Definition 2 and Lemma 1 in the following. Figure 3 reports the matrices A and $\tilde{U}$ for a toy example with four poses.

Definition 2 (Unit Gain Graphs). A *unit gain graph* (see, e.g., [61]) is a graph in which to each orientation of an edge (i,j) is assigned a complex number $\tilde{z}_{ij}$ (with $|\tilde{z}_{ij}| = 1$), which is the inverse of the complex number $\frac{1}{\tilde{z}_{ij}}$ assigned to the opposite orientation (j,i). Moreover, a *complex incidence matrix* of a unit gain graph is a matrix in which each row corresponds to an edge and the row corresponding to edge $e = (i,j)$ has $-\tilde{z}_{ij}$ on the i-th column, $+1$ on the j-th column, and zero elsewhere.

Roughly speaking, a unit gain graph describes a problem in which we can "flip" the orientation of an edge by inverting the corresponding complex weight. To understand what this property means in our context, recall the definition (12), and consider the following chain of equalities:

$$\|\bar{U}r\|_2^2 = \sum_{(i,j)\in\mathscr{E}} \|r_j - R_{ij}r_i\|_2^2 = \sum_{(i,j)\in\mathscr{E}} \|r_i - R_{ij}^\top r_j\|_2^2 \tag{26}$$

which, written in the complex domain, becomes:

$$\|\tilde{U}\tilde{r}\|_2^2 = \sum_{(i,j)\in\mathscr{E}} |\tilde{r}_j - \mathrm{e}^{\mathrm{j}\theta_{ij}}\tilde{r}_i|^2 = \sum_{(i,j)\in\mathscr{E}} |\tilde{r}_i - \mathrm{e}^{-\mathrm{j}\theta_{ij}}\tilde{r}_j|^2 = \sum_{(i,j)\in\mathscr{E}} |\tilde{r}_i - \frac{1}{\mathrm{e}^{\mathrm{j}\theta_{ij}}}\tilde{r}_j|^2 \tag{27}$$

Eq. (27) essentially says that the term $\|\tilde{U}\tilde{r}\|_2^2$ does not change if we flip the orientation of an edge and invert the relative rotation measurement. The proof of the following lemma is straightforward from (27).

Lemma 1 (Properties of $\tilde{U}$). *Matrix $\tilde{U}$ is a complex incidence matrix of a unit gain graph with weights $R_{ij}^{\vee} = \mathrm{e}^{\mathrm{j}\theta_{ji}}$ associated to each edge (i, j).*

Our interest towards unit gain graphs is motivated by the recent results in [61] on the spectrum of the incidence matrix of those graphs. Using these results, we can characterize the presence of eigenvalues in zero for the matrix $\tilde{W}$, as specified in the following proposition (proof in section "Proof of Proposition 6: Zero Eigenvalues in $\tilde{W}$" in Appendix).

Proposition 6 (Zero Eigenvalues in $\tilde{W}$). *The complex anchored pose graph matrix $\tilde{W}$ has a single eigenvalue in zero if and only if the pose graph is balanced or is a tree.*

Besides analyzing the spectrum of $\tilde{W}$, it is of interest to understand how the complex matrix $\tilde{W}$ relates to the real matrix W. The following proposition states that there is a tight correspondence between the eigenvalues of the real pose graph matrix W and its complex counterpart $\tilde{W}$.

Proposition 7 (Spectrum of Complex Graph Matrices). *The $2(2n-1)$ eigenvalues of W are the $2n-1$ eigenvalues of $\tilde{W}$, repeated twice.*

See section "Proof of Proposition 7: Spectrum of Complex and Real Pose Graph Matrices" in Appendix for a proof.

4 Lagrangian Duality in PGO

In the previous section we wrote the PGO problem in complex variables as per Eq. (23). In the following, we refer to this problem as the *primal* PGO problem, that, defining $\tilde{x} \doteq [\tilde{\rho}^* \; \tilde{r}^*]^*$, can be written in compact form as

$$\begin{aligned} f^\star = \min_{\tilde{x}} \quad & \tilde{x}^* \tilde{W} \tilde{x} & \text{(Primal problem)} \\ \text{s.t.:} \quad & |\tilde{x}_i|^2 = 1, \; i = n, \ldots, 2n-1, & \end{aligned} \tag{28}$$

In this section we derive the Lagrangian dual of (28), which is given in Sect. 4.1. Then, in Sect. 4.2, we discuss an SDP relaxation of (28), that can be interpreted as the dual of the dual problem. Finally, in Sect. 4.3 we analyze the properties of the dual problem, and discuss how it relates with the primal PGO problem.

4.1 The Dual Problem

The Lagrangian of the primal problem (28) is

$$\mathbb{L}(\tilde{x},\lambda) = \tilde{x}^* \tilde{W} \tilde{x} + \sum_{i=1}^{n} \lambda_i (1 - |\tilde{x}_{n+i-1}|^2)$$

where $\lambda_i \in \mathbb{R}$, $i = 1,\ldots,n$, are the *Lagrange multipliers* (or *dual variables*). Recalling the structure of $\tilde{W}$ from (24), the Lagrangian becomes:

$$\mathbb{L}(\tilde{x},\lambda) \;=\; \tilde{x}^* \begin{bmatrix} L & \tilde{S} \\ \tilde{S}^* & \tilde{Q}(\lambda) \end{bmatrix} \tilde{x} + \sum_{i=1}^{n} \lambda_i \;=\; \tilde{x}^* \tilde{W}(\lambda) \tilde{x} + \sum_{i=1}^{n} \lambda_i,$$

where for notational convenience we defined

$$\tilde{Q}(\lambda) \doteq \tilde{Q} - \mathrm{diag}(\lambda_1,\ldots,\lambda_n), \qquad \tilde{W}(\lambda) \doteq \begin{bmatrix} L & \tilde{S} \\ \tilde{S}^* & \tilde{Q}(\lambda) \end{bmatrix}$$

The *dual function* $d : \mathbb{R}^n \to \mathbb{R}$ is the infimum of the Lagrangian with respect to $\tilde{x}$:

$$d(\lambda) = \inf_{\tilde{x}} \mathbb{L}(\tilde{x},\lambda) = \inf_{\tilde{x}} \tilde{x}^* \tilde{W}(\lambda) \tilde{x} + \sum_{i=1}^{n} \lambda_i, \tag{29}$$

For any choice of λ the dual function provides a lower bound on the optimal value of the primal problem [7, Sect. 5.1.3]. Therefore, the *Lagrangian dual problem* looks for a *maximum* of the dual function over λ:

$$d^\star \doteq \max_{\lambda} d(\lambda) = \max_{\lambda} \inf_{\tilde{x}} \tilde{x}^* \tilde{W}(\lambda) \tilde{x} + \sum_{i=1}^{n} \lambda_i, \tag{30}$$

The infimum over $\tilde{x}$ of $\mathbb{L}(\tilde{x},\lambda)$ drifts to $-\infty$ unless $\tilde{W}(\lambda) \succeq 0$. Therefore we can safely restrict the maximization to vectors λ that are such that $\tilde{W}(\lambda) \succeq 0$; these are called *dual-feasible*. Moreover, at any dual-feasible λ, the $\tilde{x}$ minimizing the Lagrangian are those that make $\tilde{x}^* \tilde{W}(\lambda) \tilde{x} = 0$. Therefore, (30) reduces to the following *dual problem*

$$\begin{aligned} d^\star = \max_{\lambda} \quad & \textstyle\sum_i \lambda_i, \qquad\qquad \text{(Dual problem)} \\ \text{s.t.:} \quad & \tilde{W}(\lambda) \succeq 0. \end{aligned} \tag{31}$$

The importance of the dual problem is twofold. First, it holds that

$$d^{\star} \leq f^{\star} \tag{32}$$

This property is called *weak duality*, see, e.g., [7, Sect. 5.2.2]. For particular problems the inequality (32) becomes an equality, and in such cases we say that *strong duality* holds. Second, since $d(\lambda)$ is concave (minimum of affine functions), the dual problem (31) is always convex in λ, regardless the convexity properties of the primal problem. The dual PGO problem (31) is a semidefinite program (SDP).

For a given λ, we denote by $\mathscr{X}(\lambda)$ the set of $\tilde{x}$ that attain the optimal value in problem (29), if any:

$$\mathscr{X}(\lambda) \doteq \{\tilde{x}_\lambda \in \mathbb{C}^{2n-1} \; : \; \tilde{x}_\lambda = \arg\min_{\tilde{x}} \; \mathbb{L}(\tilde{x}, \lambda) = \arg\min_{\tilde{x}} \; \tilde{x}^* \tilde{W}(\lambda) \tilde{x}\}$$

Since we already observed that for any dual-feasible λ the points $\tilde{x}$ that minimize the Lagrangian are such that $\tilde{x}^* \tilde{W}(\lambda) \tilde{x} = 0$, it follows that:

$$\mathscr{X}(\lambda) = \{\tilde{x} \in \mathbb{C}^{2n-1} : \tilde{W}(\lambda)\tilde{x} = 0\} = \text{Kernel}(\tilde{W}(\lambda)), \quad \text{for } \lambda \text{ dual-feasible.} \tag{33}$$

The following result ensures that if a vector in $\mathscr{X}(\lambda)$ is feasible for the primal problem, then it is also an optimal solution for the PGO problem.

Theorem 1. *Given $\lambda \in \mathbb{R}^n$, if an $\tilde{x}_\lambda \in \mathscr{X}(\lambda)$ is primal feasible, then $\tilde{x}_\lambda$ is primal optimal; moreover, λ is dual optimal, and the duality gap is zero.*

A proof of this theorem is given in section "Proof of Theorem 1: Primal-dual Optimal Pairs" in Appendix.

4.2 SDP Relaxation and the Dual of the Dual

We have seen that a lower bound $d^{\star}$ on the optimal value $f^{\star}$ of the primal (28) can be obtained by solving the Lagrangian dual problem (31). Here, we outline another, direct, relaxation method to obtain such bound.

Observing that $\tilde{x}^* \tilde{W} \tilde{x} = \text{Tr}(\tilde{W} \tilde{x} \tilde{x}^*)$, we rewrite (28) equivalently as

$$\begin{aligned} f^{\star} = \min_{\tilde{X}, \tilde{x}} \; & \text{Tr } \tilde{W} \tilde{X} \\ \text{s.t.: } & \text{Tr } E_i \tilde{X} = 1, \quad i = n, \ldots, 2n-1, \\ & \tilde{X} = \tilde{x}\tilde{x}^*. \end{aligned} \tag{34}$$

where E_i is a matrix that is zero everywhere, except for the i-th diagonal element, which is one. The condition $\tilde{X} = \tilde{x}\tilde{x}^*$ is equivalent to (1) $\tilde{X} \succeq 0$ and (2) $\tilde{X}$ has rank one. Thus, (34) is rewritten by eliminating $\tilde{x}$ as

$$
\begin{aligned}
f^{\star} = \min_{\tilde{X}} \ & \operatorname{Tr} \tilde{W}\tilde{X} \\
\text{s.t.:} \ & \operatorname{Tr} E_i\tilde{X} = 1, \quad i = n, \ldots, 2n-1, \\
& \tilde{X} \succeq 0 \\
& \operatorname{rank}(\tilde{X}) = 1.
\end{aligned} \tag{35}
$$

Dropping the rank constraint, which is non-convex, we obtain the following SDP relaxation (see, e.g., [80]) of the primal problem:

$$
\begin{aligned}
s^{\star} = \min_{\tilde{X}} \quad & \operatorname{Tr} \tilde{W}\tilde{X} \\
\text{s.t.:} \ & \operatorname{Tr} E_i\tilde{X} = 1, i = n, \ldots, 2n-1, \\
& \tilde{X} \succeq 0
\end{aligned} \tag{36}
$$

which we can also rewrite as

$$
\begin{aligned}
s^{\star} = \min_{\tilde{X}} \ & \operatorname{Tr} \tilde{W}\tilde{X} \qquad \text{(SDP relaxation)} \\
\text{s.t.:} \ & \tilde{X}_{ii} = 1, i = n, \ldots, 2n-1, \\
& \tilde{X} \succeq 0
\end{aligned} \tag{37}
$$

where $\tilde{X}_{ii}$ denotes the i-th diagonal entry in $\tilde{X}$. Obviously, $s^{\star} \leq f^{\star}$, since the feasible set of (37) contains that of (35). One may then ask what is the relation between the Lagrangian dual and the SDP relaxation of problem (37): the answer is that the former is the dual of the latter hence, under constraint qualification, it holds that $s^{\star} = d^{\star}$, i.e., the SDP relaxation and the Lagrangian dual approach yield the *same* lower bound on $f^{\star}$. This is formalized in the following proposition.

Proposition 8. *The Lagrangian dual of problem* (37) *is problem* (31), *and vice-versa. Strong duality holds between these two problems, i.e.,* $d^{\star} = s^{\star}$. *Moreover, if the optimal solution* $\tilde{X}^{\star}$ *of* (37) *has rank one, then* $s^{\star} = f^{\star}$, *and hence* $d^{\star} = f^{\star}$.

Proof. The fact that the SDPs (37) and (31) are related by duality can be found in standard textbooks (e.g. [7, Example 5.13]); moreover, since these are convex programs, under constraint qualification, the duality gap is zero, i.e., $d^{\star} = s^{\star}$. To prove that $\operatorname{rank}(\tilde{X}^{\star}) = 1 \Rightarrow s^{\star} = d^{\star} = f^{\star}$, we observe that (i) $\operatorname{Tr}\tilde{W}\tilde{X}^{\star} \doteq s^{\star} \leq f^{\star}$ since (37) is a relaxation of (35). However, when $\operatorname{rank}(\tilde{X}^{\star}) = 1$, $\tilde{X}^{\star}$ is feasible for problem (37), hence, by optimality of $f^{\star}$, it holds (ii) $f^{\star} \leq f(\tilde{X}^{\star}) = \operatorname{Tr}\tilde{W}\tilde{X}^{\star}$. Combining (i) and (ii) we prove that, when $\operatorname{rank}(\tilde{X}^{\star}) = 1$, then $f^{\star} = s^{\star}$, which also implies $f^{\star} = d^{\star}$. □

To the best of our knowledge this is the first time in which this SDP relaxation has been proposed to solve PGO; in the context of SLAM, anther SDP relaxation has been proposed by Liu et al. [53], but it does not use the chordal distance and approximates the expression of the relative rotation measurements. For the rotation subproblem, SDP relaxations have been proposed in [32, 67, 72]. According to

Proposition 8, one advantage of the SDP relaxation approach is that we can a-posteriori check if the duality (or, in this case, the relaxation) gap is zero, from the optimal solution $\tilde{X}^\star$. Indeed, if one solves (37) and finds that the optimal $\tilde{X}^\star$ has rank one, then we actually solved (28), hence the relaxation gap is zero. Moreover, in this case, from spectral decomposition of $\tilde{X}^\star$ we can get a vector $\tilde{x}^\star$ such that $\tilde{X}^\star = (\tilde{x}^\star)(\tilde{x}^\star)^*$, and this vector is an optimal solution to the primal problem.

In the following section we derive similar a-posteriori conditions for the dual problem (31). These conditions enable the computation of a primal optimal solution. Moreover, they allow discussing the uniqueness of such solution. Furthermore, we prove that in special cases we can provide *a-priori* conditions that guarantee that the duality gap is zero.

4.3 Analysis of the Dual Problem

In this section we provide conditions under which the duality gap is zero. These conditions depend on the spectrum of $\tilde{W}(\lambda^\star)$, which arises from the solution of (31). We refer to $\tilde{W}(\lambda^\star)$ as the *penalized pose graph matrix*. A first proposition establishes that (31) attains an optimal solution.

Proposition 9. *The optimal value $d^\star$ in* (31) *is attained at a finite $\lambda^\star$. Moreover, the penalized pose graph matrix $\tilde{W}(\lambda^\star)$ has an eigenvalue in 0.*

Proof. Since $\tilde{W}(\lambda) \succeq 0$ implies that the diagonal entries are nonnegative, the feasible set of (31) is contained in the set $\{\lambda : \tilde{W}_{ii} - \lambda_i \geq 0,\ i = 1, \ldots, 2n-1\}$ (recall that $\tilde{W}_{ii}$ are reals according to Remark 2). On the other hand, $\lambda_l = 0_{2n-1}$ is feasible and all points in the set $\{\lambda : \lambda_i \geq 0$ yield an objective that is at least as good as the objective value at λ_l. Therefore, the problem is equivalent to $\max_\lambda \sum_i \lambda_i$ subject to the original constraint, plus a box constraint on $\lambda \in \{0 \leq \lambda_i \leq \tilde{W}_{ii},\ i = 1, \ldots, n\}$. Thus we maximize a linear function over a compact set, hence a finite optimal solution $\lambda^\star$ must be attained.

Now let us prove that $\tilde{W}(\lambda^\star)$ has an eigenvalue in zero. Assume by contradiction that $\tilde{W}(\lambda^\star) \succ 0$. From the Schur complement rule we know:

$$\tilde{W}(\lambda^\star) \succ 0 \Leftrightarrow \begin{cases} L \succ 0 \\ \tilde{Q}(\lambda^\star) - \tilde{S}^* L^{-1} \tilde{S} \succ 0 \end{cases} \tag{38}$$

The condition $L \succ 0$ is always satisfied for a connected graph, since $L = A^\top A$, and the anchored incidence matrix A, obtained by removing a node from the original incidence matrix, is always full-rank for connected graphs [68, Sect. 19.3]. Therefore, our assumption $\tilde{W}(\lambda^\star) \succ 0$ implies that

$$\tilde{Q}(\lambda^\star) - \tilde{S}^* L^{-1} \tilde{S} = \tilde{Q} - \tilde{S}^* L^{-1} \tilde{S} - \mathrm{diag}(\lambda^\star) \succ 0 \tag{39}$$

Now, let

$$\epsilon = \lambda_{\min}(\tilde{Q}(\lambda^\star) - \tilde{S}^* L^{-1}\tilde{S}) > 0.$$

which is positive by the assumption $\tilde{W}(\lambda^\star) \succ 0$. Consider $\lambda = \lambda^\star + \epsilon\mathbf{1}$, then

$$\tilde{Q}(\lambda) - \tilde{S}^* L^{-1}\tilde{S} = \tilde{Q}(\lambda) - \tilde{S}^* L^{-1}\tilde{S} - \epsilon I \succeq 0,$$

thus λ is dual feasible, and $\sum_i \lambda_i > \sum_i \lambda_i^\star$, which would contradict optimality of $\lambda^\star$. We thus proved that $\tilde{Q}(\lambda^\star)$ must have a zero eigenvalue. □

Proposition 10 (No Duality Gap). *If the zero eigenvalue of the penalized pose graph matrix* $\tilde{W}(\lambda^\star)$ *is* simple *then the duality gap is zero, i.e.,* $d^\star = f^\star$.

Proof. We have already observed in Proposition 8 that (37) is the dual problem of (31), therefore, we can interpret $\tilde{X}$ as a Lagrange multiplier for the constraint $\tilde{W}(\lambda) \succeq 0$. If we consider the optimal solutions $\tilde{X}^\star$ and $\lambda^\star$ of (37) and (31), respectively, the *complementary slackness* condition ensures that $\text{Tr}(\tilde{W}(\lambda^\star)\tilde{X}^\star) = 0$ (see [7, Example 5.13]). Let us parametrize $\tilde{X}^\star \succeq 0$ as

$$\tilde{X}^\star = \sum_{i=1}^{2n-1} \mu_i \tilde{v}_i \tilde{v}_i^*,$$

where $0 \le \mu_1 \le \mu_2 \le \cdots \le \mu_{2n-1}$ are the eigenvalues of $\tilde{X}$, and $\tilde{v}_i$ form a unitary set of eigenvectors. Then, the complementary slackness condition becomes

$$\begin{aligned}
\text{Tr}(\tilde{W}(\lambda^\star)\tilde{X}^\star) &= \text{Tr}\left(\tilde{W}(\lambda^\star) \sum_{i=1}^{2n-1} \mu_i \tilde{v}_i \tilde{v}_i^*\right) \\
&= \sum_{i=1}^{2n-1} \mu_i \text{Tr}\left(\tilde{W}(\lambda^\star)\tilde{v}_i\tilde{v}_i^*\right) \\
&= \sum_{i=1}^{2n-1} \mu_i\, \tilde{v}_i^*\, \tilde{W}(\lambda^\star)\tilde{v}_i \;\; = \;\; 0.
\end{aligned}$$

Since $\tilde{W}(\lambda^\star) \succeq 0$, the above quantity is zero at a nonzero $\tilde{X}^\star$ ($\tilde{X}^\star$ cannot be zero since it needs to satisfy the constraints $\tilde{X}_{ii} = 1$) if and only if $\mu_i = 0$ for $i = m+1, \ldots, 2n-1$, and $\tilde{W}(\lambda^\star)\tilde{v}_i = 0$ for $i = 1, \ldots, m$, where m is the multiplicity of 0 as an eigenvalue of $\tilde{W}(\lambda^\star)$. Hence $\tilde{X}^\star$ has the form

$$\tilde{X}^\star = \sum_{i=1}^{m} \mu_i \tilde{v}_i \tilde{v}_i^*, \tag{40}$$

where $\tilde{v}_i$, $i = 1, \ldots, m$, form a unitary basis of the null-space of $\tilde{W}(\lambda^\star)$. Now, if $m = 1$, then the solution $\tilde{X}^\star$ to problem (37) has rank one, but according to Proposition 8 this implies $d^\star = f^\star$, proving the claim. □

In the following we say that $\tilde{W}(\lambda^\star)$ satisfies the *single zero eigenvalue property* (SZEP) if its zero eigenvalue is simple. The following corollary provides a more explicit relation between the solution of the primal and the dual problem when $\tilde{W}(\lambda^\star)$ satisfies the SZEP.

Corollary 1 (SZEP $\Rightarrow \tilde{x}^\star \in \mathscr{X}(\lambda^\star)$). *If the zero eigenvalue of* $\tilde{W}(\lambda^\star)$ *is* simple, *then the set* $\mathscr{X}(\lambda^\star)$ *contains a primal optimal solution. Moreover, the primal optimal solution is unique, up to an arbitrary rotation.*

Proof. Let $\tilde{x}^\star$ be a primal optimal solution, and let $f^\star = (\tilde{x}^\star)^* \tilde{W}(\tilde{x}^\star)$ be the corresponding optimal value. From Proposition 10 we know that the SZEP implies that the duality gap is zero, i.e., $d^\star = f^\star$, hence

$$\sum_{i=1}^{n} \lambda_i^\star = (\tilde{x}^\star)^* \tilde{W}(\tilde{x}^\star). \tag{41}$$

Since $\tilde{x}^\star$ is a solution of the primal, it must be feasible, hence $|\tilde{x}_i^\star|^2 = 1$, $i = n, \ldots, 2n-1$. Therefore, the following equalities holds:

$$\sum_{i=1}^{n} \lambda_i^\star = \sum_{i=1}^{n} \lambda_i^\star |\tilde{x}_{n+i-1}^\star|^2 = (\tilde{x}^\star)^* \begin{bmatrix} 0 & 0 \\ 0 & \mathrm{diag}(\lambda^\star) \end{bmatrix} (\tilde{x}^\star) \tag{42}$$

Plugging (42) back into (41):

$$(\tilde{x}^\star)^* \left[\tilde{W} - \begin{bmatrix} 0 & 0 \\ 0 & \mathrm{diag}(\lambda^\star) \end{bmatrix} \right] (\tilde{x}^\star) = 0 \Leftrightarrow (\tilde{x}^\star)^* \tilde{W}(\lambda^\star)(\tilde{x}^\star) = 0 \tag{43}$$

which proves that $\tilde{x}^\star$ belongs to the null space of $\tilde{W}(\lambda^\star)$, which coincides with our definition of $\mathscr{X}(\lambda^\star)$ in (33), proving the first claim.

Let us prove the second claim. From the first claim we know that the SZEP implies that any primal optimal solution is in $\mathscr{X}(\lambda^\star)$. Moreover, when $\tilde{W}(\lambda^\star)$ has a single eigenvalue in zero, then $\mathscr{X}(\lambda^\star) = \mathrm{Kernel}(\tilde{W}(\lambda^\star))$ is 1-dimensional and can be written as $\mathscr{X}(\lambda^\star) = \{\tilde{\gamma} \tilde{x}^\star : \tilde{\gamma} \in \mathbb{C}\}$, or, using the polar form for $\tilde{\gamma}$:

$$\mathscr{X}(\lambda^\star) = \{\eta \mathrm{e}^{\mathrm{j}\varphi} \, \tilde{x}^\star : \eta, \varphi \in \mathbb{R}\} \tag{44}$$

From (44) it's easy to see that any $\eta \neq 1$ would alter the norm of $\tilde{x}^\star$, leading to a solution that it's not primal feasible. On the other hand, any $\mathrm{e}^{\mathrm{j}\varphi}\tilde{x}^\star$ belongs to $\mathscr{X}(\lambda^\star)$, and it's primal feasible ($|\mathrm{e}^{\mathrm{j}\varphi}\tilde{x}_i^\star| = |\tilde{x}_i^\star|$), hence by Theorem 1, any $\mathrm{e}^{\mathrm{j}\varphi}\tilde{x}^\star$ is primal optimal. We conclude the proof by noting that the multiplication by $\mathrm{e}^{\mathrm{j}\varphi}$ corresponds to a global rotation of the pose estimate $\tilde{x}^\star$: this can be easily understood from the relation (18). □

Proposition 10 provides an *a-posteriori* condition on the duality gap, that requires solving the dual problem; while Sect. 6 will show that this condition is very useful in practice, it is also interesting to devise a-priori conditions, that can be assessed from the pose graph matrix $\tilde{W}$, without solving the dual problem. A first step in this direction is the following proposition.

Proposition 11 (Strong Duality in Trees and Balanced Pose Graphs). *Strong duality holds for any balanced pose graph optimization problem, and for any pose graph whose underlying graph is a tree.*

Proof. Balanced pose graphs and trees have in common the fact that they attain $f^\star = 0$ (Propositions 1 and 2). By weak duality we know that $d^\star \leq 0$. However, $\lambda = 0_n$ is feasible (as $\tilde{W} \succeq 0$) and attains $d(\lambda) = 0$, hence $\lambda = 0_n$ is feasible and dual optimal, proving $d^\star = f^\star$. □

5 Algorithms

In this section we exploit the results presented so far to devise an algorithm to solve PGO. The idea is to solve the dual problem, and use $\lambda^\star$ and $\tilde{W}(\lambda^\star)$ to compute a solution for the primal PGO problem. We split the presentation into two sections: Sect. 5.1 discusses the case in which $\tilde{W}(\lambda^\star)$ satisfies the SZEP, while Sect. 5.2 discusses the case in which $\tilde{W}(\lambda^\star)$ has multiple eigenvalues in zero. This distinction is important as in the former case (which is the most common in practice) we can compute a provably optimal solution for PGO, while in the latter case our algorithm returns an estimate that is not necessarily optimal. Finally, in Sect. 5.3 we summarize our algorithm and present the corresponding pseudocode.

5.1 *Case 1: $\tilde{W}(\lambda^\star)$ Satisfies the SZEP*

According to Corollary 1, if $\tilde{W}(\lambda^\star)$ has a single zero eigenvalue, then the optimal solution of the primal problem $\tilde{x}^\star$ is in $\mathcal{X}(\lambda^\star)$, where $\mathcal{X}(\lambda^\star)$ coincides with the null space of $\tilde{W}(\lambda^\star)$, as per (33). Moreover, this null space is 1-dimensional, hence it can be written explicitly as:

$$\mathcal{X}(\lambda^\star) = \text{Kernel}(\tilde{W}(\lambda^\star)) = \{\tilde{v} \in \mathbb{C}^{2n-1} : \tilde{v} = \gamma \tilde{x}^\star\}, \tag{45}$$

which means that any vector in the null space is a scalar multiple of the primal optimal solution $\tilde{x}^\star$. This observation suggests a computational approach to compute $\tilde{x}^\star$. We can first compute an eigenvector $\tilde{v}$ corresponding to the single zero eigenvalue of $\tilde{W}(\lambda^\star)$ (this is a vector in the null space of $\tilde{W}(\lambda^\star)$). Then, since $\tilde{x}^\star$ must be primal feasible (i.e., $|\tilde{x}_n| = \ldots = |\tilde{x}_{2n-1}| = 1$), we compute a suitable scalar γ that makes $\frac{1}{\gamma}\tilde{v}$ primal feasible. This scalar is clearly $\gamma = |\tilde{v}_n| = \ldots = |\tilde{v}_{2n-1}|$ (we essentially need to normalize the norm of the last n entries of $\tilde{v}$). The existence of a suitable γ, and hence the fact that $|\tilde{v}_n| = \ldots = |\tilde{v}_{2n-1}| > 0$, is guaranteed by Corollary 1. As a result we get the optimal solution $\tilde{x}^\star = \frac{1}{\gamma}\tilde{v}$. The pseudocode of our approach is given in Algorithm 1, and further discussed in Sect. 5.3.

5.2 *Case 2: $\tilde{W}(\lambda^\star)$ does* **not** *Satisfy the SZEP*

Currently we are not able to compute a guaranteed optimal solution for PGO, when $\tilde{W}(\lambda^\star)$ has multiple eigenvalues in zero. Nevertheless, it is interesting to exploit the solution of the dual problem for finding a (possibly suboptimal) estimate, which can be used, for instance, as initial guess for an iterative technique.

Eigenvector Method One idea to compute a suboptimal solution from the dual problem is to follow the same approach of Sect. 5.1: we compute an eigenvector of $\tilde{W}(\lambda^\star)$, corresponding to one of the zero eigenvalues, and we normalize it to make it feasible. In this case, we are not guaranteed that $|\tilde{v}_n| = \ldots = |\tilde{v}_{2n-1}| > 0$ (as in the previous section), hence the normalization has to be done component-wise, for each of the last n entries of $\tilde{v}$. In the following, we consider an alternative approach, which we have seen to perform better in practice (see experiments in Sect. 6).

Null Space Method This approach is based on the insight of Theorem 1: if there is a primal feasible $\tilde{x} \in \mathscr{X}(\lambda^\star)$, then $\tilde{x}$ must be primal optimal. Therefore we look for a vector $\tilde{x} \in \mathscr{X}(\lambda^\star)$ that is "close" to the feasible set. According to (33), $\mathscr{X}(\lambda^\star)$ coincides with the null space of $\tilde{W}(\lambda^\star)$. Let us denote with $\tilde{V} \in \mathbb{C}^{(2n-1)\times q}$ a basis of the null space of $\tilde{W}(\lambda^\star)$, where q is the number of zero eigenvalues of $\tilde{W}(\lambda^\star)$.[4] Any vector $\tilde{x}$ in the null space of $\tilde{W}(\lambda^\star)$ can be written as $\tilde{x} = \tilde{V}\tilde{z}$, for some vector $\tilde{z} \in \mathbb{C}^q$. Therefore we propose to compute a possibly suboptimal estimate $\tilde{x} = \tilde{V}\tilde{z}^\star$, where $\tilde{z}^\star$ solves the following optimization problem:

$$\max_{\tilde{z}} \sum_{i=1}^{2n-1} \mathrm{real}(\tilde{V}_i\tilde{z}) + \mathrm{imag}(\tilde{V}_i\tilde{z}) \qquad (46)$$

$$\text{s.t.:} \quad |\tilde{V}_i\tilde{z}|^2 \leq 1, \qquad i = n, \ldots, 2n-1$$

where $\tilde{V}_i$ denotes the i-th row of $\tilde{V}$, and real(·) and imag(·) return the real and the imaginary part of a complex number, respectively. For an intuitive explanation of problem (46), we notice that the feasible set of the primal problem (28) is described by $|\tilde{x}_i|^2 = 1$, for $i = n, \ldots, 2n-1$. In problem (46) we relax the equality constraints

[4] $\tilde{V}$ can be computed from singular value decomposition of $\tilde{W}(\lambda^\star)$.

to convex inequality constraints $|\tilde{x}_i|^2 \leq 1$, for $i = n, \ldots, 2n-1$; these can be written as $|\tilde{V}_i \tilde{z}|^2 \leq 1$, recalling that we are searching in the null space of $\tilde{W}(\lambda^\star)$, which is spanned by $\tilde{V}\tilde{z}$. Then, the objective function in (46) encourages "large" elements $\tilde{V}_i \tilde{z}$, hence pushing the inequality $|\tilde{V}_i \tilde{z}|^2 \leq 1$ to be tight. While other metrics can force large entries $\tilde{V}_i \tilde{z}$, we preferred the linear metric (46) to preserve convexity.

Note that $\tilde{x} = \tilde{V}\tilde{z}^\star$, in general, is neither optimal nor feasible for our PGO problem (28), hence we need to normalize it to get a feasible estimate. The experimental section provides empirical evidence that, despite being heuristic in nature, this method performs well in practice, outperforming—among the others—the eigenvector method presented earlier in this section.

5.3 Pseudocode and Implementation Details

The pseudocode of our algorithm is given in Algorithm 1. The first step is to solve the dual problem, and check the a-posteriori condition of Proposition 10. If the SZEP is satisfied, then we can compute the optimal solution by scaling the eigenvector of $\tilde{W}(\lambda^\star)$ corresponding to the zero eigenvalue μ_1. This is the case described in Sect. 5.1 and is the most relevant in practice, since the vast majority of robotics problems falls in this case.

The "else" condition corresponds to the case in which $\tilde{W}(\lambda^\star)$ has multiple eigenvalue in zero. The pseudocode implements the null space approach of Sect. 5.2. The algorithm computes a basis for the null space of $\tilde{W}(\lambda^\star)$ and solves (46) to find a vector belonging to the null space (i.e., in the form $\tilde{x} = \tilde{V}\tilde{z}$) that is close to the feasible set. Since such vector is not guaranteed to be primal feasible (and it is not in general), the algorithm normalizes the last n entries of $\tilde{x}^\star = \tilde{V}\tilde{z}^\star$, so to satisfy the unit norm constraints in (28). Besides returning the estimate $\tilde{x}^\star$, the algorithm also provides an optimality certificate when $\tilde{W}(\lambda^\star)$ has a single eigenvalue in zero.

Algorithm 1 Solving PGO using Lagrangian duality

Input: complex PGO matrix $\tilde{W}$
Output: primal solution $\tilde{x}^\star$ and optimality certificate `isOpt`

solve the dual problem (31) and get $\lambda^\star$
if $\tilde{W}(\lambda^\star)$ has a single eigenvalue μ_1 at zero **then**
  compute the eigenvector $\tilde{v}$ of $\tilde{W}(\lambda^\star)$ corresponding to μ_1
  compute $\tilde{x}^\star = \frac{1}{\gamma}\tilde{v}$, where $\gamma = |\tilde{v}_j|$, for any $j \in \{n, \ldots, 2n-1\}$
  set `isOpt` = `true`
else
  compute a basis $\tilde{V}$ for the null space of $\tilde{W}(\lambda^\star)$ using SVD
  compute $\tilde{z}^\star$ by solving the convex problem (46)
  set $\tilde{x}^\star = \tilde{V}\tilde{z}^\star$ and normalize $|\tilde{x}_i|$ to 1, for all $i = n, \ldots, 2n-1$
  set `isOpt` = `unknown`
end if
return ($\tilde{x}^\star$, `isOpt`)

---

6 Numerical Analysis and Discussion

The objective of this section is four-fold. First, we validate our theoretical derivation, providing experimental evidence that supports the claims. Second, we show that the duality gap is zero in a vast amount or practical problems. Third, we confirm the effectiveness of Algorithm 1 to solve PGO. Fourth, we provide toy examples in which the duality gap is greater than zero, hoping that this can stimulate further investigation towards *a-priori* conditions that ensure zero duality gap.

Simulation Setup For each run we generate a random graph with $n = 10$ nodes, unless specified otherwise. We draw the position of each pose by a uniform distribution in a $10\,\text{m} \times 10\,\text{m}$ square. Similarly, ground truth node orientations are randomly selected in $(-\pi, +\pi]$. Then we create set of edges defining a spanning *path* of the graph (these are usually called *odometric edges*); moreover, we add further edges to the edge set, by connecting random pairs of nodes with probability $P_c = 0.1$ (these are usually called *loop closures*). From the randomly selected *true* poses, and for each edge (i,j) in the edge set, we generate the relative pose measurement using the following model:

$$\begin{aligned} \Delta_{ij} &= R_i^{\top}\left(p_j - p_i\right) + \epsilon_{\Delta}, \qquad && \epsilon_{\Delta} \sim N(0_2, \sigma_{\Delta}^2) \\ R_{ij} &= R_i^{\top}\, R_j\, R(\epsilon_R), && \epsilon_R \sim N(0, \sigma_R^2) \end{aligned} \tag{47}$$

where $\epsilon_{\Delta} \in \mathbb{R}^2$ and $\epsilon_R \in \mathbb{R}$ are zero-mean Normally distributed random variables, with standard deviation σ_{Δ} and σ_R, respectively, and $R(\epsilon_R)$ is a random planar rotation of an angle ϵ_R. Unless specified otherwise, all statistics are computed over 100 runs.

Spectrum of $\tilde{W}$ In Proposition 6, we showed that the complex anchored pose graph matrix $\tilde{W}$ has at most one eigenvalue in zero, and the zero eigenvalue only appears when the pose graph is balanced or is a tree.

Figure 4a reports the value of the smallest eigenvalue of $\tilde{W}$ (in log scale) for different σ_R, with fixed $\sigma_{\Delta} = 0\,\text{m}$. When also σ_R is zero, the pose graph is balanced, hence the smallest eigenvalue of $\tilde{W}$ is (numerically) zero. For increasing levels of noise, the smallest eigenvalue increases and stays away from zero. Similarly, Fig. 4b reports the value of the smallest observed eigenvalue of $\tilde{W}$ (in log scale) for different σ_{Δ}, with fixed $\sigma_R = 0\,\text{rad}$.

Duality Gap is Zero in Many Cases This section shows that for the levels of measurement noise of practical interest, the matrix $\tilde{W}(\lambda^{\star})$ satisfies the Single Zero Eigenvalue Property (SZEP), hence the duality gap is zero (Proposition 10). We consider the same measurement model of Eq. (47), and we analyze the percentage of tests in which $\tilde{W}(\lambda^{\star})$ satisfies the SZEP.

Figure 5a shows the percentage of the experiments in which the penalized pose graph matrix $\tilde{W}(\lambda^{\star})$ has a single zero eigenvalue, for different values of rotation noise σ_R, and keeping fixed the translation noise to $\sigma_{\Delta} = 0.1\,\text{m}$ (this is a typical

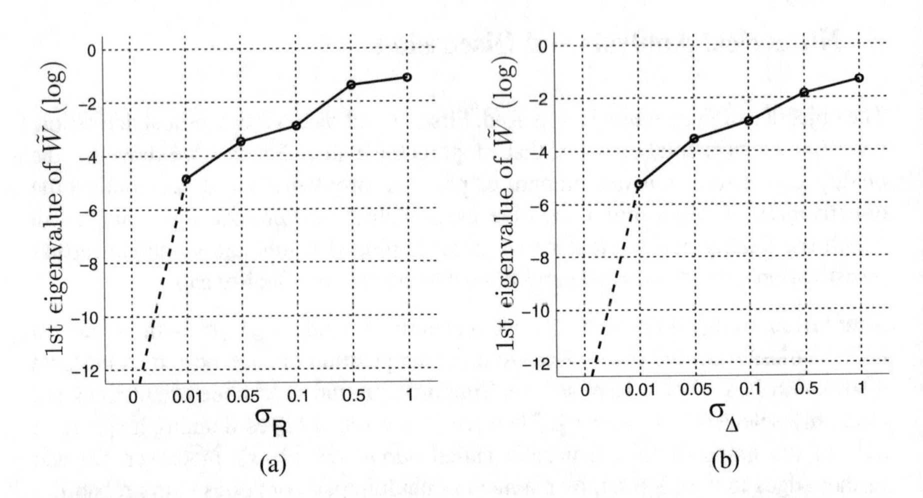

Fig. 4 Smallest eigenvalue of $\tilde{W}$ (in logarithmic scale) for different levels of (**a**) rotation noise (std: σ_R), and (**b**) translation noise (std: σ_Δ). The figure show the minimum observed value over 100 Monte Carlo runs, for non-tree graphs. The minimum eigenvalue is zero only if the graph is balanced

value in mobile robotics applications). For $\sigma_R \leq 0.5$ rad, $\tilde{W}(\lambda^\star)$ satisfies the SZEP in all tests. This means that in this range of operation, Algorithm 1 is guaranteed to compute a globally-optimal solution for PGO. For $\sigma_R = 1$ rad, the percentage of successful experiments drops, while still remaining larger than 90 %. Note that $\sigma_R = 1$ rad is a very large rotation noise (in robotics, typically $\sigma_R \leq 0.3$ rad [11]), and it is not far from the case in which rotation measurements are uninformative (uniformly distributed in $(-\pi, +\pi]$). To push our evaluation further we also tested this extreme case. When rotation noise is uniformly distributed in $(-\pi, +\pi]$, we obtained a percentage of successful tests (single zero eigenvalue) of 69 %, which confirms that the number of cases in which we can compute a globally optimal solution drops gracefully when increasing the noise levels.

Figure 5b shows the percentage of the experiments in which $\tilde{W}(\lambda^\star)$ has a single zero eigenvalue, for different values of translation noise σ_Δ, and keeping fixed the rotation noise to $\sigma_R = 0.1$ rad. Also in this case, for practical noise regimes, our approach can compute a global solution in all cases. The percentage of successful tests drops to 98 % when the translation noise has standard deviation 1m. We also tested the case of uniform noise on translation measurements. When we draw the measurement noise from a uniform distribution in $[-5, 5]^2$ (recall that the poses are deployed in a 10×10 square), the percentage of successful experiments is 68 %.

We also tested the percentage of experiments satisfying the SZEP for different levels of connectivity of the graph, controlled by the parameter P_c. We observed 100 % successful experiments, independently on the choice of P_c, for $\sigma_R = \sigma_\Delta = 0.1$ and $\sigma_R = \sigma_\Delta = 0.5$. A more interesting case if shown in Fig. 5c and corresponds to the case $\sigma_R = \sigma_\Delta = 1$. The SZEP is always satisfied for $P_c = 0$: this is natural

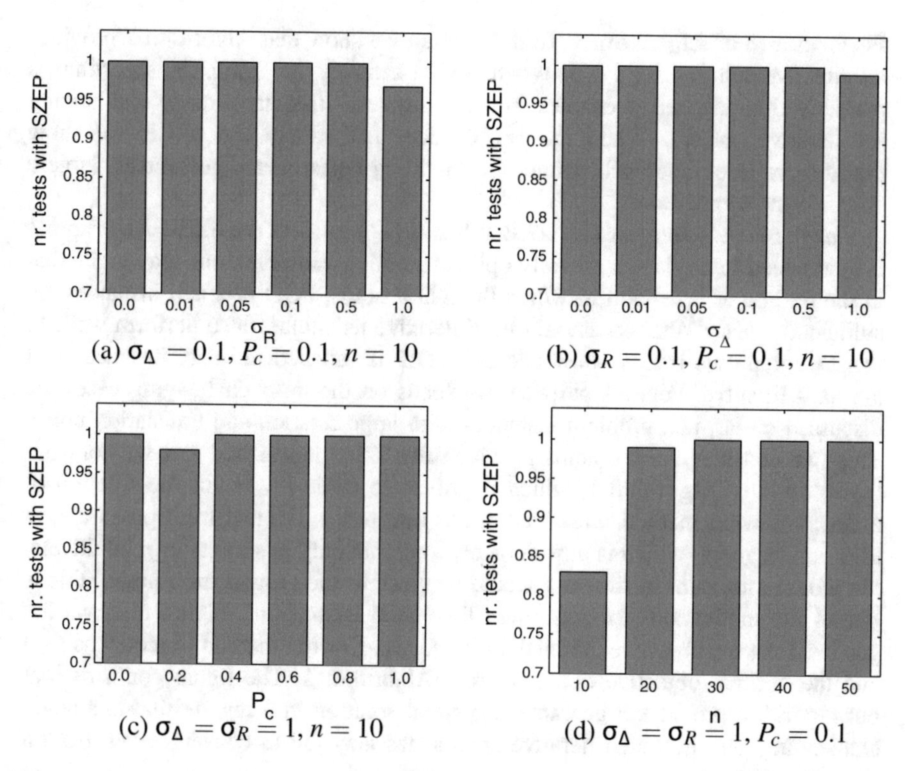

Fig. 5 Percentage of problems in which $\widetilde{W}(\lambda^{\star})$ satisfied the SZEP property, for different (**a**) rotation measurement noise σ_R, (**b**) translation measurement noises σ_Δ, (**c**) loop closure probability P_c, (**d**) number of nodes n

as $P_c = 0$ always produces trees, for which we are guaranteed to satisfy the SZEP (Proposition 11). For $P_c = 0.2$ the SZEP fails in few runs. Finally, increasing the connectivity beyond $P_c = 0.4$ re-establishes 100 % of successful tests. This would suggest that the connectivity level of the graph influences the duality gap, and better connected graphs have more changes to have zero duality gap.

Finally, we tested the percentage of experiments satisfying the SZEP for different number of nodes n. We tested the following number of nodes: $n = \{10, 20, 30, 40, 50\}$. For $\sigma_R = \sigma_\Delta = 0.1$ and $\sigma_R = \sigma_\Delta = 0.5$ the SZEP was satisfied in 100 % of the tests, and we omit the results for brevity. The more challenging case $\sigma_R = \sigma_\Delta = 1$ is shown in Fig. 5d. The percentage of successful tests increases for larger number of poses. We remark that current SDP solvers do not scale well to large problems, hence a Monte Carlo analysis over larger problems becomes prohibitive. We refer the reader to [12] for single-run experiments on larger PGO problems, which confirm that the duality gap is zero in problems arising in real-world robotics applications.

Performance of Algorithm 1 In this section we show that Algorithm 1 provides an effective solution for PGO. When $\tilde{W}(\lambda^\star)$ satisfies the SZEP, the algorithm is provably optimal, and it enables to solve problems that are already challenging for iterative solvers. When the $\tilde{W}(\lambda^\star)$ does *not* satisfy the SZEP, we show that the proposed approach, while not providing performance guarantees, largely outperforms competitors.

Case 1: $\tilde{W}(\lambda^\star)$ satisfies the SZEP. When $\tilde{W}(\lambda^\star)$ satisfies the SZEP, Algorithm 1 is guaranteed to produce a globally optimal solution. However, one may argue that in the regime of operation in which the SZEP holds, PGO problem instances are sufficiently "easy" that commonly used iterative techniques also perform well. In this paragraph we briefly show that the SZEP is satisfied in many instances that are hard to solve. For this purpose, we focus on the most challenging cases we discussed so far, i.e., problem instances with large rotation and translation noise. Then we consider the problems in which the SZEP is satisfied and we compare the solution of Algorithm 1, which is proven to attain $f^\star$, versus the solution of a Gauss-Newton method initialized at the *true* poses. Ground truth poses are an ideal initial guess (which is unfortunately available only in simulation): intuitively, the global minimum of the cost should be close to the ground truth poses (this is one of the motivations for maximum likelihood estimation). Figure 6 shows the gap between the objective attained by the Gauss-Newton method (denoted as f_{GN}) and the optimal objective obtained from Algorithm 1. The figure confirms that our algorithm provides a guaranteed optimal solution in a regime that is already challenging, and in which iterative approaches may fail to converge even from a good initialization.

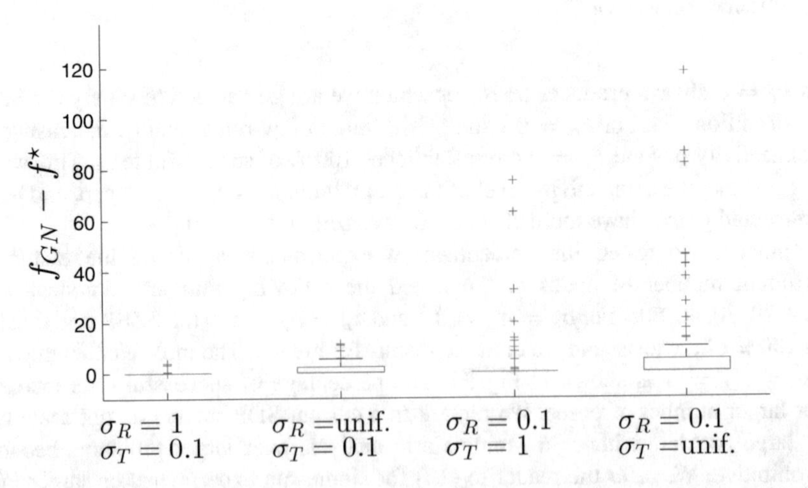

Fig. 6 Statistics on tests in which the SZEP is satisfied: the figure reports the gap between the optimal objective $f^\star$ attained by Algorithm 1 and the objective f_{GN} attained by a Gauss-Newton method initialized at the true poses. We simulate different combinations of noise (see x-axis labels), keeping fixed $n = 10$ and $P_c = 0.1$. The label "unif." denotes uniform noise for rotations (in $(-\pi, +\pi]$) or translations (in $[-5, +5]$)

Case 2: $\tilde{W}(\lambda^\star)$ *does* not *satisfy the SZEP.* In this case, Algorithm 1 computes an estimate, according to the *null space approach* proposed in Sect. 5.2; we denote this approach with the label NS. To evaluate the performance of the proposed approach, we considered 100 instances in which the SZEP was *not* satisfied and we compared our approach against the following methods: a Gauss-Newton method initialized at the ground truth poses (GN), the eigenvector method described at the beginning of Sect. 5.2 (Eig), and the SDP relaxation of Sect. 4.2 (SDP). For the SDP approach, we compute the solution $\tilde{X}^\star$ of the relaxed problem (37). If $\tilde{X}^\star$ has rank larger than 1, we find the closest rank-1 matrix $\tilde{X}_{\text{rank-1}}$ from singular value decomposition [28]. Then we factorize $\tilde{X}_{\text{rank-1}}$ as $\tilde{X}_{\text{rank-1}} = \tilde{x}\tilde{x}^*$ ($\tilde{x}$ can be computed via Cholesky factorization of $\tilde{X}_{\text{rank-1}}$ [71]). We report the results of our comparison in the first row of Fig. 7, where we show for different noise setups (sub-figures a1–a4), the cost of the estimate produced by the four approaches. The proposed null space approach (NS) largely outperforms the Eig and the SDP approaches, and has comparable performance with an "oracle" GN approach which knows the ground truth poses.

One may also compare the performance of the approaches NS, Eig, SDP after refining the corresponding estimates with a Gauss-Newton method, which tops off residual errors. The cost obtained by the different techniques, with the Gauss-Newton refinement, are shown in the second row of Fig. 7. For this case we also added one more initialization technique in the comparison: we consider an approach that solves for rotations first, using the eigenvalue method in [71], and

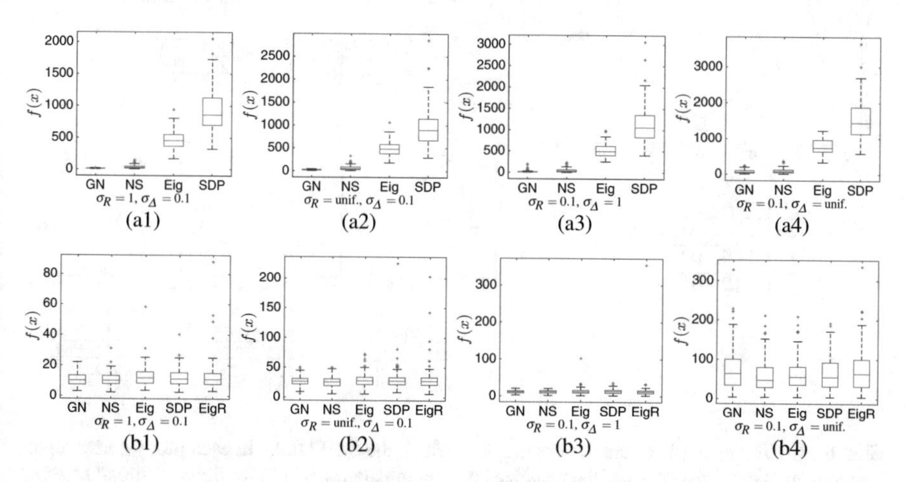

Fig. 7 Statistics on tests in which the SZEP is *not* satisfied: **(a1)**–**(a4)** Comparison of different PGO solvers for different levels of noise. The compared approaches are: a Gauss-Newton method initialized at the ground truth poses (GN), the proposed null space approach (NS), the eigenvector method (Eig), the SDP relaxation (SDP). **(b1)**–**(b4)** Comparison of the techniques GN, NS, Eig, SDP, refined with a Gauss-Newton method, and an alternative approach which solves for rotations first (EigR)

then applies the Gauss-Newton method from the rotation guess.[5] Figure 7b1–b4 show less differences (in average) among the techniques, as in most cases the Gauss-Newton refinement is able to converge starting from all the compared initializations. However, for the techniques `Eig`, `SDP`, and `EigR` we see many red sample points, which denote cases in which the error is larger than the 75th percentile; these are the cases in which the techniques failed to converge and produced a large cost. On the other hand, the proposed `NS` approach is less prone to converge to a bad minimum (fewer and lower red samples).

Chain Graph Counterexample and Discussion In this section we consider a simple graph topology: the *chain graph*. A chain graph is a graph with edges $(1,2), (2,3), \ldots, (n-1, n), (n, 1)$. Removing the last edge we obtain a tree (or, more specifically, a *path*), for which the SZEP is always satisfied. Therefore the question is: *is the SZEP always satisfied in PGO whose underlying graph is a chain?* The answer, unfortunately, is no. Figure 8a provides an example of a very simple chain graph with five nodes that fails to meet the SZEP property. The figure reports the 4 smallest eigenvalues of $\tilde{W}(\lambda^\star)$ $(\mu_1, \ldots, \mu_4)$, and the first two are numerically zero.

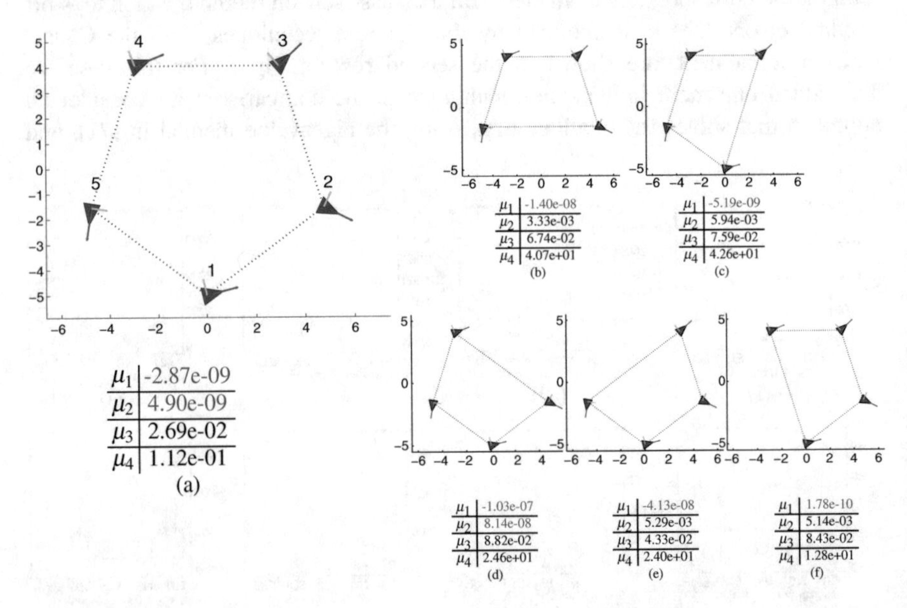

Fig. 8 **(a)** Toy example of chain pose graph in which the SZEP fails. In each plot we also report the four smallest eigenvalues of the penalized pose graph matrix $\tilde{W}(\lambda^\star)$ for the corresponding PGO problem. Removing a node from the original graph may change the duality properties of the graph. In **(b)**, **(c)**, **(d)**, **(e)**, **(f)** we remove nodes 1, 2, 3, 4, 5, respectively. Removing any node, except node 3, leads to a graph that satisfied the SZEP

[5]This was not included in the first row of Fig. 7 as it does not provide a guess for the positions of the nodes.

If the chain graph were balanced, Proposition 11 says that the SZEP needs to be satisfied. Therefore, one may argue that failure to meet the SZEP depends on the amount of error accumulated along the loop in the graph. Surprisingly, also this intuition fails. In Fig. 8b–f we show the pose graphs obtained by removing a single node from the pose graph in Fig. 8a. When removing a node, say k, we introduce a relative measurement between nodes $k-1$ and $k+1$, that is equal to the composition of the relative measurements associated to the edges $(k-1, k)$ and $(k, k+1)$ in the original graph. By constructions, the resulting graphs have the same accumulated errors (along each loop) as the original graph. However, interestingly, they do not necessarily share the same duality properties of the original graph. The graphs obtained by removing nodes $1, 2, 4, 5$ (shown in figures b, c, e, f, respectively), in fact, satisfy the SZEP. On the other hand, the graph in Fig. 8c still has 2 eigenvalues in zero. The data to reproduce these toy examples are reported in section "Numerical Data for the Toy Examples in Sect. 6" in Appendix.

We conclude with a test showing that the SZEP is not only dictated by the underlying rotation subproblem but also depends heavily on the translation part of the optimization problem. To show this we consider variations of the PGO problem in Fig. 8a, in which we "scale" all translation measurements by a constant factor. When the scale factor is smaller than one we obtain a PGO problem in which nodes are closer to each other; for scale > 1 we obtain larger inter-nodal measurements; the scale equal to 1 coincides with the original problem. Figure 9 shows the second eigenvalue of $\tilde{W}(\lambda^\star)$ for different scaling of the original graphs. Scaling down the measurements in the graph of Fig. 8a can re-establish the SZEP. Interestingly, this is in agreement with the convergence analysis of [10], which shows that the basin of convergence becomes larger when scaling down the inter-nodal distances.

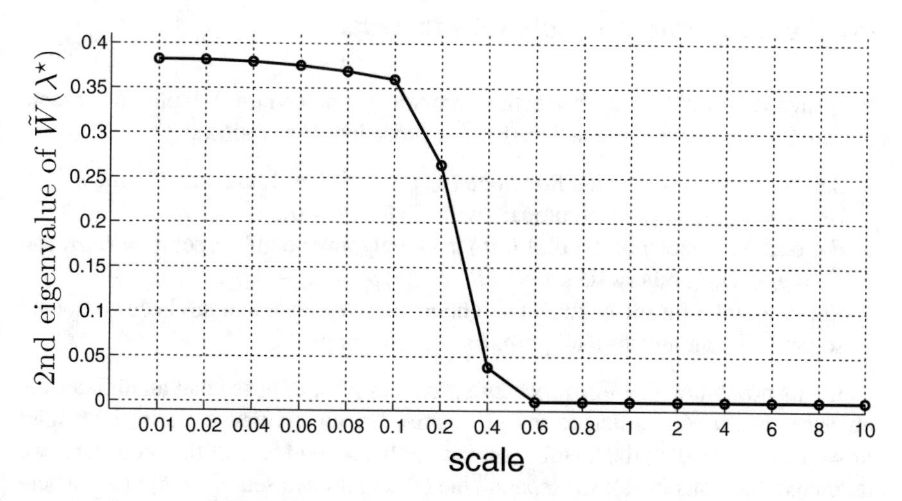

Fig. 9 Second eigenvalue of the matrix $\tilde{W}(\lambda^\star)$ for different variations of the toy graph of Fig. 8a. Each variation is obtained by scaling the translation measurements of the original graph by the amount specified on the x-axis of this figure. When the scale of the measurement is ≤ 0.4 the second eigenvalue of $\tilde{W}(\lambda^\star)$ is larger than zero, hence the SZEP is satisfied

7 Conclusion

We show that the application of Lagrangian duality in PGO provides an appealing approach to compute a globally optimal solution. More specifically, we propose four contributions. First, we rephrase PGO as a problem in complex variables. This allows drawing connection with the recent literature on *unit gain* graphs, and enables results on the spectrum of the pose graph matrix. Second, we formulate the Lagrangian dual problem and we analyze the relations between the primal and the dual solutions. Our key result proves that the duality gap is connected to the number of eigenvalues of the *penalized pose graph matrix*, which arises from the solution of the dual problem. In particular, if this matrix has a *single eigenvalue in zero* (SZEP), then (1) the duality gap is zero, (2) the primal PGO problem has a unique solution (up to an arbitrary roto-translation), and (3) the primal solution can be computed by *scaling* an eigenvector of the penalized pose graph matrix. The third contribution is an algorithm that returns a guaranteed optimal solution when the SZEP is satisfied, and (empirically) provides a very good estimate when the SZEP fails. Finally, we report numerical results, that show that (1) the SZEP holds for noise levels of practical robotics applications, (2) the proposed algorithm outperforms several existing approaches, (3) the satisfaction of the SZEP depends on multiple factors, including graph connectivity, number of poses, and measurement noise.

Appendix

Proof of Proposition 1: Zero Cost in Trees

We prove Proposition 1 by inspection, providing a procedure to build an estimate that annihilates every summand in (8). The procedure is as follows:

1. Select a root node, say the first node (p_i, r_i), with $i = 1$, and set it to the origin, i.e., $p_i = 0_2$, $r_i = [1\ 0]^\top$ (compare with (7) for $\theta_i = 0$);
2. For each neighbor j of the root i, if j is an outgoing neighbor, set $r_j = R_{ij} r_i$, and $p_j = p_i + D_{ij} r_i$, otherwise set $r_j = R_{ji}^\top r_i$, and $p_j = p_i + D_{ji} r_j$;
3. Repeat point 2 for the unknown neighbors of every node that has been computed so far, and continue until all poses have been computed.

Let us now show that this procedure produces a set of poses that annihilates the objective in (8). According to the procedure, we set the first node to the origin: $p_1 = 0_2$, $r_1 = [1\ 0]^\top$; then, before moving to the second step of the procedure, we rearrange the terms in (8): we separate the edges into two sets $\mathcal{E} = \mathcal{E}_1 \cup \bar{\mathcal{E}}_1$, where $\mathcal{E}_1$ is the set of edges incident on node 1 (the root), and $\bar{\mathcal{E}}_1$ are the remaining edges. Then the cost can be written as:

$$f(p,r) = \sum_{(i,j)\in\mathscr{E}_1} \|p_j - p_i - D_{ij}r_i\|_2^2 + \|r_j - R_{ij}r_i\|_2^2 +$$
$$+ \sum_{(i,j)\in\bar{\mathscr{E}}_1} \|p_j - p_i - D_{ij}r_i\|_2^2 + \|r_j - R_{ij}r_i\|_2^2 \tag{48}$$

We can further split the set $\mathscr{E}_1$ into edges that have node 1 as a tail (i.e., edges in the form $(1,j)$) and edges that have node 1 as head (i.e., $(j,1)$):

$$f(p,r) = \sum_{(1,j),j\in\mathscr{N}_1^{\text{out}}} \|p_j - p_1 - D_{1j}r_1\|_2^2 + \|r_j - R_{1j}r_1\|_2^2 +$$
$$+ \sum_{(j,1),j\in\mathscr{N}_1^{\text{in}}} \|p_1 - p_j - D_{j1}r_j\|_2^2 + \|r_1 - R_{j1}r_j\|_2^2 +$$
$$+ \sum_{(i,j)\in\bar{\mathscr{E}}_1} \|p_j - p_i - D_{ij}r_i\|_2^2 + \|r_j - R_{ij}r_i\|_2^2 \tag{49}$$

Now, we set each node j in the first two summands as prescribed in step 2 of the procedure. By inspection one can verify that this choice annihilates the first two summands and the cost becomes:

$$f(p,r) = \sum_{(i,j)\in\bar{\mathscr{E}}_1} \|p_j - p_i - D_{ij}r_i\|_2^2 + \|r_j - R_{ij}r_i\|_2^2 \tag{50}$$

Now we select a node k that has been computed at the previous step, but has some neighbor that is still unknown. As done previously, we split the set $\bar{\mathscr{E}}_1$ into two disjoint subsets: $\bar{\mathscr{E}}_1 = \mathscr{E}_k \cup \bar{\mathscr{E}}_k$, where the set $\mathscr{E}_k$ contains the edges in $\bar{\mathscr{E}}_1$ that are incident on k, and $\bar{\mathscr{E}}_k$ contains the remaining edges:

$$f(p,r) = \sum_{\{(k,j),j\in\mathscr{N}_k^{\text{out}}\}\cap\bar{\mathscr{E}}_1} \|p_j - p_k - D_{kj}r_k\|_2^2 + \|r_j - R_{kj}r_k\|_2^2 +$$
$$+ \sum_{\{(j,k),j\in\mathscr{N}_k^{\text{in}}\}\cap\bar{\mathscr{E}}_1} \|p_k - p_j - D_{jk}r_j\|_2^2 + \|r_k - R_{jk}r_j\|_2^2 +$$
$$+ \sum_{(i,j)\in\bar{\mathscr{E}}_k} \|p_j - p_i - D_{ij}r_i\|_2^2 + \|r_j - R_{ij}r_i\|_2^2 \tag{51}$$

Again, setting neighbors j as prescribed in step 2 of the procedure, annihilates the first two summands in (51). Repeating the same reasoning for all nodes that have been computed, but still have unknown neighbors, we can easily show that all terms in (51) become zero (the assumption of graph connectivity ensures that we can reach all nodes), proving the claim.

Proof of Proposition 2: Zero Cost in Balanced Graphs

Similarly to section "Proof of Proposition 1: Zero Cost in Trees" in this Appendix, we prove Proposition 2 by showing that in balanced graphs one can always build a solution that attains zero cost.

For the assumption of connectivity, we can find a spanning tree $\mathscr{T}$ of the graph, and split the terms in the cost function accordingly:

$$f(p,r) = \sum_{(i,j)\in\mathscr{T}} \|p_j - p_i - D_{ij}r_i\|_2^2 + \|r_j - R_{ij}r_i\|_2^2 + \\ + \sum_{(i,j)\in\bar{\mathscr{T}}} \|p_j - p_i - D_{ij}r_i\|_2^2 + \|r_j - R_{ij}r_i\|_2^2 \tag{52}$$

where $\bar{\mathscr{T}} \doteq \mathscr{E} \setminus \mathscr{T}$ are the *chords* of the graph w.r.t. $\mathscr{T}$.

Then, using the procedure in section "Proof of Proposition 1: Zero Cost in Trees" in this Appendix we construct a solution $\{r_i^\star, p_i^\star\}$ that attains zero cost for the measurements in the spanning tree $\mathscr{T}$. Therefore, our claim only requires to demonstrate that the solution built from the spanning tree also annihilates the terms in $\bar{\mathscr{T}}$:

$$f(p^\star, r^\star) = \sum_{(i,j)\in\bar{\mathscr{T}}} \|p_j^\star - p_i^\star - D_{ij}r_i^\star\|_2^2 + \|r_j^\star - R_{ij}r_i^\star\|_2^2 \tag{53}$$

To prove the claim, we consider one of the chords in $\bar{\mathscr{T}}$ and we show that the cost at $\{r_i^\star, p_i^\star\}$ is zero. The cost associated to a chord $(i,j) \in \bar{\mathscr{T}}$ is:

$$\|p_j^\star - p_i^\star - D_{ij}r_i^\star\|_2^2 + \|r_j^\star - R_{ij}r_i^\star\|_2^2 \tag{54}$$

Now consider the unique path $\mathscr{P}_{ij}$ in the spanning tree $\mathscr{T}$ that connects i to j, and number the nodes along this path as $i, i+1, \ldots, j-1, j$.

Let us start by analyzing the second summand in (54), which corresponds to the rotation measurements. According to the procedure in section "Proof of Proposition 1: Zero Cost in Trees" in this Appendix to build the solution for $\mathscr{T}$, we propagate the estimate from the root of the tree. Then it is easy to see that:

$$r_j^\star = R_{j-1j} \cdots R_{i+1i+2} R_{ii+1} r_i^\star \tag{55}$$

where R_{ii+1} is the rotation associated to the edge $(i, i+1)$, or its transpose if the edge is in the form $(i+1, i)$ (i.e., it is traversed backwards along $\mathscr{P}_{ij}$). Now we notice that the assumption of balanced graph implies that the measurements compose to the identity along every cycle in the graph. Since the chord (i,j) and the path $\mathscr{P}_{ij}$ form a cycle in the graph, it holds:

$$R_{j-1j} \cdots R_{i+1i+2} R_{ii+1} = R_{ij} \tag{56}$$

Substituting (56) back into (55) we get:

$$r_j^\star = R_{ij} r_i^\star \tag{57}$$

which can be easily seen to annihilate the second summand in (54).

Now we only need to demonstrate that also the first summand in (54) is zero. The procedure in section "Proof of Proposition 1: Zero Cost in Trees" in this Appendix leads to the following estimate for the position of node j:

$$\begin{aligned} p_j^\star &= p_i^\star + D_{ii+1} r_i^\star + D_{i+1i+2} r_{i+1}^\star + \cdots + D_{j-1j} r_{j-1}^\star \\ &= p_i^\star + D_{ii+1} r_i^\star + D_{i+1i+2} R_{ii+1} r_i^\star + \cdots + D_{j-1j} R_{j-2j-1} \cdots R_{i+1i+2} R_{ii+1} r_i^\star \\ &= p_i^\star + \left(D_{ii+1} + D_{i+1i+2} R_{ii+1} + \cdots + D_{j-1j} R_{j-2j-1} \cdots R_{i+1i+2} R_{ii+1}\right) r_i^\star \end{aligned} \tag{58}$$

The assumption of balanced graph implies that position measurements compose to zero along every cycle, hence:

$$\begin{aligned} \Delta_{ij} = \Delta_{ii+1} &+ R_{ii+1}\Delta_{i+1i+2} + R_{i+1i+2} R_{ii+1} \Delta_{i+2i+3} + \cdots \\ &+ R_{j-2j-1} \cdots R_{i+1i+2} R_{ii+1} \Delta_{j-1j} \end{aligned} \tag{59}$$

or equivalently:

$$\begin{aligned} D_{ij} = D_{ii+1} &+ D_{i+1i+2} R_{ii+1} + \cdots \\ &+ D_{j-1j} R_{j-2j-1} \cdots R_{i+1i+2} R_{ii+1} \end{aligned} \tag{60}$$

Substituting (60) back into (58) we obtain:

$$p_j^\star = p_i^\star + D_{ij} r_i^\star$$

which annihilates the first summand in (54), concluding the proof.

Proof of Proposition 4: Properties of $\mathcal{W}$

Let us prove that $\mathcal{W}$ has (at least) two eigenvalues in zero. We already observed that the top-left block of $\mathcal{W}$ is $\bar{\mathcal{L}} = \mathcal{L} \otimes I_2$, where $\mathcal{L}$ is the Laplacian matrix of the graph underlying the PGO problem. The Laplacian $\mathcal{L}$ of a connected graph has a single eigenvalue in zero, and the corresponding eigenvector is 1_n (see, e.g., [18, Sects. 1.2–1.3]), i.e., $\mathcal{L} \cdot 1_n = 0$. Using this property, it is easy to show that the matrix $N \doteq [0_n^\top \; 1_n^\top]^\top \otimes I_2$ is in the nullspace of $\mathcal{W}$, i.e., $\mathcal{W} N = 0$. Since N has rank 2, this implies that the nullspace of $\mathcal{W}$ has at least dimension 2, which proves the first claim.

Let us now prove that the matrix $\mathscr{W}$ is composed by 2×2 blocks $[\mathscr{W}]_{ij}$, with $[\mathscr{W}]_{ij} \in \alpha SO(2)$, $\forall i,j = 1,\dots,2n$, and $[\mathscr{W}]_{ii} = \alpha_{ii} I_2$ with $\alpha_{ii} \geq 0$. We prove this by direct inspection of the blocks of $\mathscr{W}$. Given the structure of $\mathscr{W}$ in (14), the claim reduces to proving that the matrices $\bar{\mathscr{L}}$, $\bar{Q}$, and $\bar{A}^\top \bar{D}$ are composed by 2×2 blocks in $\alpha SO(2)$, and the diagonal blocks of $\bar{\mathscr{L}}$ and $\bar{Q}$ are multiples of the identity matrix. To this end, we start by observing that $\bar{\mathscr{L}} = \mathscr{L} \otimes I_2$, hence all blocks in $\bar{\mathscr{L}}$ are multiples of the 2×2 identity matrix, which also implies that they belong to $\alpha SO(2)$. Consider next the matrix $\bar{Q} \doteq \bar{D}^\top \bar{D} + \bar{U}^\top \bar{U}$. From the definition of $\bar{D}$ it follows that $\bar{D}^\top \bar{D}$ is zero everywhere, except the 2×2 diagonal blocks:

$$[\bar{D}^\top \bar{D}]_{ii} = \sum_{j \in \mathscr{N}_i^{\text{out}}} \|\Delta_{ij}\|_2^2 I_2, \qquad i = 1,\dots,n. \tag{61}$$

Similarly, from simple matrix manipulation we obtain the following block structure of $\bar{U}^\top \bar{U}$:

$$\begin{aligned} [\bar{U}^\top \bar{U}]_{ii} &= d_i I_2, & i &= 1,\dots,n; \\ [\bar{U}^\top \bar{U}]_{ij} &= -R_{ij}, & (i,j) &\in \mathscr{E}; \\ [\bar{U}^\top \bar{U}]_{ij} &= -R_{ji}^\top, & (j,i) &\in \mathscr{E}; \\ [\bar{U}^\top \bar{U}]_{ij} &= 0_{2\times 2}, & &\text{otherwise}. \end{aligned} \tag{62}$$

where d_i is the degree (number of neighbours) of node i. Combining (61) and (62) we get the following structure for $\bar{Q}$:

$$\begin{aligned} [\bar{Q}]_{ii} &= \beta_i I_2, & i &= 1,\dots,n; \\ [\bar{Q}]_{ij} &= -R_{ij}, & (i,j) &\in \mathscr{E}; \\ [\bar{Q}]_{ij} &= -R_{ji}^\top, & (j,i) &\in \mathscr{E}; \\ [\bar{Q}]_{ij} &= 0_{2\times 2}, & &\text{otherwise}. \end{aligned} \tag{63}$$

where we defined $\beta_i \doteq d_i + \sum_{j \in \mathscr{N}_i^{\text{out}}} \|\Delta_{ij}\|_2^2$. Clearly, $\bar{Q}$ has blocks in $\alpha SO(2)$ and the diagonal blocks are nonnegative multiples of I_2.

Now, it only remains to inspect the structure of $\bar{A}^\top \bar{D}$. The matrix $\bar{A}^\top \bar{D}$ has the following structure:

$$\begin{aligned} [\bar{A}^\top \bar{D}]_{ii} &= \textstyle\sum_{j \in \mathscr{N}_i^{\text{out}}} D_{ij}, & i &= 1,\dots,n; \\ [\bar{A}^\top \bar{D}]_{ij} &= -D_{ji}, & (j,i) &\in \mathscr{E}; \\ [\bar{A}^\top \bar{D}]_{ij} &= 0_{2\times 2}, & &\text{otherwise}. \end{aligned} \tag{64}$$

Note that $\sum_{j \in \mathscr{N}_i^{\text{out}}} D_{ij}$ is the sum of matrices in $\alpha SO(2)$, hence it also belongs to $\alpha SO(2)$. Therefore, also all blocks of $\bar{A}^\top \bar{D}$ are in $\alpha SO(2)$, thus concluding the proof.

Proof of Proposition 5: Cost in the Complex Domain

Let us prove the equivalence between the complex cost and its real counterpart, as stated in Proposition 5.

We first observe that the dot product between two 2-vectors $x_1, x_2 \in \mathbb{R}^2$, can be written in terms of their complex representation $\tilde{x}_1 \doteq x_1^\vee$, and $\tilde{x}_2 \doteq x_2^\vee$, as follows:

$$x_1^\top x_2 = \frac{\tilde{x}_1^* \tilde{x}_2 + \tilde{x}_1 \tilde{x}_2^*}{2} \tag{65}$$

Moreover, we know that the action of a matrix $Z \in \alpha SO(2)$ can be written as the product of complex numbers, see (18).

Combining (65) and (18) we get:

$$x_1^\top Z x_2 \sim \frac{\tilde{x}_1^* \ \tilde{z} \ \tilde{x}_2 + \tilde{x}_1 \ \tilde{z}^* \ \tilde{x}_2^*}{2} \tag{66}$$

where $\tilde{z} = Z^\vee$. Furthermore, when Z is multiple of the identity matrix, it easy to see that $z = Z^\vee$ is actually a real number, and Eq. (66) becomes:

$$x_1^\top Z x_1 \sim \tilde{x}_1^* \ z \ \tilde{x}_1 \tag{67}$$

With the machinery introduced so far, we are ready to rewrite the cost $x^\top W x$ in complex form. Since W is symmetric, the product becomes:

$$x^\top W x = \sum_{i=1}^{2n-1} \left[x_i^\top [W]_{ii} x_i + \sum_{j=i+1}^{2n-1} 2 \ x_i^\top [W]_{ij} x_j \right] \tag{68}$$

Using the fact that $[W]_{ii}$ is a multiple of the identity matrix, $\tilde{W}_{ii} \doteq [W]_{ii}^\vee \in \mathbb{R}$, and using (67) we conclude $x_i^\top [W]_{ii} x_i = \tilde{x}_i^* \tilde{W}_{ii} \tilde{x}_i$. Moreover, defining $\tilde{W}_{ij} \doteq [W]_{ij}^\vee$ (these will be complex numbers, in general), and using (66), Eq. (68) becomes:

$$\begin{aligned} x^\top W x &= \sum_{i=1}^{2n-1} \left[\tilde{x}_i^* \tilde{W}_{ii} \tilde{x}_i + \sum_{j=i+1}^{2n-1} (\tilde{x}_i^* \tilde{W}_{ij} \tilde{x}_j + \tilde{x}_i \tilde{W}_{ij}^* \tilde{x}_j^*) \right] \\ &= \sum_{i=1}^{2n-1} \left[\tilde{x}_i^* \tilde{W}_{ii} \tilde{x}_i + \sum_{j \neq i} \tilde{x}_i^* \tilde{W}_{ij} \tilde{x}_j \right] = \tilde{x}^* \tilde{W} \tilde{x} \end{aligned} \tag{69}$$

where we completed the lower triangular part of $\tilde{W}$ as $\tilde{W}_{ji} = \tilde{W}_{ij}^*$.

Proof of Proposition 6: Zero Eigenvalues in $\tilde{W}$

Let us denote with N_0 the number of zero eigenvalues of the pose graph matrix $\tilde{W}$. N_0 can be written in terms of the dimension of the matrix ($\tilde{W} \in \mathbb{C}^{(2n-1)\times(2n-1)}$) and the rank of the matrix:

$$N_0 = (2n-1) - \text{rank}(\tilde{W}) \tag{70}$$

Now, recalling the factorization of $\tilde{W}$ given in (25), we note that:

$$\text{rank}(\tilde{W}) = \text{rank}\left(\begin{bmatrix} A & \tilde{D} \\ 0 & \tilde{U} \end{bmatrix}\right) = \text{rank}(A) + \text{rank}(\tilde{U}) \tag{71}$$

where the second relation follows from the upper triangular structure of the matrix. Now, we know from [68, Sect. 19.3] that the anchored incidence matrix A, obtained by removing a row from the the incidence matrix of a connected graph, is full rank:

$$\text{rank}(A) = n - 1 \tag{72}$$

Therefore:

$$N_0 = n - \text{rank}(\tilde{U}) \tag{73}$$

Now, since we recognized that $\tilde{U}$ is the complex incidence matrix of a unit gain graph (Lemma 1), we can use the result of Lemma 2.3 in [61], which says that:

$$\text{rank}(\tilde{U}) = n - b, \tag{74}$$

where b is the number of connected components in the graph that are balanced. Since we are working on a connected graph (Assumption 1), b can be either one (balanced graph or tree), or zero otherwise. Using (73) and (74), we obtain $N_0 = b$, which implies that $N_0 = 1$ for balanced graphs or trees, or $N_0 = 0$, otherwise.

Proof of Proposition 7: Spectrum of Complex and Real Pose Graph Matrices

Recall that any Hermitian matrix has real eigenvalues, and possibly complex eigenvectors. Let $\mu \in \mathbb{R}$ be an eigenvalue of $\tilde{W}$, associated with an eigenvector $\tilde{v} \in \mathbb{C}^{2n-1}$, i.e.,

$$\tilde{W}\tilde{v} = \mu\tilde{v} \tag{75}$$

From Eq. (75) we have, for $i = 1, \ldots, 2n-1$,

$$\sum_{j=1}^{2n-1} \tilde{W}_{ij}\tilde{v}_j = \mu\tilde{v}_i \Leftrightarrow \sum_{j=1}^{2n-1} [W]_{ij} v_j = \mu v_i \tag{76}$$

where v_i is such that $v_i^\vee = \tilde{v}_i$. Since Eq. (76) holds for all $i = 1, \ldots, 2n-1$, it can be written in compact form as:

$$Wv = \mu v \tag{77}$$

hence v is an eigenvector of the real anchored pose graph matrix W, associated with the eigenvalue μ. This proves that any eigenvalue of $\tilde{W}$ is also an eigenvalue of W.

To prove that the eigenvalue μ is actually repeated twice in W, consider now Eq. (75) and multiply both members by the complex number $\mathrm{e}^{\mathrm{j}\frac{\pi}{2}}$:

$$\tilde{W}\tilde{v}\mathrm{e}^{\mathrm{j}\frac{\pi}{2}} = \mu\tilde{v}\mathrm{e}^{\mathrm{j}\frac{\pi}{2}} \tag{78}$$

For $i = 1, \ldots, 2n-1$, we have:

$$\sum_{j=1}^{2n-1} \tilde{W}_{ij}^* \tilde{v}_j \mathrm{e}^{\mathrm{j}\frac{\pi}{2}} = \mu\tilde{v}_i\mathrm{e}^{\mathrm{j}\frac{\pi}{2}} \Leftrightarrow \sum_{j=1}^{2n-1} [W]_{ij} w_j = \mu w_i \tag{79}$$

where w_i is such that $w_i^\vee = \tilde{v}_j\mathrm{e}^{\mathrm{j}\frac{\pi}{2}}$. Since Eq. (79) holds for all $i = 1, \ldots, 2n-1$, it can be written in compact form as:

$$Ww = \mu w \tag{80}$$

hence also w is an eigenvector of W associated with the eigenvalue μ.

Now it only remains to demonstrate that v and w are linearly independent. One can readily check that, if $\tilde{v}_i$ is in the form $\tilde{v}_i = \eta_i \mathrm{e}^{\mathrm{j}\theta_i}$, then

$$v_i = \eta_i \begin{bmatrix} \cos(\theta_i) \\ \sin(\theta_i) \end{bmatrix}. \tag{81}$$

Moreover, observing that $\tilde{v}_j\mathrm{e}^{\mathrm{j}\frac{\pi}{2}} = \eta_i\mathrm{e}^{\mathrm{j}(\theta_i+\pi/2)}$, then

$$w_i = \eta_i \begin{bmatrix} \cos(\theta_i + \pi/2) \\ \sin(\theta_i + \pi/2) \end{bmatrix} = \eta_i \begin{bmatrix} -\sin(\theta_i) \\ \cos(\theta_i) \end{bmatrix} \tag{82}$$

From (81) and (82) is it easy to see that $v^\top w = 0$, thus v, w are orthogonal, hence independent. To each eigenvalue μ of $\tilde{W}$ there thus correspond an identical eigenvalue of W, of geometric multiplicity at least two. Since $\tilde{W}$ has $2n-1$ eigenvalues and W has $2(2n-1)$ eigenvalues, we conclude that to each eigenvalue

μ of $\tilde{W}$ there correspond exactly two eigenvalues of W in μ. The previous proof also shows how the set of orthogonal eigenvectors of W is related to the set of eigenvectors of $\tilde{W}$.

Proof of Theorem 1: Primal-dual Optimal Pairs

We prove that, given $\lambda \in \mathbb{R}^n$, if an $\tilde{x}_\lambda \in \mathscr{X}(\lambda)$ is primal feasible, then x_λ is primal optimal; moreover, λ is dual optimal, and the duality gap is zero.

By weak duality we know that for any λ:

$$\mathscr{L}(x_\lambda, \lambda) \leq f^\star \tag{83}$$

However, if x_λ is primal feasible, by optimality of $f^\star$, it must also hold

$$f^\star \leq f(x_\lambda) \tag{84}$$

Now we observe that for a feasible x_λ, the terms in the Lagrangian associated to the constraints disappear and $\mathscr{L}(x_\lambda, \lambda) = f(x_\lambda)$. Using the latter equality and the inequalities (83) and (84) we get:

$$f^\star \leq f(x_\lambda) = \mathscr{L}(x_\lambda, \lambda) \leq f^\star \tag{85}$$

which implies $f(x_\lambda) = f^\star$, i.e., x_λ is primal optimal.

Further, we have that

$$d^\star \geq \min_x \mathscr{L}(x, \lambda) = \mathscr{L}(x_\lambda, \lambda) = f(x_\lambda) = f^\star,$$

which, combined with weak duality ($d^\star \leq f^\star$), implies that $d^\star = f^\star$ and that λ attains the dual optimal value.

Numerical Data for the Toy Examples in Sect. 6

Ground truth nodes poses, written as $x_i = [p_i^\top, \theta_i]$:

$$\begin{aligned}
x_1 &= [\ 0.0000\ \ -5.0000\ \ 0.2451\ ] \\
x_2 &= [\ 4.7553\ \ -1.5451\ \ -0.4496\ ] \\
x_3 &= [\ 2.9389\ \ 4.0451\ \ 0.7361\ ] \\
x_4 &= [\ -2.9389\ \ 4.0451\ \ 0.3699\ ] \\
x_5 &= [\ -4.7553\ \ -1.5451\ \ -1.7225\ ]
\end{aligned} \tag{86}$$

Relative measurements, for each edge (i,j), written as $(i,j) : [\Delta_{ij}^{\top}, \theta_{ij}]$:

$$\begin{aligned}
(1,2) &: [\ 4.6606 \quad 1.2177 \quad 2.8186\ ] \\
(2,3) &: [\ -4.4199 \quad 4.8043 \quad 0.1519\ ] \\
(3,4) &: [\ -4.1169 \quad 4.9322 \quad 0.5638\ ] \\
(4,5) &: [\ -3.6351 \quad -5.0908 \quad -0.5855\ ] \\
(5,1) &: [\ 3.4744 \quad 5.9425 \quad 2.5775\ ]
\end{aligned} \tag{87}$$

References

1. Aragues, R., Carlone, L., Calafiore, G., Sagues, C.: Multi-agent localization from noisy relative pose measurements. In: IEEE International Conference on Robotics and Automation (ICRA), Shanghai, pp. 364–369 (2011)
2. Arie-Nachimson, M., Kovalsky, S.Z., Kemelmacher-Shlizerman, I., Singer, A., Basri, R.: Global motion estimation from point matches. In: 3DIMPVT, 2012
3. Bandeira, A.S., Singer, A., Spielman, D.A.: A Cheeger inequality for the graph connection Laplacian. SIAM. J. Matrix Anal. Appl. **34**(4), 1611–1630 (2013)
4. Barooah, P., Hespanha, J.P.: Estimation on graphs from relative measurements. Control Syst. Mag. **27**(4), 57–74 (2007)
5. Biswas, P., Lian, T., Wang, T., Ye, Y.: Semidefinite programming based algorithms for sensor network localization. ACM Trans. Sensor Netw. **2**(2), 188–220 (2006)
6. Borra, D., Carli, R., Lovisari, E., Fagnani, F., Zampieri, S.: Autonomous calibration algorithms for planar networks of cameras. In Proceedings of the 2012 American Control Conference, pp. 5126–5131 (2012)
7. Boyd, S., Vandenberghe, L.: Convex Optimization. Cambridge University Press, Cambridge (2004)
8. Calafiore, G.C., Carlone, L., Wei, M.: Distributed optimization techniques for range localization in networked systems. In: IEEE Conference on Decision and Control, pp. 2221–2226 (2010)
9. Carlevaris-Bianco, N., Eustice, R.M.: Generic factor-based node marginalization and edge sparsification for pose-graph SLAM. In: IEEE International Conference on Robotics and Automation (ICRA), pp. 5728–5735 (2013)
10. Carlone, L.: Convergence analysis of pose graph optimization via Gauss-Newton methods. In: IEEE International Conference on Robotics and Automation (ICRA), pp. 965–972 (2013)
11. Carlone, L., Censi, A.: From angular manifolds to the integer lattice: guaranteed orientation estimation with application to pose graph optimization. IEEE Trans. Robot. **30**(2), 475–492 (2014)
12. Carlone, L., Dellaert, F.: Duality-based verification techniques for 2D SLAM. In: International Conference on Robotics and Automation (ICRA), pp. 4589–4596 (2014)
13. Carlone, L., Aragues, R., Castellanos, J.A., Bona, B.: A linear approximation for graph-based simultaneous localization and mapping. In: Robotics: Science and Systems (RSS) (2011)
14. Carlone, L., Aragues, R., Castellanos, J.A., Bona, B.: A fast and accurate approximation for planar pose graph optimization. Intl. J. Robot. Res. **33**(7), 965–987 (2014)
15. Carlone, L., Tron, R., Daniilidis, K., Dellaert, F.: Initialization techniques for 3D SLAM: a survey on rotation estimation and its use in pose graph optimization. In: IEEE International Conference on Robotics and Automation (ICRA) (2015)
16. Chirikjian, G.S.: Stochastic Models, Information Theory, and Lie Groups, Volume 2: Analytic Methods and Modern Applications (Applied and Numerical Harmonic Analysis). Birkhauser, Basel (2012)

17. Chiuso, A., Picci, G., Soatto, S.: Wide-sense estimation on the special orthogonal group. Commun. Inf. Syst. **8**, 185–200 (2008)
18. Chung, F.R.K.: Spectral Graph Theory. CBMS Regional Conference Series in Mathematics, vol. 92 American Mathematical Society, Providence, RI (1996)
19. Costa, J., Patwari, N., Hero, A.: Distributed weighted-multidimensional scaling for node localization in sensor networks. ACM Trans. Sensor Netw. **2**(1), 39–64 (2006)
20. Cucuringu, M., Lipman, Y., Singer, A.: Sensor network localization by eigenvector synchronization over the Euclidean group. ACM Trans. Sensor Netw. **8**(3), 19:1–19:42 (2012)
21. Cucuringu, M., Singer, A., Cowburn, D.: Eigenvector synchronization, graph rigidity and the molecule problem. Inf. Infer.: J. IMA **1**(1), 21–67 (2012)
22. Dellaert, F., Carlson, J., Ila, V., Ni, K., Thorpe, C.E.: Subgraph-preconditioned conjugate gradient for large scale slam. In: IEEE/RSJ International Conference on Intelligent Robots and Systems (IROS) (2010)
23. Doherty, L., Pister, K., El Ghaoui, L.: Convex position estimation in wireless sensor networks. In: IEEE INFOCOM, vol. 3, pp. 1655–1663 (2001)
24. Dubbelman, G., Browning, B.: Closed-form online pose-chain slam. In: IEEE International Conference on Robotics and Automation (ICRA) (2013)
25. Dubbelman, G., Esteban, I., Schutte, K.: Efficient trajectory bending with applications to loop closure. In: IEEE/RSJ International Conference on Intelligent Robots and Systems (IROS), pp. 1–7 (2010)
26. Dubbelman, G., Hansen, P., Browning, B., Dias, M.B.: Orientation only loop-closing with closed-form trajectory bending. In: IEEE International Conference on Robotics and Automation (ICRA) (2012)
27. Duckett, T., Marsland, S., Shapiro, J.: Fast, on-line learning of globally consistent maps. Auton. Robot. **12**(3), 287–300 (2002)
28. Eckart, C., Young, G.: The approximation of one matrix by another low rank. Psychometrika **1**, 211–218 (1936)
29. Eren, T., Goldenberg, O.K., Whiteley, W., Yang, Y.R.: Rigidity, computation, and randomization in network localization. In: INFOCOM 2004. Twenty-third Annual Joint Conference of the IEEE Computer and Communications Societies, vol. 4, pp. 2673–2684. IEEE, Hong Kong (2004)
30. Eren, T., Whiteley, W., Belhumeur, P.N.: Using angle of arrival (bearing) information in network localization. In: IEEE Conference on Decision and Control, pp. 4676–4681 (2006)
31. Eustice, R.M., Singh, H., Leonard, J.J., Walter, M.R.: Visually mapping the RMS Titanic: conservative covariance estimates for SLAM information filters. Int. J. Robot. Res. **25**(12), 1223–1242 (2006)
32. Fredriksson, J., Olsson, C.: Simultaneous multiple rotation averaging using Lagrangian duality. In: Asian Conference on Computer Vision (ACCV) (2012)
33. Frese, U., Larsson, P., Duckett, T.: A multilevel relaxation algorithm for simultaneous localisation and mapping. IEEE Trans. Robot. **21**(2), 196–207 (2005)
34. Govindu, V.M.: Combining two-view constraints for motion estimation. In: IEEE Conference on Computer Vision and Pattern Recognition (CVPR), pp. 218–225 (2001)
35. Govindu, V.M.: Lie-algebraic averaging for globally consistent motion estimation. In: IEEE Conference on Computer Vision and Pattern Recognition (CVPR) (2004)
36. Grisetti, G., Stachniss, C., Burgard, W.: Non-linear constraint network optimization for efficient map learning. Trans. Intell. Transport. Syst. **10**(3), 428–439 (2009)
37. Hartley, R., Trumpf, J., Dai, Y., Li, H.: Rotation averaging. IJCV **103**(3), 267–305 (2013)
38. Hatanaka, T., Fujita, M., Bullo, F.: Vision-based cooperative estimation via multi-agent optimization. In: IEEE Conference on Decision and Control (2010)
39. Hazewinkel, M. (ed.) Complex number. In: Encyclopedia of Mathematics. Springer, New York (2001)
40. Huang, S., Lai, Y., Frese, U., Dissanayake, G.: How far is SLAM from a linear least squares problem? In: IEEE/RSJ International Conference on Intelligent Robots and Systems (IROS), pp. 3011–3016 (2010)

41. Huang, S., Wang, H., Frese, U., Dissanayake, G.: On the number of local minima to the point feature based SLAM problem. In: IEEE International Conference on Robotics and Automation (ICRA), pp. 2074–2079 (2012)
42. Indelman, V., Nelson, E., Michael, N., Dellaert, F.: Multi-robot pose graph localization and data association from unknown initial relative poses via expectation maximization. In: IEEE International Conference on Robotics and Automation (ICRA) (2014)
43. Johannsson, H., Kaess, M., Fallon, M., Leonard, J.J.: Temporally scalable visual SLAM using a reduced pose graph. In: IEEE International Conference on Robotics and Automation (ICRA), pp. 54–61 (2013)
44. Kaess, M., Ranganathan, A., Dellaert, F.: iSAM: incremental smoothing and mapping. IEEE Trans. Robot. **24**(6), 1365–1378 (2008)
45. Kaess, M., Johannsson, H., Roberts, R., Ila, V., Leonard, J., Dellaert, F.: iSAM2: incremental smoothing and mapping using the Bayes tree. Int. J. Robot. Res. **31**, 217–236 (2012)
46. Kim, B., Kaess, M., Fletcher, L., Leonard, J., Bachrach, A., Roy, N., Teller, S.: Multiple relative pose graphs for robust cooperative mapping. In: IEEE International Conference on Robotics and Automation (ICRA), Anchorage, Alaska, May 2010, pp. 3185–3192
47. Knuth, J., Barooah, P.: Collaborative 3D localization of robots from relative pose measurements using gradient descent on manifolds. In: IEEE International Conference on Robotics and Automation (ICRA), pp. 1101–1106 (2012)
48. Knuth, J., Barooah, P.: Collaborative localization with heterogeneous inter-robot measurements by Riemannian optimization. In: IEEE International Conference on Robotics and Automation (ICRA) (2013)
49. Knuth, J., Barooah, P.: Error growth in position estimation from noisy relative pose measurements. Robot. Auton. Syst. **61**(3), 229–224 (2013)
50. Konolige, K.: Large-scale map-making. In: Proceedings of the 21th AAAI National Conference on AI, San Jose, CA (2004)
51. Kümmerle, R., Grisetti, G., Strasdat, H., Konolige, K., Burgard, W.: g2o: A general framework for graph optimization. In: Proceedings of the IEEE International Conference on Robotics and Automation (ICRA), Shanghai, May 2011
52. Levenberg, K.: A method for the solution of certain nonlinear problems in least squares. Quart. Appl. Math **2**(2), 164–168 (1944)
53. Liu, M., Huang, S., Dissanayake, G., Wang, H.: A convex optimization based approach for pose SLAM problems. In: IEEE/RSJ International Conference on Intelligent Robots and Systems (IROS), pp. 1898–1903 (2012)
54. Lu, F., Milios, E.: Globally consistent range scan alignment for environment mapping. Auton. Robots **4**, 333–349 (1997)
55. Mao, G., Fidan, B., Anderson, B.: Wireless sensor network localization techniques. Comput. Networks **51**(10), 2529–2553 (2007)
56. Martinec, D., Pajdla, T.: Robust rotation and translation estimation in multiview reconstruction. In: IEEE Conference on Computer Vision and Pattern Recognition (CVPR), pp. 1–8 (2007)
57. Olfati-Saber, R.: Swarms on sphere: a programmable swarm with synchronous behaviors like oscillator networks. In: IEEE Conference on Decision and Control, pp. 5060–5066 (2006)
58. Olson, E., Leonard, J., Teller, S.: Fast iterative alignment of pose graphs with poor initial estimates. In: IEEE International Conference on Robotics and Automation (ICRA), May 2006, pp. 2262–2269
59. Peters, J.R., Borra, D., Paden, B., Bullo, F.: Sensor network localization on the group of 3D displacements. SIAM J. Control Optim. (2014, submitted)
60. Piovan, G., Shames, I., Fidan, B., Bullo, F., Anderson, B.: On frame and orientation localization for relative sensing networks. Automatica **49**(1), 206–213 (2013)
61. Reff, N.: Spectral properties of complex unit graphs (2011). arXiv 1110.4554
62. Rosen, D.M., Kaess, M., Leonard, J.J.: An incremental trust-region method for robust online sparse least-squares estimation. In: IEEE International Conference on Robotics and Automation (ICRA), St. Paul, MN, May 2012, pp. 1262–1269

63. Rosen, D.M., Kaess, M., Leonard, J.J.: RISE: an incremental trust-region method for robust online sparse least-squares estimation. IEEE Trans. Robot. **30**(5), 1091–1108 (2014)
64. Russell, W.J., Klein, D.J., Hespanha, J.P.: Optimal estimation on the graph cycle space. IEEE Trans. Signal Process. **59**(6), 2834–2846 (2011)
65. Sarlette, A., Sepulchre, R.: Consensus optimization on manifolds. SIAM J. Control Optim. **48**(1), 56–76 (2009)
66. Saunderson, J., Parrilo, P.A., Willsky, A.: Semidefinite descriptions of the convex hull of rotation matrices (2014). arXiv preprint: http://arxiv.org/abs/1403.4914
67. Saunderson, J., Parrilo, P.A., Willsky, A.: Semidefinite relaxations for optimization problems over rotation matrices. In: IEEE Conference on Decision and Control, May 2014
68. Schrijver, A.: *Theory of Linear and Integer Programming*. Wiley, New York (1998)
69. Sharp, G.C., Lee, S.W., Wehe, D.K.: Multiview registration of 3D scenes by minimizing error between coordinate frames. IEEE Trans. Pattern Anal. Mach. Intell. **26**(8), 1037–1050 (2004)
70. Singer, A.: A remark on global positioning from local distances. Proc. Natl. Acad. Sci. **105**(28), 9507–9511 (2008)
71. Singer, A.: Angular synchronization by eigenvectors and semidefinite programming. Appl. Comput. Harmon. Anal. **30**, 20–36 (2010)
72. Singer, A., Shkolnisky, Y.: Three-dimensional structure determination from common lines in Cryo-EM by eigenvectors and semidefinite programming. SIAM J. Imag. Sci. **4**(2), 543–572 (2011)
73. Stanfield, R.: Statistical theory of DF finding. J. IEE **94**(5), 762–770 (1947)
74. Thunberg, J., Montijano, E., Hu, X.: Distributed attitude synchronization control. In: IEEE Conference on Decision and Control (2011)
75. Tron, R., Afsari, B., Vidal, R.: Intrinsic consensus on SO(3) with almost global convergence. In: IEEE Conference on Decision and Control (2012)
76. Tron, R., Afsari, B., Vidal, R.: Riemannian consensus for manifolds with bounded curvature. IEEE Trans. Autom. Control **58**(4), 921–934 (2012)
77. Tron, R., Carlone, L., Dellaert, F., Daniilidis, K.: Rigid components identification and rigidity enforcement in bearing-only localization using the graph cycle basis. In: American Control Conference (2015)
78. Wang, H., Hu, G., Huang, S., Dissanayake, G.: On the structure of nonlinearities in pose graph SLAM. In: Robotics: Science and Systems (RSS) (2012)
79. Wang, L., Singer, A.: Exact and stable recovery of rotations for robust synchronization. Inf. Infer.: J. IMA pp. 1–53 (2013) doi: 10.1093/imaiai/drn000
80. Zhang, S.: Quadratic maximization and semidefinite relaxation. Math. Programm. Ser. A **87**, 453–465 (2000)

State-Feedback Control of Positive Switching Systems with Markovian Jumps

Patrizio Colaneri, Paolo Bolzern, José C. Geromel, and Grace S. Deaecto

Abstract This chapter deals with positive linear systems in continuous-time affected by a switching signal representing a disturbance driven by a Markov chain. A state-feedback control law has to be designed in order to ensure mean stability and input–output $\mathscr{L}_\infty$-induced or $\mathscr{L}_1$-induced mean performance. The chapter is divided into two parts. In the first, the control action is based on the knowledge of both the state of the system and the sample path of the Markovian process (mode-dependent control). In the second, instead, only the state-variable is known (mode-independent control). In the mode-dependent case, as well as in the single-input mode-independent case, necessary and sufficient conditions for the existence of feasible feedback gains are provided based on linear programming tools, also yielding a full parametrization of feasible solutions. In the multi-input mode-independent case, sufficient conditions are worked out in terms of convex programming. Some numerical examples illustrate the theory.

Keywords Positive systems • Markov Jumps • Stabilization • Input-output performance

P. Colaneri (✉)
Politecnico di Milano, DEIB, IEIIT-CNR, Milano, Italy
e-mail: patrizio.colaneri@polimi.it

P. Bolzern
Politecnico di Milano, DEIB, Milano, Italy
e-mail: paolo.bolzern@polimi.it

J.C. Geromel
School of Electrical and Computer Engineering, UNICAMP, Brazil
e-mail: geromel@dsce.fee.unicamp.br

G.S. Deaecto
School of Mechanical Engineering, UNICAMP, Brazil
e-mail: grace@fem.unicamp.br

B. Goldengorin (ed.), *Optimization and Its Applications in Control and Data Sciences*, Springer Optimization and Its Applications 115,
DOI 10.1007/978-3-319-42056-1_6

1 Introduction

This chapter deals with stabilization and control of (continuous-time) positive Markov jump linear systems (PMJLS). The switching signal θ is a Markov process associated with a given transition rate matrix.

The class of positive systems in the deterministic setting has been widely studied in the past years. Relevant general textbooks are available, see [9, 11, 12], and more specific contributions dealing with Lyapunov functions and input–output norms can be found in [2, 6, 14–16]. As for the class of Markov Jump Linear Systems (MJLS), a wide corpus of results is available, see the textbooks [5, 7].

On the other hand, only a few papers on PMJLS (in continuous-time) are available up to now. To the best of the authors knowledge, the first contribution pointing out the usefulness of the linear programming (LP) approach to the study of PMJLS is [2]. More recently, in [4], various notions of stability and their relationships are studied, while results on stochastic stabilization are provided in [17]. An application to an epidemiological model can be found in [1]. A very recent survey on analysis and design of PMJLS is available in [3].

The chapter is divided into two parts. In the first, the attention is concentrated on mode-dependent state feedback laws $\mathbf{u}(t) = K_{\theta(t)}\mathbf{x}(t)$, whereas in the second the focus is on mode-independent state feedback laws $\mathbf{u}(t) = K\mathbf{x}(t)$. The addressed problems deal with mean stabilization, $\mathscr{L}_\infty$-induced optimal control and $\mathscr{L}_1$-induced optimal control with a deterministic disturbance $\mathbf{w}(t)$. Notably, the first two problems admit a complete parametrization of the mode-dependent feedback gains in terms of linear programming. The same is true for the third when the disturbance $\mathbf{w}(t)$ is scalar. The results in the mode-dependent case are inspired by similar results for time-invariant deterministic positive systems provided in [6] for $\mathscr{L}_\infty$-induced control and [8] for $\mathscr{L}_1$-induced control. The $\mathscr{L}_2$-induced optimal control design problem is treated in [16] for the deterministic case, for which diagonal positive solutions of the design constraints exist. This is not the case of PMJLS, so that this point constitutes a subject of future research. A rather complete exposition of the mode-dependent control for PMJLS can be found in [3].

The part concerning mode-independent control is inspired by [10]. Its originality for PMJLS stems from the fact that, in the single-input case, a complete parametrization of the state-feedback laws $\mathbf{u}(t) = K\mathbf{x}(t)$ can be worked out using standard LP tools. For multi-input systems, a sufficient condition for the existence of a feasible gain K is developed using convex programming.

1.1 Notation

The semiring of nonnegative real numbers is $\mathbb{R}_+$. The (i,j)-th entry of a matrix M will be denoted by $[M]_{ij}$ and The i-th entry of a vector $\mathbf{v}$ will be denoted by $[\mathbf{v}]_i$. A matrix M (in particular, a vector) with entries in $\mathbb{R}_+$ is a *nonnegative* matrix (or vector). The matrix (or vector) M is *positive* ($M > 0$) if nonnegative and nonzero, and *strictly positive* ($M \gg 0$) if all its entries are positive.

The symbol $\mathbf{1}_n$ denotes the n-dimensional vector with all entries equal to 1. The symbol $\mathbf{e}_k$ will denote the k-th column of the identity matrix (the size will be clear from the context).

The convex polytope of the nonnegative m-tuples of real numbers that sum up to 1 will be denoted by

$$\mathscr{P}_m := \left\{ \alpha \in \mathbb{R}_+^m : \sum_{i=1}^m [\alpha]_i = 1 \right\} = \left\{ \alpha \in \mathbb{R}_+^m : \mathbf{1}_m' \alpha = 1 \right\}.$$

A square matrix $A = [a_{ij}]$ is said to be *Metzler* if its off-diagonal entries are nonnegative, namely $a_{ij} \geq 0$ for every $i \neq j$. An $n \times n$ Metzler matrix A, with $n > 1$, is *reducible* if there exists a permutation matrix P such that

$$P'AP = \begin{bmatrix} A_{11} & A_{12} \\ 0 & A_{22} \end{bmatrix},$$

where A_{11} is a $k \times k$ matrix, $1 \leq k \leq n-1$. A Metzler matrix which is not reducible is called *irreducible*. For an irreducible $n \times n$ Metzler matrix A it is known that its eigenvalue with maximum real part is real with multiplicity 1, and is called the *Perron-Frobenius eigenvalue*. The corresponding eigenspace is generated by a strictly positive eigenvector in $\mathscr{P}_n$, called *Perron-Frobenius eigenvector*. For further details on positive systems the reader is referred to [9]. A symmetric matrix Q is negative definite if and only if all its eigenvalues are negative. A negative definite matrix is indicated by the expression $Q \prec 0$.

The symbol $\mathscr{G}_m$ indicates the set of $m \times m$ Metzler matrices A such that $A\mathbf{1}_m = 0$. Therefore matrices in $\mathscr{G}_m$ represent infinitesimal transition rate matrices associated with a continuous-time Markov chain.

The 1-norm of a matrix $M \in \mathbb{R}^{n \times m}$ is defined as $\|M\|_1 = \max_j \sum_{i=1}^n |[M]_{ij}|$. The ∞-norm of a matrix $M \in \mathbb{R}^{n \times m}$ is defined as $\|M\|_\infty = \max_i \sum_{j=1}^m |[M]_{ij}|$. For a positive matrix M, the absolute value is obviously irrelevant, and these norms can be computed as $\|M\|_1 = \max_j \mathbf{1}_n^\top M \mathbf{e}_j$ and $\|M\|_\infty = \max_i \mathbf{e}_i^\top M \mathbf{1}_m$, respectively.

If $\{M_i \in \mathbb{R}^{n \times m}, i = 1, 2, \ldots, N\}$ is a set of matrices, the symbol $\operatorname{col}_i\{M_i\}$ will denote the matrix in $\mathbb{R}^{nN \times m}$ obtained by orderly stacking all the matrices M_i in a single block column. Analogously, the symbol $\operatorname{row}_i\{M_i\}$ will denote the matrix in $\mathbb{R}^{m \times nN}$ obtained by orderly stacking all the matrices M_i in a single block row. Finally, the symbol $\operatorname{diag}_i\{M_i\}$ will denote the block diagonal matrix in $\mathbb{R}^{nN \times mN}$ obtained by orderly putting M_i on the diagonal. The range for the index i will be omitted, if clear from the context. The symbol $\otimes$ will be used to indicate the Kronecker product.

The set $\mathscr{L}_\infty$ includes all the (nonnegative) deterministic signals with finite ∞-norm, i.e. bounded on the time interval $[0, \infty)$. The set $\mathscr{L}_1$ includes all the (nonnegative) deterministic signals with finite 1-norm, i.e. integrable over the time interval $[0, \infty)$. The symbol $\delta(t)$ denotes the unit impulse.

The expectation of a stochastic variable v will be denoted as $E[v]$. The conditional expectation of v given the event $\mathscr{A}$ will be indicated by $E[v|\mathscr{A}]$. If $\mathscr{A}_1, \mathscr{A}_2, \ldots, \mathscr{A}_N$ are mutually exclusive events whose union covers the entire event space, then the well-known law of total expectation claims that $E[v] = \sum_{i=1}^{N} E[v|\mathscr{A}_i]\Pr[\mathscr{A}_i]$, where $\Pr[\mathscr{A}_i]$ denotes the probability of event $\mathscr{A}_i$.

2 Stability and Induced Norms of a PMJLS

This section is devoted to the analysis of mean stability (M-stability), $\mathscr{L}_\infty$–induced and $\mathscr{L}_1$–induced norms of a PMJLS. Define $\mathscr{N} = \{1, 2, \ldots, N\}$ and let $(\Omega, \mathscr{F}, P)$ be a complete probability space equipped with a right-continuous filtration $\{\mathscr{F}_t, t \in \mathbb{R}_+\}$. In the sequel, reference will be made to a time-homogeneous Markov process $\theta(t)$ adapted to $\{\mathscr{F}_t\}$ and taking values in $\mathscr{N}$. The PMJLS is described by

$$\dot{\mathbf{x}}(t) = F_{\theta(t)}\mathbf{x}(t) + B_{\theta(t)}\mathbf{w}(t) \tag{1}$$

$$\mathbf{z}(t) = L_{\theta(t)}\mathbf{x}(t) + D_{\theta(t)}\mathbf{w}(t) \tag{2}$$

where $\mathbf{x}$ is the n-dimensional state, $\mathbf{w}$ is a n_w-dimensional disturbance input and $\mathbf{z}$ is the n_z-dimensional performance output. The time evolution for $t \geq 0$ of $\theta(t) \in \mathscr{N} = \{1, 2, \cdots, N\}$ is completely characterized by its initial probability distribution and the constant transition rates λ_{ij} from mode i to mode $j \neq i$. More precisely, define the transition probabilities $\Pr\{\theta(t+h) = j|\theta(t) = i\} = \lambda_{ij}h + o(h), i \neq j$, where $h > 0$, and $\lambda_{ij} \geq 0$ is the transition rate from mode i at time t to mode j at time $t + h$. Letting

$$\lambda_{ii} = -\sum_{j=1, j\neq i}^{N} \lambda_{ij}$$

and defining $\Lambda = [\lambda_{ij}]$, the matrix $\Lambda \in \mathscr{G}_N$ is called the transition rate matrix (or infinitesimal generator) of the Markov process. From now on we assume that the matrix Λ is irreducible. Letting $\boldsymbol{\pi}(t) \in \mathscr{P}_N$ represent the probability distribution of $\theta(t)$, this assumption implies that the stationary probability distribution $\bar{\boldsymbol{\pi}}$ is the unique solution in $\mathscr{P}_N$ of the equation $\Lambda^\top \bar{\boldsymbol{\pi}} = 0$, i.e. the equilibrium point of the Kolmogorov differential equation

$$\dot{\boldsymbol{\pi}}(t) = \Lambda^\top \boldsymbol{\pi}(t) \tag{3}$$

If $\boldsymbol{\pi}(0) = \bar{\boldsymbol{\pi}}$, the process $\theta(t)$ is stationary and ergodic.

The matrices F_i appearing in the model are $n \times n$ Metzler matrices, while B_i, L_i, D_i are nonnegative matrices for all $i \in \mathscr{N}$. It is assumed that the disturbance signal $\mathbf{w}$ is nonnegative for $t \geq 0$. In view of these assumptions, if the initial state at time

$t = 0$ is nonnegative, the state vector $\mathbf{x}$ remains in the positive orthant and the output signal $\mathbf{z}$ is nonnegative as well, for all $t \geq 0$.

Our analysis will concentrate on the $\mathscr{L}_\infty$–induced and $\mathscr{L}_1$–induced norms of system (1), (2) under the standing assumption that the input $\mathbf{w}$ is a *deterministic* disturbance. However, for such a study, it will be useful to assume in passing that the input $\mathbf{w}$ is a stochastic signal adapted to the filtration $\mathscr{F}_t$, i.e. it may depend on the current and past values of $\theta(t)$. We therefore define

$$\mathbf{m}_i(t) := E[\mathbf{x}(t)|\theta(t) = i][\boldsymbol{\pi}(t)]_i, \quad \mathbf{m} = \underset{i}{\mathrm{col}}\{\mathbf{m}_i\}$$

$$\mathbf{v}_i(t) := E[\mathbf{w}(t)|\theta(t) = i][\boldsymbol{\pi}(t)]_i, \quad \mathbf{v} = \underset{i}{\mathrm{col}}\{\mathbf{v}_i\}$$

Observing that $E[\mathbf{x}(t)] = \sum_{i=1}^{N} \mathbf{m}_i(t)$ and $E[\mathbf{w}(t)] = \sum_{i=1}^{N} \mathbf{v}_i(t)$, an elementary computation shows that the PMJLS can be given a representation in the mean values in terms of the following deterministic system of order nN:

$$\dot{\mathbf{m}}(t) = \tilde{F}\mathbf{m}(t) + \breve{B}\mathbf{v}(t) \tag{4}$$

$$E[\mathbf{z}(t)] = \tilde{L}\tilde{\mathbf{m}}(t) + \breve{D}\mathbf{v}(t) \tag{5}$$

where

$$\tilde{F} = \underset{i}{\mathrm{diag}}\{F_i\} + \Lambda^\top \otimes I_n \tag{6}$$

$$\breve{B} = \underset{i}{\mathrm{diag}}\{B_i\} \tag{7}$$

$$\tilde{L} = \underset{i}{\mathrm{row}}\{L_i\} \tag{8}$$

$$\breve{D} = \underset{i}{\mathrm{row}}\{D_i\} \tag{9}$$

It is important to note that $\tilde{F}$ is a Metzler matrix, while $\breve{B}$, $\tilde{L}$ and $\breve{D}$ are nonnegative matrices. System (4), (5) along with (6)–(9) will be called stochastic-input mean (SIM) system and the associated transfer function

$$\breve{G}(s) = \tilde{L}(sI - \tilde{F})^{-1}\breve{B} + \breve{D} \tag{10}$$

will be referred to as stochastic-input mean (SIM) transfer function.

As said, we are interested in the case where the input $\mathbf{w}$ is deterministic. In such a case, it results that

$$\mathbf{v}(t) = \boldsymbol{\pi}(t) \otimes \mathbf{w}(t) \tag{11}$$

The dynamical system from $\mathbf{w}$ to $E[\mathbf{z}]$ can be easily written as follows:

$$\dot{\mathbf{m}}(t) = \tilde{F}\mathbf{m}(t) + \tilde{B}(t)\mathbf{w}(t) \tag{12}$$

$$E[\mathbf{z}(t)] = \tilde{L}\mathbf{m}(t) + \tilde{D}(t)\mathbf{w}(t) \tag{13}$$

where

$$\tilde{B}(t) = \operatorname*{col}_i\{B_i[\pi(t)]_i\} \tag{14}$$

$$\tilde{D}(t) = \sum_{i=1}^{N} D_i[\pi(t)]_i \tag{15}$$

with $\pi(t)$ being the solution of (3) under a given initial condition $\pi(0)$. Notice that system (12), (13) is time-varying, and becomes time-invariant if $\pi(0) = \bar{\pi}$, so that $\pi(t) = \bar{\pi}$, $t \geq 0$. Such a case corresponds to taking the expectations $E[\mathbf{x}(t)|\theta(t) = i]$ and $E[\mathbf{z}(t)]$ with respect to the stationary probability distribution $\bar{\pi}$. Therefore, we can define the deterministic-input mean (DIM) system associated to the given PMJLS as

$$\dot{\mathbf{m}}(t) = \tilde{F}\mathbf{m}(t) + \tilde{B}\mathbf{w}(t) \tag{16}$$

$$E[\mathbf{z}(t)] = \tilde{L}\mathbf{m}(t) + \tilde{D}\mathbf{w}(t) \tag{17}$$

with

$$\tilde{B} = \operatorname*{col}_i\{B_i[\bar{\pi}]_i\} \tag{18}$$

$$\tilde{D} = \sum_{i=1}^{N} D_i[\bar{\pi}]_i \tag{19}$$

Note that $\tilde{B}$ and $\tilde{D}$ are nonnegative matrices. The deterministic-input mean (DIM) transfer function is then defined as

$$\tilde{G}(s) = \tilde{L}(sI - \tilde{F})^{-1}\tilde{B} + \tilde{D} \tag{20}$$

and the following relation between the DIM and SIM mean transfer functions holds:

$$\tilde{G}(s) = \breve{G}(s)\,(\bar{\pi} \otimes I_{n_w})$$

The matrix $\tilde{F}$, along with the DIM and SIM transfer functions, play a fundamental role in the characterization of M-stability of a PMJLS and the computation of its $\mathscr{L}_\infty$-induced and $\mathscr{L}_1$-induced input–output norms. This is discussed in the next subsections.

2.1 M-Stability

System (1) is said to be mean stable (M-stable) if, for any nonnegative initial state $\mathbf{x}(0)$ and any initial probability distribution $\pi(0)$, the expected value of the free motion of the state vector $\mathbf{x}(t)$ asymptotically converges to zero. This characterization of stability for PMJLS's is equivalent to first-moment stability, namely to the convergence to zero of the free motion of any norm $\|\mathbf{x}(t)\|$ of the state vector, see [3]. An M-stable system is also exponentially stable and, thanks to monotonicity of positive systems, M-stability can be ascertained by only checking the convergence to zero of the expected value of the free motion of the state vector for a single initial state in the strictly positive orthant.

Recalling the definition of $\mathbf{m}$ and (16), it is clear that M-stability is equivalent to Hurwitz stability of the Metzler matrix $\tilde{F}$. Checking Hurwitz stability of a Metzler matrix can be done via LP. Precisely, stability is equivalent to the existence of a strictly positive vector $\mathbf{s} \in \mathbb{R}_+^{nN}$ such that $\tilde{F}\mathbf{s} \ll 0$ (equivalently, a strictly positive vector $\mathbf{p} \in \mathbb{R}_+^{nN}$ such that $\tilde{F}^\top \mathbf{p} \ll 0$). Recalling the structure of $\tilde{F}$, the necessary and sufficient condition of M-stability of system (1) is formulated as follows, see [2–4].

Proposition 1. *The following statements are equivalent:*

(i) System (1) is M-stable

(ii) There exist strictly positive vectors $\mathbf{s}_i \in \mathbb{R}_+^n$, $i \in \mathcal{N}$, such that

$$F_i\mathbf{s}_i + \sum_{j=1}^{N} \lambda_{ji}\mathbf{s}_j \ll 0, \quad i \in \mathcal{N} \tag{21}$$

(iii) There exist strictly positive vectors $\mathbf{p}_i \in \mathbb{R}_+^n$, $i \in \mathcal{N}$, such that

$$F_i^\top\mathbf{p}_i + \sum_{j=1}^{N} \lambda_{ij}\mathbf{p}_j \ll 0, \quad i \in \mathcal{N} \tag{22}$$

2.2 $\mathcal{L}_\infty$-Induced Norm

Assume that the system (1), (2) is M-stable and let $\mathbf{x}(0) = 0$, $\boldsymbol{\pi}(0) = \bar{\boldsymbol{\pi}}$. Moreover, let $\mathbf{w} > 0$ be a deterministic bounded disturbance, i.e. $\mathbf{w} \in \mathcal{L}_\infty$. Therefore, it makes sense to compute the $\mathcal{L}_\infty$–induced norm, defined as

$$J_\infty := \sup_{\mathbf{w}\in\mathcal{L}_\infty, \mathbf{w}>0} \frac{\sup_{k,t\geq 0} E[[\mathbf{z}(t)]_k]}{\sup_{k,t\geq 0}[\mathbf{w}(t)]_k}$$

Such a performance index provides a measure of average disturbance attenuation in terms of peak-to-peak worst-case gain. Letting $\tilde{g}(t)$ be the impulse response of the DIM system (16), (17), and taking, without loss of generality, $\mathbf{w}(t) \leq \mathbf{1}_{n_w}$, $t \geq 0$, it follows that

$$\begin{aligned}
\sup_{k,t\geq 0} E[[\mathbf{z}(t)]_k] &= \sup_{k,t\geq 0} \int_0^t \mathbf{e}_k^\top \tilde{g}(t-\tau)\mathbf{w}(\tau)d\tau \\
&\leq \sup_{k,t\geq 0} \int_0^t \mathbf{e}_k^\top \tilde{g}(\tau)\mathbf{1}_{n_w}d\tau
\end{aligned}$$

$$\leq \max_k \int_0^\infty \mathbf{e}_k^\top \tilde{g}(\tau)\mathbf{1}_{n_w} d\tau$$
$$\leq \max_k \mathbf{e}_k^\top \tilde{G}(0)\mathbf{1}_{n_w} = \|\tilde{G}(0)\|_\infty$$

and it is clear that the supremum value is reached for $\mathbf{w}(t) = \mathbf{1}_{n_w}$, $t \geq 0$. Therefore $J_\infty = \|\tilde{G}(0)\|_\infty$. Checking whether $\|\tilde{G}(0)\|_\infty < \rho$, for a given $\rho > 0$ is an LP problem, precisely stated in the following proposition, whose proof can be found in [3].

Proposition 2. *System (1), (2) is M-stable and $J_\infty < \rho$ if and only if there exist strictly positive vectors $\mathbf{s}_i \in \mathbb{R}_+^n$, $i \in \mathcal{N}$, such that*

$$F_i \mathbf{s}_i + \sum_{j=1}^N \lambda_{ji} \mathbf{s}_j + [\bar{\boldsymbol{\pi}}]_i B_i \mathbf{1}_{n_w} \ll 0, \quad i \in \mathcal{N} \tag{23}$$

$$\sum_{i=1}^N (L_i \mathbf{s}_i + [\bar{\boldsymbol{\pi}}]_i D_i \mathbf{1}_{n_w}) \ll \mathbf{1}_{n_z} \rho \tag{24}$$

As apparent from the definition and the positivity of the system, the worst disturbance for J_∞ is a constant vector with equal entries, namely $\mathbf{w}(t) = \mathbf{1}_{n_w}$.

2.3 $\mathcal{L}_1$-Induced Norm

For an M-stable system (1), (2), with $\mathbf{x}(0) = 0$, $\boldsymbol{\pi}(0) = \bar{\boldsymbol{\pi}}$, and an integrable deterministic disturbance $\mathbf{w} > 0 \in \mathcal{L}_1$, the $\mathcal{L}_1$–induced norm is defined as

$$J_1 := \sup_{\mathbf{w} \in \mathcal{L}_1, \mathbf{w} > 0} \frac{E[\int_0^\infty \mathbf{1}_{n_z}^\top \mathbf{z}(t) dt]}{\int_0^\infty \mathbf{1}_{n_w}^\top \mathbf{w}(t) dt}$$

It provides an alternative measure of disturbance attenuation in terms of average integral worst-case gain. It results that

$$E\left[\int_0^\infty \mathbf{1}_{n_z}^\top \mathbf{z}(t) dt\right] = \int_0^\infty \mathbf{1}_{n_z}^\top \left(\int_0^t \tilde{g}(t-\tau)\mathbf{w}(\tau) d\tau\right) dt$$
$$= \int_0^\infty \mathbf{1}_{n_z}^\top \left(\int_\tau^\infty \tilde{g}(t-\tau) dt\right) \mathbf{w}(\tau) d\tau$$
$$= \int_0^\infty \mathbf{1}_{n_z}^\top \left(\int_0^\infty \tilde{g}(t) dt\right) \mathbf{w}(\tau) d\tau$$

$$= \mathbf{1}_{n_z}^\top \tilde{G}(0) \int_0^\infty \mathbf{w}(\tau)d\tau$$

$$= \sum_{k=1}^{n_w} \mathbf{1}_{n_z}^\top \tilde{G}(0)\mathbf{e}_k \int_0^\infty [\mathbf{w}(t)]_k dt$$

and hence

$$\frac{E[\int_0^\infty \mathbf{1}_{n_z}^\top \mathbf{z}(t)dt]}{\int_0^\infty \mathbf{1}_{n_w}^\top \mathbf{w}(t)dt} = \sum_{k=1}^{n_w} \mathbf{1}_{n_z}^\top \tilde{G}(0)\mathbf{e}_k \beta_k$$

with

$$\beta_k = \frac{\int_0^\infty [\mathbf{w}(t)]_k dt}{\int_0^\infty \mathbf{1}_{n_w}^\top \mathbf{w}(t)dt}$$

Since $\beta_k \geq 0$ and $\sum_{k=1}^{n_w} \beta_k = 1$, it is clear that $J_1 = \max_k \mathbf{1}_{n_z}^\top \tilde{G}(0)\mathbf{e}_k = \|\tilde{G}(0)\|_1$, and the worst disturbance is the (non $\mathscr{L}_1$–integrable) impulsive signal $\mathbf{w}(t) = \delta(t)\mathbf{e}_k$, where k is the index maximizing $\mathbf{1}_{n_z}^\top \tilde{G}(0)\mathbf{e}_k$. Checking whether $\|\tilde{G}(0)\|_1 < \rho$, for a given $\rho > 0$ is an LP problem, precisely stated in the following proposition, proven in [3].

Proposition 3. *System (1), (2) is M-stable and* $J_1 < \rho$ *if and only if there exist strictly positive vectors* $\mathbf{p}_i \in \mathbb{R}_+^n$, $i \in \mathscr{N}$, *such that*

$$F_i^\top \mathbf{p}_i + \sum_{j=1}^{N} \lambda_{ij}\mathbf{p}_j + L_i^\top \mathbf{1}_{n_z} \ll 0, \quad i \in \mathscr{N} \tag{25}$$

$$\sum_{i=1}^{N} (B_i^\top \mathbf{p}_i + D_i^\top \mathbf{1}_{n_z})[\bar{\pi}]_i \ll \rho \mathbf{1}_{n_w} \tag{26}$$

As apparent from the positivity of the system, the worst disturbance is an impulse applied to the worst input channel, namely $\mathbf{w}(t) = \delta(t)\mathbf{e}_k$, where $k = \arg \max_{i=1,\ldots,n_w} \mathbf{1}_{n_z}^\top \tilde{G}(0)\mathbf{e}_i$.

Remark 1. Notice that the conditions (23), (24) and (25), (26) can be interpreted as dual inequalities, reflecting the duality of the two norms considered, i.e.

$$\|\tilde{G}(0)\|_\infty = \max_{k=1,2,\ldots,n_z} \mathbf{e}_k^\top \tilde{G}(0)\mathbf{1}_{n_w}, \quad \|\tilde{G}(0)\|_1 = \max_{k=1,2,\ldots,n_w} \mathbf{1}_{n_z}^\top \tilde{G}(0)\mathbf{e}_k$$

Inequalities (23), (24) are generally used to cope with state-feedback problems, whereas inequalities (25), (26) are used for output injection problems. In the sequel, we will deal only with the first problem. In this regard, notice that in order to check whether $\|\tilde{G}(0)\|_1 < \rho$ one can use inequalities (23), (24) applied to the SISO systems with mean transfer functions $\mathbf{1}_{n_z}^\top \tilde{G}(0)\mathbf{e}_k$, and check whether they are satisfied with a certain ρ, for each $k = 1, 2, \ldots, n_w$.

2.4 Transient Analysis

The computation of the $\mathscr{L}_\infty$ and $\mathscr{L}_1$-induced norms has been carried out under the assumption $\boldsymbol{\pi}(0) = \bar{\boldsymbol{\pi}}$, which has allowed to consider the time-invariant DIM system (16), (17) and the associated DIM transfer function $\tilde{G}(s)$. Now, we address the problem of computing the induced gains for the time-varying system (12), (13). This corresponds to the case when the initial probability distribution is different from the stationary one, so that the dynamics of the Kolmogorov equation (3) has to be taken into account. Since $\boldsymbol{\pi}(0)$ is a generic element of $\mathscr{P}_N$, we are well advised to maximize the gain also with respect to the elements of this set. We start from the worst $\mathscr{L}_1$–induced gain under arbitrary initial distribution $\boldsymbol{\pi}(0)$, defined as

$$\tilde{J}_1 := \sup_{\mathbf{w}\in\mathscr{L}_1, \mathbf{w}>0, \boldsymbol{\pi}(0)\in\mathscr{P}_N} \frac{E[\int_0^\infty \mathbf{1}_{n_z}^\top \mathbf{z}(t)dt]}{\int_0^\infty \mathbf{1}_{n_w}^\top \mathbf{w}(t)dt}$$

Notice that the time-varying system (12), (13) is obtained from the time-invariant system (4), (5) using (11) and observing that $\tilde{B}(t) = \breve{B}\,(\boldsymbol{\pi}(t)\otimes I_{n_w})$ and $\tilde{D}(t) = \breve{D}\,(\boldsymbol{\pi}(t)\otimes I_{n_w})$. Therefore, letting $\breve{g}(t)$ denote the impulse response of system (4), (5), it results that

$$\begin{aligned}
E\left[\int_0^\infty \mathbf{1}_{n_z}^\top \mathbf{z}(t)dt\right] &= \int_0^\infty \mathbf{1}_{n_z}^\top \left(\int_0^t \breve{g}(t-\tau)\,(\boldsymbol{\pi}(\tau)\otimes\mathbf{w}(\tau))\,d\tau\right) dt \\
&= \int_0^\infty \mathbf{1}_{n_z}^\top \left(\int_\tau^\infty \breve{g}(t-\tau)dt\right)(\boldsymbol{\pi}(\tau)\otimes\mathbf{w}(\tau))\,d\tau \\
&= \int_0^\infty \mathbf{1}_{n_z}^\top \left(\int_0^\infty \breve{g}(t)dt\right)(\boldsymbol{\pi}(\tau)\otimes\mathbf{w}(\tau))\,d\tau \\
&= \mathbf{1}_{n_z}^\top \breve{G}(0)\int_0^\infty (\boldsymbol{\pi}(t)\otimes\mathbf{w}(t))\,dt
\end{aligned}$$

Moreover $\mathbf{1}_{Nn_w}^\top(\boldsymbol{\pi}(t)\otimes\mathbf{w}(t)) = \mathbf{1}_{n_w}^\top\mathbf{w}(t)$. Therefore, for any assigned vector $\boldsymbol{\pi}(0)\in\mathscr{P}_N$, the $\mathscr{L}_1$-induced norm is less than ρ if and only if the inequalities (25), (26) are feasible with $[\bar{\boldsymbol{\pi}}]_i$ replaced by $[\boldsymbol{\pi}(0)]_i$. Concerning the computation of $\tilde{J}_1$, it is clear that $\tilde{J}_1 = \|\breve{G}(0)\|_1$, where the worst disturbance is $\mathbf{w}(t) = \delta(t)\mathbf{e}_k$ and the worst initial probability vector is $\boldsymbol{\pi}(0) = \mathbf{e}_i$, where $k\in\{1,2,\ldots,n_w\}$ and $i\in\mathscr{N}$ are the maximizing indices of $\mathbf{1}_{n_z}^\top\breve{G}(0)\,(\mathbf{e}_i\otimes I_{n_w})\,\mathbf{e}_k$. The linear program for assessing that $\tilde{J}_1 < \rho$ is still given by Proposition 3 by replacing inequality (26) with

$$B_i^\top\mathbf{p}_i + D_i^\top\mathbf{1}_{n_z} \ll \rho\mathbf{1}_{n_w}, \quad i\in\mathscr{N} \tag{27}$$

Note that this condition coincides with that provided in Theorem 4 of [13], where the $\mathscr{L}_1$ norm is computed considering an arbitrary $\pi(0)$. The above considerations could be used in the sequel to derive variants of Theorems 3, 11, when one is interested in considering $\pi(0) \neq \bar{\pi}$.

Next, consider the worst $\mathscr{L}_\infty$–induced gain under arbitrary initial distribution $\pi(0)$, defined as

$$\tilde{J}_\infty = \sup_{\mathbf{w}\in\mathscr{L}_\infty, \mathbf{w}>0, \pi(0)\in\mathscr{P}_N} \frac{\sup_{k,t\geq 0} E[[\mathbf{z}(t)]_k]}{\sup_{k,t\geq 0}[\mathbf{w}(t)]_k}$$

The computation of the $\mathscr{L}_\infty$-induced norm for an assigned $\pi(0)$ is hard since it depends on the whole trajectory $\pi(t)$. The formulation of algebraic necessary and sufficient conditions for guaranteeing such a norm to be less than ρ (and consequently to evaluate $\tilde{J}_\infty$) is still an open issue. However, suitable bounds can be easily derived. In this respect notice that, for $\mathbf{w}(t) \leq \mathbf{1}_{n_w}$, $t \geq 0$, and any $\pi(0) \in \mathscr{P}_N$,

$$\begin{aligned}
\sup_{k,t\geq 0} E\left[[\mathbf{z}(t)]_k\right] &= \sup_{k,t\geq 0} \int_0^t \mathbf{e}_k^\top \breve{g}(t-\tau)\left(\pi(\tau)\otimes \mathbf{w}(\tau)\right) d\tau \\
&\leq \sup_{k,t\geq 0} \int_0^t \mathbf{e}_k^\top \breve{g}(\tau)\left(\pi(t-\tau)\otimes \mathbf{1}_{n_w}\right) d\tau \\
&\leq \max_k \int_0^\infty \mathbf{e}_k^\top \breve{g}(\tau)\mathbf{1}_{Nn_w} d\tau = \|\breve{G}(0)\|_\infty
\end{aligned}$$

It is then clear that the worst disturbance is $\mathbf{w}(t) = \mathbf{1}_{n_w}$ and $\tilde{J}_\infty \leq \|\breve{G}(0)\|_\infty$.

On the other hand, $\tilde{J}_\infty$ is greater than the gain obtained with $\pi(0) = \bar{\pi}$, which coincides with J_∞ computed in Sect. 2.2. In conclusion, it turns out that

$$J_\infty = \|\tilde{G}(0)\|_\infty \leq \tilde{J}_\infty \leq \|\breve{G}(0)\|_\infty$$

Since $\tilde{J}_\infty \leq \|\breve{G}(0)\|_\infty$, a sufficient condition ensuring $\tilde{J}_\infty < \rho$ is given by Proposition 2, where the scalars $[\bar{\pi}]_i$ in the linear inequalities (23), (24) are replaced by 1.

For a more detailed characterization of the induced norms of a PMJLS and their relations with induced norms under stochastic disturbances, the interested reader is referred to [3].

2.5 *Stability and Norms Under Positive Perturbations*

From (21), it is apparent that an M-stable system cannot be destabilized by nonpositive perturbations of the system matrices F_i. Conversely, nonnegative perturbations of F_i cannot result in an M-stable system if the original system is not M-stable.

Moreover, it can be shown that the Perron-Frobenius eigenvalue of $\tilde{F}$ is a monotonic nondecreasing function of positive perturbations of any entry of matrices F_i.

For M-stable systems, matrix $\tilde{G}(0) = \tilde{D} - \tilde{L}(\tilde{F})^{-1}\tilde{B}$ is nonnegative and its entries are monotonically nondecreasing functions of the entries of F_i, L_i, B_i, D_i as long as M-stability is preserved. This conclusion is straightforward for nonnegative perturbations of the parameters in L_i, B_i, D_i in view of formulas (8), (18) and (19). Now consider a nonnegative perturbation Δ_i of the matrix F_i and, according to (6), let $\hat{F} = \operatorname{diag}_i(F_i + \Delta_i) + \Lambda^\top \otimes I_n$. It is apparent that $\hat{F} \geq \tilde{F}$. Assuming that $\hat{F}$ is Hurwitz, then $-\hat{F}^{-1}$ (as well as $-\tilde{F}^{-1}$) is a nonnegative matrix. Moreover $\hat{F}^{-1}\hat{F}\tilde{F}^{-1} \geq \hat{F}^{-1}\tilde{F}\tilde{F}^{-1}$, which implies $-\tilde{F}^{-1} \geq -\hat{F}^{-1}$. In conclusion, being the entries of $\tilde{G}(0)$ nonnegative and monotonically nondecreasing functions of the system matrix parameters F_i, L_i, B_i, D_i, its norm (in any specification) is nondecreasing as well. In other words, no positive perturbation of any system matrix can lead to an improvement in the values of J_∞ or J_1.

All what said for the DIM transfer function in $s = 0$, namely $\tilde{G}(0)$, also holds for the SIM transfer function in $s = 0$, i.e. $\breve{G}(0) = \breve{D} - \tilde{L}(\tilde{F})^{-1}\breve{B}$.

3 Stabilization and Norm Minimization via Mode-Dependent State-Feedback

Here we discuss the effect of a memoryless state-feedback law applied to a PMJLS described by

$$\dot{\mathbf{x}}(t) = A_{\theta(t)}\mathbf{x}(t) + B_{\theta(t)}\mathbf{w}(t) + G_{\theta(t)}\mathbf{u}(t) \tag{28}$$

$$\mathbf{z}(t) = C_{\theta(t)}\mathbf{x}(t) + D_{\theta(t)}\mathbf{w}(t) + H_{\theta(t)}\mathbf{u}(t) \tag{29}$$

where the input signal $\mathbf{u} \in \mathbb{R}^{n_u}$ has been added together with the relevant matrices G_i and H_i, assumed to be nonnegative for all $i \in \mathscr{N}$. We consider a mode-dependent state-feedback control of the form

$$\mathbf{u}(t) = K_{\theta(t)}\mathbf{x}(t)$$

where $K_i \in R^{n_u \times n}$ for all i. Notice that both the stochastic process $\theta(t)$ and the state variable $\mathbf{x}(t)$ are considered measurable. For brevity, we only focus on design problems where M-stability is concerned. In particular, the addressed problems concern mean stabilization, $\mathscr{L}_\infty$-induced and $\mathscr{L}_1$-induced control with deterministic disturbance $\mathbf{w}(t)$. Notably, the first two control problems admit a complete parametrization of the feedback gains in terms of linear programming. For similar results on time-invariant deterministic positive systems, see [6].

3.1 Mean Stabilization

Assume $\mathbf{w}(t) = 0$, $t \geq 0$, and consider the state equation only, i.e.

$$\dot{\mathbf{x}}(t) = A_{\theta(t)}\mathbf{x}(t) + G_{\theta(t)}\mathbf{u}(t) \tag{30}$$

The problem of mean stabilization (M-stabilization) can be cast as follows.

Problem 1. Parameterize the set $\mathscr{K}$ of all K_i, $i \in \mathscr{N}$, such that $F_i = A_i + G_i K_i$ are Metzler matrices, $i \in \mathscr{N}$, and the closed-loop system

$$\dot{\mathbf{x}}(t) = \left(A_{\theta(t)} + G_{\theta(t)}K_{\theta(t)}\right)\mathbf{x}(t) \tag{31}$$

is M-stable.

A little thought reveals that this problem is ill-posed if one requires that $\mathbf{u}(t)$ is nonnegative for all $t \geq 0$ and all initial states $\mathbf{x}(0) \in \mathbb{R}^n_+$. Indeed, such a condition is equivalent to requiring that K_i, $i \in \mathscr{N}$, are nonnegative matrices. Therefore, if system (31) is M-stable, in view of Proposition 1 there exist strictly positive vectors $\mathbf{s}_i \in \mathbb{R}^n_+$ such that

$$(A_i + G_i K_i)\,\mathbf{s}_i + \sum_{j=1}^{N} \lambda_{ji}\mathbf{s}_j \ll 0$$

Being both K_i and G_i nonnegative for all $i \in \mathscr{N}$, it turns out that

$$A_i\mathbf{s}_i + \sum_{j=1}^{N} \lambda_{ji}\mathbf{s}_j \ll 0$$

This means that the open-loop system is already M-stable. Therefore, it is not possible to stabilize an unstable PMJLS through a state-feedback law with nonnegative gains. In the same vein, recalling the effect of positive perturbations discussed in Sect. 2.5, if lower bounds for the entries of K_i are known, i.e. $K_i \geq \underline{K}_i$, $i \in \mathscr{N}$, then Problem 1 is solvable if and only if the limiting gains $\underline{K}_i$ are such that the closed-loop matrices $A_i + G_i\underline{K}_i$ are Metzler for all $i \in \mathscr{N}$ and the closed-loop system with $\mathbf{u}(t) = K_{\theta(t)}\mathbf{x}(t)$ is mean-stable.

One can relax the positivity constraints on $\mathbf{u}(t)$, by allowing matrices K_i, $i \in \mathscr{N}$ to have nonpositive entries, but requiring at the same time that the state vector remains in the nonnegative orthant for any initial state. This requirement corresponds to select the gains K_i, $i \in \mathscr{N}$, in such a way that the closed-loop matrices $A_i + G_i K_i$ are Metzler for each $i \in \mathscr{N}$. This leads to the following linear constraints for the entries of K_i:

$$\mathbf{e}_r^\top G_i K_i \mathbf{e}_p \geq -[A_i]_{rp}, \quad i \in \mathscr{N}, \quad r \neq p = 1, 2, \ldots, n \tag{32}$$

Therefore, the following result can be formulated.

Theorem 1. *There exist K_i, $i \in \mathcal{N}$, such that $A_i + G_iK_i$, $i \in \mathcal{N}$, are Metzler matrices and the closed-loop system (31) is M-stable if and only if there exist strictly positive vectors $\mathbf{s}_i \in \mathbb{R}^n_+$ and vectors $\mathbf{h}_i^p \in \mathbb{R}^{n_u}$, $i \in \mathcal{N}$, $p = 1, 2, \ldots, n$, such that*

$$A_i\mathbf{s}_i + G_i\sum_{p=1}^{n}\mathbf{h}_i^p + \sum_{j=1}^{N}\lambda_{ji}\mathbf{s}_j \ll 0 \tag{33}$$

$$\mathbf{e}_r^\top G_i\mathbf{h}_i^p + [A_i]_{rp}\mathbf{e}_p^\top\mathbf{s}_i \geq 0, \quad r \neq p = 1, 2, \ldots, n \tag{34}$$

for all $i \in \mathcal{N}$. Matrices K_i, $i \in \mathcal{N}$, are then obtained from

$$K_i\mathbf{e}_p = (\mathbf{e}_p^\top\mathbf{s}_i)^{-1}\mathbf{h}_i^p, \quad p = 1, 2, \ldots, n \tag{35}$$

Proof. Assume that (33), (34) are feasible. Then, construct matrices K_i according to (35). Therefore, for $r \neq p$, it must hold that

$$\begin{aligned}
[A_i + G_iK_i]_{rp} &= \mathbf{e}_r^\top\,(A_i + G_iK_i)\,\mathbf{e}_p \\
&= [A_i]_{rp} + \mathbf{e}_r^\top\left(G_i\sum_{q=1}^{n}K_i\mathbf{e}_q\mathbf{e}_q^\top\right)\mathbf{e}_p \\
&= (\mathbf{e}_p^\top\mathbf{s}_i)^{-1}\left([A_i]_{rp}(\mathbf{e}_p^\top\mathbf{s}_i) + \mathbf{e}_r^\top G_i\mathbf{h}_i^p\right)
\end{aligned}$$

Thanks to (34), it turns out that $[A_i + G_iK_i]_{rp} \geq 0$ for $r \neq p$. This means that matrices $A_i + G_iK_i$ are Metzler, for all i. Moreover,

$$\begin{aligned}
0 \gg A_i\mathbf{s}_i + G_i\sum_{p=1}^{n}\mathbf{h}_i^p + \sum_{j=1}^{N}\lambda_{ji}\mathbf{s}_j \\
= A_i\mathbf{s}_i + G_i\sum_{p=1}^{n}K_i\mathbf{e}_p\mathbf{e}_p^\top\mathbf{s}_i + \sum_{j=1}^{N}\lambda_{ji}\mathbf{s}_j \\
= (A_i + G_iK_i)\,\mathbf{s}_i + \sum_{j=1}^{N}\lambda_{ji}\mathbf{s}_j
\end{aligned}$$

so that the closed-loop system (31) is M-stable.

Viceversa, assume that there exist matrices K_i such that $A_i + G_iK_i$ are Metzler matrices for all i and the closed-loop system (31) is M-stable. Then, using Proposition 1, there exist strictly positive vectors $\mathbf{s}_i \in \mathbb{R}^n_+$ such that

$$(A_i + G_iK_i)\,\mathbf{s}_i + \sum_{j=1}^{N}\lambda_{ji}\mathbf{s}_j \ll 0$$

Letting

$$\mathbf{h}_i^p = \mathbf{e}_p^\top \mathbf{s}_i K_i \mathbf{e}_p, \quad p = 1, 2, \ldots, n$$

and reversing the arguments of the sufficiency part, it follows that both inequalities (33) and (34) are satisfied.

Remark 2. It is worth noticing that Theorem 1 provides a parametrization of all the state-feedback control gains ensuring positivity and M-stability of the closed-loop system. Letting $\mathscr{K}$ be the set of the N-tuple gains $\mathbb{K}$ solving the problem, the parametrization is indeed given by

$$\mathscr{K} = \{\mathbb{K} : K_i \mathbf{e}_p = (\mathbf{e}_p^\top \mathbf{s}_i)^{-1} \mathbf{h}_i^p, \quad p = 1, 2, \cdots, n, \quad i \in \mathscr{N}\} \tag{36}$$

where $\mathbf{h}_i^p$ and $\mathbf{s}_i$ solve (33), (34). This parametrization is useful if one wants to include in the design further properties besides stability, e.g. minimization of specified closed-loop performance indices.

Remark 3. Assume that the system is single input ($n_u = 1$) and make the simplifying assumption that $G_i \gg 0$ for each $i \in \mathscr{N}$ (otherwise $G_i \to G_i + \epsilon \mathbf{1}_n$). Notice from (32) that $A_i + G_i K_i$ is Metzler if and only if $K_i \geq \underline{K}_i^{[A,G]}$, where

$$\underline{K}_i^{[A,G]} \mathbf{e}_p = \max_{r \neq p} -\frac{[A_i]_{rp}}{\mathbf{e}_r^\top G_i}, \quad i \in \mathscr{N}, \quad p = 1, 2, \ldots, n \tag{37}$$

In other words, there exist matrices $\underline{K}_i^{[A,G]}$ such that the Metzler conditions are verified for any $K_i \geq \underline{K}_i^{[A,G]}$. In view of the discussion presented in Sect. 2.5, if the feedback law $u(t) = \underline{K}_{\theta(t)}^{[A,G]} \mathbf{x}(t)$ is not M-stabilizable, then Problem 1 has no solution. When $n_u = 1$, the parametrization of all gains solving Problem 1 can be written as

$$\mathscr{K} = \{\mathbb{K} : K_i \mathbf{e}_p = \underline{K}_i^{[A,G]} \mathbf{e}_p + (\mathbf{e}_p^\top \mathbf{s}_i)^{-1} \hat{h}_i^p, \quad p = 1, 2, \cdots, n, \quad i \in \mathscr{N}\} \tag{38}$$

where $\hat{h}_i^p$ are nonnegative scalars and $\mathbf{s}_i$ are strictly positive vectors satisfying

$$(A_i + G_i \underline{K}_i^{[A,G]}) \mathbf{s}_i + G_i \sum_{p=1}^{n} \hat{h}_i^p + \sum_{j=1}^{N} \lambda_{ji} \mathbf{s}_j \ll 0 \tag{39}$$

for all $i \in \mathscr{N}$.

On the other hand, a necessary condition for M-stability is that $A_i + G_i K_i + \lambda_{ii} I_n$ are Hurwitz stable. Being such matrices Metzler, this implies negativity of their diagonal entries, i.e. $K_i \leq \overline{K}_i$ where

$$\overline{K}_i \mathbf{e}_p = -\frac{[A_i]_{pp} + \lambda_{ii} + \epsilon}{\mathbf{e}_p^\top G_i}, \quad i \in \mathscr{N}, \quad p = 1, 2, \ldots, n \tag{40}$$

with ϵ being an arbitrarily small positive number. In conclusion, in the scalar input case ($n_u = 1$), the gains K_i such that the closed-loop system is positive and M-stable should lie necessarily inside the interval

$$\underline{K}_i^{[A,G]} \leq K_i \leq \overline{K}_i \tag{41}$$

where $\underline{K}_i^{[A,G]}$ and $\overline{K}_i$ are defined in (37) and (40), respectively. Notice that it may happen that this vector interval is void. This would entail that Problem 1 does not admit any solution.

3.2 $\mathscr{L}_\infty$-Induced Mode-Dependent Control

Assume that $\mathbf{w}(t), t \geq 0$, is a nonnegative bounded disturbance for system (28), (29), that $\mathbf{x}(0) = 0$ and $\boldsymbol{\pi}(0) = \bar{\boldsymbol{\pi}}$. We aim at finding K_i, $i \in \mathscr{N}$, such that $A_i + B_i K_i$, $i \in \mathscr{N}$, are Metzler matrices and the $\mathscr{L}_\infty$-induced norm for the closed-loop system

$$\dot{\mathbf{x}}(t) = \left(A_{\theta(t)} + G_{\theta(t)} K_{\theta(t)}\right) \mathbf{x}(t) + B_{\theta(t)} \mathbf{w}(t) \tag{42}$$

$$\mathbf{z}(t) = \left(C_{\theta(t)} + H_{\theta(t)} K_{\theta(t)}\right) \mathbf{x}(t) + D_{\theta(t)} \mathbf{w}(t) \tag{43}$$

is minimized. Here, we consider only the $\mathscr{L}_\infty$-induced norm associated with deterministic positive bounded signals $\mathbf{w}(t)$, i.e.

$$J_\infty(\mathbb{K}) = \sup_{\mathbf{w} \in \mathscr{L}_\infty, \mathbf{w} > 0} \frac{\sup_{k,t \geq 0} E[[\mathbf{z}(t)]_k]}{\sup_{k,t \geq 0} [\mathbf{w}(t)]_k}$$

where $\mathbb{K} = \{K_1, K_2, \ldots, K_N\}$. Letting $\tilde{G}_{\mathbb{K}}(s)$ be the DIM transfer function of the PMJLS (42), (43) for a certain $\mathbb{K}$, we have seen in Sect. 2.2 that $J_\infty(\mathbb{K}) = \|\tilde{G}_{\mathbb{K}}(0)\|_\infty$.

Problem 2. Parameterize the set $\mathscr{K}_\infty$ of all gains in $\mathscr{K}$ such that (i) the closed-loop system (42), (43) is positive and M-stable, (ii) $J_\infty(\mathbb{K}) < \rho$ for a given positive scalar ρ.

Notice that point (i) requires that $A_i + G_i K_i$, $i \in \mathscr{N}$, are Metzler matrices and $C_i + H_i K_i$, $i \in \mathscr{N}$, are nonnegative matrices. The following parametrization result can be proven.

Theorem 2. *There exist K_i, $i \in \mathscr{N}$, such that the closed-loop system (42), (43) is positive and M-stable with $\mathscr{L}_\infty$-induced norm less than $\rho > 0$ if and only if there exist strictly positive vectors $\mathbf{s}_i \in \mathbb{R}^n_+$ and vectors $\mathbf{h}_i^p \in \mathbb{R}^{n_u}$, $i \in \mathscr{N}, p = 1, 2, \ldots, n$, such that*

$$A_i\mathbf{s}_i + G_i\sum_{p=1}^{n}\mathbf{h}_i^p + \sum_{j=1}^{N}\lambda_{ji}\mathbf{s}_j + [\bar{\pi}]_i B_i \mathbf{1}_{n_w} \ll 0 \tag{44}$$

$$\sum_{i=1}^{N}\left(C_i\mathbf{s}_i + H_i\sum_{p=1}^{n}\mathbf{h}_i^p + [\bar{\pi}]_i D_i \mathbf{1}_{n_w}\right) \ll \rho \mathbf{1}_{n_z} \tag{45}$$

$$\mathbf{e}_r^\top G_i \mathbf{h}_i^p + [A]_{rp}\mathbf{e}_p^\top \mathbf{s}_i \geq 0, \quad r \neq p \tag{46}$$

$$\mathbf{e}_q^\top H_i \mathbf{h}_i^p + [C]_{qp}\mathbf{e}_p^\top \mathbf{s}_i \geq 0 \tag{47}$$

for all $i \in \mathscr{N}$, $r = 1,2,\ldots,n$, $p = 1,2,\ldots,n$, $q = 1,2,\ldots,n_z$. Matrices K_i, $i \in \mathscr{N}$, are then given as follows:

$$K_i\mathbf{e}_p = (\mathbf{e}_p^\top \mathbf{s}_i)^{-1}\mathbf{h}_i^p, \quad p = 1,2,\ldots,n \tag{48}$$

Proof. The proof is similar to the one of Theorem 1 and hence is only sketched. Conditions (44), (45) with position (48) are equivalent to

$$(A_i + G_iK_i)\mathbf{s}_i + \sum_{j=1}^{N}\lambda_{ji}\mathbf{s}_j + [\bar{\pi}]_i B_i \mathbf{1}_{n_w} \ll 0$$

$$\sum_{i=1}^{N}((C_i + H_iK_i)\mathbf{s}_i + [\bar{\pi}]_i D_i \mathbf{1}_{n_w}) \ll \rho \mathbf{1}_{n_z}$$

for all $i \in \mathscr{N}$. In view of Proposition 2, they correspond to a closed-loop system having $\mathscr{L}_\infty$-induced norm less than ρ. Moreover conditions (46), (47) with (48) are equivalent to say that $A_i + B_iK_i$, $i \in \mathscr{N}$, are Metzler matrices and $C_i + H_iK_i$ are nonnegative matrices, $i \in \mathscr{N}$.

Remark 4. It is worth noticing that Theorem 2 provides a parametrization of all the state-feedback control laws ensuring positivity, M-stability and $\mathscr{L}_\infty$-induced norm less that ρ of the closed-loop system. Letting $\mathscr{K}_\infty$ be the set of N-tuple gains K_i solving Problem 2, the parametrization is indeed given by

$$\mathscr{K}_\infty = \{\mathbb{K} : K_i\mathbf{e}_p = (\mathbf{e}_p^\top \mathbf{s}_i)^{-1}\mathbf{h}_i^p, \quad p = 1,2,\cdots,n, \quad i \in \mathscr{N}\} \tag{49}$$

where $\mathbf{h}_i^p$ and $\mathbf{s}_i$ solve (44)–(47).

Remark 5. The conditions of Theorem 2 are expressed in terms of inequalities that are linear in all the decision variables and the performance parameter ρ. Then, the minimization of ρ can be carried out through routine LP methods, so obtaining the set of $\mathscr{L}_\infty$-optimal gains.

Remark 6. In the scalar input case ($n_u = 1$), and assuming $G_i \gg 0$, $H_i \gg 0$, it is possible to conclude that the gains K_i are necessarily included in the intervals

$$\underline{K}_i^\star \le K_i \le \overline{K}_i, \quad i \in \mathcal{N} \tag{50}$$

where $\overline{K}_i$ is defined in (40) and

$$\underline{K}_i^\star = \max\{\underline{K}_i^{[A,G]}, \underline{K}_i^{[C,H]}\} \tag{51}$$

with $\underline{K}_i^{[A,G]}$ defined in (37) and

$$\underline{K}_i^{[C,H]}\mathbf{e}_p = \max_r -\frac{[C_i]_{rp}}{\mathbf{e}_r^\top H_i}, \quad p = 1, 2, \dots, n$$

Note that the max operator in (51) is to be intended elementwise.

3.3 $\mathcal{L}_1$-Induced Mode-Dependent Control

In this section we assume that $\mathbf{w}(t)$, $t \geq 0$, is a nonnegative integrable deterministic disturbance for system (28), (29). We aim at finding K_i, $i \in \mathcal{N}$, such that $A_i + G_iK_i$, $i \in \mathcal{N}$, are Metzler matrices and the $\mathcal{L}_1$-induced norm

$$J_1(\mathbb{K}) = \sup_{\mathbf{w}\in\mathcal{L}_1, \mathbf{w}>0} \frac{E[\int_0^\infty \mathbf{1}_{n_z}^\top \mathbf{z}(t)dt]}{\int_0^\infty \mathbf{1}_{n_w}^\top \mathbf{w}(t)dt}$$

for the closed-loop system (42), (43) is minimized, where $\mathbb{K} = \{K_1, K_2, \dots, K_N\}$. We have seen in Sect. 2.3 that $J_1(\mathbb{K}) = \|\tilde{G}_{\mathbb{K}}(0)\|_1$. A further objective is to parameterize the set $\mathcal{K}_1$ of all gains in $\mathcal{K}$ such that (1) the closed-loop system is M-stable, (2) the closed-loop system has the state-space description of a positive system, (3) $J_1(\mathbb{K}) < \rho$ for a given positive scalar ρ.

The following result does not provide a complete parametrization but only a sufficient condition for the existence of the feedback gain matrices ensuring positivity of the closed-loop system, M-stability and guaranteed $\mathcal{L}_1$-induced norm.

Theorem 3. *There exist K_i, $i \in \mathcal{N}$, such that the closed-loop system (42), (43) is positive and M-stable with $\mathcal{L}_1$-induced norm less than $\rho > 0$ if there exist strictly positive vectors $\mathbf{s}_i \in \mathbb{R}_+^n$ and vectors $\mathbf{h}_i^p \in \mathbb{R}^{n_u}$, $i \in \mathcal{N}$, $p = 1, 2, \dots, n$, such that*

$$A_i\mathbf{s}_i + G_i \sum_{p=1}^n \mathbf{h}_i^p + \sum_{j=1}^N \lambda_{ji}\mathbf{s}_j + [\bar{\pi}]_i B_i \mathbf{e}_k \ll 0 \tag{52}$$

$$\mathbf{1}_{n_z}^\top \sum_{i=1}^N \left(C_i\mathbf{s}_i + H_i \sum_{p=1}^n \mathbf{h}_i^p + [\bar{\pi}]_i D_i \mathbf{e}_k \right) < \rho \tag{53}$$

$$\mathbf{e}_r^\top G_i \mathbf{h}_i^p + [A]_{rp} \mathbf{e}_p^\top \mathbf{s}_i \geq 0, \quad r \neq p \tag{54}$$

$$\mathbf{e}_q^\top H_i \mathbf{h}_i^p + [C]_{qp} \mathbf{e}_p^\top \mathbf{s}_i \geq 0 \tag{55}$$

for all $i \in \mathcal{N}$, $r = 1, 2, \ldots, n$, $p = 1, 2, \ldots, n$, $q = 1, 2, \ldots, n_z$, $k = 1, 2, \ldots, n_w$. Matrices K_i, $i \in \mathcal{N}$, are then given as follows:

$$K_i \mathbf{e}_p = (\mathbf{e}_p^\top \mathbf{s}_i)^{-1} \mathbf{h}_i^p, \quad p = 1, 2, \ldots, n \tag{56}$$

Proof. Conditions (52), (53) with position (56) are equivalent to

$$(A_i + G_i K_i)\mathbf{s}_i + \sum_{j=1}^{N} \lambda_{ji} \mathbf{s}_j + [\bar{\pi}]_i B_i \mathbf{e}_k \ll 0 \tag{57}$$

$$\mathbf{1}_{n_z}^\top \sum_{i=1}^{N} ((C_i + H_i K_i)\mathbf{s}_i + [\bar{\pi}]_i D_i \mathbf{e}_k) < \rho \tag{58}$$

for all $i \in \mathcal{N}$. Moreover conditions (54), (55) with (56) are equivalent to say that $A_i + G_i K_i$, $i \in \mathcal{N}$, are Metzler matrices and $C_i + H_i K_i$ are nonnegative matrices, $i \in \mathcal{N}$. Therefore the closed-loop system is positive. Let $\tilde{G}_{\mathbb{K}}(s)$ be the mean transfer function of the PMJLS with $\mathbf{u} = K_\theta \mathbf{x}$. We have to prove that its $\mathcal{L}_1$-induced norm is less than ρ. To this aim, recall that such a norm is given by $\|\tilde{G}_{\mathbb{K}}(0)\|_1$. Then,

$$\|\tilde{G}_{\mathbb{K}}(0)\|_1 = \max_{k=1,2,\ldots,n_w} \|\mathbf{1}_{n_z}^\top \tilde{G}_{\mathbb{K}}(0)\mathbf{e}_k\|_1 = \max_{k=1,2,\ldots,n_w} \|\mathbf{1}_{n_z}^\top \tilde{G}_{\mathbb{K}}(0)\mathbf{e}_k\|_\infty$$

The last equality holds true since $\mathbf{1}_{n_z}^\top \tilde{G}_{\mathbb{K}}(0)\mathbf{e}_k$ is a scalar. Thanks to Proposition 2, inequalities (57), (58) imply that $\|\mathbf{1}_{n_z}^\top \tilde{G}_{\mathbb{K}}(0)\mathbf{e}_k\|_\infty < \rho$, for any $k = 1, 2, \ldots, n_w$, so that the conclusion $\|\tilde{G}_{\mathbb{K}}(0)\|_1 < \rho$ is proven.

Remark 7. The conditions of Theorem 3 are also necessary (and hence provide a complete parametrization of the set $\mathcal{K}_1$) in the case $n_w = 1$, i.e. for single disturbance systems. As a matter of fact, in such a case it results that

$$J_1(\mathbb{K}) = \|\tilde{G}_{\mathbb{K}}(0)\|_1 = \mathbf{1}_{n_z}^\top \tilde{G}_{\mathbb{K}}(0) = \|\mathbf{1}_{n_z}^\top \tilde{G}_{\mathbb{K}}(0)\|_\infty$$

Then, imposing $J_1(\mathbb{K}) < \rho$ is equivalent to imposing $\|\mathbf{1}_{n_z}^\top \tilde{G}_{\mathbb{K}}(0)\|_\infty < \rho$, which can be performed by means of Theorem 2 applied to a system with scalar output $\mathbf{1}_{n_z}^\top \mathbf{z}(t)$. For this system, the inequalities (44), (45) of Theorem 2 coincide with the inequalities (52), (53) of Theorem 3 with $n_w = 1$. The reason why the conditions of Theorem 3 are not necessary in the general case $n_w > 1$ is the requirement of the existence of common vectors $\mathbf{s}_i$ and $\mathbf{h}_i^r$ satisfying (52), (53) for all input channels $k = 1, 2, \ldots, n_w$. On the other hand, relaxing this requirement would lead to channel dependent gains, in view of (56). With similar arguments as those in [8], it is argued that necessity of the conditions of Theorem 3 holds for robust performance if matrices $B_1, B_2, \ldots, B_N$ and $D_1, D_2, \ldots, D_N$ belong to a suitable uncertainty set. Moreover notice that, when the system is single-input single-output ($n_w = n_z = 1$), the conditions of Theorems 2 and 3 do coincide. As a matter of fact, for a scalar transfer function all induced norms are equal. Finally, for $n_u = 1$, all admissible gains $\mathbb{K}$ must satisfy the constraint (50).

4 Stabilization and Norm Minimization via Mode-Independent State-Feedback

This section is devoted to the design of a single state-feedback gain K, independent of the current mode θ, such that the closed-loop system is M-stable and satisfies performance requirements in terms of $\mathscr{L}_\infty$–induced or $\mathscr{L}_1$–induced norms.

4.1 Mode-Independent M-Stabilization

First we study the problem of mode-independent M-stabilization. To be precise, we aim at finding a single K such that

$$\dot{\mathbf{x}} = (A_\theta + G_\theta K)\mathbf{x} \tag{59}$$

is an M-stable PMJLS. From Theorem 1 above we can derive a necessary and sufficient condition for mode-independent M-stabilization by imposing that

$$(\mathbf{e}_r^\top s_j)\mathbf{h}_i^r = (\mathbf{e}_r^\top s_i)\mathbf{h}_j^r, \quad i \in \mathscr{N}, j \in \mathscr{N}, \quad r = 1, 2, \cdots, n \tag{60}$$

These are bilinear constraints associated to (33), (34), which are difficult to handle. However, a full parametrization of the stabilizing gains K can be obtained as shown in the following theorem.

Theorem 4. *There exists K such that $A_i + G_i K$, $i \in \mathscr{N}$, are Metzler matrices and the closed-loop system (59) is M-stable if and only if there exist strictly positive vectors $\mathbf{s}_i \in \mathbb{R}_+^n$ and vectors $\underline{\mathbf{h}}_i^p \in \mathbb{R}^{n_u}$, $\overline{\mathbf{h}}_i^p \in \mathbb{R}^{n_u}$, $i \in \mathscr{N}$, $p = 1, 2, \ldots, n$, such that*

$$A_i \mathbf{s}_i + G_i \sum_{p=1}^{n} \overline{\mathbf{h}}_i^p + \sum_{j=1}^{N} \lambda_{ji} \mathbf{s}_j \ll 0, \tag{61}$$

$$\mathbf{e}_r^\top G_i \underline{\mathbf{h}}_i^p + [A_i]_{rp} \mathbf{e}_p^\top \mathbf{s}_i \geq 0, \quad r \neq p = 1, 2, \ldots, n \tag{62}$$

$$\underline{\mathbf{h}}_i^p \leq \overline{\mathbf{h}}_i^p, \quad p = 1, 2, \ldots, n \tag{63}$$

for all $i \in \mathcal{N}$. All admissible gains K are then obtained from

$$(\mathbf{e}_p^\top \mathbf{s}_i)^{-1}\underline{\mathbf{h}}_i^p \leq K\mathbf{e}_p \leq (\mathbf{e}_p^\top \mathbf{s}_i)^{-1}\overline{\mathbf{h}}_i^p, \quad i \in \mathcal{N}, p = 1, 2, \ldots, n \tag{64}$$

Proof. First note that inequalities (63) are necessary in order to ensure that all the intervals in (64) are not void. Now, assume that (61)–(62) are feasible and take K satisfying (64). First, we show that $A_i + G_iK$ are Metzler matrices. Indeed, from (63) and the left inequality of (64) it holds that, for $r \neq p$,

$$\begin{aligned}
[A_i + G_iK]_{rp} &= [A_i]_{rp} + \mathbf{e}_r^\top G_iK\mathbf{e}_p \\
&\geq (\mathbf{e}_p^\top \mathbf{s}_i)^{-1}\left(\mathbf{e}_r^\top G_i\underline{\mathbf{h}}_i^p + [A_i]_{rp}\mathbf{e}_p^\top \mathbf{s}_i\right) \\
&\geq 0
\end{aligned}$$

As for stability, taking into account (61) and the right inequality of (64), it follows that

$$\begin{aligned}
(A_i + G_iK)\mathbf{s}_i + \sum_{j=1}^{N} \lambda_{ji}\mathbf{s}_j &= A_i\mathbf{s}_i + G_i \sum_{p=1}^{n} K\mathbf{e}_p\mathbf{e}_p^\top \mathbf{s}_i + \sum_{j=1}^{N} \lambda_{ji}\mathbf{s}_j \\
&\leq A_i\mathbf{s}_i + G_i \sum_{p=1}^{n} \overline{\mathbf{h}}_i^p + \sum_{j=1}^{N} \lambda_{ji}\mathbf{s}_j \\
&\ll 0
\end{aligned}$$

implying M-stability of the closed-loop system.

Viceversa, suppose that K is an admissible gain. Taking $\underline{\mathbf{h}}_i^p = \overline{\mathbf{h}}_i^p = K\mathbf{e}_p\mathbf{e}_p^\top \mathbf{s}_i$ for all i and p, the gain is consistent with inequalities (64) and conditions (62), (63) are trivially verified. M-stability of the closed-loop system is equivalent to the existence of strictly positive vectors $\mathbf{s}_i$ satisfying

$$(A_i + G_iK)\mathbf{s}_i + \sum_{j=1}^{N} \lambda_{ji}\mathbf{s}_j \ll 0$$

With the aforementioned definition of $\overline{\mathbf{h}}_i^p$, (61) directly follows.

Remark 8. The above theorem provides a full parametrization of the set of mode-independent M-stabilizing gains. However, conditions (64) are not linear in the unknowns. Moreover, feasibility of the linear constraints (61)–(63) does not ensure that a solution K exists in the intersection of the vector intervals defined in (64).

If $n_u = 1$, a necessary and sufficient condition for the existence of a mode-independent M-stabilizing gain, also providing a full parametrization of the set of M-stabilizing gains, is now presented. Preliminarily define the matrices

$$\underline{K}^{[A,G]} = \max_{i \in \mathcal{N}} \underline{K}_i^{[A,G]}, \quad \overline{K} = \min_{i \in \mathcal{N}} \overline{K}_i \tag{65}$$

where the max and min operators are to be intended elementwise. From Remark 3, it appears that all M-stabilizing gains K must satisfy $\underline{K}^{[A,G]} \leq K \leq \overline{K}$. Thus, we have the following result.

Theorem 5. *Let $n_u = 1$, $G_i \gg 0$, $i \in \mathcal{N}$, and recall (65). There exists K such that $A_i + G_i K$, $i \in \mathcal{N}$, are Metzler matrices and the closed-loop system (59) is M-stable if and only if it is M-stable with $u(t) = \underline{K}^{[A,G]}\mathbf{x}(t)$. Moreover, all admissible K satisfy*

$$\underline{K}^{[A,G]}\mathbf{e}_p \leq K\mathbf{e}_p \leq \underline{K}^{[A,G]}\mathbf{e}_p + (\mathbf{e}_p^\top \mathbf{s}_i)^{-1}\hat{h}_i^p, \quad i \in \mathcal{N}, p = 1, 2, \ldots, n \tag{66}$$

where the strictly positive vectors $\mathbf{s}_i \in \mathbb{R}_+^n$ and the nonnegative scalars $\hat{h}_i^p$, $i \in \mathcal{N}$, $p = 1, 2, \ldots, n$, solve the inequalities

$$\left(A_i + G_i \underline{K}^{[A,G]}\right)\mathbf{s}_i + G_i \sum_{p=1}^{n} \hat{h}_i^p + \sum_{j=1}^{N} \lambda_{ji}\mathbf{s}_j \ll 0 \tag{67}$$

for all $i \in \mathcal{N}$.

Proof. The proof straightforwardly derives from the definition of $\underline{K}^{[A,G]}$ and the nonnegativity of the scalars $\hat{h}_i^p$.

An alternative approach to design a mode-independent feedback gain K in the multi-input case $n_u > 1$ is based on the so-called minimax theory, see [10], which allows to find an admissible solution in the convex hull of a number of given gain matrices. To this end, the following result is important.

Lemma 1. *Let a $N \times M$ real matrix Q be given. The following statements are equivalent:*

(i)

$$\exists \mu \in \mathcal{P}_M : Q\mu \ll 0$$

(ii)

$$\min_{\mu \in \mathcal{P}_N} \max_{\nu \in \mathcal{P}_M} \nu^\top Q\mu < 0$$

If $N = M$, then (i) and (ii) are equivalent to

(iii)

$$\mathbf{x}^\top (Q + Q^\top)\mathbf{x} < 0, \; \forall \mathbf{x} > 0$$

Let K_i, $i \in \mathcal{N}$, be constructed as in (35) of Theorem 1, where $\mathbf{s}_i$, $\mathbf{h}_i$ solve (33), (34). Such gains solve the mode-dependent M-stabilization problem. One can look for a mode-independent gain K in the convex hull of the given mode-dependent gains K_i.

Theorem 6. *Let K_i, $i \in \mathscr{N}$, be given. There exists K in the convex hull of K_i such that the closed-loop system (59) is positive and M-stable if there exist strictly positive vectors $\mathbf{s}_i \in \mathbb{R}^n_+$ and a matrix $Q \in \mathbb{R}^{N \times N}$ such that*

$$(A_i + G_i K_j)\mathbf{s}_i + \sum_{k=1}^{N} \lambda_{ki}\mathbf{s}_k - [Q]_{ij}\mathbf{1}_n \ll 0, \quad j \in \mathscr{N} \tag{68}$$

$$\mathbf{e}_r^\top \left(A_i + G_i K_j\right) \mathbf{e}_p + [Q]_{ij} \geq 0, \quad j \in \mathscr{N}, \; r \neq p = 1, 2, \ldots, n \tag{69}$$

$$Q + Q^\top \prec 0 \tag{70}$$

for all $i \in \mathscr{N}$. An admissible mode independent gain matrix K is then obtained from

$$K = \sum_{j=1}^{N} [\boldsymbol{\mu}]_j K_j \tag{71}$$

where $\boldsymbol{\mu}$ is any solution in $\mathscr{P}_N$ of $Q\boldsymbol{\mu} \ll 0$.

Proof. In view of Lemma 1, condition (70) guarantees the existence of $\boldsymbol{\mu} \in \mathscr{P}_N$ satisfying $Q\boldsymbol{\mu} \ll 0$. Multiplying inequalities (68), (69) by $[\boldsymbol{\mu}]_j$ and summing up, it results that

$$(A_i + G_i K)\mathbf{s}_i + \sum_{k=1}^{N} \lambda_{ki}\mathbf{s}_k \ll 0, \quad i \in \mathscr{N}$$

$$\mathbf{e}_r^\top (A_i + G_i K)\, \mathbf{e}_p \geq 0, \quad i \in \mathscr{N}, \; r \neq p = 1, 2, \ldots, n$$

with K given by (71). Therefore all matrices $A_i + G_i K$ are Metzler and the closed-loop system with $\mathbf{u}(t) = K\mathbf{x}(t)$ is M-stable.

A similar design method can be worked out if nN gains K_j are given. The following result is in order.

Theorem 7. *Let K_j, $j = 1, 2, \cdots, nN$ be given. There exists K in the convex hull of K_j such that the closed-loop system is positive and M-stable if there exist vectors $\mathbf{q}_{ij} \in \mathbb{R}^n$, $i \in \mathscr{N}$, $j = 1, 2, \ldots, nN$, strictly positive vectors $\mathbf{s}_i \in \mathbb{R}^n_+$, $i \in \mathscr{N}$, and $\boldsymbol{\mu} \in \mathscr{P}_{nN}$ such that*

$$(A_i + G_i K_j)\mathbf{s}_i + \sum_{k=1}^{N} \lambda_{ki}\mathbf{s}_k - \mathbf{q}_{ij} \ll 0, \quad j = 1, 2, \ldots, nN \tag{72}$$

$$\mathbf{e}_r^\top \left(A_i + G_i K_j\right) \mathbf{e}_p + \mathbf{e}_r^\top \mathbf{q}_{ij} \geq 0, \quad j = 1, 2, \ldots, nN, \; r \neq p = 1, 2, \ldots, n \tag{73}$$

$$Q + Q^\top \prec 0 \tag{74}$$

for all $i \in \mathscr{N}$, *where the square matrix* $Q \in \mathbb{R}^{nN \times nN}$ *is given by* $Q = \mathrm{row}_j\{\mathrm{col}_i\{\mathbf{q}_{ij}\}\}$. *An admissible mode independent gain matrix* K *is then obtained from*

$$K = \sum_{j=1}^{nN} [\mu]_j K_j \tag{75}$$

where μ *is any solution in* $\mathscr{P}_{nN}$ *of* $Q\mu \ll 0$.

Proof. The proof is similar to that of Theorem 6 and is therefore omitted.

Remark 9. The previous results only offer sufficient conditions for mode-independent M-stabilization via convex programming. The gains K_j that are given in advance can be chosen by solving the mode-dependent M-stabilization problem. A necessary condition for the existence of an M-stabilizing gain K is indeed the existence of mode-dependent gains in $\mathscr{K}$ solving Problem 1.

4.2 $\mathscr{L}_\infty$ Mode-Independent Control

Assume that $\mathbf{w}(t)$, $t \geq 0$, is a nonnegative bounded disturbance for system (28), (29). We are interested in finding K such that $A_i + G_i K$, $i \in \mathscr{N}$, are Metzler matrices and the $\mathscr{L}_\infty$-induced norm of the closed-loop system

$$\dot{\mathbf{x}}(t) = \left(A_{\theta(t)} + G_{\theta(t)} K\right) \mathbf{x}(t) + B_{\theta(t)} \mathbf{w}(t) \tag{76}$$

$$\mathbf{z}(t) = \left(C_{\theta(t)} + H_{\theta(t)} K\right) \mathbf{x}(t) + D_{\theta(t)} \mathbf{w}(t) \tag{77}$$

is less than a prescribed bound ρ. Here, we consider only the $\mathscr{L}_\infty$-induced norm associated with deterministic positive bounded signals $\mathbf{w}(t)$, i.e.

$$J_\infty(K) = \sup_{\mathbf{w} \in \mathscr{L}_\infty, \mathbf{w} > 0} \frac{\sup_{k,t \geq 0} E[[\mathbf{z}(t)]_k]}{\sup_{k,t \geq 0} [\mathbf{w}(t)]_k}$$

The aim is to find a gain K (and possibly a parameterization of all gains K) such that (1) the closed-loop system is M-stable, (2) the closed-loop system has the state-space description of a positive system, (3) $J_\infty(K) < \rho$ for a given positive scalar ρ. A first result in this direction is provided by the following theorem.

Theorem 8. *There exists* K *such that the closed-loop system (76), (77) is positive and M-stable with* $\mathscr{L}_\infty$*-induced norm less than* $\rho > 0$ *if and only if there exist strictly positive vectors* $\mathbf{s}_i \in \mathbb{R}^n_+$ *and vectors* $\underline{\mathbf{h}}_i^p \in \mathbb{R}^{n_u}$, $\overline{\mathbf{h}}_i^p \in \mathbb{R}^{n_u}$, $i \in \mathscr{N}$, $p = 1, 2, \ldots, n$, *such that*

$$A_i \mathbf{s}_i + G_i \sum_{p=1}^{n} \overline{\mathbf{h}}_i^p + \sum_{j=1}^{N} \lambda_{ji} \mathbf{s}_j + [\bar{\pi}]_i B_i \mathbf{1}_{n_w} \ll 0 \tag{78}$$

$$\sum_{i=1}^{N} \left(C_i \mathbf{s}_i + H_i \sum_{p=1}^{n} \overline{\mathbf{h}}_i^p + [\bar{\pi}]_i D_i \mathbf{1}_{n_w} \right) \ll \rho \mathbf{1}_{n_z} \tag{79}$$

$$\mathbf{e}_r^\top G_i \underline{\mathbf{h}}_i^p + [A]_{rp} \mathbf{e}_p^\top \mathbf{s}_i \geq 0, \quad r \neq p \tag{80}$$

$$\mathbf{e}_q^\top H_i \underline{\mathbf{h}}_i^p + [C]_{qp} \mathbf{e}_p^\top \mathbf{s}_i \geq 0 \tag{81}$$

$$\underline{\mathbf{h}}_i^p \leq \overline{\mathbf{h}}_i^p \tag{82}$$

for all $i \in \mathcal{N}$, $r = 1, 2, \ldots, n$, $p = 1, 2, \ldots, n$, $q = 1, 2, \ldots, n_z$. *All admissible gains K are then obtained from*

$$(\mathbf{e}_p^\top \mathbf{s}_i)^{-1} \underline{\mathbf{h}}_i^p \leq K \mathbf{e}_p \leq (\mathbf{e}_p^\top \mathbf{s}_i)^{-1} \overline{\mathbf{h}}_i^p, \quad i \in \mathcal{N}, p = 1, 2, \ldots, n \tag{83}$$

Proof. The proof is similar to the one of Theorem 4 and hence is only sketched. Assume that (78)–(82) are feasible and take a gain K satisfying (83). Positivity of the closed-loop system can be proven thanks to inequalities (80), (81) and the left inequality of (83). Moreover, inequalities (78), (79) and the right inequality of (83) entail that

$$(A_i + G_i K) \mathbf{s}_i + \sum_{j=1}^{N} \lambda_{ji} \mathbf{s}_j + [\bar{\pi}]_i B_i \mathbf{1}_{n_w} \ll 0$$

$$\sum_{i=1}^{N} ((C_i + H_i K) \mathbf{s}_i + [\bar{\pi}]_i D_i \mathbf{1}_{n_w}) \ll \rho \mathbf{1}_{n_z}$$

for all $i \in \mathcal{N}$. Hence the closed-loop $\mathcal{L}_\infty$-induced norm is less than ρ, in view of Proposition 2.

Viceversa, if an admissible gain K exists, taking $\underline{\mathbf{h}}_i^p = \overline{\mathbf{h}}_i^p = K \mathbf{e}_p \mathbf{e}_p^\top \mathbf{s}_i$ for all i and p, the gain is consistent with inequalities (83) and conditions (80)–(82) are trivially verified. M-stability of the closed-loop system with $\mathcal{L}_\infty$-induced norm less than ρ is equivalent to the existence of strictly positive vectors $\mathbf{s}_i$ satisfying

$$(A_i + G_i K) \mathbf{s}_i + \sum_{j=1}^{N} \lambda_{ji} \mathbf{s}_j + [\bar{\pi}]_i B_i \mathbf{1}_{n_w} \ll 0$$

$$\sum_{i=1}^{N} ((C_i + H_i K) \mathbf{s}_i + [\bar{\pi}]_i D_i \mathbf{1}_{n_w}) \ll \rho \mathbf{1}_{n_z}$$

With the aforementioned definition of $\overline{\mathbf{h}}_i^p$, the inequalities (78), (79) directly follow.

Remark 10. A similar observation as that done in Remark 8 holds for the above theorem. In particular, feasibility of the linear constraints (78)–(82) does not guarantee the existence of a feasible mode-independent gain K.

Remarkably, for the single input case ($n_u = 1$) a necessary and sufficient condition holds, based on the definition of

$$\underline{K}^\star = \max\{\underline{K}^{[A,G]}, \underline{K}^{[C,H]}\} \tag{84}$$

where the maximum is to be taken elementwise, $\underline{K}^{[A,G]}$ is defined in (65), while

$$\underline{K}^{[C,H]} = \max_{i \in \mathscr{N}} \underline{K}_i^{[C,H]}$$

It is apparent that any admissible gain K must satisfy the constraint

$$\underline{K}^\star \leq K \leq \overline{K} \tag{85}$$

Theorem 9. *Let $n_u = 1$, $G_i \gg 0$, $i \in \mathscr{N}$ and recall (84). There exists K such that the closed-loop system (76), (77) is positive and M-stable with $\mathscr{L}_\infty$-induced norm less than $\rho > 0$ if and only if it is so with $u(t) = \underline{K}^\star \mathbf{x}(t)$. Moreover, all admissible K satisfy*

$$\underline{K}^\star \mathbf{e}_p \leq K\mathbf{e}_p \leq \underline{K}^\star \mathbf{e}_p + (\mathbf{e}_p^\top \mathbf{s}_i)^{-1} \hat{h}_i^p, \quad i \in \mathscr{N}, p = 1, 2, \ldots, n \tag{86}$$

where the strictly positive vectors $\mathbf{s}_i \in \mathbb{R}_+^n$ and the nonnegative scalars $\hat{h}_i^p$, $i \in \mathscr{N}$, $p = 1, 2, \ldots, n$, solve the inequalities

$$(A_i + G_i\underline{K}^\star)\,\mathbf{s}_i + G_i \sum_{p=1}^{n} \hat{h}_i^p + \sum_{j=1}^{N} \lambda_{ji}\mathbf{s}_j \ll 0 \tag{87}$$

$$\sum_{i=1}^{N} \left((C_i + H_i\underline{K}^\star)\,\mathbf{s}_i + H_i \sum_{p=1}^{n} \hat{h}_i^p + [\bar{\pi}]_i D_i \mathbf{1}_{n_w} \right) \ll \rho \mathbf{1}_{n_z} \tag{88}$$

for all $i \in \mathscr{N}$.

Proof. The proof straightforwardly derives from the definition of $\underline{K}^\star$ and the nonnegativity of the scalars $\hat{h}_i^p$.

Remark 11. Recalling Remarks 3 and 6, in the scalar input case ($n_u = 1$) and $G_i \gg 0$, it is possible to conclude that the gain K necessarily satisfies the constraint (85). Notice also that a solution to the mode-independent $\mathscr{L}_\infty$-induced control exists if and only if the closed-loop system with $u(t) = \underline{K}^\star \mathbf{x}(t)$ is M-stable and $J_\infty(\underline{K}^\star) < \rho$. This can be checked via LP by means of Proposition 2.

A minimax approach, similar to the one provided for mode-independent M-stabilization, can be worked out. The aim is to search for a mode-independent gain K in the convex hull of pre-computed mode-dependent gains K_i, $i \in \mathscr{N}$. In this respect, we have the following sufficient condition, whose proof is omitted as similar to the ones presented previously.

Theorem 10. *Let K_i, $i \in \mathscr{N}$, be given. There exists K in the convex hull of K_i such that the closed-loop system (76), (77) is positive and M-stable with $\mathscr{L}_\infty$-induced norm less than $\rho > 0$ if there exist strictly positive vectors $\mathbf{s}_i \in \mathbb{R}^n_+$ and a matrix $Q \in \mathbb{R}^{N\times N}$ such that*

$$(A_i + G_iK_j)\mathbf{s}_i + \sum_{k=1}^{N} \lambda_{ki}\mathbf{s}_k + [\bar{\boldsymbol{\pi}}]_i B_i \mathbf{1}_{n_w} - [Q]_{ij}\mathbf{1}_n \ll 0 \tag{89}$$

$$\sum_{i=1}^{N} \left((C_i + H_iK_j)\, \mathbf{s}_i + [\bar{\boldsymbol{\pi}}]_i D_i \mathbf{1}_{n_w} - [Q]_{ij}\mathbf{1}_{n_z} \right) \ll \rho \mathbf{1}_{n_z} \tag{90}$$

$$\mathbf{e}_r^\top (A_i + G_iK_j)\mathbf{e}_p + [Q]_{ij} \geq 0, \quad r \neq p \tag{91}$$

$$\mathbf{e}_q^\top (C_i + H_iK_j)\mathbf{e}_p + [Q]_{ij} \geq 0 \tag{92}$$

for all $i \in \mathscr{N}$, $r = 1, 2, \ldots, n$, $p = 1, 2, \ldots, n$, $q = 1, 2, \ldots, n_z$. An admissible mode independent gain matrix K is then obtained from

$$K = \sum_{j=1}^{N} [\boldsymbol{\mu}]_j K_j \tag{93}$$

where $\boldsymbol{\mu}$ is any solution in $\mathscr{P}_N$ of $Q\boldsymbol{\mu} \ll 0$.

It goes without saying that an extension of Theorem 7 to cope with mode-independent gain design with guaranteed $\mathscr{L}_\infty$-induced performance could be also worked out, searching K in the convex hull of given nN control gain matrices K_i.

4.3 Mode-Independent $\mathscr{L}_1$-Induced Optimal Control

In this section we assume that $\mathbf{w}(t)$, $t \geq 0$, is a nonnegative integrable deterministic disturbance for system (28), (29). With reference to the closed-loop system (76), (77), we are interested in finding K such that $A_i + G_iK$, $i \in \mathscr{N}$, are Metzler matrices and the $\mathscr{L}_1$-induced norm

$$J_1(K) = \sup_{\mathbf{w} \in \mathscr{L}_1, \mathbf{w} > 0} \frac{E[\int_0^\infty \mathbf{1}_{n_z}^\top \mathbf{z}(t)dt]}{\int_0^\infty \mathbf{1}_{n_w}^\top \mathbf{w}(t)dt}$$

is minimized. More precisely, the aim is to find a gain K (and possibly a parameterization of all gains K) such that (1) the closed-loop system is M-stable, (2) the closed-loop system has the state-space description of a positive system, (3) $J_1(K) < \rho$ for a given positive scalar ρ.

The following result does not provide a complete parametrization but only a sufficient condition for the existence of a set of feedback gain matrices ensuring positivity of the closed-loop system, M-stability and guaranteed $\mathscr{L}_1$-induced norm. The proof is omitted since similar to the ones provided so far.

Theorem 11. *There exists K such that the closed-loop system (76), (77) is positive and M-stable with $\mathscr{L}_1$-induced norm less than $\rho > 0$ if there exist strictly positive vectors $\mathbf{s}_i \in \mathbb{R}^n_+$ and vectors $\underline{\mathbf{h}}_i^p \in \mathbb{R}^{n_u}$, $\overline{\mathbf{h}}_i^p \in \mathbb{R}^{n_u}$, $i \in \mathscr{N}$, $p = 1, 2, \ldots, n$, such that*

$$A_i\mathbf{s}_i + G_i\sum_{p=1}^{n}\overline{\mathbf{h}}_i^p + \sum_{j=1}^{N}\lambda_{ji}\mathbf{s}_j + [\bar{\pi}]_i B_i\mathbf{e}_k \ll 0 \tag{94}$$

$$\mathbf{1}_p^\top\sum_{i=1}^{N}\left(C_i\mathbf{s}_i + H_i\sum_{p=1}^{n}\overline{\mathbf{h}}_i^p + [\bar{\pi}]_i D_i\mathbf{e}_k\right) < \rho \tag{95}$$

$$\mathbf{e}_r^\top G_i\underline{\mathbf{h}}_i^p + [A]_{rp}\mathbf{e}_p^\top\mathbf{s}_i \geq 0, \quad r \neq p \tag{96}$$

$$\mathbf{e}_q^\top H_i\underline{\mathbf{h}}_i^p + [C]_{qp}\mathbf{e}_p^\top\mathbf{s}_i \geq 0 \tag{97}$$

$$\underline{\mathbf{h}}_i^p \leq \overline{\mathbf{h}}_i^p \tag{98}$$

for all $i \in \mathscr{N}$, $r = 1, 2, \ldots, n$, $p = 1, 2, \ldots, n$, $q = 1, 2, \ldots, n_z$, $k = 1, 2, \ldots, n_w$. Admissible gains K are then given as follows:

$$(\mathbf{e}_p^\top\mathbf{s}_i)^{-1}\underline{\mathbf{h}}_i^p \leq K\mathbf{e}_p \leq (\mathbf{e}_p^\top\mathbf{s}_i)^{-1}\overline{\mathbf{h}}_i^p, \quad p = 1, 2, \ldots, n \tag{99}$$

Remark 12. Following Remark 7, the conditions of Theorem 11 are also necessary (and hence provide a complete parametrization of the feedback gains) in the case $n_w = 1$, i.e. for single disturbance systems.

Remark 13. In the scalar input case ($n_u = 1$) and $G_i \gg 0$, the gain K necessarily satisfies the constraint (85) and the mode-independent $\mathscr{L}_1$-induced control problem has a solution if and only if the closed-loop system with $u(t) = \underline{K}^\star\mathbf{x}(t)$ is M-stable and $J_1(\underline{K}^\star) < \rho$. This can be checked via LP by means of Proposition 3.

Remark 14. A minimax approach for the computation of a mode-independent gain K with guaranteed $\mathscr{L}_1$-induced performance can be pursued along the rationale used in the previous sections for M-stabilization and $\mathscr{L}_\infty$-induced performance, see Theorems 6, 7 and 10. For brevity, the details are omitted.

5 Examples

This section contains three examples. The third is an extension of the one presented in [3], and consists of a fourth-order compartmental model with four reservoirs.

Example 1. Consider the second-order PMJLS with $N = 2$

$$\dot{\mathbf{x}}(t) = A_{\theta(t)}\mathbf{x}(t) + \begin{bmatrix} 1 \\ 1 \end{bmatrix} (\mathbf{u}(t) + \mathbf{w}(t))$$

$$\mathbf{z}(t) = \begin{bmatrix} 1 & 1 \end{bmatrix} \mathbf{x}(t)$$

where

$$A_1 = \begin{bmatrix} 2 & 1 \\ 3 & -3 \end{bmatrix}, \quad A_2 = \begin{bmatrix} -2 & 8 \\ 1 & 3 \end{bmatrix}, \quad \Lambda = \begin{bmatrix} -1 & 1 \\ 5 & -5 \end{bmatrix}$$

We first aim at parameterizing the set of gains $\mathbb{K} = \{K_1, K_2\}$ such that the closed-loop system matrices $A_1 + G_1 K_1$ and $A_2 + G_2 K_2$ are Metzler and the associated closed-loop system is M-stable. For being Metzler, it is necessary and sufficient that

$$K_1 \geq \underline{K}_1^{[A,G]} = \begin{bmatrix} -3 & -1 \end{bmatrix}, \quad K_2 \geq \underline{K}_2^{[A,G]} = \begin{bmatrix} -1 & -8 \end{bmatrix}$$

Moreover, for stability it is necessary that

$$K_1 \leq \overline{K}_1 = \begin{bmatrix} -1-\epsilon & 4-\epsilon \end{bmatrix}, \quad K_2 \leq \overline{K}_2 = \begin{bmatrix} 7-\epsilon & 2-\epsilon \end{bmatrix}$$

Thanks to Remark 3, it can be concluded that Problem 1 is solvable if and only if $u(t) = \underline{K}_{\theta(t)}^{[A,G]} x(t)$ is a mode-dependent M-stabilizing feedback law. This is indeed the case. Notice also that, being $n_w = n_u = n_z = 1$, the $\mathscr{L}_\infty$-induced norm and the $\mathscr{L}_1$-induced norm do coincide and are minimized by taking the "smallest" possible gains $K_i = \underline{K}_i^\star = \underline{K}_i^{[A,G]}$, $i = 1, 2$. Therefore, the optimal value is $\rho = 1.0286$ and the optimal gains are

$$K_1 = \begin{bmatrix} -3 & -1 \end{bmatrix}, \quad K_2 = \begin{bmatrix} -1 & -8 \end{bmatrix}$$

Finally, for mode-independent M-stabilization, one has first to construct a feedback matrix $\underline{K}^{[A,G]} = \underline{K}^\star$ taking the maximum elements of K_i, namely

$$\underline{K}^\star = \begin{bmatrix} -1 & -1 \end{bmatrix}$$

Any matrix K ensuring that $A_i + G_i K$ are Metzler matrices should be such that $K \geq \underline{K}^\star$. The PMJLS with $u(t) = \underline{K}^\star x(t)$ is not M-stable, as witnessed by the violation of the second constraint of (85). In conclusion, no mode-independent M-stabilizing feedback gain exists.

Example 2. Consider the PMJLS (28), (29) with $N = 2$,

$$A_1 = \begin{bmatrix} 2 & 0 \\ 0 & -1 \end{bmatrix}, \quad A_2 = \begin{bmatrix} -15 & 1 \\ 0 & -2 \end{bmatrix}, \quad \Lambda = \begin{bmatrix} -1 & 1 \\ 5 & -5 \end{bmatrix}$$

$$B_1 = B_2 = \begin{bmatrix} 1 \\ 1 \end{bmatrix}, \quad G_1 = G_2 = \begin{bmatrix} 1 \\ 0 \end{bmatrix}$$

$$C_1 = C_2 = \begin{bmatrix} 1 & 1 \end{bmatrix}, \quad D_1 = D_2 = 0, \quad H_1 = H_2 = \alpha$$

As for the closed-loop system being positive, a necessary and sufficient condition is that $K_1\mathbf{e}_2 \geq \bar{\gamma}_1 = 0$, $K_2\mathbf{e}_2 \geq \bar{\gamma}_2 = -1$. A simple analysis of Hurwitz stability of the matrix $\tilde{F}$, defined in (6) with $F_i = A_i + G_iK_i$, reveals that the closed-loop system with $K_i = [\beta_i \quad \gamma_i]$ is M-stable for any $\gamma_i \geq \bar{\gamma}_i$ and β_i such that the matrix

$$\begin{bmatrix} 1 + \beta_1 & 1 \\ 5 & -20 + \beta_2 \end{bmatrix}$$

is Hurwitz stable. Notice also that a stabilizing mode-independent law exists. All such feedback matrices are given by

$$K = \begin{bmatrix} \beta & \gamma \end{bmatrix}, \quad \beta < -1.2355, \quad \gamma \geq 0$$

Now consider the scalar output z and consider first the case $\alpha = 0$. The optimization of $J_1(\mathbb{K}) = J_\infty(\mathbb{K})$ returns unbounded gains associated with the optimal cost $\rho = 0.8704$. The optimal mode-dependent gains are

$$K_1 = \begin{bmatrix} -\infty & 0 \end{bmatrix}, \quad K_2 = \begin{bmatrix} -\infty & -1 \end{bmatrix}$$

The fact that the optimal gains are unbounded can be easily explained by observing that the first entry of both gains can be made arbitrarily negative without destroying stability and positivity of the closed-loop system. Since with high feedback gain the expectation of the first state-variable decays to zero arbitrarily fast and the second is not affected by feedback, the optimal cost represents the optimal performance of the open-loop scalar subsystem

$$\dot{y}(t) = a_{\theta(t)}y(t) + w(t), \quad z(t) = y(t), \quad a_1 = -1, \qquad a_2 = -2$$

Similarly, it can be shown that the optimal mode-independent gain is

$$K = \max\{K_1, K_2\} = \begin{bmatrix} -\infty & 0 \end{bmatrix}$$

with the same value of the performance index.

Now let $\alpha > 0$, so as to weigh the input in the cost and obtain bounded gains. Note that, for a fixed value α, all the entries of the gains K_1, K_2 must be greater than

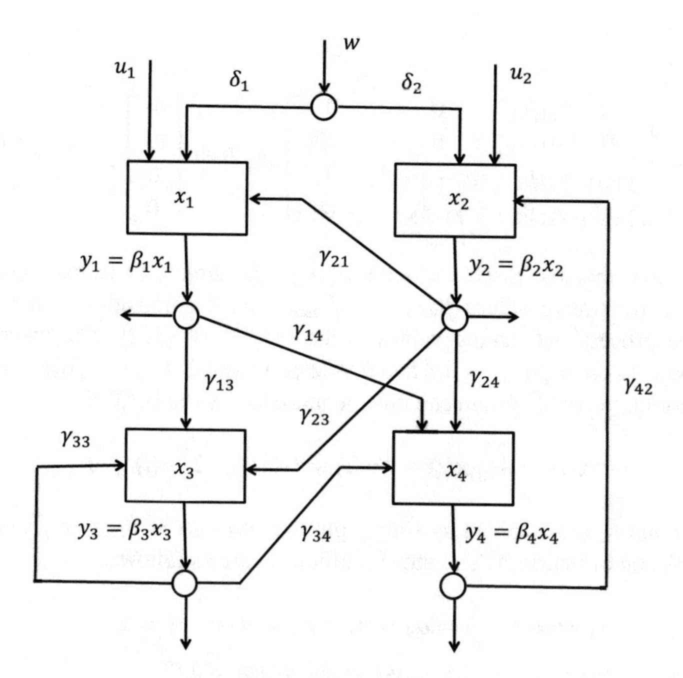

Fig. 1 The compartmental system considered in Example 3

$-\alpha^{-1}$ in order to guarantee positivity of $z(t)$. Considering also the requirement on the Metzler property of the closed-loop matrices, the mode-dependent gains must satisfy

$$K_1 \geq K_1^\star = \begin{bmatrix} -\alpha^{-1} & 0 \end{bmatrix}, \quad K_2 \geq K_2^\star = \begin{bmatrix} -\alpha^{-1} & \max\{-\alpha^{-1}, -1\} \end{bmatrix}$$

These constraints may prevent the existence of stabilizing gains. Indeed, it can be observed that closed-loop M-stability is lost for $\alpha \geq \alpha^* = 0.81$. The optimal value of the cost $J_1(\mathbb{K}) = J_\infty(\mathbb{K})$ is monotonically increasing with respect to α, from a minimum of 0.8704 (when $\alpha = 0$) to infinity (when α tends to α^*). The same occurs for the optimal mode-independent gain $K = \begin{bmatrix} -\alpha^{-1} & 0 \end{bmatrix}$.

Example 3. This example is concerned with a compartmental model with 4 reservoirs connected as in Fig. 1. The system state $\mathbf{x}$ is a 4-dimensional vector corresponding to the storage of water in the four reservoirs. The signal w is a scalar disturbance, while the inflows $u_1 = [\mathbf{u}]_1$ and $u_2 = [\mathbf{u}]_2$ are control inputs constrained by the assumption that the total control inflow is nonnegative, i.e. $u_1 + u_2 \geq 0$. The model takes the form

$$\dot{\mathbf{x}}(t) = A\mathbf{x}(t) + Bw(t) + G\mathbf{u}(t), \quad \mathbf{1}_2^\top \mathbf{u}(t) \geq 0$$

with

$$A = \begin{bmatrix} -\alpha_1 & \gamma_{21}\beta_2 & 0 & 0 \\ 0 & -\alpha_2 & 0 & \gamma_{42}\beta_4 \\ \gamma_{13}\beta_1 & \gamma_{23}\beta_2 & -\alpha_3 + \gamma_{33}\beta_3 & 0 \\ \gamma_{14}\beta_1 & \gamma_{24}\beta_2 & \gamma_{34}\beta_3 & -\alpha_4 \end{bmatrix}, \quad B = \begin{bmatrix} \delta_1 \\ \delta_2 \\ 0 \\ 0 \end{bmatrix}, \quad G = \begin{bmatrix} I_2 \\ 0 \end{bmatrix}$$

Suppose now that the discharge parameter β_2 is subject to sudden fluctuations between two extreme values β_{2min} and β_{2max}, and the transition is governed by a Markov process $\theta(t)$ taking values in the set $\mathcal{N} = \{1, 2\}$. The value $\theta = 1$ corresponds to $\beta_2 = \beta_{2max} = 0.8$ and $\theta = 2$ corresponds to $\beta_2 = \beta_{2min} = 0$.

The compartmental system can now be modeled by the PMJLS

$$\dot{\mathbf{x}}(t) = A_{\theta(t)}\mathbf{x}(t) + Bw(t) + G\mathbf{u}(t), \quad \mathbf{1}_2^\top \mathbf{u}(t) \geq 0 \tag{100}$$

where A_1 and A_2 are obtained by simply putting either $\beta_2 = \beta_{2max}$ or $\beta_2 = \beta_{2min}$ in the expression of matrix A. The other coefficients are as follows:

$$\alpha_1 = \alpha_2 = \alpha_3 = \alpha_4 = 1, \quad \beta_1 = \beta_3 = \beta_4 = 1$$

$$\gamma_{13} = \gamma_{14} = 0.5, \quad \gamma_{21} = \gamma_{23} = \gamma_{24} = 1/3$$

$$\gamma_{33} = 0.2, \quad \gamma_{34} = 0.4, \quad \gamma_{42} = 0.5, \quad \delta_1 = \delta_2 = 0.5$$

Assume that the transition rate matrix of the Markov process $\theta(t)$ is

$$\Lambda = \begin{bmatrix} -0.1 & 0.1 \\ 0.5 & -0.5 \end{bmatrix}$$

We will focus on the time evolution of the entire state, i.e. $\mathbf{z}(t) = \mathbf{x}(t)$. We first aim at finding the mode-dependent control law $\mathbf{u}(t) = K_{\theta(t)}\mathbf{x}(t)$ that minimizes the $\mathcal{L}_\infty$-induced norm from w to $\mathbf{z}$, under the constraint $u_1 + u_2 \geq 0$, that can be enforced by

$$\mathbf{1}_2^\top K_i \geq 0, \quad i = 1, 2 \tag{101}$$

As discussed in Sect. 2.2, the closed-loop $\mathcal{L}_\infty$-induced norm is

$$J_\infty(\mathbb{K}) = \sup_{t \geq 0} \max_k E[[\mathbf{x}(t)]_k]$$

obtained for $w(t) = 1,\ t \geq 0$. Theorem 2 has been applied along with the minimization of the guaranteed cost ρ, by also adding to (44)–(47) the additional linear constraint (101) in the form

$$\mathbf{1}_2^\top \mathbf{h}_i^p \geq 0, \quad i = 1, 2, \quad p = 1, 2, 3, 4$$

The result is given by (48), namely

$$K_1 = \begin{bmatrix} -367.61 & 519.95 & 0 & 0.083 \\ 367.61 & -519.95 & 0 & -0.083 \end{bmatrix}, \quad K_2 = \begin{bmatrix} -247190.12 & 1250.79 & 0 & 0.21 \\ 247190.12 & -1250.79 & 0 & -0.21 \end{bmatrix}$$

and the optimal attenuation level is $\rho = 0.67$. Notice that the open-loop system ($K_1 = K_2 = 0$) is M-stable and its $\mathscr{L}_\infty$-induced norm is $\rho = 0.74$. Hence, through a mode-dependent feedback a 10 % improvement has been achieved.

For mode-independent control, one can resort to Theorem 10, by searching K in the convex hull of matrices K_1 and K_2 computed above and minimizing the $\mathscr{L}_\infty$-induced norm. The optimal K in this set is $K = K_1$ yielding $\rho = 0.81$. As apparent, this value is greater than the one provided by the null gain. Thus, in this example, the minimax approach with the given K_i does not lead to an improvement with respect to open-loop.

Turn now to the $\mathscr{L}_1$-induced norm. The system disturbance is scalar and hence the optimal value of the attenuation level attainable via mode-dependent control can be found by minimizing ρ using Theorem 3. The worst disturbance, as discussed in Sect. 2.3, is $w(t) = \delta(t)$ and

$$J_1(\mathbb{K}) = E\left[\int_0^\infty \mathbf{1}_4^\top \mathbf{x}(t)dt\right]$$

is minimized with

$$K_1 = \begin{bmatrix} -298.39 & 310042.80 & 0 & 0.02 \\ 298.39 & -310042.80 & 0 & -0.02 \end{bmatrix}, \quad K_2 = \begin{bmatrix} -310117.93 & 1624.47 & 0 & 0.02 \\ 310117.93 & -1624.47 & 0 & -0.02 \end{bmatrix}$$

yielding the optimal attenuation level is $\rho = 2.30$. In Fig. 2, 20 realizations of the closed-loop impulse response under mode-dependent $\mathscr{L}_1$ optimal control are plotted, together with the transient of the expected value.

As for the mode-independent control, again one can look at the convex hull of matrices K_1 and K_2 and minimize the $\mathscr{L}_1$-induced norm. The result is $K = K_2$, to which corresponds $\rho = 2.48$. Notice that in open-loop the system is M-stable and the $\mathscr{L}_1$-induced norm is $\rho = 2.52$. Therefore in this case the minimax approach leads to a slight improvement with respect to open-loop.

6 Concluding Remarks

In this paper, several issues concerning state-feedback design for M-stabilization and norm optimization for the class of Positive Markov Jump Linear Systems have been addressed. In particular, full parametrization of the feedback gains for both mode-dependent and mode-independent stabilization and $\mathscr{L}_\infty$ control as been provided. Remarkably, in the mode-dependent and in the single-input

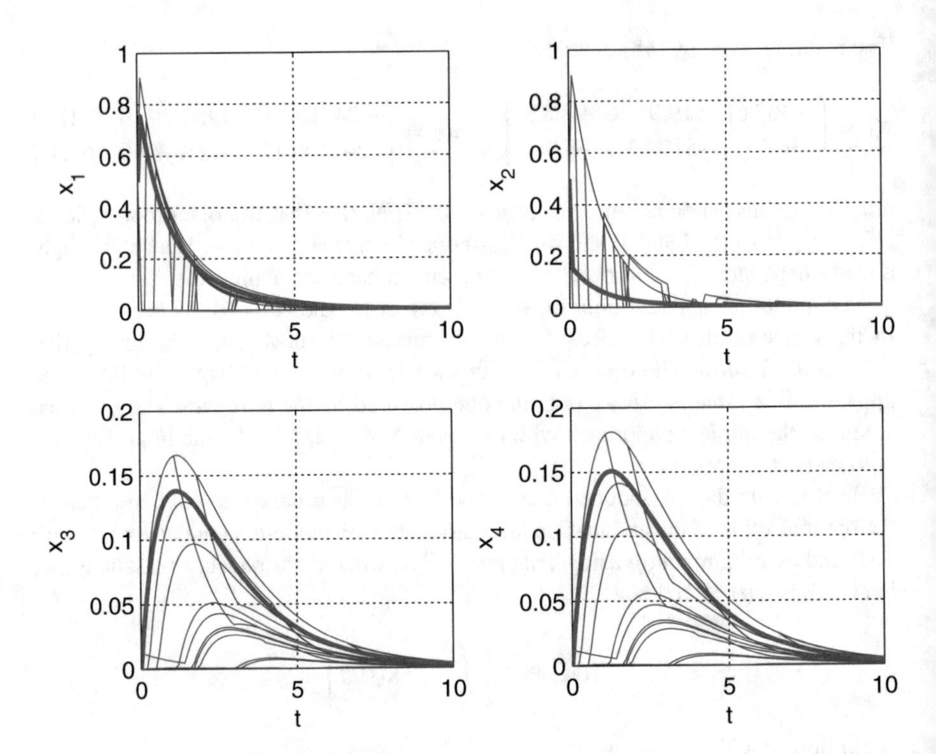

Fig. 2 Plot of 20 realizations of the mode-dependent closed-loop impulse response. The *red curves* represent the transient of the expected value

mode-independent cases, the parametrization is obtained through Linear Programming. Various problems are still to be more deeply understood, e.g. the multi-disturbance $\mathscr{L}_1$ control and the multi-input mode-independent $\mathscr{L}_\infty$ and $\mathscr{L}_1$ design. Moreover, how to extend, via convex programming, the above treatment to mean square stability and $\mathscr{L}_2$-induced norm, while preserving positivity, is still an open problem that deserves further investigation. A main difficulty in extending the theory for deterministic positive systems developed in [16] is the fact that the existence of diagonal solutions of the coupled Lyapunov-Metzler inequalities is not a necessary condition for mean square stability. Finally, following [15] in the deterministic case, an interesting open problem is to design output-feedback controllers for PMJLS forcing positivity and prescribed stochastic stability requirements.

References

1. Ait Rami, M., Bokharaie, V.S., Mason, O., Wirth, F.R.: Stability criteria for SIS epidemiological models under switching policies. Discrete Cont. Dyn. Ser. B **19**, 2865–2887 (2014)
2. Ait Rami, M., Shamma, J.: Hybrid positive systems subject to Markovian switching. In: Proceedings of the 3rd IFAC Conference on Analysis and Design of Hybrid Systems, Zaragoza (2009)

3. Bolzern, P., Colaneri, P.: Positive Markov jump linear systems. Found. Trends Syst. Control **2**(3–4), 275–427. http://dx.doi.org/10.1561/2600000006
4. Bolzern, P., Colaneri, P., De Nicolao, G.: Stochastic stability of positive Markov jump linear systems. Automatica **50**, 1181–1187 (2014)
5. Boukas, E.K.: Stochastic Switching Systems. Birkhäuser, Basel (2005)
6. Briat, C.: Robust stability and stabilization of uncertain linear positive systems via integral linear constraints: L_1-gain and L_∞-gain characterization. Int. J. Robust Nonlinear Control **23**, 1932–1954 (2013)
7. Costa, O.L.V., Fragoso, M.D., Todorov, M.G.: Continuous-time Markov Jump Linear Systems. Springer, Berlin (2013)
8. Ebihara, Y., Peaucelle, D., Arzelier, D.: Optimal L_1 controller synthesis for positive systems and its robustness properties. In: Proceedings of the American Control Conference, Montreal, pp. 5992–5997 (2012)
9. Farina, L., Rinaldi, S.: Positive Linear Systems - Theory and Applications. Wiley, New York, NY (2000)
10. Geromel, J.C., Deaecto G.S., Colaneri, P.: Minimax control of Markov jump linear systems. Adaptive Control and Signal Processing (2015). doi: 10.1002/acs.2575
11. Haddad, W., Chellaboina, V., Hui, Q.: Nonnegative and Compartmental Dynamical Systems. Princeton University Press, Princeton (2010)
12. Kaczorek, T.: Positive 1D and 2D Systems. Springer, London (2002)
13. Ogura, M., Preciado, V.M.: Optimal design of switched networks of positive linear systems via geometric programming. IEEE Trans. Control Netw. Syst. (2015). doi: 10.1109/TCNS.2015.2489339
14. Rantzer A.: Distributed control of positive systems. In: Proceedings of the 50th IEEE Conference on Decision and Control, Orlando, FL, pp. 6608–6611 (2011)
15. Rantzer, A.: Scalable control of positive systems. Eur. J. Control **24**, 72–80 (2015)
16. Tanaka, T., Langbort, C.: The bounded real lemma for internally positive systems and H-infinity structured static state feedback. IEEE Trans. Autom. Control **56**, 2218–2223 (2011)
17. Zhang, J., Han, Z., Zhu, F.: Stochastic stability and stabilization of positive systems with Markovian jump parameters. Nonlinear Anal.: Hybrid Syst. **12**, 147–155 (2014)

Matrix-Free Convex Optimization Modeling

Steven Diamond and Stephen Boyd

Abstract We introduce a convex optimization modeling framework that transforms a convex optimization problem expressed in a form natural and convenient for the user into an equivalent cone program in a way that preserves fast linear transforms in the original problem. By representing linear functions in the transformation process not as matrices, but as graphs that encode composition of linear operators, we arrive at a matrix-free cone program, i.e., one whose data matrix is represented by a linear operator and its adjoint. This cone program can then be solved by a matrix-free cone solver. By combining the matrix-free modeling framework and cone solver, we obtain a general method for efficiently solving convex optimization problems involving fast linear transforms.

Keywords Convex optimization • Matrix-free optimization • Conic programming • Optimization modeling

1 Introduction

Convex optimization modeling systems like YALMIP [83], CVX [57], CVXPY [36], and Convex.jl [106] provide an automated framework for converting a convex optimization problem expressed in a natural human-readable form into the standard form required by a solver, calling the solver, and transforming the solution back to the human-readable form. This allows users to form and solve convex optimization problems quickly and efficiently. These systems easily handle problems with a few thousand variables, as well as much larger problems (say, with hundreds of thousands of variables) with enough sparsity structure, which generic solvers can exploit.

S. Diamond (✉)
Department of Computer Science, Stanford University, Stanford, CA 94305, USA
e-mail: diamond@cs.stanford.edu

S. Boyd
Department of Electrical Engineering, Stanford University, Stanford, CA 94305, USA
e-mail: boyd@stanford.edu

B. Goldengorin (ed.), *Optimization and Its Applications in Control and Data Sciences*, Springer Optimization and Its Applications 115,
DOI 10.1007/978-3-319-42056-1_7

The overhead of the problem transformation, and the additional variables and constraints introduced in the transformation process, result in longer solve times than can be obtained with a custom algorithm tailored specifically for the particular problem. Perhaps surprisingly, the additional solve time (compared to a custom solver) for a modeling system coupled to a generic solver is often not as much as one might imagine, at least for modest sized problems. In many cases the convenience of easily expressing the problem makes up for the increased solve time using a convex optimization modeling system.

Many convex optimization problems in applications like signal and image processing, or medical imaging, involve hundreds of thousands or many millions of variables, and so are well out of the range that current modeling systems can handle. There are two reasons for this. First, the standard form problem that would be created is too large to store on a single machine, and second, even if it could be stored, standard interior-point solvers would be too slow to solve it. Yet many of these problems are readily solved on a single machine by custom solvers, which exploit fast linear transforms in the problems. The key to these custom solvers is to directly use the fast transforms, never forming the associated matrix. For this reason these algorithms are sometimes referred to as *matrix-free* solvers.

The literature on matrix-free solvers in signal and image processing is extensive; see, e.g., [9, 10, 22, 23, 51, 97, 117]. There has been particular interest in matrix-free solvers for LASSO and basis pursuit denoising problems [10, 24, 42, 46, 74, 108]. Matrix-free solvers have also been developed for specialized control problems [109, 110]. The most general matrix-free solvers target semidefinite programs [75] or quadratic programs and related problems [52, 99]. The software closest to a convex optimization modeling system for matrix-free problems is TFOCS, which allows users to specify many types of convex problems and solve them using a variety of matrix-free first-order methods [11].

To better understand the advantages of matrix-free solvers, consider the nonnegative deconvolution problem

$$\begin{array}{ll} \text{minimize} & \|c * x - b\|_2 \\ \text{subject to} & x \geq 0, \end{array} \tag{1}$$

where $x \in \mathbf{R}^n$ is the optimization variable, $c \in \mathbf{R}^n$ and $b \in \mathbf{R}^{2n-1}$ are problem data, and $*$ denotes convolution. Note that the problem data has size $O(n)$. There are many custom matrix-free methods for efficiently solving this problem, with $O(n)$ memory and a few hundred iterations, each of which costs $O(n \log n)$ floating point operations (flops). It is entirely practical to solve instances of this problem of size $n = 10^7$ on a single computer [77, 81].

Existing convex optimization modeling systems fall far short of the efficiency of matrix-free solvers on problem (1). These modeling systems target a standard form in which a problem's linear structure is represented as a sparse matrix. As a result, linear functions must be converted into explicit matrix multiplication. In particular, the operation of convolving by c will be represented as multiplication by a $(2n-1) \times n$ Toeplitz matrix C. A modeling system will thus transform problem (1) into the problem

$$\begin{array}{ll} \text{minimize} & \|Cx - b\|_2 \\ \text{subject to} & x \geq 0, \end{array} \tag{2}$$

as part of the conversion into standard form.

Once the transformation from (1) to (2) has taken place, there is no hope of solving the problem efficiently. The explicit matrix representation of C requires $O(n^2)$ memory. A typical interior-point method for solving the transformed problem will take a few tens of iterations, each requiring $O(n^3)$ flops. For this reason existing convex optimization modeling systems will struggle to solve instances of problem (1) with $n = 10^4$, and when they are able to solve the problem, they will be dramatically slower than custom matrix-free methods.

The key to matrix-free methods is to exploit fast algorithms for evaluating a linear function and its adjoint. We call an implementation of a linear function that allows us to evaluate the function and its adjoint a *forward-adjoint oracle* (FAO). In this paper we describe a new algorithm for converting convex optimization problems into standard form while preserving fast linear functions. (A preliminary version of this paper appeared in [35].) This yields a convex optimization modeling system that can take advantage of fast linear transforms, and can be used to solve large problems such as those arising in image and signal processing and other areas, with millions of variables. This allows users to rapidly prototype and implement new convex optimization based methods for large-scale problems. As with current modeling systems, the goal is not to attain (or beat) the performance of a custom solver tuned for the specific problem; rather it is to make the specification of the problem straightforward, while increasing solve times only moderately.

The outline of our paper is as follows. In Sect. 2 we give many examples of useful FAOs. In Sect. 3 we explain how to compose FAOs so that we can efficiently evaluate the composition and its adjoint. In Sect. 4 we describe cone programs, the standard intermediate-form representation of a convex problem, and solvers for cone programs. In Sect. 5 we describe our algorithm for converting convex optimization problems into equivalent cone programs while preserving fast linear transforms. In Sect. 6 we report numerical results for the nonnegative deconvolution problem (1) and a special type of linear program, for our implementation of the abstract ideas in the paper, using versions of the existing cone solvers SCS [94] and POGS [45] modified to be matrix-free. (The main modification was using the matrix-free equilibration described in [37].) Even with our simple, far from optimized matrix-free cone solvers, we demonstrate scaling to problems far larger than those that can be solved by generic methods (based on sparse matrices), with acceptable performance loss compared to specialized custom algorithms tuned to the problems.

We reserve certain details of our matrix-free canonicalization algorithm for the appendix. In "Equivalence of the Cone Program" we explain the precise sense in which the cone program output by our algorithm is equivalent to the original convex optimization problem. In "Sparse Matrix Representation" we describe how existing modeling systems generate a sparse matrix representation of the cone program. The details of this process have never been published, and it is interesting to compare with our algorithm.

2 Forward-Adjoint Oracles

A general linear function $f : \mathbf{R}^n \to \mathbf{R}^m$ can be represented on a computer as a dense matrix $A \in \mathbf{R}^{m\times n}$ using $O(mn)$ bytes. We can evaluate $f(x)$ on an input $x \in \mathbf{R}^n$ in $O(mn)$ flops by computing the matrix-vector multiplication Ax. We can likewise evaluate the adjoint $f^*(y) = A^T y$ on an input $y \in \mathbf{R}^m$ in $O(mn)$ flops by computing $A^T y$.

Many linear functions arising in applications have structure that allows the function and its adjoint to be evaluated in fewer than $O(mn)$ flops or using fewer than $O(mn)$ bytes of data. The algorithms and data structures used to evaluate such a function and its adjoint can differ wildly. It is thus useful to abstract away the details and view linear functions as *forward-adjoint oracles* (FAOs), i.e., a tuple $\Gamma = (f, \Phi_f, \Phi_{f^*})$ where f is a linear function, Φ_f is an algorithm for evaluating f, and Φ_{f^*} is an algorithm for evaluating f^*. We use n to denote the size of f's input and m to denote the size of f's output.

While we focus on linear functions from $\mathbf{R}^n$ into $\mathbf{R}^m$, the same techniques can be used to handle linear functions involving complex arguments or values, i.e., from $\mathbf{C}^n$ into $\mathbf{C}^m$, from $\mathbf{R}^n$ into $\mathbf{C}^m$, or from $\mathbf{C}^n$ into $\mathbf{R}^m$, using the standard embedding of complex n-vectors into real $2n$-vectors. This is useful for problems in which complex data arise naturally (e.g., in signal processing and communications), and also in some cases that involve only real data, where complex intermediate results appear (typically via an FFT).

2.1 Vector Mappings

We present a variety of FAOs for functions that take as argument, and return, vectors.

Scalar Multiplication Scalar multiplication by $\alpha \in \mathbf{R}$ is represented by the FAO $\Gamma = (f, \Phi_f, \Phi_{f^*})$, where $f : \mathbf{R}^n \to \mathbf{R}^n$ is given by $f(x) = \alpha x$. The adjoint f^* is the same as f. The algorithms Φ_f and Φ_{f^*} simply scale the input, which requires $O(m + n)$ flops and $O(1)$ bytes of data to store α. Here $m = n$.

Multiplication by a Dense Matrix Multiplication by a dense matrix $A \in \mathbf{R}^{m\times n}$ is represented by the FAO $\Gamma = (f, \Phi_f, \Phi_{f^*})$, where $f(x) = Ax$. The adjoint $f^*(u) = A^T u$ is also multiplication by a dense matrix. The algorithms Φ_f and Φ_{f^*} are the standard dense matrix multiplication algorithm. Evaluating Φ_f and Φ_{f^*} requires $O(mn)$ flops and $O(mn)$ bytes of data to store A and A^T.

Multiplication by a Sparse Matrix Multiplication by a sparse matrix $A \in \mathbf{R}^{m\times n}$, i.e., a matrix with many zero entries, is represented by the FAO $\Gamma = (f, \Phi_f, \Phi_{f^*})$, where $f(x) = Ax$. The adjoint $f^*(u) = A^T u$ is also multiplication by a sparse matrix. The algorithms Φ_f and Φ_{f^*} are the standard algorithm for multiplying by a sparse matrix in (for example) compressed sparse row format. Evaluating Φ_f and Φ_{f^*} requires $O(\mathbf{nnz}(A))$ flops and $O(\mathbf{nnz}(A))$ bytes of data to store A and A^T, where $\mathbf{nnz}$ is the number of nonzero elements in a sparse matrix [34, Chap. 2].

Multiplication by a Low-Rank Matrix Multiplication by a matrix $A \in \mathbf{R}^{m\times n}$ with rank k, where $k \ll m$ and $k \ll n$, is represented by the FAO $\Gamma = (f, \Phi_f, \Phi_{f^*})$, where $f(x) = Ax$. The matrix A can be factored as $A = BC$, where $B \in \mathbf{R}^{m\times k}$ and $C \in \mathbf{R}^{k\times n}$. The adjoint $f^*(u) = C^T B^T u$ is also multiplication by a rank k matrix. The algorithm Φ_f evaluates $f(x)$ by first evaluating $z = Cx$ and then evaluating $f(x) = Bz$. Similarly, Φ_{f^*} multiplies by B^T and then C^T. The algorithms Φ_f and Φ_{f^*} require $O(k(m+n))$ flops and use $O(k(m+n))$ bytes of data to store B and C and their transposes. Multiplication by a low-rank matrix occurs in many applications, and it is often possible to approximate multiplication by a full rank matrix with multiplication by a low-rank one, using the singular value decomposition or methods such as sketching [79].

Discrete Fourier Transform The discrete Fourier transform (DFT) is represented by the FAO $\Gamma = (f, \Phi_f, \Phi_{f^*})$, where $f : \mathbf{R}^{2p} \to \mathbf{R}^{2p}$ is given by

$$\begin{aligned} f(x)_k &= \tfrac{1}{\sqrt{p}} \textstyle\sum_{j=1}^{p} \operatorname{Re}\left(\omega_p^{(j-1)(k-1)}\right) x_j - \operatorname{Im}\left(\omega_p^{(j-1)(k-1)}\right) x_{j+p} \\ f(x)_{k+p} &= \tfrac{1}{\sqrt{p}} \textstyle\sum_{j=1}^{p} \operatorname{Im}\left(\omega_p^{(j-1)(k-1)}\right) x_j + \operatorname{Re}\left(\omega_p^{(j-1)(k-1)}\right) x_{j+p} \end{aligned}$$

for $k = 1, \ldots, p$. Here $\omega_p = e^{-2\pi i/p}$. The adjoint f^* is the inverse DFT. The algorithm Φ_f is the fast Fourier transform (FFT), while Φ_{f^*} is the inverse FFT. The algorithms can be evaluated in $O((m+n)\log(m+n))$ flops, using only $O(1)$ bytes of data to store the dimensions of f's input and output [31, 82]. Here $m = n = 2p$. There are many fast transforms derived from the DFT, such as the discrete Hartley transform [17] and the discrete sine and cosine transforms [2, 86], with the same computational complexity as the FFT.

Convolution Convolution with a kernel $c \in \mathbf{R}^p$ is defined as $f : \mathbf{R}^n \to \mathbf{R}^m$, where

$$f(x)_k = \sum_{i+j=k+1} c_i x_j, \quad k = 1, \ldots, m. \tag{3}$$

Different variants of convolution restrict the indices i, j to different ranges, or interpret vector elements outside their natural ranges as zero or using periodic (circular) indexing.

Standard (column) convolution takes $m = n+p-1$, and defines c_i and x_j in (3) as zero when the index is outside its range. In this case the associated matrix $\mathbf{Col}(c) \in \mathbf{R}^{n+p-1\times n}$ is Toeplitz, with each column a shifted version of c:

$$\mathbf{Col}(c) = \begin{bmatrix} c_1 & & \\ c_2 & \ddots & \\ \vdots & \ddots & c_1 \\ c_p & & c_2 \\ & \ddots & \vdots \\ & & c_p \end{bmatrix}.$$

Another standard form, row convolution, restricts the indices in (3) to the range $k = p, \ldots, n$. For simplicity we assume that $n \geq p$. In this case the associated matrix $\mathbf{Row}(c) \in \mathbf{R}^{n-p+1 \times n}$ is Toeplitz, with each row a shifted version of c, in reverse order:

$$\mathbf{Row}(c) = \begin{bmatrix} c_p & c_{p-1} & \cdots & c_1 & & \\ & \ddots & \ddots & & \ddots & \\ & & c_p & c_{p-1} & \cdots & c_1 \end{bmatrix}.$$

The matrices $\mathbf{Col}(c)$ and $\mathbf{Row}(c)$ are related by the equalities

$$\mathbf{Col}(c)^T = \mathbf{Row}(\mathbf{rev}(c)), \qquad \mathbf{Row}(c)^T = \mathbf{Col}(\mathbf{rev}(c)),$$

where $\mathbf{rev}(c)_k = c_{p-k+1}$ reverses the order of the entries of c.

Yet another variant on convolution is circular convolution, where we take $p = n$ and interpret the entries of vectors outside their range modulo n. In this case the associated matrix $\mathbf{Circ}(c) \in \mathbf{R}^{n \times n}$ is Toeplitz, with each column and row a (circularly) shifted version of c:

$$\mathbf{Circ}(c) = \begin{bmatrix} c_1 & c_n & c_{n-1} & \cdots & \cdots & c_2 \\ c_2 & c_1 & c_n & \ddots & & \vdots \\ c_3 & c_2 & \ddots & \ddots & \ddots & \vdots \\ \vdots & \ddots & \ddots & \ddots & c_n & c_{n-1} \\ \vdots & & \ddots & c_2 & c_1 & c_n \\ c_n & \cdots & \cdots & c_3 & c_2 & c_1 \end{bmatrix}.$$

Column convolution with $c \in \mathbf{R}^p$ is represented by the FAO $\Gamma = (f, \Phi_f, \Phi_{f^*})$, where $f : \mathbf{R}^n \to \mathbf{R}^{n+p-1}$ is given by $f(x) = \mathbf{Col}(c)x$. The adjoint f^* is row convolution with $\mathbf{rev}(c)$, i.e., $f^*(u) = \mathbf{Row}(\mathbf{rev}(c))u$. The algorithms Φ_f and Φ_{f^*} are given in Algorithms 1 and 2, and require $O((m + n + p)\log(m + n + p))$ flops. Here $m = n + p - 1$. If the kernel is small (i.e., $p \ll n$), Φ_f and Φ_{f^*} instead evaluate (3) directly in $O(np)$ flops. In either case, the algorithms Φ_f and Φ_{f^*} use $O(p)$ bytes of data to store c and $\mathbf{rev}(c)$ [30, 82].

Circular convolution with $c \in \mathbf{R}^n$ is represented by the FAO $\Gamma = (f, \Phi_f, \Phi_{f^*})$, where $f : \mathbf{R}^n \to \mathbf{R}^n$ is given by $f(x) = \mathbf{Circ}(c)x$. The adjoint f^* is circular convolution with

$$\tilde{c} = \begin{bmatrix} c_1 \\ c_n \\ c_{n-1} \\ \vdots \\ c_2 \end{bmatrix}.$$

Algorithm 1 Column convolution $c * x$

Precondition: $c \in \mathbf{R}^p$ is a length p array. $x \in \mathbf{R}^n$ is a length n array. $y \in \mathbf{R}^{n+p-1}$ is a length $n+p-1$ array.

Extend c and x into length $n+p-1$ arrays by appending zeros.
$\hat{c} \leftarrow$ FFT of c.
$\hat{x} \leftarrow$ FFT of x.
for $i = 1, \ldots, n+p-1$ **do**
  $y_i \leftarrow \hat{c}_i \hat{x}_i$.
end for
$y \leftarrow$ inverse FFT of y.

Postcondition: $y = c * x$.

Algorithm 2 Row convolution $c * u$

Precondition: $c \in \mathbf{R}^p$ is a length p array. $u \in \mathbf{R}^{n+p-1}$ is a length $n+p-1$ array. $v \in \mathbf{R}^n$ is a length n array.

Extend $\mathbf{rev}(c)$ and v into length $n+p-1$ arrays by appending zeros.
$\hat{c} \leftarrow$ inverse FFT of zero-padded $\mathbf{rev}(c)$.
$\hat{u} \leftarrow$ FFT of u.
for $i = 1, \ldots, n+p-1$ **do**
  $v_i \leftarrow \hat{c}_i \hat{u}_i$.
end for
$v \leftarrow$ inverse FFT of v.
Reduce v to a length n array by removing the last $p-1$ entries.

Postcondition: $v = c * u$.

Algorithm 3 Circular convolution $c * x$

Precondition: $c \in \mathbf{R}^n$ is a length n array. $x \in \mathbf{R}^n$ is a length n array. $y \in \mathbf{R}^n$ is a length n array.

$\hat{c} \leftarrow$ FFT of c.
$\hat{x} \leftarrow$ FFT of x.
for $i = 1, \ldots, n$ **do**
  $y_i \leftarrow \hat{c}_i \hat{x}_i$.
end for
$y \leftarrow$ inverse FFT of y.

Postcondition: $y = c * x$.

The algorithms Φ_f and Φ_{f^*} are given in Algorithm 3, and require $O((m+n)\log(m+n))$ flops. The algorithms Φ_f and Φ_{f^*} use $O(m+n)$ bytes of data to store c and $\tilde{c}$ [30, 82]. Here $m=n$.

Discrete Wavelet Transform The discrete wavelet transform (DWT) for orthogonal wavelets is represented by the FAO $\Gamma = (f, \Phi_f, \Phi_{f^*})$, where the function $f : \mathbf{R}^{2^p} \to \mathbf{R}^{2^p}$ is given by

$$f(x) = \begin{bmatrix} D_1 G_1 & \\ D_1 H_1 & \\ & I_{2^p-2} \end{bmatrix} \cdots \begin{bmatrix} D_{p-1} G_{p-1} & \\ D_{p-1} H_{p-1} & \\ & I_{2^p-1} \end{bmatrix} \begin{bmatrix} D_p G_p \\ D_p H_p \end{bmatrix} x, \tag{4}$$

where $D_k \in \mathbf{R}^{2^{k-1} \times 2^k}$ is defined such that $(D_k x)_i = x_{2i}$ and the matrices $G_k \in \mathbf{R}^{2^k \times 2^k}$ and $H_k \in \mathbf{R}^{2^k \times 2^k}$ are given by

$$G_k = \mathbf{Circ}\left(\begin{bmatrix} g \\ 0 \end{bmatrix}\right), \qquad H_k = \mathbf{Circ}\left(\begin{bmatrix} h \\ 0 \end{bmatrix}\right).$$

Here $g \in \mathbf{R}^q$ and $h \in \mathbf{R}^q$ are low and high pass filters, respectively, that parameterize the DWT. The adjoint f^* is the inverse DWT. The algorithms Φ_f and Φ_f^* repeatedly convolve by g and h, which requires $O(q(m+n))$ flops and uses $O(q)$ bytes to store h and g [85]. Here $m = n = 2^p$. Common orthogonal wavelets include the Haar wavelet and the Daubechies wavelets [32, 33]. There are many variants on the particular DWT described here. For instance, the product in (4) can be terminated after fewer than p multiplications by G_k and H_k [70], G_k and H_k can be defined as a different type of convolution matrix, or the filters g and h can be different lengths, as in biorthogonal wavelets [28].

Discrete Gauss Transform The discrete Gauss transform (DGT) is represented by the FAO $\Gamma = (f_{Y,Z,h}, \Phi_f, \Phi_{f^*})$, where the function $f_{Y,Z,h} : \mathbf{R}^n \to \mathbf{R}^m$ is parameterized by $Y \in \mathbf{R}^{m\times d}$, $Z \in \mathbf{R}^{n\times d}$, and $h > 0$. The function $f_{Y,Z,h}$ is given by

$$f_{Y,Z,h}(x)_i = \sum_{j=1}^{n} \exp(-\|y_i - z_j\|^2/h^2) x_j, \quad i = 1, \ldots, m,$$

where $y_i \in \mathbf{R}^d$ is the ith column of Y and $z_j \in \mathbf{R}^d$ is the jth column of Z. The adjoint of $f_{Y,Z,h}$ is the DGT $f_{Z,Y,h}$. The algorithms Φ_f and Φ_{f^*} are the improved fast Gauss transform, which evaluates $f(x)$ and $f^*(u)$ to a given accuracy in $O(d^p(m+n))$ flops. Here p is a parameter that depends on the accuracy desired. The algorithms Φ_f and Φ_{f^*} use $O(d(m+n))$ bytes of data to store Y, Z, and h [114]. An interesting application of the DGT is efficient multiplication by a Gaussian kernel [113].

Multiplication by the Inverse of a Sparse Triangular Matrix Multiplication by the inverse of a sparse lower triangular matrix $L \in \mathbf{R}^{n\times n}$ with nonzero elements on its diagonal is represented by the FAO $\Gamma = (f, \Phi_f, \Phi_{f^*})$, where $f(x) = L^{-1}x$.

The adjoint $f^*(u) = (L^T)^{-1}u$ is multiplication by the inverse of a sparse upper triangular matrix. The algorithms Φ_f and Φ_{f^*} are forward and backward substitution, respectively, which require $O(\mathbf{nnz}(L))$ flops and use $O(\mathbf{nnz}(L))$ bytes of data to store L and L^T [34, Chap. 3].

Multiplication by a Pseudo-Random Matrix Multiplication by a matrix $A \in \mathbf{R}^{m\times n}$ whose columns are given by a pseudo-random sequence (i.e., the first m values of the sequence are the first column of A, the next m values are the second column of A, etc.) is represented by the FAO $\Gamma = (f, \Phi_f, \Phi_{f^*})$, where $f(x) = Ax$. The adjoint $f^*(u) = A^T u$ is multiplication by a matrix whose rows are given by a pseudo-random sequence (i.e., the first m values of the sequence are the first row of A^T, the next m values are the second row of A^T, etc.). The algorithms Φ_f and Φ_{f^*} are the standard dense matrix multiplication algorithm, iterating once over the pseudo-random sequence without storing any of its values. The algorithms require $O(mn)$ flops and use $O(1)$ bytes of data to store the seed for the pseudo-random sequence. Multiplication by a pseudo-random matrix might appear, for example, as a measurement ensemble in compressed sensing [50].

Multiplication by the Pseudo-Inverse of a Graph Laplacian Multiplication by the pseudo-inverse of a graph Laplacian matrix $L \in \mathbf{R}^{n\times n}$ is represented by the FAO $\Gamma = (f, \Phi_f, \Phi_{f^*})$, where $f(x) = L^\dagger x$. A graph Laplacian is a symmetric matrix with nonpositive off diagonal entries and the property $L\mathbf{1} = 0$, i.e., the diagonal entry in a row is the negative sum of the off-diagonal entries in that row. (This implies that it is positive semidefinite.) The adjoint f^* is the same as f, since $L = L^T$. The algorithms Φ_f and Φ_{f^*} are one of the fast solvers for graph Laplacian systems that evaluate $f(x) = f^*(x)$ to a given accuracy in around $O(\mathbf{nnz}(L))$ flops [73, 101, 111]. (The details of the computational complexity are much more involved.) The algorithms use $O(\mathbf{nnz}(L))$ bytes of data to store L.

2.2 Matrix Mappings

We now consider linear functions that take as argument, or return, matrices. We take the standard inner product on matrices $X, Y \in \mathbf{R}^{p\times q}$,

$$\langle X, Y\rangle = \sum_{i=1,\ldots,p,\ j=1,\ldots,q} X_{ij}Y_{ij} = \mathbf{Tr}(X^T Y).$$

The adjoint of a linear function $f : \mathbf{R}^{p\times q} \to \mathbf{R}^{s\times t}$ is then the function $f^* : \mathbf{R}^{s\times t} \to \mathbf{R}^{p\times q}$ for which

$$\mathbf{Tr}(f(X)^T Y) = \mathbf{Tr}(X^T f^*(Y)),$$

holds for all $X \in \mathbf{R}^{p\times q}$ and $Y \in \mathbf{R}^{s\times t}$.

Vec and Mat The function **vec** : $\mathbf{R}^{p\times q} \to \mathbf{R}^{pq}$ is represented by the FAO $\Gamma = (f, \Phi_f, \Phi_{f^*})$, where $f(X)$ converts the matrix $X \in \mathbf{R}^{p\times q}$ into a vector $y \in \mathbf{R}^{pq}$ by stacking the columns. The adjoint f^* is the function **mat** : $\mathbf{R}^{pq} \to \mathbf{R}^{p\times q}$, which outputs a matrix whose columns are successive slices of its vector argument. The algorithms Φ_f and Φ_{f^*} simply reinterpret their input as a differently shaped output in $O(1)$ flops, using only $O(1)$ bytes of data to store the dimensions of f's input and output.

Sparse Matrix Mappings Many common linear functions on and to matrices are given by a sparse matrix multiplication of the vectorized argument, reshaped as the output matrix. For $X \in \mathbf{R}^{p\times q}$ and $f(X) = Y \in \mathbf{R}^{s\times t}$,

$$Y = \mathbf{mat}(A\,\mathbf{vec}(X)).$$

The form above describes the general linear mapping from $\mathbf{R}^{p\times q}$ to $\mathbf{R}^{s\times t}$; we are interested in cases when A is sparse, i.e., has far fewer than $pqst$ nonzero entries. Examples include extracting a submatrix, extracting the diagonal, forming a diagonal matrix, summing the rows or columns of a matrix, transposing a matrix, scaling its rows or columns, and so on. The FAO representation of each such function is $\Gamma = (f, \Phi_f, \Phi_{f^*})$, where f is given above and the adjoint is given by

$$f^*(U) = \mathbf{mat}(A^T\,\mathbf{vec}(U)).$$

The algorithms Φ_f and Φ_{f^*} are the standard algorithms for multiplying a vector by a sparse matrix in (for example) compressed sparse row format. The algorithms require $O(\mathbf{nnz}(A))$ flops and use $O(\mathbf{nnz}(A))$ bytes of data to store A and A^T [34, Chap. 2].

Matrix Product Multiplication on the left by a matrix $A \in \mathbf{R}^{s\times p}$ and on the right by a matrix $B \in \mathbf{R}^{q\times t}$ is represented by the FAO $\Gamma = (f, \Phi_f, \Phi_{f^*})$, where $f : \mathbf{R}^{p\times q} \to \mathbf{R}^{s\times t}$ is given by $f(X) = AXB$. The adjoint $f^*(U) = A^T UB^T$ is also a matrix product. There are two ways to implement Φ_f efficiently, corresponding to different orders of operations in multiplying out AXB. In one method we multiply by A first and B second, for a total of $O(s(pq + qt))$ flops (assuming that A and B are dense). In the other method we multiply by B first and A second, for a total of $O(p(qt + st))$ flops. The former method is more efficient if

$$\frac{1}{t} + \frac{1}{p} < \frac{1}{s} + \frac{1}{q}.$$

Similarly, there are two ways to implement Φ_{f^*}, one requiring $O(s(pq + qt))$ flops and the other requiring $O(p(qt+st))$ flops. The algorithms Φ_f and Φ_{f^*} use $O(sp+qt)$ bytes of data to store A and B and their transposes. When $p = q = s = t$, the flop count for Φ_f and Φ_{f^*} simplifies to $O\left((m + n)^{1.5}\right)$ flops. Here $m = n = pq$. (When the matrices A or B are sparse, evaluating $f(X)$ and $f^*(U)$ can be done even more efficiently.) The matrix product function is used in Lyapunov and algebraic Riccati inequalities and Sylvester equations, which appear in many problems from control theory [49, 110].

2-D Discrete Fourier Transform The 2-D DFT is represented by the FAO $\Gamma = (f, \Phi_f, \Phi_{f^*})$, where $f : \mathbf{R}^{2p\times q} \to \mathbf{R}^{2p\times q}$ is given by

$$\begin{aligned}
f(X)_{k\ell} &= \tfrac{1}{\sqrt{pq}} \textstyle\sum_{s=1}^{p} \sum_{t=1}^{q} \mathrm{Re}\left(\omega_p^{(s-1)(k-1)} \omega_q^{(t-1)(\ell-1)}\right) X_{st} \\
&\qquad - \mathrm{Im}\left(\omega_p^{(s-1)(k-1)} \omega_q^{(t-1)(\ell-1)}\right) X_{s+p,t} \\
f(X)_{k+p,\ell} &= \tfrac{1}{\sqrt{pq}} \textstyle\sum_{s=1}^{p} \sum_{t=1}^{q} \mathrm{Im}\left(\omega_p^{(s-1)(k-1)} \omega_q^{(t-1)(\ell-1)}\right) X_{st} \\
&\qquad + \mathrm{Re}\left(\omega_p^{(s-1)(k-1)} \omega_q^{(t-1)(\ell-1)}\right) X_{s+p,t},
\end{aligned}$$

for $k = 1, \dots, p$ and $\ell = 1, \dots, q$. Here $\omega_p = e^{-2\pi i/p}$ and $\omega_q = e^{-2\pi i/q}$. The adjoint f^* is the inverse 2-D DFT. The algorithm Φ_f evaluates $f(X)$ by first applying the FFT to each row of X, replacing the row with its DFT, and then applying the FFT to each column, replacing the column with its DFT. The algorithm Φ_{f^*} is analogous, but with the inverse FFT and inverse DFT taking the role of the FFT and DFT. The algorithms Φ_f and Φ_{f^*} require $O((m+n)\log(m+n))$ flops, using only $O(1)$ bytes of data to store the dimensions of f's input and output [80, 82]. Here $m = n = 2pq$.

2-D Convolution 2-D convolution with a kernel $C \in \mathbf{R}^{p\times q}$ is defined as $f : \mathbf{R}^{s\times t} \to \mathbf{R}^{m_1\times m_2}$, where

$$f(X)_{k\ell} = \sum_{i_1+i_2=k+1, j_1+j_2=\ell+1} C_{i_1 j_1} X_{i_2 j_2}, \quad k = 1, \dots, m_1, \quad \ell = 1, \dots, m_2. \tag{5}$$

Different variants of 2-D convolution restrict the indices i_1, j_1 and i_2, j_2 to different ranges, or interpret matrix elements outside their natural ranges as zero or using periodic (circular) indexing. There are 2-D analogues of 1-D column, row, and circular convolution.

Standard 2-D (column) convolution, the analogue of 1-D column convolution, takes $m_1 = s + p - 1$ and $m_2 = t + q - 1$, and defines $C_{i_1 j_1}$ and $X_{i_2 j_2}$ in (5) as zero when the indices are outside their range. We can represent the 2-D column convolution $Y = C * X$ as the matrix multiplication

$$Y = \mathbf{mat}(\mathbf{Col}(C)\,\mathbf{vec}(X)),$$

where $\mathbf{Col}(C) \in \mathbf{R}^{(s+p-1)(t+q-1)\times st}$ is given by:

$$\mathbf{Col}(C) = \begin{bmatrix} \mathbf{Col}(c_1) & & \\ \mathbf{Col}(c_2) & \ddots & \\ \vdots & \ddots & \mathbf{Col}(c_1) \\ \mathbf{Col}(c_q) & & \mathbf{Col}(c_2) \\ & \ddots & \vdots \\ & & \mathbf{Col}(c_q) \end{bmatrix}.$$

Here $c_1, \ldots, c_q \in \mathbf{R}^p$ are the columns of C and $\mathbf{Col}(c_1), \ldots, \mathbf{Col}(c_q) \in \mathbf{R}^{s+p-1 \times s}$ are 1-D column convolution matrices.

The 2-D analogue of 1-D row convolution restricts the indices in (5) to the range $k = p, \ldots, s$ and $\ell = q, \ldots, t$. For simplicity we assume $s \geq p$ and $t \geq q$. The output dimensions are $m_1 = s - p + 1$ and $m_2 = t - q + 1$. We can represent the 2-D row convolution $Y = C * X$ as the matrix multiplication

$$Y = \mathbf{mat}(\mathbf{Row}(C)\,\mathbf{vec}(X)),$$

where $\mathbf{Row}(C) \in \mathbf{R}^{(s-p+1)(t-q+1) \times st}$ is given by:

$$\mathbf{Row}(C) = \begin{bmatrix} \mathbf{Row}(c_q) & \mathbf{Row}(c_{q-1}) & \ldots & \mathbf{Row}(c_1) & & \\ & \ddots & \ddots & & \ddots & \\ & & \mathbf{Row}(c_q) & \mathbf{Row}(c_{q-1}) & \ldots & \mathbf{Row}(c_1) \end{bmatrix}.$$

Here $\mathbf{Row}(c_1), \ldots, \mathbf{Row}(c_q) \in \mathbf{R}^{s-p+1 \times s}$ are 1-D row convolution matrices. The matrices $\mathbf{Col}(C)$ and $\mathbf{Row}(C)$ are related by the equalities

$$\mathbf{Col}(C)^T = \mathbf{Row}(\mathbf{rev}(C)), \qquad \mathbf{Row}(C)^T = \mathbf{Col}(\mathbf{rev}(C)),$$

where $\mathbf{rev}(C)_{k\ell} = C_{p-k+1, q-\ell+1}$ reverses the order of the columns of C and of the entries in each row.

In the 2-D analogue of 1-D circular convolution, we take $p = s$ and $q = t$ and interpret the entries of matrices outside their range modulo s for the row index and modulo t for the column index. We can represent the 2-D circular convolution $Y = C * X$ as the matrix multiplication

$$Y = \mathbf{mat}(\mathbf{Circ}(C)\,\mathbf{vec}(X)),$$

where $\mathbf{Circ}(C) \in \mathbf{R}^{st \times st}$ is given by:

$$\mathbf{Circ}(C) = \begin{bmatrix} \mathbf{Circ}(c_1) & \mathbf{Circ}(c_t) & \mathbf{Circ}(c_{t-1}) & \ldots & \ldots & \mathbf{Circ}(c_2) \\ \mathbf{Circ}(c_2) & \mathbf{Circ}(c_1) & \mathbf{Circ}(c_t) & \ddots & & \vdots \\ \mathbf{Circ}(c_3) & \mathbf{Circ}(c_2) & \ddots & \ddots & \ddots & \vdots \\ \vdots & \ddots & \ddots & \ddots & \mathbf{Circ}(c_t) & \mathbf{Circ}(c_{t-1}) \\ \vdots & & \ddots & \mathbf{Circ}(c_2) & \mathbf{Circ}(c_1) & \mathbf{Circ}(c_t) \\ \mathbf{Circ}(c_t) & \ldots & \ldots & \mathbf{Circ}(c_3) & \mathbf{Circ}(c_2) & \mathbf{Circ}(c_1) \end{bmatrix}.$$

Here $\mathbf{Circ}(c_1), \ldots, \mathbf{Circ}(c_t) \in \mathbf{R}^{s \times s}$ are 1-D circular convolution matrices.

2-D column convolution with $C \in \mathbf{R}^{p \times q}$ is represented by the FAO $\Gamma = (f, \Phi_f, \Phi_{f^*})$, where $f : \mathbf{R}^{s \times t} \to \mathbf{R}^{s+p-1 \times t+q-1}$ is given by

$$f(X) = \mathbf{mat}(\mathbf{Col}(C)\,\mathbf{vec}(X)).$$

The adjoint f^* is 2-D row convolution with $\mathbf{rev}(C)$, i.e.,

$$f^*(U) = \mathbf{mat}(\mathbf{Row}(\mathbf{rev}(C))\,\mathbf{vec}(U)).$$

The algorithms Φ_f and Φ_{f^*} are given in Algorithms 4 and 5, and require $O((m+n)\log(m+n))$ flops. Here $m = (s+p-1)(t+q-1)$ and $n = st$. If the kernel is small (i.e., $p \ll s$ and $q \ll t$), Φ_f and Φ_{f^*} instead evaluate (5) directly in $O(pqst)$ flops. In either case, the algorithms Φ_f and Φ_{f^*} use $O(pq)$ bytes of data to store C and $\mathbf{rev}(C)$ [82, Chap. 4]. Often the kernel is parameterized (e.g., a Gaussian kernel), in which case more compact representations of C and $\mathbf{rev}(C)$ are possible [44, Chap. 7].

Algorithm 4 2-D column convolution $C * X$

Precondition: $C \in \mathbf{R}^{p\times q}$ is a length pq array. $X \in \mathbf{R}^{s\times t}$ is a length st array. $Y \in \mathbf{R}^{s+p-1\times t+q-1}$ is a length $(s+p-1)(t+q-1)$ array.

Extend the columns and rows of C and X with zeros so $C, X \in \mathbf{R}^{s+p-1\times t+q-1}$.
$\hat{C} \leftarrow$ 2-D DFT of C.
$\hat{X} \leftarrow$ 2-D DFT of X.
for $i = 1, \ldots, s+p-1$ **do**
  for $j = 1, \ldots, t+q-1$ **do**
    $Y_{ij} \leftarrow \hat{C}_{ij}\hat{X}_{ij}$.
  end for
end for
$Y \leftarrow$ inverse 2-D DFT of Y.

Postcondition: $Y = C * X$.

Algorithm 5 2-D row convolution $C * U$

Precondition: $C \in \mathbf{R}^{p\times q}$ is a length pq array. $U \in \mathbf{R}^{s+p-1\times t+q-1}$ is a length $(s+p-1)(t+q-1)$ array. $V \in \mathbf{R}^{s\times t}$ is a length st array.

Extend the columns and rows of $\mathbf{rev}(C)$ and V with zeros so $\mathbf{rev}(C), V \in \mathbf{R}^{s+p-1\times t+q-1}$.
$\hat{C} \leftarrow$ inverse 2-D DFT of zero-padded $\mathbf{rev}(C)$.
$\hat{U} \leftarrow$ 2-D DFT of U.
for $i = 1, \ldots, s+p-1$ **do**
  for $j = 1, \ldots, t+q-1$ **do**
    $V_{ij} \leftarrow \hat{C}_{ij}\hat{U}_{ij}$.
  end for
end for
$V \leftarrow$ inverse 2-D DFT of V.
Truncate the rows and columns of V so that $V \in \mathbf{R}^{s\times t}$.

Postcondition: $V = C * U$.

Algorithm 6 2-D circular convolution $C * X$

Precondition: $C \in \mathbf{R}^{s\times t}$ is a length st array. $X \in \mathbf{R}^{s\times t}$ is a length st array. $Y \in \mathbf{R}^{s\times t}$ is a length st array.

$\hat{C} \leftarrow$ 2-D DFT of C.
$\hat{X} \leftarrow$ 2-D DFT of X.
for $i = 1, \ldots, s$ **do**
  for $j = 1, \ldots, t$ **do**
    $Y_{ij} \leftarrow \hat{C}_{ij}\hat{X}_{ij}$.
  end for
end for
$Y \leftarrow$ inverse 2-D DFT of Y.

Postcondition: $Y = C * X$.

2-D circular convolution with $C \in \mathbf{R}^{s\times t}$ is represented by the FAO $\Gamma = (f, \Phi_f, \Phi_{f*})$, where $f : \mathbf{R}^{s\times t} \rightarrow \mathbf{R}^{s\times t}$ is given by

$$f(X) = \mathbf{mat}(\mathbf{Circ}(C)\,\mathbf{vec}(X)).$$

The adjoint f^* is 2-D circular convolution with

$$\tilde{C} = \begin{bmatrix} C_{1,1} & C_{1,t} & C_{1,t-1} & \ldots & C_{1,2} \\ C_{s,1} & C_{s,t} & C_{s,t-1} & \ldots & C_{s,2} \\ C_{s-1,1} & C_{s-1,t} & C_{s-1,t-1} & \ldots & C_{s-1,2} \\ \vdots & \vdots & \vdots & \ddots & \vdots \\ C_{2,1} & C_{2,t} & C_{2,t-1} & \ldots & C_{2,2} \end{bmatrix}.$$

The algorithms Φ_f and Φ_{f*} are given in Algorithm 6, and require $O((m+n)\log(m+n))$ flops. The algorithms Φ_f and Φ_{f*} use $O(m+n)$ bytes of data to store C and $\tilde{C}$ [82, Chap. 4]. Here $m = n = st$.

2-D Discrete Wavelet Transform The 2-D DWT for separable, orthogonal wavelets is represented by the FAO $\Gamma = (f, \Phi_f, \Phi_{f*})$, where $f : \mathbf{R}^{2^p\times 2^p} \rightarrow \mathbf{R}^{2^p\times 2^p}$ is given by

$$f(X)_{ij} = W_k \cdots W_{p-1} W_p X W_p^T W_{p-1}^T \cdots W_k^T,$$

where $k = \max\{\lceil \log_2(i)\rceil, \lceil \log_2(j)\rceil, 1\}$ and $W_k \in \mathbf{R}^{2^p\times 2^p}$ is given by

$$W_k = \begin{bmatrix} D_k G_k & \\ D_k H_k & \\ & I \end{bmatrix}.$$

Here D_k, G_k, and H_k are defined as for the 1-D DWT. The adjoint f^* is the inverse 2-D DWT. As in the 1-D DWT, the algorithms Φ_f and Φ_{f^*} repeatedly convolve by the filters $g \in \mathbf{R}^q$ and $h \in \mathbf{R}^q$, which requires $O(q(m+n))$ flops and uses $O(q)$ bytes of data to store g and h [70]. Here $m = n = 2^p$. There are many alternative wavelet transforms for 2-D data; see, e.g., [20, 38, 69, 102].

2.3 *Multiple Vector Mappings*

In this section we consider linear functions that take as argument, or return, multiple vectors. (The idea is readily extended to the case when the arguments or return values are matrices.) The adjoint is defined by the inner product

$$\langle (x_1, \ldots, x_k), (y_1, \ldots, y_k) \rangle = \sum_{i=1}^{k} \langle x_i, y_i \rangle = \sum_{i=1}^{k} x_i^T y_i.$$

The adjoint of a linear function $f : \mathbf{R}^{n_1} \times \cdots \times \mathbf{R}^{n_k} \to \mathbf{R}^{m_1} \times \cdots \times \mathbf{R}^{m_\ell}$ is then the function $f^* : \mathbf{R}^{m_1} \times \cdots \times \mathbf{R}^{m_\ell} \to \mathbf{R}^{n_1} \times \cdots \times \mathbf{R}^{n_k}$ for which

$$\sum_{i=1}^{\ell} f(x_1, \ldots, x_k)_i^T y_i = \sum_{i=1}^{k} x_i^T f^*(y_1, \ldots, y_\ell)_i,$$

holds for all $(x_1, \ldots, x_k) \in \mathbf{R}^{n_1} \times \cdots \times \mathbf{R}^{n_k}$ and $(y_1, \ldots, y_\ell) \in \mathbf{R}^{m_1} \times \cdots \times \mathbf{R}^{m_\ell}$. Here $f(x_1, \ldots, x_k)_i$ and $f^*(y_1, \ldots, y_\ell)_i$ refer to the ith output of f and f^*, respectively.

Sum and Copy The function **sum** : $\mathbf{R}^m \times \cdots \times \mathbf{R}^m \to \mathbf{R}^m$ with k inputs is represented by the FAO $\Gamma = (f, \Phi_f, \Phi_{f^*})$, where $f(x_1, \ldots, x_k) = x_1 + \cdots + x_k$. The adjoint f^* is the function **copy** : $\mathbf{R}^m \to \mathbf{R}^m \times \cdots \times \mathbf{R}^m$, which outputs k copies of its input. The algorithms Φ_f and Φ_{f^*} require $O(m + n)$ flops to sum and copy their input, respectively, using only $O(1)$ bytes of data to store the dimensions of f's input and output. Here $n = km$.

Vstack and Split The function **vstack** : $\mathbf{R}^{m_1} \times \cdots \times \mathbf{R}^{m_k} \to \mathbf{R}^n$ is represented by the FAO $\Gamma = (f, \Phi_f, \Phi_{f^*})$, where $f(x_1, \ldots, x_k)$ concatenates its k inputs into a single vector output. The adjoint f^* is the function **split** : $\mathbf{R}^n \to \mathbf{R}^{m_1} \times \cdots \times \mathbf{R}^{m_k}$, which divides a single vector into k separate components. The algorithms Φ_f and Φ_{f^*} simply reinterpret their input as a differently sized output in $O(1)$ flops, using only $O(1)$ bytes of data to store the dimensions of f's input and output. Here $n = m = m_1 + \cdots + m_k$.

2.4 Additional Examples

The literature on fast linear transforms goes far beyond the preceding examples. In this section we highlight a few notable omissions. Many methods have been developed for matrices derived from physical systems. The multigrid [62] and algebraic multigrid [18] methods efficiently apply the inverse of a matrix representing discretized partial differential equations (PDEs). The fast multipole method accelerates multiplication by matrices representing pairwise interactions [21, 59], much like the fast Gauss transform [60]. Hierarchical matrices are a matrix format that allows fast multiplication by the matrix and its inverse, with applications to discretized integral operators and PDEs [14, 63, 64].

Many approaches exist for factoring an invertible sparse matrix into a product of components whose inverses can be applied efficiently, yielding a fast method for applying the inverse of the matrix [34, 41]. A sparse LU factorization, for instance, decomposes an invertible sparse matrix $A \in \mathbf{R}^{n\times n}$ into the product $A = LU$ of a lower triangular matrix $L \in \mathbf{R}^{n\times n}$ and an upper triangular matrix $U \in \mathbf{R}^{n\times n}$. The relationship between $\mathbf{nnz}(A)$, $\mathbf{nnz}(L)$, and $\mathbf{nnz}(U)$ is complex and depends on the factorization algorithm [34, Chap. 6].

We only discussed 1-D and 2-D DFTs and convolutions, but these and related transforms can be extended to arbitrarily many dimensions [40, 82]. Similarly, many wavelet transforms naturally operate on data indexed by more than two dimensions [76, 84, 116].

3 Compositions

In this section we consider compositions of FAOs. In fact we have already discussed several linear functions that are naturally and efficiently represented as compositions, such as multiplication by a low-rank matrix and sparse matrix mappings. Here though we present a data structure and algorithm for efficiently evaluating any composition and its adjoint, which gives us an FAO representing the composition.

A composition of FAOs can be represented using a directed acyclic graph (DAG) with exactly one node with no incoming edges (the start node) and exactly one node with no outgoing edges (the end node). We call such a representation an *FAO DAG*.

Each node in the FAO DAG stores the following attributes:

- An FAO $\Gamma = (f, \Phi_f, \Phi_{f^*})$. Concretely, f is a symbol identifying the function, and Φ_f and Φ_{f^*} are executable code.
- The data needed to evaluate Φ_f and Φ_{f^*}.
- A list E_{in} of incoming edges.
- A list E_{out} of outgoing edges.

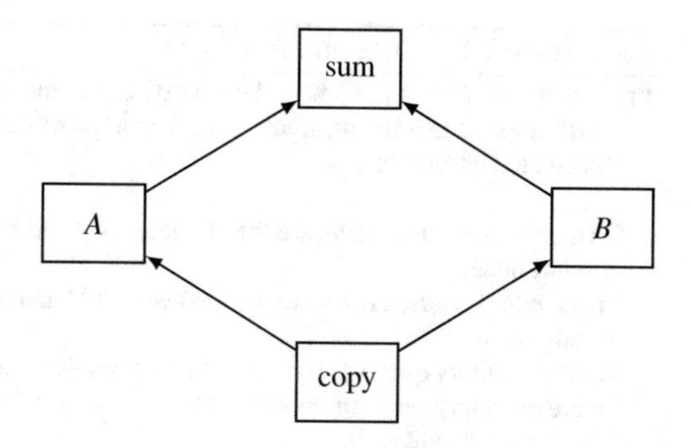

Fig. 1 The FAO DAG for $f(x) = Ax + Bx$

Each edge has an associated array. The incoming edges to a node store the arguments to the node's FAO. When the FAO is evaluated, it writes the result to the node's outgoing edges. Matrix arguments and outputs are stored in column-major order on the edge arrays.

As an example, Fig. 1 shows the FAO DAG for the composition $f(x) = Ax + Bx$, where $A \in \mathbf{R}^{m\times n}$ and $B \in \mathbf{R}^{m\times n}$ are dense matrices. The **copy** node duplicates the input $x \in \mathbf{R}^n$ into the multi-argument output $(x, x) \in \mathbf{R}^n \times \mathbf{R}^n$. The A and B nodes multiply by A and B, respectively. The **sum** node sums two vectors together. The **copy** node is the start node, and the **sum** node is the end node. The FAO DAG requires $O(mn)$ bytes to store, since the A and B nodes store the matrices A and B and their transposes. The edge arrays also require $O(mn)$ bytes of memory.

3.1 *Forward Evaluation*

To evaluate the composition $f(x) = Ax + Bx$ using the FAO DAG in Fig. 1, we first evaluate the start node on the input $x \in \mathbf{R}^n$, which copies x onto both outgoing edges. We evaluate the A and B nodes (serially or in parallel) on their incoming edges, and write the results (Ax and Bx) to their outgoing edges. Finally, we evaluate the end node on its incoming edges to obtain the result $Ax + Bx$.

The general procedure for evaluating an FAO DAG is given in Algorithm 7. The algorithm evaluates the nodes in a topological order. The total flop count is the sum of the flops from evaluating the algorithm Φ_f on each node. If we allocate all scratch space needed by the FAO algorithms in advance, then no memory is allocated during the algorithm.

3.2 *Adjoint Evaluation*

Given an FAO DAG G representing a function f, we can easily generate an FAO DAG G^* representing the adjoint f^*. We modify each node in G, replacing the node's FAO (f, Φ_f, Φ_{f^*}) with the FAO $(f^*, \Phi_{f^*}, \Phi_f)$ and swapping E_{in} and E_{out}.

Algorithm 7 Evaluate an FAO DAG

Precondition: $G = (V, E)$ is an FAO DAG representing a function f. V is a list of nodes. E is a list of edges. I is a list of inputs to f. O is a list of outputs from f. Each element of I and O is represented as an array.

Create edges whose arrays are the elements of I and save them as the list of incoming edges for the start node.
Create edges whose arrays are the elements of O and save them as the list of outgoing edges for the end node.
Create an empty queue Q for nodes that are ready to evaluate.
Create an empty set S for nodes that have been evaluated.
Add G's start node to Q.
while Q is not empty **do**
  $u \leftarrow$ pop the front node of Q.
  Evaluate u's algorithm Φ_f on u's incoming edges, writing the result to u's outgoing edges.
  Add u to S.
  for each edge $e = (u, v)$ in u's E_{out} **do**
    if for all edges (p, v) in v's E_{in}, p is in S **then**
      Add v to the end of Q.
    end if
  end for
end while

Postcondition: O contains the outputs of f applied to inputs I.

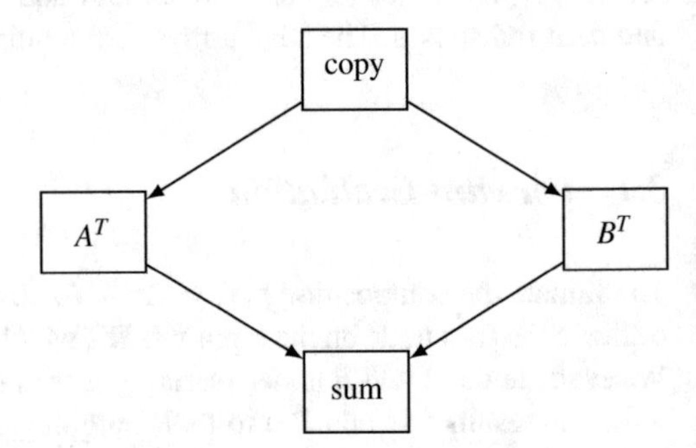

Fig. 2 The FAO DAG for $f^*(u) = A^T u + B^T u$ obtained by transforming the FAO DAG in Fig. 1

We also reverse the orientation of each edge in G. We can apply Algorithm 7 to the resulting graph G^* to evaluate f^*. Figure 2 shows the FAO DAG in Fig. 1 transformed into an FAO DAG for the adjoint.

3.3 *Parallelism*

Algorithm 7 can be easily parallelized, since the nodes in the ready queue Q can be evaluated in any order. A simple parallel implementation could use a thread pool with t threads to evaluate up to t nodes in the ready queue at a time. The evaluation of individual nodes can also be parallelized by replacing a node's algorithm Φ_f with a parallel variant. For example, the standard algorithms for dense and sparse matrix multiplication have simple parallel variants.

The extent to which parallelism speeds up evaluation of an FAO DAG is difficult to predict. Naive parallel evaluation may be slower than serial evaluation due to communication costs and other overhead. Achieving a perfect parallel speed-up would require sophisticated analysis of the DAG to determine which aspects of the algorithm to parallelize, and may only be possible for highly structured DAGs like one describing a block matrix [54].

3.4 *Optimizing the DAG*

The FAO DAG can often be transformed so that the output of Algorithm 7 is the same but the algorithm is executed more efficiently. Such optimizations are especially important when the FAO DAG will be evaluated on many different inputs (as will be the case for matrix-free solvers, to be discussed later). For example, the FAO DAG representing $f(x) = ABx + ACx$ where $A, B, C \in \mathbf{R}^{n \times n}$, shown in Fig. 3, can be transformed into the FAO DAG in Fig. 4, which requires one fewer multiplication by A. The transformation is equivalent to rewriting $f(x) = ABx + ACx$ as $f(x) = A(Bx + Cx)$. Many other useful graph transformations can be derived from the rewriting rules used in program analysis and code generation [3].

Sometimes graph transformations will involve pre-computation. For example, if two nodes representing the composition $f(x) = b^T cx$, where $b, c \in \mathbf{R}^n$, appear in an FAO DAG, the DAG can be made more efficient by evaluating $\alpha = b^T c$ and replacing the two nodes with a single node for scalar multiplication by α.

The optimal rewriting of a DAG will depend on the hardware and overall architecture on which the multiplication algorithm is being run. For example, if the

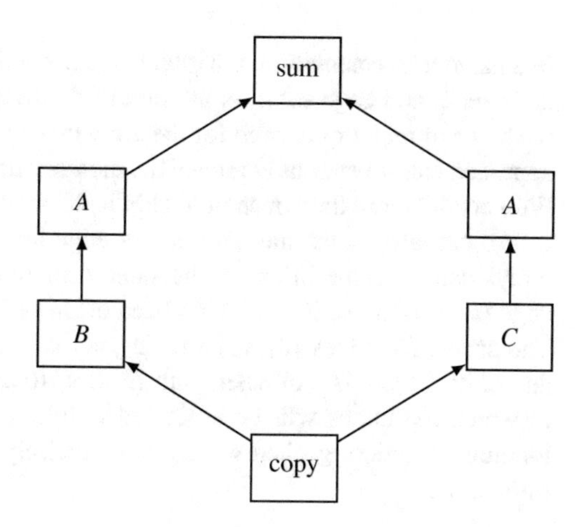

Fig. 3 The FAO DAG for $f(x) = ABx + ACx$

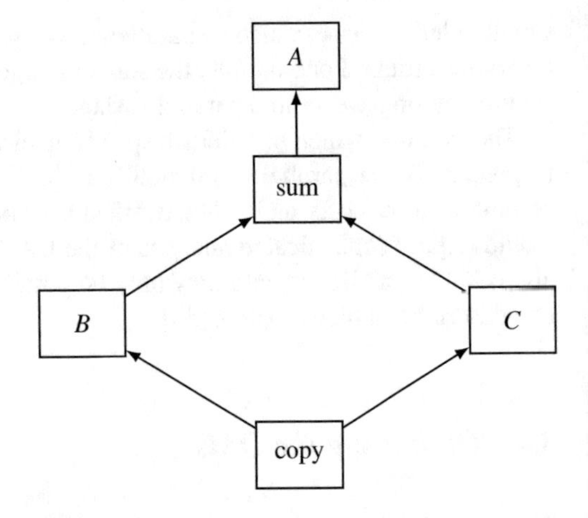

Fig. 4 The FAO DAG for $f(x) = A(Bx + Cx)$

algorithm is being run on a distributed computing cluster then a node representing multiplication by a large matrix

$$A = \begin{bmatrix} A_{11} & A_{12} \\ A_{21} & A_{22} \end{bmatrix},$$

could be split into separate nodes for each block, with the nodes stored on different computers. This rewriting would be necessary if the matrix A is so large it cannot be stored on a single machine. The literature on optimizing compilers suggests many approaches to optimizing an FAO DAG for evaluation on a particular architecture [3].

3.5 Reducing the Memory Footprint

In a naive implementation, the total bytes needed to represent an FAO DAG G, with node set V and edge set E, is the sum of the bytes of data on each node $u \in V$ and the bytes of memory needed for the array on each edge $e \in E$. A more sophisticated approach can substantially reduce the memory needed. For example, when the same FAO occurs more than once in V, duplicate nodes can share data.

We can also reuse memory across edge arrays. The key is determining which arrays can never be in use at the same time during Algorithm 7. An array for an edge (u, v) is in use if node u has been evaluated but node v has not been evaluated. The arrays for edges (u_1, v_1) and (u_2, v_2) can never be in use at the same time if and only if there is a directed path from v_1 to u_2 or from v_2 to u_1. If the sequence in which the nodes will be evaluated is fixed, rather than following an unknown topological ordering, then we can say precisely which arrays will be in use at the same time.

After we determine which edge arrays may be in use at the same time, the next step is to map the edge arrays onto a global array, keeping the global array as small as possible. Let $L(e)$ denote the length of edge e's array and $U \subseteq E \times E$ denote the set of pairs of edges whose arrays may be in use at the same time. Formally, we want to solve the optimization problem

$$\begin{array}{ll} \text{minimize} & \max_{e \in E}\{z_e + L(e)\} \\ \text{subject to} & [z_e, z_e + L(e) - 1] \cap [z_f, z_f + L(f) - 1] = \emptyset, \quad (e,f) \in U \\ & z_e \in \{1, 2, \ldots\}, \quad e \in E, \end{array} \tag{6}$$

where the z_e are the optimization variables and represent the index in the global array where edge e's array begins.

When all the edge arrays are the same length, problem (6) is equivalent to finding the chromatic number of the graph with vertices E and edges U. Problem (6) is thus NP-hard in general [72]. A reasonable heuristic for problem (6) is to first find a graph coloring of (E, U) using one of the many efficient algorithms for finding graph colorings that use a small number of colors; see, e.g., [19, 65]. We then have a mapping ϕ from colors to sets of edges assigned to the same color. We order the colors arbitrarily as $c_1, \ldots, c_k$ and assign the z_e as follows:

$$z_e = \begin{cases} 1, & e \in \phi(c_1) \\ \max\limits_{f \in \phi(c_{i-1})} \{z_f + L(f)\}, & e \in \phi(c_i), \quad i > 1. \end{cases}$$

Additional optimizations can be made based on the unique characteristics of different FAOs. For example, the outgoing edges from a **copy** node can share the incoming edge's array until the outgoing edges' arrays are written to (i.e., copy-on-write). Another example is that the outgoing edges from a **split** node can point to segments of the array on the incoming edge. Similarly, the incoming edges on a **vstack** node can point to segments of the array on the outgoing edge.

3.6 Software Implementations

Several software packages have been developed for constructing and evaluating compositions of linear functions. The MATLAB toolbox SPOT allows users to construct expressions involving both fast transforms, like convolution and the DFT, and standard matrix multiplication [66]. TFOCS, a framework in MATLAB for solving convex problems using a variety of first-order algorithms, provides functionality for constructing and composing FAOs [11]. The Python package `linop` provides methods for constructing FAOs and combining them into linear expressions [107]. Halide is a domain specific language for image processing that makes it easy to optimize compositions of fast transforms for a variety of architectures [98].

Our approach to representing and evaluating compositions of functions is similar to the approach taken by autodifferentiation tools. These tools represent a composite function $f : \mathbf{R}^n \to \mathbf{R}^m$ as a DAG [61], and multiply by the Jacobian $J \in \mathbf{R}^{m\times n}$ and its adjoint efficiently through graph traversal. Forward mode autodifferentiation computes $x \to Jx$ efficiently by traversing the DAG in topological order. Reverse mode autodifferentiation, or backpropagation, computes $u \to J^T u$ efficiently by traversing the DAG once in topological order and once in reverse topological order [8]. An enormous variety of software packages have been developed for autodifferentiation; see [8] for a survey. Autodifferentiation in the form of backpropagation plays a central role in deep learning frameworks such as TensorFlow [1], Theano [7, 13], Caffe [71], and Torch [29].

4 Cone Programs and Solvers

In this section we describe cone programs, the standard intermediate-form representation of a convex problem, and solvers for cone programs.

4.1 Cone Programs

A cone program is a convex optimization problem of the form

$$\begin{array}{ll} \text{minimize} & c^T x \\ \text{subject to} & Ax + b \in \mathcal{K}, \end{array} \tag{7}$$

where $x \in \mathbf{R}^n$ is the optimization variable, $\mathcal{K}$ is a convex cone, and $A \in \mathbf{R}^{m\times n}$, $c \in \mathbf{R}^n$, and $b \in \mathbf{R}^m$ are problem data. Cone programs are a broad class that include linear programs, second-order cone programs, and semidefinite programs as special cases [16, 92]. We call the cone program *matrix-free* if A is represented implicitly as an FAO, rather than explicitly as a dense or sparse matrix.

The convex cone $\mathcal{K}$ is typically a Cartesian product of simple convex cones from the following list:

- Zero cone: $\mathcal{K}_0 = \{0\}$.
- Free cone: $\mathcal{K}_{\text{free}} = \mathbf{R}$.
- Nonnegative cone: $\mathcal{K}_+ = \{x \in \mathbf{R} \mid x \geq 0\}$.
- Second-order cone: $\mathcal{K}_{\text{soc}} = \{(x, t) \in \mathbf{R}^{n+1} \mid x \in \mathbf{R}^n,\ t \in \mathbf{R},\ \|x\|_2 \leq t\}$.
- Positive semidefinite cone: $\mathcal{K}_{\text{psd}} = \{\mathbf{vec}(X) \mid X \in \mathbf{S}^n,\ z^T X z \geq 0 \text{ for all } z \in \mathbf{R}^n\}$.
- Exponential cone ([95, Sect. 6.3.4]):

$$\mathcal{K}_{\text{exp}} = \{(x, y, z) \in \mathbf{R}^3 \mid y > 0,\ ye^{x/y} \leq z\} \cup \{(x, y, z) \in \mathbf{R}^3 \mid x \leq 0,\ y = 0,\ z \geq 0\}.$$

- Power cone [68, 90, 100]:

$$\mathcal{K}_{\text{pwr}}^{a} = \{(x, y, z) \in \mathbf{R}^3 \mid x^a y^{(1-a)} \geq |z|, \ x \geq 0, \ y \geq 0\},$$

  where $a \in [0, 1]$.

These cones are useful in expressing common problems (via canonicalization), and can be handled by various solvers (as discussed below). Note that all the cones are subsets of $\mathbf{R}^n$, i.e., real vectors. It might be more natural to view the elements of a cone as matrices or tuples, but viewing the elements as vectors simplifies the matrix-free canonicalization algorithm in Sect. 5.

Cone programs that include only cones from certain subsets of the list above have special names. For example, if the only cones are zero, free, and nonnegative cones, the cone program is a linear program; if in addition it includes the second-order cone, it is called a second-order cone program. A well studied special case is so-called symmetric cone programs, which include the zero, free, nonnegative, second-order, and positive semidefinite cones. Semidefinite programs, where the cone constraint consists of a single positive semidefinite cone, are another common case.

4.2 *Cone Solvers*

Many methods have been developed to solve cone programs, the most widely used being interior-point methods; see, e.g., [16, 91, 93, 112, 115].

Interior-Point A large number of interior-point cone solvers have been implemented. Most support symmetric cone programs. SDPT3 [105] and SeDuMi [103] are open-source solvers implemented in MATLAB; CVXOPT [6] is an open-source solver implemented in Python; MOSEK [89] is a commercial solver with interfaces to many languages. ECOS is an open-source cone solver written in library-free C that supports second-order cone programs [39]; Akle extended ECOS to support the exponential cone [4]. DSDP5 [12] and SDPA [47] are open-source solvers for semidefinite programs implemented in C and C++, respectively.

First-Order First-order methods are an alternative to interior-point methods that scale more easily to large cone programs, at the cost of lower accuracy. PDOS [26] is a first-order cone solver based on the alternating direction method of multipliers (ADMM) [15]. PDOS supports second-order cone programs. POGS [45] is an ADMM based solver that runs on a GPU, with a version that is similar to PDOS and targets second-order cone programs. SCS is another ADMM-based cone solver, which supports symmetric cone programs as well as the exponential and power cones [94]. Many other first-order algorithms can be applied to cone programs (e.g., [22, 78, 96]), but none have been implemented as a robust, general purpose cone solver.

Matrix-Free Matrix-free cone solvers are an area of active research, and a small number have been developed. PENNON is a matrix-free semidefinite program (SDP) solver [75]. PENNON solves a series of unconstrained optimization problems using Newton's method. The Newton step is computed using a preconditioned conjugate gradient method, rather than by factoring the Hessian directly. Many other matrix-free algorithms for solving SDPs have been proposed (e.g., [25, 48, 104, 118]). CVXOPT can be used as a matrix-free cone solver, as it allows users to specify linear functions as Python functions for evaluating matrix-vector products, rather than as explicit matrices [5].

Several matrix-free solvers have been developed for quadratic programs (QPs), which are a superset of linear programs and a subset of second-order cone programs. Gondzio developed a matrix-free interior-point method for QPs that solves linear systems using a preconditioned conjugate gradient method [52, 53, 67]. PDCO is a matrix-free interior-point solver that can solve QPs [99], using LSMR to solve linear systems [43].

5 Matrix-Free Canonicalization

Canonicalization is an algorithm that takes as input a data structure representing a general convex optimization problem and outputs a data structure representing an equivalent cone program. By solving the cone program, we recover the solution to the original optimization problem. This approach is used by convex optimization modeling systems such as YALMIP [83], CVX [57], CVXPY [36], and Convex.jl [106]. The same technique is used in the code generators CVXGEN [87] and QCML [27].

The downside of canonicalization's generality is that special structure in the original problem may be lost during the transformation into a cone program. In particular, current methods of canonicalization convert fast linear transforms in the original problem into multiplication by a dense or sparse matrix, which makes the final cone program far more costly to solve than the original problem.

The canonicalization algorithm can be modified, however, so that fast linear transforms are preserved. The key is to represent all linear functions arising during the canonicalization process as FAO DAGs instead of as sparse matrices. The FAO DAG representation of the final cone program can be used by a matrix-free cone solver to solve the cone program. The modified canonicalization algorithm never forms explicit matrix representations of linear functions. Hence we call the algorithm *matrix-free canonicalization*.

The remainder of this section has the following outline: In Sect. 5.1 we give an informal overview of the matrix-free canonicalization algorithm. In Sect. 5.2 we define the expression DAG data structure, which is used throughout the matrix-free canonicalization algorithm. In Sect. 5.3 we define the data structure used to represent convex optimization problems as input to the algorithm. In Sect. 5.4 we define the representation of a cone program output by the matrix-free canonicalization algorithm. In Sect. 5.5 we present the matrix-free canonicalization algorithm itself.

For clarity, we move some details of canonicalization to the appendix. In "Equivalence of the Cone Program" we give a precise definition of the equivalence between the cone program output by the canonicalization algorithm and the original convex optimization problem given as input. In "Sparse Matrix Representation" we explain how the standard canonicalization algorithm generates a sparse matrix representation of a cone program.

5.1 *Informal Overview*

In this section we give an informal overview of the matrix-free canonicalization algorithm. Later sections define the data structures used in the algorithm and make the procedure described in this section formal and explicit.

We are given an optimization problem

$$\begin{array}{ll} \text{minimize} & f_0(x) \\ \text{subject to} & f_i(x) \leq 0, \quad i = 1, \ldots, p \\ & h_i(x) + d_i = 0, \quad i = 1, \ldots, q, \end{array} \tag{8}$$

where $x \in \mathbf{R}^n$ is the optimization variable, $f_0 : \mathbf{R}^n \to \mathbf{R}, \ldots, f_p : \mathbf{R}^n \to \mathbf{R}$ are convex functions, $h_1 : \mathbf{R}^n \to \mathbf{R}^{m_1}, \ldots, h_q : \mathbf{R}^n \to \mathbf{R}^{m_q}$ are linear functions, and $d_1 \in \mathbf{R}^{m_1}, \ldots, d_q \in \mathbf{R}^{m_q}$ are vector constants. Our goal is to convert the problem into an equivalent matrix-free cone program, so that we can solve it using a matrix-free cone solver.

We assume that the problem satisfies a set of requirements known as *disciplined convex programming* [55, 58]. The requirements ensure that each of the $f_0, \ldots, f_p$ can be represented as partial minimization over a cone program. Let each function f_i have the cone program representation

$$\begin{array}{lll} f_i(x) = \text{minimize (over } t^{(i)}) & g_0^{(i)}(x, t^{(i)}) + e_0^{(i)} & \\ \text{subject to} & g_j^{(i)}(x, t^{(i)}) + e_j^{(i)} \in \mathcal{K}_j^{(i)}, & j = 1, \ldots, r^{(i)}, \end{array}$$

where $t^{(i)} \in \mathbf{R}^{s^{(i)}}$ is the optimization variable, $g_0^{(i)}, \ldots, g_{r^{(i)}}^{(i)}$ are linear functions, $e_0^{(i)}, \ldots, e_{r^{(i)}}^{(i)}$ are vector constants, and $\mathcal{K}_1^{(i)}, \ldots, \mathcal{K}_{r^{(i)}}^{(i)}$ are convex cones.

We rewrite problem (8) as the equivalent cone program

$$\begin{array}{ll} \text{minimize} & g_0^{(0)}(x, t^{(0)}) + e_0^{(0)} \\ \text{subject to} & -g_0^{(i)}(x, t^{(i)}) - e_0^{(i)} \in \mathcal{K}_+, \quad i = 1, \ldots, p, \\ & g_j^{(i)}(x, t^{(i)}) + e_j^{(i)} \in \mathcal{K}_j^{(i)} \quad i = 1, \ldots, p, \quad j = 1, \ldots, r^{(i)} \\ & h_i(x) + d_i \in \mathcal{K}_0^{m_i}, \quad i = 1, \ldots, q. \end{array} \tag{9}$$

We convert problem (9) into the standard form for a matrix-free cone program given in (7) by representing $g_0^{(0)}$ as the inner product with a vector $c \in \mathbf{R}^{n+s^{(0)}}$,

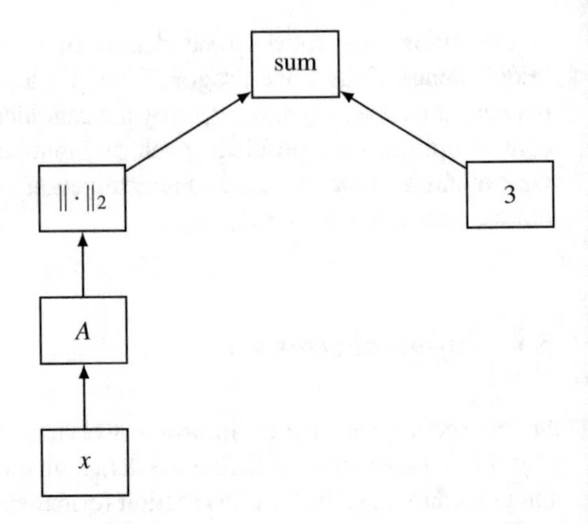

Fig. 5 The expression DAG for $f(x) = \|Ax\|_2 + 3$

concatenating the d_i and $e_j^{(i)}$ vectors into a single vector b, and representing the matrix A implicitly as the linear function that stacks the outputs of all the h_i and $g_j^{(i)}$ (excluding the objective $g_0^{(0)}$) into a single vector.

5.2 *Expression DAGs*

The canonicalization algorithm uses a data structure called an *expression DAG* to represent functions in an optimization problem. Like the FAO DAG defined in Sect. 3, an expression DAG encodes a composition of functions as a DAG where a node represents a function and an edge from a node u to a node v signifies that an output of u is an input to v. Figure 5 shows an expression DAG for the composition $f(x) = \|Ax\|_2 + 3$, where $x \in \mathbf{R}^n$ and $A \in \mathbf{R}^{m \times n}$.

Formally, an expression DAG is a connected DAG with one node with no outgoing edges (the end node) and one or more nodes with no incoming edges (start nodes). Each node in an expression DAG has the following attributes:

- A symbol representing a function f.
- The data needed to parameterize the function, such as the power p for the function $f(x) = x^p$.
- A list E_{in} of incoming edges.
- A list E_{out} of outgoing edges.

Each start node in an expression DAG is either a constant function or a variable. A variable is a symbol that labels a node input. If two nodes u and v both have incoming edges from variable nodes with symbol t, then the inputs to u and v are the same.

We say an expression DAG is affine if every non-start node represents a linear function. If in addition every start node is a variable, we say the expression DAG is linear. We say an expression DAG is constant if it contains no variables, i.e., every start node is a constant.

5.3 Optimization Problem Representation

An *optimization problem representation* (OPR) is a data structure that represents a convex optimization problem. The input to the matrix-free canonicalization algorithm is an OPR. An OPR can encode any mathematical optimization problem of the form

$$\begin{array}{ll} \text{minimize (over } y \text{ w.r.t. } \mathcal{K}_0) & f_0(x,y) \\ \text{subject to} & f_i(x,y) \in \mathcal{K}_i, \quad i = 1, \ldots, \ell, \end{array} \tag{10}$$

where $x \in \mathbf{R}^n$ and $y \in \mathbf{R}^m$ are the optimization variables, $\mathcal{K}_0$ is a proper cone, $\mathcal{K}_1, \ldots, \mathcal{K}_\ell$ are convex cones, and for $i = 0, \ldots, \ell$, we have $f_i : \mathbf{R}^n \times \mathbf{R}^m \to \mathbf{R}^{m_i}$ where $\mathcal{K}_i \subseteq \mathbf{R}^{m_i}$. (For background on convex optimization with respect to a cone, see, e.g., [16, Sect. 4.7].)

Problem (10) is more complicated than the standard definition of a convex optimization problem given in (8). The additional complexity is necessary so that OPRs can encode partial minimization over cone programs, which can involve minimization with respect to a cone and constraints other than equalities and inequalities. These partial minimization problems play a major role in the canonicalization algorithm. Note that we can easily represent equality and inequality constraints using the zero and nonnegative cones.

Concretely, an OPR is a tuple (s, o, C) where

- The element s is a tuple $(V, \mathcal{K})$ representing the problem's objective sense. The element V is a set of symbols encoding the variables being minimized over. The element $\mathcal{K}$ is a symbol encoding the proper cone the problem objective is being minimized with respect to.
- The element o is an expression DAG representing the problem's objective function.
- The element C is a set representing the problem's constraints. Each element $c_i \in C$ is a tuple $(e_i, \mathcal{K}_i)$ representing a constraint of the form $f(x,y) \in \mathcal{K}$. The element e_i is an expression DAG representing the function f and $\mathcal{K}_i$ is a symbol encoding the convex cone $\mathcal{K}$.

The matrix-free canonicalization algorithm can only operate on OPRs that satisfy the two DCP requirements [55, 58]. The first requirement is that each nonlinear function in the OPR have a known representation as partial minimization over a cone program. See [56] for many examples of such representations.

The second requirement is that the objective o be verifiable as convex with respect to the cone $\mathcal{K}$ in the objective sense s by the DCP composition rule. Similarly, for each element $(e_i, \mathcal{K}_i) \in C$, the constraint that the function represented by e_i lie in the convex cone represented by $\mathcal{K}_i$ must be verifiable as convex by the composition rule. The DCP composition rule determines the curvature of a composition $f(g_1(x), \ldots, g_k(x))$ from the curvatures and ranges of the arguments $g_1, \ldots, g_k$, the curvature of the function f, and the monotonicity of f on the range of its arguments. See [55] and [106] for a full discussion of the DCP composition rule. Additional rules are used to determine the range of a composition from the range of its arguments.

Note that it is not enough for the objective and constraints to be convex. They must also be structured so that the DCP composition rule can verify their convexity. Otherwise the cone program output by the matrix-free canonicalization algorithm is not guaranteed to be equivalent to the original problem.

To simplify the exposition of the canonicalization algorithm, we will also require that the objective sense s represent minimization over all the variables in the problem with respect to the nonnegative cone, i.e., the standard definition of minimization. The most general implementation of canonicalization would also accept OPRs that can be transformed into an equivalent OPR with an objective sense that meets this requirement.

5.4 Cone Program Representation

The matrix-free canonicalization algorithm outputs a tuple $(c_{\text{arr}}, d_{\text{arr}}, b_{\text{arr}}, G, \mathcal{K}_{\text{list}})$ where

- The element c_{arr} is a length n array representing a vector $c \in \mathbf{R}^n$.
- The element d_{arr} is a length one array representing a scalar $d \in \mathbf{R}$.
- The element b_{arr} is a length m array representing a vector $b \in \mathbf{R}^m$.
- The element G is an FAO DAG representing a linear function $f(x) = Ax$, where $A \in \mathbf{R}^{m \times n}$.
- The element $\mathcal{K}_{\text{list}}$ is a list of symbols representing the convex cones $(\mathcal{K}_1, \ldots, \mathcal{K}_\ell)$.

The tuple represents the matrix-free cone program

$$\begin{array}{ll} \text{minimize} & c^T x + d \\ \text{subject to} & Ax + b \in \mathcal{K}, \end{array} \tag{11}$$

where $\mathcal{K} = \mathcal{K}_1 \times \cdots \times \mathcal{K}_\ell$.

We can use the FAO DAG G and Algorithm 7 to represent A as an FAO, i.e., export methods for multiplying by A and A^T. These two methods are all a matrix-free cone solver needs to efficiently solve problem (11).

5.5 Algorithm

The matrix-free canonicalization algorithm can be broken down into subroutines. We describe these subroutines before presenting the overall algorithm.

Conic-Form The Conic-Form subroutine takes an OPR as input and returns an equivalent OPR where every non-start node in the objective and constraint expression DAGs represents a linear function. The output of the Conic-Form subroutine represents a cone program, but the output must still be transformed into a data structure that a cone solver can use, e.g., the cone program representation described in Sect. 5.4.

The general idea of the Conic-Form algorithm is to replace each nonlinear function in the OPR with an OPR representing partial minimization over a cone program. Recall that the canonicalization algorithm requires that all nonlinear functions in the problem be representable as partial minimization over a cone program. The OPR for each nonlinear function is spliced into the full OPR. We refer the reader to [56] and [106] for a full discussion of the Conic-Form algorithm.

The Conic-Form subroutine preserves fast linear transforms in the problem. All linear functions in the original OPR are present in the OPR output by Conic-Form. The only linear functions added are ones like **sum** and scalar multiplication that are very efficient to evaluate. Thus, evaluating the FAO DAG representing the final cone program will be as efficient as evaluating all the linear functions in the original problem (8).

Linear and Constant The Linear and Constant subroutines take an affine expression DAG as input and return the DAG's linear and constant components, respectively. Concretely, the Linear subroutine returns a copy of the input DAG where every constant start node is replaced with a variable start node and a node mapping the variable output to a vector (or matrix) of zeros with the same dimensions as the constant. The Constant subroutine returns a copy of the input DAG where every variable start node is replaced with a zero-valued constant node of the same dimensions. Figures 7 and 8 show the results of applying the Linear and Constant subroutines to an expression DAG representing $f(x) = x + 2$, as depicted in Fig. 6.

Evaluate The Evaluate subroutine takes a constant expression DAG as input and returns an array. The array contains the value of the function represented by the expression DAG. If the DAG evaluates to a matrix $A \in \mathbf{R}^{m \times n}$, the array represents $\mathbf{vec}(A)$. Similarly, if the DAG evaluates to multiple output vectors $(b_1, \ldots, b_k) \in \mathbf{R}^{n_1} \times \cdots \times \mathbf{R}^{n_k}$, the array represents $\mathbf{vstack}(b_1, \ldots, b_k)$. For example, the output of the Evaluate subroutine on the expression DAG in Fig. 8 is a length one array with first entry equal to 2.

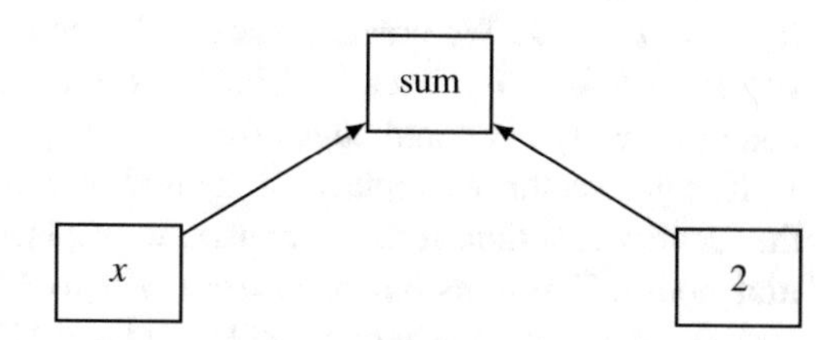

Fig. 6 The expression DAG for $f(x) = x + 2$

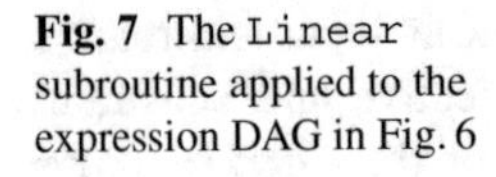

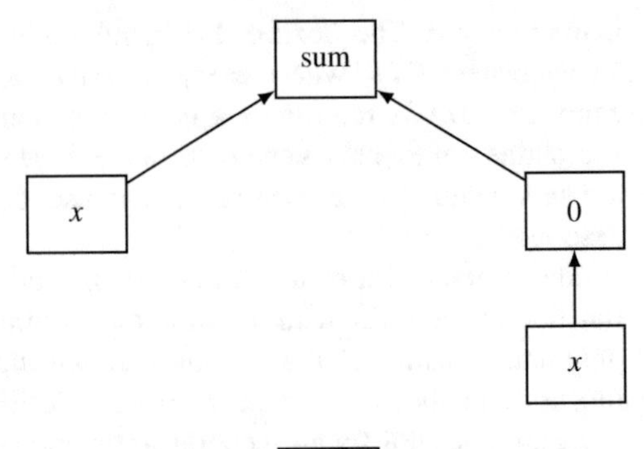

Fig. 7 The `Linear` subroutine applied to the expression DAG in Fig. 6

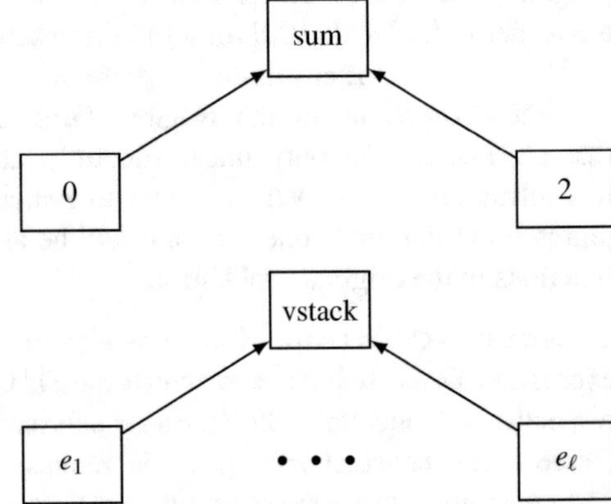

Fig. 8 The `Constant` subroutine applied to the expression DAG in Fig. 6

Fig. 9 The expression DAG for $\mathbf{vstack}(e_1, \ldots, e_\ell)$

`Graph-Repr` The `Graph-Repr` subroutine takes a list of linear expression DAGs, $(e_1, \ldots, e_\ell)$, and an ordering over the variables in the expression DAGs, $<_V$, as input and outputs an FAO DAG G. We require that the end node of each expression DAG represent a function with a single vector as output.

We construct the FAO DAG G in three steps. In the first step, we combine the expression DAGs into a single expression DAG $H^{(1)}$ by creating a **vstack** node and adding an edge from the end node of each expression DAG to the new node. The expression DAG $H^{(1)}$ is shown in Fig. 9.

In the second step, we transform $H^{(1)}$ into an expression DAG $H^{(2)}$ with a single start node. Let $x_1, \ldots, x_k$ be the variables in $(e_1, \ldots, e_\ell)$ ordered by $<_V$. Let n_i be the length of x_i if the variable is a vector and of $\mathbf{vec}(x_i)$ if the variable is a matrix, for $i = 1, \ldots, k$. We create a start node representing the function $\mathbf{split} : \mathbf{R}^n \to \mathbf{R}^{n_1} \times \cdots \times \mathbf{R}^{n_k}$. For each variable x_i, we add an edge from output i of the start node to a **copy** node and edges from that **copy** node to all the nodes representing x_i. If x_i is a vector, we replace all the nodes representing x_i with nodes representing the identity function. If x_i is a matrix, we replace all the nodes representing x_i with **mat** nodes. The transformation from $H^{(1)}$ to $H^{(2)}$ when $\ell = 1$ and e_1 represents $f(x) = x + A(x + y)$, where $x, y \in \mathbf{R}^n$ and $A \in \mathbf{R}^{n \times n}$, are depicted in Figs. 10 and 11.

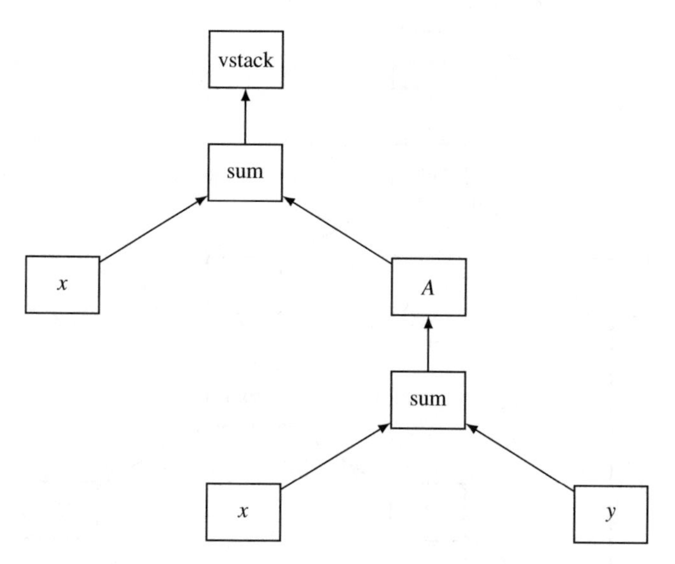

Fig. 10 The expression DAG $H^{(1)}$ when $\ell = 1$ and e_1 represents $f(x, y) = x + A(x + y)$

In the third and final step, we transform $H^{(2)}$ from an expression DAG into an FAO DAG G. $H^{(2)}$ is almost an FAO DAG, since each node represents a linear function and the DAG has a single start and end node. To obtain G we simply add the node and edge attributes needed in an FAO DAG. For each node u in $H^{(2)}$ representing the function f, we add to u an FAO (f, Φ_f, Φ_{f^*}) and the data needed to evaluate Φ_f and Φ_{f^*}. The node already has the required lists of incoming and outgoing edges. We also add an array to each of $H^{(2)}$'s edges.

`Optimize-Graph` The `Optimize-Graph` subroutine takes an FAO DAG G as input and outputs an equivalent FAO DAG G^{opt}, meaning that the output of Algorithm 7 is the same for G and G^{opt}. We choose G^{opt} by optimizing G so that the runtime of Algorithm 7 is as short as possible (see Sect. 3.4). We also compress the FAO data and edge arrays to reduce the graph's memory footprint (see Sect. 3.5). We could optimize the graph for the adjoint, G^*, as well, but asymptotically at least the flop count and memory footprint for G^* will be the same as for G, meaning optimizing G is the same as jointly optimizing G and G^*.

`Matrix-Repr` The `Matrix-Repr` subroutine takes a list of linear expression DAGs, $(e_1, \ldots, e_\ell)$, and an ordering over the variables in the expression DAGs, $<_V$, as input and outputs a sparse matrix. Note that the input types are the same as in the `Graph-Repr` subroutine. In fact, for a given input the sparse matrix output by `Matrix-Repr` represents the same linear function as the FAO DAG output by `Graph-Repr`. The `Matrix-Repr` subroutine is used by the standard canonicalization algorithm to produce a sparse matrix representation of a cone program. The implementation of `Matrix-Repr` is described in the appendix in "Sparse Matrix Representation".

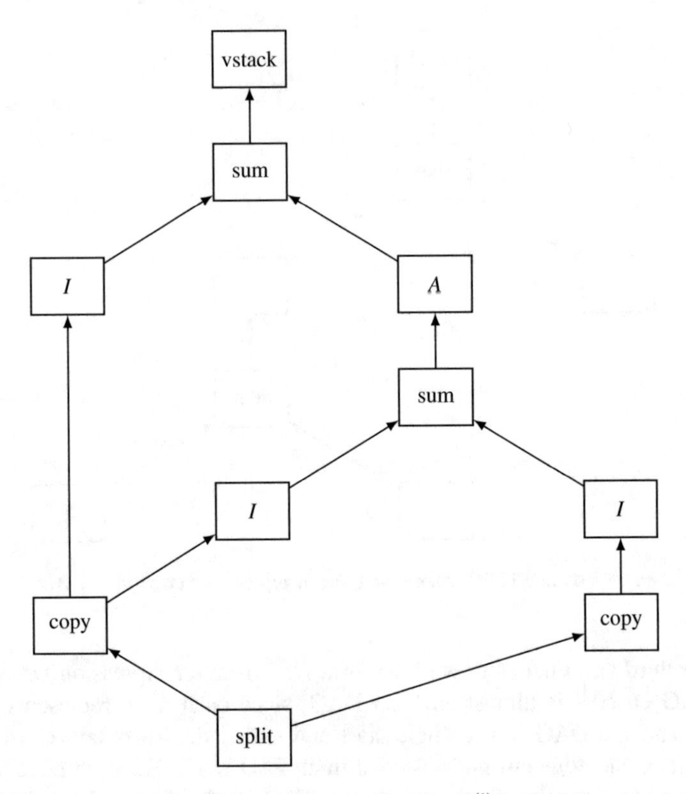

Fig. 11 The expression DAG $H^{(2)}$ obtained by transforming $H^{(1)}$ in Fig. 10

Overall Algorithm With all the subroutines in place, the matrix-free canonicalization algorithm is straightforward. The implementation is given in Algorithm 8.

Algorithm 8 Matrix-free canonicalization

Precondition: p is an OPR that satisfies the requirements of DCP.

$(s, o, C) \leftarrow$ `Conic-Form`(p).
Choose any ordering $<_V$ on the variables in (s, o, C).
Choose any ordering $<_C$ on the constraints in C.
$((e_1, \mathcal{K}_1), \ldots, (e_\ell, \mathcal{K}_\ell)) \leftarrow$ the constraints in C ordered according to $<_C$.
$c_{\text{mat}} \leftarrow$ `Matrix-Repr`((`Linear`(o)), $<_V$).
Convert c_{mat} from a 1-by-n sparse matrix into a length n array c_{arr}.
$d_{\text{arr}} \leftarrow$ `Evaluate`(`Constant`(o)).
$b_{\text{arr}} \leftarrow$ **vstack**(`Evaluate`(`Constant`(e_1)), ..., `Evaluate`(`Constant`(e_ℓ))).
$G \leftarrow$ `Graph-Repr`((`Linear`(e_1), ..., `Linear`(e_ℓ)), $<_V$).
$G^{\text{opt}} \leftarrow$ `Optimize-Graph`(G)
$\mathcal{K}_{\text{list}} \leftarrow (\mathcal{K}_1, \ldots, \mathcal{K}_\ell)$.
return $(c_{\text{arr}}, d_{\text{arr}}, b_{\text{arr}}, G^{\text{opt}}, \mathcal{K}_{\text{list}})$.

6 Numerical Results

We have implemented the matrix-free canonicalization algorithm as an extension of CVXPY [36], available at

https://github.com/mfopt/mf_cvxpy.

To solve the resulting matrix-free cone programs, we implemented modified versions of SCS [94] and POGS [45] that are truly matrix-free, available at

https://github.com/mfopt/mf_scs,
https://github.com/mfopt/mf_pogs.

(The main modification was using the matrix-free equilibration described in [37].) Our implementations are still preliminary and can be improved in many ways. We also emphasize that the canonicalization is independent of the particular matrix-free cone solver used.

In this section we benchmark our implementation of matrix-free canonicalization and of matrix-free SCS and POGS on several convex optimization problems involving fast linear transforms. We compare the performance of our matrix-free convex optimization modeling system with that of the current CVXPY modeling system, which represents the matrix A in a cone program as a sparse matrix and uses standard cone solvers. The standard cone solvers and matrix-free SCS were run serially on a single Intel Xeon processor, while matrix-free POGS was run on a Titan X GPU.

6.1 Nonnegative Deconvolution

We applied our matrix-free convex optimization modeling system to the nonnegative deconvolution problem (1). The Python code below constructs and solves problem (1). The constants c and b and problem size n are defined elsewhere. The code is only a few lines, and it could be easily modified to add regularization on x or apply a different cost function to $c * x - b$. The modeling system would automatically adapt to solve the modified problem.

```
# Construct the optimization problem.
x = Variable(n)
cost = norm2(conv(c, x) - b)
prob = Problem(Minimize(cost),
               [x >= 0])
# Solve using matrix-free SCS.
prob.solve(solver=MF_SCS)
```

Problem Instances We used the following procedure to generate interesting (nontrivial) instances of problem (1). For all instances the vector $c \in \mathbf{R}^n$ was

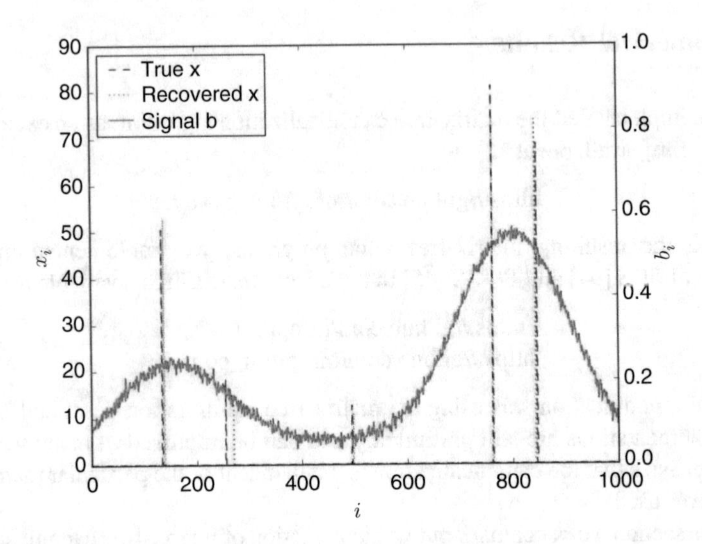

Fig. 12 Results for a problem instance with $n = 1000$

a Gaussian kernel with standard deviation $n/10$. All entries of c less than 10^{-6} were set to 10^{-6}, so that no entries were too close to zero. The vector $b \in \mathbf{R}^{2n-1}$ was generated by picking a solution $\tilde{x}$ with five entries randomly chosen to be nonzero. The values of the nonzero entries were chosen uniformly at random from the interval $[0, n/10]$. We set $b = c * \tilde{x} + v$, where the entries of the noise vector $v \in \mathbf{R}^{2n-1}$ were drawn from a normal distribution with mean zero and variance $\|c * \tilde{x}\|^2/(400(2n-1))$. Our choice of v yielded a signal-to-noise ratio near 20.

While not relevant to solving the optimization problem, the solution of the nonnegative deconvolution problem often, but not always, (approximately) recovers the original vector $\tilde{x}$. Figure 12 shows the solution recovered by ECOS [39] for a problem instance with $n = 1000$. The ECOS solution $x^\star$ had a cluster of 3–5 adjacent nonzero entries around each spike in $\tilde{x}$. The sum of the entries was close to the value of the spike. The recovered x in Fig. 12 shows only the largest entry in each cluster, with value set to the sum of the cluster's entries.

Results Figure 13 compares the performance on problem (1) of the interior-point solver ECOS [39] and matrix-free versions of SCS and POGS as the size n of the optimization variable increases. We limited the solvers to 10^4 s.

For each variable size n we generated ten different problem instances and recorded the average solve time for each solver. ECOS and matrix-free SCS were run with an absolute and relative tolerance of 10^{-3} for the duality gap, ℓ_2-norm of the primal residual, and ℓ_2-norm of the dual residual. Matrix-free POGS was run with an absolute tolerance of 10^{-4} and a relative tolerance of 10^{-3}.

For each solver, we plot the solve times and the least-squares linear fit to those solve times (the dotted line). The slopes of the lines show how the solvers scale. The least-squares linear fit for the ECOS solve times has slope 3.1, which indicates that

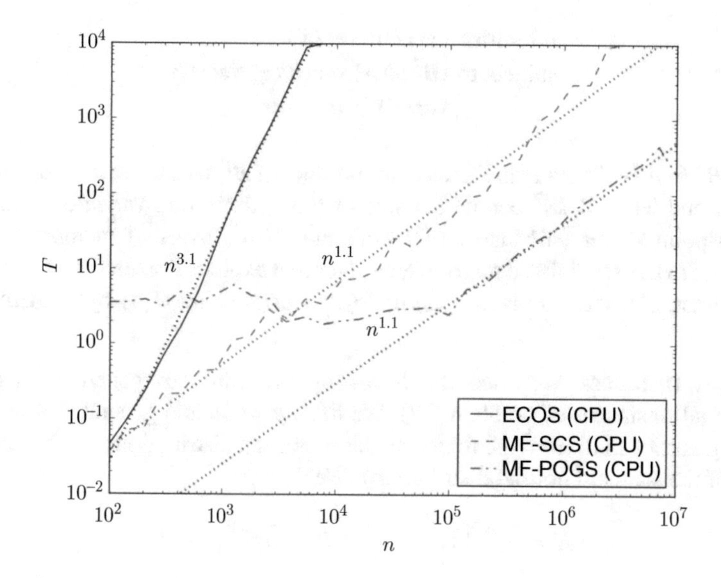

Fig. 13 Solve time in seconds T versus variable size n

the solve time scales like n^3, as expected. The least-squares linear fit for the matrix-free SCS solve times has slope 1.1, which indicates that the solve time scales like the expected $n \log n$. The least-squares linear fit for the matrix-free POGS solve times in the range $n \in [10^5, 10^7]$ has slope 1.1, which indicates that the solve time scales like the expected $n \log n$. For $n < 10^5$, the GPU is not saturated, so increasing n barely increases the solve time.

6.2 Sylvester LP

We applied our matrix-free convex optimization modeling system to Sylvester LPs, or convex optimization problems of the form

$$\begin{array}{ll} \text{minimize} & \mathbf{Tr}(D^T X) \\ \text{subject to} & AXB \leq C \\ & X \geq 0, \end{array} \tag{12}$$

where $X \in \mathbf{R}^{p\times q}$ is the optimization variable, and $A \in \mathbf{R}^{p\times p}$, $B \in \mathbf{R}^{q\times q}$, $C \in \mathbf{R}^{p\times q}$, and $D \in \mathbf{R}^{p\times q}$ are problem data. The inequality $AXB \leq C$ is a variant of the Sylvester equation $AXB = C$ [49].

Existing convex optimization modeling systems will convert problem (12) into the vectorized format

$$\begin{array}{ll} \text{minimize} & \mathbf{vec}(D)^T\,\mathbf{vec}(X) \\ \text{subject to} & (B^T \otimes A)\,\mathbf{vec}(X) \leq \mathbf{vec}(C) \\ & \mathbf{vec}(X) \geq 0, \end{array} \tag{13}$$

where $B^T \otimes A \in \mathbf{R}^{pq \times pq}$ is the Kronecker product of B^T and A. Let $p = kq$ for some fixed k, and let $n = kq^2$ denote the size of the optimization variable. A standard interior-point solver will take $O(n^3)$ flops and $O(n^2)$ bytes of memory to solve problem (13). A specialized matrix-free solver that exploits the matrix product AXB, by contrast, can solve problem (12) in $O(n^{1.5})$ flops using $O(n)$ bytes of memory [110].

Problem Instances We used the following procedure to generate interesting (nontrivial) instances of problem (12). We fixed $p = 5q$ and generated $\tilde{A}$ and $\tilde{B}$ by drawing entries i.i.d. from the folded standard normal distribution (i.e., the absolute value of the standard normal distribution). We then set

$$A = \tilde{A}/\|\tilde{A}\|_2 + I, \qquad B = \tilde{B}/\|\tilde{B}\|_2 + I,$$

so that A and B had positive entries and bounded condition number. We generated D by drawing entries i.i.d. from a standard normal distribution. We fixed $C = \mathbf{1}\mathbf{1}^T$. Our method of generating the problem data ensured the problem was feasible and bounded.

Results Figure 14 compares the performance on problem (12) of the interior-point solver ECOS [39] and matrix-free versions of SCS and POGS as the size $n = 5q^2$ of

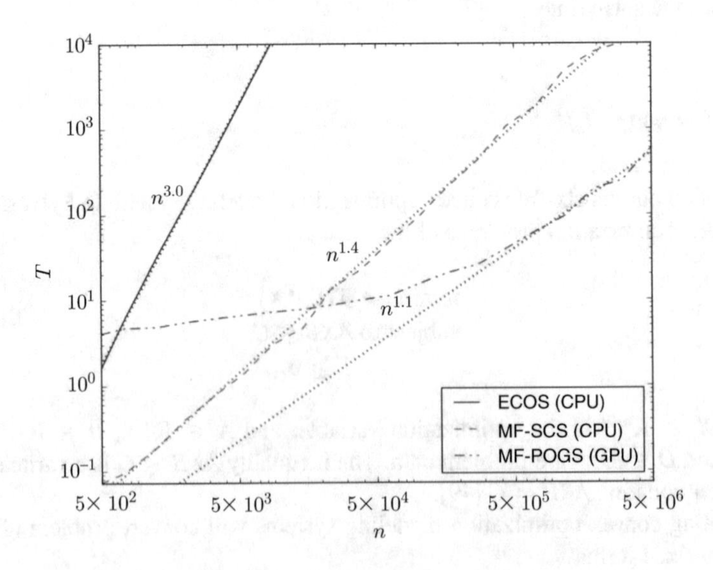

Fig. 14 Solve time in seconds T versus variable size n

the optimization variable increases. We limited the solvers to 10^4 s. For each variable size n we generated ten different problem instances and recorded the average solve time for each solver. ECOS and matrix-free SCS were run with an absolute and relative tolerance of 10^{-3} for the duality gap, ℓ_2-norm of the primal residual, and ℓ_2-norm of the dual residual. Matrix-free POGS was run with an absolute tolerance of 10^{-4} and a relative tolerance of 10^{-3}.

For each solver, we plot the solve times and the least-squares linear fit to those solve times (the dotted line). The slopes of the lines show how the solvers scale. The least-squares linear fit for the ECOS solve times has slope 3.0, which indicates that the solve time scales like n^3, as expected. The least-squares linear fit for the matrix-free SCS solve times has slope 1.4, which indicates that the solve time scales like the expected $n^{1.5}$. The least-squares linear fit for the matrix-free POGS solve times in the range $n \in [5 \times 10^5, 5 \times 10^6]$ has slope 1.1. The solve time scales more slowly than the expected $n^{1.5}$, likely because the GPU was not fully saturated even on the largest problem instances. For $n < 5 \times 10^5$, the GPU was far from saturated, so increasing n barely increases the solve time.

Acknowledgements We thank Eric Chu, Michal Kočvara, and Alex Aiken for helpful comments on earlier versions of this work, and Chris Fougner, John Miller, Jack Zhu, and Paul Quigley for their work on the POGS cone solver and CVXcanon [88], which both contributed to the implementation of matrix-free CVXPY. We also thank the anonymous reviewers for useful feedback. This material is based upon work supported by the National Science Foundation Graduate Research Fellowship under Grant No. DGE-114747 and by the DARPA XDATA program.

Appendix

Equivalence of the Cone Program

In this section we explain the precise sense in which the cone program output by the matrix-free canonicalization algorithm is equivalent to the original convex optimization problem.

Theorem 1. *Let p be a convex optimization problem whose OPR is a valid input to the matrix-free canonicalization algorithm. Let $\Phi(p)$ be the cone program represented by the output of the algorithm given p's OPR as input. All the variables in p are present in $\Phi(p)$, along with new variables introduced during the canonicalization process [55, 106]. Let $x \in \mathbf{R}^n$ represent the variables in p stacked into a vector and $t \in \mathbf{R}^m$ represent the new variables in $\Phi(p)$ stacked into a vector.*

The problems p and $\Phi(p)$ are equivalent in the following sense:

1. *For all x feasible in p, there exists $t^\star$ such that $(x, t^\star)$ is feasible in $\Phi(p)$ and $p(x) = \Phi(p)(x, t^\star)$.*
2. *For all (x, t) feasible in $\Phi(p)$, x is feasible in p and $p(x) \leq \Phi(p)(x, t)$.*

For a point x feasible in p, by $p(x)$ we mean the value of p's objective evaluated at x. The notation $\Phi(p)(x, t)$ is similarly defined.

Proof. See [55].

Theorem 1 implies that p and $\Phi(p)$ have the same optimal value. Moreover, p is infeasible if and only if $\Phi(p)$ is infeasible, and p is unbounded if and only if $\Phi(p)$ is unbounded. The theorem also implies that any solution $x^\star$ to p is part of a solution $(x^\star, t^\star)$ to $\Phi(p)$ and vice versa.

A similar equivalence holds between the Lagrange duals of p and $\Phi(p)$, but the details are beyond the scope of this paper. See [55] for a discussion of the dual of the cone program output by the canonicalization algorithm.

Sparse Matrix Representation

In this section we explain the `Matrix-Repr` subroutine used in the standard canonicalization algorithm to obtain a sparse matrix representation of a cone program. Recall that the subroutine takes a list of linear expression DAGs, $(e_1, \ldots, e_\ell)$, and an ordering over the variables in the expression DAGs, $<_V$, as input and outputs a sparse matrix A.

The algorithm to carry out the subroutine is not discussed anywhere in the literature, so we present here the version used by CVXPY [36]. The algorithm first converts each expression DAG into a map from variables to sparse matrices, representing a sum of terms. For example, if the map ϕ maps the variable $x \in \mathbf{R}^n$ to the sparse matrix coefficient $B \in \mathbf{R}^{m\times n}$ and the variable $y \in \mathbf{R}^n$ to the sparse matrix coefficient $C \in \mathbf{R}^{m\times n}$, then ϕ represents the sum $Bx + Cy$.

The conversion from expression DAG to map of variables to sparse matrices is done using Algorithm 9. The algorithm uses the subroutine `Matrix-Coeff`, which takes a node representing a linear function f and indices i and j as inputs and outputs a sparse matrix D. Let $\tilde{f}$ be a function defined on the range of f's ith input such that $\tilde{f}(x)$ is equal to f's jth output when f is evaluated on ith input x and zero-valued matrices (of the appropriate dimensions) for all other inputs. The output of `Matrix-Coeff` is the sparse matrix D such that for any value x in the domain of $\tilde{f}$,

$$D\,\mathbf{vec}(x) = \mathbf{vec}(\tilde{f}(x)).$$

The sparse matrix coefficients in the maps of variables to sparse matrices are assembled into a single sparse matrix A, as follows: Let $x_1, \ldots, x_k$ be the variables in the expression DAGs, ordered according to $<_V$. Let n_i be the length of x_i if the variable is a vector and of $\mathbf{vec}(x_i)$ if the variable is a matrix, for $i = 1, \ldots, k$. Let m_j be the length of expression DAG e_j's output, for $j = 1, \ldots, \ell$. The coefficients for x_1 are placed in the first n_1 columns in A, the coefficients for x_2 in the next n_2 columns, etc. Similarly, the coefficients from e_1 are placed in the first m_1 rows of A, the coefficients from e_2 in the next m_2 rows, etc.

Algorithm 9 Convert an expression DAG into a map from variables to sparse matrices

Precondition: e is a linear expression DAG that outputs a single vector.

Create an empty queue Q for nodes that are ready to evaluate.
Create an empty set S for nodes that have been evaluated.
Create a map M from (node, output index) tuples to maps of variables to sparse matrices.
for every start node u in e **do**
  $x \leftarrow$ the variable represented by node u.
  $n \leftarrow$ the length of x if the variable is a vector and of $\mathbf{vec}(x)$ if the variable is a matrix.
  $M[(u,1)] \leftarrow$ a map with key x and value the n-by-n identity matrix.
  Add u to S.
end for
Add all nodes in e to Q whose only incoming edges are from start nodes.
while Q is not empty **do**
  $u \leftarrow$ pop the front node of Q.
  Add u to S.
  for edge (u,p) in u's E_{out}, with index j **do**
    Create an empty map M_j from variables to sparse matrices.
    for edge (v,u) in u's E_{in}, with index i **do**
      $A^{(ij)} \leftarrow \texttt{Matrix-Coeff}(u,i,j)$.
      $k \leftarrow$ the index of (v,u) in v's E_{out}.
      for key x and value C in $M[(v,k)]$ **do**
        if M_j has an entry for x **then**
          $M_j[x] \leftarrow M_j[x] + A^{(ij)}C$.
        else
          $M_j[x] \leftarrow A^{(ij)}C$.
        end if
      end for
    end for
    $M[(u,j)] \leftarrow M_j$.
    if for all edges (q,p) in p's E_{in}, q is in S **then**
      Add p to the end of Q.
    end if
  end for
end while
$u_{\text{end}} \leftarrow$ the end node of e.
return $M[(u_{\text{end}},1)]$.

References

1. Abadi, M., Agarwal, A., Barham, P., Brevdo, E., Chen, Z., Citro, C., Corrado, G., Davis, A., Dean, J., Devin, M., Ghemawat, S., Goodfellow, I., Harp, A., Irving, G., Isard, M., Jia, Y., Jozefowicz, R., Kaiser, L., Kudlur, M., Levenberg, J., Mané, D., Monga, R., Moore, S., Murray, D., Olah, C., Schuster, M., Shlens, J., Steiner, B., Sutskever, I., Talwar, K., Tucker, P., Vanhoucke, V., Vasudevan, V., Viégas, F., Vinyals, O., Warden, P., Wattenberg, M., Wicke, M., Yu, Y., Zheng, X.: TensorFlow: large-scale machine learning on heterogeneous systems. (2015) http://tensorflow.org/. Cited 2 March 2016

2. Ahmed, N., Natarajan, T., Rao, K.: Discrete cosine transform. IEEE Trans. Comput. **C-23**(1), 90–93 (1974)
3. Aho, A., Lam, M., Sethi, R., Ullman, J.: Compilers: Principles, Techniques, and Tools, 2nd edn. Addison-Wesley Longman, Boston (2006)
4. Akle, S.: Algorithms for unsymmetric cone optimization and an implementation for problems with the exponential cone. Ph.D. thesis, Stanford University (2015)
5. Andersen, M., Dahl, J., Liu, Z., Vandenberghe, L.: Interior-point methods for large-scale cone programming. In: Sra, S., Nowozin, S., Wright, S. (eds.) Optimization for Machine Learning, pp. 55–83. MIT Press, Cambridge (2012)
6. Andersen, M., Dahl, J., Vandenberghe, L.: CVXOPT: Python software for convex optimization, version 1.1 (2015). http://cvxopt.org/. Cited 2 March 2016
7. Bastien, F., Lamblin, P., Pascanu, R., Bergstra, J., Goodfellow, I., Bergeron, A., Bouchard, N., Bengio, Y.: Theano: new features and speed improvements. In: Deep Learning and Unsupervised Feature Learning, Neural Information Processing Systems Workshop (2012)
8. Baydin, A., Pearlmutter, B., Radul, A., Siskind, J.: Automatic differentiation in machine learning: a survey. Preprint (2015). http://arxiv.org/abs/1502.05767. Cited 2 March 2016
9. Beck, A., Teboulle, M.: Fast gradient-based algorithms for constrained total variation image denoising and deblurring problems. IEEE Trans. Image Process. **18**(11), 2419–2434 (2009)
10. Beck, A., Teboulle, M.: A fast iterative shrinkage-thresholding algorithm for linear inverse problems. SIAM J. Imag. Sci. **2**(1), 183–202 (2009)
11. Becker, S., Candès, E., Grant, M.: Templates for convex cone problems with applications to sparse signal recovery. Math. Program. Comput. **3**(3), 165–218 (2011)
12. Benson, S., Ye, Y.: Algorithm 875: DSDP5—software for semidefinite programming. ACM Trans. Math. Software **34**(3), (2008)
13. Bergstra, J., Breuleux, O., Bastien, F., Lamblin, P., Pascanu, R., Desjardins, G., Turian, J., Warde-Farley, D., Bengio, Y.: Theano: a CPU and GPU math expression compiler. In: Proceedings of the Python for Scientific Computing Conference (2010)
14. Börm, S., Grasedyck, L., Hackbusch, W.: Introduction to hierarchical matrices with applications. Eng. Anal. Bound. Elem. **27**(5), 405–422 (2003)
15. Boyd, S., Parikh, N., Chu, E., Peleato, B., Eckstein, J.: Distributed optimization and statistical learning via the alternating direction method of multipliers. Found. Trends Mach. Learn. **3**, 1–122 (2011)
16. Boyd, S., Vandenberghe, L.: Convex Optimization. Cambridge University Press, Cambridge (2004)
17. Bracewell, R.: The fast Hartley transform. In: Proceedings of the IEEE, vol. 72, pp. 1010–1018 (1984)
18. Brandt, A., McCormick, S., Ruge, J.: Algebraic multigrid (AMG) for sparse matrix equations. In: D. Evans (ed.) Sparsity and its Applications, pp. 257–284. Cambridge University Press, Cambridge (1985)
19. Brélaz, D.: New methods to color the vertices of a graph. Commun. ACM **22**(4), 251–256 (1979)
20. Candès, E., Demanet, L., Donoho, D., Ying, L.: Fast discrete curvelet transforms. Multiscale Model. Simul. **5**(3), 861–899 (2006)
21. Carrier, J., Greengard, L., Rokhlin, V.: A fast adaptive multipole algorithm for particle simulations. SIAM J. Sci. Stat. Comput. **9**(4), 669–686 (1988)
22. Chambolle, A., Pock, T.: A first-order primal-dual algorithm for convex problems with applications to imaging. J. Math. Imaging Vision **40**(1), 120–145 (2011)
23. Chan, T., Esedoglu, S., Nikolova, M.: Algorithms for finding global minimizers of image segmentation and denoising models. SIAM J. Appl. Math. **66**(5), 1632–1648 (2006)
24. Chen, S., Donoho, D., Saunders, M.: Atomic decomposition by basis pursuit. SIAM J. Sci. Comput. **20**(1), 33–61 (1998)
25. Choi, C., Ye, Y.: Solving sparse semidefinite programs using the dual scaling algorithm with an iterative solver. Working paper, Department of Management Sciences, University of Iowa (2000)

26. Chu, E., O'Donoghue, B., Parikh, N., Boyd, S.: A primal-dual operator splitting method for conic optimization. Preprint (2013). http://stanford.edu/~boyd/papers/pdf/pdos.pdf. Cited 2 March 2016
27. Chu, E., Parikh, N., Domahidi, A., Boyd, S.: Code generation for embedded second-order cone programming. In: Proceedings of the European Control Conference, pp. 1547–1552 (2013)
28. Cohen, A., Daubechies, I., Feauveau, J.C.: Biorthogonal bases of compactly supported wavelets. Commun. Pure Appl. Math. **45**(5), 485–560 (1992)
29. Collobert, R., Kavukcuoglu, K., Farabet, C.: Torch7: a MATLAB-like environment for machine learning. In: BigLearn, Neural Information Processing Systems Workshop (2011)
30. Cooley, J., Lewis, P., Welch, P.: The fast Fourier transform and its applications. IEEE Trans. Educ. **12**(1), 27–34 (1969)
31. Cooley, J., Tukey, J.: An algorithm for the machine calculation of complex Fourier series. Math. Comput. **19**(90), 297–301 (1965)
32. Daubechies, I.: Orthonormal bases of compactly supported wavelets. Commun. Pure Appl. Math. **41**(7), 909–996 (1988)
33. Daubechies, I.: Ten lectures on wavelets. SIAM, Philadelphia (1992)
34. Davis, T.: Direct Methods for Sparse Linear Systems (Fundamentals of Algorithms 2). SIAM, Philadelphia (2006)
35. Diamond, S., Boyd, S.: Convex optimization with abstract linear operators. In: Proceedings of the IEEE International Conference on Computer Vision, pp. 675–683 (2015)
36. Diamond, S., Boyd, S.: CVXPY: A Python-embedded modeling language for convex optimization. J. Mach. Learn. Res. **17**(83), 1–5 (2016)
37. Diamond, S., Boyd, S.: Stochastic matrix-free equilibration. J. Optim. Theory Appl. (2016, to appear)
38. Do, M., Vetterli, M.: The finite ridgelet transform for image representation. IEEE Trans. Image Process. **12**(1), 16–28 (2003)
39. Domahidi, A., Chu, E., Boyd, S.: ECOS: an SOCP solver for embedded systems. In: Proceedings of the European Control Conference, pp. 3071–3076 (2013)
40. Dudgeon, D., Mersereau, R.: Multidimensional Digital Signal Processing. Prentice-Hall, Englewood Cliffs (1984)
41. Duff, I., Erisman, A., Reid, J.: Direct Methods for Sparse Matrices. Oxford University Press, New York (1986)
42. Figueiredo, M., Nowak, R., Wright, S.: Gradient projection for sparse reconstruction: application to compressed sensing and other inverse problems. IEEE J. Sel. Top. Signal Process. **1**(4), 586–597 (2007)
43. Fong, D., Saunders, M.: LSMR: an iterative algorithm for sparse least-squares problems. SIAM J. Sci. Comput. **33**(5), 2950–2971 (2011)
44. Forsyth, D., Ponce, J.: Computer Vision: A Modern Approach. Prentice Hall, Upper Saddle River (2002)
45. Fougner, C., Boyd, S.: Parameter selection and pre-conditioning for a graph form solver. (2015, preprint). http://arxiv.org/pdf/1503.08366v1.pdf. Cited 2 March 2016
46. Fountoulakis, K., Gondzio, J., Zhlobich, P.: Matrix-free interior point method for compressed sensing problems. Math. Program. Comput. **6**(1), 1–31 (2013)
47. Fujisawa, K., Fukuda, M., Kobayashi, K., Kojima, M., Nakata, K., Nakata, M., Yamashita, M.: SDPA (semidefinite programming algorithm) user's manual – version 7.0.5. Tech. rep. (2008)
48. Fukuda, M., Kojima, M., Shida, M.: Lagrangian dual interior-point methods for semidefinite programs. SIAM J. Optim. **12**(4), 1007–1031 (2002)
49. Gardiner, J., Laub, A., Amato, J., Moler, C.: Solution of the Sylvester matrix equation $AXB^T + CXD^T = E$. ACM Trans. Math. Software **18**(2), 223–231 (1992)
50. Gilbert, A., Strauss, M., Tropp, J., Vershynin, R.: One sketch for all: fast algorithms for compressed sensing. In: Proceedings of the ACM Symposium on Theory of Computing, pp. 237–246 (2007)

51. Goldstein, T., Osher, S.: The split Bregman method for ℓ_1-regularized problems. SIAM J. Imag. Sci. **2**(2), 323–343 (2009)
52. Gondzio, J.: Matrix-free interior point method. Comput. Optim. Appl. **51**(2), 457–480 (2012)
53. Gondzio, J.: Convergence analysis of an inexact feasible interior point method for convex quadratic programming. SIAM J. Optim. **23**(3), 1510–1527 (2013)
54. Gondzio, J., Grothey, A.: Parallel interior-point solver for structured quadratic programs: application to financial planning problems. Ann. Oper. Res. **152**(1), 319–339 (2007)
55. Grant, M.: Disciplined convex programming. Ph.D. thesis, Stanford University (2004)
56. Grant, M., Boyd, S.: Graph implementations for nonsmooth convex programs. In: Blondel, V., Boyd, S., Kimura, H. (eds.) Recent Advances in Learning and Control. Lecture Notes in Control and Information Sciences, pp. 95–110. Springer, London (2008)
57. Grant, M., Boyd, S.: CVX: MATLAB software for disciplined convex programming, version 2.1 (2014). http://cvxr.com/cvx. Cited 2 March 2016
58. Grant, M., Boyd, S., Ye, Y.: Disciplined convex programming. In: Liberti, L., Maculan, N. (eds.) Global Optimization: From Theory to Implementation, Nonconvex Optimization and its Applications, pp. 155–210. Springer, New York (2006)
59. Greengard, L., Rokhlin, V.: A fast algorithm for particle simulations. J. Comput. Phys. **73**(2), 325–348 (1987)
60. Greengard, L., Strain, J.: The fast Gauss transform. SIAM J. Sci. Stat. Comput. **12**(1), 79–94 (1991)
61. Griewank, A.: On automatic differentiation. In: Iri, M., Tanabe, K. (eds.) Mathematical Programming: Recent Developments and Applications, pp. 83–108. Kluwer Academic, Tokyo (1989)
62. Hackbusch, W.: Multi-Grid Methods and Applications. Springer, Heidelberg (1985)
63. Hackbusch, W.: A sparse matrix arithmetic based on $\mathscr{H}$-matrices. Part I: introduction to $\mathscr{H}$-matrices. Computing **62**(2), 89–108 (1999)
64. Hackbusch, W., Khoromskij, B., Sauter, S.: On $\mathscr{H}^2$-matrices. In: Bungartz, H.J., Hoppe, R., Zenger, C. (eds.) Lectures on Applied Mathematics, pp. 9–29. Springer, Heidelberg (2000)
65. Halldórsson, M.: A still better performance guarantee for approximate graph coloring. Inf. Process. Lett. **45**(1), 19–23 (1993)
66. Hennenfent, G., Herrmann, F., Saab, R., Yilmaz, O., Pajean, C.: SPOT: a linear operator toolbox, version 1.2 (2014). http://www.cs.ubc.ca/labs/scl/spot/index.html. Cited 2 March 2016
67. Hestenes, M., Stiefel, E.: Methods of conjugate gradients for solving linear systems. J. Res. Nat. Bur. Stand. **49**(6), 409–436 (1952)
68. Hien, L.: Differential properties of Euclidean projection onto power cone. Math. Methods Oper. Res. **82**(3), 265–284 (2015)
69. Jacques, L., Duval, L., Chaux, C., Peyré, G.: A panorama on multiscale geometric representations, intertwining spatial, directional and frequency selectivity. IEEE Trans. Signal Process. **91**(12), 2699–2730 (2011)
70. Jensen, A., la Cour-Harbo, A.: Ripples in Mathematics. Springer, Berlin (2001)
71. Jia, Y., Shelhamer, E., Donahue, J., Karayev, S., Long, J., Girshick, R., Guadarrama, S., Darrell, T.: Caffe: Convolutional architecture for fast feature embedding. (2014, preprint). http://arxiv.org/abs/1408.5093. Cited 2 March 2016
72. Karp, R.: Reducibility among combinatorial problems. In: Miller, R., Thatcher, J., Bohlinger, J. (eds.) Complexity of Computer Computations, The IBM Research Symposia Series, pp. 85–103. Springer, New York (1972)
73. Kelner, J., Orecchia, L., Sidford, A., Zhu, A.: A simple, combinatorial algorithm for solving SDD systems in nearly-linear time. In: Proceedings of the ACM Symposium on Theory of Computing, pp. 911–920 (2013)
74. Kim, S.J., Koh, K., Lustig, M., Boyd, S., Gorinevsky, D.: An interior-point method for large-scale ℓ_1-regularized least squares. IEEE J. Sel. Top. Signal Process. **1**(4), 606–617 (2007)
75. Kočvara, M., Stingl, M.: On the solution of large-scale SDP problems by the modified barrier method using iterative solvers. Math. Program. **120**(1), 285–287 (2009)

76. Kovacevic, J., Vetterli, M.: Nonseparable multidimensional perfect reconstruction filter banks and wavelet bases for $\mathscr{R}^n$. IEEE Trans. Inf. Theory **38**(2), 533–555 (1992)
77. Krishnaprasad, P., Barakat, R.: A descent approach to a class of inverse problems. J. Comput. Phys. **24**(4), 339–347 (1977)
78. Lan, G., Lu, Z., Monteiro, R.: Primal-dual first-order methods with $O(1/\epsilon)$ iteration-complexity for cone programming. Math. Program. **126**(1), 1–29 (2011)
79. Liberty, E.: Simple and deterministic matrix sketching. In: Proceedings of the ACM SIGKDD International Conference on Knowledge Discovery and Data Mining, pp. 581–588 (2013)
80. Lim, J.: Two-dimensional Signal and Image Processing. Prentice-Hall, Upper Saddle River (1990)
81. Lin, Y., Lee, D., Saul, L.: Nonnegative deconvolution for time of arrival estimation. In: Proceedings of the IEEE International Conference on Acoustics, Speech, and Signal Processing, vol. 2, pp. 377–380 (2004)
82. Loan, C.V.: Computational Frameworks for the Fast Fourier Transform. SIAM, Philadelphia (1992)
83. Lofberg, J.: YALMIP: A toolbox for modeling and optimization in MATLAB. In: Proceedings of the IEEE International Symposium on Computed Aided Control Systems Design, pp. 294–289 (2004)
84. Lu, Y., Do, M.: Multidimensional directional filter banks and surfacelets. IEEE Trans. Image Process. **16**(4), 918–931 (2007)
85. Mallat, S.: A theory for multiresolution signal decomposition: the wavelet representation. IEEE Trans. Pattern Anal. Mach. Intell. **11**(7), 674–693 (1989)
86. Martucci, S.: Symmetric convolution and the discrete sine and cosine transforms. IEEE Trans. Signal Process. **42**(5), 1038–1051 (1994)
87. Mattingley, J., Boyd, S.: CVXGEN: A code generator for embedded convex optimization. Optim. Eng. **13**(1), 1–27 (2012)
88. Miller, J., Zhu, J., Quigley, P.: CVXcanon, version 0.0.22 (2015). https://github.com/cvxgrp/CVXcanon. Cited 2 March 2016
89. MOSEK optimization software, version 7 (2015). https://mosek.com/. Cited 2 March 2016
90. Nesterov, Y.: Towards nonsymmetric conic optimization. Optim. Methods Software **27**(4–5), 893–917 (2012)
91. Nesterov, Y., Nemirovskii, A.: Interior-Point Polynomial Algorithms in Convex Programming. SIAM, Philadelphia (1994)
92. Nesterov, Y., Nemirovsky, A.: Conic formulation of a convex programming problem and duality. Optim. Methods Softw. **1**(2), 95–115 (1992)
93. Nocedal, J., Wright, S.: Numerical Optimization. Springer, New York (2006)
94. O'Donoghue, B., Chu, E., Parikh, N., Boyd, S.: Conic optimization via operator splitting and homogeneous self-dual embedding. J. Optim. Theory Appl. **169**(3), 1042–1068 (2016)
95. Parikh, N., Boyd, S.: Proximal algorithms. Found. Trends Optim. **1**(3), 123–231 (2014)
96. Pock, T., Chambolle, A.: Diagonal preconditioning for first order primal-dual algorithms in convex optimization. In: Proceedings of the IEEE International Conference on Computer Vision, pp. 1762–1769 (2011)
97. Pock, T., Cremers, D., Bischof, H., Chambolle, A.: An algorithm for minimizing the Mumford-Shah functional. In: Proceedings of the IEEE International Conference on Computer Vision, pp. 1133–1140 (2009)
98. Ragan-Kelley, J., Barnes, C., Adams, A., Paris, S., Durand, F., Amarasinghe, S.: Halide: A language and compiler for optimizing parallelism, locality, and recomputation in image processing pipelines. In: Proceedings of the ACM SIGPLAN Conference on Programming Language Design and Implementation, pp. 519–530 (2013)
99. Saunders, M., Kim, B., Maes, C., Akle, S., Zahr, M.: PDCO: Primal-dual interior method for convex objectives (2013). http://web.stanford.edu/group/SOL/software/pdco/. Cited 2 March 2016
100. Skajaa, A., Ye, Y.: A homogeneous interior-point algorithm for nonsymmetric convex conic optimization. Math. Program. **150**(2), 391–422 (2014)

101. Spielman, D., Teng, S.H.: Nearly-linear time algorithms for graph partitioning, graph sparsification, and solving linear systems. In: Proceedings of the ACM Symposium on Theory of Computing, pp. 81–90 (2004)
102. Starck, J.L., Candès, E., Donoho, D.: The curvelet transform for image denoising. IEEE Trans. Image Process. **11**(6), 670–684 (2002)
103. Sturm, J.: Using SeDuMi 1.02, a MATLAB toolbox for optimization over symmetric cones. Optim. Methods Softw. **11**(1–4), 625–653 (1999)
104. Toh, K.C.: Solving large scale semidefinite programs via an iterative solver on the augmented systems. SIAM J. Optim. **14**(3), 670–698 (2004)
105. Toh, K.C., Todd, M., Tütüncü, R.: SDPT3 — a MATLAB software package for semidefinite programming, version 4.0. Optim. Methods Softw. **11**, 545–581 (1999)
106. Udell, M., Mohan, K., Zeng, D., Hong, J., Diamond, S., Boyd, S.: Convex optimization in Julia. In: Proceedings of the Workshop for High Performance Technical Computing in Dynamic Languages, pp. 18–28 (2014)
107. Vaillant, G.: linop, version 0.7 (2013). http://pythonhosted.org//linop/. Cited 2 March 2016
108. van den Berg, E., Friedlander, M.: Probing the Pareto frontier for basis pursuit solutions. SIAM J. Sci. Comput. **31**(2), 890–912 (2009)
109. Vandenberghe, L., Boyd, S.: A polynomial-time algorithm for determining quadratic Lyapunov functions for nonlinear systems. In: Proceedings of the European Conference on Circuit Theory and Design, pp. 1065–1068 (1993)
110. Vandenberghe, L., Boyd, S.: A primal-dual potential reduction method for problems involving matrix inequalities. Math. Program. **69**(1–3), 205–236 (1995)
111. Vishnoi, K.: Laplacian solvers and their algorithmic applications. Theor. Comput. Sci. **8**(1–2), 1–141 (2012)
112. Wright, S.: Primal-Dual Interior-Point Methods. SIAM, Philadelphia (1987)
113. Yang, C., Duraiswami, R., Davis, L.: Efficient kernel machines using the improved fast Gauss transform. In: Saul, L., Weiss, Y., Bottou, L. (eds.) Advances in Neural Information Processing Systems 17, pp. 1561–1568. MIT Press, Cambridge (2005)
114. Yang, C., Duraiswami, R., Gumerov, N., Davis, L.: Improved fast Gauss transform and efficient kernel density estimation. In: Proceedings of the IEEE International Conference on Computer Vision, vol. 1, pp. 664–671 (2003)
115. Ye, Y.: Interior Point Algorithms: Theory and Analysis. Wiley-Interscience, New York (2011)
116. Ying, L., Demanet, L., Candès, E.: 3D discrete curvelet transform. In: Proceedings of SPIE: Wavelets XI, vol. 5914, pp. 351–361 (2005)
117. Zach, C., Pock, T., Bischof, H.: A duality based approach for realtime TV-ℓ_1 optical flow. In: Hamprecht, F., Schnörr, C., Jähne, B. (eds.) Pattern Recognition. Lecture Notes in Computer Science, vol. 4713, pp. 214–223. Springer, Heidelberg (2007)
118. Zhao, X.Y., Sun, D., Toh, K.C.: A Newton-CG augmented Lagrangian method for semidefinite programming. SIAM J. Optim. **20**(4), 1737–1765 (2010)

Invariance Conditions for Nonlinear Dynamical Systems

Zoltán Horváth, Yunfei Song, and Tamás Terlaky

Abstract Recently, Horváth et al. (Appl Math Comput, submitted) proposed a novel unified approach to study, i.e., invariance conditions, sufficient and necessary conditions, under which some convex sets are invariant sets for linear dynamical systems. In this paper, by utilizing analogous methodology, we generalize the results for nonlinear dynamical systems. First, the Theorems of Alternatives, i.e., the nonlinear Farkas lemma and the S-lemma, together with Nagumo's Theorem are utilized to derive invariance conditions for discrete and continuous systems. Only standard assumptions are needed to establish invariance of broadly used convex sets, including polyhedral and ellipsoidal sets. Second, we establish an optimization framework to computationally verify the derived invariance conditions. Finally, we derive analogous invariance conditions without any conditions.

Keywords Invariant set • Nonlinear dynamical system • Polyhedral set • Ellipsoid • Convex set

1 Introduction

Positively invariant set is an important concept, and it has a wide range of applications in dynamical systems and control theory, see, e.g., [6, 7, 17, 24]. Let a state space and a dynamical system be given. A subset $\mathcal{S}$ in the state space is called a *positively invariant set* of the dynamical system if any forward trajectory originated from $\mathcal{S}$ stays in $\mathcal{S}$. For the sake of simplicity, throughout the paper we use *invariant set* to refer to positively invariant set. Some classical examples of invariant sets are

Z. Horváth
Department of Mathematics and Computational Sciences, Széchenyi István University, 9026 Győr, Egyetem tér 1, Hungary
e-mail: horvathz@sze.hu

Y. Song • T. Terlaky (✉)
Department of Industrial and Systems Engineering, Lehigh University, 200 West Packer Avenue, Bethlehem, PA 18015-1582, USA
e-mail: yus210@lehigh.edu; terlaky@lehigh.edu

B. Goldengorin (ed.), *Optimization and Its Applications in Control and Data Sciences*, Springer Optimization and Its Applications 115,
DOI 10.1007/978-3-319-42056-1_8

equilibria, limit cycles, etc. (see [26]). In higher dimensional spaces, examples of invariant sets are e.g., invariant torus and chaotic attractor, see, e.g., [26]. A similar concept in dynamical system is stability, which is one of the most commonly studied property of invariant sets. Intuitively, an invariant set is referred to as stable if any trajectories starting close to it remain close to it, and unstable if they do not. Positively invariant set is an important concept and an efficient tool for the design of controllers of constrained systems. For example, for a given closed-loop control system, the state and control constraints hold when the initial state is chosen from a certain positively invariant set, see, e.g., [31].

A fundamental question is to develop efficient tools to verify if a given set is an invariant set for a given (discrete or continuous) dynamical system. Sufficient and necessary conditions under which a set is an invariant set for a dynamical system are important both from the theoretical and practical aspects. Such sufficient and necessary conditions are usually referred to as *invariance conditions*, see, e.g., [12]. Invariance conditions can be considered as special tools to study the relationship between the invariant set and the dynamical system. They also provide alternative ways to design efficient algorithms to construct invariant sets. Linear discrete and continuous dynamical systems have been extensively studied in recent decades, since such systems have a wide range of applications in control theory, see, e.g., [1, 5, 16]. Invariance condition for linear systems are relatively easy to derive while analogous conditions for nonlinear systems are more difficult to derive. Convex sets are often chosen as candidates for invariant sets of linear dynamical systems. These sets include polyhedron, see, e.g., [4–6], polyhedral cone, see, e.g., [10, 27], ellipsoid, see, e.g., [7, 31], and Lorenz cone, see, e.g., [3, 23, 25]. Recently, a novel unified method is presented in [12] to derive invariance conditions for these classical sets for both linear discrete and linear continuous dynamical systems. Invariant sets for nonlinear dynamical systems are more complicated to study. The localization problem of compact invariant sets for discrete nonlinear system is studied in [14]. A simple criteria to verify if a general convex set is a robust control invariant set for a nonlinear uncertain system is presented in [8]. Invariant set for discrete system is studied in [15], and an application to model predictive control is provided. The steplength threshold for preserving invariance of a set when applying a discretization method to continuous systems is studied in [11, 13].

In this paper, we present invariance conditions for some classical sets for nonlinear discrete and continuous dynamical systems. This is motivated by the fact that most problems in the real world are often described by nonlinear systems rather than linear systems. Therefore there is a need to investigate efficient invariance condition to verify sets to be invariant sets for a nonlinear dynamical system. The main tools used to derive invariance conditions for discrete and continuous dynamical systems are the so called Theorems of Alternatives, e.g., Farkas lemma [20, 22], S-Lemma [20, 30], and Nagumo Theorem [6, 18], respectively. For each invariance condition, we also present an equivalent optimization problem, which provides the possibility to use current advanced optimization algorithms or software to verify the invariance property of given sets.

The main contribution of this paper is that we propose novel invariance conditions for general discrete and continuous systems using a novel and simple approach. Our novel approach establishes a close connection between the theory of invariant sets and optimization theory, as well as provides the possibility of using current advanced optimization algorithms and methodology to solve invariant set problems.

Notation and Conventions. To avoid unnecessary repetitions, the following notations and conventions are used in this paper. The i-th row of a matrix G is denoted by G_i^T. The interior and the boundary of a set $\mathcal{S}$ is denoted by int($\mathcal{S}$) and $\partial\mathcal{S}$, respectively. The index set $\{1, 2, \ldots, n\}$ is denoted by $\mathcal{I}(n)$.

2 Preliminaries

In this paper, we consider the following discrete and continuous dynamical systems:

$$x_{k+1} = f_d(x_k), \tag{1}$$

$$\dot{x}(t) = f_c(x(t)), \tag{2}$$

where $x_k, x \in \mathbb{R}^n$ are *state variables*, and $f_d, f_c : \mathbb{R}^n \to \mathbb{R}^n$ are continuous differentiable functions. When $f_d(x) = Ax$ (or $f_c(x) = Ax$) with A being an n by n matrix, then (1) (or (2)) is a linear discrete (or continuous) dynamical system.

Definition 1. A set $\mathcal{S}$ is an invariant set for the discrete system (1) if $x_k \in \mathcal{S}$ implies $x_{k+1} \in \mathcal{S}$ for all $k \in \mathbb{N}$. A set $\mathcal{S}$ is an invariant set for the continuous system (2) if $x(0) \in \mathcal{S}$ implies $x(t) \in \mathcal{S}$ for all $t \geq 0$.

A *polyhedron*[1], denoted by $\mathcal{P} \in \mathbb{R}^n$, is represented as $\mathcal{P} = \{x \in \mathbb{R}^n \mid Gx \leq b\}$, where $G \in \mathbb{R}^{m \times n}$ and $b \in \mathbb{R}^m$. An *ellipsoid*, denoted by $\mathcal{E} \in \mathbb{R}^n$, centered at the origin is defined as $\mathcal{E} = \{x \in \mathbb{R}^n \mid x^T Q x \leq 1\}$, where $Q \in \mathbb{R}^{n \times n}$ and $Q \succ 0$. Note that any ellipsoid with nonzero center can be transformed to an ellipsoid centered at the origin, see, e.g., [9]. A set $\mathcal{S}$ is said to be *convex* if $\alpha x + (1 - \alpha)y \in \mathcal{S}$ for any $x, y \in \mathcal{S}$ and $\alpha \in [0, 1]$. One can show that any polyhedra and ellipsoids are both convex sets.

The following nonlinear Farkas lemma [20] and S-lemma [20, 30], which are also refereed to as the Theorems of Alternatives, are extensively studied in the optimization community. In this paper, we apply these two lemmas as our tools to derive invariance conditions of sets for discrete systems.

Theorem 1 (Nonlinear Farkas lemma[2] [20]). *Let $h(y), g_1(y), \ldots, g_m(y) : \mathbb{R}^n \to \mathbb{R}$ be convex functions. Assume that the Slater condition is satisfied. Then the following two statements are equivalent:*

---

[1]For the sake of simplicity, we assume that there exists an interior point in the polyhedron.

[2]The Slater condition means that there exists a $\hat{y} \in \mathbb{R}^n$, such that $g_j(\hat{y}) \leq 0$ for all j when $g_j(x)$ is linear, and $g_j(\hat{y}) < 0$ for all j when $g_j(x)$ is nonlinear.

- *The inequality systems* $h(y) < 0,\ g_j(y) \le 0, j = 1, 2, \ldots, m$ *have no solution.*
- *There exist* $\beta_1, \beta_2, \ldots, \beta_m \ge 0$, *such that* $h(y) + \sum_{j=1}^{m} \beta_j g_j(y) \ge 0$ *for all* $y \in \mathbb{R}^n$.

Theorem 2 ($\mathcal{S}$-Lemma [20, 30]). *Let* $h(y), g(y) : \mathbb{R}^n \to \mathbb{R}$ *be quadratic functions, and suppose that there is a* $\hat{y} \in \mathbb{R}^n$ *such that* $g(\hat{y}) < 0$. *Then the following two statements are equivalent:*

- *The inequality system* $h(y) < 0, g(y) \le 0$ *has no solution.*
- *There exists a scalar* $\beta \ge 0$, *such that* $h(y) + \beta g(y) \ge 0$, *for all* $y \in \mathbb{R}^n$.

The following Nagumo Theorem [18] is a general theoretical result which can be considered as invariance condition of a closed and convex set for continuous systems. This theorem is chosen as a tool to derive the invariance condition of sets for continuous systems.

Theorem 3 (Nagumo [6, 18]). *Let* $\mathcal{S} \subseteq \mathbb{R}^n$ *be a closed convex set, and assume that* $\dot{x}(t) = f(x(t))$, *where* $f : \mathbb{R}^n \to \mathbb{R}^n$ *is a continuous function, has a unique solution for every* $x(0) \in \mathcal{S}$. *Then* $\mathcal{S}$ *is an invariant set for this system if and only if*

$$f(x) \in \mathcal{T}_{\mathcal{S}}(x), \text{ for all } x \in \partial\mathcal{S}, \tag{3}$$

where $\mathcal{T}_{\mathcal{S}}(x)$ *is the tangent cone*[3] *of* $\mathcal{S}$ *at* x.

The geometric interpretation of Theorem 3 is clear, i.e., a set $\mathcal{S}$ is an invariant set for the continuous system if and only if the tangent line of the trajectory $x(t)$ cannot point out of its tangent cone. According to [6], we have that the Nagumo Theorem cannot be extended to discrete systems.

3 Invariance Conditions for Discrete Systems

In this section, under certain assumptions, we present invariance conditions of polyhedral sets, ellipsoids, and convex sets for discrete systems. The introduction of these assumptions ensures that the Theorems of Alternatives can be applied to derive invariance conditions. First, an invariance condition of polyhedral sets for discrete systems is presented as follows.

Theorem 4. *Let a polyhedron* $\mathcal{P} = \{x \mid Gx \le b\}$, *where* $G \in \mathbb{R}^{m\times n}$ *and* $b \in \mathbb{R}^m$, *and the discrete system be given as in (1). Assume that* $b_i - G_i^T f_d(x)$ *are convex functions for all* $i \in \mathcal{I}(m)$. *Then* $\mathcal{P}$ *is an invariant set for the discrete system (1) if and only if there exists a matrix* $H \ge 0$, *such that*

$$HGx - Gf_d(x) \ge Hb - b, \quad \text{for all } x \in \mathbb{R}^n. \tag{4}$$

[3] The tangent cone $\mathcal{T}_{\mathcal{S}}(x)$ is denoted as follows: $\mathcal{T}_{\mathcal{S}}(x) = \{y \in \mathbb{R}^n \mid \liminf_{t\to 0+} \frac{\text{dist}(x+ty,\mathcal{S})}{t} = 0\}$, where $\text{dist}(x, \mathcal{S}) = \inf_{s\in\mathcal{S}} \|x - s\|$.

Proof. We have that $\mathcal{P}$ is an invariant set for the discrete system (1) if and only if $\mathcal{P} \subseteq \mathcal{P}' = \{x \mid Gf_d(x) \leq b\}$. The latter one means that for every $i \in \mathcal{I}(m)$, the system $G_i^T f_d(x) > b_i$ and $Gx \leq b$ has no solution. Let us assume to the contrary that there exists an x^* and i^*, such that $G_{i^*}^T f_d(x^*) > b_{i^*}$ and $Gx^* \leq b$. Then we have $x^* \in \mathcal{P}$ but $x^* \notin \mathcal{P}'$, which contradicts to $\mathcal{P} \subseteq \mathcal{P}'$. Also, since $b_i - G_i^T f_d(x)$ is a convex function, then, according to the convex Farkas Lemma 1, we have that there exists a vector $H_i \geq 0$ and $H_i \in \mathbb{R}^n$, such that

$$b_i - G_i^T f_d(x) + H_i^T (Gx - b) \geq 0, \quad \text{for all } x \in \mathbb{R}^n.$$

Writing H_i^T for all $i \in \mathcal{I}(m)$ together into a matrix H, we have $H \geq 0$ and

$$b - Gf_d(x) + H(Gx - b) \geq 0, \quad \text{for all } x \in \mathbb{R}^n,$$

which is the same as (4).

One can use algebraic method to verify if condition (4) holds when $f_d(x)$ is given. The algebraic method may be very challenge. Here we present the following optimization methodology to equivalently solve condition (4).

Remark 1. Consider the following m optimization problems

$$\max_{H_i \geq 0} \min_{x \in \mathbb{R}^n} \{H_i^T Gx - G_i^T f_d(x) - H_i^T b + b_i\} \text{ for all } i \in \mathcal{I}(m). \tag{5}$$

If the global optimal objective values of the m optimization problems in (5) are all nonnegative, then we can claim that condition (4) holds.

In Theorem 4, we do not specifically assume that the system is a linear or a nonlinear system. If the system in Theorem 4 is a linear dynamical system, then we have the following corollary, which is an invariance condition of polyhedral sets for linear systems. Note that Corollary 1 can also be referred to [12]. An alternative proof for Corollary 1, using optimality conditions, is presented in Appendix.

Corollary 1 ([5, 12]). *Let a polyhedron $\mathcal{P} = \{x \mid Gx \leq b\}$, where $G \in \mathbb{R}^{m \times n}$ and $b \in \mathbb{R}^m$ be given, and the discrete system given in (1) be linear, i.e., $f_d(x) = A_d x$, where $A_d \in \mathbb{R}^{n \times n}$. Then $\mathcal{P}$ is an invariant set of the discrete system (1) if and only if there exists a matrix $H \in \mathbb{R}^{m \times m}$ and*[4] *$H \geq 0$, such that $HG = GA$ and $Hb \leq b$.*

Proof. Since the system is linear, $b_i - G_i^T Ax$ are convex functions for all $i \in \mathcal{I}(m)$. According to Theorem 4, there exists a matrix $H \geq 0$, such that condition (4) holds, i.e.,

$$(HG - GA)x \geq Hb - b, \quad \text{for all } x \in \mathbb{R}^n. \tag{6}$$

[4] Here $H \geq 0$ means that all the entries of H are nonnegative.

Note that (6) holds for all $x \in \mathbb{R}^n$. One can easily show that (6) is equivalent to $HG = GA$ and $Hb \leq b$. The proof is complete.

In Theorem 4, we have the condition that $b_i - G_i^T f_d(x)$ are convex function for all i. Recall that a function, which is twice differentiable, is convex if and only if its Hessian is positive semidefinite for all x. Thus, to verify if the functions $b_i - G_i^T f_d(x)$ are convex, it is sufficient to verify if $G_i^T \nabla^2 f(x) \preceq 0$ for all $x \in \mathbb{R}^n$. We now give an example to illustrate Theorem 4.

Example 1. Let the discrete system be given as $\xi_{k+1} = -\xi_k + 2\eta_k - \xi_k^2, \eta_{k+1} = -2\xi_k - \eta_k + \eta_k^2$, and the polyhedron be given as $\mathcal{P} = \{(\xi, \eta) \mid \xi - \eta \leq -10, 2\xi - \eta \leq 10, \xi - 2\eta \leq -20\}$.

We first show that $\mathcal{P}$ is an invariant set for the discrete system, i.e., $(\xi_{k+1}, \eta_{k+1}) \in \mathcal{P}$ for all $(\xi_k, \eta_k) \in \mathcal{P}$. For simplicity, we only prove the first constraint, i.e., $\xi_{k+1} - \eta_{k+1} \leq -10$. In fact, we have $\xi_{k+1} - \eta_{k+1} = -\xi_k^2 - \eta_k^2 + \xi_k + 3\eta_k = -\xi_k^2 - (\eta_k - 2.5)^2 + \xi_k - 2\eta_k + 6.25 \leq \xi_k - 2\eta_k + 6.25 \leq -20 + 6.25 \leq -10$. The other two constraints can be proved in a similar manner. On the other hand, one can show that the assumption in Theorem 4 is satisfied for this example. Then we can find a suitable $H \geq 0$ such that condition (4) holds. One can easily verify that $H = [0, 0, 1; 0, 0, 0; 1, 0, 1]$ satisfies condition (4). Then according to Theorem 4, we have that $\mathcal{P}$ is an invariant set for the discrete system.

We now consider an invariance condition for ellipsoids for the discrete system (1).

Theorem 5. *Let an ellipsoid $\mathcal{E} = \{x \mid x^T Q x \leq 1\}$, where $Q \in \mathbb{R}^{n \times n}$ and $Q \succ 0$, and the discrete system be given as in (1). Assume that $(f_d(x))^T Q f_d(x)$ is a concave function. Then $\mathcal{E}$ is an invariant set for the discrete system (1) if and only if there exists a $\beta \geq 0$, such that*

$$\beta x^T Q x - (f_d(x))^T Q f_d(x) \geq \beta - 1, \quad \text{for all } x \in \mathbb{R}^n. \tag{7}$$

Proof. The ellipsoid $\mathcal{E}$ is an invariant set for the discrete system if and only if $\mathcal{E} \subseteq \mathcal{E}'$, where $\mathcal{E}' = \{x \mid (f_d(x))^T Q f_d(x) \leq 1\}$. We also note that $\mathcal{E} \subseteq \mathcal{E}'$ is equivalent to $(\mathbb{R}^n \setminus \mathcal{E}') \cap \mathcal{E} = \emptyset$, i.e., the inequality system $1 - (f_d(x))^T Q f_d(x) < 0$ and $x^T Q x - 1 \leq 0$ has no solution. Since $(f_d(x))^T Q f_d(x)$ is a concave function, we have that $1 - (f_d(x))^T Q f_d(x)$ is a convex function. Note that $x^T Q x - 1$ is also a convex function, according to Theorem 1, there exists a $\beta \geq 0$, such that

$$-(f_d(x))^T Q f_d(x) + 1 + \beta(x^T Q x - 1) \geq 0, \quad \text{for all } x \in \mathbb{R}^n,$$

which is the same as (7).

Remark 2. If we choose $x = 0$ in condition (7), then we have $\beta \leq 1 - (f_d(0))^T Q f_d(0)$, which can be considered as an upper bound of β.

Similarly, we present the following optimization problem which is equivalent to condition (7).

Remark 3. Consider the following optimization problem

$$\max_{\beta \geq 0} \min_{x \in \mathbb{R}^n} \{\beta x^T Qx - (f_d(x))^T Qf_d(x) - \beta + 1\}. \tag{8}$$

If the optimal objective value of optimization problem (8) is nonnegative, then condition (7) holds.

If the system in Theorem 5 is a linear dynamical system, then we have the following corollary, which is an invariance condition of ellipsoids for linear system. Note that Corollary 2 can also be referred to [12].

Corollary 2 ([12]). *Let an ellipsoid $\mathcal{E} = \{x \,|\, x^T Qx \leq 1\}$, where $Q \in \mathbb{R}^{n \times n}$ and $Q \succ 0$, and a linear discrete system be given as in (1), i.e., $f_d(x) = A_d x$, where $A_d \in \mathbb{R}^{n \times n}$. Then $\mathcal{E}$ is an invariant set for the discrete system (1), if and only if there exists a $\beta \in [0, 1]$, such that $A_d^T QA_d - \beta Q \preceq 0$.*

Proof. According to Theorem 5, we have that there exists a $\beta \geq 0$, such that

$$x^T(\beta Q - A_d^T QA_d)x \geq \beta - 1, \text{ for all } x \in \mathbb{R}^n. \tag{9}$$

If we choose $x = 0$, then we have $\beta \leq 1$. Assume that $A_d^T QA_d - \beta Q \npreceq 0$, then there exists a negative eigenvalue λ and the corresponding eigenvector $x^* \neq 0$ such that $(\beta Q - A_d^T QA_d)x^* = \lambda x^*$, where $\lambda < 0$. Let $y^* = \alpha x^*$ with $\alpha < \sqrt{\frac{\beta-1}{\lambda}} \frac{1}{\|x^*\|}$, then we have $(y^*)^T(\beta Q - A_d^T QA_d)y^* < \beta - 1$, which contradicts (9). Thus, we have $A_d^T QA_d - \beta Q \preceq 0$.

Observe that parameter β presented in Corollary 2 can be eliminated. In fact, one can show that $A_d^T QA_d - \beta Q \preceq 0$ for $\beta \in [0, 1]$ and $Q \succ 0$ is equivalent to $A_d^T QA_d - Q \preceq 0$, see [12].

In Theorem 5, we have the condition that $(f_d(x))^T Qf_d(x)$ is a concave function. In fact, this is equivalent to verify if the Hessian of $(f_d(x))^T Qf_d(x)$ is negative semidefinite for all $x \in \mathbb{R}^n$. We now give an example to illustrate Theorem 5.

Example 2. Let the discrete system be $\xi_{k+1} = \frac{\sqrt{\xi_k + \eta_k}}{2}, \eta_{k+1} = \frac{\sqrt{\xi_k - 3\eta_k}}{2}$, and the ellipsoid be given as $\mathcal{E} = \{(\xi, \eta) \,|\, \xi^2 + \eta^2 \leq 1\}$.

For any $(\xi_k, \eta_k) \in \mathcal{E}$, we have $\xi_{k+1}^2 + \eta_{k+1}^2 = \frac{\xi_k - \eta_k}{2} \leq \frac{\sqrt{2}}{2}\sqrt{\xi_k^2 + \eta_k^2} < 1$, which shows that $\mathcal{E}$ is an invariant set for the discrete system. On the other hand, let $f(x) = (f_1(x), f_2(x))^T = (\frac{\sqrt{\xi_k + \eta_k}}{2}, \frac{\sqrt{\xi_k - 3\eta_k}}{2})^T$ and $Q = [1, 0; 0, 1]$. Then we have that $f(x)^T Qf(x)$ is a concave function. If we choose $\beta = \frac{1}{4}$, then condition (7) yields $(\xi_k - 1)^2 + (\eta_k - 1)^2 + 1 \geq 0$ for any $(\xi_k, \eta_k) \in \mathbb{R}^2$. This, according to Theorem 5, also shows that $\mathcal{E}$ is an invariant set for the discrete system.

We now consider invariance conditions for more general convex sets for discrete system (1). Let a convex set be given as:

$$\mathcal{S} = \{x \in \mathbb{R}^n \,|\, g(x) \leq 0\}, \tag{10}$$

where $g : \mathbb{R}^n \to \mathbb{R}$ is a convex function. Then we have the following theorem, which gives invariance condition for the convex set (10) for discrete system (1).

Theorem 6. *Let the convex set $\mathcal{S}$ be given as in (10), and the discrete system be given as in (1). Assume that there exists $x^0 \in \mathbb{R}^n$ such that $g(x) < 0$, and that $g(f_d(x))$ is a concave function. Then $\mathcal{S}$ is an invariant set for the discrete system if and only if there exists an $\alpha \geq 0$, such that*

$$\alpha g(x) - g(f_d(x)) \geq 0, \quad \textit{for all } x \in \mathbb{R}^n. \tag{11}$$

Moreover, if $g(x)$ and $g(f_d(x))$ are quadratic functions, then the assumption that $g(f_d(x))$ is a concave function is not required.

Proof. The major tool used in this proof is the convex Farkas Lemma, i.e., Theorem 1. Note that to ensure $\mathcal{S}$ is an invariant set for the discrete system, we need to prove $\mathcal{S} \subseteq \mathcal{S}' = \{x \mid g(f_d(x)) \leq 0\}$, i.e., $(\mathbb{R}^n \setminus \mathcal{S}') \cap \mathcal{S} = \emptyset$. Then the following inequality system has no solution:

$$-g(f_d(x)) < 0, \quad g(x) \leq 0.$$

According to Theorem 1, there exists an $\alpha \geq 0$, such that

$$-g(f_d(x)) + \alpha g(x) \geq 0, \quad \text{for } x \in \mathbb{R}^n,$$

which is the same as (11). For the case of quadratic functions, we can use a similar argument and the S-Lemma to prove the last statement.

Remark 4. The set $\mathcal{S}$ given as in (10) is represented by only a single convex function. In fact, the first statement in Theorem 6 can be easily extended to the set which is presented by several convex functions, e.g., polyhedral sets.

The first statement in Theorem 6 requires that $g(f_d(x))$ is a concave function given that $g(x)$ is a convex function. Let us consider x defined in a one dimensional space as an example[5] to illustrate this case is indeed possible. Since $f_d(x)$ is a convex function, we have $f_d''(x) \geq 0$ for all $x \in \mathbb{R}$. For simplicity, we denote $h(x) = -g(f_d(x))$. Then we have

$$h''(x) = -g''(f_d(x))(f_d(x))^2 - g'(f_d(x))f_d''(x). \tag{12}$$

If $h''(x) > 0$ for all $x \in \mathbb{R}$, then $h(x)$ is a convex function, i.e., $g(f_d(x))$ is a concave function. We now find a sufficient condition such that $h'(x) > 0$ for all $x \in \mathbb{R}$. Assume that $g(x)$ is a decreasing convex nonlinear function and $g(x)$ has no lower

[5]The example uses the following theorem: If $\tilde{g}(x)$ is a nondecreasing function, and $\tilde{f}(x)$ is a convex function, then $\tilde{g}(\tilde{f}(x))$ is a convex function.

bound, we have $g'(x) < 0$ and $g''(x) > 0$ for all $x \in \mathbb{R}$. Assume $f_d(x)$ is a concave function, we have $f_d''(x) < 0$. This yields $-\frac{g'(f_d(x))}{g_d''(f(x))} > 0 \geq \frac{(f'(x))^2}{f''(x)}$, i.e., $h''(x) > 0$.

Remark 5. Consider the following optimization problem:

$$\max_{\alpha \geq 0} \min_{x \in \mathbb{R}^n} \{\alpha g(x) - g(f_d(x))\}. \tag{13}$$

If the optimal objective value of optimization problem (13) is nonnegative, we can claim that condition (11) holds.

Thus far we have three "max-min" optimization problems shown as in (5), (8), and (13). It is usually not easy to solve a "max-min" problem. In fact, these three problems can be transformed into a nonlinear optimization problem. Here we consider (13) as an example to illustrate this idea. From here, we assume that $g(x)$ in (10) is continuously differentiable.

Theorem 7. *Optimization problem (13) is equivalent to the nonlinear optimization problem*

$$\max_{x,\alpha} \{\alpha g(x) - g(f_d(x)) \mid \alpha \nabla_x g(x) - \nabla_x g(f_d(x)) = 0, \alpha \geq 0\}. \tag{14}$$

Proof. Since $\alpha \geq 0$, and the functions $g(x)$ and $-g(f_d(x))$ are both convex functions, we have that $\alpha g(x) - g(f_d(x))$ is also a convex function. Also, for $\alpha \geq 0$, the optimization problem

$$\min_{x \in \mathbb{R}^n} \{\alpha g(x) - g(f_d(x))\}, \tag{15}$$

is a convex optimization problem in $\mathbb{R}^n$, thus problem (15) has a Wolfe dual, see, e.g., [21, 29] given as follows:

$$\max_{x \in \mathbb{R}^n} \{\alpha g(x) - g(f_d(x)) \mid \alpha \nabla_x g(x) - \nabla_x (g(f_d(x))) = 0\}. \tag{16}$$

Consequently, problem (13) is equivalent to the nonlinear optimization problem (14).

Remark 6. One can use a proof similar to the one presented in Theorem 7 to derive equivalent nonlinear optimization problems for the optimization problems presented in (5) and (8).

We now consider an alternative way to investigate invariance conditions for discrete systems. The following lemma is easy to prove.

Lemma 1. *Let $\phi(x), \psi(x) : \mathbb{R}^n \to \mathbb{R}$. The following two statements are equivalent:*

- *The inequality system $\phi(x) \leq 0, \psi(x) > 0$ has no solution.*

- *The optimal objective value of the optimization problem*

$$\max\{\epsilon \mid \phi(x) \le 0, -\psi(x) + \epsilon \le 0\} \tag{17}$$

is nonpositive.

According to Lemma 1, we have the following lemma.

Lemma 2. *Let the discrete system be given as in (1). Let $\mathcal{S}_1 = \{x \in \mathbb{R}^n \mid \phi(x) \le 0\}$ and $\mathcal{S}_2 = \{x \in \mathbb{R}^n \mid \psi(x) \le 0\}$ be two closed sets*[6]*, where $\phi(x), \psi(x) : \mathbb{R}^n \to \mathbb{R}$. Then $x \in \mathcal{S}_1$ implies $f_d(x) \in \mathcal{S}_2$ if and only if the optimal objective value of the following optimization problem*

$$\max\{\epsilon \mid \phi(x) \le 0, -\psi(f_d(x)) + \epsilon \le 0\}, \tag{18}$$

is nonpositive.

Proof. We have that $x \in \mathcal{S}_1$ implies $f_d(x) \in \mathcal{S}_2$ if and only if $\mathcal{S}_1 \subseteq \tilde{\mathcal{S}}_2 = \{x \mid \psi(f_d(x)) \le 0\}$. This is equivalent to $(\mathbb{R}^n \setminus \tilde{\mathcal{S}}_2) \cap \mathcal{S}_1 = \emptyset$, i.e., the systems $\phi(x) \le 0$ and $\psi(f_d(x)) > 0$ have no solution. Then, according to Lemma 1, the lemma is immediate.

According to Lemma 2, we have the following theorem.

Theorem 8. *Let the discrete system be given as in (1), and let $\bar{\mathcal{S}} = \{x \in \mathbb{R}^n \mid \phi(x) \le 0\}$ be a set, where $\phi(x) : \mathbb{R}^n \to \mathbb{R}$. Then $\bar{\mathcal{S}}$ is an invariant set for the discrete system if and only if the optimal objective value of the following optimization problem*

$$\max\{\epsilon \mid \phi(x) \le 0, -\phi(f_d(x)) + \epsilon \le 0\} \tag{19}$$

is nonpositive.

Proof. The set is an invariant set for the discrete system if and only if $\bar{\mathcal{S}} \subseteq \tilde{\mathcal{S}} = \{x \mid \phi(f_d(x)) \le 0\}$. According to Lemma 2, the theorem is immediate.

4 Invariance Conditions for Continuous Systems

In this section, we consider invariance conditions for continuous systems in the form of (2). For discrete systems, in Sect. 3, we transformed the invariance conditions into "max-min" optimization problems, which were later proved to be equivalent to traditional nonlinear optimization problems. For the continuous systems, we transform the invariance conditions into nonlinear optimization problems, too.

First, we consider an invariance condition for continuous system (2) and for polyhedral sets $\mathcal{P} = \{x \; Gx \le b\}$, where $G \in \mathbb{R}^{m\times n}$ and $b \in \mathbb{R}^m$. For simplicity

[6] It is not necessary to assume that the two sets are convex sets.

we assume that the origin is in the interior of the polyhedral set, thus we have $\mathcal{P} = \{x \in \mathbb{R}^n \mid Gx \leq b\} = \{x \in \mathbb{R}^n \mid g_i^T x \leq b_i, i \in \mathcal{I}(m)\}$, where $b > 0$.

Theorem 9. *Let a polyhedral set be given as $\mathcal{P} = \{x \in \mathbb{R}^n \mid g_i^T x \leq b_i, i \in \mathcal{I}(m)\}$, where $b > 0$, and let $\mathcal{P}^i = \{x \in \mathcal{P} \mid g_i^T x = b_i\}$ for $i \in \mathcal{I}(m)$. Then $\mathcal{P}$ is an invariant set for the continuous system (2) if and only if for all $i \in \mathcal{I}(m)$*

$$g_i^T f_c(x) \leq 0 \text{ holds for all } x \in \mathcal{P}^i. \tag{20}$$

Proof. Let $x \in \partial\mathcal{P}$. Then we have that x is in the relative interior of a face, on the relative boundary, or a vertex of $\mathcal{P}$. There exists a maximal index set $\mathcal{I}_x$ such that $x \in \bigcap_{i \in \mathcal{I}_x} \mathcal{P}^i$. We note that $\mathcal{T}_{\mathcal{P}}(x) = \{y \in \mathbb{R}^n \mid g_i^T y \leq 0, i \in \mathcal{I}_x\}$, then, according to Nagumo Theorem 3, the theorem is immediate.

Remark 7. Let us assume a polyhedral set $\mathcal{P}$ be given as in the statement of Theorem 9. Consider the following m optimization problems:

$$\max\{g_i^T f_c(x) \mid g_i^T x = b_i \text{ and } x \in \mathcal{P}\}, i \in \mathcal{I}(m). \tag{21}$$

If the optimal objective values of all the m optimization problems in (21) are nonpositive, then we can claim that (20) holds.

Clearly, when $g_i^T f_c(x)$ is a concave function, problem (21) is a convex problem, which can be solved efficiently by using nonlinear convex optimization solvers, like MOSEK [2]. Otherwise, this problem is a nonconvex problem, which may need special nonlinear algorithms to solve [19, 28].

Invariance conditions for continuous system (2) and for ellipsoids or Lorenz cones is presented in the following theorem.

Theorem 10. *Let the ellipsoid $\mathcal{E} = \{x \mid x^T Q x \leq 1\}$, where $Q \in \mathbb{R}^{n \times n}$ and $Q \succ 0$, and the continuous system be given as in (2). Then $\mathcal{E}$ is an invariant set for the continuous system (2) if and only if*

$$(f_c(x))^T Q x \leq 0, \text{ for all } x \in \partial\mathcal{E}. \tag{22}$$

Proof. Note that $\partial\mathcal{E} = \{x \mid x^T Q x = 1\}$, thus the outer normal vector of $\mathcal{E}$ at $x \in \partial\mathcal{E}$ is $f_d(x)$. Then we have that the tangent cone at $x \in \partial\mathcal{E}$ is given as $\mathcal{T}_{\mathcal{E}}(x) = \{y \mid y^T Q x \leq 0\}$, thus this theorem follows by the Nagumo Theorem 3.

Note that Theorem 10 can be applied to a Lorenz cone $\mathcal{C}_{\mathcal{L}}$, see, e.g., [12].

Remark 8. Let us consider an ellipsoid $\mathcal{E}$ and the following optimization problem:

$$\max\{(f_c(x))^T Q x \mid x^T Q x = 1\}. \tag{23}$$

If the global optimal objective value of optimization problem (23) is nonpositive, then condition (22) holds.

We note that problem (23) is not a convex optimization problem since the set of feasible solutions $\{x \mid x^T Qx = 1\}$ is not a convex region. Thus, nonconvex optimization algorithms, such as the ones implemented in [28] are required to solve this problem.

Theorem 11. *Let the convex set $\mathcal{S}$ be given as in (10) and let function $g(x)$ be continuously differentiable. Then $\mathcal{S}$ is an invariant set for the continuous system (2) if and only if*

$$(\nabla g(x))^T f_c(x) \leq 0, \quad \text{for all } x \in \partial\mathcal{S}. \tag{24}$$

Proof. The outer normal vector at $x \in \partial\mathcal{S}$ is $\nabla g(x)$. Since $\mathcal{S}$ is a convex set, we have

$$\mathcal{T}_{\mathcal{S}}(x) = \{y \mid (\nabla g(x))^T y \leq 0\}. \tag{25}$$

The proof is immediate by applying Nagumo's Theorem 3.

Remark 9. Consider the following optimization problem:

$$\max\{\alpha \mid \alpha = (\nabla g(x))^T f_c(x),\ g(x) = 0\}. \tag{26}$$

If the optimal objective value of optimization problem (26) is nonpositive, then we can claim that condition (24) holds.

We note that when problem (26) is not a convex optimization problem, thus we may need a nonconvex optimization algorithm to solve this problem.

5 General Results for Discrete Systems

In Sect. 3, invariance conditions for polyhedral sets, ellipsoids, and convex sets are presented under certain assumptions. In this section, invariance conditions for these sets for discrete systems are presented without any assumption. First let us consider polyhedral sets.

Theorem 12. *Let the polyhedron $\mathcal{P} = \{x \mid Gx \leq b\}$, where $G \in \mathbb{R}^{m\times n}$ and $b \in \mathbb{R}^m$, and the discrete system be given as in (1). Then $\mathcal{P}$ is an invariant set for the discrete system (1) if and only if there exists a matrix $H \geq 0$, such that*

$$HGx - Gf_d(x) \geq Hb - b, \quad \text{for all } x \in \mathcal{P}. \tag{27}$$

Proof. Sufficiency: Condition (27) can be reformulated as $b - Gf_d(x) \geq H(b - Gx)$, where $x \in \mathcal{P}$, i.e., $b - Gx \geq 0$. Since $H \geq 0$, we have $b - Gf_d(x) \geq 0$, i.e., $f_d(x) \in \mathcal{P}$ for all $x \in \mathcal{P}$. Thus $\mathcal{P}$ is an invariant set for the discrete system. *Necessity:* Assume $\mathcal{P}$ is an invariant set for the discrete system. Then for any $x_k \in \mathcal{P}$, we have

$x_{k+1} = f_d(x_k) \in \mathcal{P}$, i.e., we have that $b - Gx \geq 0$ implies $b - Gf_d(x) \geq 0$. Thus, we can choose $H = 0$.

Note that the difference between conditions (4) and (27) is that the same inequality holds, for $x \in \mathbb{R}^n$ in (4), and for $x \in \mathcal{P}$ in (27), respectively. Similarly, we also have the following remark.

Remark 10. Consider the following m optimization problems

$$\max_{H_i \geq 0} \min_{x} \{H_i^T Gx - G_i^T f_d(x) - H_i^T b + b_i \mid Gx \leq b\} \quad i \in \mathcal{I}(m). \tag{28}$$

If the global optimal objective values of the m optimization problems in (28) are all nonnegative, then condition (27) holds.

We now present an invariance condition for ellipsoids for discrete systems. In this invariance condition, for ellipsoids no assumption is needed.

Theorem 13. *Let the ellipsoid $\mathcal{E} = \{x \mid x^T Qx \leq 1\}$, where $Q \in \mathbb{R}^{n\times n}$ and $Q \succ 0$, and let the discrete system be given as in (1). Then $\mathcal{E}$ is an invariant set for the discrete system if and only if there exists a $\beta \geq 0$, such that*

$$\beta x^T Qx - (f_d(x))^T Qf_d(x) \geq \beta - 1, \quad \text{for all } x \in \mathcal{E}. \tag{29}$$

Proof. Sufficiency: Condition (29) can be reformulated as $1 - (f_d(x))^T Qf_d(x) \geq \beta(1 - x^T Qx)$, where $x \in \mathcal{E}$. Thus we have $1 - (f_d(x))^T Qf_d(x) \geq 0$, i.e., $f_d(x) \in \mathcal{E}$. Thus $\mathcal{E}$ is an invariant set for the discrete system. *Necessity:* It is immediate by choosing $\beta = 0$.

Remark 11. Consider the following optimization problem

$$\max_{\beta \geq 0} \min_{x} \{\beta x^T Qx - (f_d(x))^T Qf_d(x) - \beta + 1 \mid x^T Qx \leq 1\}. \tag{30}$$

If the optimal objective value of optimization problem (30) is nonnegative, then condition (29) holds.

We now present an invariance condition for convex sets and for discrete systems. In this invariance condition, no assumption is needed for convex sets.

Theorem 14. *Let the convex set $\mathcal{S}$ be given as in (10) and let the discrete system be given as in (1). Then $\mathcal{S}$ is an invariant set for the discrete system if and only if there exists an $\alpha \geq 0$, such that*

$$\alpha g(x) - g(f_d(x)) \geq 0, \quad \text{for all } x \in \mathcal{S}. \tag{31}$$

Proof. Sufficiency: Condition (31) can be reformulated as $\alpha g(x) \geq g(f_d(x))$, where $x \in \mathcal{S}$, i.e., $g(x) \leq 0$. According to $\alpha \geq 0$, we have $g(f_d(x)) \leq 0$, i.e., $f_d(x) \in \mathcal{S}$.

Thus $\mathcal{S}$ is an invariant set for the discrete system. *Necessity:* It is immediate by choosing $\alpha = 0$.

Remark 12. Consider the following optimization problem:

$$\max_{\alpha \geq 0} \min_{x \in \mathbb{R}^n} \{\alpha g(x) - g(f_d(x))\}. \quad (32)$$

If the optimal objective value of optimization problem (32) is nonnegative, then condition (31) holds.

We note that there are no assumptions in Theorems 12, 13, and 14, which means we cannot use the Wolfe duality theory. Thus we cannot transform the "max-min" optimization problems in Remark 10, 11, and 12 into nonlinear maximization problems. The absence of convexity assumptions makes the theorems more broadly applicable, however the nonlinear feasibility problems (27), (29), and (31) are nonconvex, thus their verification is significantly harder than solving convex feasibility problems. We pointed out in the introduction that there are very few papers studying invariance conditions for nonlinear systems. The nonlinear feasibility problems (27), (29), and (31) provide us a novel perspective to consider invariance conditions. They also bring the possibility of applying state-of-the-art optimization algorithms to solve the nonlinear problems related to invariance conditions.

6 Conclusions

In this paper we derived invariance conditions for some classical sets for nonlinear dynamical systems by utilizing a methodology analogous to the one presented in [12]. This is motivated by the fact that most problems in the real world are modeled by nonlinear dynamical systems, because they often show nonlinear characteristics. The Theorems of Alternatives, i.e., the nonlinear Farkas lemma and the S-lemma, together with Nagumo's Theorem are our main tools to derive invariance conditions for discrete and continuous systems. We derive the invariance conditions for these classic sets for nonlinear systems with some, and without any, conditions. We also propose an optimization problem for each invariance condition. Then to verify the invariance condition is equivalent to solve the corresponding optimization problem. These invariance conditions provide potential ways to design algorithms to establish invariant sets for a system. The introduction of the associated optimization problem opens new avenues to use advanced optimization algorithms and software to solve invariant set problems.

Acknowledgements This research is supported by a Start-up grant of Lehigh University, and by TAMOP-4.2.2.A-11/1KONV-2012-0012: Basic research for the development of hybrid and electric vehicles. The TAMOP Project is supported by the European Union and co-financed by the European Regional Development Fund.

Appendix

Theorem 15 ([5, 12]). *Let $\mathcal{P} = \{x \mid Gx \leq b\}$ be a polyhedron, where $G \in \mathbb{R}^{m\times n}$ and $b \in \mathbb{R}^m$. Let the discrete system, given as in (1), be linear, i.e., $f(x) = Ax$. Then $\mathcal{P}$ is an invariant set for the discrete system (1) if and only if there exists a matrix $H \geq 0$, such that $HG = GA$ and $Hb \leq b$.*

Proof. We have that $\mathcal{P}$ is an invariant set for the linear system if and only if the optimal objective values of the following m linear optimization problems are all nonnegative:

$$\min\{b_i - G_i^T Ax \mid Gx \leq b\} \quad i \in \mathcal{I}(m). \tag{33}$$

Problems (33) are equivalent to

$$-\, b_i + \max\{G_i^T Ax \mid Gx \leq b\} \quad i \in \mathcal{I}(m). \tag{34}$$

The duals of these linear optimization problems presented in (34) are for all $i \in \mathcal{I}(m)$

$$\begin{aligned} -b_i + \min\ & b^T H_i \\ \text{s.t. } & G^T H_i = A^T G_i \\ & H_i \geq 0. \end{aligned} \tag{35}$$

Due to the Strong Duality Theorem of linear optimization, see, e.g., [22], the primal and dual objective function values are equal at optimal solutions, thus $G_i^T Ax = b^T H_i$. As the optimal value of (33) is nonnegative for all $i \in \mathcal{I}(m)$, we have $b_i - b^T H_i \geq 0$. Thus $b \geq Hb$. The proof is complete.

References

1. Aliluiko, A.M., Mazko, O.H.: Invariant cones and stability of linear dynamical systems. Ukr. Math. J. **58**(11), 1635–1655 (2006)
2. Andersen, E.D.: MOSEK. https://www.mosek.com/ November (2015)
3. Birkhoff, G.: Linear transformations with invariant cones. Am. Math. Mon. **74**(3), 274–276 (1967)
4. Bitsoris, G.: On the positive invariance of polyhedral sets for discrete-time systems. Syst. Control Lett. **11**(3), 243–248 (1988)
5. Bitsoris, G.: Positively invariant polyhedral sets of discrete-time linear systems. Int. J. Control. **47**(6), 1713–1726 (1988)
6. Blanchini, F.: Set invariance in control. Automatica **35**(11), 1747–1767 (1999)
7. Boyd, S., Ghaoui, L.E., Feron, E., Balakrishnan, V.: Linear Matrix Inequalities in System and Control Theory. SIAM Studies in Applied Mathematics, SIAM, Philadelphia (1994)

8. Fiacchini, M., Alamo, T., Camacho, E.F.: On the computation of convex robust control invariant sets for nonlinear systems. Automatica **46**(8), 1334–1338 (2010)
9. Harris, J.: Algebraic Geometry: A First Course. Springer Science, New York (1992)
10. Horváth, Z.: Positively invariant cones of dynamical systems under Runge-Kutta and Rosenbrock-type discretization. Proc. Appl. Math. Mech. **4** 688–689 (2004)
11. Horváth, Z., Song, Y., Terlaky, T.: Invariance preserving discretizations of dynamical systems. J. Optim. Theory Appl. (2015, submitted)
12. Horváth, Z., Song, Y., Terlaky, T.: A novel unified approach to invariance in dynamical systems. Appl. Math. Comput. (2015, submitted)
13. Horváth, Z., Song, Y., Terlaky, T.: Steplength thresholds for invariance preserving of discretization methods of dynamical systems on a polyhedron. Discr. Cont. Dynamical System - A **35**(7), 2997–2013 (2015)
14. Kanatnikov, A.N., Krishchenko, A.P.: Localization of compact invariant sets of nonlinear systems. Int. J. Bifurcation Chaos **21**, 2057–2065 (2011)
15. Kerrigan, E.C., Maciejowski, J.M.: Invariant sets for constrained nonlinear discrete-time systems with application to feasibility in model predictive control. In: Proceedings of the 39th IEEE Conference on Decision and Control (2000)
16. Krein, M.G., Rutman, M.A.: Linear operators leaving invariant a cone in a Banach space. Uspehi Matem. Nauk (N. S.), 3**1**(23), 1–95 (1948) (in Russian). Am. Math. Soc. Transl. **26**, 128 pp. (1950)
17. Luenberger, D.: Introduction to Dynamic Systems: Theory, Models, and Applications, 1st edn. Wiley, New York (1979)
18. Nagumo, M.: Über die Lage der Integralkurven gewöhnlicher Differentialgleichungen. Proc. Phys.-Math. Soc. Jpn. **24**(3), 551–559 (1942)
19. Pintér, J.D.: LGO. http://www.pinterconsulting.com/l_s_d.html, November (2015)
20. Pólik, I., Terlaky, T.: A survey of the *S*-lemma. SIAM Rev. **49**(3), 371–418 (2007)
21. Pólik, I., Terlaky, T.: Interior point methods for nonlinear optimization. In: Di Pillo, G., Schoen, F. (ed.) Nonlinear Optimization, Springer Lecture Notes in Mathematics, Chap. 4, pp. 215–27. Springer Science, Heidelberg (2010)
22. Roos, C., Terlaky, T., Vial, J.-P.: Interior Point Methods for Linear Optimization. Springer Science, Heidelberg (2006)
23. Schneider, H., Vidyasagar, M.: Cross-positive matrices. SIAM J. Numer. Anal. **7**(4), 508–519 (1970)
24. Shen, J.: Positive invariance of constrained affine dynamics and its applications to hybrid systems and safety verification. IEEE Trans. Autom. Control **57**(1), 3–18 (2012)
25. Stern, R., Wolkowicz, H.: Exponential nonnegativity on the ice cream cone. SIAM J. Matrix Anal. Appl. **12**(1), 160–165 (1991)
26. Teschl, G.: Ordinary Differential Equations and Dynamical Systems. American Mathematical Society, Providence (2012)
27. Tiwari, A., Fung, J., Bhattacharya, R., Murray, R.M.: Polyhedral cone invariance applied to rendezvous of multiple agents. In: 43th IEEE Conference on Decision and Control, vol. 1, pp. 165–170 (2004)
28. Waechter, A.: IPOPT. https://projects.coin-or.org/Ipopt, November (2015)
29. Wolfe, P.: A duality theorem for non-linear programming. Quart. Appl. Math. **19**, 239–244 (1961)
30. Yakubovich, V.: *S*-procedure in nonlinear control theory. Vestn. Leningr. Univ. **1**, 62–77 (1971)
31. Zhou, K., Doyle, J., Glover, K.: Robust and Optimal Control, 1st edn. Prentice Hall, Upper Saddle River (1995)

Modeling of Stationary Periodic Time Series by ARMA Representations

Anders Lindquist and Giorgio Picci

Dedicated to Boris Teodorovich Polyak on the occasion of his 80th birthday

Abstract This is a survey of some recent results on the rational circulant covariance extension problem: Given a partial sequence $(c_0, c_1, \ldots, c_n)$ of covariance lags $c_k = \mathbb{E}\{y(t+k)\overline{y(t)}\}$ emanating from a stationary periodic process $\{y(t)\}$ with period $2N > 2n$, find all possible rational spectral functions of $\{y(t)\}$ of degree at most $2n$ or, equivalently, all bilateral and unilateral ARMA models of order at most n, having this partial covariance sequence. Each representation is obtained as the solution of a pair of dual convex optimization problems. This theory is then reformulated in terms of circulant matrices and the connections to reciprocal processes and the covariance selection problem is explained. Next it is shown how the theory can be extended to the multivariate case. Finally, an application to image processing is presented.

Keywords Discrete moment problem • Periodic processes • Circulant covariance extension • Bilateral ARMA models • Image processing

1 Introduction

The rational covariance extension problem to determine a rational spectral density given a finite number of covariance lags has been studied in great detail [2, 5–7, 9, 10, 17, 19, 20, 24, 35], and it can be formulated as a (truncated) trigonometric moment problem with a degree constraint. Among other things, it is the basic problem in partial stochastic realization theory [2] and certain Toeplitz

A. Lindquist (✉)
Shanghai Jiao Tong University, Shanghai, China

Royal Institute of Technology, Stockholm, Sweden
e-mail: alq@kth.se

G. Picci
University of Padova, Padova, Italy
e-mail: picci@dei.unipd.it

B. Goldengorin (ed.), *Optimization and Its Applications in Control and Data Sciences*, Springer Optimization and Its Applications 115,
DOI 10.1007/978-3-319-42056-1_9

matrix completion problems. In particular, it provides a parameterization of the family of (unilateral) autoregressive moving-average (ARMA) models of stationary stochastic processes with the same finite sequence of covariance lags. We also refer the reader to the recent monograph [31], in which this problem is discussed in the context of stochastic realization theory.

Covariance extension for *periodic* stochastic processes, on the other hand, leads to matrix completion of Toeplitz matrices with circulant structure and to partial stochastic realizations in the form of *bilateral* ARMA models

$$\sum_{k=-n}^{n} q_k y(t-k) = \sum_{k=-n}^{n} p_k e(t-k)$$

for a stochastic processes $\{y(t)\}$, where $\{e(t)\}$ is the corresponding conjugate process. This connects up to a rich realization theory for reciprocal processes [26–29]. As we shall see there are also (forward and backward) unilateral ARMA representations for periodic processes.

In [12] a maximum-entropy approach to this circulant covariance extension problem was presented, providing a procedure for determining the unique bilateral AR model matching the covariance sequence. However, more recently it was discovered that the circulant covariance extension problem can be recast in the context of the optimization-based theory of moment problems with rational measures developed in [1, 3, 4, 6, 8–10, 21, 22] allowing for a complete parameterization of all bilateral ARMA realizations. This led to a complete theory for the scalar case [30], which was then extended to the multivariable case in [32]. Also see [38] for modifications of this theory to skew periodic processes and [37] for fast numerical procedures.

The AR theory of [12] has been successfully applied to image processing of textures [13, 36], and we anticipate an enhancement of such methods by allowing for more general ARMA realizations.

The present survey paper is to a large extent based on [30, 32] and [12]. In Sect. 2 we begin by characterizing stationary periodic processes. In Sect. 3 we formulate the rational covariance extension problem for periodic processes as a moment problem with atomic measure and present the solution in the context of the convex optimization approach of [1, 3, 4, 6, 8–10]. These results are then reformulated in terms of circulant matrices in Sect. 4 and interpreted in term of bilateral ARMA models in Sect. 5 and in terms of unilateral ARMA models in Sect. 6. In Sect. 7 we investigate the connections to reciprocal processes of order n [12] and the covariance selection problem of Dempster [15]. In Sect. 8 we consider the situation when both partial covariance data and logarithmic moment (cepstral) data is available. To simplify the exposition the theory has so far been developed in the context of scalar processes, but in Sect. 9 we show how it can be extended to the multivariable case. All of these results are illustrated by examples taken from [30] and [32]. Section 10 is devoted to applications in image processing.

2 Periodic Stationary Processes

Consider a zero-mean full-rank stationary process $\{y(t)\}$, in general complex-valued, defined on a finite interval $[-N+1, N]$ of the integer line $\mathbb{Z}$ and extended to all of $\mathbb{Z}$ as a periodic stationary process with period $2N$ so that

$$y(t + 2kN) = y(t) \tag{1}$$

almost surely. By stationarity there is a representation

$$y(t) = \int_{-\pi}^{\pi} e^{it\theta} d\hat{y}(\theta), \quad \text{where } \mathbb{E}\{|d\hat{y}|^2\} = dF(\theta), \tag{2}$$

(see, e.g., [31, p. 74]), and therefore

$$c_k := \mathbb{E}\{y(t+k)\overline{y(t)}\} = \int_{-\pi}^{\pi} e^{ik\theta} dF(\theta). \tag{3}$$

Also, in view of (1),

$$\int_{-\pi}^{\pi} e^{it\theta} \left(e^{i2N\theta} - 1\right) d\hat{y} = 0,$$

and hence

$$\int_{-\pi}^{\pi} \left|e^{i2N\theta} - 1\right|^2 dF = 0,$$

which shows that the support of dF must be contained in $\{k\pi/N;\, k = -N+1, \dots, N\}$. Consequently the spectral density of $\{y(t)\}$ consists of point masses on the discrete unit circle $\mathbb{T}_{2N} := \{\zeta_{-N+1}, \zeta_{-n+2}, \dots, \zeta_N\}$, where

$$\zeta_k = e^{ik\pi/N}. \tag{4}$$

More precisely, define the function

$$\Phi(\zeta) = \sum_{k=-N+1}^{N} c_k \zeta^{-k} \tag{5}$$

on $\mathbb{T}_{2N}$. This is the discrete Fourier transform (DFT) of the sequence $(c_{-N+1}, \dots, c_N)$, which can be recovered by the inverse DFT

$$c_k = \frac{1}{2N} \sum_{j=-N+1}^{N} \Phi(\zeta_j)\zeta_j^k = \int_{-\pi}^{\pi} e^{ik\theta} \Phi(e^{i\theta}) d\nu, \tag{6}$$

where ν is a step function with steps $\frac{1}{2N}$ at each ζ_k; i.e.,

$$d\nu(\theta) = \sum_{j=-N+1}^{N} \delta(e^{i\theta} - \zeta_j)\frac{d\theta}{2N}. \tag{7}$$

Consequently, by (3), $dF(\theta) = \Phi(e^{i\theta})d\nu(\theta)$. We note in passing that

$$\int_{-\pi}^{\pi} e^{ik\theta} d\nu(\theta) = \delta_{k0}, \tag{8}$$

where δ_{k0} equals one for $k = 0$ and zero otherwise. To see this, note that, for $k \neq 0$,

$$\begin{aligned}(1 - \zeta_k)\int_{-\pi}^{\pi} e^{ik\theta} d\nu &= \frac{1}{2N}\sum_{j=-N+1}^{N}\left(\zeta_k^j - \zeta_k^{j+1}\right) \\ &= \frac{1}{2N}\left(\zeta_k^{-N+1} - \zeta_k^{N+1}\right) = 0.\end{aligned}$$

Since $\{y(t)\}$ is stationary and full rank, the Toeplitz matrix

$$\mathbf{T}_n = \begin{bmatrix} c_0 & \bar{c}_1 & \bar{c}_2 & \cdots & \bar{c}_n \\ c_1 & c_0 & \bar{c}_1 & \cdots & \bar{c}_{n-1} \\ c_2 & c_1 & c_0 & \cdots & \bar{c}_{n-2} \\ \vdots & \vdots & \vdots & \ddots & \vdots \\ c_n & c_{n-1} & c_{n-2} & \cdots & c_0 \end{bmatrix} \tag{9}$$

is positive definite for all $n \in \mathbb{Z}$. However, this condition is not sufficient for $c_0, c_1, \ldots, c_n$ to be a bona-fide covariance sequence of a periodic process, as can be seen from the following simple example. Consider a real-valued periodic stationary process y of period four. Then

$$\mathbb{E}\left\{\begin{bmatrix} y(1) \\ y(2) \\ y(3) \\ y(4) \end{bmatrix} \left[y(1)\; y(2)\; y(3)\; y(4)\right]\right\} = \begin{bmatrix} c_0 & c_1 & c_2 & c_3 \\ c_1 & c_0 & c_1 & c_2 \\ c_2 & c_1 & c_0 & c_1 \\ c_3 & c_2 & c_1 & c_0 \end{bmatrix}.$$

Then looking at the covariance matrix for two periods, we obtain

$$\mathbb{E}\left\{\begin{bmatrix} y(1) \\ y(2) \\ \vdots \\ y(8) \end{bmatrix} \begin{bmatrix} y(1) \; y(2) \cdots y(8) \end{bmatrix}\right\} = \begin{bmatrix} c_0 & c_1 & c_2 & c_3 & c_0 & c_1 & c_2 & c_3 \\ c_1 & c_0 & c_1 & c_2 & c_1 & c_0 & c_1 & c_2 \\ c_2 & c_1 & c_0 & c_1 & c_2 & c_1 & c_0 & c_1 \\ c_3 & c_2 & c_1 & c_0 & c_3 & c_2 & c_1 & c_0 \\ c_0 & c_1 & c_2 & c_3 & c_0 & c_1 & c_2 & c_3 \\ c_1 & c_0 & c_1 & c_2 & c_1 & c_0 & c_1 & c_2 \\ c_2 & c_1 & c_0 & c_1 & c_2 & c_1 & c_0 & c_1 \\ c_3 & c_2 & c_1 & c_0 & c_3 & c_2 & c_1 & c_0 \end{bmatrix},$$

which is a Toeplitz matrix only when $c_3 = c_1$. Therefore the condition $c_3 = c_1$ is necessary. Consequently

$$\mathbf{T}_8 = \begin{bmatrix} c_0 & c_1 & c_2 & c_1 & c_0 & c_1 & c_2 & c_1 \\ c_1 & c_0 & c_1 & c_2 & c_1 & c_0 & c_1 & c_2 \\ c_2 & c_1 & c_0 & c_1 & c_2 & c_1 & c_0 & c_1 \\ c_1 & c_2 & c_1 & c_0 & c_1 & c_2 & c_1 & c_0 \\ c_0 & c_1 & c_2 & c_1 & c_0 & c_1 & c_2 & c_1 \\ c_1 & c_0 & c_1 & c_2 & c_1 & c_0 & c_1 & c_2 \\ c_2 & c_1 & c_0 & c_1 & c_2 & c_1 & c_0 & c_1 \\ c_1 & c_2 & c_1 & c_0 & c_1 & c_2 & c_1 & c_0 \end{bmatrix}$$

is a *circulant matrix*, where the columns are shifted cyclically, the last component moved to the top. Circulant matrices will play a key role in the following.

3 The Covariance Extension Problem for Periodic Processes

Suppose that we are given a partial covariance sequence $c_0, c_1, \ldots, c_n$ with $n < N$ such that the Toeplitz matrix $\mathbf{T}_n$ is positive definite. Consider the problem of finding and extension $c_{n+1}, c_{n+2}, \ldots, c_N$ so that the corresponding sequence $c_0, c_1, \ldots, c_N$ is the covariance sequence of a stationary process of period $2N$.

In general this problem will have infinitely many solutions, and, for reasons that will become clear later, we shall restrict our attention to spectral functions (5) which are rational in the sense that

$$\Phi(\zeta) = \frac{P(\zeta)}{Q(\zeta)}, \tag{10}$$

where P and Q are Hermitian pseudo-polynomials of degree at most n, that is of the form

$$P(\zeta) = \sum_{k=-n}^{n} p_k \zeta^{-k}, \quad p_{-k} = \bar{p}_k. \tag{11}$$

Let $\mathfrak{P}_+(N)$ be the cone of all pseudo-polynomials (11) that are positive on the discrete unit circle $\mathbb{T}_{2N}$, and let $\mathfrak{P}_+ \subset \mathfrak{P}_+(N)$ be the subset of pseudo-polynomials (11) such that $P(e^{i\theta}) > 0$ for all $\theta \in [-\pi, \pi]$. Moreover let $\mathfrak{C}_+(N)$ be the dual cone of all partial covariance sequences $\mathbf{c} = (c_0, c_1, \dots, c_n)$ such that

$$\langle \mathbf{c}, \mathbf{p} \rangle := \sum_{k=-n}^{n} c_k \bar{p}_k > 0 \quad \text{for all } P \in \overline{\mathfrak{P}_+(N)} \setminus \{0\},$$

and let $\mathfrak{C}_+$ be defined in the same way as the dual cone of $\mathfrak{P}_+$. It can be shown [25] that $\mathbf{c} \in \mathfrak{C}_+$ is equivalent to the Toeplitz condition $\mathbf{T}_n > 0$. Since $\mathfrak{P}_+ \subset \mathfrak{P}_+(N)$, we have $\mathfrak{C}_+(N) \subset \mathfrak{C}_+$, so in general $\mathbf{c} \in \mathfrak{C}_+(N)$ is a stricter condition than $\mathbf{T}_n > 0$.

The proof of the following theorem can be found in [30].

Theorem 1. *Let $\mathbf{c} \in \mathfrak{C}_+(N)$. Then, for each $P \in \mathfrak{P}_+(N)$, there is a unique $Q \in \mathfrak{P}_+(N)$ such that*

$$\Phi = \frac{P}{Q}$$

satisfies the moment conditions

$$\int_{-\pi}^{\pi} e^{ik\theta} \Phi(e^{i\theta}) d\nu(\theta) = c_k, \quad k = 0, 1, \dots, n. \tag{12}$$

Consequently the family of solutions (10) of the covariance extension problem stated above are parameterized by $P \in \mathfrak{P}_+(N)$ in a bijective fashion. From the following theorem we see that, for any $P \in \mathfrak{P}_+(N)$, the corresponding unique $Q \in \mathfrak{P}_+(N)$ can be obtained by convex optimization. We refer the reader to [30] for the proofs.

Theorem 2. *Let $\mathbf{c} \in \mathfrak{C}_+(N)$ and $P \in \mathfrak{P}_+(N)$. Then the problem to maximize*

$$\mathbb{I}_P(\Phi) = \int_{-\pi}^{\pi} P(e^{i\theta}) \log \Phi(e^{i\theta}) d\nu \tag{13}$$

subject to the moment conditions (12) *has a unique solution, namely* (10)*, where Q is the unique optimal solution of the problem to minimize*

$$\mathbb{J}_P(Q) = \langle \mathbf{c}, \mathbf{q} \rangle - \int_{-\pi}^{\pi} P(e^{i\theta}) \log Q(e^{i\theta}) d\nu \tag{14}$$

over all $Q \in \mathfrak{P}_+(N)$, where $\mathbf{q} := (q_0, q_1, \dots, q_n)$. The functional $\mathbb{J}_P$ is strictly convex.

Theorems 1 and 2 are discrete versions of corresponding results in [6, 9]. The solution corresponding to $P = 1$ is called the *maximum-entropy solution* by virtue of (13).

Remark 3. As $N \to \infty$ the process y looses it periodic character, and its spectral density Φ_∞ becomes continuous and defined on the whole unit circle so that

$$\int_{-\pi}^{\pi} e^{ik\theta}\Phi_\infty(e^{i\theta})\frac{d\theta}{2\pi} = c_k, \quad k = 0, 1, \dots, n. \tag{15}$$

In fact, denoting by Q_N the solution of Theorem 1, it was shown in [30] that $\Phi_\infty = P/Q_\infty$, where, for each fixed P,

$$Q_\infty = \lim_{N\to\infty} Q_N$$

is the unique Q such that $\Phi_\infty = P/Q$ satisfies the moment conditions (15).

4 Reformulation in Terms of Circulant Matrices

Circulant matrices [14] are Toeplitz matrices with a special circulant structure

$$\mathrm{Circ}\{\gamma_0, \gamma_1, \dots, \gamma_\nu\} = \begin{bmatrix} \gamma_0 & \gamma_\nu & \gamma_{\nu-1} & \cdots & \gamma_1 \\ \gamma_1 & \gamma_0 & \gamma_\nu & \cdots & \gamma_2 \\ \gamma_2 & \gamma_1 & \gamma_0 & \cdots & \gamma_3 \\ \vdots & \vdots & \vdots & \ddots & \vdots \\ \gamma_\nu & \gamma_{\nu-1} & \gamma_{\nu-2} & \cdots & \gamma_0 \end{bmatrix}, \tag{16}$$

where the columns (or, equivalently, rows) are shifted cyclically, and where $\gamma_0, \gamma_1, \dots, \gamma_\nu$ here are taken to be complex numbers. In our present covariance extension problem we consider *Hermitian* circulant matrices

$$\mathbf{M} := \mathrm{Circ}\{m_0, m_1, m_2, \dots, m_N, \bar{m}_{N-1}, \dots, \bar{m}_2, \bar{m}_1\}, \tag{17}$$

which can be represented in form

$$\mathbf{M} = \sum_{k=-N+1}^{N} m_k \mathbf{S}^{-k}, \quad m_{-k} = \bar{m}_k \tag{18}$$

where $\mathbf{S}$ is the nonsingular $2N \times 2N$ cyclic shift matrix

$$\mathbf{S} := \begin{bmatrix} 0 & 1 & 0 & 0 & \dots & 0 \\ 0 & 0 & 1 & 0 & \dots & 0 \\ 0 & 0 & 0 & 1 & \dots & 0 \\ \vdots & \vdots & \vdots & \ddots & \ddots & \vdots \\ 0 & 0 & 0 & 0 & 0 & 1 \\ 1 & 0 & 0 & 0 & 0 & 0 \end{bmatrix}. \tag{19}$$

The pseudo-polynomial

$$M(\zeta) = \sum_{k=-N+1}^{N} m_k \zeta^{-k}, \quad m_{-k} = \bar{m}_k \tag{20}$$

is called the *symbol* of $\mathbf{M}$. Clearly $\mathbf{S}$ is itself a circulant matrix (although not Hermitian) with symbol $S(\zeta) = \zeta$. A necessary and sufficient condition for a matrix $\mathbf{M}$ to be circulant is that

$$\mathbf{S}\mathbf{M}\mathbf{S}^{\mathsf{T}} = \mathbf{M}. \tag{21}$$

Hence, since $\mathbf{S}^{-1} = \mathbf{S}^{\mathsf{T}}$, the inverse of a circulant matrix is also circulant. More generally, if $\mathbf{A}$ and $\mathbf{B}$ are circulant matrices of the same dimension with symbols $A(\zeta)$ and $B(\zeta)$ respectively, then $\mathbf{AB}$ and $\mathbf{A} + \mathbf{B}$ are circulant matrices with symbols $A(\zeta)B(\zeta)$ and $A(\zeta) + B(\zeta)$, respectively. In fact, the circulant matrices of a fixed dimension form an algebra—more precisely, a commutative *-algebra with the involution * being the conjugate transpose—and the DFT is an *algebra homomorphism* of the set of circulant matrices onto the pseudo-polynomials of degree at most N in the variable $\zeta \in \mathbb{T}_{2N}$. Consequently, circulant matrices commute, and, if $\mathbf{M}$ is a circulant matrix with symbol $M(\zeta)$, then $\mathbf{M}^{-1}$ is circulant with symbol $M(\zeta)^{-1}$.

The proof of the following proposition is immediate.

Proposition 4. *Let $\{y(t);\, t = -N+1, \dots, N\}$ be a stationary process with period $2N$ and covariance lags* (3)*, and let $\mathbf{y}$ be the $2N$-dimensional stochastic vector $\mathbf{y} = [y(-N+1), y(-N+2), \cdots, y(N)]^{\mathsf{T}}$. Then, with * denoting conjugate transpose,*

$$\boldsymbol{\Sigma} := \mathbb{E}\{\mathbf{y}\mathbf{y}^*\} = \operatorname{Circ}\{c_0, c_1, c_2, \dots, c_N, \bar{c}_{N-1}, \dots, \bar{c}_2, \bar{c}_1\} \tag{22}$$

is a $2N \times 2N$ Hermitian circulant matrix with symbol $\Phi(\zeta)$ given by (5)*.*

The covariance extension problem of Sect. 3, called the *circulant rational covariance extension problem*, can now be reformulated as a matrix extension problem. The given covariance data $\mathbf{c} = (c_0, c_1, \dots, c_n)$ can be represented as a circulant matrix

$$\mathbf{C} = \operatorname{Circ}\{c_0, c_1, \dots, c_n, 0, \dots, 0, \bar{c}_n, \bar{c}_{n-1}, \dots, \bar{c}_1\} \tag{23}$$

with symbol

$$C(\zeta) = \sum_{k=-n}^{n} c_k \zeta^{-k}, \tag{24}$$

where the unknown covariance lags $c_{n+1}, c_{n+2}, \ldots, c_N$ in (22), to be determined, here are replaced by zeros. A circulant matrix of type (23) is called *banded of order* n. We recall that $n < N$. From now one we drop the attribute 'Hermitian' since we shall only consider such circulant matrices in the sequel. A banded circulant matrix of order n will thus be determined by $n + 1$ (complex) parameters.

The next lemma establishes the connection between circulant matrices and their symbols.

Lemma 5. *Let* **M** *be a circulant matrix with symbol* $M(\zeta)$. *Then*

$$\mathbf{M} = \mathbf{F}^* \text{diag}\big(M(\zeta_{-N+1}), M(\zeta_{-N+2}), \ldots, M(\zeta_N)\big)\mathbf{F}, \tag{25}$$

where **F** *is the unitary matrix*

$$\mathbf{F} = \frac{1}{\sqrt{2N}} \begin{bmatrix} \zeta_{-N+1}^{N-1} & \zeta_{-N+1}^{N-2} & \cdots & \zeta_{-N+1}^{-N} \\ \vdots & \vdots & \cdots & \vdots \\ \zeta_0^{N-1} & \zeta_0^{N-2} & \cdots & \zeta_0^{-N} \\ \vdots & \vdots & \cdots & \vdots \\ \zeta_N^{N-1} & \zeta_N^{N-2} & \cdots & \zeta_N^{-N} \end{bmatrix}. \tag{26}$$

Moreover, if $M(\zeta_k) > 0$ *for all k, then*

$$\log \mathbf{M} = \mathbf{F}^* \text{diag}\big(\log M(\zeta_{-N+1}), \log M(\zeta_{-N+2}), \ldots, \log M(\zeta_N)\big)\mathbf{F}. \tag{27}$$

Proof. The discrete Fourier transform $\mathfrak{F}$ maps a sequence $(g_{-N+1}, g_{-N+2}, \ldots, g_N)$ into the sequence of complex numbers

$$G(\zeta_j) := \sum_{k=-N+1}^{N} g_k \zeta_j^{-k}, \qquad j = -N+1, -N+2, \ldots, N. \tag{28}$$

The sequence **g** can be recovered from G by the inverse transform

$$g_k = \int_{-\pi}^{\pi} e^{ik\theta} G(e^{i\theta}) d\nu(\theta), \quad k = -N+1, -N+2, \ldots, N. \tag{29}$$

This correspondence can be written

$$\hat{\mathbf{g}} = \mathbf{F}\mathbf{g}, \tag{30}$$

where $\hat{\mathbf{g}} := (2N)^{-\frac{1}{2}}\big(G(\zeta_{-N+1}), \dots, G(\zeta_N)\big)^{\mathsf{T}}$, $\mathbf{g} := (g_{-N+1}, \dots, g_N)^{\mathsf{T}}$, and $\mathbf{F}$ is the nonsingular $2N \times 2N$ Vandermonde matrix (26). Clearly $\mathbf{F}$ is unitary. Since

$$\mathbf{Mg} = \sum_{k=-N+1}^{N} m_k \mathbf{S}^{-k}$$

and $[\mathbf{S}^{-k}\mathbf{g}]_j = g_{j-k}$, where $g_{k+2N} = g_k$, we have

$$\begin{aligned}\mathfrak{F}(\mathbf{Mg}) &= \sum_{j=-N+1}^{N} \zeta^{-j} \sum_{k=-N+1}^{N} m_k g_{j-k} \\ &= \sum_{k=-N+1}^{N} m_k \zeta^{-k} \sum_{j=-N+1}^{N} g_{j-k}\zeta^{-(j-k)} = M(\zeta)\mathfrak{F}\mathbf{g},\end{aligned}$$

which yields

$$\sqrt{2N}(\mathbf{FMg})_j = M(\zeta_j)\sqrt{2N}(\mathbf{Fg})_j, \quad j = -N+1, -N+2, \dots, N,$$

from which (25) follows. Finally, since, as a function of $z \in \mathbb{C}$, $\log M(z)$ is analytic in the neighborhood of each $M(\zeta_k) > 0$, the eigenvalues of $\log \mathbf{M}$ are just the real numbers $\log M(\zeta_k)$, $k = -N+1, \dots, N$, by the spectral mapping theorem [16, p. 557], and hence (27) follows.

We are now in a position to reformulate Theorems 1 and 2 in terms of circulant matrices. To this end first note that, in view of Lemma 5, the cone $\mathfrak{P}_+(N)$ corresponds to the class of positive-definite banded $2N \times 2N$ circulant matrices $\mathbf{P}$ of order n. Moreover, by Plancherel's Theorem for DFT, which is a simple consequence of (8), we have

$$\sum_{k=-n}^{n} c_k \bar{p}_k = \frac{1}{2N} \sum_{j=-N+1}^{N} C(\zeta_j)P(\zeta_j),$$

and hence, by Lemma 5,

$$\langle \mathbf{c}, \mathbf{p} \rangle = \frac{1}{2N}\operatorname{tr}(\mathbf{CP}), \tag{31}$$

where tr denotes trace.

Consequently, $\mathbf{c} \in \mathfrak{C}_+(N)$ if and only if $\operatorname{tr}(\mathbf{CP}) > 0$ for all nonzero, positive-semidefinite, banded $2N \times 2N$ circulant matrices $\mathbf{P}$ of order n. Moreover, if $\mathbf{Q}$ and $\mathbf{P}$ are circulant matrices with symbols $P(\zeta)$ and $Q(\zeta)$, respectively, then, by Lemma 5, $P(\zeta)/Q(\zeta)$ is the symbol of $\mathbf{Q}^{-1}\mathbf{P}$. Therefore Theorem 1 has the following matrix version.

Theorem 6. *Let* $\mathbf{c} \in \mathfrak{C}_+(N)$*, and let* $\mathbf{C}$ *be the corresponding circulant matrix* (23)*. Then, for each positive-definite banded* $2N \times 2N$ *circulant matrices* $\mathbf{P}$ *of order* n*, there is unique positive-definite banded* $2N \times 2N$ *circulant matrices* $\mathbf{Q}$ *of order* n *such that*

$$\boldsymbol{\Sigma} = \mathbf{Q}^{-1}\mathbf{P} \tag{32}$$

is a circulant extension (22) *of* $\mathbf{C}$.

In the same way, Theorem 2 has the following matrix version, as can be seen by applying Lemma 5.

Theorem 7. *Let* $\mathbf{c} \in \mathfrak{C}_+(N)$*, and let* $\mathbf{C}$ *be the corresponding circulant matrix* (23)*. Moreover, let* $\mathbf{P}$ *be a positive-definite banded* $2N \times 2N$ *circulant matrix of order* n*. Then the problem to maximize*

$$\mathscr{I}_{\mathbf{P}}(\boldsymbol{\Sigma}) = \operatorname{tr}(\mathbf{P} \log \boldsymbol{\Sigma}) \tag{33}$$

subject to

$$\mathbf{E}_n{}^T \boldsymbol{\Sigma} \mathbf{E}_n = \mathbf{T}_n, \quad \text{where } \mathbf{E}_n = \begin{bmatrix} \mathbf{I}_n \\ \mathbf{0} \end{bmatrix} \tag{34}$$

has a unique solution, namely (32)*, where* $\mathbf{Q}$ *is the unique optimal solution of the problem to minimize*

$$\mathscr{J}_{\mathbf{P}}(\mathbf{q}) = \operatorname{tr}(\mathbf{C}\mathbf{Q}) - \operatorname{tr}(\mathbf{P} \log \mathbf{Q}) \tag{35}$$

over all positive-definite banded $2N \times 2N$ *circulant matrices* $\mathbf{Q}$ *of order* n*, where* $\mathbf{q} := (q_0, q_1, \ldots, q_n)$*. The functional* $\mathscr{J}_{\mathbf{P}}$ *is strictly convex.*

5 Bilateral ARMA Models

Suppose now that we have determined a circulant matrix extension (32). Then there is a stochastic vector $\mathbf{y}$ formed from the a stationary periodic process with corresponding covariance lags (3) so that

$$\boldsymbol{\Sigma} := \mathbb{E}\{\mathbf{y}\mathbf{y}^*\} = \operatorname{Circ}\{c_0, c_1, c_2, \ldots, c_N, \bar{c}_{N-1}, \ldots, \bar{c}_2, \bar{c}_1\}.$$

Let $\hat{\mathbb{E}}\{y(t) \mid y(s),\, s \neq t\}$ be the wide sense conditional mean of $y(t)$ given all $\{y(s),\, s \neq t\}$. Then the error process

$$d(t) := y(t) - \hat{\mathbb{E}}\{y(t) \mid y(s),\, s \neq t\} \tag{36}$$

is orthogonal to all random variables $\{y(s),\ s \neq t\}$, i.e., $\mathbb{E}\{y(t)\,\overline{d(s)}\} = \sigma^2\,\delta_{ts},\ t,s \in \mathbb{Z}_{2N} := \{-N+1, -N+2, \ldots, N\}$, where σ^2 is a positive number. Equivalently, $\mathbb{E}\{\mathbf{y}\mathbf{d}^*\} = \sigma^2\mathbf{I}$, where $\mathbf{I}$ is the $2N \times 2N$ identity matrix. Setting $\mathbf{e} := \mathbf{d}/\sigma^2$, we then have

$$\mathbb{E}\{\mathbf{e}\mathbf{y}^*\} = \mathbf{I}, \tag{37}$$

i.e., the corresponding process e is the *conjugate process* of y [33]. Interpreting (36) in the mod $2N$ arithmetics of $\mathbb{Z}_{2N}$, $\mathbf{y}$ admits a linear representation of the form

$$\mathbf{G}\mathbf{y} = \mathbf{e}, \tag{38}$$

where $\mathbf{G}$ is a $2N \times 2N$ Hermitian circulant matrix with ones on the main diagonal. Since $\mathbf{G}\mathbb{E}\{\mathbf{y}\mathbf{y}^*\} = \mathbb{E}\{\mathbf{e}\mathbf{y}^*\} = \mathbf{I}$, $\mathbf{G}$ is also positive definite and the covariance matrix $\boldsymbol{\Sigma}$ is given by

$$\boldsymbol{\Sigma} = \mathbf{G}^{-1}, \tag{39}$$

which is circulant, since the inverse of a circulant matrix is itself circulant. In fact, a stationary process $\mathbf{y}$ is full-rank periodic in $\mathbb{Z}_{2N}$, if and only if $\boldsymbol{\Sigma}$ is a Hermitian positive definite circulant matrix [12].

Since $\mathbf{G}$ is a Hermitian circulant matrix, it has a symbol

$$G(\zeta) = \sum_{k=-N+1}^{N} g_k \zeta^{-k}, \quad g_{-k} = \bar{g}_k,$$

and the linear equation can be written in the autoregressive (AR) form

$$\sum_{k=-N+1}^{N} g_k y(t-k) = e(t). \tag{40}$$

However, in general $\mathbf{G}$ is not banded and $n << N$, and therefore (40) is not a parsimonious representation. Instead using the solution (32), we have $\mathbf{G} = \mathbf{P}^{-1}\mathbf{Q}$, where $\mathbf{P}$ and $\mathbf{Q}$ are banded of order n with symbols

$$P(\zeta) = \sum_{k=-n}^{n} p_k \zeta^{-k} \quad \text{and} \quad Q(\zeta) = \sum_{k=-n}^{n} q_k \zeta^{-k},$$

and hence (38) can be written

$$\mathbf{Q}\mathbf{y} = \mathbf{P}\mathbf{e},$$

or equivalently in the ARMA form

$$\sum_{k=-n}^{n} q_k y(t-k) = \sum_{k=-n}^{n} p_k e(t-k). \tag{41}$$

Consequently, by Theorem 6, there is a unique bilateral ARMA model (41) for each banded positive-definite Hermitian circulant matrix $\mathbf{P}$ of order n, provided $\mathbf{c} \in \mathfrak{C}_+(N)$. Of course, we could use the maximum-entropy solution with $\mathbf{P} = \mathbf{I}$ leading to an AR model

$$\sum_{k=-n}^{n} q_k y(t-k) = e(t). \tag{42}$$

Next, to illustrate the accuracy of bilateral AR modeling by the methods described so far we give some simulations from [30], provided by Chiara Masiero. Given an AR model of order $n = 8$ with poles as depicted in Fig. 1, we compute a covariance sequence $\mathbf{c} = (c_0, c_1, \ldots, c_n)$ with $n = 8$, which is then used to solve the optimization problem (35) with $\mathbf{P} = \mathbf{I}$ to obtain a bilateral AR approximations of degree eight for various choices of N. In Fig. 2, the top picture depicts the spectral density for $N = 128$ together with the true spectral density (dashed line), and the bottom picture illustrates how the estimation error decreases with increasing N.

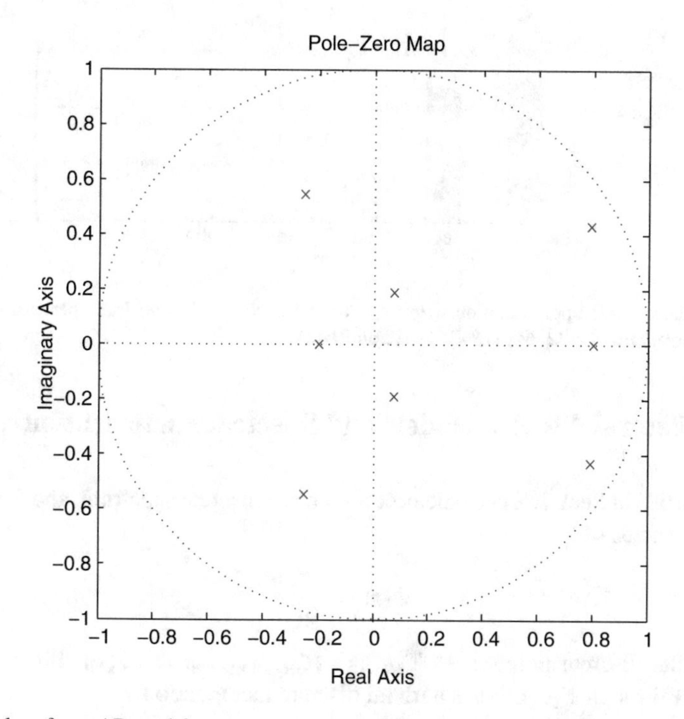

Fig. 1 Poles of true AR model

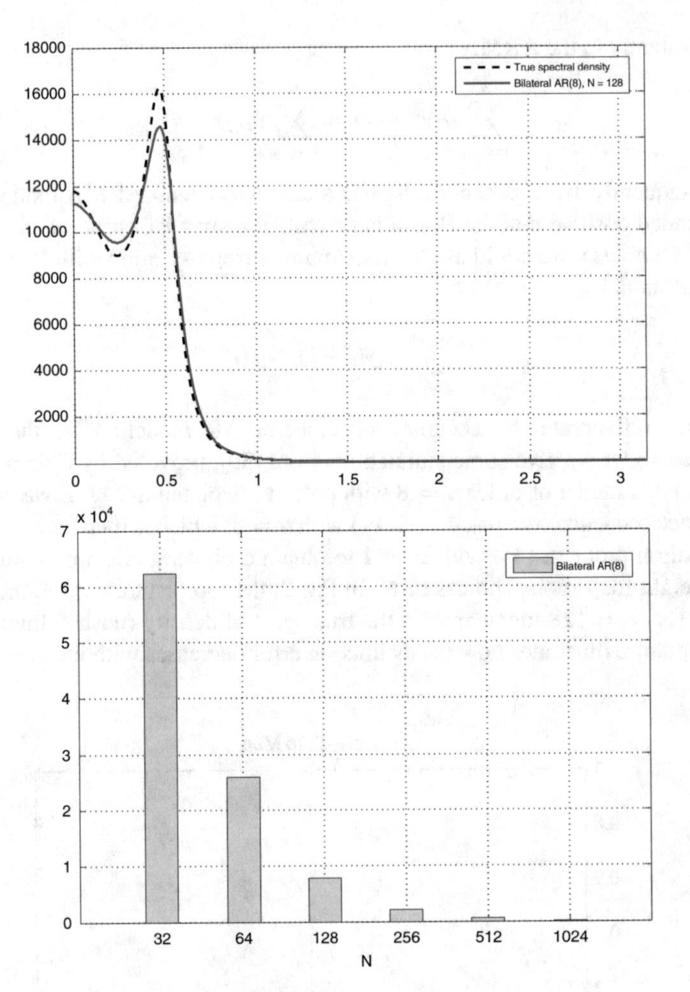

Fig. 2 Bilateral AR approximation: (*top*) spectrum for N = 128 and true spectrum (*dashed*); (*bottom*) errors for N=32, 64, 128, 256, 512 and 1024

6 Unilateral ARMA Models and Spectral Factorization

As explained in Sect. 2, a periodic process y has a discrete spectrum, and Theorem 1 provides values of

$$\Phi(z) = \frac{P(z)}{Q(z)}$$

only in the discrete points $z \in \mathbb{T}_{2N} := \{\zeta_{-N+1}, \zeta_{-n+2}, \dots, \zeta_N\}$. Since Φ takes positive values on $\mathbb{T}_{2N}$, there is a trivial discrete factorization

$$\Phi(\zeta_k) = W(\zeta_k)W(\zeta_k)^* \quad k = -N+1, \dots, N. \tag{43}$$

Defining

$$W_k = \frac{1}{2N} \sum_{j=-N+1}^{N} W(\zeta_j)\zeta_j^k, \quad k = -N+1, \dots, N,$$

we can write (43) in the form

$$\Phi(\zeta) = W(\zeta)W(\zeta)^*, \tag{44}$$

where $W(\zeta)$ is the discrete Fourier transform

$$W(\zeta) = \sum_{k=-N+1}^{N} W_k \zeta^{-k}.$$

Formally substituting the variable $z \in \mathbb{T}$ in place of ζ in W, we obtain a spectral factorization equation

$$\tilde{\Phi}(z) = W(z)W(z)^*, \quad z \in \mathbb{T}, \tag{45}$$

defined on the whole unit circle, where the continuous spectral density $\tilde{\Phi}(z)$, frequency sampled with sampling interval $\frac{\pi}{N}$, satisfies $\tilde{\Phi}(\zeta) = \Phi(\zeta)$ on $\mathbb{T}_{2N}$. This is a spectral density of a non-periodic stationary process but should not be confused with Φ_∞ in Remark 3, which is the unique continuous Φ with numerator polynomial P and the same first $n+1$ covariance lags as the periodic process y, i.e.,

$$\int_{-\pi}^{\pi} e^{ik\theta} \Phi_\infty(e^{i\theta}) \frac{d\theta}{2\pi} = c_k, \quad k = 0, 1, \dots, n.$$

In fact, although

$$\int_{-\pi}^{\pi} e^{ik\theta} \tilde{\Phi}(e^{i\theta}) d\nu(\theta) = c_k, \quad k = 0, 1, \dots, n, \tag{46}$$

the non-periodic process with spectral density $\tilde{\Phi}$ has the covariance lags

$$\tilde{c}_k = \int_{-\pi}^{\pi} e^{ik\theta} \tilde{\Phi}(e^{i\theta}) \frac{d\theta}{2\pi}, \quad k = 0, 1, \dots, n,$$

which differ from $c_0, c_1, \dots, c_n$. However, setting $\Delta\theta_j := \theta_j - \theta_{j-1}$ where $e^{i\theta_j} = \zeta_j$, we see from (4) that $\Delta\theta_j = \pi/N$ and that the integral (46) with $\tilde{\Phi}$ fixed is the Riemann sum

$$\sum_{j=-N+1}^{N} e^{ik\theta_j} \tilde{\Phi}(\zeta_j) \frac{\Delta\theta_j}{2\pi}$$

converging to $\tilde{c}_k$ for $k = 0, 1, \dots, n$ as $N \to \infty$.

By Proposition 4, $\Phi(\zeta)$ is the symbol of the circulant covariance matrix $\boldsymbol{\Sigma}$, and hence (44) can be written in the matrix form

$$\boldsymbol{\Sigma} = \mathbf{W}\mathbf{W}^*, \tag{47}$$

where $\mathbf{W}$ is the circulant matrix with symbol $W(\zeta)$. The spectral density (45) has a unique outer spectral factor $W(z)$; see, e.g., [31]. As explained in detail in [11], this corresponds in the discrete setting to $W(\zeta)$ taking the form

$$W(\zeta) = \sum_{k=0}^{N} W_k \zeta^{-k}, \tag{48}$$

which in turn corresponds to $\mathbf{W}$ being *lower-triangular circulant*, i.e.,

$$\mathbf{W} = \text{Circ}\{W_0, W_1, \dots, W_N, 0, \dots, 0\}. \tag{49}$$

Note that a lower-triangular circulant matrix is not lower triangular as the circulant structure has to be preserved. Since $\boldsymbol{\Sigma}$ is invertible, then so is $\mathbf{W}$.

Next define the periodic stochastic process $\{w(t),\ t = -N+1\dots, N\}$ for which $\mathbf{w} = [w(-N+1), w(-N+2), \dots, w(N)]^\mathsf{T}$ is given by

$$\mathbf{w} = \mathbf{W}^{-1}\mathbf{y}. \tag{50}$$

Then, in view of (47), we obtain $\mathbb{E}\{\mathbf{w}\mathbf{w}^*\} = \mathbf{I}$, i.e., the process w is a white noise process. Consequently we have the unilateral representation

$$y(t) = \sum_{k=0}^{N} W_k w(t-k)$$

in terms of white noise.

To construct an ARMA model we appeal to the following result, which is easy to verify in terms of symbols but, as demonstrated in [11], also holds for block circulant matrices considered in Sect. 9.

Lemma 8. *There exists an integer N_0 such that the following holds for $N \geq N_0$. A positive definite, Hermitian, circulant matrix $\mathbf{M}$ admits a factorization $\mathbf{M} = \mathbf{V}\mathbf{V}^*$, where $\mathbf{V}$ is of a banded lower-diagonal circulant matrix of order $n < N$, if and only if $\mathbf{M}$ is bilaterally banded of order n.*

By Theorem 6, $\boldsymbol{\Sigma} = \mathbf{Q}^{-1}\mathbf{P}$, where $\mathbf{Q}$ and $\mathbf{P}$ are banded, positive definite, Hermitian, circulant matrices of order n. Hence, for N sufficiently large, by Lemma 8 there are factorizations

$$\mathbf{Q} = \mathbf{A}\mathbf{A}^* \quad \text{and} \quad \mathbf{P} = \mathbf{B}\mathbf{B}^*,$$

where $\mathbf{A}$ and $\mathbf{B}$ are banded lower-diagonal circulant matrices of order n. Consequently, $\boldsymbol{\Sigma} = \mathbf{A}^{-1}\mathbf{B}(\mathbf{A}^{-1}\mathbf{B})^*$, i.e.,

$$\mathbf{W} = \mathbf{A}^{-1}\mathbf{B}, \tag{51}$$

which together with (50) yields $\mathbf{Ay} = \mathbf{Bw}$, i.e., the unilateral ARMA model

$$\sum_{k=0}^{n} a_k y(t-k) = \sum_{k=0}^{n} b_k w(t-k). \tag{52}$$

Since $\mathbf{A}$ is nonsingular, $a_0 \neq 0$, and hence we can normalize by setting $a_0 = 1$. In particular, if $\mathbf{P} = \mathbf{I}$, we obtain the AR representation

$$\sum_{k=0}^{n} a_k y(t-k) = b_0 w(t). \tag{53}$$

Symmetrically, there is factorization

$$\boldsymbol{\Sigma} = \bar{\mathbf{W}}\bar{\mathbf{W}}^*, \tag{54}$$

where $\bar{\mathbf{W}}$ is upper-diagonal circulant, i.e. the transpose of a lower-diagonal circulant matrix, and a white-noise process

$$\bar{\mathbf{w}} = \bar{\mathbf{W}}^{-1}\mathbf{y}. \tag{55}$$

Likewise there are factorizations

$$\mathbf{Q} = \bar{\mathbf{A}}\bar{\mathbf{A}}^* \quad \text{and} \quad \mathbf{P} = \bar{\mathbf{B}}\bar{\mathbf{B}}^*,$$

where $\bar{\mathbf{A}}$ and $\bar{\mathbf{B}}$ are banded upper-diagonal circulant matrices of order n. This yields a backward unilateral ARMA model

$$\sum_{k=-n}^{0} \bar{a}_k y(t-k) = \sum_{k=-n}^{0} \bar{b}_k \bar{w}(t-k). \tag{56}$$

These representations should be useful in the smoothing problem for periodic systems [29].

7 Reciprocal Processes and the Covariance Selection Problem

Let $\mathbf{A}$, $\mathbf{B}$ and $\mathbf{X}$ be subspaces in a certain common ambient Hilbert space of zero mean second order random variables. We say that $\mathbf{A}$ and $\mathbf{B}$ are *conditionally orthogonal* given $\mathbf{X}$ if

$$\alpha - \hat{\mathbb{E}}\{\alpha \mid \mathbf{X}\} \perp \beta - \hat{\mathbb{E}}\{\beta \mid \mathbf{X}\}, \quad \forall \alpha \in \mathbf{A}, \forall \beta \in \mathbf{B} \tag{57}$$

(see, e.g., [31]), which we denote $\mathbf{A} \perp \mathbf{B} \mid \mathbf{X}$, and which clearly is equivalent to

$$\mathbb{E}\left\{\hat{\mathbb{E}}\{\alpha \mid \mathbf{X}\}\overline{\hat{\mathbb{E}}\{\beta \mid \mathbf{X}\}}\right\} = \mathbb{E}\{\alpha\overline{\beta}\}, \quad \forall \alpha \in \mathbf{A}, \forall \beta \in \mathbf{B}. \tag{58}$$

Conditional orthogonality is the same as conditional uncorrelatedness, and hence conditional independence in the Gaussian case.

Let $\mathbf{y}_{[t-n,t)}$ and $\mathbf{y}_{(t,t+n]}$ be the n-dimensional random column vectors obtained by stacking $y(t-n), y(t-n+1)\dots, y(t-1)$ and $y(t+1), y(t+2)\dots, y(t+n)$, respectively, in that order. In the same way, $\mathbf{y}_{[t-n,t]}$ is obtained by appending $y(t)$ to $\mathbf{y}_{[t-n,t)}$ as the last element, etc. Here and in the following the sums $t-k$ and $t+k$ are to be understood modulo $2N$. For any interval $(t_1, t_2) \subset [-N+1, N]$, we denote by $(t_1, t_2)^c$ the complementary set in $[1, 2N]$.

Definition 9. A *reciprocal process of order n* on $(-N, N]$ is a process $\{y(t);\, t = -N+1, \dots, N\}$ such that

$$\hat{\mathbb{E}}\{y(t) \mid y(s),\ s \neq t\} = \hat{\mathbb{E}}\{y(t) \mid \mathbf{y}_{[t-n,t)} \vee \mathbf{y}_{(t,t+n]}\} \tag{59}$$

for $t \in (-N, N]$.

This is a generalization introduced in [12] of the concept of *reciprocal process* [23], which can be trivially extended to vector processes. In fact, a reciprocal process in the original sense is here a reciprocal process of order one. This concept does not require stationarity, although here it will always be assumed.

It follows from [31, Proposition 2.4.2 (iii)] that $\{y(t)\}$ is reciprocal of order n if and only if

$$\hat{\mathbb{E}}\{y(t) \mid y(s),\ s \in [t-n,\ t+n]^c\} = \hat{\mathbb{E}}\{y(t) \mid \mathbf{y}_{[t-n,t)} \vee \mathbf{y}_{(t,t+n]}\} \tag{60}$$

for $t \in [-N+1, N]$. In particular, the estimation error

$$\begin{aligned} d(t) &:= y(t) - \hat{\mathbb{E}}\{y(t) \mid y(s),\ s \neq t\} \\ &= y(t) - \hat{\mathbb{E}}\{y(t) \mid \mathbf{y}_{[t-n,t)} \vee \mathbf{y}_{(t,t+n]}\} \end{aligned} \tag{61}$$

must clearly be orthogonal to all random variables $\{y(s),\ s \neq t\}$; i.e. $\mathbb{E}\{d(t)\overline{y(s)}\} = \sigma^2\delta_{st}$, where σ^2 is the variance of $d(t)$. Then $e(t) := d(t)/\sigma^2$ is the (normalized) conjugate process of y satisfying (37), i.e.,

$$\mathbb{E}\{e(t)\overline{y(s)}\} = \delta_{ts}. \tag{62}$$

Since $e(t+k)$ is a linear combination of the components of the random vector $\mathbf{y}_{[t+k-n,t+k+n]}$, it follows from (62) that both $e(t+k)$ and $e(t-k)$ are orthogonal to $e(t)$ for $k > n$. Hence the process $\{e(t)\}$ has correlation bandwidth n, i.e.,

$$\mathbb{E}\{e(t+k)\,e(t)^*\} = 0 \quad \text{for } n < |k| < 2N - n,\ k \in [-N+1, N], \tag{63}$$

and consequently $(\mathbf{y}, \mathbf{e})$ satisfies (38), where $\mathbf{G}$ is banded of order n, which corresponds to an AR representation (42).

Consequently, the AR solutions of the rational circulant covariance extension problem are precisely the ones corresponding to a reciprocal process $\{y(t)\}$ of order n. Next we demonstrate how this representation is connected to the *covariance selection problem* of Dempster [15] by deriving a generalization of this seminal result.

Let $J := \{j_1, \ldots, j_p\}$ and $K := \{k_1, \ldots, k_q\}$ be two subsets of $\{-N+1, -N+2, \ldots, N\}$, and define $\mathbf{y}_J$ and $\mathbf{y}_K$ as the subvectors of $\mathbf{y} = (y_{-N+1}, y_{-N+2}, \cdots, y_N)^{\mathsf{T}}$ with indices in J and K, respectively. Moreover, let

$$\check{\mathbf{Y}}_{J,K} := \text{span}\{y(t);\ t \notin J,\ t \notin K\} = \check{\mathbf{Y}}_J \cap \check{\mathbf{Y}}_K,$$

where $\check{\mathbf{Y}}_J := \text{span}\{y(t);\ t \notin J\}$. With a slight misuse of notation, we shall write

$$\mathbf{y}_J \perp \mathbf{y}_K \mid \check{\mathbf{Y}}_{J,K}, \tag{64}$$

to mean that the subspaces spanned by the components of $\mathbf{y}_J$ and $\mathbf{y}_K$, respectively, are conditionally orthogonal given $\check{\mathbf{Y}}_{J,K}$. This condition can be characterized in terms of the inverse of the covariance matrix $\boldsymbol{\Sigma} := \mathbb{E}\{\mathbf{y}\mathbf{y}^*\} = \left[\sigma_{ij}\right]_{i,j=-N+1}^{N}$ of y.

Theorem 10. *Let* $\mathbf{G} := \boldsymbol{\Sigma}^{-1} = \left[g_{ij}\right]_{i,j=1}^{N}$ *be the concentration matrix of the random vector* y*. Then the conditional orthogonality relation* (64) *holds if and only if* $g_{jk} = 0$ *for all* $(j,k) \in J \times K$.

Proof. Let E_J be the $2N \times 2N$ diagonal matrix with ones in the positions $(j_1, j_1), \ldots, (j_m, j_m)$ and zeros elsewhere and let E_K be defined similarly in terms of index set K. Then $\check{\mathbf{Y}}_J$ is spanned by the components of $\mathbf{y} - E_J\mathbf{y}$ and $\check{\mathbf{Y}}_K$ by the components of $\mathbf{y} - E_K\mathbf{y}$. Let

$$\tilde{\mathbf{y}}_K := \mathbf{y}_K - \hat{\mathbb{E}}\{\mathbf{y}_K \mid \check{\mathbf{Y}}_K\},$$

and note that its $q \times q$ covariance matrix

$$\tilde{\boldsymbol{\Sigma}}_K := \mathbb{E}\{\tilde{\mathbf{y}}_K \tilde{\mathbf{y}}_K^*\}$$

must be positive definite, for otherwise some linear combination of the components of $\mathbf{y}_K$ would belong to $\check{\mathbf{Y}}_K$. Let $\tilde{\mathbf{y}}_K = G_K\mathbf{y}$ for some $q \times 2N$ matrix G_K. Since $\tilde{\mathbf{y}}_K \perp \check{\mathbf{Y}}_K$,

$$\mathbb{E}\{\tilde{\mathbf{y}}_K(\mathbf{y} - E_K\mathbf{y})^*\} = 0$$

and therefore $\mathbb{E}\{\tilde{\mathbf{y}}_K\mathbf{y}^*\} = G_K\boldsymbol{\Sigma}$ must be equal to $\mathbb{E}\{\tilde{\mathbf{y}}_K(E_K\mathbf{y})^*\}$, which, by $\tilde{\mathbf{y}}_K \in \check{\mathbf{Y}}_K^{\perp}$, in turn equals

$$\mathbb{E}\{\tilde{\mathbf{y}}_K(E_K\mathbf{y})^*\} = \mathbb{E}\{\tilde{\mathbf{y}}_K\hat{\mathbb{E}}\{(E_K\mathbf{y})^* \mid \check{\mathbf{Y}}_K^{\perp}\}\}.$$

However, since the nonzero components of $\hat{\mathbb{E}}\{E_K\mathbf{y} \mid \check{\mathbf{Y}}_K^{\perp}\}$ are those of $\tilde{\mathbf{y}}_K$, there is an $2N \times q$ matrix Π_K with the unit vectors e'_{k_i}, $i = 1, \dots, q$, as the rows such that

$$\hat{\mathbb{E}}\{E_K\mathbf{y} \mid \check{\mathbf{Y}}_K^{\perp}\} = \Pi_K\tilde{\mathbf{y}}_K,$$

and hence

$$\mathbb{E}\{\tilde{\mathbf{y}}_K(E_K\mathbf{y})^*\} = \mathbb{E}\{\tilde{\mathbf{y}}_K\tilde{\mathbf{y}}_K^*\}\Pi_K^* = \tilde{\boldsymbol{\Sigma}}_K\Pi_K^*.$$

Consequently, $G_K\boldsymbol{\Sigma} = \tilde{\boldsymbol{\Sigma}}_K\Pi_K^*$, i.e.,

$$G_K = \tilde{\boldsymbol{\Sigma}}_K\Pi_K^*\boldsymbol{\Sigma}^{-1}.$$

In the same way, $\tilde{\mathbf{y}}_J = G_J\mathbf{y}$, where G_J is the $q \times 2N$ matrix

$$G_J = \tilde{\boldsymbol{\Sigma}}_J\Pi_J^*\boldsymbol{\Sigma}^{-1},$$

and therefore

$$\mathbb{E}\{\tilde{\mathbf{y}}_J\tilde{\mathbf{y}}_K^*\} = \tilde{\boldsymbol{\Sigma}}_J\Pi_J^*\boldsymbol{\Sigma}^{-1}\Pi_K\tilde{\boldsymbol{\Sigma}}_K,$$

which is zero if and only if $\Pi_J^*\boldsymbol{\Sigma}^{-1}\Pi_K = 0$, i.e., $g_{jk} = 0$ for all $(j,k) \in J \times K$.

It remains to show that $\mathbb{E}\{\tilde{\mathbf{y}}_J\tilde{\mathbf{y}}_K^*\} = 0$ is equivalent to (64), which in view of (58), can be written

$$\mathbb{E}\left\{\hat{\mathbb{E}}\{\mathbf{y}_J \mid \check{\mathbf{Y}}_{J,K}\}\hat{\mathbb{E}}\{\mathbf{y}_K \mid \check{\mathbf{Y}}_{J,K}\}^*\right\} = \mathbb{E}\{\mathbf{y}_J\mathbf{y}_K^*\}.$$

However,

$$\mathbb{E}\{\tilde{\mathbf{y}}_J\tilde{\mathbf{y}}_K^*\} = \mathbb{E}\{\mathbf{y}_J\mathbf{y}_K^*\} - \mathbb{E}\left\{\hat{\mathbb{E}}\{\mathbf{y}_J \mid \check{\mathbf{Y}}_J\}\hat{\mathbb{E}}\{\mathbf{y}_K \mid \check{\mathbf{Y}}_K\}^*\right\},$$

so the proof will complete if we show that

$$\mathbb{E}\left\{\hat{\mathbb{E}}\{\mathbf{y}_J \mid \check{\mathbf{Y}}_J\}\hat{\mathbb{E}}\{\mathbf{y}_K \mid \check{\mathbf{Y}}_K\}^*\right\} = \mathbb{E}\left\{\hat{\mathbb{E}}\{\mathbf{y}_J \mid \check{\mathbf{Y}}_{J,K}\}\hat{\mathbb{E}}\{\mathbf{y}_K \mid \check{\mathbf{Y}}_{J,K}\}^*\right\} \tag{65}$$

the proof of which follows precisely the lines of Lemma 2.6.9 in [31, p. 56].

Taking J and K to be singletons we recover as a special case Dempster's original result [15].

To connect back to Definition 9 of a reciprocal process of order n, use the equivalent condition (60) so that, with $J = \{t\}$ and $K = [t-n,\, t+n]^c$, $\mathbf{y}_J = y(t)$ and $\mathbf{y}_K$ are conditionally orthogonal given $\check{\mathbf{Y}}_{J,K} = \mathbf{y}_{[t-n,t)} \vee \mathbf{y}_{(t,t+n]}$. Then $J \times K$ is the set $\{t \times [t-n,\, t+n]^c\,;\, t \in (-N, N]\}$, and hence Theorem 10 states precisely that the circulant matrix $\mathbf{G}$ is banded of order n. We stress that in general $\mathbf{G} = \boldsymbol{\Sigma}^{-1}$ is not banded, as the underlying process $\{y(t)\}$ is not reciprocal of degree n, and we then have an ARMA representation as explained in Sect. 5.

8 Determining P with the Help of Logarithmical Moments

We have shown that the solutions of the circulant rational covariance extension problem, as well as the corresponding bilateral ARMA models, are completely parameterized by $P \in \mathfrak{P}_+(N)$, or, equivalently, by their corresponding banded circulant matrices $\mathbf{P}$. This leads to the question of how to determine the $\mathbf{P}$ from given data.

To this end, suppose that we are also given the logarithmic moments

$$\gamma_k = \int_{-\pi}^{\pi} e^{ik\theta} \log \Phi(e^{i\theta}) d\nu, \quad k = 1, 2, \dots, n. \tag{66}$$

In the setting of the classical trigonometric moment problem such moments are known as *cepstral coefficients*, and in speech processing, for example, they are estimated from observed data for purposes of design.

Following [30] and, in the context of the trigonometric moment problem, [7, 10, 18, 34], we normalize the elements in $\mathfrak{P}_+(N)$ to define $\tilde{\mathfrak{P}}_+(N) := \{P \in \mathfrak{P}_+(N) \mid p_0 = 1\}$ and consider the problem to find a nonnegative integrable Φ maximizing

$$\mathbb{I}(\Phi) = \int_{-\pi}^{\pi} \log \Phi(e^{i\theta}) d\nu = \frac{1}{2N} \sum_{j=-N+1}^{N} \log \Phi(\zeta_j) \tag{67}$$

subject to the moment constraints (6) and (66). It is shown in [30] that if there is a maximal Φ that is positive on the unit circle, it is given by

$$\Phi(\zeta) = \frac{P(\zeta)}{Q(\zeta)}, \tag{68}$$

where (P, Q) is the unique solution of the dual problem to minimize

$$\mathbb{J}(P, Q) = \langle \mathbf{c}, \mathbf{q} \rangle - \langle \boldsymbol{\gamma}, \mathbf{p} \rangle + \int_{-\pi}^{\pi} P(e^{i\theta}) \log \left(\frac{P(e^{i\theta})}{Q(e^{i\theta})} \right) d\nu \tag{69}$$

over all $(P, Q) \in \tilde{\mathfrak{P}}_+(N) \times \mathfrak{P}_+(N)$, where $\boldsymbol{\gamma} = (\gamma_0, \gamma_1, \dots, \gamma_n)$ and $\mathbf{p} = (p_0, p_1, \dots, p_n)$ with $\gamma_0 = 0$ and $p_0 = 1$.

The problem is that the dual problem might have a minimizer on the boundary so that there is no stationery point in the interior, and then the constraints will in general not be satisfied [30]. Therefore the problem needs to be regularized in the style of [17]. More precisely, we consider the regularized problem to minimize

$$\mathbb{J}_\lambda(P,Q) = \mathbb{J}(P,Q) - \lambda \int_{-\pi}^{\pi} \log P(e^{i\theta}) d\nu \tag{70}$$

for some suitable $\lambda > 0$ over all $(P,Q) \in \tilde{\mathfrak{P}}_+(N) \times \mathfrak{P}_+(N)$. Setting $\mathbf{J}_\lambda(\mathbf{P},\mathbf{Q}) := 2N\mathbb{J}_\lambda(P,Q)$, (70) can be written

$$\mathbf{J}_\lambda(\mathbf{P},\mathbf{Q}) = \operatorname{tr}\{\mathbf{C}\mathbf{Q}\} - \operatorname{tr}\{\boldsymbol{\Gamma}\mathbf{P}\} + \operatorname{tr}\{\mathbf{P}\log \mathbf{P}\mathbf{Q}^{-1}\} - \lambda \operatorname{tr}\{\log \mathbf{P}\}, \tag{71}$$

where $\boldsymbol{\Gamma}$ is the Hermitian circulant matrix with symbol

$$\Gamma(\zeta) = \sum_{k=-n}^{n} \gamma_k \zeta^{-k}, \quad \gamma_{-k} = \bar{\gamma}_k. \tag{72}$$

Therefore, in the circulant matrix form, the regularized dual problem amounts to minimizing (71) over all banded Hermitian circulant matrices $\mathbf{P}$ and $\mathbf{Q}$ of order n subject to $p_0 = 1$. It is shown in [30] that

$$\boldsymbol{\Sigma} = \mathbf{Q}^{-1}\mathbf{P}, \tag{73}$$

or, equivalently in symbol form (68), maximizes

$$\mathbf{I}(\boldsymbol{\Sigma}) = \operatorname{tr}\{\log \boldsymbol{\Sigma}\} = \log\det \boldsymbol{\Sigma}, \tag{74}$$

or, equivalently (67), subject to (6) and (66), the latter constraint modified so that the logarithmic moment γ_k is exchanged for $\gamma_k + \varepsilon_k$, $k = 1, 2, \dots, n$, where

$$\varepsilon_k = \int_{-\pi}^{\pi} e^{ik\theta} \frac{\lambda}{\hat{P}(e^{i\theta})} d\nu = \frac{\lambda}{2N} \operatorname{tr}\{\mathbf{S}^k \hat{\mathbf{P}}^{-1}\}, \tag{75}$$

$\hat{P}$ being the optimal P.

The following example from [30], provided by Chiara Masiero, illustrates the advantages of this procedure. We start from an ARMA model with $n = 8$ poles and three zeros distributed as in Fig. 3, from which we compute $\mathbf{c} = (c_0, c_1, \dots, c_n)$ and $\boldsymbol{\gamma} = (\gamma_1, \dots, \gamma_n)$ for various choices of the order n. First we determine the maximum entropy solution from $\mathbf{c}$ with $n = 12$ and $N = 1024$. The resulting spectral function Φ is depicted in the top plot of Fig. 4 together with the true spectrum. Next we compute Φ by the procedure in this section using $\mathbf{c}$ and $\boldsymbol{\gamma}$ with $n = 8$ and $N = 128$. The result is depicted in the bottom plot of Fig. 4 again together with the true spectrum. This illustrates the advantage of bilateral ARMA modeling as compared to bilateral AR modeling, as a much lower value on N provides a better approximation, although n is smaller.

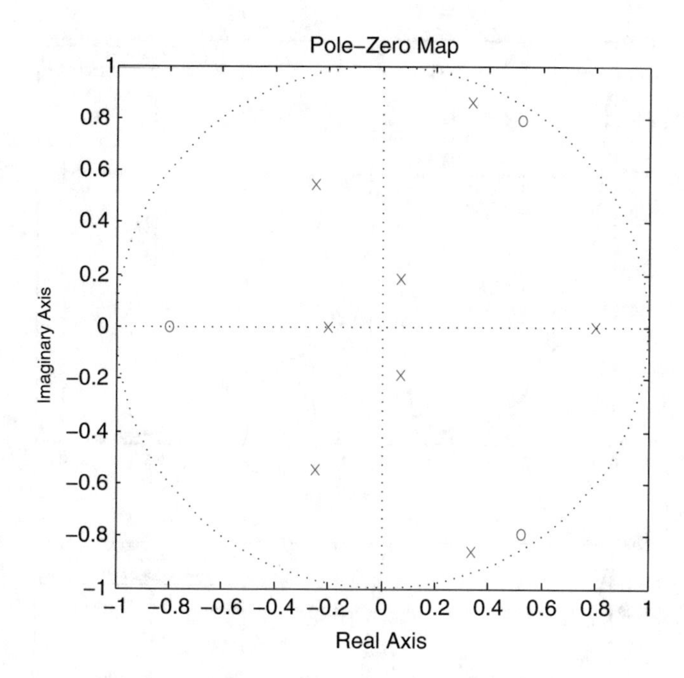

Fig. 3 Poles and zeros of true ARMA model

9 Extensions to the Multivariate Case

To simplify notation we have so far restricted our attention to scalar stationary periodic processes. We shall now demonstrate that most of the results can be simply extended to the multivariate case, provided we restrict the analysis to scalar pseudo-polynomials $P(\zeta)$. In fact, most of the equations in the previous section will remain intact if we allow ourselves to interpret the scalar quantities as matrix-valued ones.

Let $\{y(t)\}$ be a zero-mean stationary m-dimensional process defined on $\mathbb{Z}_{2N}$; i.e., a stationary process defined on a finite interval $[-N+1, N]$ of the integer line $\mathbb{Z}$ and extended to all of $\mathbb{Z}$ as a periodic stationary process with period $2N$. Moreover, let $C_{-N+1}, C_{-N+2}, \dots, C_N$ be the $m \times m$ covariance lags $C_k := \mathbb{E}\{y(t+k)y(t)^*\}$, and define its discrete Fourier transformation

$$\Phi(\zeta_j) := \sum_{k=-N+1}^{N} C_k \zeta_j^{-k}, \qquad j = -N+1, \dots, N, \tag{76}$$

which is a positive, Hermitian matrix-valued function of ζ. Then, by the inverse discrete Fourier transformation,

$$C_k = \frac{1}{2N} \sum_{j=-N+1}^{N} \zeta_j^k \Phi(\zeta_j) = \int_{-\pi}^{\pi} e^{ik\theta} \Phi(e^{i\theta}) d\nu, \quad k = -N+1, \dots, N, \tag{77}$$

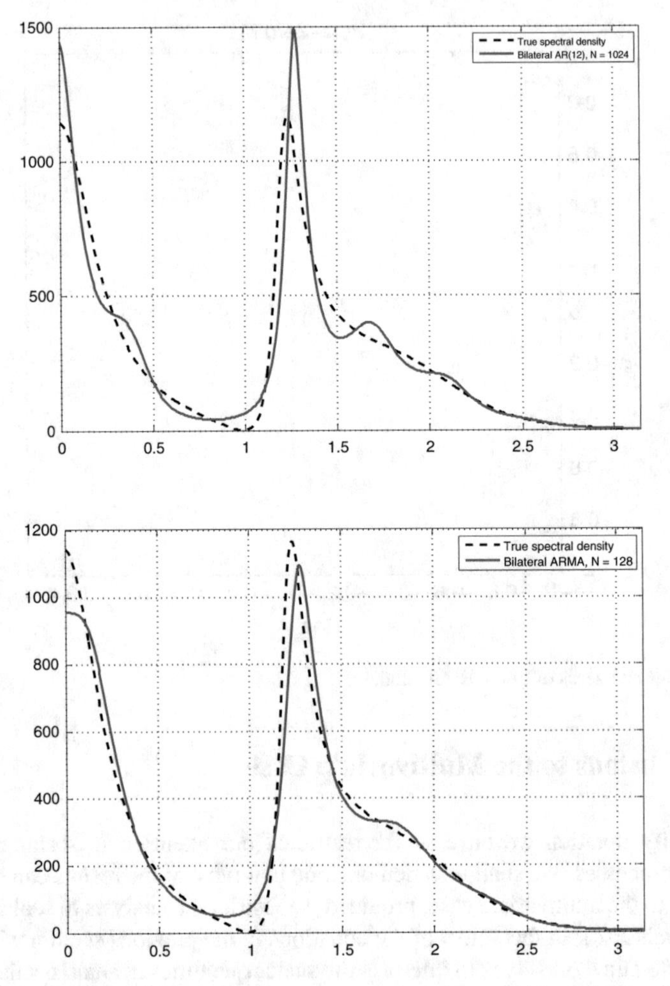

Fig. 4 Bilateral approximations with true spectrum (*dashed*): (*top*) bilateral AR with $n = 12$ and $N = 1024$; (*bottom*) bilateral ARMA with $n = 8$ and $N = 128$ using both covariance and logarithmic moment estimates

where the Stieljes measure $d\nu$ is given by (7). The $m \times m$ matrix function Φ is the *spectral density* of the vector process y. In fact, let

$$\hat{y}(\zeta_k) := \sum_{t=-N+1}^{N} y(t)\zeta_k^{-t}, \quad k = -N+1, \ldots, N, \tag{78}$$

be the discrete Fourier transformation of the process y. Since

$$\frac{1}{2N} \sum_{t=-N+1}^{N} (\zeta_k \zeta_\ell^*)^t = \delta_{k\ell}$$

by (8), the random variables (78) are uncorrelated, and

$$\frac{1}{2N}\mathbb{E}\{\hat{y}(\zeta_k)\hat{y}(\zeta_\ell)^*\} = \Phi(\zeta_k)\delta_{k\ell}. \tag{79}$$

This yields a spectral representation of y analogous to the usual one, namely

$$y(t) = \frac{1}{2N}\sum_{k=-N+1}^{N} \zeta_k^t\,\hat{y}(\zeta_k) = \int_{-\pi}^{\pi} e^{ik\theta}\,d\hat{y}(\theta), \tag{80}$$

where $d\hat{y} := \hat{y}(e^{i\theta})d\nu$.

Next, we define the class $\mathfrak{P}_+^{(m,n)}(N)$ of $m \times m$ Hermitian pseudo-polynomials

$$Q(\zeta) = \sum_{k=-n}^{n} Q_k\zeta^{-k}, \quad Q_{-k} = Q_k^* \tag{81}$$

of degree at most n that are positive definite on the discrete unit circle $\mathbb{T}_{2N}$, and let $\mathfrak{P}_+^{(m,n)} \subset \mathfrak{P}_+^{(m,n)}(N)$ be the subset of all (81) such that $Q(e^{i\theta})$ is positive define for all $\theta \in [-\pi, \pi]$. Moreover let $\mathfrak{C}_+^{(m,n)}(N)$ be the dual cone of all $C = (C_0, C_1, \ldots, C_n)$ such that

$$\langle C, Q\rangle := \sum_{k=-n}^{n} \operatorname{tr}\{C_k Q_k^*\} > 0 \quad \text{for all } Q \in \overline{\mathfrak{P}_+^{(m,n)}(N)} \setminus \{0\},$$

and let $\mathfrak{C}_+^{(m,n)} \supset \mathfrak{C}_+^{(m,n)}(N)$ be defined as the dual cone of $\mathfrak{P}_+^{(m,n)}$. Analogously to the scalar case it can be shown that $C \in \mathfrak{C}_+^{(m,n)}$ if and only if the block-Toeplitz matrix

$$\mathbf{T}_n = \begin{bmatrix} C_0 & C_1^* & C_2^* & \cdots & C_n^* \\ C_1 & C_0 & C_1^* & \cdots & C_{n-1}^* \\ C_2 & C_1 & C_0 & \cdots & C_{n-2}^* \\ \vdots & \vdots & \vdots & \ddots & \vdots \\ C_n & C_{n-1} & C_{n-2} & \cdots & C_0 \end{bmatrix} \tag{82}$$

is positive definite [32], a condition that is necessary, but in general not sufficient, for $C \in \mathfrak{C}_+^{(m,n)}(N)$ to hold.

The basic problem is the following. Given the sequence $C = (C_0, C_1, \ldots, C_n) \in \mathfrak{C}_+^{(m,n)}(N)$ of $m \times m$ covariance lags, find an extension $C_{n+1}, C_{n+2}, \ldots, C_N$ with $C_{-k} = C_k^*$ such that the spectral function Φ defined by (76) has the rational form

$$\Phi(\zeta) = P(\zeta)Q(\zeta)^{-1}, \quad P \in \mathfrak{P}_+^{(1,n)}(N),\ Q \in \mathfrak{P}_+^{(m,n)}(N). \tag{83}$$

Theorem 11. *Let $C \in \mathfrak{C}_+^{(m,n)}(N)$. Then, for each $P \in \mathfrak{P}_+^{(1,n)}(N)$, there is a unique $Q \in \mathfrak{P}_+^{(m,n)}(N)$ such that*

$$\Phi = PQ^{-1} \tag{84}$$

satisfies the moment conditions

$$\int_{-\pi}^{\pi} e^{ik\theta} \Phi(e^{i\theta}) d\nu = C_k, \quad k = 0, 1, \ldots, n. \tag{85}$$

Theorem 11 is a direct consequence of the following theorem, which also provides an algorithm for computing the solution.

Theorem 12. *For each $(C, P) \in \mathfrak{C}_+^{(m,n)}(N) \times \mathfrak{P}_+^{(1,n)}(N)$, the problem to maximize the functional*

$$\mathbb{I}_P(\Phi) = \int_{-\pi}^{\pi} P(e^{i\theta}) \log\det \Phi(e^{i\theta}) d\nu \tag{86}$$

subject to the moment conditions (85) *has a unique solution $\hat{\Phi}$, and it has the form*

$$\hat{\Phi}(\zeta) = P(\zeta)\hat{Q}(\zeta)^{-1}, \tag{87}$$

where $\hat{Q} \in \mathfrak{P}_+^{(m,n)}(N)$ is the unique solution to the dual problem to minimize

$$\mathbb{J}_P(Q) = \langle C, Q \rangle - \int_{-\pi}^{\pi} P(e^{i\theta}) \log\det Q(e^{i\theta}) d\nu \tag{88}$$

over all $Q \in \mathfrak{P}_+^{(m,n)}(N)$.

The proofs of Theorems 11 and 12 follow the lines of [32]. It can also be shown that the moment map sending $Q \in \mathfrak{P}_+^{(m,n)}(N)$ to $C \in \mathfrak{C}_+^{(m,n)}(N)$ is a diffeomorphism.

To formulate a matrix version of Theorems 11 and 12 we need to introduce (Hermitian) block-circulant matrices

$$\mathbf{M} = \sum_{k=-N+1}^{N} S^{-k} \otimes M_k, \quad M_{-k} = M_k^* \tag{89}$$

where $\otimes$ is the Kronecker product and S is the nonsingular $2N \times 2N$ cyclic shift matrix (19). The notation $\mathbf{S}$ will now be reserved for the $2mN \times 2mN$ block-shift matrix

$$\mathbf{S} = S \otimes I_m = \begin{bmatrix} 0 & I_m & 0 & \ldots & 0 \\ 0 & 0 & I_m & \ldots & 0 \\ \vdots & \vdots & \vdots & \ddots & \vdots \\ 0 & 0 & 0 & 0 & I_m \\ I_m & 0 & 0 & 0 & 0 \end{bmatrix}. \tag{90}$$

As before $\mathbf{S}^{2N} = \mathbf{S}^0 = \mathbf{I} := I_{2mN}$, $\mathbf{S}^{k+2N} = \mathbf{S}^k$, and $\mathbf{S}^{2N-k} = \mathbf{S}^{-k} = (\mathbf{S}^k)^{\mathsf{T}}$. Moreover

$$\mathbf{SMS}^* = \mathbf{M} \tag{91}$$

is both necessary and sufficient for $\mathbf{M}$ to be $m \times m$ block-circulant. The symbol of $\mathbf{M}$ is the $m \times m$ pseudo-polynomial

$$M(\zeta) = \sum_{k=-N+1}^{N} M_k \zeta^{-k}, \quad M_{-k} = M_k^*. \tag{92}$$

We shall continue using the notation

$$\mathbf{M} := \text{Circ}\{M_0, M_1, M_2, \ldots, M_N, M_{N-1}^*, \ldots, M_1^*\} \tag{93}$$

also for (Hermitain) block-circulant matrices.

The problem can now be reformulated in the following way. Given the banded block-circulant matrix

$$\mathbf{C} = \sum_{k=-n}^{n} S^{-k} \otimes C_k, \quad C_{-k} = C_k^* \tag{94}$$

of order n, find an extension $C_{n+1}, C_{n+2}, \ldots, C_N$ such that the block-circulant matrix

$$\boldsymbol{\Sigma} = \sum_{k=-N+1}^{N} S^{-k} \otimes C_k, \quad C_{-k} = C_k^* \tag{95}$$

has the symbol (83).

To proceed we need a block-circulant version of Lemma 5.

Lemma 13. *Let* $\mathbf{M}$ *be a block-circulant matrix with symbol* $M(\zeta)$. *Then*

$$\mathbf{M} = \mathbf{F}^* \text{diag}\big(M(\zeta_{-N+1}), M(\zeta_{-N+2}), \ldots, M(\zeta_N)\big)\mathbf{F}, \tag{96}$$

where $\mathbf{F}$ *is the unitary* $2mN \times 2mN$ *matrix*

$$\mathbf{F} = \frac{1}{\sqrt{2N}} \begin{bmatrix} \zeta_{-N+1}^{N-1} I_m & \zeta_{-N+1}^{N-2} I_m & \cdots & \zeta_{-N+1}^{-N} I_m \\ \vdots & \vdots & \cdots & \vdots \\ \zeta_0^{N-1} I_m & \zeta_0^{N-2} I_m & \cdots & \zeta_0^{-N} I_m \\ \vdots & \vdots & \cdots & \vdots \\ \zeta_N^{N-1} I_m & \zeta_N^{N-2} I_m & \cdots & \zeta_N^{-N} I_m \end{bmatrix}. \tag{97}$$

Moreover, if $M(\zeta_k)$ is positive definite for all k, then

$$\log \mathbf{M} = \mathbf{F}^* \text{diag}\big(\log M(\zeta_{-N+1}), \log M(\zeta_{-N+2}), \dots, \log M(\zeta_N)\big)\mathbf{F}, \tag{98}$$

where diag stands for block diagonal.

The proof of Lemma 13 will be omitted, as it follows the same lines as that of Lemma 5 with straight-forward modification to the multivariate case. Clearly the inverse

$$\mathbf{M}^{-1} = \mathbf{F}^* \text{diag}\big(M(\zeta_{-N+1})^{-1}, M(\zeta_{-N+2})^{-1}, \dots, M(\zeta_N)^{-1}\big)\mathbf{F} \tag{99}$$

is also block-circulant, and

$$\mathbf{S} = \mathbf{F}^* \text{diag}\big(\zeta_{-N+1} I_m, \zeta_{-N+2} I_m, \dots, \zeta_N I_m\big)\mathbf{F}. \tag{100}$$

However, unlike the scalar case, block-circulant matrices do not commute in general.

Given Lemma 13, we are now in a position to reformulate Theorems 11 and 12 in matrix from.

Theorem 14. *Let $C \in \mathfrak{C}_+^{(m,n)}(N)$, and let* $\mathbf{C}$ *be the corresponding block-circulant matrix* (94) *and* (82) *the corresponding block-Toeplitz matrix. Then, for each positive-definite banded $2mN \times 2mN$ block-circulant matrices*

$$\mathbf{P} = \sum_{k=-n}^{n} S^{-k} \otimes p_k I_m, \quad p_{-k} = \bar{p}_k \tag{101}$$

of order n, where $P(\zeta) = \sum_{k=-n}^{n} p_k \zeta^{-k} \in \mathfrak{P}_+^{(1,n)}(N)$, there is a unique sequence $Q = (Q_0, Q_1, \dots, Q_n)$ of $m \times m$ matrices defining a positive-definite banded $2mN \times 2mN$ block-circulant matrix

$$\mathbf{Q} = \sum_{k=-n}^{n} S^{-k} \otimes Q_k, \quad Q_{-k} = Q_k^* \tag{102}$$

of order n such that

$$\boldsymbol{\Sigma} = \mathbf{Q}^{-1}\mathbf{P} \tag{103}$$

is a block-circulant extension (95) *of* $\mathbf{C}$. *The block-circulant matrix* (103) *is the unique maximizer of the function*

$$\mathscr{I}_{\mathbf{P}}(\boldsymbol{\Sigma}) = \text{tr}(\mathbf{P} \log \boldsymbol{\Sigma}) \tag{104}$$

subject to

$$\mathbf{E}_n{}^T \boldsymbol{\Sigma} \mathbf{E}_n = \mathbf{T}_n, \quad \textit{where } \mathbf{E}_n = \begin{bmatrix} \mathbf{I}_{mn} \\ \mathbf{0} \end{bmatrix}. \tag{105}$$

Moreover, $\mathbf{Q}$ *is the unique optimal solution of the problem to minimize*

$$\mathscr{J}_{\mathbf{P}}(\mathbf{Q}) = \operatorname{tr}(\mathbf{CQ}) - \operatorname{tr}(\mathbf{P} \log \mathbf{Q}) \tag{106}$$

over all positive-definite banded $2mN \times 2mN$ *block-circulant matrices* (102) *of order* n. *The functional* $\mathscr{J}_{\mathbf{P}}$ *is strictly convex.*

For $\mathbf{P} = \mathbf{I}$ we obtain the maximum-entropy solution considered in [12], where the primal problem to maximize $\mathscr{I}_{\mathbf{I}}$ subject to (105) was presented. In [12] there was also an extra constraint (91), which, as we can see, is not needed, since it is automatically fulfilled. For this reason the dual problem presented in [12] is more complicated than merely minimizing $\mathscr{J}_{\mathbf{I}}$.

Next suppose we are also given the (scalar) logarithmic moments (66) and that $C \in \mathfrak{C}_+^{(m,n)}(N)$. Then, if the problem to maximize $\operatorname{tr}\{\log \boldsymbol{\Sigma}\}$ subject to (105) and (66) over all positive-definite block-circulant matrices (95) has a solution, then it has the form

$$\boldsymbol{\Sigma} = \mathbf{Q}^{-1}\mathbf{P} \tag{107}$$

where the $(\mathbf{P}, \mathbf{Q})$ is a solution of the dual problem to minimize

$$\mathbf{J}(\mathbf{P}, \mathbf{Q}) = \operatorname{tr}\{\mathbf{CQ}\} - \operatorname{tr}\{\boldsymbol{\Gamma}\mathbf{P}\} + \operatorname{tr}\{\mathbf{P} \log \mathbf{P}\mathbf{Q}^{-1}\}, \tag{108}$$

over all positive-definite block-circulant matrices of the type (101) and (102) with the extra constrain $p_0 = 1$, where $\boldsymbol{\Gamma}$ is the block-circulant matrix formed in the style of (102) from

$$\Gamma(\zeta) = \sum_{k=-n}^{n} \gamma_k \zeta^{-k}, \quad \gamma_{-k} = \bar{\gamma}_k. \tag{109}$$

However, the minimum of (108) may end up on the boundary, in which case the constraint (66) may fail to be satisfied. Therefore, as in the scalar case, we need to regularize the problem by instead minimizing

$$\mathbf{J}_\lambda(\mathbf{P}, \mathbf{Q}) = \operatorname{tr}\{\mathbf{CQ}\} - \operatorname{tr}\{\boldsymbol{\Gamma}\mathbf{P}\} + \operatorname{tr}\{\mathbf{P} \log \mathbf{P}\mathbf{Q}^{-1}\} - \lambda \operatorname{tr}\{\log \mathbf{P}\}. \tag{110}$$

This problem has a unique optimal solution (107) satisfying (105), but not (66). The appropriate logarithmic moment constraint is obtained as in the scalar case by exchanging γ_k for $\gamma_k + \varepsilon_k$ for each $k = 1, 2, \ldots, n$, where ε_k is given by (75).

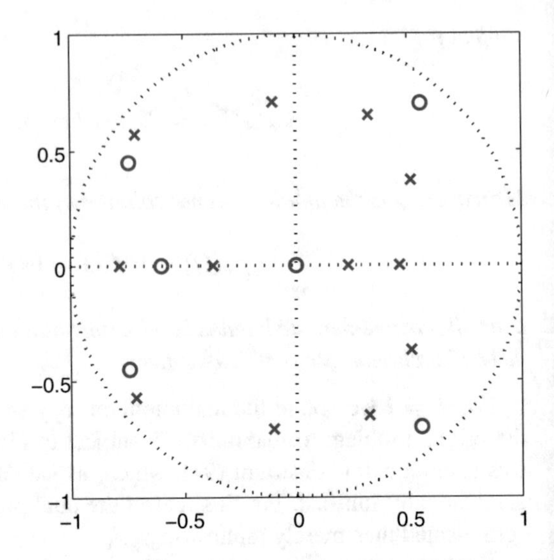

Fig. 5 Poles and zeros of an ARMA 2×2 model of order $n = 6$

Again each solution leads to an ARMA model

$$\sum_{k=-n}^{n} Q_k y(t-k) = \sum_{k=-n}^{n} p_k e(t-k), \tag{111}$$

where $\{e(t)\}$ is the conjugate process of $\{y(t)\}$, $Q_0, Q_1, \dots, Q_n$ are $m \times m$ matrices, whereas $p_0, p_1, \dots, p_n$ are scalar with $p_0 = 1$.

We illustrate this theory with a simple example from [32], where a covariance sequence $C := (C_0, C_1, \dots C_n)$ and a cepstral sequence $\boldsymbol{\gamma} := (\gamma_1, \gamma_2, \dots, \gamma_n)$ have been computed from a two-dimensional ARMA process with a spectral density $\Phi := PQ^{-1}$, where P is a scalar pseudo-polynomial of degree three and Q is a 2×2 matrix-valued pseudo-polynomial of degree $n = 6$. Its zero and poles are illustrated in Fig. 5.

Given C and $\boldsymbol{\gamma}$, we apply the procedure in this section to determine a pair $(\mathbf{P}, \mathbf{Q})$ of order $n = 6$. For comparison we also compute an bilateral AR approximation with $n = 12$ fixing $\mathbf{P} = \mathbf{I}$. As illustrated in Fig. 6, the bilateral ARMA model of order $n = 6$ computed with $N = 32$ outperforms the bilateral AR model of order $n = 12$ with $N = 64$.

The results of Sect. 5 can also be generalized to the multivariate case along the lines described in [11].

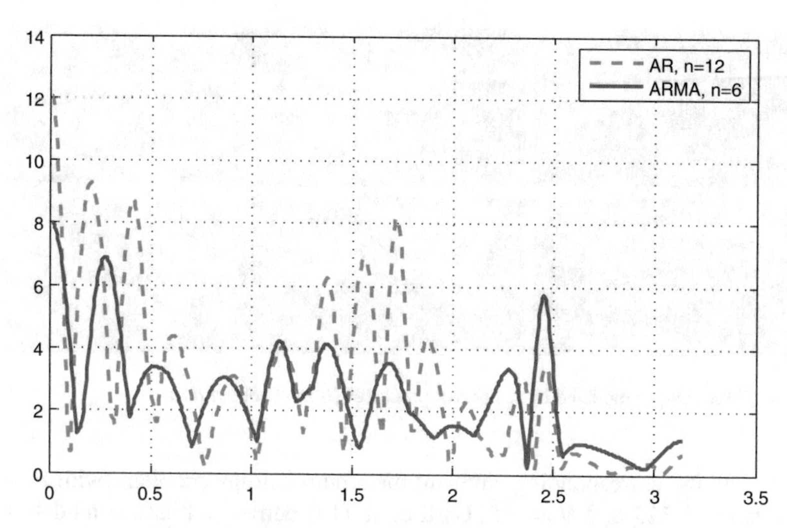

Fig. 6 The norm of the approximation error for a bilateral AR of order 12 for $N = 64$ and a bilateral ARMA of order 6 for $N = 32$

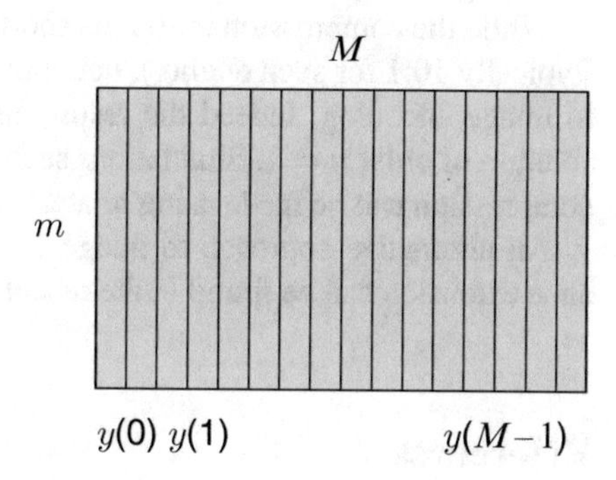

Fig. 7 An image modeled as a reciprocal vector process

10 Application to Image Processing

In [12] the circulant maximum-entropy solution has been used to model spatially stationary images (*textures*) [40] in terms of (vector-valued) stationary periodic processes. The image could be thought of as an $m \times M$ matrix of pixels where the columns form a m-dimensional reciprocal process $\{y(t)\}$, which can extended to a periodic process with period $M > N$ outside the interval $[0, N]$; see Fig. 7.

This imposes the constraint $C_{M-k} = C_k{}^{\mathsf{T}}$ on the covariance lags $C_k := E\{y(t+k)y(t)^{\mathsf{T}}\}$, leading to a circulant Toeplitz matrix. The problem considered in [12] is to model the process $\{y(t)\}$ given (estimated) $C_0, C_1, \dots, C_n$, where $n < N$ with an efficient low-dimensional model. This is precisely a problem of the type considered in Sect. 9.

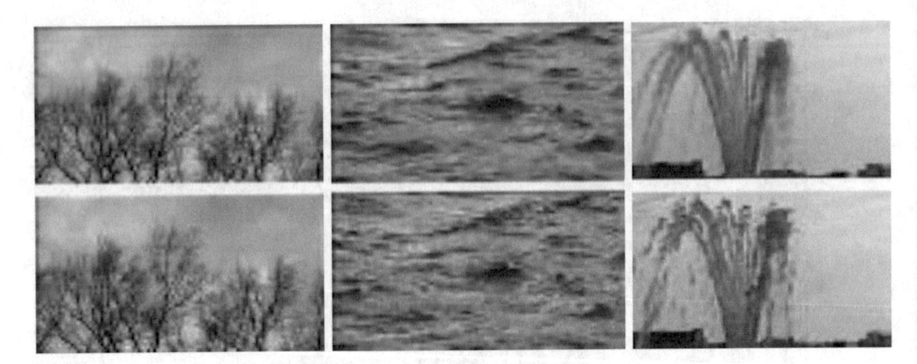

Fig. 8 Three images modeled by reciprocal processes (original at *bottom*)

Solving the corresponding circulant maximum-entropy problem (with $\mathbf{P} = \mathbf{I}$), $n = 1$, $m = 125$ and $N = 88$, Carli et al. [12] derived a bilateral model of the images at the bottom row of Fig. 8 to compress the images in the top row, thereby achieving a compression of 5:1.

While the compression ratio falls short of competing with current jpeg standards (typically 10:1 for such quality), our approach suggests a new stochastic alternative to image encoding. Indeed the results in Fig. 8 apply just the maximum entropy solution of order $n = 1$. Simulations such as those in Fig. 4 suggest that much better compression can be made using bilateral ARMA modeling.

An alternative approach to image compression using multidimensional covariance extension can be found in the recent paper [39].

References

1. Blomqvist, A., Lindquist, A., Nagamune, R.: Matrix-valued Nevanlinna-Pick interpolation with complexity constraint: an optimization approach. IEEE Trans. Autom. Control **48**, 2172–2190 (2003)
2. Byrnes, C.I., Lindquist, A.: On the partial stochastic realization problem. IEEE Trans. Autom. Control **42**, 1049–1069 (1997)
3. Byrnes, C.I., Lindquist, A.: The generalized moment problem with complexity constraint. Integr. Equ. Oper. Theory **56**, 163–180 (2006)
4. Byrnes, C.I., Lindquist, A.: The moment problem for rational measures: convexity in the spirit of Krein. In: Modern Analysis and Application: Mark Krein Centenary Conference, Vol. I: Operator Theory and Related Topics, Book Series: Operator Theory Advances and Applications, vol. 190, pp. 157–169. Birkhäuser, Basel (2009)
5. Byrnes, C.I., Lindquist, A, Gusev, S.V., Matveev, A.V.: A complete parameterization of all positive rational extensions of a covariance sequence. IEEE Trans. Autom. Control **40**, 1841–1857 (1995)
6. Byrnes, C.I., Gusev, S.V., Lindquist, A.: A convex optimization approach to the rational covariance extension problem. SIAM J. Contr. Opt. **37**, 211–229 (1999)
7. Byrnes, C.I., Enqvist, P., Lindquist, A.: Cepstral coefficients, covariance lags and pole-zero models for finite datastrings. IEEE Trans. Signal Process **50**, 677–693 (2001)

8. Byrnes, C.I., Georgiou, T.T., Lindquist, A.: A generalized entropy criterion for Nevanlinna-Pick interpolation with degree constraint. IEEE Trans. Autom. Control **45**, 822–839 (2001)
9. Byrnes, C.I., Gusev, S.V., Lindquist, A.: From finite covariance windows to modeling filters: a convex optimization approach. SIAM Rev. **43**, 645–675 (2001)
10. Byrnes, C.I., Enqvist, P., Lindquist, A.: Identifiability and well-posedness of shaping-filter parameterizations: a global analysis approach. SIAM J. Control Optim. **41**, 23–59 (2002)
11. Picci, G.: A new approach to circulant band extension: Proc. 22nd International Symposium on the Mathematical Theory of Networks and Systems (MTNS), July 11–15, Minneapolis, MN, USA, pp 123–130 (2016)
12. Carli, F.P., Ferrante, A., Pavon, M., Picci, G.: A maximum entropy solution of the covariance extension problem for reciprocal processes. IEEE Trans. Autom. Control **56**, 1999–2012 (2011)
13. Chiuso, A., Ferrante, A., Picci, G.: Reciprocal realization and modeling of textured images. In: Proceedings of the 44rd IEEE Conference on Decision and Control (2005)
14. Davis, P.: Circulant Matrices. Wiley, New York (1979)
15. Dempster, A.P.: Covariance selection. Biometrics **28**(1), 157–175 (1972)
16. Dunford, N., Schwartz, J.T.: Linear Operators, Part I: General Theory. Wiley, New York (1958)
17. Enqvist, P.: Spectral estimation by Geometric, Topological and Optimization Methods, PhD thesis, Optimization and Systems Theory, KTH, Stockholm, (2001)
18. Enqvist, P.: A convex optimization approach to ARMA(n,m) model design from covariance and cepstrum data. SIAM J. Control Optim. **43**(3), 1011–1036 (2004)
19. Georgiou, T.T.: Partial Realization of Covariance Sequences, Ph.D. thesis, CMST, University of Florida, Gainesville (1983)
20. Georgiou, T.T.: Realization of power spectra from partial covariances, IEEE Trans. Acoust. Speech Signal Process. **35**, 438–449 (1987)
21. Georgiou, T.T.: Solution of the general moment problem via a one-parameter imbedding. IEEE Trans. Autom. Control **50**, 811–826 (2005)
22. Georgiou, T.T., Lindquist, A.: Kullback-Leibler approximation of spectral density functions. IEEE Trans. Inform. Theory **49**, 2910–2917 (2003)
23. Jamison, B.: Reciprocal processes. Zeitschrift. Wahrsch. Verw. Gebiete **30**, 65–86 (1974)
24. Kalman, R.E.: Realization of covariance sequences. In: Proceedings of the Toeplitz Memorial Conference, Tel Aviv (1981)
25. Krein, M.G., Nudelman,A.A.: The Markov Moment Problem and Extremal Problems. American Mathematical Society, Providence, RI (1977)
26. Krener, A.J.: Reciprocal processes and the stochastic realization problem for acausal systems, In: Byrnes, C.I., Lindquist, A. (eds.) Modeling Identification and Robust Control, pp. 197–211. North-Holland, Amsterdam (1986)
27. Krener, A.J., Frezza, R., Levy, B.C.: Gaussian reciprocal processes and self-adjoint differential equations of second order. Stoch. Stoch. Rep. **34**, 29–56 (1991)
28. Levy, B.C., Ferrante, A.: Characterization of stationary discrete-time Gaussian reciprocal processes over a finite interval, SIAM J. Matrix Anal. Appl. **24**, 334–355 (2002)
29. Levy, B.C., Frezza, R., Krener, A.J.: Modeling and estimation of discrete-time Gaussian reciprocal processes. IEEE Trans. Autom. Contr. **35**, 1013–1023 (1990)
30. Lindquist, A., Picci, G.: The circulant rational covariance extension problem: the complete solution. IEEE Trans. Autom. Control **58**, 2848–2861 (2013)
31. Lindquist, A. Picci, G.: Linear Stochastic Systems: A Geometric Approach to Modeling, Estimation and Identification. Springer, Heidelberg, New York, Dordrecht, London (2015)
32. Lindquist, A. Masiero, C., Picci, G.: On the multivariate circulant rational covariance extension problem. In: Proceeding of the 52st IEEE Conference on Decision and Control (2013)
33. Masani, P.: The prediction theory of multivariate stochastic processes, III. Acta Math. **104**, 141–162 (1960)
34. Musicus, B.R., Kabel, A.M.: Maximum entropy pole-zero estimation, Technical Report 510, MIT Research Lab. Electronics, Aug. 1985; now available on the internet at http://dspace.mit.edu/bitstream/handle/1721.1/4233/RLE-TR-510-17684936.pdf

35. Pavon, M., Ferrante, A.: On the geometry of maximum entropy problems. SIAM Rev. **55**(3), 415–439 (2013)
36. Picci, G., Carli, F.: Modelling and simulation of images by reciprocal processes. In: Proceedings of the Tenth International Conference on Computer Modeling and Simulation UKSIM, pp. 513–518 (2008)
37. Ringh, A., Karlsson, J.: A fast solver for the circulant rational covariance extension problem. In: European Control Conference (ECC), July 2015, pp. 727–733 (2015)
38. Ringh, A., Lindquist, A.: Spectral estimation of periodic and skew periodic random signals and approximation of spectral densities. In: 33rd Chinese Control Conference (CCC), pp. 5322–5327 (2014)
39. Ringh A., Karlsson, J., Lindquist, A: Multidimensional rational covariance extension with applications to spectral estimation and image compression. SIAM J. Control Optim. **54**(4), 1950–1982 (2016)
40. Soatto, S., Doretto, G., Wu, Y.: Dynamic textures. In: Proceedings of the International Conference Computer Vision, July 2001, pp. 439–446

A New Two-Step Proximal Algorithm of Solving the Problem of Equilibrium Programming

Sergey I. Lyashko and Vladimir V. Semenov

Dedicated to Boris Polyak on the occasion of his 80th Birthday

Abstract We propose a new iterative two-step proximal algorithm for solving the problem of equilibrium programming in a Hilbert space. This method is a result of extension of L.D. Popov's modification of Arrow-Hurwicz scheme for approximation of saddle points of convex-concave functions. The convergence of the algorithm is proved under the assumption that the solution exists and the bifunction is pseudo-monotone and Lipschitz-type.

Keywords Equilibrium problem • Variational inequality • Two-step proximal algorithm • Bifunction • Pseudomonotonicity • Lipschitz condition • Convergence

1 Introduction

Throughout this chapter, we assume that H is a real Hilbert space with inner product $(\cdot,\cdot)$ and norm $\|\cdot\|$. The symbol $\rightharpoonup$ denote weak convergence.

Let C be a nonempty closed convex subset of H and $F : C \times C \to \mathbb{R}$ be a bifunction with $F(x,x) = 0$ for all $x \in C$. Consider the following equilibrium problem in the sense of Blum and Oettli [12]:

$$\text{find } x \in C \text{ such that } \quad F(x,y) \geq 0 \quad \forall\, y \in C. \tag{1}$$

The equilibrium problem (1) (problem of equilibrium programming, Ky Fan inequality) is very general in the sense that it includes, as special cases, many applied mathematical models such as: variational inequalities, fixed point problems, optimization problems, saddle point problems, Nash equilibrium point problems in

S.I. Lyashko • V.V. Semenov (✉)
Department of Computational Mathematics, Kiev National Taras Shevchenko University, 64, Vladimirskaya Str., Kiev 01601, Ukraine
e-mail: lyashko.serg@gmail.com; semenov.volodya@gmail.com

B. Goldengorin (ed.), *Optimization and Its Applications in Control and Data Sciences*, Springer Optimization and Its Applications 115,
DOI 10.1007/978-3-319-42056-1_10

non-cooperative games, complementarity problems, see [3, 5, 6, 10, 12, 14, 17, 25] and the references therein. This problem is interesting because it allows to unify all these particular problems in a convenient way. In recent years, many methods have been proposed for solving equilibrium and related problems [2–10, 14, 16, 26, 28, 32, 35–37]. The solution approximation methods for the equilibrium problem are often based on the resolvent of equilibrium bifunction (see, for instance [14]) where at each iterative step a strongly monotone regularization equilibrium problem is solved. It is also called the proximal point method [16, 18, 20, 26, 37].

The variational inequality problem is a special case of the equilibrium problem. For solving the variational inequality in Euclidean space, Korpelevich [21] introduced the extragradient method where two metric projections onto feasible sets must be found at each iterative step. This method was setted in Hilbert spaces by Nadezhkina and Takahashi [27]. Some extragradient-like algorithms proposed for solving variational inequality problems can be found in [19, 33, 34, 38]. In 2011, the authors in [13, 22] have replaced the second projection onto any closed convex set in the extragradient method by one onto a half-space and proposed the subgradient extragradient method for variational inequalities in Hilbert spaces, see also [15, 39].

In recent years, the extragradient method has been extended to equilibrium problems for monotone (more general, pseudomonotone) and Lipschitz-type continuous bifunctions and studied both theoretically and algorithmically [1, 31, 40]. In this methods we must solve two strongly convex minimization problems on a closed convex constrained set at each iterative step. We note that similar methods have been previously proposed and studied by Antipin [2–4].

In 1980, Russian mathematician Popov [30] introduced very interesting modification of Arrow-Hurwicz scheme for approximation of saddle points of convex-concave functions in Euclidean space. Let X and Y are closed convex subset of Euclidean spaces $\mathbb{R}^d$ and $\mathbb{R}^p$, respectively, and $L : X \times Y \to \mathbb{R}$ be a differentiable convex-concave function. Then, the method [30] approximation of saddle points of L on $X \times Y$ can be written as

$$\begin{cases} x_1, \bar{x}_1 \in X, \ \ y_1, \bar{y}_1 \in Y, \ \ \lambda > 0, \\ x_{n+1} = P_X\left(x_n - \lambda L_1'(\bar{x}_n, \bar{y}_n)\right), \ y_{n+1} = P_Y\left(y_n + \lambda L_2'(\bar{x}_n, \bar{y}_n)\right), \\ \bar{x}_{n+1} = P_X\left(x_{n+1} - \lambda L_1'(\bar{x}_n, \bar{y}_n)\right), \ \bar{y}_{n+1} = P_Y\left(y_{n+1} + \lambda L_2'(\bar{x}_n, \bar{y}_n)\right), \end{cases}$$

where P_X and P_Y are metric projection onto X and Y, respectively, L_1' and L_2' are partial derivatives. Under some suitable assumptions, Popov proved the convergence of this method.

In this chapter, we have been motivated and inspired by the results of the authors in [30, 31], proposed a new two-step proximal algorithm for solving equilibrium problems. This algorithm is the extension of Popov method [30].

The set of solutions of the equilibrium problem (1) is denoted $EP(F, C)$. Further, we assume that the solution set $EP(F, C)$ is nonempty.

Here, for solving equilibrium problem (1), we assume that the bifunction F satisfies the following conditions:

(A1) $F(x,x) = 0$ for all $x \in C$;
(A2) for all $x, y \in C$ from $F(x,y) \geq 0$ it follows that $F(y,x) \leq 0$ (pseudo-monotonicity);
(A3) for all $x \in C$ the function $F(x,\cdot)$ is convex and lower semicontinuous on C;
(A4) for all $y \in C$ the function $F(\cdot,y)$ is weakly upper semicontinuous on C;
(A5) for all $x, y, z \in C$ the next inequality holds

$$F(x,y) \leq F(x,z) + F(z,y) + a\,\|x-z\|^2 + b\,\|z-y\|^2,$$

where a, b are positive constants (Lipschitz-type continuity);
(A6) for all bounded sequences (x_n), (y_n) from C we have

$$\|x_n - y_n\| \to 0 \;\Rightarrow\; F(x_n, y_n) \to 0.$$

It is easy to show that under the assumptions (A1)–(A4), we have

$$x \in EP(F,C) \quad \Leftrightarrow \quad x \in C: \; F(y,x) \leq 0 \;\; \forall\, y \in C.$$

In particular, the set $EP(F,C)$ is convex and closed (see, for instance [31]).

The hypothesis (A5) was introduced by Mastroeni [25]. It is necessary to imply the convergence of the auxiliary principle method for equilibrium problems. For example, the bifunction $F(x,y) = (Ax, y-x)$ with k-Lipschitz operator $A: C \to H$ satisfies (A5). Actually,

$$\begin{aligned} F(x,y) - F(x,z) - F(z,y) &= (Ax, y-x) - (Ax, z-x) - (Az, y-z) = \\ &= (Ax - Az, y-z) \leq \|Ax - Az\|\,\|y-z\| \leq k\,\|x-z\|\,\|y-z\| \leq \\ &\leq \frac{k}{2}\,\|x-z\|^2 + \frac{k}{2}\,\|y-z\|^2. \end{aligned}$$

This implies that F satisfies the condition (A5) with $a = b = k/2$.

The condition (A6) is satisfied by bifunction $F(x,y) = (Ax, y-x)$ with Lipschitz operator $A: C \to H$.

2 The Algorithm

Let $g: H \to \mathbb{R} \cup \{+\infty\}$ be a convex, lower semicontinuous, and proper. The proximity operator of a function g is the operator $\mathrm{prox}_g : H \to \mathrm{dom}\, g \subseteq H$ ($\mathrm{dom}\, g$ denotes the effective domain of g) which maps every $x \in H$ to the unique minimizer of the function $g + \|\cdot - x\|^2/2$, i.e.,

$$\forall x \in H \quad \mathrm{prox}_g x = \mathrm{argmin}_{y \in \mathrm{dom}\, g} \left\{ g(y) + \frac{1}{2}\|y-x\|^2 \right\}.$$

We have

$$z = \mathrm{prox}_g x \quad \Leftrightarrow \quad g(y) - g(z) + (z - x, y - z) \geq 0 \quad \forall y \in \mathrm{dom}\, g.$$

Proximity operators have attractive properties that make them particularly well suited for iterative minimization algorithms. For instance, prox_g is firmly nonexpansive and its fixed point set is precisely the set of minimizers of g. For detailed accounts of the proximity operators theory, see [11].

Now we extend the Popov method [30] to an equilibrium problem (1). In Algorithm 1 we are going to describe, in order to be able to obtain its convergence, the parameter λ must satisfy some condition (see convergence Theorem 1).

Algorithm 1. For $x_1, y_1 \in C$ generate the sequences $x_n, y_n \in C$ with the iterative scheme

$$\begin{cases} x_{n+1} = \mathrm{prox}_{\lambda F(y_n,\cdot)} x_n = \mathrm{argmin}_{y \in C} \left\{ \lambda F(y_n, y) + \frac{1}{2}\|y - x_n\|^2 \right\}, \\ y_{n+1} = \mathrm{prox}_{\lambda F(y_n,\cdot)} x_{n+1} = \mathrm{argmin}_{y \in C} \left\{ \lambda F(y_n, y) + \frac{1}{2}\|y - x_{n+1}\|^2 \right\}, \end{cases}$$

where $\lambda > 0$.

Extragradient method for the equilibrium problem (1) has the form

$$\begin{cases} y_n = \mathrm{prox}_{\lambda F(x_n,\cdot)} x_n, \\ x_{n+1} = \mathrm{prox}_{\lambda F(y_n,\cdot)} x_n, \end{cases}$$

where $\lambda > 0$ [31]. A distinctive and attractive feature of the Algorithm 1 consists in the fact that in the iterative step is used only one function $F(y_n, \cdot)$.

Remark 1. If $F(x, y) = (Ax, y - x)$, then Algorithm 1 takes the form:

$$\begin{cases} x_1 \in C, \; y_1 \in C, \\ x_{n+1} = P_C(x_n - \lambda A y_n), \\ y_{n+1} = P_C(x_{n+1} - \lambda A y_n), \end{cases}$$

where P_C is the operator of metric projection onto the set C.

A particular case of the scheme from the Remark 1 was proposed by Popov [30] for search of saddle points of convex-concave functions, which are defined on finite-dimensional Euclidean space. In recent works Malitsky and Semenov [23, 24] proved the convergence of this algorithm for variational inequalities with monotone and Lipschitz operators in infinite-dimensional Hilbert space, and proposed some modifications of this algorithm.

For substantiation of the iterative Algorithm 1 we note first, that if for some number $n \in \mathbb{N}$ next equalities are satisfied

$$x_{n+1} = x_n = y_n \tag{2}$$

than $y_n \in EP(F, C)$ and the following stationarity condition holds

$$y_k = x_k = y_n \quad \forall\, k \geq n.$$

Actually, the equality

$$x_{n+1} = \text{prox}_{\lambda F(y_n, \cdot)} x_n$$

means that

$$F(y_n, y) - F(y_n, x_{n+1}) + \frac{(x_{n+1} - x_n, y - x_{n+1})}{\lambda} \geq 0 \quad \forall y \in C.$$

From (2) it follows that

$$F(y_n, y) \geq 0 \ \ \forall y \in C,$$

i.e. $y_n \in EP(F, C)$.

Taking this into account the practical variant of the Algorithm 1 can be written as

Algorithm 2. Choose $x_1 \in C$, $y_1 \in C$, $\lambda > 0$, and $\varepsilon > 0$.

Step 1. For x_n and y_n compute

$$x_{n+1} = \text{prox}_{\lambda F(y_n, \cdot)} x_n.$$

Step 2. If $\max\{\|x_{n+1} - x_n\|, \|y_n - x_n\|\} \leq \varepsilon$, then STOP, else compute

$$y_{n+1} = \text{prox}_{\lambda F(y_n, \cdot)} x_{n+1}.$$

Step 3. Set $n := n + 1$ and go to Step 1.

Further, we assume that for all numbers $n \in \mathbb{N}$ the condition (2) doesn't hold. In the following section the weak convergence of the sequences (x_n), (y_n) generated by the Algorithm 1 is proved.

3 Convergence Results

To prove the convergence we need next facts.

Lemma 1. *Let non-negative sequences* (a_n), (b_n) *such that*

$$a_{n+1} \le a_n - b_n.$$

Then exists the limit $\lim_{n\to\infty} a_n \in \mathbb{R}$ *and* $\sum_{n=1}^{\infty} b_n < +\infty$.

Lemma 2 (Opial [29]). *Let the sequence* (x_n) *of elements from Hilbert space* H *converges weakly to* $x \in H$. *Then for all* $y \in H \setminus \{x\}$ *we have*

$$\liminf_{n\to\infty} \|x_n - x\| < \liminf_{n\to\infty} \|x_n - y\|.$$

We start the analysis of the convergence with the proof of important inequality for sequences (x_n) and (y_n), generated by the Algorithm 1.

Lemma 3. *Let sequences* (x_n), (y_n) *be generated by the Algorithm 1, and let* $z \in EP(F, C)$. *Then, we have*

$$\|x_{n+1} - z\|^2 \le \|x_n - z\|^2 - (1 - 2\lambda b)\|x_{n+1} - y_n\|^2 - \\ - (1 - 4\lambda a)\|y_n - x_n\|^2 + 4\lambda a \|x_n - y_{n-1}\|^2. \quad (3)$$

Proof. We have

$$\|x_{n+1} - z\|^2 = \|x_n - z\|^2 - \|x_n - x_{n+1}\|^2 + 2(x_{n+1} - x_n, x_{n+1} - z) = \\ = \|x_n - z\|^2 - \|x_n - y_n\|^2 - \|y_n - x_{n+1}\|^2 - \\ -2(x_n - y_n, y_n - x_{n+1}) + 2(x_{n+1} - x_n, x_{n+1} - z). \quad (4)$$

From the definition of points x_{n+1} and y_n it follows that

$$\lambda F(y_n, z) - \lambda F(y_n, x_{n+1}) \ge (x_{n+1} - x_n, x_{n+1} - z), \quad (5)$$

$$\lambda F(y_{n-1}, x_{n+1}) - \lambda F(y_{n-1}, y_n) \ge -(x_n - y_n, y_n - x_{n+1}). \quad (6)$$

Using inequalities (5), (6) to estimate inner products in (4), we get

$$\|x_{n+1} - z\|^2 \le \|x_n - z\|^2 - \|x_n - y_n\|^2 - \|y_n - x_{n+1}\|^2 + \\ +2\lambda \{F(y_n, z) - F(y_n, x_{n+1}) + F(y_{n-1}, x_{n+1}) - F(y_{n-1}, y_n)\}. \quad (7)$$

From pseudomonotonicity of the bifunction F and $z \in EP(F, C)$ it follows that

$$F(y_n, z) \le 0,$$

and Lipschitz-type continuity F guaranties the satisfying of inequality

$$-F(y_n, x_{n+1}) + F(y_{n-1}, x_{n+1}) - F(y_{n-1}, y_n) \leq$$
$$\leq a \|y_{n-1} - y_n\|^2 + b \|y_n - x_{n+1}\|^2 .$$

Using the above estimations (7), we get

$$\|x_{n+1} - z\|^2 \leq \|x_n - z\|^2 - \|x_n - y_n\|^2 - \|y_n - x_{n+1}\|^2 +$$
$$+2\lambda a \|y_{n-1} - y_n\|^2 + 2\lambda b \|y_n - x_{n+1}\|^2 . \tag{8}$$

The term $\|y_{n-1} - y_n\|^2$ we estimate in the next way

$$\|y_{n-1} - y_n\|^2 \leq 2 \|y_{n-1} - x_n\|^2 + 2 \|y_n - x_n\|^2 .$$

Taking this into account (8), we get the inequality

$$\|x_{n+1} - z\|^2 \leq \|x_n - z\|^2 - \|x_n - y_n\|^2 - \|y_n - x_{n+1}\|^2 +$$
$$+4\lambda a \|y_{n-1} - x_n\|^2 + 4\lambda a \|y_n - x_n\|^2 + 2\lambda b \|y_n - x_{n+1}\|^2 ,$$

i.e. the inequality (3). □

Proceed directly to proof of the convergence of the algorithm. Let $z \in EP(F, C)$. Assume

$$a_n = \|x_n - z\|^2 + 4\lambda a \|y_{n-1} - x_n\|^2 ,$$
$$b_n = (1 - 4\lambda a) \|y_n - x_n\|^2 + (1 - 4\lambda a - 2\lambda b) \|y_n - x_{n+1}\|^2 .$$

Then inequality (3) takes form

$$a_{n+1} \leq a_n - b_n .$$

The following condition are required

$$0 < \lambda < \frac{1}{2(2a + b)} .$$

Then from Lemma 1 we can conclude that exists the limit

$$\lim_{n\to\infty} \left(\|x_n - z\|^2 + 4\lambda a \|y_{n-1} - x_n\|^2 \right)$$

and

$$\sum_{n=1}^{\infty} \left((1 - 4\lambda a) \|y_n - x_n\|^2 + (1 - 4\lambda a - 2\lambda b) \|y_n - x_{n+1}\|^2 \right) < +\infty .$$

Whence we obtain

$$\lim_{n\to\infty} \|y_n - x_n\| = \lim_{n\to\infty} \|y_n - x_{n+1}\| = \lim_{n\to\infty} \|x_n - x_{n+1}\| = 0 \tag{9}$$

and convergence of the sequence $(\|x_n - z\|)$ for all $z \in EP(F, C)$. In particular, sequences (x_n), (y_n) are bounded.

Now we consider the subsequence (x_{n_k}), which converges weakly to the point $\bar{z} \in C$. Then from (9) it follows that $y_{n_k} \rightharpoonup \bar{z}$. Show that $\bar{z} \in EP(F, C)$. We have

$$F(y_n, y) \geq F(y_n, x_{n+1}) + \frac{(x_{n+1} - x_n, x_{n+1} - y)}{\lambda} \quad \forall y \in C. \tag{10}$$

Passing to the limit (10) taking into account (9) and conditions (A4), (A6), we get

$$F(\bar{z}, y) \geq \limsup_{k\to\infty} F(y_{n_k}, y) \geq \lim_{k\to\infty} \{F(y_{n_k}, x_{n_k+1}) + \\ + \frac{(x_{n_k+1} - x_{n_k}, x_{n_k+1} - y)}{\lambda}\Bigg\} = 0 \quad \forall y \in C,$$

i.e. $\bar{z} \in EP(F, C)$.

Now we show that $x_n \rightharpoonup \bar{z}$. Then from (9) it follows that $y_n \rightharpoonup \bar{z}$. Assume the converse. Let exists the subsequence (x_{m_k}) such that $x_{m_k} \rightharpoonup \tilde{z}$ and $\tilde{z} \neq \bar{z}$. It is clear that $\tilde{z} \in EP(F, C)$. Use the Lemma 2 twice. We have

$$\lim_{n\to\infty} \|x_n - \bar{z}\| = \lim_{k\to\infty} \|x_{n_k} - \bar{z}\| < \lim_{k\to\infty} \|x_{n_k} - \tilde{z}\| = \lim_{n\to\infty} \|x_n - \tilde{z}\| = \\ = \lim_{k\to\infty} \|x_{m_k} - \tilde{z}\| < \lim_{k\to\infty} \|x_{m_k} - \bar{z}\| = \lim_{n\to\infty} \|x_n - \bar{z}\|,$$

it is impossible. So, sequence (x_n) converges weakly to $\bar{z} \in EP(F, C)$.

Thus, we obtain the following result.

Theorem 1. *Let H be a Hilbert space, $C \subseteq H$ is nonempty convex closed set, for bifunction $F : C \times C \to \mathbb{R}$ conditions (A1)–(A6) are satisfied and $EP(F, C) \neq \emptyset$. Assume that $\lambda \in \left(0, \frac{1}{2(2a+b)}\right)$. Then sequences (x_n), (y_n) generated by the Algorithm 1 converge weakly to the solution $\bar{z} \in EP(F, C)$ of the equilibrium problem (1), and $\lim_{n\to\infty} \|x_n - y_n\| = 0$.*

Remark 2. The asymptotics $\lim_{n\to\infty} \|x_n - y_n\| = 0$ can be specified up to the following:

$$\liminf_{n\to\infty} \sqrt{n}\|x_n - y_n\| = 0. \tag{11}$$

Indeed, if (11) does not hold, then $\|x_n - y_n\| \geq \mu n^{-1/2}$ for some $\mu > 0$ and all sufficiently large n. Hence, the series $\sum \|x_n - y_n\|^2$ diverges. We have obtained an contradiction.

4 Conclusion and Future Work

In this work we have proposed a new iterative two-step proximal algorithm for solving the equilibrium programming problem in the Hilbert space. The method is the extension of Popov's modification [30] for Arrow-Hurwitz scheme for search of saddle points of convex-concave functions. The convergence of the algorithm is proved under the assumption that the solution exists and the bifunction is pseudo-monotone and Lipschitz-type.

In one of a forthcoming work we'll consider the next regularized variant of the algorithm that converges strongly

$$\begin{cases} x_{n+1} = \operatorname{prox}_{\lambda F(y_n,\cdot)} (1-\alpha_n)\, x_n, \\ y_{n+1} = \operatorname{prox}_{\lambda F(y_n,\cdot)} (1-\alpha_{n+1})\, x_{n+1}, \end{cases}$$

where $\lambda > 0$, (α_n) is infinitesimal sequence of positive numbers. Also we plan to study the variant of the method using Bregman's distance instead of Euclidean.

The interesting question is the substantiation of using Algorithm 1 as the element of an iterative method for equilibrium problem with a priori information, described in the form of inclusion to the fixed points set of quasi-nonexpansive operator.

Another promising area is the development of Algorithm 1 variants for solving stochastic equilibrium problems.

Acknowledgements We are grateful to Yura Malitsky, Yana Vedel for discussions. We are very grateful to the referees for their really helpful and constructive comments. Vladimir Semenov thanks the State Fund for Fundamental Researches of Ukraine for support.

References

1. Anh, P.N.: Strong convergence theorems for nonexpansive mappings and Ky Fan inequalities. J. Optim. Theory Appl. **154**, 303–320 (2012)
2. Antipin, A.S.: Equilibrium programming: gradient methods. Autom. Remote Control **58**(8), Part 2, 1337–1347 (1997)
3. Antipin, A.S.: Equilibrium programming: proximal methods. Comput. Math. Math. Phys. **37**, 1285–1296 (1997)
4. Antipin, A.: Equilibrium programming problems: prox-regularization and prox-methods. In: Recent Advances in Optimization. Lecture Notes in Economics and Mathematical Systems, vol. 452, pp. 1–18. Springer, Heidelberg (1997)
5. Antipin, A.S.: Extraproximal approach to calculating equilibriums in pure exchange models. Comput. Math. Math. Phys. **46**, 1687–1998 (2006)
6. Antipin, A.S.: Multicriteria equilibrium programming: extraproximal methods. Comput. Math. Math. Phys. **47**, 1912–1927 (2007)
7. Antipin, A.S., Vasil'ev, F.P., Shpirko, S.V.: A regularized extragradient method for solving equilibrium programming problems. Comput. Math. Math. Phys. **43**(10), 1394–1401 (2003)
8. Antipin, A.S., Artem'eva, L.A., Vasil'ev, F.P.: Multicriteria equilibrium programming: extragradient method. Comput. Math. Math. Phys. **50**(2), 224–230 (2010)

9. Antipin, A.S., Jacimovic, M., Mijailovic, N.: A second-order continuous method for solving quasi-variational inequalities. Comput. Math. Math. Phys. **51**(11), 1856–1863 (2011)
10. Antipin, A.S., Artem'eva, L.A., Vasil'ev, F.P.: Extraproximal method for solving two-person saddle-point games. Comput. Math. Math. Phys. **51**(9), 1472–1482 (2011)
11. Bauschke, H.H., Combettes, P.L.: Convex Analysis and Monotone Operator Theory in Hilbert Spaces. Springer, New York (2011)
12. Blum, E., Oettli, W.: From optimization and variational inequalities to equilibrium problems. Math. Stud. **63**, 123–145 (1994)
13. Censor, Y., Gibali, A., Reich, S.: The subgradient extragradient method for solving variational inequalities in Hilbert space. J. Optim. Theory Appl. **148**(2), 318–335 (2011)
14. Combettes, P.L., Hirstoaga, S.A.: Equilibrium programming in Hilbert spaces. J. Nonlinear Convex Anal. **6**(1), 117–136 (2005)
15. Denisov, S.V., Semenov, V.V., Chabak, L.M.: Convergence of the modified extragradient method for variational inequalities with non-Lipschitz operators. Cybern. Syst. Anal. **51**, 757–765 (2015)
16. Flam, S.D., Antipin, A.S.: Equilibrium programming using proximal-like algorithms. Math. Program. **78**, 29–41 (1997)
17. Giannessi, F., Maugeri, A., Pardalos, P.M.: Equilibrium Problems: Nonsmooth Optimization and Variational Inequality Models. Kluwer Academic, New York (2004)
18. Iusem, A.N., Sosa, W.: On the proximal point method for equilibrium problems in Hilbert spaces. Optimization **59**, 1259–1274 (2010)
19. Khobotov, E.N.: Modification of the extra-gradient method for solving variational inequalities and certain optimization problems. USSR Comput. Math. Math. Phys. **27**(5), 120–127 (1987)
20. Konnov, I.V.: Application of the proximal point method to nonmonotone equilibrium problems. J. Optim. Theory Appl. **119**, 317–333 (2003)
21. Korpelevich, G.M.: The extragradient method for finding saddle points and other problems. Ekonomika i Matematicheskie Metody **12**, 747–756 (1976) (In Russian)
22. Lyashko, S.I., Semenov, V.V., Voitova, T.A.: Low-cost modification of Korpelevich's methods for monotone equilibrium problems. Cybern. Syst. Anal. **47**, 631–639 (2011)
23. Malitsky, Yu.V., Semenov, V.V.: An extragradient algorithm for monotone variational inequalities. Cybern. Syst. Anal. **50**, 271–277 (2014)
24. Malitsky, Yu.V., Semenov, V.V.: A hybrid method without extrapolation step for solving variational inequality problems. J. Glob. Optim. **61**, 193–202 (2015)
25. Mastroeni, G.: On auxiliary principle for equilibrium problems. In: Daniele, P., et al. (eds.) Equilibrium Problems and Variational Models, pp. 289–298. Kluwer Academic, Dordrecht (2003)
26. Moudafi, A.: Proximal point methods extended to equilibrium problems. J. Nat. Geom. **15**, 91–100 (1999)
27. Nadezhkina, N., Takahashi, W.: Weak convergence theorem by an extragradient method for nonexpansive mappings and monotone mappings. J. Optim. Theory Appl. **128**, 191–201 (2006)
28. Nurminski, E.A.: The use of additional diminishing disturbances in Fejer models of iterative algorithms. Comput. Math. Math. Phys. **48**, 2154–2161 (2008)
29. Opial, Z.: Weak convergence of the sequence of successive approximations for nonexpansive mappings. Bull. Am. Math. Soc. **73**, 591–597 (1967)
30. Popov, L.D.: A modification of the Arrow-Hurwicz method for search of saddle points. Math. Notes Acad. Sci. USSR **28**(5), 845–848 (1980)
31. Quoc, T.D., Muu, L.D., Hien, N.V.: Extragradient algorithms extended to equilibrium problems. Optimization **57**, 749–776 (2008)
32. Semenov, V.V.: On the parallel proximal decomposition method for solving the problems of convex optimization. J. Autom. Inf. Sci. **42**(4), 13–18 (2010)
33. Semenov, V.V.: A strongly convergent splitting method for systems of operator inclusions with monotone operators. J. Autom. Inf. Sci. **46**(5), 45–56 (2014)
34. Semenov, V.V.: Hybrid splitting methods for the system of operator inclusions with monotone operators. Cybern. Syst. Anal. **50**, 741–749 (2014)

35. Semenov, V.V.: Strongly convergent algorithms for variational inequality problem over the set of solutions the equilibrium problems. In: Zgurovsky, M.Z., Sadovnichiy, V.A. (eds.) Continuous and Distributed Systems, pp. 131–146. Springer, Heidelberg (2014)
36. Stukalov, A.S.: An extraproximal method for solving equilibrium programming problems in a Hilbert space. Comput. Math. Math. Phys. **46**, 743–761 (2006)
37. Takahashi, S., Takahashi, W.: Viscosity approximation methods for equilibrium problems and fixed point problems in Hilbert spaces. J. Math. Anal. Appl. **331**, 506–515 (2007)
38. Tseng, P.: A modified forward-backward splitting method for maximal monotone mappings. SIAM J. Control Optim. **38**(2), 431–446 (2000)
39. Verlan, D.A., Semenov, V.V., Chabak, L.M.: A strongly convergent modified extragradient method for variational inequalities with non-Lipschitz operators. J. Autom. Inf. Sci. **47**(7), 31–46 (2015)
40. Vuong, P.T., Strodiot, J.J, Nguyen, V.H.: Extragradient methods and linesearch algorithms for solving Ky Fan inequalities and fixed point problems. J. Optim. Theory Appl. **155**, 605–627 (2012)

Nonparametric Ellipsoidal Approximation of Compact Sets of Random Points

Sergey I. Lyashko, Dmitry A. Klyushin, Vladimir V. Semenov, Maryna V. Prysiazhna, and Maksym P. Shlykov

Abstract One of the main problems of stochastic control theory is the estimation of attainability sets, or information sets. The most popular and natural approximations of such sets are ellipsoids. B.T. Polyak and his disciples use two kinds of ellipsoids covering a set of points—minimal volume ellipsoids and minimal trace ellipsoids. We propose a way to construct an ellipsoidal approximation of an attainability set using nonparametric estimations. These ellipsoids can be considered as an approximation of minimal volume ellipsoids and minimal trace ellipsoids. Their significance level depends only on the number of points and only one point from the set lays on a bound of such ellipsoid. This unique feature allows to construct a statistical depth function, rank multivariate samples and identify extreme points. Such ellipsoids in combination with traditional methods of estimation allow to increase accuracy of outer ellipsoidal approximations and estimate the probability of attaining a target set of states.

Keywords Ellipsoidal approximation • Attainability set • Information set • Nonparametric estimation • Extreme point • Confidence ellipse

1 Introduction

Ellipsoidal estimation of parameters and states of systems is one of the most popular tools [1]. In many problems the volume or trace of an ellipsoid are used as an optimality criteria [2]. However, if the initial states of a system are random it is more natural to use some confidence ellipsoids that contain the initial states and have required statistical properties. This problem is closely related with the problem of detection of outliers in random sets and ordering random points.

The problem of determining the initial ellipsoid may be reduced to the ordering of multidimensional random samples. The standard systematization of methods

S.I. Lyashko • D.A. Klyushin • V.V. Semenov (✉) • M.V. Prysiazhna • M.P. Shlykov
Kiev National Taras Shevchenko University, Vladimirskaya Str. 64, Kiev, Ukraine
e-mail: lyashko.serg@gmail.com; dokmed5@gmail.com; semenov.volodya@gmail.com; marrinep@gmail.com; maksym.shlykov@gmail.com

B. Goldengorin (ed.), *Optimization and Its Applications in Control and Data Sciences*, Springer Optimization and Its Applications 115,
DOI 10.1007/978-3-319-42056-1_11

of multidimensional ordering was suggested by Barnett [3]. According to this approach, methods of multidimensional ordering are subdivided on marginal, reduced, partial and conditional. Marginal methods order samples by separate components. The reduced methods calculate distance of each samples from distribution center. Partial ordering means the subdividing of samples into groups of identical samples. In conditional methods ordering of samples with respect to the chosen component influencing the others is made.

Now, the most popular approach to the multidimensional ordering is the approach based on the concept of a statistical depth of samples concerning distribution center and corresponding peeling methods. These methods were proposed by Tukey [4], Titterington [5], Oja [6], Liu [7], Zuo and Serfling [8] and others authors. They allow to consider geometrical properties of multidimensional distributions. We offer a new method of ordering multidimensional data based on Petunin's ellipses and ellipsoids [9]. Note that unlike the method proposed in [10], this method does not mean peeling, i.e. performance of the repeating iterations of the same procedure applied to the decreasing set of points, and it orders all points at once.

Without loss of generality, we will consider the algorithm of construction of Petunin's ellipse at a plane, and then we will go to R^m, where $m > 2$. The initial data for the algorithm is a set of points $M_n = \{\mathbf{x}_1, \ldots, \mathbf{x}_n\}$, where $\mathbf{x}_n = (x_n, y_n)$.

2 Petunin's Ellipses

At the first stage, we search for the most distant points (x_k, y_k) and (x_l, y_l) in the set

$$M_n = \{(x_1, y_1), \ldots, (x_n, y_n)\}.$$

Then, we connect the points (x_k, y_k) and (x_l, y_l) by the line segment L. Next, we search for the points (x_r, y_r) and $\left(x_q, y_q\right)$, which are most distant from L. Further, we connect the points (x_r, y_r) and $\left(x_q, y_q\right)$ by the segments L_1 and L_2, which are parallel to L. Then, we connect the points (x_k, y_k) and (x_l, y_l) by the line segments L_3 and L_4, which are perpendicular to the line segment L. Intersections of the line segment L_1, L_2, L_3, and L_4 form a rectangle Π with the side length a and b (Fig. 1).

For definiteness, put $a \leq b$. Transport the left lower corner of the rectangle to the origin of a new system of axes Ox' and Oy' using the rotation and parallel translation. The points (x_1, y_1), (x_2, y_2), ..., (x_n, y_n) become the points $\left(x'_1, y'_1\right)$, $\left(x'_2, y'_2\right)$, ..., $\left(x'_n, y'_n\right)$. Map the points $\left(x'_1, y'_1\right)$, $\left(x'_2, y'_2\right)$, ..., $\left(x'_n, y'_n\right)$ to the points $\left(\alpha x'_1, y'_1\right)$, $\left(\alpha x'_2, y'_2\right)$, ..., $\left(\alpha x'_n, y'_n\right)$, where $\alpha = \frac{a}{b}$. As a result, we have the set of points covered by the square S.

Find the center $\left(x'_0, y'_0\right)$ of the square S and distances $r_1, r_2, \ldots, r_n$ between the center and every points $\left(\alpha x'_1, y'_1\right)$, $\left(\alpha x'_2, y'_2\right)$, ..., $\left(\alpha x'_n, y'_n\right)$. The greatest number $R = \max(r_1, r_2, \ldots, r_n)$ defines the circle with the center at the point $\left(x'_0, y'_0\right)$ with radius R. Finally, all the points $\left(\alpha x'_1, y'_1\right)$, $\left(\alpha x'_2, y'_2\right)$, ..., $\left(\alpha x'_n, y'_n\right)$ are laying in the circle with the radius R. Stretching this circle along the Ox' with the coefficient

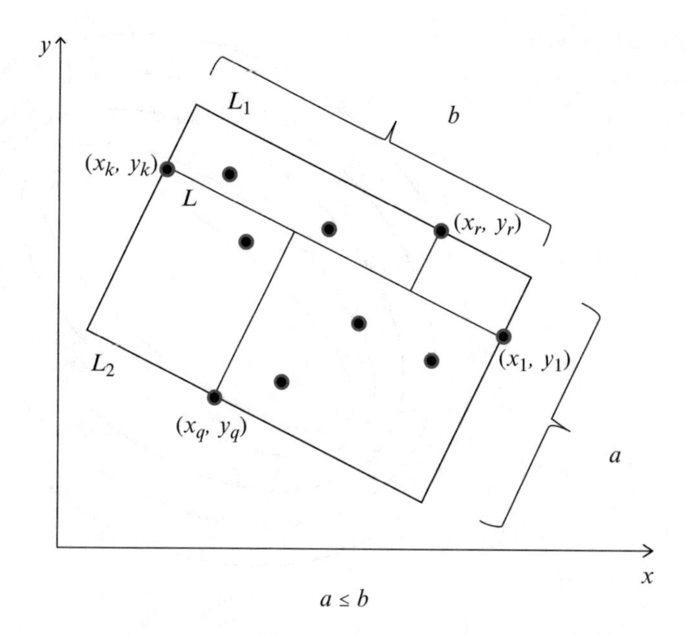

Fig. 1 The Petunin's rectangle

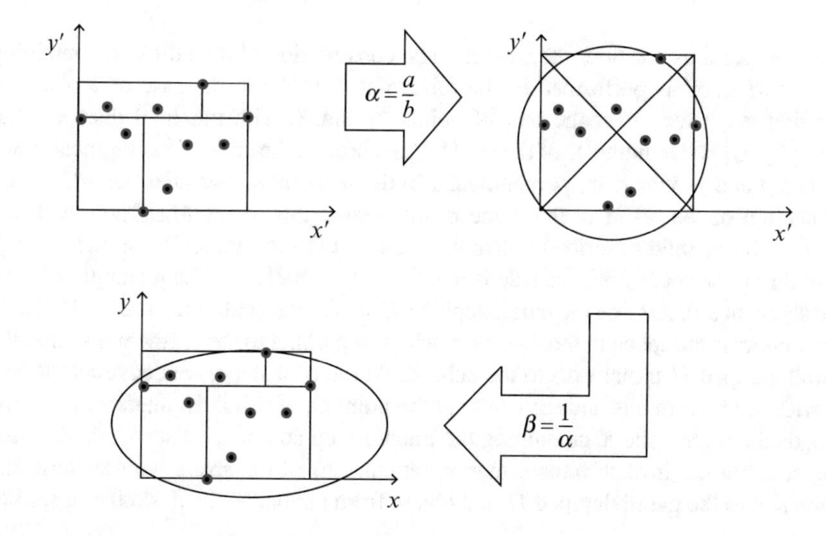

Fig. 2 Construction of Petunin's ellipse

$\beta = \frac{1}{\alpha}$ and using inverse transformations of rotation and parallel translation we obtain a Petunin's ellipse (Fig. 2).

As a result, only one point from the sample is located on every embedded ellipsoid, so we arrange them (Fig. 3).

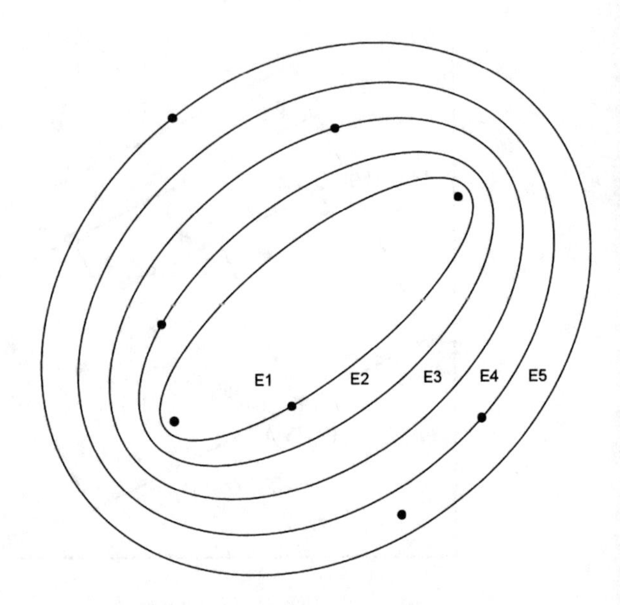

Fig. 3 Embedded Petunin's ellipses

3 Classic Method of Construction of Petunin's Ellipsoids

For clarity, let us consider the case R^3. The construction of the ellipsoid containing the set M in R^3 is performed in the following way. As in the case of a plane, at the first stage we select the pair of points X_i and X_j with maximal distance. Let $a = \left[X_i, X_j\right]$ is the diameter of the set M. Pass through the ends of the segment a two planes β and γ, which are perpendicular to the segment a. Consider the orthogonal projection of the set M at the plane β and denote this set by M_β. Then, with the help of the method described above we construct the rectangle Π_β on the plane β containing the set M_β, whose side is parallel to this diameter. The rectangle Π_β and the segment a determine the parallelepiped $\Pi = \Pi_\beta \times a$ containing the set M. Then, we compress the space in the direction, which is parallel to the segment a so that the parallelepiped Π transforms to the cube K. At the next stage, we construct the ball C with minimal radius and the center at the point O, which is an intersection of the diagonals of the cube K containing the transformed compressed set M. At the final stage, using the inverse transformation (extension) of the space we transform the cube K into the parallelepiped Π and obtain from the ball C the desired ellipsoid E.

4 Statistical Properties of Petunin's Ellipsoids

Let $G = (M, F)$, $M \subset R$ be the population of random variables with an unknown distribution function F. The bulk of population G is a subset $B \subseteq M$ such that $P\{x \in B\} = 1 - \alpha$, where x as an arbitrary element from the sample retrieved from the population G using simple sampling, and α is given significance level (for example, $\alpha = 0,05$). If sample values $x_1, x_2, \ldots, x_n$ are exchangeable random variables with the same absolutely continuous distribution function, that according to Hill's assumption [11],

$$P\left(x_{n+1} = x \in \left(x_{(i)}, x_{(j)}\right)\right) = \frac{j-i}{n+1},$$

where x_{n+1} is the next sample value from G, and $x_{(i)}$ and $x_{(j)}$ are order statistics. The following facts take place [10].

Theorem 1. *If $\eta_1, \eta_2, \ldots, \eta_{n+1}$ are exchangeable identically distributed random variables with absolutely continuous joint distribution function, such that*

$$P\{\eta_k = \eta_m\} = 0$$

for $k \neq m$, then

$$P\{\eta_k \geq \eta_1, \ldots, \eta_k \geq \eta_{k-1}, \eta_k \geq \eta_{k+1}, \ldots, \eta_k \geq \eta_{n+1}\} = \frac{1}{n+1}.$$

Proof. Let $F_k(x_1, x_2, \ldots x_k)$ be the joint distribution function of any k values from $\eta_1, \eta_2, \ldots, \eta_{n+1}$. For any $k = 1, 2, \ldots, n+1$ we have:

$$\begin{aligned}&P\{\eta_k \geq \eta_1, \ldots, \eta_k \geq \eta_{k-1}, \eta_k \geq \eta_{k+1}, \ldots, \eta_k \geq \eta_{n+1}\} = \\ &= \int\limits_{R^1} P\{x \geq \eta_1, \ldots, x \geq \eta_{k-1}, x \geq \eta_{k+1}, \ldots, x \geq \eta_{n+1}/\eta_k = x\} f(x)\, dx,\end{aligned}$$

where $f(x) = F'(x)$ is the probability density of one random value.

Note that

$$\begin{aligned}&P\{x \geq \eta_1, \ldots, x \geq \eta_{k-1}, x \geq \eta_{k+1}, \ldots, x \geq \eta_{n+1}/\eta_k = x\} = \\ &\lim_{\varepsilon \to 0} \frac{P\{x \geq \eta_1, \ldots, x \geq \eta_{k-1}, x \geq \eta_{k+1}, \ldots, x \geq \eta_{n+1}, x - \varepsilon < \eta_k\}}{P\{x - \varepsilon < \eta_k\}} = \\ &= \lim_{\varepsilon \to 0} \frac{F_{n+1}(x, x, \ldots, x, x) - F_{n+1}(x, x, \ldots, x, x - \varepsilon)}{F(x) - F(x - \varepsilon)} \\ &= \lim_{\varepsilon \to 0} \frac{\left(F_{n+1}(x, x, \ldots, x, x) - F_{n+1}(x, x, \ldots, x, x - \varepsilon)\right)/_{\varepsilon}}{\left(F(x) - F(x - \varepsilon)\right)/_{\varepsilon}} = \\ &= \frac{\frac{\partial}{\partial y} F_{n+1}\left(\overbrace{x, \ldots, x}^{n}, y\right)}{f(x)}\end{aligned}$$

Hence,

$$P\{\eta_k \geq \eta_1, \ldots, \eta_k \geq \eta_{k-1}, \eta_k \geq \eta_{k+1}, \ldots, \eta_k \geq \eta_{n+1}\} = \\ = \int_{R^1} \frac{\partial}{\partial y} F_{n+1}(x, \ldots x, y)dx.$$

Thus, the values $\int_{R^1} \frac{\partial}{\partial y} F_{n+1}(x, \ldots x, y)$ are equal for $k = 1, 2, \ldots, n+1$. The event $\{\eta_k \geq \eta_1, \ldots, \eta_k \geq \eta_{k-1}, \eta_k \geq \eta_{k+1}, \ldots, \eta_k \geq \eta_{n+1}\}$ means that a random variable attains the greatest value. These events form a complete group, i.e.

$$\sum_{k=1}^{n+1} P\{\eta_k \geq \eta_1, \ldots, \eta_k \geq \eta_{k-1}, \eta_k \geq \eta_{k+1}, \ldots, \eta_k \geq \eta_{n+1}\} =$$

$$(n+1) \int_{R^1} \frac{\partial}{\partial y} F_{n+1}(x, \ldots x, y)dx = 1.$$

Thus,

$$P\{\eta_k \geq \eta_1, \ldots, \eta_k \geq \eta_{k-1}, \eta_k \geq \eta_{k+1}, \ldots, \eta_k \geq \eta_{n+1}\} = \frac{1}{n+1}.$$

Theorem 2. *If $\eta_1, \eta_2, \ldots, \eta_{n+1}$ are exchangeable random variables with absolutely continuous distribution function such that $P\{\eta_k = \eta_{(j)}\} = 0$ for $k \neq m$, and $\eta_{(1)}\eta_{(2)} \ldots \eta_{(n)}$ is a variational series constructed on the first n values, then*

$$P\{\eta_{n+1} \in (\eta_{(j)}, \eta_{(j+1)}]\} = \frac{1}{n+1}.$$

Proof. Divide the half-interval $(\eta_{(i)}, \eta_{(j+1)}]$ into intervals:

$$P\{\eta_{n+1} \in (\eta_{(j)}, \eta_{(j+1)}]\} = P\{\eta_{n+1} > \eta_{(j)}, \eta_{n+1} \geq \eta_{(j+1)}\} = \\ = P\{\eta_{n+1} \in (\eta_{(j)}, \eta_{(j+1)})\} + P\{\eta_{n+1} = \eta_{(j+1)}\} =$$

$$= \sum_{i_1, i_2, \ldots, i_n} P\{\eta_{n+1} > \eta_{i_1}, \ldots, \eta_{n+1} > \eta_{i_j}, \eta_{n+1} < \eta_{i_{j+1}}, \ldots, \eta_{n+1} < \eta_{i_n}\}.$$

The first term contains C_n^j terms, i.e. all the combinations $\{i_1, i_2, \ldots, i_n\}$ of the numbers $\{1, 2, \ldots, n\}$ such that j values among $\eta_1, \eta_2, \ldots, \eta_n$ are less that η_{n+1}, and other values $n - j$ are not less than η_{n+1}. Using inclusion-elimination principle we have that

$$P\left\{\eta_{n+1} \in \left(\eta_{(j)}, \eta_{(j+1)}\right]\right\} =$$

$$= \sum_{i_1,i_2,\ldots,i_n} P\left\{\underbrace{\eta_{n+1} > \eta_{i_1}, \ldots, \eta_{n+1} > \eta_{i_j}}_{B}, \underbrace{\eta_{n+1} \geq \eta_{i_{j+1}}}_{A_1}, \ldots, \underbrace{\eta_{n+1} \geq \eta_{i_n}}_{A_{n-j}}\right\} =$$

$$= \sum_{i_1,i_2,\ldots,i_n} \sum_{k=0}^{n-j} (-1)^k \sum_{\substack{\{s_1,s_2,\ldots,s_k\}\subseteq \\ \subseteq\{i_{j+1},i_{j+2},\ldots,i_n\}}} P\left\{\eta_{n+1} > \eta_{i_1}, \ldots, \eta_{n+1} > \eta_{s_k}\right\}.$$

Since the distribution function F is absolutely continuous on the interval $\left(\eta_{(j)}, \eta_{(j+1)}\right)$, by Theorem 1 we have

$$P\left\{\eta_{n+1} \in \left(\eta_{(j)}, \eta_{(j+1)}\right]\right\} =$$

$$= \sum_{i_1,i_2,\ldots,i_n} \sum_{k=0}^{n-j} (-1)^k \sum_{\substack{\{s_1,s_2,\ldots,s_k\}\subseteq \\ \subseteq\{i_{j+1},i_{j+2},\ldots,i_n\}}} \frac{1}{j+k+1} =$$

$$= C_n^j \sum_{k=0}^{n-j} (-1)^k C_{n-j}^k \frac{1}{j+k+1} = C_n^j \sum_{k=0}^{n-j} (-1)^k C_{n-j}^k \int_0^1 t^{j+k} dt =$$

$$= C_n^j \int_0^1 t^j (1-t)^{n-j} dt =$$

$$= C_n^j \int_0^1 \sum_{k=0}^{n-j} (-1)^k C_{n-j}^k t^{j+k} dt.$$

It follows from the definition of the beta function that

$$P\left\{\eta_{n+1} \in \left(\eta_{(j)}, \eta_{(j+1)}\right]\right\} = C_n^j \mathrm{B}\,(j+1, n-j+1) =$$

$$= C_n^j \frac{\Gamma\,(j+1)\,\Gamma\,(n-j+1)}{\Gamma\,(n+2)} =$$

$$= C_n^j \frac{j!\,(n-j)!}{(n+1)!} = \frac{n!}{j!\,(n-j)!} \frac{j!\,(n-j)!}{(n+1)!} = \frac{1}{n+1}.$$

Corollary 1. *It follows from Theorem 2 that*

$$P\left\{\eta_{n+1} \in \left(\eta_{(i)}, \eta_{(j)}\right)\right\} = \frac{j-i}{n+1}$$

$$\forall 0 i < j n+1, where \eta_0 = -\infty, \eta_{n+1} = +\infty.$$

Proof. It is easy to see that

$$P\{\eta_{n+1} \in (\eta_{(i)}, \eta_{(j)})\} = P\{\tilde{x} \in [x_{(i)}, x_{(i+1)}) \dots [x_{(j-1)}, x_{(j)})\} =$$
$$= P\{\eta_{n+1} \in [\eta_{(i)}, \eta_{(i+1)})\} + \dots + P\{\eta_{n+1} \in [\eta_{(j-1)}, \eta_{(j)})\} = \frac{j-i}{n+1}.$$

Lemma 1. *If two random values* $\mathbf{x}_1 = (\eta_1^1, \eta_2^1, \dots, \eta_n^1)$ *and* $\mathbf{x}_2 = (\eta_1^2, \eta_2^2, \dots, \eta_n^2)$, *have the same probability density* f_G, *then for any point* $X_0 = (x_1^0, x_2^0, \dots, x_n^0)$ *the Euclidean distanced* $r_1 = \sqrt{(\eta_1^1 - x_1^0)^2 + (\eta_2^1 - x_2^0)^2 + \dots + (\eta_n^1 - x_n^0)^2}$ *and* $r_2 = \sqrt{(\eta_1^2 - x_1^0)^2 + (\eta_2^2 - x_2^0)^2 + \dots + (\eta_n^2 - x_n^0)^2}$ *are identically distributed random variables.*

Proof. Let us construct the distribution function for random variable $r_1 = r(\mathbf{x}_1)$. $F_{r_1}(x) = P\{r_1 x\}$.For $x < 0$ the probability is zero 0 since r_1 is the a nonnegative value. Consider the case $x0$:

$$P\{r_1 x\} = P\left\{\sqrt{(\eta_1^1 - x_1^0)^2 + (\eta_2^1 - x_2^0)^2 + \dots + (\eta_n^1 - x_n^0)^2} x\right\}.$$

The domain for which

$$\sqrt{(\eta_1^1 - x_1^0)^2 + (\eta_2^1 - x_2^0)^2 + \dots + (\eta_n^1 - x_n^0)^2} x$$

is a ball in n-dimensional space centered at X_0 with radius x. Denote this domain by B_0. Then

$$P\{r_1 x\} = P\{\mathbf{x}_1 \in O_{X_0}^x\} = \int_{B_0} f_G(t_1, t_2, \dots, t_n)\, dt_1 dt_2 \dots dt_n.$$

So, we have shown that the distribution function of the random variable r_1 has the form:

$$F_{r_1}(x) = \begin{cases} 0, \ x < 0, \\ \int_{B_0} f_G(t_1, t_2, \dots, t_n)\, dt_1 dt_2 \dots dt_n, \ x \geq x0 \end{cases}$$

Since the random vectors $\mathbf{x}_1$ and $\mathbf{x}_2$ have the same probability density, repeating the arguments for r_2 we obtain:

$$F_{r_2}(x) = \begin{cases} 0, \ x < 0, \\ \int_{B_0} f_G(t_1, t_2, \dots, t_n)\, dt_1 dt_2 \dots dt_n, \ x \geq x0, \end{cases}$$

In other words,

$$F_{r_1}(x) = F_{r_2}(x),$$

This means that r_1 and r_2 are identically distributed random variables.

Theorem 3. *If the vectors* $\mathbf{x}_1, \mathbf{x}_2, \ldots, \mathbf{x}_n$ *are exchangeable and identically distributed random variables from population G,* E_n *is Petunin's ellipsoid containing the point* $\mathbf{x}_1, \mathbf{x}_2, \ldots, \mathbf{x}_n$ *and* $\vec{x}_{n+1} \in G$*, then*

$$P(\mathbf{x}_{n+1} \in E_n) = \frac{n}{n+1}.$$

Proof. After affine transformations, distribution functions of all random vectors change equally. Respectively, $\mathbf{x}'_1, \mathbf{x}'_2, \ldots, \mathbf{x}'_n$ also are identically distributed random variables from G', which contains the points transformed by the same transformation, including $\vec{x}'_{n+1}$. Therefore, the points $\mathbf{x}'_1, \mathbf{x}'_2, \ldots, \mathbf{x}'_n$ remain exchangeable. By Lemma 1, the random variables $r_1, r_2, \ldots, r_n$ are identically distributed. Since their order depends only on the order of exchangeable points $\mathbf{x}'_1, \mathbf{x}'_2, \ldots, \mathbf{x}'_n$, then they are exchangeable also. If to calculate the distance r_{n+1} to the center $\vec{x}_{n+1}$ of the hypercube, then by Theorem 2

$$P(r_{n+1} < r_{(n)}) = \frac{n}{n+1} = P(\mathbf{x}_{n+1} \in E_n).$$

Corollary 2. *The significance level of Petunin's ellipsoid no greater than 0,05, is attained for* $n > 39$.

Corollary 3. *Since the point* $X_0 = (x_1^0, x_2^0, \ldots, x_n^0)$ *is an arbitrary, the both the square center and centroid of the point set can be taker as the starting point.*

The described algorithm have several special features:

1. Only one point always is located on the ellipsoid boundary.
2. The ellipsoid contains n points with the probability $\frac{n}{n+1}$ independently from the distribution function.
3. It is always possible to limit the point set by 40 samples, guaranteeing that the significance level is no greater than 0.05.

5 Effective Method of Construction of Petunin's Ellipsoids

In this section we describe the modification of the classical algorithm for an arbitrary finite-dimensional space. Note, that the main goal of the classical algorithm is to construct a linear transformation that at each step maps diameter vectors to axes. All we have to do after that is to make scaling and to find the most distant point from the center of a hypercube. Our goal is to describe the construction of the linear transformation introduced above. Assume that we do not perform rotation at each step, but only make projections and translations. This means that we have all the diameters of sets. By construction, these diameters form an orthogonal system. Let us normalize it and denote by $B_m = \{\overrightarrow{a}_1, \overrightarrow{a}_2, \ldots, \overrightarrow{a}_m\}$.

Then, we must find an unitary transformation that maps this orthonormalized system to basis. This problem becomes trivial after taking into account the fact that the inverse transformation maps basis vectors to our system. Thus, we know the matrix of the inverse transformation:

$$U^{-1} = (\overrightarrow{a}_1 | \overrightarrow{a}_2 | \ldots | \overrightarrow{a}_m)$$

or, equivalently,

$$U = (\overrightarrow{a}_1 | \overrightarrow{a}_2 | \ldots | \overrightarrow{a}_m)^{-1}.$$

Another convenient fact about unitary operators is $U^{-1} = U^T$, which means that the only thing to make for finding the inverse transformation is transposition. Our next goal is to simplify the process of moving to the space of lower dimension and make only one translation. Consider the first step of the algorithm for that. Let we found the diameter vectors of our input set (denote them by $\overrightarrow{x}_k$ and $\overrightarrow{x}_l$).

As we do not make a rotation and translation, we must project our points onto the affine subspace $\overrightarrow{x}_k + L$. Here L is an orthogonal complement of a line determined by $\overrightarrow{x}_k - \overrightarrow{x}_l$. In fact it is a hyperplane which contains $\overrightarrow{x_k}$ and is orthogonal to $\overrightarrow{x}_k - \overrightarrow{x}_l$.

Denote the points obtained after projection by $M_n^{(1)} = \{\mathbf{x}_i^{(1)}, i = \overline{1, n-1}\}$. It is worth to say that after projection the number of points will decrease by one, because the projection of $\overrightarrow{x_k}$ is $\overrightarrow{x_l}$ by construction. Then, we show that on the next steps we do not have to move to the spaces with lower dimension. Instead, it is enough to perform projection onto the hyperplane that is orthogonal to the diameter of the corresponding set. Let at some step we have $M_n^{(k)} = \{\mathbf{x}_i^{(k)}, i = \overline{1, n-k}\}$—the set of points in R^m that lie in some affine subspace $y + L_1$, whose dimension is p.

Assume that we found the diameter of this set and it equals to $\mathbf{x}_l^{(k)} - \mathbf{x}_t^{(k)}$ and consider the hyperplane that is orthogonal to this vector and contains $\mathbf{x}_t^{(k)}$. Its equation looks as follows:

$$(\mathbf{x}, \mathbf{x}_l^{(k)} - \mathbf{x}_t^{(k)}) = (\mathbf{x}_t^{(k)}, \mathbf{x}_l^{(k)} - \mathbf{x}_t^{(k)}).$$

The projection operator for this hyperplane is:

$$P\mathbf{x} = \mathbf{x} - \frac{(\mathbf{x}, \mathbf{x}_l^{(k)} - \mathbf{x}_t^{(k)}) - (\mathbf{x}_t^{(k)}, \mathbf{x}_l^{(k)} - \mathbf{x}_t^{(k)})}{||\mathbf{x}_l^{(k)} - \mathbf{x}_t^{(k)}||^2} (\mathbf{x}_l^{(k)} - \mathbf{x}_t^{(k)}).$$

As $\mathbf{x}_t^{(k)}$ lies in L_1 we have $\mathbf{y} + L_1 = \mathbf{x}_l^{(k)} + L_1$. This implies that $\mathbf{x} = \overrightarrow{x}_t^{(k)} + \mathbf{y}, \mathbf{y} \in L_1$ and

$$P\mathbf{x} = \mathbf{x} - \alpha(\mathbf{x}_l^{(k)} - \mathbf{x}_t^{(k)}) = \overrightarrow{x}_t^{(k)} + \mathbf{y} - \alpha(\mathbf{x}_l^{(k)} - \mathbf{x}_t^{(k)}),$$

where

$$\alpha = \frac{(\mathbf{x}, \mathbf{x}_l^{(k)} - \mathbf{x}_t^{(k)}) - (\mathbf{x}_t^{(k)}, \mathbf{x}_l^{(k)} - \mathbf{x}_t^{(k)})}{||\mathbf{x}_l^{(k)} - \mathbf{x}_t^{(k)}||^2}.$$

Also we know that $\alpha(\mathbf{x}_l^{(k)} - \mathbf{x}_t^{(k)}) \in L_1$ and this implies that $P\mathbf{x} \in \overrightarrow{x}_t^{(k)} + L_1$. At the same time, by the definition of projection Px belongs to the hyperplane $(\mathbf{x}, \mathbf{x}_l^{(k)} - \mathbf{x}_t^{(k)}) = (\mathbf{x}_t^{(k)}, \mathbf{x}_l^{(k)} - \mathbf{x}_t^{(k)})$ which is also an affine subspace. Thus, the projections of our points will lie at the intersection of two affine subspaces one of which is our hyperplane. Denote it by $\mathbf{x}_t^{(k)} + L_2$, where L_2 is a $(m-1)$-dimensional subspace. The intersection of these affine subspaces is affine subspace $\mathbf{x}_t^{(k)} + L_1 \cap L_2$. At the same the fact that $\mathbf{x}_l^{(k)} - \mathbf{x}_t^{(k)}$ is orthogonal to L_2 means that $L_1 + L_2 = R^m$. To find the dimension of $L_1 \cap L_2$ consider the Grassmann's formula for our subspaces:

$$m = m - 1 + p - \dim L_1 \cap L_2$$

and we have

$$\dim L_1 \cap L_2 = p - 1.$$

Knowing this, we can perform projection onto the hyperplane which is orthogonal to the diameter and contains one of its points. But in fact the dimension of the affine subspace, which contains our projected points will decrease. In the original algorithm we move to the orthogonal complement instead of this. Finally, we have the following algorithm:

Input data: $M_n = \{\mathbf{x}_1, \ldots, \mathbf{x}_n\}$—vectors from R^m

Algorithm

1. $M_n^{(0)} \leftarrow M_n, B = \emptyset, k = 0$.
2. While $k < m$ do:
   - 2.1. Find $\overrightarrow{x}_l^{(k)}$ and $\overrightarrow{x}_t^{(k)}$—the diameter points of $M_n^{(k)}$
   - 2.2. $B \leftarrow B \cup \left\{ \frac{\mathbf{x}_l^{(k)} - \mathbf{x}_t^{(k)}}{||\mathbf{x}_l^{(k)} - \mathbf{x}_t^{(k)}||} \right\}$.
   - 2.3. $M_n^{(k+1)} \leftarrow P_L M_n^{(k)}$, where L is a hyperplane $(\mathbf{x}, \mathbf{x}_l^{(k)} - \mathbf{x}_t^{(k)}) = (\mathbf{x}_t^{(k)}, \mathbf{x}_l^{(k)} - \mathbf{x}_t^{(k)})$.
   - 2.4. $k \leftarrow k + 1$.
3. Build matrix U whose columns are the vectors of B.
4. $M_n^{'} \leftarrow U^T M_n$.
5. Find minimal axis-aligned bounding box for the points of $M_n^{'}$. Denote by $\mathbf{c}$ the center of this bounding box.
6. Build scaling transformation S which maps this bounding box into hypercube.
7. $M_n^{''} \leftarrow SM_n^{'}, \mathbf{c}^{'} \leftarrow S\mathbf{c}$;

8. $R \leftarrow \max\left\{||\mathbf{x} - \mathbf{c}'||, x \in M''_n\right\}$.
9. $E \leftarrow \frac{1}{R^2} I$, where I is a unit matrix of size m.
10. $E \leftarrow US^T ESU^T$, $\mathbf{c} \leftarrow U\mathbf{c}$.

Output data: E is the matrix of our ellipsoid, $\mathbf{c}$ is its center.

The only constraint for the application of our algorithm is that the ellipsoid must exist and be non-degenerate. In other words, there must not exist a subspace of the input space R^m that contains the input set of points. At each step of both algorithms we are computing the diameter of a finite set of points. This procedure requires $O(mn(n-k))$ operations for the corresponding step k. After that we project our set to the affine subspace. The complexity of this operation is $O(m(n-k))$. As we have to make it m times the total complexity of all projections and diameters searches will be

$$T_1 = T_{diameter} + T_{projection} = O\left(\sum_{k=1}^{m}(mn(n-k) + m(n-k))\right) = O(m^2n^2 + m^2n)$$

After that we provide linear transformations of the input sample which take

$$T_{linear\ transformation} = O(m^2 n)$$

Another step we perform is the search of the most distant point from the center of rectangle. This operation has complexity $O(mn)$ and we won't include it to the result as it will be absorbed by other summands of our estimation. But we also have the last step of our algorithm which computes matrices product. This action's complexity is

$$T_{matrices\ multiplication} = O(m^3)$$

Summing all the parts of our algorithm we will obtain

$$T_{total} = T_1 + T_{linear\ transformation} + T_{matrices\ multiplication} = O(m^2n^2 + m^2n + m^3)$$

This result may be improved for two- and three-dimensional spaces with convex hull construction. We may build a convex hull for our set and perform search on its vertices. Asymptotically this algorithm will require $O(n \lg n + z(z-1)/2)$ operations, where z is the number of vertices of the convex hull. In the worst case the complexity of the whole algorithm will be the same, but for most samples the number of vertices in convex hull is small in comparison with the sample size. This means that the complexity for such samples will be $O(n \lg n)$.

Memory complexity of our algorithm will be $O(mn + m^2)$ as we need to keep the input sample, projected sample (it changes at each step), the matrices for our ellipsoid and transformations.

Table 1 Comparison results for two and three dimensions

	Original algorithm	Modified algorithm
R^2	1	2,0082
R^3	1,9951	2,9404

We compared original and modified algorithms in two- and three-dimensional spaces. Both of them were implemented in R programming language. There were generated 100 samples of 200 vectors each. All the vectors were from Gaussian multivariate distribution with the same covariance matrix. Every sample consisted of two groups of 100 vectors each with different mean vectors. For each algorithm the total time of its work on all the 100 samples was computed. Taking as a unit the time of work for original algorithm in two-dimensional space the results of our comparison are displayed in Table 1. It is worth saying that for three-dimensional space after the first projection in the original algorithm we obtain the case m=2. This means that after rotation we can use the algorithm for m=2 and build the bounding box after that. It has been taken into account while implementing the original algorithm. This optimization decreases the number of diameters search by one and we have to perform them $m - 1$ times. At the same time the modified algorithm performs diameters search m times. All the other actions made by algorithms are analogous. Thus the ratio between the times of work of the two algorithms on the same data sets must be close to $(m - 1)/m$.

6 Conclusion

The proposed ellipsoidal approximation of an attainability set using nonparametric estimations may be considered as an adequate approximation of minimal volume ellipsoids and minimal trace ellipsoids. Its significance level depends only on the number of points and only one point from the set lays at the bound of such ellipsoid. This unique feature allows to construct a statistical depth function, order multivariate vectors and identify outliers. As a result, it is possible to increase accuracy of outer ellipsoidal approximations and estimate the probability of attaining a target set of states.

References

1. Polyak B.T., Nazin S.A., Durieub C., Walterc E.: Ellipsoidal parameter or state estimation under model uncertainty. Automatica **40**, 1171–1179 (2004)
2. Kiselev, O.N., Polyak, B.T.: Ellipsoidal estimation based on a generalized criterion. Remote Control **52**, 1281–1292 (1991)
3. Barnett, V.: The ordering of multivariate data. J. R. Stat. Soc. Ser. A (General) **139**, 318–355 (1976)

4. Tukey, J.W.: Mathematics and the picturing of data. In: Proceedings of the International Congress of Mathematicians, pp. 523–531. Montreal, Canada (1975)
5. Titterington, D.M.: Estimation of correlation coefficients by ellipsoidal trimming. Appl. Stat. **27**, 227–234 (1978)
6. Oja, H.: Descriptive statistics for multivariate distributions. Stat. Probab. Lett. **1**, 327–332 (1983)
7. Liu, R.J.: On a notion of data depth based on random simplices. Ann. Stat. **18**, 405–414 (1990)
8. Zuo, Y., Serfling, R.: General notions of statistical depth function. Ann. Stat. **28**, 461–482 (2000)
9. Petunin, Yu.I., Rublev, B.V.: Pattern recognition with the help quadratic discriminant function. J. Math. Sci. **97**, 3959–3967 (1999)
10. Lyashko, S.I., Klyushin, D.A., Alexeenko, V.V.: Multivariate ranking using elliptical peeling. Cybern. Syst. Anal. **49**, 511–516 (2013)
11. Hill, B.: Posteriori distribution of percentiles: Bayes' theorem for sampling from a population. J. Am. Stat. Assoc. **63**:677–691 (1968).

Extremal Results for Algebraic Linear Interval Systems

Daniel N. Mohsenizadeh, Vilma A. Oliveira, Lee H. Keel, and Shankar P. Bhattacharyya

Abstract This chapter explores some important characteristics of algebraic linear systems containing interval parameters. Applying the Cramer's rule, a parametrized solution of a linear system can be expressed as the ratio of two determinants. We show that these determinants can be expanded as multivariate polynomial functions of the parameters. In many practical problems, the parameters in the system characteristic matrix appear with rank one, resulting in a rational multilinear form for the parametrized solutions. These rational multilinear functions are monotonic with respect to each parameter. This monotonic characteristic plays an important role in the analysis and design of algebraic linear interval systems in which the parameters appear with rank one. In particular, the extremal values of the parametrized solutions over the box of interval parameters occur at the vertices of the box.

Keywords Algebraic linear interval systems • Parametrized solutions • Extremal results

1 Introduction

Linear interval systems arise in many branches of science and engineering, such as control systems, communications, economics, sociology, and genomics. The

---

D.N. Mohsenizadeh • S.P. Bhattacharyya (✉)
Department of Electrical and Computer Engineering, Texas A&M University, College Station, TX 77843, USA
e-mail: danielmz@tamu.edu; bhatt@ece.tamu.edu

V.A. Oliveira
Department of Electrical and Computer Engineering, University of Sao Paulo at Sao Carlos, Sao Carlos, SP, Brazil
e-mail: vilma@sc.usp.br

L.H. Keel
Department of Electrical and Computer Engineering, Tennessee State University, Nashville, TN 37203, USA
e-mail: keel@gauss.tsuniv.edu

B. Goldengorin (ed.), *Optimization and Its Applications in Control and Data Sciences*, Springer Optimization and Its Applications 115,
DOI 10.1007/978-3-319-42056-1_12

problem of analyzing and designing linear interval systems has theoretical and practical importance and has been open for the last few decades. Several results concerning the analysis of systems with real parametric interval can be found in the early works in [1, 4, 6, 12–14, 16]. In [19], a method is proposed to calculate the exact bounds of the solution set which is based on solving special boundary problems of the linear interval system. The bounds of the solution set can be obtained by applying a linear programming algorithm as explained in [15], and followed up in [7], while a branch-and-bound scheme is presented in [20]. The results developed in [18] can be used to obtain outer estimations of parametric AE solution sets for linear systems with interval parameters. The sign-definite decomposition method can be used to decide the robust positivity (or negativity) of a polynomial over a box of interval parameters by evaluating the sign of the decomposed polynomials at the vertices of the box [2, 5, 8]. This chapter concentrates on the class of algebraic linear systems containing interval parameters and takes a novel approach to determine the exact extremal values of the solution set over a box in the parameter space. We show that this can be accomplished by finding the general functional form of the solution set in terms of the interval parameters and then using its properties. Furthermore, in the case of unknown linear systems, these properties allow us to determine the unknown parameters of the function by a small set of measurements made on the system.

A parametrized solution of a linear system containing parameters can be expressed as the ratio of two determinants. These determinants can be expanded as polynomial functions of the parameters, resulting in a rational polynomial form for the parametrized solutions [3, 9, 11, 17]. If the interval parameters in the system characteristic matrix appear with rank one, which is the case in many practical applications, then the rational polynomial form reduces to a rational multilinear function, being monotonic in each parameter [10]. This monotonic characteristic leads us to extract the extremal results for linear interval systems. In particular, we show that if the rank one condition holds, then the extremal values of the solution set over a box in the parameter space occur at the vertices of that box. This result enables us to evaluate the performance of a linear interval system over a box of parameters by checking the respective performance index at the vertices.

This chapter is organized as follows. In Sect. 2 we provide some mathematical preliminaries on the parametrized solutions of linear systems. Section 3 presents the extremal results for linear systems with interval parameters appearing with rank one. Section 4 provides some examples. Finally, we summarize with our concluding remarks in Sect. 5.

2 Linear Systems with Parameters

Consider the linear system

$$\begin{aligned} 0 &= A(p)x + Bu, \\ y &= C(p)x + Du, \end{aligned} \tag{1}$$

where $A_{n\times n}(p)$ is the system characteristic matrix, $p = [p_1, p_2, \ldots, p_l]^T$ is the vector of system parameters, $u_{r\times 1}$ is the vector of input variables, $x_{n\times 1}$ is the vector of system state variables, $y_{m\times 1}$ is the vector of system outputs, and $B_{n\times r}$ and $D_{m\times r}$ are system matrices with scalar entries. Let

$$z := \begin{pmatrix} x \\ y \end{pmatrix}, \tag{2}$$

then, (1) can be rewritten as

$$\underbrace{\begin{pmatrix} A(p) & 0 \\ -C(p) & I \end{pmatrix}}_{\tilde{A}(p)} \underbrace{\begin{pmatrix} x \\ y \end{pmatrix}}_{z} + \underbrace{\begin{pmatrix} B \\ -D \end{pmatrix}}_{\tilde{B}} u = \begin{pmatrix} 0 \\ 0 \end{pmatrix}. \tag{3}$$

It is clear that $|\tilde{A}(p)| = |A(p)|$. Let us define

$$T_{ij}(p) := \begin{pmatrix} A(p) & b_j \\ -c_i(p) & d_{ij} \end{pmatrix}, \; i = 1, \ldots, m, \; j = 1, \ldots, r, \tag{4}$$

with $c_i(p), i = 1, \ldots, m$ the i-th row of $C(p)$, $b_j, j = 1, \ldots, r$ the j-th column of B, and d_{ij} the corresponding (i, j) element of D. We also define

$$\begin{aligned} \alpha(p) &:= |A(p)|, \\ \beta_{ij}(p) &:= |T_{ij}(p)|. \end{aligned} \tag{5}$$

In a linear interval system the parameters p and inputs u vary in intervals. Suppose that $\mathcal{B}$ denotes the box of intervals and is characterized as $\mathcal{B} = \mathcal{P} \times \mathcal{U}$ where

$$\begin{aligned} \mathcal{P} &= \{p : p_k^- \leq p_k \leq p_k^+, \; k = 1, \ldots, l\}, \\ \mathcal{U} &= \{u : u_j^- \leq u_j \leq u_j^+, \; j = 1, \ldots, r\}. \end{aligned} \tag{6}$$

We make the following assumption regarding the system in (3).

Assumption 1. $|A(p)| \neq 0, \; \forall p \in \mathcal{P}.$

This assumption is true for many physical systems, because if there exists a vector $p_0 \in \mathcal{P}$ so that $A(p_0)$ becomes a singular matrix, then the vector of system state variables x will not have a unique value which is not the case for physical systems.

Theorem 1. *For the system described in* (1)*, and under the Assumption 1, the input-output relationship is*

$$y_i = \sum_{j=1}^{r} \frac{\beta_{ij}(p)}{\alpha(p)} u_j, \; i = 1, \ldots, m, \tag{7}$$

with $\beta_{ij}(p)$ and $\alpha(p)$ as defined in (5).

Proof. Let $T_{ij}(p)$ and $\beta_{ij}(p)$, for $i = 1, \ldots, m$, $j = 1, \ldots, r$, be as defined in (4) and (5), respectively. Also, let $A(p)$ satisfy the statement of Assumption 1. Applying the Cramer's rule to (3) and using the fact that $|\tilde{A}(p)| = |A(p)|$, the i-th output y_i can be expressed as

$$y_i = \sum_{j=1}^{r} \frac{|T_{ij}(p)|}{|A(p)|} u_j, \; i = 1, \ldots, m, \tag{8}$$

and the result follows. □

Remark 1. Suppose that p is fixed at p^*. Then, the form in (7) states the well-known Superposition Principle.

The form of the multivariate polynomials $\alpha(p)$ and $\beta_{ij}(p)$ in (7) can be determined using the following assumption and lemma.

Assumption 2. *p appears affinely in $A(p)$ and $C(p)$, that is*

$$A(p) = A_0 + p_1 A_1 + p_2 A_2 + \cdots + p_l A_l, \tag{9}$$

$$C(p) = C_0 + p_1 C_1 + p_2 C_2 + \cdots + p_l C_l. \tag{10}$$

Based on the statement of Assumption 2, $T_{ij}(p)$ defined in (4) can be written as

$$T_{ij}(p) = T_{ij0} + p_1 T_{ij1} + \cdots + p_l T_{ijl}. \tag{11}$$

Lemma 1. *If p appears affinely in $A(p)$, and*

$$r_k = \text{rank}(A_k), \; k = 1, 2, \ldots, l, \tag{12}$$

then, $\alpha(p) = |A(p)|$ is a multivariate polynomial in p of degree at most r_k in p_k, $k = 1, 2, \ldots, l$:

$$\alpha(p) = \sum_{k_l=0}^{r_l} \cdots \sum_{k_2=0}^{r_2} \sum_{k_1=0}^{r_1} \alpha_{k_1 k_2 \cdots k_l} p_1^{k_1} p_2^{k_2} \cdots p_l^{k_l}. \tag{13}$$

Proof. The proof follows easily from the properties of determinants. □

Remark 2. The number of coefficients $\alpha_{k_1 k_2 \cdots k_l}$ in (13) is $\prod_{k=1}^{l}(r_k + 1)$.

Remark 3. Based on the form (9) and the rank conditions (12), we say that p_k appears in $A(p)$ with rank r_k.

Applying Lemma 1 to $T_{ij}(p)$, then $\beta_{ij}(p) = |T_{ij}(p)|$ will be a multivariate polynomial in p of degree at most r_{ijk} in p_k where

$$r_{ijk} = \text{rank}(T_{ijk}),\ i = 1, \ldots, m,\ j = 1, \ldots, r,\ k = 1, \ldots, l. \tag{14}$$

3 Extremal Results

In many physical systems, the parameters p appear in $A(p)$ with rank one. For instance, resistors, impedances and dependent sources in an electrical circuit, mechanical properties of links in a truss structure, pipe resistances in a linear hydraulic network, and blocks in a control system block diagram, all appear with rank one in the characteristic matrix $A(p)$ of the system. Likewise, p appears with rank one in matrices T_{ij}, $i = 1, \ldots, m$, $j = 1, \ldots, r$. Based on this rank condition, we state the following lemma which is helpful in establishing the extremal results that will be discussed shortly.

Lemma 2. *For the system in* (1), *if Assumptions 1 and 2 hold, and*

$$\begin{aligned} &\text{rank}(A_k) = 1,\ k = 1, 2, \ldots, l, \\ &\text{rank}(T_{ijk}) = 1,\ i = 1, \ldots, m,\ j = 1, \ldots, r,\ k = 1, \ldots, l, \end{aligned} \tag{15}$$

then, $\frac{\partial y_i}{\partial p_k}$ *and* $\frac{\partial y_i}{\partial u_j}$ *are sign-invariant functions of* p_k *and* u_j*, respectively, over* $\mathcal{P} \times \mathcal{U}$ *defined in* (6).

Proof. Consider y_i as a function of p_k with p_t, $t \neq k$ fixed at p_t^*, and u_j, $j = 1, \ldots, r$ fixed at u_j^*. Then,

$$y_i = \sum_{j=1}^{r} \frac{\beta_{ij0} + \beta_{ij1} p_k}{\alpha_0 + \alpha_1 p_k} u_j^*, \tag{16}$$

where we used Lemma 1, for the case $p = p_k$, to expand the polynomials $\beta_{ij}(p)$ and $\alpha(p)$ in (7) according to the rank conditions in (15). Thus,

$$\frac{\partial y_i}{\partial p_k} = \sum_{j=1}^{r} \frac{\beta_{ij1}\alpha_0 - \beta_{ij0}\alpha_1}{(\alpha_0 + \alpha_1 p_k)^2} u_j^*, \tag{17}$$

which is sign-invariant over $p_k \in [p_k^-, p_k^+]$. Now, if p is fixed at p^*, then

$$\frac{\partial y_i}{\partial u_j} = \frac{\beta_{ij}(p^*)}{\alpha(p^*)}, \tag{18}$$

which is sign-invariant over $u_j \in [u_j^-, u_j^+]$. This argument is true for each p_k, $k = 1, \ldots, l$ and each u_j, $j = 1, \ldots, r$. This completes the proof. □

We state the extremal results for linear interval systems as the following theorem.

Theorem 2. *Consider the system in* (1) *under the Assumptions 1 and 2, and the rank conditions in* (15). *Suppose that p and u are varying in the box* $\mathcal{B} = \mathcal{P} \times \mathcal{U}$, *defined in* (6), *with* $v := 2^{l+r}$ *vertices, labeled* $V_1, V_2, \cdots, V_v$. *Then, the extremal values of* y_i, $i = 1, \ldots, m$ *occur at the vertices of* $\mathcal{B}$:

$$\min_{p,u \in \mathcal{B}} y_i(p,u) = \min\{y_i(V_1), y_i(V_2), \ldots, y_i(V_v)\},$$

$$\max_{p,u \in \mathcal{B}} y_i(p,u) = \max\{y_i(V_1), y_i(V_2), \ldots, y_i(V_v)\}.$$

Proof. Proof follows immediately from Lemma 2. □

4 Example

4.1 A Linear Interval System

Consider the following linear system

$$0 = \underbrace{\begin{pmatrix} p_1 & 3 & 0 & 1 & -2p_1 \\ -5 & 0 & 3 & 1 & 2 \\ -2p_1 & 0 & 1 & 2 & 4p_1 \\ 0 & -1 & 0 & 0 & -1 \\ 1 & -4 & 0 & -3 & 0 \end{pmatrix}}_{A(p_1)} \underbrace{\begin{pmatrix} x_1 \\ x_2 \\ x_3 \\ x_4 \\ x_5 \end{pmatrix}}_{x} + \underbrace{\begin{pmatrix} 1 \\ 0 \\ 0 \\ 0 \\ 0 \end{pmatrix}}_{B} u_1,$$

$$y_1 = \underbrace{\begin{pmatrix} -p_1 & 0 & 0 & 0 & 2p_1 \end{pmatrix}}_{C(p_1)} \underbrace{\begin{pmatrix} x_1 \\ x_2 \\ x_3 \\ x_4 \\ x_5 \end{pmatrix}}_{x} + \underbrace{2}_{D} u_1. \tag{19}$$

where p_1 and u_1 are varying in the rectangle,

$$\mathcal{B} = \{(p_1, u_1) \mid 1 \le p_1 \le 4,\ 2 \le u_1 \le 3\}, \tag{20}$$

with vertices:

$$\begin{aligned} A &= (1,2), \quad B = (1,3), \\ C &= (4,3), \quad D = (4,2). \end{aligned} \tag{21}$$

Suppose that the extremal values of the output y_1 over the rectangle $\mathcal{B}$ are to be evaluated.

For the above system, Assumptions 1 and 2 hold. The matrix $A(p_1)$ in (19) can be decomposed as

$$A(p_1) = \underbrace{\begin{pmatrix} 0 & 3 & 0 & 1 & 0 \\ -5 & 0 & 3 & 1 & 2 \\ 0 & 0 & 1 & 2 & 0 \\ 0 & -1 & 0 & 0 & -1 \\ 1 & -4 & 0 & -3 & 0 \end{pmatrix}}_{A_0} + \underbrace{\begin{pmatrix} 1 & 0 & 0 & 0 & -2 \\ 0 & 0 & 0 & 0 & 0 \\ -2 & 0 & 0 & 0 & 4 \\ 0 & 0 & 0 & 0 & 0 \\ 0 & 0 & 0 & 0 & 0 \end{pmatrix}}_{A_1} p_1, \tag{22}$$

where $\text{rank}(A_1) = 1$. The matrix $T_{11}(p_1)$ will be

$$T_{11}(p_1) = \begin{pmatrix} p_1 & 3 & 0 & 1 & -2p_1 & 1 \\ -5 & 0 & 3 & 1 & 2 & 0 \\ -2p_1 & 0 & 1 & 2 & 4p_1 & 0 \\ 0 & -1 & 0 & 0 & -1 & 0 \\ 1 & -4 & 0 & -3 & 0 & 0 \\ -p_1 & 0 & 0 & 0 & 2p_1 & 2 \end{pmatrix}, \tag{23}$$

which can be written as

$$T_{11}(p_1) = \underbrace{\begin{pmatrix} 0 & 3 & 0 & 1 & 0 & 1 \\ -5 & 0 & 3 & 1 & 2 & 0 \\ 0 & 0 & 1 & 2 & 0 & 0 \\ 0 & -1 & 0 & 0 & -1 & 0 \\ 1 & -4 & 0 & -3 & 0 & 0 \\ 0 & 0 & 0 & 0 & 0 & 2 \end{pmatrix}}_{T_{110}} + \underbrace{\begin{pmatrix} 1 & 0 & 0 & 0 & -2 & 0 \\ 0 & 0 & 0 & 0 & 0 & 0 \\ -2 & 0 & 0 & 0 & 4 & 0 \\ 0 & 0 & 0 & 0 & 0 & 0 \\ 0 & 0 & 0 & 0 & 0 & 0 \\ -1 & 0 & 0 & 0 & 2 & 0 \end{pmatrix}}_{T_{111}} p_1, \tag{24}$$

with $\text{rank}(T_{111}) = 1$. Based on these rank conditions, the statement of Theorem 2 can be applied. Thus, setting (p_1, u_1) to the values corresponding to the vertices A, B, C, D in (21) and solving (19) for y_1, gives

$$\begin{aligned} \min_{p_1,u_1 \in \mathcal{B}} y_1 &= 5.5 \quad \text{at vertex A}, \\ \max_{p_1,u_1 \in \mathcal{B}} y_1 &= 9.6 \quad \text{at vertex C}. \end{aligned} \tag{25}$$

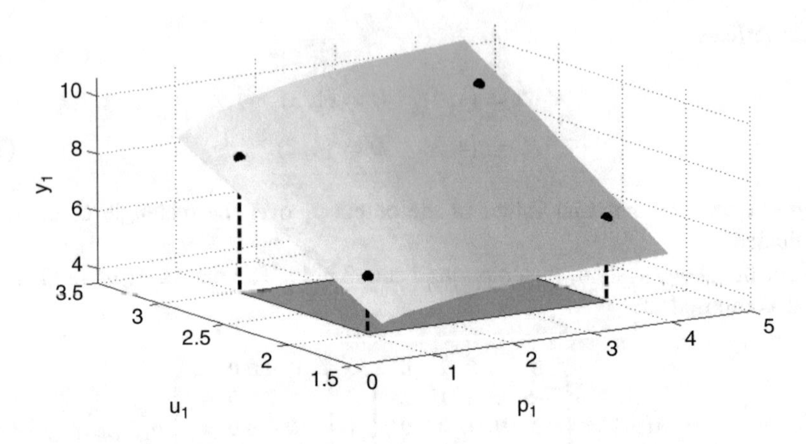

Fig. 1 Plot of $y_1(p_1, u_1)$

Alternatively, using Theorem 1, the function $y_1(p_1, u_1)$ can be calculated as

$$y_1(p_1, u_1) = \frac{38 + 63p_1}{19 + 18p_1} u_1, \tag{26}$$

which is plotted in Fig. 1 (the rectangle $\mathcal{B}$, defined in (20), is also shown). It can be easily seen that the extremal values of y_1 occur at the vertices of $\mathcal{B}$ and are the same as those obtained in (25).

4.2 A Linear DC Circuit

Consider the linear DC circuit shown in Fig. 2 where $R_1, R_2, \ldots, R_5$ are resistors, V is an independent voltage source, J is an independent current source, and $V_{\text{amp},1}, V_{\text{amp},2}$ are dependent voltage sources with amplification factors K_1, K_2, respectively. Applying the Kirchhoff's laws, the governing equations can be written as

$$\begin{pmatrix} 1 & -1 & -1 & 0 & 0 \\ 0 & 0 & 1 & 1 & 0 \\ 0 & 1 & 0 & -1 & -1 \\ R_1 & 0 & R_3 & 0 & 0 \\ K_1 & R_2 & -R_3 & K_2 & R_5 \end{pmatrix} \begin{pmatrix} I_1 \\ I_2 \\ I_3 \\ I_4 \\ I_5 \end{pmatrix} = \begin{pmatrix} 0 & 0 \\ 0 & 1 \\ 0 & 0 \\ 1 & 0 \\ 0 & 0 \end{pmatrix} \begin{pmatrix} V \\ J \end{pmatrix}. \tag{27}$$

Suppose that $R_2 = 1, R_3 = 2, R_4 = 5, R_5 = 3, K_1 = 1, K_2 = 4$, and the resistance R_1 and the sources V and J are the interval parameters varying in the box,

$$\mathcal{B} = \{(R_1, V, J) \mid 2 \leq R_1 \leq 5,\ 1 \leq V \leq 3,\ 2 \leq J \leq 4\}. \tag{28}$$

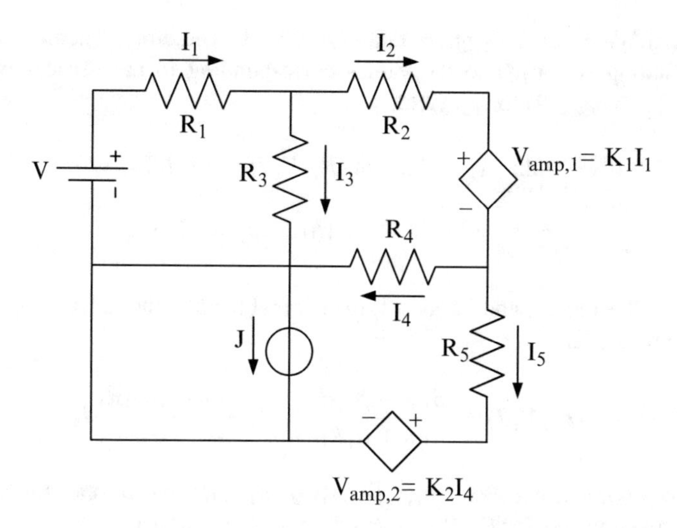

Fig. 2 A linear DC circuit

Assuming that the output is $y_1 = -R_1 I_1 + I_3 + V + 2J$, the set of system equations will be

$$0 = \underbrace{\begin{pmatrix} 1 & -1 & -1 & 0 & 0 \\ 0 & 0 & 1 & 1 & 0 \\ 0 & 1 & 0 & -1 & -1 \\ R_1 & 0 & 2 & 0 & 0 \\ 1 & 1 & -2 & 4 & 3 \end{pmatrix}}_{A(R_1)} \underbrace{\begin{pmatrix} I_1 \\ I_2 \\ I_3 \\ I_4 \\ I_5 \end{pmatrix}}_{I} + \underbrace{\begin{pmatrix} 0 & 0 \\ 0 & 1 \\ 0 & 0 \\ 1 & 0 \\ 0 & 0 \end{pmatrix}}_{B} \underbrace{\begin{pmatrix} V \\ J \end{pmatrix}}_{u},$$

$$y_1 = \underbrace{\begin{pmatrix} -R_1 & 0 & 1 & 0 & 0 \end{pmatrix}}_{C(R_1)} \underbrace{\begin{pmatrix} I_1 \\ I_2 \\ I_3 \\ I_4 \\ I_5 \end{pmatrix}}_{I} + \underbrace{\begin{pmatrix} 1 & 2 \end{pmatrix}}_{D} \underbrace{\begin{pmatrix} V \\ J \end{pmatrix}}_{u}. \tag{29}$$

Suppose that it is of interest to find the extremal values of y_1 over the box $\mathcal{B}$ in (28).

It can be easily verified that the Assumptions 1 and 2 hold. The matrices $A(R_1), T_{11}(R_1)$ and $T_{12}(R_1)$ can be written as

$$\begin{aligned} A(R_1) &= A_0 + A_1 R_1, \\ T_{11}(R_1) &= T_{110} + T_{111} R_1, \\ T_{12}(R_1) &= T_{120} + T_{121} R_1, \end{aligned} \tag{30}$$

where $\text{rank}(A_1) = \text{rank}(T_{111}) = \text{rank}(T_{121}) = 1$. Therefore, Theorem 2 can be applied. Setting (R_1, V, J) to the values corresponding to the vertices of box $\mathcal{B}$ in (28) and solving (29) for y_1, yields

$$\min_{R_1,V,J\in\mathcal{B}} y_1 = 4.9 \quad \text{at } (R_1, V, J) = (2, 1, 2),$$

$$\max_{R_1,V,J\in\mathcal{B}} y_1 = 11.7 \quad \text{at } (R_1, V, J) = (5, 3, 4). \tag{31}$$

Based on Theorem 1 and Lemma 1, the general form of the function $y_1(R_1, V, J)$ can be expressed as

$$y_1(R_1, V, J) = \frac{\beta_{110} + \beta_{111}R_1}{\alpha_0 + \alpha_1 R_1}V + \frac{\beta_{120} + \beta_{121}R_1}{\alpha_0 + \alpha_1 R_1}J, \tag{32}$$

where the unknown constants $\beta_{110}, \beta_{111}, \beta_{120}, \beta_{121}, \alpha_0$ and α_1 can be calculated using the given values for $R_2, R_3, R_4, R_5, K_1, K_2$. Thus, we get

$$y_1(R_1, V, J) = \frac{5 + 14R_1}{10 + 7R_1}V + \frac{20 + 11R_1}{10 + 7R_1}J. \tag{33}$$

It can be easily verified that the extremal values of y_1 occur at the vertices of $\mathcal{B}$ and are the same as those obtained in (31).

Remark 4. In the case of an unknown circuit, that is the values of R_2, R_3, R_4, R_5, K_1, K_2 are not known, the unknown constants $\beta_{110}, \beta_{111}, \beta_{120}, \beta_{121}, \alpha_0$ and α_1 in (32) can be determined through measurements, which is the generalization of the well-known Thevenin's Theorem [3, 11, 17].

5 Conclusions

In this chapter we described some important characteristics of the parametrized solutions of linear interval systems. We showed that if the interval parameters in the characteristic matrix of the system appear with rank one, which is the case in many practical applications, then the parametrized solutions will be monotonic in each parameter. We also showed how the Superposition Principle can be obtained from the general functional form developed in this chapter. Furthermore, we described that the extremal values of the parametrized solutions over a box in the parameter space occur at the vertices of the box.

Acknowledgements Dedicated to Professor Boris Polyak in honor of his 80th birthday.

References

1. Alefeld, G., Herzberger, J.: Introduction to Interval Computations. Academic Press, New York (1983)
2. Bhattacharyya, S.P., Datta, A., Keel, L.H.: Linear Control Theory: Structure, Robustness, and Optimization. CRC, Boca Raton (2009)
3. Bhattacharyya, S.P., Keel, L.H., Mohsenizadeh, D.N.: Linear Systems: A Measurement Based Approach. Springer, Berlin (2013)
4. Deif, A.: Sensitivity Analysis in Linear Systems. Springer, New York (1986)
5. Elizondo-Gonzalez, C.: Necessary and sufficient conditions for robust positivity of polynomic functions via sign decomposition. In: 3rd IFAC Symposium on Robust Control Design. Prague, Czech Republic (2000)
6. Horowitz, I.M.: Synthesis of Feedback Systems. Academic, New York (1963)
7. Jansson, C.: Calculation of exact bounds for the solution set of linear interval systems. Linear Algebra Appl. **251**, 321–340 (1997)
8. Knap, M.J., Keel, L.H., Bhattacharyya, S.P.: Robust Hurwitz stability via sign-definite decomposition. Linear Algebra Appl. **434**(7), 1663–1676 (2011)
9. Mohsenizadeh, D.N., Keel, L.H., Bhattacharyya, S.P.: An equivalent plant representation for unknown control systems. In: 7th ASME Dynamic Systems and Control Conference. San Antonio (2014)
10. Mohsenizadeh, D.N., Keel, L.H., Bhattacharyya, S.P.: An extremal result for unknown interval linear systems. In: 19th IFAC World Congress, pp. 6502–6507. Cape Town (2014)
11. Mohsenizadeh, N., Nounou, H., Nounou, M., Datta, A., Bhattacharyya, S.P.: Linear circuits: a measurement based approach. Int. J. Circuit Theory Appl. **43**(2), 205–232 (2015)
12. Moore, R.E.: Interval Analysis. Prentice-Hall, Englewood Cliffs (1966)
13. Neumaier, A.: New techniques for the analysis of linear interval equations. Linear Algebra Appl. **58**, 273–325 (1984)
14. Neumaier, A.: Linear interval equations. In: Nickel, K. (ed.) Interval Mathematics. Lecture Notes in Computer Science. Springer, Berlin (1985)
15. Oettli, W.: On the solution set of a linear system with inaccurate coefficients. SIAM J. Numer. Anal. **2**, 115–118 (1965)
16. Oettli, W., Prager, W.: Compatibility of approximate solution of linear equations with given error bounds for coefficients and right-hand sides. Numer. Math. **6**, 405–409 (1964)
17. Oliveira, V.A., Felizardo, K.R., Bhattacharyya, S.P.: A model-free measurement based approach to circuit analysis and synthesis based on linear interval systems. In: 24th IEEE International Symposium on Industrial Electronics, pp. 38–43 (2015)
18. Popova, E.D., Hladik, M.: Outer enclosures to the parametric AE solution set. Soft Comput. **17**(8), 1403–1414 (2013)
19. Rohn, J.: Systems of linear interval equations. Linear Algebra Appl. **126**, 39–78 (1989)
20. Shary, S.P.: Optimal solutions of interval linear algebraic systems. Interval Comput. **2**, 7–30 (1991)

Applying the Gradient Projection Method to a Model of Proportional Membership for Fuzzy Cluster Analysis

Susana Nascimento

Abstract This paper presents a fuzzy proportional membership model for clustering (FCPM). Unlike the other clustering models, FCPM requires that each entity may express an extent of each prototype, which makes its criterion to loose the conventional prototype-additive structure. The methods for fitting the model at different fuzziness parameter values are presented. Because of the complexity of the clustering criterion, minimization of the errors requires the gradient projection method (GPM). We discuss how to find the projection of a vector on the simplex of the fuzzy membership vectors and how the stepsize length of the GPM had been fixed. The properties of the clusters found with the FCPM are discussed. Especially appealing seems the property to keep the extremal cluster prototypes stable even after addition of many entities around the grand mean.

Keywords Fuzzy proportional membership • Gradient projection method • Extremal cluster prototype

1 Introduction

Since the introduction of fuzzy partitions by Ruspini [49], fuzzy clustering has grown and become quite prominent in cluster analysis and modeling. The conceptual definition of fuzzy membership is particularly appealing to quantify a grade of membership of entities to clusters. The most known approach to fuzzy clustering is the method of Fuzzy c-Means (FCM) introduced by Dunn [16] and Bezdek [7, 8]. The FCM has received much attention from the scientific community and has been extensively applied in diverse tasks of pattern recognition and image processing [9, 10, 24], data mining and engineering systems design [1, 41, 48], bioinformatics [2], as well as to handle distributed data without resorting to approximations [54].

S. Nascimento (✉)
Department of Computer Science and NOVA Laboratory for Computer Science and Informatics (NOVA LINCS), Faculdade de Ciências e Tecnologia, Universidade Nova de Lisboa, 2829-516 Caparica, Portugal
e-mail: snt@fct.unl.pt

B. Goldengorin (ed.), *Optimization and Its Applications in Control and Data Sciences*, Springer Optimization and Its Applications 115,
DOI 10.1007/978-3-319-42056-1_13

The FCM method and its extensions show how a cluster structure is derived from the data. Yet the method provides no feedback on reconstruction of the data from the found cluster structure, that is, on representing the data as a function of clusters. Specifically, the fuzzy memberships, though expressing the similarity between entities and prototypes indeed, are not involved in the reconstruction of the observations from the clusters.

In this paper we present a fuzzy clustering framework, within the data recovery paradigm advocated by Mirkin [33, 34], for mining for typological structures from data. The model assumes the existence of some prototypes which serve as "ideal" patterns to data entities. To relate the prototypes to the observations, we assume that the observed entities share parts of the prototypes, such that an entity may bear 70 % of a prototype $V1$ and 30 % of prototype $V2$, which simultaneously expresses the entity's membership to the respective clusters. The underlying structure of this model can be described by a fuzzy K-partition defined in such a way that the membership of an entity to a cluster expresses the proportion of the cluster's prototype present in the entity. Specifically, we assume that any entity may independently relate to any prototype, up to the condition that the memberships for any entity must sum to unity, which is similar to the assumption in the FCM criterion. The model, *proportional membership fuzzy clustering* (FCPM) [38–40] provides a family of clustering criteria, FCPM-m, with fuzziness parameter m ($m = 0, 1, 2$), leading to cluster structures with central prototypes (FCPM-0, FCPM-1), closely matching the FCM, as well as cluster structures with extremal prototypes (FCPM-2), close to the concept of ideal, or archetypal, type [15, 35].

To minimize the FCPM criteria, we developed an alternating optimization (AO) algorithm which involves two iterative steps. First, given a set of K prototypes,[1] find the optimal fuzzy membership values satisfying the fuzzy constraints. Second, given the solution from the first step, obtain the optimal set of prototypes. The latter step considers the first-order condition of optimality. To minimize this constrained minimization problem, we adapt the gradient projection method (GPM) by Levitin and Polyak [28], Polyak [45, 46]. The GPM is a powerful and simple method to solve bound-constrained optimization problems and has been successfully applied in various machine learning and data mining algorithms. In [30] the GPM has been applied to non-negative matrix factorization (NMF), a minimization problem with bound constraints, with a method of alternating least-squares using projected gradients leading to faster convergence than the popular multiplicative update method. In [55] the authors used the GPM to determine the hyper-parameters of radial basis kernel functions (RBF) in a multiple kernel learning algorithm by incorporating sequential minimization with the GPM. The approach was applied to multi-class classification problems. In [53] the GPM was applied to a convex programming problem of automatically learning the feature weights to obtain the feature order preferences of a clustering algorithm. In [50] the authors introduce a more broad approach of projected Newton-type methods for solving large-scale

[1] In the follow up text the number of clusters is denoted by K instead of c as in the FCM.

optimization problems present in machine learning. Yet, in [37] fuzzy decision tree is combined with the GPM to control the growth of the parameters during the tuning optimization process. To the best of our knowledge, no previous works on fuzzy clustering took advantage of the GPM. We describe the adaptation of the GPM to the FCPM model in the following aspects: (i) computation of the gradient for the clustering criteria; (ii) choice of a constant step-size length; (iii) method for finding the projection of a vector on the simplex of the membership vectors.

Increasing attention has been given to create synthetic data generators (DG's) aimed at evaluation or demonstration of clustering methods properties. The most popular DG's (e.g. ones integrated in the WEKA system [23]) assume that data are typically generated from traditional multivariate normal distributions ranging from simple to complex covariance structures [25, 32], or based on selected density and a certain "difficulty level" for outlier detection [44]. Yet, more general data generators have been designed, like KNIME [4] a modular platform for data generation for general purpose study of algorithms, a data generator to create clusters from several different distributional families with multidimensional overlap between clusters to compare different initialization strategies of K-means [52], or a flexible data generator producing Gaussian clusters with controlled parameters of between- and within-cluster spread (to model cluster intermix) to study properties of K-means clustering algorithm [14]. However, none of these approaches focus on fuzzy clustering structures. To analyse the data recovery properties of FCPM we constructed a DG according to the FCPM model, and conducted a simulation study with data sets with distinct dimensionalities, exhibiting a typological structure.

Another contribution of this study is to show how an ideal type structure can be found with the FCPM. Ideal types are extreme points that synthesize data representing "pure individual types", and are assigned by the most discriminating features of data points. Recent works on archetypal analysis [11, 13, 17–19, 21, 29, 36, 47] show the importance of the concept of 'archetypal type' in machine learning and data mining applications. The model of archetypal analysis postulates that entities are convex combinations of the archetypes to be found, and that the magnitudes of the K coefficients for each case reflect the relative proximity to each archetype. However, in FCPM the membership values measure the proportions of the prototypes present in the entity and the entities share proportions of the set of prototypes. We further explore the tendency of FCPM to find prototypes with extreme values contrasting to the central prototypes found by FCM. We take a case study from the field of psychiatry [31] presenting extreme cluster prototypes. When augmenting the original data with patients bearing less severe syndromes and running the algorithms, the prototypes found by the FCM move towards the more moderate characteristics of the data while the FCPM prototypes are almost extreme, highlighting the suitability of the FCPM to model the concept of ideal type.

The rest of the paper is organized as follows. The first section presents the FCPM model along with its properties. Section 2 describes a version of alternating minimization algorithm for the model; this involves 'major' and 'minor' iterations with the former involving application of the GPM. Section 3 describes the simulation study involving the stage of data generation as well as a set of simulation experiments to analyse the data recovery properties of the FCPM and FCM methods. It appears the FCPM methods have some properties that allow it to choose the number of clusters by moving some prototypes together or out of the data area. Section 4 describes our experiment in exhibiting the 'extremal' properties of the FCPM types making the FCPM prototypes akin to archetypes.

2 The Model of Fuzzy Proportional Membership

Consider an entity-to-feature data matrix X preprocessed into Y by shifting the origin to the gravity center of all the entities (rows) in Y and rescaling features (columns) by their ranges with $s_h = \max_i (x_{ih}) - \min_i (x_{ih})$. Thus, $Y = [y_{ih}]$ is a $n \times p$ entity-to-feature data table where each entity, described by p features, is defined by the row-vector $\mathbf{y}_i = [y_{ih}] \in \mathrm{Re}^p$ $(i = 1 \cdots n\,;\, h = 1 \cdots p)$.

In the follow-up model in (1) we assume that each entity $\mathbf{y}_i = [y_{ih}]$ of Y is related to each prototype $\mathbf{v}_k = [v_{kh}]$ $(k = 1, \cdots, K)$, as it is in the FCM. We further assume that the membership value u_{ki} is not just a weight, but it expresses the proportion of $\mathbf{v}_k$ which is present in $\mathbf{y}_i$. That is, we consider that approximately $y_{ih} = u_{ki} v_{kh}$ for every feature h. More formally, we suppose that

$$y_{ih} = u_{ki} v_{kh} + e_{kih}, \tag{1}$$

where the residual values e_{kih} are as small as possible.

2.1 *Generic FCPM Square-Error Criterion*

According to model (1), we define the clustering criterion as fitting of each data point to a share of each of the prototypes, represented by the degree of membership. This goal is achieved by minimising all the residual values in (1) via the square-error criterion

$$E_0(U, V) = \sum_{k=1}^{K} \sum_{i=1}^{n} \sum_{h=1}^{p} (y_{ih} - u_{ki} v_{kh})^2 \tag{2}$$

over all admissible u_{ki} and v_{kh}, that is, with regard to the fuzzy constraints

$$0 \le u_{kj} \le 1, \;\text{ for all } k = 1, \ldots, K, j = 1, \ldots, n \tag{3}$$

$$\sum_{k=1}^{K} u_{ki} = 1, \;\text{ for all } i = 1, \ldots, n. \tag{4}$$

on the space of fuzzy K-partitions M_{fKn}.

Equation (1) along with the square-error criterion (2) to be minimised by unknown membership matrix $U \in M_{fKn}$ and set of prototypes $V = (\mathbf{v}_1, \mathbf{v}_2, \ldots, \mathbf{v}_K) \in \mathrm{Re}^{K\times p}$ for Y given, is referred to as the generic *fuzzy clustering proportional membership model*, FCPM-0, for short.

The following properties of the FCPM model (1) have to be considered:

- Each prototype, $\mathbf{v}_k$ is a "model" or "ideal" point such that any entity, $\mathbf{y}_i$, bears a proportion of it, u_{ki}, up to the residuals. The proportion, u_{ki}, is considered as the value of membership of $\mathbf{y}_i$ to the cluster k. So, both the prototypes and memberships are reflected in the model of data generation.
- Equation (1) can be considered as a device to reconstruct the data from the model. The clustering criterion follows the square-error framework to warrant that the reconstruction is, on average, as exact as possible.
- The model (1) may be considered over-specified: any observed entity must share a proportion of each of the prototypes, which ideally may occur only if all the entities and prototypes belong to the same uni-dimensional space. Such a solution is not realistic, especially when contradictory tendencies are present in the data. This property of the generic model led to some over-estimation effects which requires to extend the criterion.
- Due to the bilinear nature of model (1), a property of the clustering criterion (2) is that it remains constant if vectors $\mathbf{v}_i$ and $\mathbf{u}_i$ are changed for $\mathbf{v}_i/\gamma$ and $\mathbf{u}_i\gamma$ for some i, where γ is an arbitrary real. In particular, tending γ to zero, the membership vector, $\mathbf{u}_i\gamma$, tends to zero while the prototype $\mathbf{v}_i/\gamma$ to infinity, without any change in corresponding differences ε in criterion (2). This way, the following phenomenon may occur in the process of adjusting solutions during alternating minimisation of criterion (2): to decrease some of the differences in (2) the membership values involved can be increased with simultaneously decreasing other membership values to zero along with moving corresponding prototypes to infinity. Tending some prototypes to infinity is a specific pattern of non-convergence of the alternating minimisation, which may occur in the generic FCPM model.
- The former two properties may make the model sensitive to the number of clusters K, which is to be specified by the user. When this number is greater than the number of prototypes fitting well in the model, some of the prototypes in a computation may be driven out to infinity in the process of alternating minimisation of the criterion (2).

2.2 Extension of FCPM Criterion

The requirement of FCPM generic criterion (2) that each entity be expressed as a part of each prototype is too strong. Sometimes it is more realistic to consider that only meaningful proportions, those expressed by high membership values, should be taken into account in Eq. (1).

In order to smooth the effect of high residual values e_{ijh} at small memberships u_{ki}, let us weight the squared residuals in (2) by a power m ($m = 1, 2$) of corresponding u_{ki}:

$$E_m(U, V) = \sum_{k=1}^{K} \sum_{i=1}^{n} \sum_{h=1}^{p} u_{ki}^m (y_{ih} - u_{ki} v_{kh})^2, \tag{5}$$

subject to the fuzzy constraints (3) and (4).

The models corresponding to these criteria at different parameter m are referred to as FCPM-1 ($m = 1$) and FCPM-2 ($m = 2$). Criterion (2) is also (5) at $m = 0$. The power m in (5) is similar to the power m in the FCM clustering criterion [8], controlling the extent of sharing among fuzzy clusters. One may assign any value $m \geq 0$ to it. To an extent, their roles are similar in FCM and FCPM: the greater the m, the greater the weight assigned to the difference between the observation and the model value at larger membership values and, in contrast, the smaller the weight of the difference at smaller membership values. Yet in the FCM the model value is just the cluster center component, v_{kh}, which does not depend on the membership value, whereas in the FCPM it is $u_{ki} v_{kh}$, which does depend on the membership value. However, the dependence is not clear-cut; for example, at $y_{ij} < v_{kh}$, the larger u_{kh}^m, the better, and it is quite opposite if $y_{ij} > v_{kh}$. In this study no other values of m are considered, a similar approach to the common practice in FCM of accepting $m = 2$ (e.g. [12, 43]). Other versions of the FCPM are considered in [38].

3 The FCPM Method and Algorithm

3.1 Alternating Minimization: Major and Minor Iterations

Let us consider the FCPM criteria above in the general format of criterion $E : M_{fKn} \times \mathrm{Re}^{K \times p} \rightarrow \mathrm{Re}^+$, to be minimized:

$$\min_{U \in M_{fKn},\, V \in \mathrm{Re}^{K \times p}} E(U, V), \tag{6}$$

following the generic alternating minimization scheme.

The alternating minimization algorithm applied to problem (6) involves two iterating steps. First, given $\widehat{V} \in \mathrm{Re}^{K \times p}$, minimize $E(U, \widehat{V})$ with respect to $U \in M_{fKn}$. Second, given the solution from the first step, $\widehat{U} \in M_{fKn}$, minimize $E(\widehat{U}, V)$ over $V \in \mathrm{Re}^{K \times p}$.

Given fuzzy membership matrix $U^{(t)}$, at tth iteration, minimization of $E(\widehat{U}, V^{(t)})$ with respect to $V \in \mathrm{Re}^{K \times p}$ can be done according to the first-order condition of optimality (i.e. $\nabla E_m(\widehat{U}, V^{(t)}) = 0$). This condition implies that

$$v_{kh}^{(t)} = \frac{\left\langle \left(\mathbf{u}_k^{(t)}\right)^{m+1}, \mathbf{y}_h \right\rangle}{\left\langle \left(\mathbf{u}_k^{(t)}\right)^{m+1}, \mathbf{u}_k^{(t)} \right\rangle}, \tag{7}$$

where parameter m takes value $m = 0, 1, 2$ for either version of FCPM-m, and $\langle \cdot \rangle$ denotes the dot product.

This equation resembles the equation for calculating the FCM prototypes, which suggests that the FCPM does capture the averaging nature of FCM. However, there is a difference as well. In formula (7), the power $m + 1$ of $\mathbf{u}$ in the numerator differs from the power $m + 2$ of $\mathbf{u}$ in the denominator, while these powers coincide in the calculation of FCM prototypes. Therefore, the FCM prototypes are convex combinations of the observed points, yet the FCPM prototypes are not.

The minimization of criterion $E(U, \widehat{V})$ with respect to $U \in M_{fKn}$ is not that straightforward as in FCM because the fuzziness constraints (3) and (4) are not automatically satisfied for the FCPM solutions. That requires an iterative process of its own. After preliminarily experimenting with several options, like the Lagrangian multipliers [6], the gradient projection method developed by Levitin and Polyak [5] has been picked up to solve the constrained minimization problem. This method works especially well for criterion E_0 in (2) as will be shown in the next section.

The gradient projection method is iterative. To distinguish between iterations of the alternating minimization process and iterations within the gradient projection method, we refer to the former ones as "major" iterations and, to the latter ones, as "minor" iterations.

3.2 *The Gradient Projection Method*

The gradient projection method (GPM) belongs to the family of steepest descent methods for constrained minimization. It is adapted here to minimize $E_m(U, \widehat{V})$ over $U \in M_{fKn}$.

Let $f : \mathrm{Re}^K \to \mathrm{Re}$ be a continuously differentiable function to be minimized over a convex subset $Q \subset \mathrm{Re}^K$. For any $\mathbf{z}$ in Re^K, let us denote its projection on Q by $\mathscr{P}_Q(\mathbf{z})$, so that $\mathscr{P}_Q(\mathbf{z})$ minimizes $\|\mathbf{x} - \mathbf{z}\|$ over all $\mathbf{x} \in Q$. Notice that $\mathscr{P}_Q(\mathbf{z})$, denotes the unique projection of vector $\mathbf{z} \in \mathrm{Re}^K$ on Q.

The GPM for solving this constraint optimization problem starts with an arbitrary $\mathbf{x}^{(0)} \in Q$ and iteratively transforms it according to the following rule:

$$\mathbf{x}^{(t+1)} = \mathscr{P}_Q(\mathbf{x}^{(t)} - \alpha \nabla f(\mathbf{x}^{(t)})), \tag{8}$$

where α is a positive constant step size and $\nabla f(\mathbf{x})$ the gradient of f at $\mathbf{x} \in Q$.

The following conditions of convergence of the gradient projection method [6, 46] have to be pointed out.

A vector function $g : \mathrm{Re}^K \rightarrow \mathrm{Re}^K$ is said to satisfy the Lipschitz continuity condition with constant L if

$$\|g(\mathbf{x}) - g(\mathbf{y})\| \leq L\,\|\mathbf{x}-\mathbf{y}\| \qquad \forall \mathbf{x}, \mathbf{y} \in \overset{K}{\mathrm{Re}}\,. \tag{9}$$

A function $f(\mathbf{x})$ is said to be strictly convex with constant $l > 0$ if

$$f(\mathbf{x}+\mathbf{y}) - f(\mathbf{x}) > \langle \nabla f(\mathbf{x}), \mathbf{y} \rangle + \frac{l}{2}\,\|\mathbf{y}\|^2 \qquad \forall \mathbf{x}, \mathbf{y} \in \overset{K}{\mathrm{Re}}\,. \tag{10}$$

For a twice differentiable function f, this is equivalent to its Hessian $\nabla^2 f$, being bounded over Re^K, that is, $\nabla^2 f(\mathbf{x}) \succeq l \cdot I$, where I is the diagonal matrix, and $A \succeq B$ means that $A - B$ is a positive semi-definite matrix.

In order for the GPM to be effective, the constraint set Q should be such that the projection operation $\mathscr{P}_Q(.)$ could be easily carried out. Polyak [46] mentioned several cases of this type including the case(s) in which Q involves upper and/or lower bounds on all the variables in the problem. However, when Q is a general polyhedron, the projection (8) requires solving a quadratic programming problem.

3.3 Applying the Gradient Projection Method at FCPM Minor Iterations

Let us denote the set of membership vectors $\mathbf{u}_j$ satisfying conditions (3) and (4) by Q, which is a convex set (cf. [8, Theorem 6.2.]). With $V^{(t)}$ pre-specified, the function $E(U, \widehat{V})$ is to be minimized over such U whose columns, $\mathbf{u}_j$, belong to Q.

The gradient projection method (8) applied to minimize $E(U, \widehat{V})$ can be stated as follows:

$$\mathbf{u}_i^{(t)} = \mathscr{P}_Q(\mathbf{u}_i^{(t-1)} - \alpha \nabla E(\mathbf{u}_i^{(t-1)}, \widehat{V})), \qquad i = 1, \cdots, n. \tag{11}$$

The possibility of translation of the problem defined over matrices in terms of separate membership vectors in (11) is due to the fact that for each $\mathbf{u}_i^{(t)}$ its components $u_{ki}^{(t)}$ only depend on $u_{ki}^{(t-1)}$.

In order to apply method (11), one needs to specify the following three parts of it: (i) Computation of $\nabla E(\mathbf{u}_i^{(t-1)}, \widehat{V})$; (ii) Choice of a constant step-size length α; (iii) Finding the projection $\mathscr{P}_Q(\mathbf{d}_i)$ for $\mathbf{d}_i = \mathbf{u}_i^{(t-1)} - \alpha \nabla E(\mathbf{u}_i^{(t-1)}, \widehat{V}) \in \mathrm{Re}^K$ $(i = 1, \cdots, n)$.

For the sake of simplicity, we start from the criterion of the generic model E_0 in (2), as the $E(U, \widehat{V})$. Then we extend the analysis to the other values of m. The demos of the mathematical properties presented may be consulted in [38, 40].

The function $E_0(U, \widehat{V})$ is convex and twice differentiable over its variables u_{ij} [38]. The elements of its gradient are

$$\nabla E_0\left([\mathbf{u}_i], \widehat{V}\right) = 2(\langle \mathbf{v}_k, \mathbf{v}_k\rangle\, u_{ki} - \langle \mathbf{y}_i, \mathbf{v}_k\rangle), \qquad k = 1, \cdots, K, \tag{12}$$

and its Hessian is a $Kn \times Kn$ diagonal matrix whose $((k,i),(k,i))$th element is $2\langle \mathbf{v}_k, \mathbf{v}_k\rangle$.

Let us denote $l = 2\min_k \langle \mathbf{v}_k, \mathbf{v}_k\rangle$ and $L = 2\max_k \langle \mathbf{v}_k, \mathbf{v}_k\rangle$. Naturally, $L \cdot I \succeq \nabla^2 E_0(U, \widehat{V}) \succeq l \cdot I$. We assume all $\mathbf{v}_k$ are nonzero which implies $L \geq l > 0$.

The gradient ∇E_0 satisfies the Lipschitz condition over Q with constant L thus defined. Indeed,

$$\begin{aligned}\nabla E_0\left([\mathbf{u}_i], \widehat{V}\right) - \nabla E_0\left([\mathbf{z}_i], \widehat{V}\right) &= 2\langle \mathbf{v}_k, \mathbf{v}_k\rangle (u_{ki} - z_{ki}) \\ &\leq L(u_{ki} - z_{ki}), \qquad \forall \mathbf{u}_i, \mathbf{z}_i \in Q.\end{aligned} \tag{13}$$

which implies the same inequality in terms of the vector norms, that is, the Lipschitz condition (9).

The situation for functions $E_m(U, \widehat{V})$ $(m = 1, 2)$ is different: neither is convex over Q, though each satisfies the Lipschitz condition.

The elements of the gradients ∇E_m are expressed as

$$\begin{aligned}\nabla E_m\left([\mathbf{u}_i], \widehat{V}\right) = {} & (m+2)\langle \mathbf{v}_k, \mathbf{v}_k\rangle\, u_{ki}^{m+1} - 2(m+1)\langle \mathbf{v}_k, \mathbf{y}_i\rangle\, u_{ki}^{m} + \\ & m\langle \mathbf{y}_i, \mathbf{y}_i\rangle\, u_{ki}^{m-1},\end{aligned} \tag{14}$$

for $m \geq 0$.

Note that (12) is a special case of (14) at $m = 0$. On the other hand, it is proven that

$$\left|\nabla E_m\left([\mathbf{u}_i], \widehat{V}\right) - \nabla E_m\left([\mathbf{z}_i], \widehat{V}\right)\right| \leq L\,|(u_{ki} - z_{ki})|, \tag{15}$$

with $|\cdot|$ the L_1 norm, and L a constant equal to

$$L = (m+2)(m+1)V + 2m(m+1)YV + m(m-1)Y, \tag{16}$$

with $V = max_k \langle \mathbf{v}_k, \mathbf{v}_k\rangle$, $YV = max_{k,i} |\langle \mathbf{v}_k, \mathbf{y}_i\rangle|$ and $Y = max_i \langle \mathbf{y}_i, \mathbf{y}_i\rangle$. This shows that ∇E_m (with $m \geq 1$) satisfies the Lipschitz condition for the L_1-norm with constant L defined above.

Although the Lipschitz continuity condition (9) is defined for the L_2-norm, it is known that the condition holds or does not hold in both L_2 and L_1-norms simultaneously, though the constant L in (9) may change [46].

When a function is not convex, yet satisfies the Lipschitz condition, the gradient projection method may converge to a local optimum only, adding this to the general local search nature of the method of alternating optimization.

According to the convergence properties of the GPM method [6, 46] we take the step-size $\alpha = \frac{2}{(1+\varepsilon)L}$, so that α spans the interval between 0 and $2/L$ when ε changes from 0 to infinity. The value $\varepsilon = 0.5$ was chosen experimentally as giving the most stable convergency rates.

Substituting L from (16) into the formula, one obtains

$$\alpha_m = \frac{1}{1.5(c_m^1 V + c_m^2 YV + c_m^3 Y)}, m = 0, 1, 2, \tag{17}$$

with V, YV and Y defined above, and coefficients c_m^j defined by:

m	0	1	2
c_m^1	1	3	6
c_m^2	0	2	6
c_m^3	0	0	1

Now we can turn to the problem (iii) of projection of the difference vectors $\mathbf{d}_i = \mathbf{u}_i^{(t-1)} - \alpha \nabla E_m(\mathbf{u}_i^{(t-1)}, \widehat{V})$ onto the set Q of vectors satisfying conditions (3) and (4).

For each criterion E_m, vectors $\mathbf{d}_i = [d_{ki}]$ to be projected onto Q are defined by equations

$$d_{ki}^{(t)} = u_{ki}^{(t-1)} - 2\alpha_m \left[(m+2) \langle \mathbf{v}_k, \mathbf{v}_k \rangle \left(u_{ki}^{(t-1)} \right)^{m+1} - 2(m+1) \langle \mathbf{v}_k, \mathbf{y}_i \rangle \left(u_{ki}^{(t-1)} \right)^m + m \langle \mathbf{y}_i, \mathbf{y}_i \rangle \left(u_{ki}^{(t-1)} \right)^{m-1} \right], \tag{18}$$

derived from (11) with ∇E_m in (14) substituted for ∇E.

We start by presenting a generic method for projecting a vector on the simplex of the membership vectors.

3.3.1 Projecting a Vector on the Simplex of the Membership Vectors

Let us consider the problem of finding a vector $\mathbf{u} = [u_k] \in Q (k = 1, \cdots, K)$, which is at the minimum distance from a pre-specified vector $\mathbf{d} = [d_k]$. This problem can be stated as follows:

$$\min_{\mathbf{u}} f(\mathbf{u}) = \|\mathbf{u} - \mathbf{d}\|^2, \tag{19}$$

subject to constraints (3) and (4).

In order to solve this problem, we assume, without any loss of generality, that $d_1 \geq d_2 \geq \cdots \geq d_K$. The optimal $\mathbf{u}^*$ of problem (19) has the same order of components, that is, $u_1^* \geq u_2^* \geq \cdots \geq u_K^*$. Let us assume that $u_1^* \geq u_2^* \geq \cdots \geq u_{k+}^* > 0$ for some $k^+ \leq K$. If $k^+ < K$, the final $K - k^+$ components are zero. For the non-zero components, the following equations hold at an optimal $\mathbf{u}^*$:

$$u_1^* - d_1 = u_2^* - d_2 = \cdots = u_{k+}^* - d_{k+}.$$

Otherwise, we could transform $\mathbf{u}^*$ as above by redistribution of values among the positive $u_1^*, \cdots, u_{k+}^*$ in such a way that its distance from $\mathbf{d}$ decreases, which would contradict the assumption that the distance had been minimized by $\mathbf{u}^*$. Thus, for the optimal $\mathbf{u}^*$, $u_1^* = d_1 - a_{k+}, u_2^* = d_2 - a_{k+}, \cdots, u_{k+}^* = d_{k+} - a_{k+}$, where a_{k+} is the common value of the differences; it can be determined from the result of summation of these equations as

$$a_{k+} = \frac{1}{k+}\sum_{k=1}^{k+} d_k - \frac{1}{k+}, \tag{20}$$

The value of k^+ is not known beforehand. To find it, the following iterative process can be applied. Start with $k^+ = K$, and at each iteration compute a_{k+} with formula (20) and take the difference $u_{k+}^* = d_{k+} - a_{k+}$. If it is less than or equal to zero, decrease k^+ by 1 and repeat the process until the difference becomes positive. Then define all the other u_k^* as follows: $u_k^* = d_k - a_{k+}$ for $k = 1, \ldots, k^+$ and $u_k^* = 0$ for $k = k^+ + 1, \cdots, K$. The process can be accelerated if, at each iteration, k^+ is decreased by the number of negative values in the set of differences $u_k^* = d_k - a_{k+}$ $(k = 1, \ldots, k^+)$. This is described in the following Algorithm A1.

Algorithm A1 *Projection*$_Q$(**d**)

---

1 *Given* **d**=$[d_k]$ $(k = 1, \cdots, K)$
2 *sort* **d**= $[d_k]$ *in the descending order;*
3 $k^+ := K$;
4 **Repeat**
5 *calculate* a_{k+} *by* (20);
6 *zeros* := *false*; $k := 0$;
7 **Repeat**
8 $k := k + 1$;
9 $u_k := d_k - a_{k+}$;
10 **If** $u_k \leq 0$ **then** *zeros* := *true*;**endIf**
11 **until** ($k = k^+$**.or.***zeros*);
12 **If** *zeros* **then**
13 **For** $j = k, \cdots, k^+$ **do** $u_j := 0$; **endFor**
14 $k^+ := k - 1$;
15 **endIf**
16 **until** ($k^+ = 0$ **.or.***notzeros*);
17 **return u** $= [u_1, \ldots, u_{k+}, 0, \ldots, 0]$;

3.4 The FCPM Alternating Optimization Algorithm

The FCPM algorithm is defined as an alternating minimization algorithm, in which each "major" iteration consists of two steps as follows. First, given prototype matrix V, the optimal membership values are found with Eq. (11) through the gradient projection iterative process (Algorithm A1). Second, given membership matrix U, the optimal prototypes are determined according to the first-degree optimality conditions (7).

Thus, the FCPM algorithm involves "major" iterations to update matrices U and V and "minor" iterations to recalculate the membership values, using the gradient projection method within each of the "major" iterations.

The algorithm starts with a set $V^{(0)}$ of K arbitrarily selected prototype points in Re^p and $U^{(0)}$ in M_{fKn}; it stops when the difference between successive prototype matrices becomes small according to an appropriate matrix norm, $|\cdot|_{err}$. The FCPM-m ($m = 0, 1, 2$) algorithm is defined in A2.

The FCPM algorithm converges only locally as the FCM does. Moreover, with a "wrong" number of pre-specified clusters, FCPM-0 may not converge at all since it may shift some prototypes to infinity. In an extensive experimental study conducted with simulated data, the number of major iterations in FCPM-0 algorithm is rather small when it converges, which is used as a stopping condition: when the number of major iterations in an FCPM-0 run goes over a large number (in our calculations, over 100), that is an indicator that the process is likely to not converge. This property is discussed in the next section.

4 Simulation Study

To analyze the data recovery properties of the FCPM model, a simulation study has been conducted with artificial generated data considering four main objectives:

O1 To compare results found by FCPM and FCM methods.
O2 To analyze the ability of FCPM to recover the original prototypes.
O3 To use the behavior of FCPM-0 as an indicator of the "right" number of clusters.
O4 To analyze the behavior of FCPM algorithms at numbers of clusters greater than those in the data.

To accomplish this, a data generator has been constructed according to the assumptions underlying the FCPM model so that each prototype is an "ideal" point such that any entity bears a proportion of it.

Algorithm A2 *FCPM-m Algorithm*

```
1   Given Y = [y_i]
2   choose K (2 ≤ K < n), (m = 0,1,2), T_1, T_2, ε > 0;
3   initialise V^(0) = {v_k^(0)}_{k=1}^K, U^(0) ∈ M_fKn, t_1 := 0;
4   Repeat
5     t_2 := 0;
6     U^(t_2) := U^(t_1);
7     Repeat
8       t_2 := t_2 + 1;
9       For i = 1, ··· n do
10        calculate d_i^(t_2) with V^(t_1), u_i^(t_2−1) by (18);
11        u_i^(t_2) := Projection_Q(d_i^(t_2))        %(Algorithm A1)
12      endFor
13    until (|U^(t_2) − U^(t_2−1)|_err < ε .or. t_2 = T_2);
14    t_1 := t_1 + 1;
15    U^(t_1) := U^(t_2);
16    calculate V^(t_1) with U^(t_1) by (7);
17  until (|V^(t_1) − V^(t_1−1)|_err < ε .or. t_1 = T_1);
18  return (V, U) := (V^(t_1), U^(t_1));
```

4.1 Generation of Data and Assessment of Results

The FCPM data generator was constructed as follows:

1. The dimension of the space (p), the number of clusters (K_0), and the number of entities generated within each cluster as numbers $n_1, n_2, \ldots, n_{K_0}$, are randomly generated within pre-specified intervals: *[min_DimP, max_DimP]* for space dimensionality p; *[min_K, max_K]* the interval for the number of clusters; *[min_PtsClt, max_PtsClt]* for the total number of points (n_k) to be generated within each cluster k ($k = 1, \cdots, K_0$). The data set cardinality is defined then as $n = \sum_{k=1}^{K_0} n_k$.
2. K_0 cluster directions are defined as follows: prototypes $\mathbf{o}_k \in \mathrm{Re}^p$ ($k = 1, \cdots, K_0$) are randomly generated within a pre-specified hyper-cube with side length between *−100.0* and *100.0*; then, their gravity center $\mathbf{o}$ is taken as the origin of the space. Each cluster direction is taken as the segment $\overrightarrow{\mathbf{o}\mathbf{o}_k}$
3. For each k ($k = 1, \cdots, K_0$), define two p-dimensional sampling boxes, one within bounds $A_k = [.9\mathbf{o}_k, 1.1\mathbf{o}_k]$ and the other within $B_k = [\mathbf{o}, \mathbf{o}_k]$. Then, generate randomly a small percentage of points, *percent_PtsOrgVs* (e.g. $0.2n_k$) in A_k and the remaining points, $(1 - \mathit{percent\_PtsOrgVs}) \cdot n_k$ (e.g. $0.8n_k$) in B_k.
4. The data generated (including the K_0 original prototypes) are normalized by centering to the origin and scaling by the ranges of features.

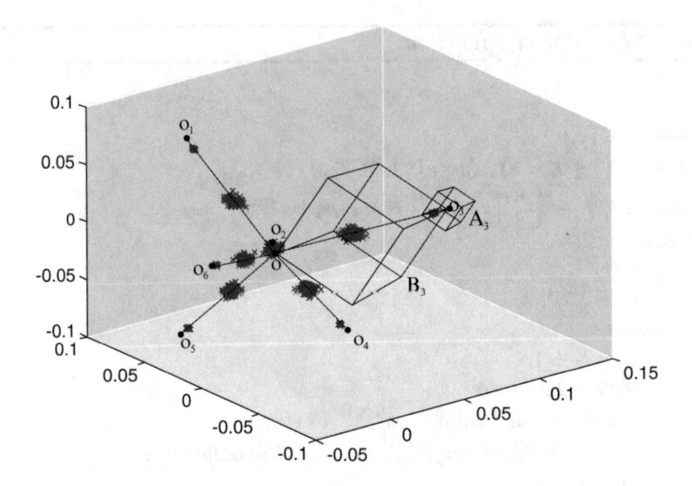

Fig. 1 Architecture of the data generated on a 3D projection of the best three principal components, with p-dimensional sampling boxes A_k and B_k for a data structure with six original prototypes

All randomly generated items come from a uniform distribution in the interval $[0, 1]$. Figure 1 illustrates the architecture of the data generator on a 3D projection of the best three principal components, with p-dimensional sampling boxes A_k and B_k, for a data structure with $K_0 = 6$ original prototypes.

Due to the characteristics of this data generator, rather complex data structures can be generated with a small number of easily interpretable parameters, something that would be much more difficult to achieve using the traditional multivariate normal distributions for data generation [3, 25, 32, 44].

A number of arbitrary decisions were made in the design of the algorithm of the FCPM data generator, which refers to the chosen parameter values, that is a common process for generating artificial data. The user can change any of these decisions by instantiating the input parameters when generating data. In the conducted simulation study these parameters have been chosen as follows: [*min_K, max_K*]=[2,8]; [*min_DimP, max_DimP*]=[2,200]; [*min_HCube, max_HCube*]=[-100.0,100.0]; *percent_DSeg= 0.1* and *percent_PtsOrgVs=0.2*.

To visualize the distribution of the data points, a 2D/3D space of the best principal components is used, into which these points are projected.

About 100 data sets have been generated. For a fixed pair, p and K_0, a group of 15 data sets were generated with different numbers of entities and different original prototypes. The experiments comprised seven such groups with p ranging from 5 to 180 and K_0 from 3 to 6.

Our criteria (5) are more complex than that of FCM and thus require more calculations. In our experiments, each of the algorithms (FCM, FCPM-0, FCPM-1, and FCPM-2) was run on the same data set and starting from the same initial prototypes (i.e. with the same initial setting) for different values of K ($K = 2, 3, 4 \ldots$).

The instantiation of the parameters of the algorithms: (i) $T_1 = 100$, $\varepsilon = 0.0001$, $|\cdot|_{err} = L_1$-norm in $\mathrm{Re}^{K\times p}$ (FCM & FCPM common parameters); (ii) $m = 2$, $\|\cdot\| = L_2$ -norm (FCM own parameters); (iii) $T_2 = 100$ (FCPM own parameters).

Our main interest on discussing the experimental results is given to the clustering results rather than the performance of the algorithms.

In order to compare the prototypes found by FCPM (V'), with "reference" prototypes V, which were either FCM found prototypes (goal O1) or generated prototypes (goal study O2), we introduced a dissimilarity coefficient to measure the squared relative quadratic mean error between corresponding prototypes of sets $V' = \left\{\mathbf{v}'_j\right\}_{j=1}^{K'}$ and $V = \{\mathbf{v}_k\}_{k=1}^{K}$, defined by:

$$D\left(V', V\right) = \frac{\sum_{k=1}^{K}\sum_{h=1}^{p}(v'_{kh} - v_{kh})^2}{\sum_{k=1}^{K}\sum_{h=1}^{p}v_{kh}^2 + \sum_{k=1}^{K}\sum_{h=1}^{p}v'^2_{kh}}. \tag{21}$$

In (21), the matching between prototypes is determined according to minimum distances. In the case in which the number of prototypes found by FCPM-0, K' is smaller than K, only K' "reference" prototypes participate in (21). Coefficient $D\left(V', V\right)$ is not negative, and it equals 0 if and only if $v_{kh} = v'_{kh}$ for all $k = 1, \cdots, K; h = 1, \cdots p$. In a typical situation, when the components of each $\mathbf{v}_k$ and $\mathbf{v}'_k$ are in the same orthants, then D is not greater than 1. Notice that the dissimilarity measure $D\left(V', V\right)$ is more or less independent of the "reference" prototypes ($\mathbf{V}$'s), their cardinality (K_0), and dimension (p); thus, it can be used to compare cluster prototypes in different settings.

In order to measure the separability of found partitions, the separability index, B_k by Backer (as defined in [8]) was used. This index is defined as follows:

$$B_k(U) = 1 - \frac{k}{k-1}\left(1 - \sum_{k=1}^{K}\sum_{i=1}^{n}\frac{u_{ki}^2}{n}\right) , \tag{22}$$

with $2 \leq k < n$. In Bezdek [8] it is proven that $0 \leq B_k(U) \leq 1$, with $B_k(U) = 1$, when U is hard; and $B_k(U) = 0$ for an entirely fuzzy k-partition, i.e. $\overline{U} = [1/k]$.

In summary, in this approach a fuzzy k-partition is induced from a hard k-partition and an affinity index. It seems that Backer's approach to clustering present greater success than the k-means family when clusters are not essentially hyper-ellipsoidal in shape.

In what follows, we refer to three types of the numbers of prototypes: (1) the number of originally generated prototypes, K_0, (2) the number of prototypes pre-specified in a run, K, and (3) the number of prototypes found by an algorithm, K'. The cluster solutions found with FCPM algorithms are characterized according to: (1) the number of clusters found, K'; (2) the dissimilarity $D(V_{FCPM}, V_{FCM})$ between the prototypes found by FCPM and those by FCM; (3) the dissimilarity

$D(V_{FCPM}, V_O)$ from the generated prototypes $O = \{\mathbf{o}_k\}_{k=1}^{K_0}$ and (4) the separability index, $B_{K'}$. The separability index was also calculated for the FCM solutions to compare different versions of the algorithm.

4.2 Discussion of the Simulation Results

Results of our experiments with FCM and FCPM algorithms lead us to distinguish between three types of data dimensionality: low, intermediate, and high, because the algorithms behave differently across these categories. With several hundred entities, the dimension p satisfying $p/K_0 \leq 5$ is considered small and p satisfying $p/K_0 \geq 25$ high. Note that the numbers of prototypes K' and K in the same computation may differ because of either of two causes:

(C1) some of the initial prototypes converge to the same stationary point;
(C2) some of the initial prototypes have been removed by the algorithm from the data cloud (this concerns FCPM-0).

In either case, $K' < K$.

Figure 2 shows the kind of cluster structure we deal with, taking a small data set generated with $K_0 = 3$ original prototypes in Re^2 ($p = 2$) with $n = 49$ points. The prototypes found by running FCM and FCPM algorithms started from the same initial setting at $K = 3$ are also displayed. The FCPM-0 algorithm moved one of the prototypes (that corresponding to cluster 2) far away from cluster 2 to the left, so that its points, in the end, share the prototype with cluster 3. Concerning the other FCPM and the FCM algorithms, all of them found their partitions with $K' = 3$ prototypes. Method FCPM-2 produced the most extremal prototypes close to the original ones, and FCPM-1 produced prototypes close to the prototypes found by FCM.

In the main series of experiments the number of prototypes looked for were taken as equal to the number of original prototypes, $K = K_0$.

The average results of running FCM and FCPM algorithms with $K = K_0$ for each of the three groups of data sets: namely, at small dimension ($p = 5$, $K_0 = 3$), intermediate dimension ($p = 50$, $K_0 = 4$), and at high dimension ($p = 180$, $K_0 = 6$), are presented in the following four tables: Table 1 shows the number K' of prototypes found, and when $K' < K$, the cause, either C1 or C2, is shown as a superscript. Table 2 shows the average dissimilarity coefficient either concerning FCM prototypes (left part), and generated prototypes (right part). These average values correspond to the average relative square error of FCPM solutions with respect to the reference prototypes. Table 3 shows the average partition coefficient, B_K. Finally, in Table 4 the number of major iterations, t_1 taken by each algorithm is shown.

Another set of experiments were carried out with $K = K_0 + 1$. The goal of study was to analyze the sensitivity of FCPM algorithms to larger numbers of pre-specified prototypes than those at which the data were generated (goal study O4). Depending

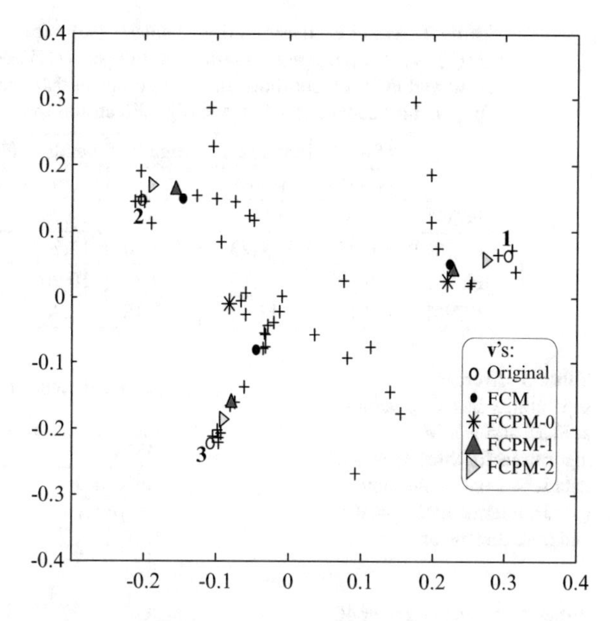

Fig. 2 Results of clustering for thc illustrative data set ($K_0 = 3, p = 2, n = 49$), with data points marked with symbol '+'. All FCPM and FCM algorithms find $K' = 3$ prototypes, except for FCPM-0, that moved prototype 2 far away to the left of its cluster. Therefore, its points share the prototype with cluster 3 according to FCPM-0

Table 1 Number of prototypes found by FCM and FCPM algorithms for the three groups of data sets: at small dimension (p=5), intermediate (p=50), and at high dimension (p=180). Upper indices C_1 and C_2 are explained in the text

	Small	Intermed.	High
k'	($K_0 = 3$)	($K_0 = 4$)	($K_0 = 6$)
FCM	3	4	1^{C1}
FCPM-0	3	4	6 or 5^{C2}
FCPM-1	3	4	6
FCPM-2	3	4	6

on the ratio p/K, the FCM and FCPM algorithms behave differently. The numbers of prototypes found by each algorithm are presented in Table 5.

The results of these experiments are the following:

Number of clusters Concerning the correctness of the number of clusters retrieved from the data, the methods fall into three groups:

(i) Methods retaining the pre-specified number of clusters: FCPM-1 and FCPM-2.

(ii) Methods which can reduce the number of clusters (especially in the high dimensional data case) due to either of the causes $C1$ (FCM) or $C2$ (FCPM-0) above.

For low and intermediate dimensions, FCPM-0 almost always finds the correct number of generated clusters (Table 1). In the high dimensional spaces, FCPM-0

Table 2 Average dissimilarity values (%) for: FCM prototypes $(D(V_{FCPM}, V_{FCM}))$ and original prototypes $(D(V_{FCPM}, V_O))$ running FCM and FCPM algorithms for three groups of data sets: small dimension (p=5), intermediate (p=50), and high dimension (p=180)

	Small	Intermed.	High	Small	Intermed.	High
	$D(V_{FCPM}, V_{FCM})$			$D(V_{FCPM}, V_O)$%		
FCM	–	–	–	14.70	17.90	96.83
FCPM-0	0.49	1.23	143.50	12.20	14.34	11.67
FCPM-1	0.89	0.16	94.20	10.20	15.31	15.82
FCPM-2	7.10	11.44	97.18	2.30	1.16	0.45

Table 3 Average separability index B_K values at FCM and FCPM algorithms for three groups of data sets: small dimension (p=5), intermediate (p=50), and high dimension (p=180)

Space dimension	Small	Intermed.	High
	B_K		
FCM	0.61	0.47	0.01
FCPM-0	0.84	0.90	0.78
FCPM-1	0.80	0.98	1.00
FCPM-2	0.43	0.36	0.30

Table 4 Average number of major iterations, t_1, for FCM and FCPM algorithms at the three groups of small, intermediate and high dimensional data sets

t_1	Small	Intermed.	High
FCM	12	15	27
FCPM-0	10	20	78/101
FCPM-1	11	9	11
FCPM-2	11	11	27

Table 5 Number of prototypes found by the FCM and FCPM algorithms run with $K = K_0 + 1$. Upper indices C_1 and C_2 are explained in the text

Space dimension	Small	Intermediate	High
k_0+1=	4	5	7
FCM	4	4^{C1}	1^{C1}
FCPM-0	3	4^{C2}	$(6;5;4)^{C2}$
FCPM-1	4	4^{C1}	6^{C1}
FCPM-2	4	4^{C2}	6^{C2}

finds the correct number of clusters in 50 % of the cases and it underestimates the number of clusters in the remaining cases removing some of the prototypes out of the data set area (as illustrated in Fig. 2). In the high dimensional spaces FCM typically leads to even smaller number of clusters, making the initial prototypes to converge to the same point. Further experiments show that this feature of FCM depends not only on the space dimension but also on the generated data structure. Indeed, for the high dimensional data, FCM views the entire data set as just one cluster around the origin of the space, because there are not that many points generated "outside" of it. However, when the proportion of points generated around the original prototypes (within the boxes A_k) is increased from 0.2 to 0.8, FCM identifies the correct number of prototypes (see [26, 27, 51] for discussions of the challenges of the curse of dimensionality). Figure 3 displays a data set

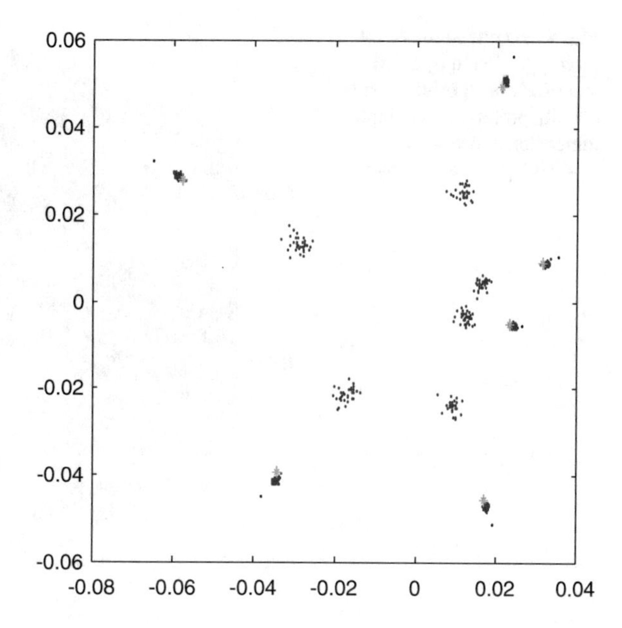

Fig. 3 A high dimension data set ($n = 975, p = 180, K_0 = 6$) with data points generated around each original prototypes in a proportion of 0.8 instead of 0.2, as in the original data generation. The original prototypes as well as the FCM and FCPM-0 found prototypes are marked

generated under this configuration, with the corresponding FCM and FCPM-0 prototypes marked. In such a case, FCM and FCPM-0 prototypes more or less coincide.

Proximity to FCM or generated Prototypes The prototypes found by FCPM-1 and FCPM-0 almost coincide with those found by FCM when the number of prototypes is correctly determined by FCM. These prototypes differ from those originally generated, since they tend to be central points in the clusters. In contrast, the FCPM-2 prototypes go to the extremes of the clusters. Indeed, FCPM-2 identifies the originally generated prototypes and, thus, yields a result differing from that of FCM (see the corresponding $D(V_{FCPM}, V_O)$ values in Table 2). This effect is especially visible when the ratio p/K increases. The prototypes found by FCPM-0 are intermediate between those found by FCM and FCPM-2. Figure 4 illustrates the relative locations of FCPM and FCM prototypes, with respect to the original prototypes, for a high dimensional data set.

Partition Separability The average values of the partition separability coefficient, B_K (Table 3) showed that the FCPM partitions are more contrasting than those by the FCM. In high dimension cases FCPM-1 led to hard clustering solutions. This is because the proportional membership is more sensitive to the discriminant attribute values characterizing a cluster, when compared with the FCM membership. The FCPM-2 gave the fuzziest partitions, typically differing from those of FCM.

Iterations On average, the number of major iterations (t_1) in FCPM-1 and FCPM-2 is smaller than in FCM, whereas in FCPM-0 this number does not differ significantly from that in FCM (for small dimensions). However, the running

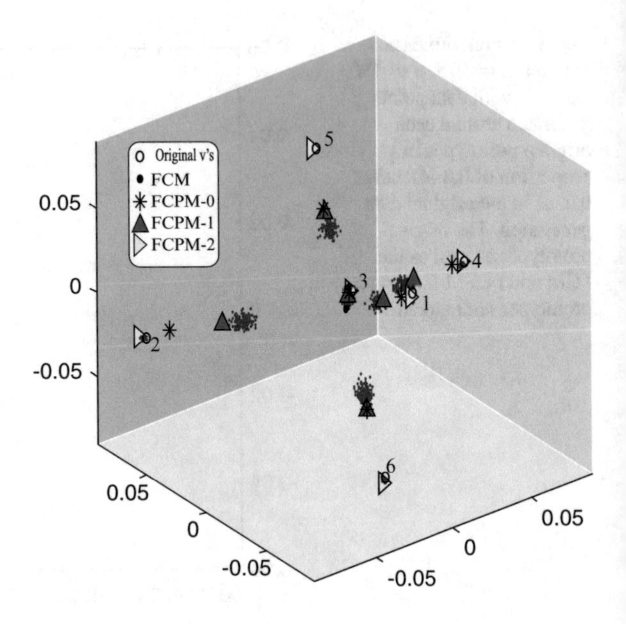

Fig. 4 3D projection of the prototypes found by FCM and FCPM's algorithms with $c = K_0$ prototypes at a high dimensional data set with $n = 887, p = 180, K_0 = 6$

time is greater for FCPM algorithms due to the time spent in the minor iterations of the gradient projection method.

Higher Number of Clusters The results presented in Table 5 show the following. For small dimensional data, FCM, FCPM-1, FCPM-2 found $K' = K_0 + 1$ distinct prototypes. The FCPM-0 removed the extra prototype out of the data space.

For intermediate dimensional data, FCM and FCPM-1 just found $K' = K_0$ distinct prototypes; the extra prototype almost always moved to coincide with one of the others. Both FCPM-2 and FCPM-0 found $K' = K_0$ prototypes by removing an extra prototype out of the data set area (Fig. 5), rather than by merging two different prototypes.

For the high dimensional data both FCM and FCPM-0 led to "degenerated" solutions: either several prototypes coincide with each other (FCM) or more than one prototype moves out of the data space preventing the algorithm to converge (FCPM-0). Still, the other methods, FCPM-1 and FCPM-2, recovered the number of prototypes that had been generated (Fig. 5). However, FCPM-1 led to hard clusters in the high dimensional cases.

Behavior of the other features ($D(V_{FCPM}, V_{FCM})$, $D(V_{FCPM}, V_O)$, and B_K), does not differ from that shown in Tables 2 and 3. Overall, in the high dimensional cases, the winner is the FCPM-2 which recovers the original "extremal" prototypes.

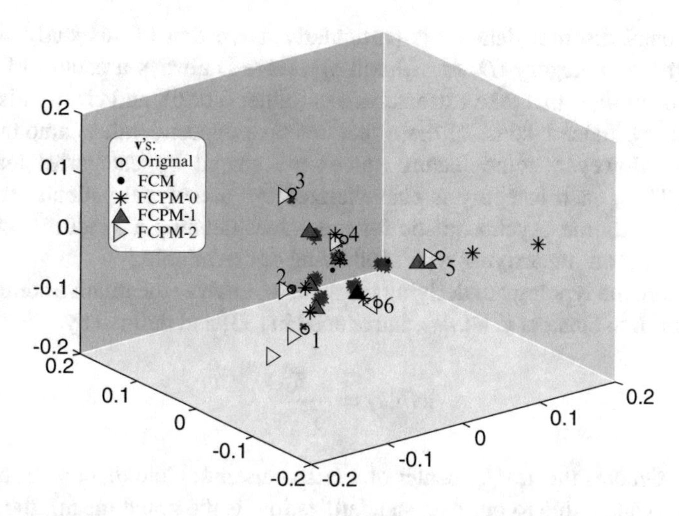

Fig. 5 3D projection of the prototypes found by FCM and FCPM with $K = K_0 + 1$ (high dimension case at $p = 180, K_0 = 6$). FCM and FCPM-0 lead to different degenerate solutions: FCM makes all the prototypes coincide with each other, whereas FCPM-0 removes more than one prototype out of the data area. FCPM-1 and FCPM-2 recover the original number of generated prototypes

5 Retrieving Extreme Types with FCPM

Archetypes are extreme points that synthesize data representing "pure individual types", and are assigned by the most discriminating features of data points in such a way that each data point is a convex combination of prototypes band, vice versa, each prototype is a convex combination of the data points [15]. In contrast to K-means or fuzzy K-means types, the archetypes are not central points but rather those extremal. In this regard we note that the FCPM models have something to do with the archetypes. Indeed, our model was introduced as an extension of the model of "ideal" fuzzy types by Mirkin and Satarov published in 1990 [35]. In that model, each data point is a convex combination of the prototypes, aka ideal types, although no reverse relation is required or maintained. Yet the extremal character of the fuzzy ideal types can be seen quite clearly. In this section, we describe an experiment demonstrating how an archetypal type structure can be recovered from data with the FCPM as a move towards FCM.

We use a data set from the psychiatric domain [31]. This data set consists of 44 patients, described by seventeen psychosomatic features (h_1–h_{17}) (see [31] for a complete description). The features are measured on a severity rating scale taking integer values between 0 and 6. The patients are partitioned into four classes of mental disorders: depressed (D), manic (M), simple schizophrenic (S_s) and paranoid schizophrenic (S_p). Each class contains eleven entities that are considered 'archetypal psychiatric patients' of that class. The parameters of this data set are: $n = 44$ ($n_1 = n_2 = n_3 = n_4 = 11$), $p = 17$ and $c = 4$.

The mental disorder data set is particularly interesting to this study since, for each diagnostic category (D, M, S_s, and S_p), there is always a pattern of features (a subset of h_1–h_{17}) that take extreme values (either 0 or 6), and clearly distinguish each category. Indeed, some of these features take opposite values among distinct categories. However, some feature values are shared by categories leading to overlaps. Thus, each category is characterized by 'archetypal patients' that show a pattern of extreme psychosomatic feature values defining a syndrome of mental conditions (i.e. an 'underlying type', following our terminology).

To capture the typology underlying the data, we analyze the relative contributions of variables h to clusters k, $w(h|k)$, introduced in [33] and defined by

$$w(h|k) = \frac{\overline{v}_{kh}^2}{\sum_h \overline{v}_{kh}^2} \tag{23}$$

where $\overline{v}_{kh}$ denotes the gravity center of a crisp cluster k. Note that the farther $\overline{v}_{kh}$ is from zero (which, due to our data standardization, is the grand mean), the easier is to separate that cluster from the other ones in terms of variable h, which is reflected in the weight values. Therefore, $w(h|k)$ can be viewed as a measure of the "degree of interestingness" of variable h in cluster k with regard to its "standard" mean value.

We consider the set of 'most contributing features within a cluster' as the features which may act as frontier values. We define the *'underlying type'* of a cluster k, $\boldsymbol{\tau}_k$, as the prototype of the cluster characterized by the set $W_{h|k}$ of its most contributing features.

Table 6 presents the contribution weights of clustering elements to the original classes of mental disorders. The mean value of the contribution weights defines a threshold, so that features contributing more than that should be considered as those characterizing the underlying type, $\boldsymbol{\tau}_\mathbf{i}$, of each of the classes/clusters. These contributions are outlined by boxes.

In order to find the underlying types of the mental disorders, algorithms FCPM-2 and FCM (with parameter $m = 2$) were run starting from the same initial setting, at $K = 4$. The headers of Tables 7 and 8 show the features characterizing each underlying type, selected from the boxed values in Table 6. The row beneath each header in these tables displays the underlying types $\boldsymbol{\tau}_i$, found by FCM and FCPM-2. The values of $\boldsymbol{\tau}_k$ are in the original data scales.

The upper rows of Table 7 display the average values of cases belonging to the corresponding diseases, whereas in Table 8 such values correspond to extreme values which are followed by the number of cases at which these values were attained. This format allows us to show the distinct cluster tendencies of clusters found by FCM and FCPM-2. It is quite clear that the underlying types found by FCPM-2 do correspond to extreme values, in contrast to the corresponding ones found by FCM—these are central ones. Concerning the membership values found, both algorithms assign the highest membership value for an entity to its original class, thus correctly grouping all entities to the their classes. The only exception occurs for entity '21' from class M, which is assigned to class S_p. This goes in line,though, with results found for other clustering algorithms such as complete linkage, K-means, in Mirkin [33].

Table 6 Relative contributions of the 17 psychosomatic variables (h_1-h_{17}) to classes D, M, S_s and S_p; The features values higher than the corresponding means are marked

	D	M	S_s	S_p
h	$w(h\|D)$	$w(h\|M)$	$w(h\|S_s)$	$w(h\|S_p)$
h_1	0.050	0.047	0.003	0.001
h_2	0.028	0.020	0.049	0.028
h_3	0.027	**0.195**	**0.096**	0.000
h_4	0.032	0.000	0.007	0.019
h_5	**0.167**	0.041	0.013	0.028
h_6	0.015	0.045	0.058	0.018
h_7	0.014	0.008	0.022	0.013
h_8	**0.117**	**0.168**	**0.167**	**0.115**
h_9	**0.183**	**0.059**	0.008	0.029
h_{10}	0.041	0.017	0.056	**0.107**
h_{11}	0.021	0.006	0.037	**0.210**
h_{12}	0.021	0.009	0.013	**0.160**
h_{13}	**0.199**	**0.075**	0.002	**0.079**
h_{14}	0.016	0.006	0.040	**0.069**
h_{15}	0.007	0.006	0.009	**0.083**
h_{16}	0.00	**0.101**	**0.274**	0.016
h_{17}	**0.064**	**0.197**	**0.146**	0.026
$\overline{w}$	0.059	0.059	0.059	0.059

5.1 *Mental Disorders Augmented Data and the Underlying Types*

In order to analyze the potential of FCPM-2 to reveal prototypes on the extreme (such as archetypes), the original data set should be modified by adding less expressed cases. Therefore, the data set was augmented by adding data of patients bearing less severe syndromes. With such an update, we can explore whether the underlying types are changed or not. To this end we added to each class data of six *mid-scale* patients and three *light-scale* patients constructed as follows. Data for each new patient, $\mathbf{x}_g = \left[x_{gh}\right]$, was generated from the data of a randomly selected original patient, $\mathbf{x}_i = [x_{ih}]$, applying the transformation

$$x_{gh} = round\,(s_F \cdot x_{ih}) + t \qquad \text{for all } h,$$

with scale-factor $s_F = 0.6$ to obtain a *mid-scale* patient and $s_F = 0.3$ to obtain a *light-scale* patient. The shift parameter t takes randomly selected value of either 0 or 1.

Tables 7 and 8 present the underlying types ($\boldsymbol{\tau}$) found in the original data set followed by the corresponding ones ($\boldsymbol{\tau}'$) for the augmented data set, for each of the algorithms under study. We can observe that the feature values of the FCM found types (in $\boldsymbol{\tau}$) have moved towards less well expressed, intermediate, values (in $\boldsymbol{\tau}'$), showing the strong tendency of the FCM to look for the central tendencies. In

Table 7 Underlying types of mental disorders (τ_D, τ_M, τ_{S_s}, τ_{S_p}), found by FCM for the original data set (τ_h^{FCM}) and for the augmented data set ($\tau_h^{\prime FCM}$), which can be compared with corresponding mean values: ($\overline{\tau}_h$) in the first rows

τ_D	h_5	h_8	h_9	h_{13}	h_{17}		
$\overline{\tau}_h$	5.18	0.09	6.00	5.64	0.73		
τ_h^{FCM}	5	0	6	5	1		
$\tau_h^{\prime FCM}$	4	0	5	4	1		

τ_M	h_3	h_8	h_9	h_{13}	h_{16}	h_{17}	
$\overline{\tau}_h$	0.18	5.82	0.82	0.09	0.00	6.00	
τ_h^{FCM}	0	6	1	0	0	6	
$\tau_h^{\prime FCM}$	1	5	1	1	1	5	

τ_{S_s}	h_3	h_8	h_{16}	h_{17}			
$\overline{\tau}_h$	5.27	0.36	5.45	0.45			
τ_h^{FCM}	5	1	5	0			
$\tau_h^{\prime FCM}$	4	1	3	1			

τ_{S_p}	h_8	h_{10}	h_{11}	h_{12}	h_{13}	h_{14}	h_{15}
$\overline{\tau}_h$	4.91	5.09	5.82	4.64	0.36	5.18	5.55
τ_h^{FCM}	5	5	5	4	1	5	5
$\tau_h^{\prime FCM}$	4	4	5	4	1	5	5

contrast, the corresponding features in FCPM-2 found types maintain their extremal nature, stressing the 'extreme' nature of FCPM-2 prototypes,and consequently their ability to identify the underlying types of clusters (and thus their stability over the most 'discriminating' features), despite the presence of many new mid- and light-scale patients. It is important to stress that the experiment has been conducted by generating six distinct augmented data sets and that the underlying types found by FCM and FCPM-2 are exactly the same for all of them.

This result shows that the FCPM keeps the knowledge of the syndromes in situations at which less severe cases are present in the data set, whereas the FCM does not. In fact, these results clearly outline the 'extremal' behavior of FCPM prototypes in contrast to the central tendency of FCM prototypes.

6 Conclusion

The fuzzy proportional membership is not just a weight, as in classical fuzzy clustering, but expresses how much the entities share the features of the prototypes. The FCPM clustering criteria provide distinct forms of pertaining the observed data points to the prototypes leading to two different cluster structures: central ones (FCPM-0, FCPM-1), like in FCM, and extremal (FCPM-2), like the ones of archetypal analysis.

Table 8 Underlying types of mental disorders (τ_D, τ_M, τ_{S_s}, τ_{S_p}), found by FCPM-2 for the original data set (τ_h^{FCPM}) and for the augmented data set ($\tau_h'^{FCPM}$); these values can be compared with corresponding extreme values/ number of cases in the first rows ($\hat{\tau}_h/\#$), followed by the cardinality of cases

τ_D	h_5	h_8	h_9	h_{13}	h_{17}		
$\hat{\tau}_h/\#$	6/#7	0/#10	6/#11	6/#7	0/#6		
τ_h^{FCPM}	6	0	6	6	0		
$\tau_h'^{FCPM}$	6	0	6	6	0		

τ_M	h_3	h_8	h_9	h_{13}	h_{16}	h_{17}	
$\hat{\tau}_h/\#$	0/#9	6/#9	0/#6	0/#10	0/#11	6/#11	
τ_h^{FCPM}	0	6	0	0	0	6	
$\tau_h'^{FCPM}$	0	6	0	0	0	6	

τ_{S_s}	h_3	h_8	h_{16}	h_{17}			
$\hat{\tau}_h/\#$	5/#8	0/#7	6/#6	0/#8			
τ_h^{FCPM}	6	0	6	0			
$\tau_h'^{FCPM}$	5	0	6	0			

τ_{S_p}	h_8	h_{10}	h_{11}	h_{12}	h_{13}	h_{14}	h_{15}
$\hat{\tau}_h/\#$	5/#6	5/#8	6/#9	5/#6	0/#7	5/#9	6/#7
τ_h^{FCPM}	6	6	6	6	0	6	6
$\tau_h'^{FCPM}$	6	6	6	6	0	6	6

The FCPM algorithm converges only locally as the FCM. The number of major iterations in FCPM algorithm is rather small when it converges. On average, the number of major iterations of versions of FCPM are smaller or of the same order of those of the FCM. However, with a "wrong" number of pre-specified clusters, FCPM-0 may not converge by shifting some prototypes to infinity.

The extremal behaviour of FCPM is explained by its sensitivity to the feature-values that are farthest from the standard (the grand mean, due to the followed standardization). This extremal behavior appears to be compatible with the notion of "interestingness" in data mining [20, 22].

The ability of FCPM to recover its own cluster structure from data with a proper designed data generator is a contribution for data-driven problem solving.

As future work, FCPM deserves to be applied to real data problems to study the type of clustering problems the fuzzy proportional membership fits.

Acknowledgements I wish to thank Professor Boris Mirkin for having introduce me to fuzzy clustering within the Data Recovery paradigm, and for all the years of discussion that we have been sharing ever since. The comments from the anonymous Reviewers are greatly acknowledged as they contributed to improve the paper.

References

1. Abonyi, J., Feil, B.: Cluster Analysis for Data Mining and System Identification. Springer, Berlin (2007)
2. Abu-Jamous, B., Fa, R., Nandi, A.K.: Integrative Cluster Analysis in Bioinformatics. Wiley, New York (2015)
3. Ada, I., Berthold, M.R.: The new iris data: modular data generators. In: KDD'10 Proceedings of the 16th ACM SIGKDD International Conference on Knowledge Discovery and Data Mining, pp. 413–422 (2010)
4. Berthold, M.R., Cebron, N., Dill, F., Gabriel, T.R., Kotter, T., Meinl, T., Ohl, P., Sieb, C., Thiel, K., Wiswedel, B.: KNIME: the Konstanz information miner. In: Studies in Classification, Data Analysis, and Knowledge Organization (GfKL 2007). Springer, Berlin (2007)
5. Bertsekas, D.: On the Goldstein-Levitin-Polyak gradient projection method. IEEE Trans. Autom. Control **21**(2), 174–184 (1976)
6. Bertsekas, D.: Nonlinear Programming. Athena Scientific, Belmont (1995)
7. Bezdek, J.: Fuzzy mathematics in pattern classification. Ph.D. Thesis, Applied Mathematics Center, Cornell University, Ithaca (1973)
8. Bezdek, J.: Pattern Recognition with Fuzzy Objective Function Algorithms. Plenum Press, New York (1981)
9. Bezdek, J.C., Pal, S.K. (eds.): Fuzzy Models for Pattern Recognition. IEEE Press, New York (1992)
10. Bezdek, J., Keller, J., Krishnapuram, R., Pal, T.: Fuzzy Models and Algorithms for Pattern Recognition and Image Processing. Kluwer Academic, Norwell (1999)
11. Bowers, C., Beale, R., Hendley, R.: Identifying archetypal perspectives in news articles. In: Proceedings of the 26th Annual BCS Interaction Specialist Group Conference on People and Computers (BCS-HCI'12), pp. 327–332 (2012)
12. Cannon, R., Dave, J., Bezdek, J.: Efficient implementation of the fuzzy c-means clustering algorithms. IEEE Trans. Pattern Anal. Mach. Intell. **8**, 248–255 (1986)
13. Chan, B., Mitchell, D., Cram, L.: Archetypal analysis of galaxy spectra. Mon. Not. R. Astron. Soc. **338**, 790–795 (2003)
14. Chiang, M.M.-T., Mirkin, B.: Intelligent choice of the number of clusters in K-means clustering: an experimental study with different cluster spreads. J. Classif. **27**(1), 3–40 (2010)
15. Cutler, A., Breiman, L.: Archetypal analysis. Technometrics **36**(4), 338–347 (1994)
16. Dunn, J.: A fuzzy relative of the ISODATA process and its use in detecting compact, well-separated clusters. J. Cybern **3**, 32–57 (1974)
17. Eisenack, K., Lüdeke, M., Kropp, J.: Construction of archetypes as a formal method to analyze social-ecological systems. In: Proceedings of the Institutional Dimensions of Global Environmental Change Synthesis Conference (2006)
18. Elder, A., Pinnel, J.: Archetypal analysis: an alternative approach to finding defining segments. In: 2003 Sawtooth Software Conference Proceedings, pp. 113–129. Sequim, WA (2003)
19. Eugster, M.: Performance profiles based on archetypal athletes. Int. J. Perform. Anal. Sport **12**(1), 166–187 (2012)
20. Fayyad, U.M., Piatetsky-Shapiro, G., Smyth, P.: From data mining to knowledge discovery: an overview. In: Fayyad, U.M., Piatetsky-Shapiro, G., Smyth, P., Uthurusamy, R. (eds.) Advances in Knowledge Discovery and Data Mining. AAAI Press/The MIT Press, Menlo Park (1996)
21. Fu, L., Medico, E.: FLAME, a novel fuzzy clustering method for the analysis of DNA microarray data. BMC Bioinf. **8**, 3 (2007)
22. Geng, L., Hamilton, H.J.: Interestingness measures for data mining: a survey. ACM Comput. Surv. **38**(3), 9 (2006)
23. Hall, M., Frank, E., Holmes, G., Pfahringer, B., Reutemann, P., Witten, I.: The WEKA data mining software: an update. ACM SIGKDD Explor. Newslett. **11**(1):10–18 (2009)
24. Höppner, F., Klawonn, F., Kruse, R., Runkler, T.: Fuzzy Cluster Analysis: Methods for Classification, Data Analysis and Image Recognition. Wiley, New York (1999)

25. Jain, A.K., Dubes, R.C.: Algorithms for Clustering Data. Prentice Hall, Englewood Cliffs (1988)
26. Jimenez, L., Landgrebe, D.: Supervised classification in high-dimensional space: geometrical, statistical and asymptotical properties of multivariate data. IEEE Trans. Syst. Man Cybern. Part C **28**(1):39–54 (1998)
27. Kriegel, H., Kröger, P., Zimek, A.: Clustering high-dimensional data: a survey on subspace clustering, pattern-based clustering, and correlation clustering. ACM Trans. Knowl. Discov. Data **3**(1), 1–58 (2009)
28. Levitin, E.S., Polyak, B.T.: Constrained minimization problems. USSR Comput. Math. Math. Phys. **6**(5), 1–50 (1966) (English transl. of paper in Zh. Vychisl. Mat. i Mat. Fiz., **6**(5), 787–823, 1965)
29. Li, S., Wang, P., Louviere, J., Carson, R.: Archetypal analysis: a new way to segment markets based on extreme individuals. In: A Celebration of Ehrenberg and Bass: Marketing Knowledge, Discoveries and Contribution, ANZMAC 2003 Conference Proceedings, pp. 1674–1679, Adelaide (2003)
30. Lin, C.-J.: Projected gradient methods for nonnegative matrix factorization. Neural Comput. **19**(10), 2756–2779 (2007)
31. Mezzich, J.E., Solomon, H.: Taxonomy and Behavioral Science. Academic Press, London (1980)
32. Milligan, G.W.: An algorithm for generating artificial test clusters. Psychometrika **50**(1), 123–127 (1985)
33. Mirkin, B.: Mathematical classification and clustering. Nonconvex Optimization and Its Applications. Kluwer Academic, Dordrecht (1996)
34. Mirkin, B.: Clustering: A Data Recovery Approach, 2nd edn. Chapman & Hall/CRC Press, London/Boca Raton (2012)
35. Mirkin, B., Satarov, G: Method of fuzzy additive types for analysis of multidimensional data. Autom. Remote. Control. **51**(5,6), 1:683–688, 2:817–821 (1990)
36. Mørup, M., Hansen, L.: Archetypal analysis for machine learning and data mining. Neurocomputing **80**, 54–63 (2012)
37. Narayanan, S.J., Bhatt, R.B., Paramasivam, I., Khalid, M., Tripathy, B.K.: Induction of fuzzy decision trees and its refinement using gradient projected-neuro-fuzzy decision tree. Int. J. Adv. Intell. Paradigms **6**(4), 346–369 (2014)
38. Nascimento, S.: Fuzzy clustering via proportional membership model. Frontiers of Artificial Intelligence and Applications, vol. 119. IOS Press, Amsterdam (2005). ISBN: 1 58603 489 8 (reprinted 2006)
39. Nascimento, S., Mirkin, B., Moura-Pires, F.: A fuzzy clustering model of data and fuzzy *c*-means. In: Langari, R. (ed.) Proceedings of The 9th IEEE International Conference on Fuzzy Systems, Fuzz-IEEE 2000. Soft Computing in the Information Age, IEEE Neural Networks Council, pp. 302–307. IEEE Press, New York (ISSN 1098-7584) (2000)
40. Nascimento, S., Mirkin, B., Moura-Pires, F.: Modeling proportional membership in fuzzy clustering. IEEE Trans. Fuzzy Syst. **11**(2), 173–186 (2003)
41. Oliveira, J.V., Pedrycz, W.: Advances in Fuzzy Clustering and Its Applications. Wiley, New York (2007)
42. Omari, A., Langer, R., Conrad, S.: TARtool: a temporal dataset generator for market basket analysis. In: Advanced Data Mining and Applications, pp. 400–410. Springer, Berlin (2008)
43. Pal, N.R., Bezdek, J.: On cluster validity for the fuzzy c-means model. IEEE Trans. Fuzzy Syst. **3**, 370–379 (1995)
44. Pei, Y., Zaiane, O.: A synthetic data generator for clustering and outlier analysis, Technical Report ID:TR06-15, 33 p. (2006)
45. Polyak, B.: A general method for solving extremum problems. Sov. Math. Dokl. **8**(3), 593–597 (1967)
46. Polyak, B.: Introduction to Optimization. Optimization Software, New York (1987)
47. Porzio, G., Ragozini, G., Vistocco, D.: On the use of archetypes as benchmarks. Appl. Stoch. Model. Bus. Ind. **24**(5), 419–437 (2008)

48. Runkler, T.A.: Data Analytics, Models and Algorithms for Intelligent Data Analysis. Springer, Berlin (2012)
49. Ruspini, E.: A new approach to clustering. Inf. Control. **15**, 22–32 (1969)
50. Schmidt, M., Kim, D., Sra, S.: Projected Newton-type methods in machine learning. In: Optimization for Machine Learning, pp. 305–330. MIT Press, Cambridge (2011)
51. Steinbach, M., Ertoz, L., Kumar, V.: The challenges of clustering high dimensional data. In: Wille, L.T. (ed.) New Directions in Statistical Physics, pp. 273–309. Springer, Berlin (2004)
52. Steinley, D., Brusco, M.: Initializing K-means batch clustering: a critical evaluation of several techniques. J. Classif. **24**(1), 99–121 (2007)
53. Sun, J., Zhao, W., Xue, J., Shen, Z. Shen, Y.: Clustering with feature order preferences. Intell. Data Anal. **14**, 479–495 (2010)
54. Vendramin, L., Naldi, M.C., Campello, R.J.G.B.: Fuzzy clustering algorithms and validity indices for distributed data. In: Emre Celebi, M. (ed.) Partitional Clustering Algorithms. Springer, Berlin (2015)
55. Yeh, C-Y, Huang, C.-W., Lee, S.-J.: Multi-Kernel support vector clustering for multi-class classification. In: Proceedings of The 3rd International Conference on Innovative Computing Information and Control (ICICIC'08). IEEE, New York (2008)

Algorithmic Principle of Least Revenue for Finding Market Equilibria

Yurii Nesterov and Vladimir Shikhman

Abstract In analogy to extremal principles in physics, we introduce the Principle of Least Revenue for treating market equilibria. It postulates that equilibrium prices minimize the total excessive revenue of market's participants. As a consequence, the necessary optimality conditions describe the clearance of markets, i.e. at equilibrium prices supply meets demand. It is crucial for our approach that the potential function of total excessive revenue be convex. This facilitates structural and algorithmic analysis of market equilibria by using convex optimization techniques. In particular, results on existence, uniqueness, and efficiency of market equilibria follow easily. The market decentralization fits into our approach by the introduction of trades or auctions. For that, Duality Theory of convex optimization applies. The computability of market equilibria is ensured by applying quasi-monotone subgradient methods for minimizing nonsmooth convex objective—total excessive revenue of the market's participants. We give an explicit implementable algorithm for finding market equilibria which corresponds to real-life activities of market's participants.

Keywords Principle of least revenue • Computation of market equilibrium • Price adjustment • Convex optimization • Subgradient methods • Decentralization of prices • Unintentional optimization

1 Introduction

We start with a celebrated quotation of Leonhard Euler (1744): "Nothing in the world takes place without optimization, and there is no doubt that all aspects of the world that have a rational basis can be explained by optimization methods." Rephrasing this in modern terms, the laws of nature can be derived by using extremal (or variational) principles. Indeed, the laws are often first-order necessary optimality

Y. Nesterov • V. Shikhman (✉)
Center for Operations Research and Econometrics (CORE), Catholic University of Louvain (UCL), 34 voie du Roman Pays, 1348 Louvain-la-Neuve, Belgium
e-mail: yurii.nesterov@uclouvain.be; vladimir.shikhman@uclouvain.be

B. Goldengorin (ed.), *Optimization and Its Applications in Control and Data Sciences*, Springer Optimization and Its Applications 115,
DOI 10.1007/978-3-319-42056-1_14

conditions for minimizing (or maximizing) a properly chosen potential function. To illustrate this idea, let us consider the Newton law

$$m\ddot{r} = F, \tag{1}$$

where $r(t)$ is the particle's position at time t, F the acting force, and m the particle's mass. The goal of classical mechanics is to solve this differential equation for various forces: gravity, electromagnetism, friction, etc. For conservative forces (gravity, electrostatics, but not friction), the force can be expressed as

$$F = -\nabla V, \tag{2}$$

where the potential energy $V(r)$ depends on the particle's position. For this case, let us define the potential function called action:

$$A(r) \stackrel{\text{def}}{=} \int_{t_1}^{t_2} \left(\frac{m\dot{r}^2}{2} - V(r) \right) dt. \tag{3}$$

The action represents the difference between the kinetic and potential energy of a particle. The integrand in (3) is called Lagrangian

$$L(r) \stackrel{\text{def}}{=} \frac{m\dot{r}^2}{2} - V(r).$$

Now, the Principle of Least Action says:

The true path taken by the particle is an extremum of the action.

In fact, it is an easy exercise from the calculus of variations to show that an extremum of the action satisfies the Newton law. Note that Newton law comes from the first-order variation of the action. Second derivative of the trajectory appears just because of the integration by parts. So, Newton law for conservative forces can be viewed as a first-order potential system. The least action principle (or Lagrangian method) became extremely fruitful in all of physics, not just mechanics. Many fundamental laws of physics can be expressed in terms of a least action principle, e.g. electromagnetism, optics, special and general relativity, particle physics etc. (see [1]). Recently in [21], the principle of least action has been applied to supply chains linking variation in production to variation in net-inventory.

From the optimization perspective, the introduction of extremal principles highlights the algorithmic aspects of laws. Instead of solving the law's systems of equations, we may consider iterative schemes for minimizing the corresponding potential function. Of course, regarding physical laws it sounds like a philosophical paradox: who minimizes the action, and by using what method? However, having in mind the application of extremal principles in social sciences, this point of view could be valuable. Namely, optimization methods for minimizing a potential function provide behavioral dynamics. They explain how a social system eventually arrives at a state which is described by the corresponding law. Although, Isaac Newton remarked in 1720: "I can calculate the motion of heavenly bodies, but not the madness of people,"—we pursue exactly this goal, at least in some particular economic setting. It will turn out that the economic behavior of people is not as mad as it may look like.

In this paper we introduce and exploit extremal principles for the economic law of supply and demand. The latter says that at a competitive market supply meets demand at an appropriate price, i.e.

$$\sum_k \tilde{y}_k = \sum_i \tilde{x}_i, \tag{4}$$

where $\tilde{y}_k$ is kth producer's and $\tilde{x}_i$ is ith consumer's bundle of goods, respectively. It is also common to refer to (4) as the market clearing condition. Here, $\tilde{y}_k \in S_k(p)$ is kth producer's supply operator, and $\tilde{x}_i \in D_i(p)$ is ith consumer's demand operator at a price p. Hence, the law of supply and demand can be equivalently written as an inclusion problem

$$0 \in \sum_k S_k(p) - \sum_i D_i(p). \tag{5}$$

Solutions of (5) are called equilibrium prices. Together with corresponding production and consumption bundles, they form market equilibrium (e.g., [8]).

The role of prices in balancing supply and demand is well-known in economics since Adam Smith. As Milton Friedman pointed out in [2]: "Adam Smith's flash of genius was his recognition that the prices ... in a free market could coordinate the activity of millions of people, each seeking his own interest, in such a way as to make everyone better off." Mathematically, this coordination may be explained by the fact that prices act as dual or Lagrange multipliers for the market clearance. In accordance with the latter, our main assumption states that supply and demand operators can be expressed as convex subdifferentials:

$$S_k(p) = \partial \mathscr{E}_k(p), \quad D_i(p) = -\partial \mathscr{E}_i(p). \tag{6}$$

$\mathscr{E}_k(p)$ is kth producer's excessive profit, and $\mathscr{E}_i(p)$ ith consumer's excessive wealth. Note that the subdifferentiability assumption (6) has the same mathematical meaning as the assumption of dealing with a conservative force (2). Let us define the potential function called total excessive revenue:

$$\mathscr{E}(p) \stackrel{\text{def}}{=} \sum_k \mathscr{E}_k(p) + \sum_i \mathscr{E}_i(p). \tag{7}$$

In our framework, the total excessive revenue represents the Lagrangian w.r.t. the market clearance. Now, we are ready to state the Principle of Least Revenue:

Equilibrium prices minimize the total excessive revenue of market's participants.

In fact, the first-order necessary optimality conditions for minimizing the total excessive revenue give us the law of supply and demand (5). The latter is due to the subdifferentiability assumption (6). Further, it is crucial for our approach that the potential function of total excessive revenue is convex. This opens up for structural and algorithmic analysis of market equilibria by using convex optimization techniques. E.g., the Walrasian tâtonnement, as suggested in [19], states that prices change proportionally to the excess demand, i.e.

$$\dot{p} \in \sum_i D_i(p) - \sum_k S_k(p).$$

Under assumption (6) the Walrasian tâtonnement becomes a subgradient system

$$\dot{p} \in -\partial \mathscr{E}(p).$$

Its discretized version is

$$p[t+1] = p[t] - h[t] \nabla \mathscr{E}(p[t]),$$

with time t, step-sizes $h[t]$, and excess supplies $\nabla \mathscr{E}(p[t]) \in \partial \mathscr{E}(p[t])$. This corresponds to the standard subgradient scheme for the convex minimization of $\mathscr{E}$. In absence of assumption (6), the Walrasian tâtonnement is fraught with severe problems. These relate, on one side, to communication, implementation, interpretation, organization and, on the other side, to stability and definiteness, see e.g. [8] for details.

The paper is organized as follows. In Sect. 2 we describe the excessive revenue model for a competitive market from [14]. We show that the least revenue principle applies to the excessive revenue model. Based on this fact, existence, uniqueness and efficiency results for market equilibrium follow. In order to minimize the total excessive revenue, quasi-monotone subgradient methods for nonsmooth convex minimization from [13] are presented in Sect. 3. They guarantee the best possible rate of convergence for the whole sequence of test points rather than of their averages. This fact allows us to prevent uncontrollable jumps of the function values at some iterations. Moreover, the sequence of record values does not enter the complexity bound. This is crucial for the applicability of quasi-monotone subgradient methods in our economic setting. Indeed, the values of the total excessive revenue are not available to market's participants. By using quasi-monotone subgradient methods, we explain in Sect. 4 how equilibrium prices can be successively adjusted. For that, we concentrate on

- *the regulation design*: regulator settles and updates prices which are taken by producers and consumers;
- *the trade design*: producers settle and update their individual prices, and consumers buy at the lowest purchase price;
- *the auction design*: consumers settle and update their individual prices, and producers sell at the highest offer price.

Notation. Our notation is quite standard. We denote by $\mathbb{R}^n$ the space of n-dimensional column vectors $x = (x^{(1)}, \dots, x^{(n)})^T$, and by $\mathbb{R}^n_+$ the set of all vectors with nonnegative components. For x and y from $\mathbb{R}^n$, we introduce the standard scalar product and the Hadamard product

$$\langle x, y \rangle = \sum_{i=1}^{n} x^{(i)} y^{(i)}, \quad x \circ y = \left(x^{(i)} y^{(i)} \right)_{i=1}^{n} \in \mathbb{R}^n.$$

Finally, a_+ denotes the positive part of the real value $a \in \mathbb{R}$: $a_+ = \max\{a, 0\}$. For $x = (x^{(1)}, \dots, x^{(n)})^T \in \mathbb{R}^n$ we denote $x_+ = \left(x_+^{(1)}, \dots, x_+^{(n)} \right)^T$. For vectors $p_1, \dots, p_K \in \mathbb{R}^n$, denote by $\min\limits_{k=1,\dots,K} p_k \in \mathbb{R}^n$ the vector with coordinates

$$\left(\min_{k=1,\dots,K} p_k \right)^{(j)} = \min_{k=1,\dots,K} p_k^{(j)}, \quad j = 1, \dots, n.$$

For the vectors $p_1, \dots, p_I \in \mathbb{R}^n$, we denote by $\max\limits_{i=1,\dots,I} p_i \in \mathbb{R}^n$ the vector with coordinates

$$\left(\max_{i=1,\dots,I} p_i \right)^{(j)} = \max_{i=1,\dots,I} p_i^{(j)}, \quad j = 1, \dots, n.$$

2 Excessive Revenue Model

Let us present the *excessive revenue model* of competitive market from [14].

2.1 *Producers and Consumers*

Consider a market with K producers, which are able to produce n different goods. Given a vector of prices $p \in \mathbb{R}^n_+$, the kth producer forms the supply operator $S_k(p)$ of production bundles $\tilde{y}_k \in \mathbb{R}^n_+$. For that, the kth producer maximizes the profit with respect to the variable cost, subsequently trying to cover the fixed cost. Namely,

- Producer $k \in \{1, \dots, K\}$ chooses first the *tentative production* bundle $y_k \in \mathbb{R}^n_+$ by solving the *profit maximization* problem:

$$\pi_k(p) \stackrel{\text{def}}{=} \max_{y_k \in \mathscr{Y}_k} \langle p, y_k \rangle - c_k(y_k). \tag{8}$$

Here, $\mathscr{Y}_k \subset \mathbb{R}^n_+$ is the production set, assumed to be nonempty, compact and convex. The producer's yield is $\langle p, y_k \rangle$. The variable cost of producing y_k is denoted by $c_k(y_k)$. We assume that c_k is a convex function on $\mathbb{R}^n_+$. Clearly, the profit $\pi_k(p)$ is convex in p as the maximum of linear functions. By $\mathscr{Y}_k^*(p)$ we denote the set of optimal solutions of (8), i.e. $y_k \in \mathscr{Y}_k^*(p)$. Note that the profit maximization problem (8) appears already in Marshallian partial equilibrium analysis, see e.g. [8].

- Secondly, the kth producer compares the profit $\pi_k(p)$ with the fixed cost of maintaining the technological set $\mathscr{Y}_k$, denoted by $\kappa_k \equiv \kappa_k(\mathscr{Y}_k) \in \mathbb{R}_+$. The latter can include the interest paid to the bank, different charges for renting the equipment, land use, etc. By this comparison, a participation level $\alpha_k \equiv \alpha_k(p) \in [0, 1]$ of kth producer is *properly* adjusted:

$$\alpha_k(p) \stackrel{\text{def}}{=} \begin{cases} 1, \text{ if } \pi_k(p) > \kappa_k, \\ 0, \text{ if } \pi_k(p) < \kappa_k. \end{cases} \tag{9}$$

In case $\pi_k(p) = \kappa_k$, $\alpha_k(p) \in [0, 1]$ is not unique and may vary. We call producers' participation levels satisfying the relation (9) *proper*.

- Finally, the *supply* operator $S_k : \mathbb{R}^n_+ \rightrightarrows \mathbb{R}^n_+$ of kth producer is given by

$$S_k(p) \stackrel{\text{def}}{=} \left\{ \tilde{y}_k = \alpha_k y_k \,\middle|\, \alpha_k \equiv \alpha_k(p) \text{ and } y_k \in \mathscr{Y}_k^*(p) \right\}. \tag{10}$$

Here, the *production* bundles are

$$\tilde{y}_k \stackrel{\text{def}}{=} \alpha_k y_k,$$

where $\alpha_k \equiv \alpha_k(p)$ is the proper participation level of the kth producer, and $y_k \in \mathscr{Y}_k^*(p)$ is the tentative production.

Let I consumers be active at the market. The ith consumer has to decide on the consumption bundle $\tilde{x}_i \in \mathbb{R}^n_+$. These consumption bundles form the demand $D_i(p)$, given the price $p \in \mathbb{R}^n_+$. The ith consumer minimizes the expenditure with an aim to guarantee the desirable utility level. Then he tries to cover this expenditure by the available wealth. Namely,

- Consumer $i \in \{1, \ldots, I\}$ decides first on the *tentative consumption* bundle $x_i \in \mathbb{R}^n_+$ by minimizing expenditure:

$$e_i(p) \stackrel{\text{def}}{=} \min_{\substack{x_i \in X_i \\ u_i(x_i) \geq \underline{u}_i}} \langle p, x_i \rangle = \min_{x_i \in \mathscr{X}_i} \langle p, x_i \rangle, \tag{11}$$

where the ith consumption set is

$$\mathscr{X}_i \stackrel{\text{def}}{=} \left\{ x_i \in X_i \,|\, u_i(x_i) \geq \underline{u}_i \right\}.$$

Here, $X_i \subset \mathbb{R}^n_+$ is assumed to be nonempty, compact and convex. By $u_i : X_i \to \mathbb{R}_+$ we denote the utility function of the ith consumer, assumed to be concave. The utility level $\underline{u}_i \in \mathbb{R}_+$ is desirable by ith consumer. The consumer's expenditure $e_i(p)$ is concave in p as the minimum of linear functions. By $\mathscr{X}_i^*(p)$ we denote the set of optimal solutions of (11), i.e. $x_i \in \mathscr{X}_i^*(p)$. The minimization of expenditure in (11) is well-known in economics as a dual problem for utility maximization. The desirable utility level $\underline{u}_i$ mainly reflects the consumer's standards on qualities of goods. In [18] the agent who faces the expenditure minimization problem (11) is called the dual consumer. The dual consumer usually acts on regular basis, thus, generating the flows of consumption. We also refer to [5, Chap. 10] and [7] for more details on the dual theory of consumption. The compactness assumption on X_i refers to the fact that the consumption is bounded. Naturally, there are physical limits to what people can consume in order to satisfy their basic needs. The unbounded desire for wealth is not an issue here, since the wealth w_i is a primitive in our model (see below and confer the discussion on this assumption in [17]).

- Secondly, the ith consumer compares the expenditure $e_i(p)$ with the available wealth $w_i \in \mathbb{R}_+$. The latter can include the budget, salary and rent payments, etc. By this comparison, a participation level $\beta_i \equiv \beta_i(p) \in [0,1]$ of ith consumer is *properly* adjusted:

$$\beta_i(p) \stackrel{\text{def}}{=} \begin{cases} 1, \text{ if } e_i(p) < w_i, \\ 0, \text{ if } e_i(p) > w_i. \end{cases} \tag{12}$$

In case $e_i(p) = w_i$, $\beta_i(p) \in [0,1]$ is not unique and may vary. We call consumers' participation levels satisfying the relation (12) *proper*.

- Finally, the *demand* operator $D_i : \mathbb{R}^n_+ \rightrightarrows \mathbb{R}^n_+$ of ith consumer is given by

$$D_i(p) \stackrel{\text{def}}{=} \left\{ \tilde{x}_i = \beta_i x_i \;\middle|\; \beta_i \equiv \beta_i(p) \text{ and } x_i \in \mathscr{X}_i^*(p) \right\}. \tag{13}$$

Here, the *consumption* bundles are

$$\tilde{x}_i \stackrel{\text{def}}{=} \beta_i x_i,$$

where $\beta_i \equiv \beta_i(p)$ is the proper participation level of the ith consumer, and $x_i \in \mathscr{X}_i^*(p)$ is the tentative consumption.

For the sake of convenient navigation along the text, we list model's data and variables:

There are two non-standard ingredients in our model which need to be explained and thoroughly justified. The first concerns the expenditure minimization problem (11) with the given level $\underline{u}_i$ of desirable utility. The second deals with the proper adjustment of participation levels α_k and β_i in (9) and (12), respectively.

(1) Expenditure minimization and responsible consumer

Data		Variables	
$\mathscr{Y}_k$	Production set	p	Prices
$c_k(\cdot)$	Variable cost function	y_k	Tentative production bundle
κ_k	Fixed cost	α_k	Producer's participation level
$\mathscr{X}_i$	Consumption set	$\tilde{y}_k \stackrel{\mathrm{def}}{=} \alpha_k y_k$	Production bundle
$u_i(\cdot)$	Utility function	x_i	Tentative consumption bundle
$\underline{u}_i$	Utility level	β_i	Consumer's participation level
w_i	Available wealth	$\tilde{x}_i \stackrel{\mathrm{def}}{=} \beta_i x_i$	Consumption bundle

The minimization of expenditure in (11) is well-known in economics as a dual problem for utility maximization (e.g. [8]):

$$v_i(p, w_i) \stackrel{\mathrm{def}}{=} \max_{\substack{x_i \in X_i \\ \langle p, x_i \rangle \leq w_i}} u_i(x). \tag{14}$$

Namely, under some technical assumptions if x_i solves (14) then it also solves (11) with $\underline{u}_i = v_i(p, w_i)$. Conversely, if x_i solves (11) then it also solves (14) with $w_i = \langle p, x_i \rangle$. In our setting the desirable utility level $\underline{u}_i$ is given, thus, it is a primitive of the model. It mainly reflects the consumer's standards on qualities of life. Hence, it does not explicitly depend on the wealth w_i as in the classical setting. Note that we model wealth effects by subsequent comparison of w_i with expenditure $e_i(p)$ rather than by the usual budget constraint $\langle p, x_i \rangle \leq w_i$ as in (14) (cf. also the discussion in (2) below). The introduction of desirable utility levels $\underline{u}_i$ as primitives is the main departure from the usual consumption model (14). This is the crucial point in our market modeling which postulates in the sense that consumer's objectives become monetary, hence transferable. As we shall see in Sect. 2.3 below, this fact implies that supply and demand operators can be expressed as convex subdifferentials, i.e. assumption (6) be valid.

Now, let us explain why in some interesting situations the desirable utility level $\underline{u}_i$ is explicitly available to consumers. For many daily goods there are physical standards to be satisfied. They constitute the so-called minimum standards of life. Additionally, some consumers often accept standards imposed by the society, e.g. through advertisement, their friends or family members. E.g., it became evident that some consumers use to shrink their usual consumption motivated by ecological reasons. Also experienced consumers, who go shopping in a supermarket say on a weekly basis, know the standards of their living. Overall, we may think of a *responsible consumer* who does care about the own expenditure. Namely, the consumer is not willing to spend more than necessary to satisfy the given standards. Thus, such a consumer tries to minimize the expenditure while guaranteeing the standards.

(2) Adjustment of participation levels and long-term behavior

Note that there is evidence from behavioral economics that consumer's choices need not be consistent with the maximization of a preference relation (see [9] and references therein). The reason for that is usually referred to as consumers' bounded rationality. Classic examples include status-quo biases, attraction, compromise and framing effects, temptation and self-control, consideration sets, and choice overload. Due to the proposed approach, the demand operator is consistent with the long-term behavior of responsible consumers. In our model, the production and consumption bundles, the consumer's wealths, and producers' costs are considered as *constant flows*. This means that we get the same amount of corresponding objects in each standard interval of time (say, 1 week). Thus, our economy can be seen as stationary. If the income of a person or a firm during this interval is greater than the expenses, then he/she can ensure a constant *rate of growth* of the own capital. In this *profitable* case, we have: $\alpha_k(p) = \beta_i(p) = 1$, i.e. producers and consumers implement their tentative bundles. If the income is strictly less than expenses, then the producer/consumer must leave the market sooner or later, i.e. $\alpha_k(p) = \beta_i(p) = 0$. This is true both for producers (bankruptcy), and for consumers (emigration from this market). We refer to those agents as being *bankrupt*. If the regular income is equal to the regular expenses, then tentative production and consumption bundles may shrink due to participation levels $\alpha_k(p), \beta_i(p)$. In this *marginal* case, producers and consumers usually have $\alpha_k(p), \beta_i(p) \in (0, 1)$. With probability one they neither fully participate in economic activities nor quit the market. In what follows, we give a behavioral explanation on how $\alpha_k(p), \beta_i(p)$ are adjusted. Note that marginal agents reach their break-even point at current prices p, hence, they make neither a profit nor a loss. For a marginal producer it means that the corresponding profit is equal to the fixed costs: $\pi_k(p) = \kappa_k$. Net saving of a marginal consumer is zero, i.e. the own wealth is equal to the minimal possible expenditure: $w_i = e_i(p)$. Hence, for $\hat{p} \approx p$ the break-even point will be mainly shifted either to profitability or bankruptcy. This reflects the existence of poverty in the society. Marginal producers face at nearby prices $\hat{p} \approx p$

$$\text{either} \quad \pi_k(\hat{p}) > \kappa_k \quad \text{or} \quad \pi_k(\hat{p}) < \kappa_k,$$

and marginal consumers face

$$\text{either} \quad w_i > e_i(\hat{p}) \quad \text{or} \quad w_i < e_i(\hat{p}).$$

Hence, sometimes marginal producers/consumers can implement their tentative bundles, i.e.

$$\alpha_k(\hat{p}) = 1, \quad \beta_i(\hat{p}) = 1,$$

and sometimes it is not possible, i.e.

$$\alpha_k(\hat{p}) = 0, \quad \beta_i(\hat{p}) = 0.$$

The particular 0–1 values of $\alpha_k(\hat{p})$ and $\beta_i(\hat{p})$ depend on the individual history of successes and failures for producers and consumers, respectively. To be more concrete, let us consider some price adjustment process $\hat{p}(t) \to p$ with discrete time $t \to \infty$. Now, the participation levels $\alpha_k(p), \beta_i(p)$ can be viewed as *frequencies* of agents' successful and failed attempts. Indeed, by averaging and taking the limit we obtain the participation levels:

$$\frac{1}{t}\sum_{s=1}^{t} \alpha_k(\hat{p}(s)) \to \alpha_k(p), \quad \frac{1}{t}\sum_{s=1}^{t} \beta_i(\hat{p}(s)) \to \beta_i(p) \quad \text{for } t \to \infty.$$

This interpretation of participation levels as frequencies is based on the *long-term behavior* of the agents. Our analysis of the price adjustment process from Sect. 4 confirms this interpretation. Namely, the limits above actually define $\alpha_k(p), \beta_i(p)$ as frequencies obtained during the price adjustment.

Let us address the question why marginal agents do not quit the market although they eventually implement only a share of their tentative production/consumption bundles. As a consequence marginal producers do not fully exploit their available capacities and cannot cover the fixed costs. Marginal consumers do not spend all their available wealths and cannot reach the desirable levels of utility. Nevertheless, these agents stay at the market since they actually *do not know* that they are marginal. During the price adjustment, the only available information is their individual history of successes and failures while attempting to produce and to consume. With above notation, at time t they know a 0–1 sequence of $\alpha_k(\hat{p}(s))$ and $\beta_i(\hat{p}(s))$, $s = 1, \ldots, t$. This particular history depends on many factors, as their luck, current prices, particular actions of other agents, etc. From time to time, marginal agents succeed to implement their tentative production/consumption bundles, but occasionally they fail. This unsure market environment causes marginal agents to temporally reduce consumption and to wait for "fair prices". Such a behavior is typical for poor people, and we can treat their fractional participation levels $\alpha_k(p)$ and $\beta_i(p)$ as a measure of *poverty*. A hidden, but very important, consequence of this marginal behavior is a possibility to clear the market as we shall see in Sect. 2.2. We conclude that the marginal agents play a crucial role in our approach to market modeling.

Overall, the participation levels $\alpha_k(p), \beta_i(p)$ are indicators of economic viability. These items account for most important, but non-standard features of our market model. Their inclusion is one of the chief novelties of the paper.

2.2 *Equilibrium Market Flows*

In accordance to the previous notations, we eventually say that the *market flow*

$$\widetilde{F} = \left((\tilde{y}_k = \alpha_k y_k)_{k=1}^{K}, (\tilde{x}_i = \beta_i x_i)_{i=1}^{I} \right)$$

is defined by the triple (p, F, γ). Here, $p \in \mathbb{R}^n_+$ is the vector of prices,

$$F = \left((y_k)_{k=1}^K, (x_i)_{i=1}^I\right) \in \prod_{k=1}^K \mathscr{Y}_k \times \prod_{i=1}^I \mathscr{X}_i$$

is the *tentative market flow*, and

$$\gamma = \left((\alpha_k)_{k=1}^K, (\beta_i)_{i=1}^I\right) \in [0,1]^{K+I}$$

is the *proper system of participation levels* (w.r.t. p and F), i.e.

$$\alpha_k = \begin{cases} 1, \text{ if } \langle p, y_k \rangle - c_k(y_k) > \kappa_k, \\ 0, \text{ if } \langle p, y_k \rangle - c_k(y_k) < \kappa_k \end{cases}, \quad \beta_i = \begin{cases} 1, \text{ if } \langle p, x_i \rangle < w_i, \\ 0, \text{ if } \langle p, x_i \rangle > w_i. \end{cases}$$

Now we define the partial market equilibrium in the standard way.

Definition 1 (Market Equilibrium). We say that $p^* \in \mathbb{R}^n$ is the *equilibrium price* if there exists a market flow

$$\widetilde{F}^* = \left(\left(\tilde{y}_k^* = \alpha_k^* y_k^*\right)_{k=1}^K, \left(\tilde{x}_i^* = \beta_i^* x_i^*\right)_{i=1}^I\right) \in \prod_{k=1}^K S_k(p^*) \times \prod_{i=1}^I D_i(p^*),$$

satisfying the *market clearing condition*

$$p^* \geq 0, \quad \sum_{k=1}^K \tilde{y}_k^* - \sum_{i=1}^I \tilde{x}_i^* \geq 0, \quad \left\langle p^*, \sum_{k=1}^K \tilde{y}_k^* - \sum_{i=1}^I \tilde{x}_i^* \right\rangle = 0. \tag{15}$$

In this case, $\widetilde{F}^*$ is called the *equilibrium market flow*. Setting

$$\gamma^* = \left(\left(\alpha_k^*\right)_{k=1}^K, \left(\beta_i^*\right)_{i=1}^I\right),$$

we call $\left(p^*, \gamma^*, \widetilde{F}^*\right)$ the *market equilibrium*.

The market clearing condition (15) states that the consumption never exceeds the production, and the markets of goods with positive prices ($p^{(j)} > 0$) are perfectly cleared:

$$\sum_{k=1}^K \tilde{y}_k^{*(j)} = \sum_{i=1}^I \tilde{x}_i^{*(j)}.$$

2.3 Principle of Least Revenue

Given a vector of prices $p \in \mathbb{R}^n_+$, producers maximize their profits and consumers minimize their expenditures. Afterwards, both properly adjust their participation levels by comparing the profits with the fixed costs, in case of producers, or by comparing the expenditures with the wealths, in case of consumers. Exactly the same behavior can be obtained by maximizing their excessive profits and excessive wealths, respectively.

The *excessive profit* of the kth producer is set as

$$\mathcal{E}_k(p) \stackrel{\text{def}}{=} (\pi_k(p) - \kappa_k)_+ = \max_{y_k \in \mathcal{Y}_k} \left(\langle p, y_k \rangle - c_k(y_k) - \kappa_k\right)_+ . \tag{16}$$

Using the substitution $\tilde{y}_k = \alpha_k y_k$, we obtain

$$\mathcal{E}_k(p) = (\pi_k(p) - \kappa_k)_+ = \max_{\alpha_k \in [0,1]} \alpha_k \left(\pi_k(p) - \kappa_k\right) =$$

$$\max_{\substack{\alpha_k \in [0,1] \\ y_k \in \mathcal{Y}_k}} \alpha_k \left(\langle p, y_k \rangle - c_k(y_k) - \kappa_k\right) = \max_{\substack{\alpha_k \in [0,1] \\ \tilde{y}_k \in \alpha_k \mathcal{Y}_k}} \langle p, \tilde{y}_k \rangle - \alpha_k c_k \left(\tilde{y}_k/\alpha_k\right) - \alpha_k \kappa_k .$$

Note that the maximization problem

$$\mathcal{E}_k(p) = \max_{\substack{\alpha_k \in [0,1] \\ \tilde{y}_k \in \alpha_k \mathcal{Y}_k}} \langle p, \tilde{y}_k \rangle - \alpha_k c_k \left(\tilde{y}_k/\alpha_k\right) - \alpha_k \kappa_k$$

is convex, and its set of optimal solutions consists of proper participation levels α_k and production bundles $\tilde{y}_k$. Moreover, $\mathcal{E}_k(p)$ is convex in p as the maximum of linear functions. Hence, the convex subdifferential of the excessive profit $\mathcal{E}_k$ gives the supply S_k of the kth producer, i.e.

$$\partial \mathcal{E}_k(p) = S_k(p). \tag{17}$$

The latter follows e.g. from [22, Theorem 2.4.18] on the convex subdifferential of a max-type function.

Analogously, we define the *excessive wealth* of the ith consumer as follows:

$$\mathcal{E}_i(p) \stackrel{\text{def}}{=} (w_i - e_i(p))_+ = \max_{x_i \in \mathcal{X}_i} \left(w_i - \langle p, x_i \rangle\right)_+ . \tag{18}$$

Using the substitution $\tilde{x}_i = \beta_i x_i$, we obtain

$$\mathcal{E}_i(p) = (w_i - e_i(p))_+ = \max_{\beta_i \in [0,1]} \beta_i \left(w_i - e_i(p)\right) =$$

$$\max_{\substack{\beta_i \in [0,1] \\ x_i \in \mathcal{X}_i}} \beta_i \left(w_i - \langle p, x_i \rangle\right) = \max_{\substack{\beta_i \in [0,1] \\ \tilde{x}_i \in \beta_i \mathcal{X}_i}} \beta_i w_i - \langle p, \tilde{x}_i \rangle .$$

Note that $\tilde{x}_i \in \beta_i \mathscr{X}_i$ means

$$\tilde{x}_i \in \beta_i X_i \text{ and } u_i\left(\tilde{x}_i/\beta_i\right) \geq \underline{u}_i.$$

In particular, the so-called perspective function $\beta_i u_i\left(\tilde{x}_i/\beta_i\right)$ is jointly concave in $(\tilde{x}_i, \beta_i)$, e.g. [4]. The maximization problem

$$\mathscr{E}_i(p) = \max_{\substack{\beta_i \in [0,1] \\ \tilde{x}_i \in \beta_i \mathscr{X}_i}} \beta_i w_i - \langle p, \tilde{x}_i \rangle$$

is convex, and its set of optimal solutions consists of proper participation levels β_i and consumption bundles $\tilde{x}_i$. Moreover, $\mathscr{E}_i(p)$ is convex in p as the maximum of linear functions. Hence, the convex subdifferential of the excessive wealth $\mathscr{E}_i$ gives the opposite demand D_i of the ith consumer, i.e.

$$\partial \mathscr{E}_i(p) = -D_i(p). \tag{19}$$

The latter follows also from [22, Theorem 2.4.18].

Overall, we define the *total excessive revenue* as the sum of excessive profits and excessive wealths:

$$\mathscr{E}(p) \stackrel{\text{def}}{=} \sum_{k=1}^{K} \mathscr{E}_k(p) + \sum_{i=1}^{I} \mathscr{E}_i(p). \tag{20}$$

Note that function $\mathscr{E}(\cdot)$ is convex since it is a sum of convex functions. Moreover, its convex subdifferential represents the excess supply due to (17) and (19).

Remark 1 (Homogeneous Case). For the homogeneous case we can give yet another explanation why marginal producers and consumers still stay at the market. Let us assume the homogeneity of the kth producer's cost function $c_k(\cdot)$, and the homogeneity of the fixed cost $\kappa_k(\cdot)$, i.e.

$$\kappa_k(\alpha \mathscr{Y}_k) = \alpha \kappa_k(\mathscr{Y}_k), \quad \alpha \in [0,1].$$

Then,

$$\mathscr{E}_k(p) = \max_{\alpha_k \in [0,1], \tilde{y}_k \in \alpha_k \mathscr{Y}_k} \langle p, \tilde{y}_k \rangle - c_k\left(\tilde{y}_k\right) - \kappa_k(\alpha_k \mathscr{Y}_k).$$

For a marginal producer with $\mathscr{E}_k(p) = 0$, this means that his activity, even within the maximal technological set $\mathscr{Y}_k$ does not generate any profit. The situation is not changing if the production activities (the set $\mathscr{Y}_k$) will be proportionally reduced by a factor $\alpha_k \in [0,1]$. Thus, it is natural to admit that in this marginal situation the producer can work with a reduced technological set $\alpha_k \mathscr{Y}_k$ by producing $\tilde{y}_k \in \alpha_k \mathscr{Y}_k$.

By doing so, he cannot cover the share $(1-\alpha_k)\kappa_k$ of the fixed cost. However, his unused capacities amounting to $(1-\alpha_k)\mathscr{Y}_k$ can be eventually exploited at other markets.

Now, we assume the homogeneity of the ith consumer's utility function $u_i(\cdot)$, and that $X_i = \mathbb{R}^n_+$. Then,

$$\mathscr{E}_i(p) = \max_{\substack{\beta_i \in [0,1],\, \tilde{x}_i \geq 0 \\ u_i(\tilde{x}_i) \geq \beta_i \underline{u}_i}} \beta_i w_i - \langle p, \tilde{x}_i \rangle .$$

If the excessive wealth of a consumer is zero, then again, there is no special reason to allocate all the wealth w_i to this expensive market. The consumer can admit to spend here only a part of it, namely $\beta_i w_i$ with some $\beta_i \in [0,1]$, which is sufficient to guarantee the share $\beta_i \underline{u}_i$ of his desirable utility level. Note that this does not change the zero level of the excessive wealth. The remaining part $(1-\beta_i)w_i$ of the wealth can be used then at other markets. □

By application of [15, Theorem 23.8] on the subdifferential of the sum of convex functions, we obtain:

Theorem 1 (Excess Supply and Total Excessive Revenue). *For $p \in \mathbb{R}^n_+$ it holds:*

$$\partial \mathscr{E}(p) = \sum_{k=1}^{K} S_k(p) - \sum_{i=1}^{I} D_i(p).$$

Proof. We apply [15, Theorem 23.8] on the subdifferential of the sum of convex functions in order to obtain

$$\partial \mathscr{E}(p) = \sum_{k=1}^{K} \partial \mathscr{E}_k(p) - \sum_{i=1}^{I} \partial \mathscr{E}_i(p).$$

Together with (17) and (19) the assertion follows. □

Theorem 1 allows us to characterize equilibrium prices as minimizers of $\mathscr{E}$.

Theorem 2 (Principle of Least Revenue). $p \in \mathbb{R}^n_+$ *is a system of equilibrium prices if and only if it solves the following convex minimization problem:*

$$\min_{p \in \mathbb{R}^n_+} \mathscr{E}(p). \tag{P}$$

Proof. **1.** Assume that $p^* \in \mathbb{R}^n$ is an equilibrium prices. Then, in view of Definition 1, there exists an equilibrium market flow

$$\widetilde{F}^* = \left(\left(\tilde{y}_k^*\right)_{k=1}^K, \left(\tilde{x}_i^*\right)_{i=1}^I \right) \in \prod_{k=1}^{K} S_k(p^*) \times \prod_{i=1}^{I} D_i(p^*),$$

satisfying the market clearing condition

$$p^* \geq 0, \quad \sum_{k=1}^{K} \tilde{y}_k^* - \sum_{i=1}^{I} \tilde{x}_i^* \geq 0, \quad \left\langle p^*, \sum_{k=1}^{K} \tilde{y}_k^* - \sum_{i=1}^{I} \tilde{x}_i^* \right\rangle = 0.$$

Denote $\zeta^* = \sum_{k=1}^{K} \tilde{y}_k - \sum_{i=1}^{I} \tilde{x}_i$. In view of Theorem 1, $\zeta^* \in \partial \mathscr{E}(p^*)$. Since $\mathscr{E}$ is convex in p, for all $p \in \mathbb{R}^n_+$ we have:

$$\mathscr{E}(p) - \mathscr{E}(p^*) \geq \langle \zeta^*, p - p^* \rangle = \langle \zeta^*, p \rangle \geq 0.$$

Thus, p^* minimizes the total excessive revenue.

2. Assume that $p^* \in \mathbb{R}^n_+$ is optimal for the minimization problem (**P**). Then there exists $\zeta^* \in \partial \mathscr{E}(p^*)$ such that

$$\langle \zeta^*, p - p^* \rangle \geq 0, \quad \text{for all } p \in \mathbb{R}^n_+.$$

Considering $p = 0$ and $p = 2p^*$, we conclude that $\langle \zeta^*, p^* \rangle = 0$. Consequently, $\zeta^* \in \mathbb{R}^n_+$. Again due to Theorem 1, there exists a market flow

$$\widetilde{F}^* = \left(\left(\tilde{y}_k^*\right)_{k=1}^{K}, \left(\tilde{x}_i^*\right)_{i=1}^{I} \right) \in \prod_{k=1}^{K} S_k(p^*) \times \prod_{i=1}^{I} D_i(p^*),$$

such that

$$\zeta^* = \sum_{k=1}^{K} \tilde{y}_k^* - \sum_{i=1}^{I} \tilde{x}_i^*.$$

Hence, $\widetilde{F}^*$ satisfies the market clearing condition, meaning that it is actually an equilibrium market flow. In view of Definition 1, p^* is an equilibrium price. □

2.4 *Existence*

Theorem 2 says that equilibrium prices correspond to optimal solutions for the minimization problem:

$$\min_{p \in \mathbb{R}^n_+} \mathscr{E}(p). \qquad \textbf{(P)}$$

This is the key to provide existence results for equilibrium prices. We denote by P^* the set of equilibrium prices. Let us introduce productive markets, at which the set of equilibrium prices P^* turns out to be nonempty and bounded.

Definition 2 (Productive Market). A market is called *productive* if there exist subsets of producers $\mathscr{K} \subset \{1,\ldots,K\}$ and consumers $\mathscr{L} \subset \{1,\ldots,L\}$, such that the corresponding production and consumption flows

$$(\{\bar{y}_k\}_{k\in\mathscr{K}}, \{\bar{x}_i\}_{i\in\mathscr{L}}) \in \prod_{k\in\mathscr{K}} \mathscr{Y}_k \times \prod_{i\in\mathscr{L}} \mathscr{X}_i$$

establish positive balances for goods:

$$\sum_{k\in\mathscr{K}} \bar{y}_k > \sum_{i\in\mathscr{L}} \bar{x}_i. \tag{21}$$

The market productivity means that there are some producers who can oversupply some consumers' needs.

Theorem 3 (Existence and Boundedness of Equilibrium Prices). *At the productive markets, the set of equilibrium prices P^* is nonempty and bounded.*

Proof. Due to Theorem 2, equilibrium prices in P^* form the set of optimal solutions of the minimization problem (**P**). We show that the latter set is bounded. For that, it is sufficient to prove that the level sets of function $\mathscr{E}(\cdot)$ are bounded. Denote $\bar{\xi} = \sum\limits_{k\in\mathscr{K}} \bar{y}_k - \sum\limits_{i\in\mathscr{L}} \bar{x}_i$. For all $p \in \mathbb{R}^n_+$ we have

$$\begin{aligned}
\mathscr{E}(p) &= \sum_{k=1}^{K} [\pi(p) - \kappa_k]_+ + \sum_{i=1}^{L} [w_i - e_i(p)]_+ \\
&\geq \sum_{k\in\mathscr{K}} [\pi(p) - \kappa_k]_+ + \sum_{i\in\mathscr{L}} [w_i - e_i(p)]_+ \\
&\geq \sum_{k\in\mathscr{K}} \pi(p) - \kappa_k + \sum_{i\in\mathscr{L}} w_i - e_i(p) \\
&\geq \sum_{k\in\mathscr{K}} (\langle p, \bar{y}_k\rangle - c_k(\bar{y}_k) - \kappa_k) + \sum_{i\in\mathscr{L}} (w_i - \langle p, \bar{x}_i\rangle) \\
&= -\sum_{k\in\mathscr{K}} (\kappa_k + c_k(\bar{y}_k)) + \sum_{i\in\mathscr{L}} w_i + \langle \bar{\xi}, p\rangle.
\end{aligned}$$

Since $\bar{\xi} > 0$, the intersection of the level sets of function $\mathscr{E}$ with $\mathbb{R}^n_+$ is bounded.

As a direct consequence of Theorem 2, we state the following result.

Theorem 4 (Convexity of Equilibrium Prices). *The set of equilibrium prices P^* is convex.*

Further, we formulate additional assumptions in order to guarantee that our market indeed works, i.e. the equilibrium prices do not vanish. Due to Theorem 2, we need to ensure that the optimal solution p^* of the minimization problem (**P**) is not at the origin. For that, we introduce the following condition rejecting the *Zero-Cost Production (ZCP)*:

$$\text{If } \alpha_k \kappa_k + \alpha_k c_k (\tilde{y}_k/\alpha_k) = 0 \text{ with } \tilde{y}_k \in \alpha_k \mathscr{Y}_k \text{ and } \alpha_k \in [0,1], \text{ then } \tilde{y}_k = 0. \tag{22}$$

This condition is automatically satisfied for $\kappa_k > 0$. If $\kappa_k = 0$, then (22) implies that for the kth producer there is no nonzero production plan with zero production cost. Recall that

$$\mathscr{E}_k(p) = \max_{\substack{\alpha_k \in [0,1] \\ \tilde{y}_k \in \alpha_k \mathscr{Y}_k}} \langle p, \tilde{y}_k \rangle - \alpha_k c_k (\tilde{y}_k/\alpha_k) - \alpha_k \kappa_k, \tag{23}$$

Therefore, condition (22) implies that $\partial \mathscr{E}_k(0) = \{0\}$. Note that $\tilde{y}_k = 0$ if $\alpha_k = 0$ in (23), hence, the term $\alpha_k c_k (\tilde{y}_k/\alpha_k)$ is set to vanish in this case.

Assume now that the wealth w_i of ith consumer is positive. Since

$$\mathscr{E}_i(p) = \max_{\substack{\beta_i \in [0,1] \\ \tilde{x}_i \in \beta_i \mathscr{X}_i}} [\beta_i w_i - \langle p, x_i \rangle ,$$

we conclude that $\partial \mathscr{E}_i(0) = -\mathscr{X}_i$. Thus, we have proved the following statement.

Lemma 1. *Let all producers satisfy ZCP-condition, and the wealths of all consumers be positive. Then,*

$$\partial \mathscr{E}(0) = -\sum_{i=1}^{L} \mathscr{X}_i. \tag{24}$$

Corollary 1 (Nonzero Equilibrium Prices). *Existence of a consumer with nonzero life standard is sufficient for having $p^* \neq 0$.*

Proof. Indeed, assume that $p^* = 0$. In view of the first-order optimality conditions for (**P**), there exists $\xi^* \in \partial \mathscr{E}(0)$ such that

$$\langle \xi^*, p \rangle \geq 0 \quad \forall p \geq 0.$$

Hence, $\xi^* = -\sum_{i=1}^{L} x_i^* \geq 0$ for some $x_i^* \in \mathscr{X}_i$. Therefore, all $x_i^* = 0$, implying zero life standards for all consumers.

It is interesting that the last statement is formulated only in terms of consumption standards. This confirms the primary role of demand in generating supply.

2.5 Efficiency

Let us present the first welfare theorem for equilibrium market flow. We are going to prove that any equilibrium market flow is efficient in the sense of Pareto optimality. This means that no producers or consumers can improve the gain (excessive profits and excessive wealths, respectively) without worsening the gain of some others. Let us start from the definition of feasible market flows.

We recall that for a given vector of prices $p \in \mathbb{R}^n_+$ and a *tentative market flow*

$$F = \left(\{y_k\}_{k=1}^K, \{x_i\}_{i=1}^L \right) \in \prod_{k=1}^K \mathscr{Y}_k \times \prod_{i=1}^L \mathscr{X}_i,$$

the system of participation levels $\gamma = \left(\{\alpha_k\}_{k=1}^K, \{\beta_i\}_{i=1}^L \right) \in [0,1]^{K+L}$ is called *proper* (with respect to π and F) if it satisfies the following conditions:

$$\alpha_k = \begin{cases} 1, \text{ if } \langle p, y_k \rangle - c_k(y_k) > \kappa_k, \\ 0, \text{ if } \langle p, y_k \rangle - c_k(y_k) < \kappa_k, \end{cases}$$

$$\beta_i = \begin{cases} 1, \text{ if } \langle p, x_i \rangle < w_i, \\ 0, \text{ if } \langle p, x_i \rangle > w_i. \end{cases}$$

Such a triple (p, F, γ) defines a *real market flow*

$$\widetilde{F} = \left(\{\tilde{y}_k = \alpha_k y_k\}_{k=1}^K, \{\tilde{x}_i = \beta_i x_i\}_{i=1}^L \right).$$

Definition 3 (Feasible Market Flow). The real market flow

$$\widetilde{F} = \left(\{\tilde{y}_k\}_{k=1}^K, \{\tilde{x}_i\}_{i=1}^L \right),$$

defined by the triple (p, F, γ), is called *feasible* if it satisfies the market clearing condition:

$$p \geq 0, \quad \sum_{k=1}^K \tilde{y}_k - \sum_{i=1}^I \tilde{x}_i \geq 0, \quad \left\langle p, \sum_{k=1}^K \tilde{y}_k - \sum_{i=1}^I \tilde{x}_i \right\rangle = 0.$$

Note that an equilibrium market flow is in particular feasible.

Definition 4 (Pareto Optimal Market Flow). A feasible market flow $\widetilde{F}$, defined by the triple

$$\left(p, F = \left(\{y_k\}_{k=1}^K, \{x_i\}_{i=1}^L \right), \gamma \right),$$

is called *Pareto optimal* if there is no feasible market flow $\widetilde{F}'$ defined by another triple

$$\left(p', F' = \left(\{y'_k\}_{k=1}^K, \{x'_i\}_{i=1}^L\right), \gamma'\right)$$

such that all inequalities

$$\begin{aligned} (\langle p', y'_k\rangle - c_k(y'_k) - \kappa_k)_+ &\geq (\langle p, y_k\rangle - c_k(y_k) - \kappa_k)_+, k = 1 \dots K, \\ (w_i - \langle p', x'_i\rangle)_+ &\geq (w_i - \langle p, x_i\rangle)_+, \; i = 1 \dots L, \end{aligned} \tag{25}$$

are satisfied, and at least one of them is strict.

Note that we define Pareto optimality with respect to excessive profits and excessive wealths. In our model they play a role of objective functions of the agents.

Theorem 5 (Efficiency of Equilibrium Market Flows). *Any equilibrium market flow is Pareto optimal.*

Proof. Using notation of Definition 4, let $\widetilde{F}^*$ be the equilibrium market flow defined by the triple (p^*, F^*, γ^*). Assume that the inequalities (25) are all valid for some feasible market flow $\widetilde{F}'$ defined by the triple (p', F', γ'). And let at least one of these inequalities be strict. For $p \in \mathbb{R}^n_+$ and $F \in \Omega \stackrel{\text{def}}{=} \prod_{k=1}^K \mathscr{Y}_k \times \prod_{i=1}^L \mathscr{X}_i$, define the function

$$\varphi(p, F) = \sum_{k=1}^K (\langle p, y_k\rangle - c_k(y_k) - \kappa_k)_+ + \sum_{i=1}^L (w_i - \langle p, x_i\rangle)_+.$$

In view of our assumption, $\varphi(p', F') > \varphi(p^*, F^*)$. Since p^* is an equilibrium price, in view of Theorem 2 and definitions (16), (18) we have:

$$\varphi(p^*, F^*) = \min_{p\geq 0} \max_{F\in\Omega} \varphi(p, F) = \max_{F\in\Omega} \min_{p\geq 0} \varphi(p, F) \geq \min_{p\geq 0} \varphi(p, F').$$

It remains to note that the market clearance condition for the flow $\widetilde{F}'$ is exactly the necessary and sufficient characterization of point p' as the optimal solution to the latter minimization problem. Therefore, $\varphi(p^*, F^*) \geq \varphi(p', F')$, a contradiction.

In view of Theorem 2, equilibrium prices minimize the total excessive revenue. Let us prove a very intuitive result that its optimal value is equal to the difference of the real consumers' wealths and the real producers' costs.

Theorem 6 (Total Excessive Revenue of the Market). *Let p^* be an equilibrium price, and*

$$\widetilde{F}^* = \left((\tilde{y}^*_k = \alpha^*_k y^*_k)_{k=1}^K, (\tilde{x}^*_i = \beta^*_i x^*_i)_{i=1}^I\right)$$

be an equilibrium market flow defined by the triple (p^*, F^*, γ^*). *Then,*

$$\mathscr{E}(p^*) = \sum_{i=1}^{I} \beta_i^* w_i - \sum_{k=1}^{K} \alpha_k^* (c_k(y_k^*) + \kappa_k) \geq 0.$$

Proof. It holds:

$$\mathscr{E}(p^*) = \sum_{k=1}^{K} \left(\langle p^*, y_k^* \rangle - c_k(u_k^*) - \kappa_k \right)_+ + \sum_{i=1}^{I} \left(w_i - \langle p^*, x_i^* \rangle \right)_+$$

$$= \sum_{k=1}^{K} \alpha_k^* \left(\langle p^*, y_k^* \rangle - c_k(u_k^*) - \kappa_k \right) + \sum_{i=1}^{I} \beta_i^* \left(w_i - \langle p^*, x_i^* \rangle \right)$$

$$= \left\langle p^*, \sum_{k=1}^{K} \alpha_k^* y_k^* - \sum_{i=1}^{I} \beta_i^* x_i^* \right\rangle - \sum_{k=1}^{K} \alpha_k^* (c_k(y_k^*) + \kappa_k) + \sum_{i=1}^{I} \beta_i^* w_i^*.$$

In view of the market clearance condition, we have

$$\left\langle p^*, \sum_{k=1}^{K} \alpha_k^* y_k^* - \sum_{i=1}^{I} \beta_i^* x_i^* \right\rangle = 0.$$

This gives us the desired expression for optimal value of $\mathscr{E}$. It is nonnegative since all terms in its definition (20) are nonnegative.

Note that the nonnegative value

$$\mathscr{E}(p^*) = \sum_{i=1}^{I} \beta_i^* w_i - \sum_{k=1}^{K} \alpha_k^* (c_k(y_k^*) + \kappa_k) \tag{26}$$

represents the total rate of accumulation of the capital within the market. In general, equilibrium prices, market flows, and participation levels are not unique. Nevertheless, all of them ensure the same value of $\mathscr{E}^* \stackrel{\text{def}}{=} \mathscr{E}(p^*)$. We call it the *total excessive revenue of the market*.

2.6 Welfare Maximization

In order to state the adjoint problem for (**P**), we set

$$\alpha \stackrel{\text{def}}{=} \{\alpha_k\}_{k=1}^K, \tilde{y} \stackrel{\text{def}}{=} \{\tilde{y}_k\}_{k=1}^K, \beta \stackrel{\text{def}}{=} \{\beta_i\}_{i=1}^I, \tilde{x} \stackrel{\text{def}}{=} \{\tilde{x}_i\}_{i=1}^I,$$

$$\mathscr{Y} \stackrel{\text{def}}{=} \prod_{k=1}^{K} \mathscr{Y}_k, \alpha\mathscr{Y} \stackrel{\text{def}}{=} \prod_{k=1}^{K} \alpha_k \mathscr{Y}_k, \mathscr{X} \stackrel{\text{def}}{=} \prod_{i=1}^{I} \mathscr{X}_i, \beta\mathscr{X} \stackrel{\text{def}}{=} \prod_{i=1}^{I} \beta_i \mathscr{X}_i.$$

Here, α, β represent participation levels, and $\tilde{y}$, $\tilde{x}$ represent production and consumption bundles, respectively. Moreover, $\alpha\mathscr{Y}$, $\beta\mathscr{X}$ represent production and consumption sets given the participation levels α, β, respectively.

The feasible set of the adjoint problem is formed by participation levels and corresponding production and consumption bundles, i.e.

$$\mathscr{A} \stackrel{\text{def}}{=} \left\{ (\alpha, \tilde{y}, \beta, \tilde{x}) \;\middle|\; \begin{array}{l} (\alpha, \tilde{y}) \in [0,1]^K \times \alpha\mathscr{Y} \\ (\beta, \tilde{x}) \in [0,1]^I \times \beta\mathscr{X} \end{array} \right\}.$$

Note that the set A is convex. Further, the following market feasibility constraint needs to be satisfied:

$$\sum_{k=1}^{K} \tilde{y}_k \geq \sum_{i=1}^{I} \tilde{x}_i, \tag{27}$$

meaning that the aggregate consumption does not exceed the aggregate production. The objective function of the adjoint problem is

$$\Phi\left(\alpha, \tilde{y}, \beta, \tilde{x}\right) \stackrel{\text{def}}{=} \sum_{i=1}^{I} \beta_i w_i - \sum_{k=1}^{K} \alpha_k c_k\left(\tilde{y}_k/\alpha_k\right) + \alpha_k \kappa_k,$$

expressing the difference between the aggregate wealth spent for consumption and producers' costs. Finally, we consider the welfare maximization problem

$$\max_{(\alpha, \tilde{y}, \beta, \tilde{x}) \in \mathscr{A}} \left\{ \Phi\left(\alpha, \tilde{y}, \beta, \tilde{x}\right) \;\middle|\; \sum_{k=1}^{K} \tilde{y}_k \geq \sum_{i=1}^{I} \tilde{x}_i \right\}. \tag{A}$$

In (**A**) the central authority assigns production and consumption bundles, as well as agents' participation levels. Moreover, it maximizes the welfare of the society while ensuring the market feasibility. In order to state (**A**), the central authority needs to know agents' cost and utility functions, production and consumption sets, etc. Obviously, this information about the agents is hardly observable to the central authority. Consequently, it cannot be justified in general that the welfare maximization problem is tackled directly. Nevertheless, note that the prices of goods play the role of Lagrange or dual multipliers for the market feasibility constraint (27), cf. already [3, 16] for similar interpretations. Hence, due to the duality theory of convex programming, the welfare maximization (**A**) is the adjoint problem for (**P**).

Theorem 7 (Adjoint for (P)). *The welfare maximization (**A**) is adjoint for the total revenue minimization (**P**):*

$$\min_{p \in \mathbb{R}^n_+} \mathscr{E}(p) = \max_{(\alpha, \tilde{y}, \beta, \tilde{x}) \in \mathscr{A}} \left\{ \Phi\left(\alpha, \tilde{y}, \beta, \tilde{x}\right) \;\middle|\; \sum_{k=1}^{K} \tilde{y}_k \geq \sum_{i=1}^{I} \tilde{x}_i \right\}.$$

We note that the productivity of the market from Definition 2 implies the standard Slater condition for the adjoint problem (**A**).

We emphasize that the adjoint problem (**A**) of the welfare maximization can hardly be solved directly. In order to construct its feasible set $\mathscr{A}$ and its objective function Φ, a central authority should acquire the knowledge on producers' costs and their production sets, on consumers' utility functions and their consumption sets. It is clear that this is implementable only within a planned economy. Even in this case, as we know e.g. from the history of communistic countries, producers and consumers are reluctant to report their market constants to the authority. In fact, they feel rather antagonistic about each other and about the authority, thus, trying to keep their information private. In turn, our approach concentrates on the total revenue minimization problem (**P**). By doing so, we explain how the free market provides a welfare maximizing solution by a decentralized price adjustment. Here, producers and consumers report only their individual supplies and demands to each other while trading or auctioning. There is no need in a central authority, since the price updates are performed by producers, in case of trade, and by consumers, in case of auction. Finally, the price adjustment balances agents' antagonistic behavior leading to a market equilibrium.

3 Quasi-Monotone Subgradient Method

We first present quasi-monotone subgradient methods for nonsmooth convex minimization from [13]. For that, we consider the following minimization problem:

$$f^* \stackrel{\text{def}}{=} \min_{x \in X} f(x), \tag{28}$$

where $X \subset \mathbb{R}^n$ is a closed convex set with nonempty interior $\operatorname{int} X$, and f is a convex function on $\mathbb{R}^n$. Moreover, let f be representable as a maximum of concave functions, i.e.

$$f(x) = \max_{a \in A} \Phi(a) + \varphi(x, a), \tag{29}$$

where $A \subset \mathbb{R}^m$ is a compact convex set, $\varphi(\cdot, a)$ is a convex function on $\mathbb{R}^n$ for every $a \in A$, and Φ, $\varphi(x, \cdot)$ are concave functions on $\mathbb{R}^m$ for every $x \in X$. Denote by $a(x)$ one of the optimal solutions of the maximization problem in (29). Then,

$$\nabla f(x) \stackrel{\text{def}}{=} \nabla_x \varphi(x, a(x)) \tag{30}$$

denotes a subgradient of f at x. This formula follows from the result on the subdifferential of a max-type function, e.g. [22, Theorem 2.4.18]. Recall that for an arbitrary subgradient $\nabla f(x)$ at $x \in X$ of a convex function f we have:

$$f(y) \geq f(x) + \langle \nabla f(x), y - x \rangle, \quad y \in X. \tag{31}$$

Using the representation (29), we also have:

$$f^* = \min_{x \in X} f(x) = \min_{x \in X} \max_{a \in A} [\Phi(a) + \varphi(x, a)] = \max_{a \in A} \left[\Phi(a) + \min_{x \in X} \varphi(x, a) \right].$$

The latter maximization problem

$$\max_{a \in A} \left[\Phi(a) + \min_{x \in X} \varphi(x, a) \right] \tag{32}$$

is called adjoint for (28) with the adjoint variable $a \in A$.

For the set X, we assume to be known a prox-function $d(x)$.

Definition 5. $d : X \mapsto \mathbb{R}$ is called a prox-function for X if the following holds:

- $d(x) \geq 0$ for all $x \in X$ and $d(x[0]) = 0$ for certain $x[0] \in X$;
- d is strongly convex on X with convexity parameter one:

$$d(y) \geq d(x) + \langle \nabla d(x), y - x \rangle + \frac{1}{2} \|y - x\|^2, \quad x, y \in X, \tag{33}$$

  where $\| \cdot \|$ is a norm on $\mathbb{R}^n$.
- Auxiliary minimization problem

$$\min_{x \in X} \{ \langle z, x \rangle + \chi d(x) \} \tag{34}$$

  is easily solvable for $z \in \mathbb{R}^n$, $\chi > 0$.

As a simple consequence of Definition 5, we have for $x \in X$:

$$d(x) \geq d(x[0]) + \langle \nabla d(x[0]), x - x[0] \rangle + \frac{1}{2} \|x - x[0]\|^2 \geq \frac{1}{2} \|x - x[0]\|^2. \tag{35}$$

For a sequence of positive parameters $\{\chi[t]\}_{t \geq 0}$, we consider the following iteration:

Quasi-monotone Subgradient Method

1. Take a current subgradient $\nabla f(x[t]) = \nabla_x \varphi(x[t], a(x[t]))$.

2. Accumulate subgradients $z[t] = z[t-1] + \nabla f(x[t]), z[-1] = 0$.

3. Compute the forecast $x^+[t] = \arg\min\limits_{x \in X} \{\langle z[t], x\rangle + \chi[t] d(x)\}$.

4. Update by combining $x[t+1] = \frac{t+1}{t+2} x[t] + \frac{1}{t+2} x^+[t]$.

(SM)

Note that from (SM) we have

$$z[t] = \sum_{r=0}^{t} \nabla f(x[r]), \quad x[t+1] = \frac{1}{t+2}\left(x[0] + \sum_{r=0}^{t} x^+[r]\right).$$

Next Lemma 2 is crucial for the convergence analysis of the quasi-monotone subgradient method (SM). It estimates the dual gap for the minimization problem (28) and its adjoint problem (32) evaluated at the historical averages.

For that, we define the penalty term δ_t and the remainder term ρ_t, $t \geq 0$, as follows:

$$\begin{aligned}\delta_t(a) &\stackrel{\text{def}}{=} -\min_{x \in X}\left\{\varphi(x, a) + \frac{\chi[t]}{t+1} d(x)\right\}, \quad a \in A,\\ \rho_t &\stackrel{\text{def}}{=} \frac{1}{t+1}\sum_{r=0}^{t} \frac{1}{2\chi[r-1]} \|\nabla f(x[r])\|_*^2, \quad \chi[-1] = \chi[0].\end{aligned}$$

Here, $\|\cdot\|_*$ is the conjugate norm to $\|\cdot\|$, i.e.

$$\|s\|_* \stackrel{\text{def}}{=} \max_{s \in \mathbb{R}^n} \{\langle s, x\rangle : \|x\| \leq 1\}, \quad s \in \mathbb{R}^n. \tag{36}$$

Further, we define the average adjoint state

$$a[t] \stackrel{\text{def}}{=} \frac{1}{t+1}\sum_{r=0}^{t} a(x[r]), \quad t \geq 0.$$

Note that $a[t] \in A$, since A is convex.

Lemma 2 is motivated by the estimate sequence technique (e.g., Sect. 2.2.1 in [10]) and is due to [13]. For the readers' convenience its proof is postponed to Appendix.

Lemma 2. *Let the sequence* $\{x[t]\}_{t\geq 0}$ *be generated by (SM) with nondecreasing parameters*

$$\chi[t+1] \geq \chi[t], \quad t \geq 0. \tag{37}$$

Then, for all $t \geq 0$ *it holds:*

$$f(x[t]) - \Phi(a[t]) + \delta_t(a[t]) \leq \rho_t. \tag{38}$$

We apply the quasi-monotone subgradient method (SM) in the following setup (S1)–(S3). Let $X = \mathbb{R}^n_+$ be equipped with the Euclidean prox-function

$$d(x) \stackrel{\text{def}}{=} \frac{1}{2} \sum_{j=1}^{n} \left(x^{(j)}\right)^2, \quad x \in X = \mathbb{R}^n_+. \tag{S1}$$

Note that the corresponding norm in Definition 5 and its conjugate according to (36) are

$$\|x\|^2 = \sum_{j=1}^{n} \left(x^{(j)}\right)^2, \quad \|s\|_*^2 = \sum_{j=1}^{n} \left(s^{(j)}\right)^2.$$

Further, we assume that φ is linear w.r.t. x:

$$\varphi(x,a) = -\sum_{j=1}^{n} x^{(j)} h_j(a), \tag{S2}$$

where $h = \left(h_j(\cdot), j = 1, \ldots, n\right)$, are convex functions on $\mathbb{R}^m$. Then, the adjoint problem (32) takes the form

$$f^* = \max_{a \in A} \left\{\Phi(a) \;\middle|\; h_j(a) \leq 0, \quad j = 1, \ldots, n\right\}. \tag{39}$$

The maximization problem (39) is assumed to satisfy the Slater condition (e.g., [15]), i.e.

$$\text{there exists } \bar{a} \in A \text{ such that } h_j(\bar{a}) < 0 \text{ for all } j = 1, \ldots, n. \tag{S3}$$

Under (S1)–(S3) we have in (SM):

$$\nabla f(x[t]) = -h(a(x[t])),$$
$$z[t] = -\sum_{r=0}^{t} h(a(x[r])),$$
$$x^+[t] = \frac{1}{\chi[t]}(-z[t])_+ = \frac{1}{\chi[t]}\left(\sum_{r=0}^{t} h(a(x[r]))\right)_+.$$

Here, the forecast $x^+[t]$ is chosen to be proportional to the historical infeasibility.

Now we are ready to proceed with the convergence analysis of the method (SM) under (S1)–(S3). Next Lemma 3 estimates the dual gap for the minimization problem (28) and its adjoint problem (39) evaluated at the historical averages.

Lemma 3. *Let the sequence* $\{x[t]\}_{t\geq 0}$ *be generated by (SM) under (S1)–(S3) with nondecreasing parameters*

$$\chi[t+1] \geq \chi[t], \quad t \geq 0.$$

Then, for all $t \geq 0$ *it holds:*

$$f(x[t]) - f^* + C_1 \tfrac{\chi[t]}{t+1} \leq f(x[t]) - \Phi(a[t]) + \tfrac{t+1}{\chi[t]} \textstyle\sum_{j=1}^{n} \left(h_j(a[t])\right)_+^2 \leq C_2 \tfrac{1}{t+1} \textstyle\sum_{r=0}^{t} \tfrac{1}{\chi[r-1]} \tag{40}$$

with positive constants $C_1, C_2 > 0$.

The proof of Lemma 3 is postponed to Appendix. For the precise dependence of constants C_1 and C_2 on the market's data see (75) and (76) in Appendix.

In order for (SM) to converge, the parameters $\{\chi[t]\}_{t\geq 0}$ need to be properly chosen. Next Lemma 4 identifies successful adjustment strategies of parameters. Namely, the parameters monotonically increase over time, but by decreasing increments.

Lemma 4. *Let nondecreasing parameters satisfy*

$$\chi[t] - \chi[t-1] \to 0, \quad \chi[t] \to \infty. \tag{41}$$

Then,

$$\frac{\chi[t]}{t+1} \to 0, \quad \text{and} \quad \frac{1}{t+1}\sum_{r=0}^{t} \frac{1}{\chi[r-1]} \to 0. \tag{42}$$

Moreover, the best achievable order of convergence in (42) is $O\left(\frac{1}{\sqrt{t}}\right)$.

For the proof of Lemma 4 see Appendix.

Remark 2. As in the proof of Lemma 4, nondecreasing parameters can be written in the cumulative form:

$$\chi[t] = \sum_{r=0}^{t} \iota[r] + \chi[-1]$$

with increments $\iota[t] \geq 0$. Then, the convergence condition (41) means that increments tend to zero and sum up to infinity, i.e.

$$\iota[t] \to 0, \quad \sum_{t=0}^{\infty} \iota[t] = \infty.$$

The latter coincides with the usual condition imposed on the step-sizes of the subgradient method for nonsmooth convex minimization (e.g., [10]). However, in our setting $\iota[t]$ play the role of incremental step-sizes. This gives rise to suppose that the parameters $\chi[t]$ can be formed by incremental learning (cf. [20]). In fact, the parameter $\chi[t]$ increases over time, however, by decreasing increments $\iota[t]$. □

Now, we are ready to prove the main convergence result for (SM) under (S1)–(S3).

Theorem 8 (Convergence of (SM)). *Let the sequence $\{x[t]\}_{t\geq 0}$ be generated by (SM) under (S1)–(S3) with nondecreasing parameters satisfying*

$$\chi[t] - \chi[t-1] \to 0, \quad \chi[t] \to \infty.$$

Then, $\{x[t]\}_{t\geq 0}$ converges to the solution set of the minimization problem (28). Moreover, the average adjoint states $\{a[t]\}_{t\geq 0}$ converge to the solution set of its adjoint problem (39). The achievable rate of convergence is of the order $O\left(\frac{1}{\sqrt{t}}\right)$.

Proof. From Lemma 3 we obtain:

$$f(x[t]) - f^* + C_1 \frac{\chi[t]}{t+1} \leq f(x[t]) - \Phi(a[t]) + \frac{t+1}{\chi[t]} \sum_{j=1}^{n} \left(h_j(a[t])\right)_+^2 \leq C_2 \frac{1}{t+1} \sum_{r=0}^{t} \frac{1}{\chi[r-1]}.$$

This inequality is composed by the objective function f of the primal problem (28), computed at the current iterates $x[t]$, objective function Φ of its adjoint problem (39), computed at historical averages $a[t]$, and the quadratic penalty $\sum_{j=1}^{n} \left(h_j(a[t])\right)_+^2$ for violation of the constraints:

$$h_j(a[t]) \leq 0, \quad j = 1, \ldots, n$$

Due to the choice of parameters $\chi[t]$, Lemma 4 provides:

$$\frac{\chi[t]}{t+1} \to 0, \text{ and } \frac{1}{t+1} \sum_{r=0}^{t} \frac{1}{\chi[r-1]} \to 0.$$

Hence, the assertion follows. □

Now, we turn our attention to the case of constant and linear parameters.

Remark 3 (Constant Parameters). Let the constant parameters be applied in (SM). Let $\varepsilon > 0$ denote the tolerance for convergence of $x[t]$ towards a solution of the primal problem (28), and $a[t]$ towards a solution of its adjoint problem (39). Our goal is to indicate the number of steps $t(\varepsilon)$ and the parameters $\chi(\varepsilon)$, in order to guarantee the tolerance ε for this primal-adjoint process. For that, we apply constant confidence parameters $\chi[t] = \chi$ to obtain

$$\frac{\chi[t]}{t+1} = \frac{\chi}{t+1}, \quad \frac{1}{t+1}\sum_{r=0}^{t}\frac{1}{\chi[r-1]} = \frac{1}{\chi}.$$

Recalling (40), the order of convergence for the primal-adjoint process is

$$\max\left\{\frac{\chi}{t+1}, \frac{1}{\chi}\right\}.$$

Choosing

$$t(\varepsilon) = O\left(\frac{1}{\varepsilon^2}\right), \chi(\varepsilon) = O\left(\frac{1}{\varepsilon}\right),$$

we have

$$\max\left\{\frac{\chi(\varepsilon)}{t(\varepsilon)+1}, \frac{1}{\chi(\varepsilon)}\right\} = O(\varepsilon).$$

□

Remark 4 (Linear Growth of Parameters). Let us define the parameters in (SM) as follows

$$\chi[t] = (t+1)\sigma,$$

where $\sigma > 0$ can be seen as a growth rate of parameters. For the forecast we have then:

$$x^+[t] = \frac{1}{\sigma}\left(\frac{1}{t+1}\sum_{r=0}^{t} h(a(x[r]))\right)_+.$$

Here, the forecast $x^+[t]$ is formed proportional to the average infeasibility. We turn our attention to the convergence of (SM) for this case. Recalling (40), the order of convergence for the primal-adjoint process is

$$\max\left\{\sigma, \frac{1}{\sigma}\cdot\frac{1}{t+1}\left(1+\sum_{r=1}^{t}\frac{1}{r}\right)\right\},$$

or equivalently,

$$\max\left\{\sigma, \frac{1}{\sigma}\cdot\frac{\ln t}{t+1}\right\}.$$

Thus, the primal-adjoint process converges up to a residuum $O(\sigma)$. □

4 Decentralization of Prices

Theorem 2 reveals the origin of equilibrium prices at the market. Namely, in order to reach an equilibrium price one needs to solve the minimization problem:

$$\min_{p\in\mathbb{R}^n_+} \mathscr{E}(p). \qquad \textbf{(P)}$$

Our goal is to explain how agents can efficiently tackle this nonsmooth convex minimization problem by successively updating prices. This can be implemented by introducing various price designs. In this paper, we focus on

- *the regulation design*: regulator settles and updates prices which are taken by producers and consumers;
- *the trade design*: producers settle and update their individual prices, and consumers buy at the lowest purchase price;
- *the auction design*: consumers settle and update their individual prices, and producers sell at the highest offer price.

4.1 Regulation

It is crucial for our approach that the updates of prices correspond to subgradient-type methods for solving (**P**). Due to Theorem 1, the subgradients $\nabla\mathscr{E}(p)$ represent the *excess supply*, i.e.

$$\nabla\mathscr{E}(p) = \sum_{k=1}^{K} y_k - \sum_{i=1}^{I} x_i, \text{ where } y_k \in S_k(p), x_i \in D_i(p). \qquad (43)$$

It can be seen from (43) that the subgradients of $\mathscr{E}$ are not known to individual agents. Indeed, $\nabla\mathscr{E}(p)$ represents the *aggregate* excess supply. For getting access to its value, one would assume the existence of a market regulator who collects the information about agents' production and consumption bundles, and aggregates them over the whole market. Here, the full information about production and consumption over the market must be available to the regulator. Besides, the prices need to be updated by the latter, thus, leading to the *price regulation*. Clearly, these assumptions can be justified within a centrally planned economy. This allows to suppose that the regulator uses the subgradients $\nabla \mathscr{E}(p)$ for updating prices. In what follows, the quasi-monotone subgradient method for solving (**P**) from Sect. 3 is applied to this end.

Let the regulator choose a sequence of positive confidence parameters $\{\chi[t]\}_{t\geq 0}$. We consider the following iteration:

Price Regulation (REG)

1. Regulator determines the aggregated excess supply $\nabla \mathscr{E}(p[t])$:

(a) kth producer computes an optimal tentative production bundle

$$y_k(p[t]) \in \mathscr{Y}_k^*(p[t]),$$

and participation level

$$\alpha_k(p[t]) = \begin{cases} 1, \text{ if } \pi_k(p[t]) \geq \kappa_k, \\ 0, \text{ if } \pi_k(p[t]) < \kappa_k, \end{cases}$$

indicating whether $y_k(p[t])$ is implemented.

The production bundle is $\alpha_k(p[t])y_k(p[t])$, i.e. either $y_k(p[t])$ or zero.

(b) ith consumer computes an optimal tentative consumption bundle

$$x_i(p[t]) \in \mathscr{X}_i^*(p[t]),$$

and participation level

$$\beta_i(p[t]) = \begin{cases} 1, \text{ if } e_i(p[t]) \leq w_i, \\ 0, \text{ if } e_i(p[t]) < w_i, \end{cases}$$

indicating whether $x_i(p[t])$ is implemented.

The consumption bundle is $\beta_i(p[t])x_i(p[t])$, i.e. either $x_i(p[t])$ or zero.

(c) regulator observes the current excess supplies

$$\nabla \mathscr{E}(p[t]) = \sum_{k=1}^{K} \alpha_k(p[t])y_k(p[t]) - \sum_{i=1}^{I} \beta_i(p[t])x_i(p[t]). \tag{44}$$

2. Regulator accumulates the excess supplies

$$z[t] = z[t-1] + \nabla \mathscr{E}(p[t]), \quad z[-1] = 0. \tag{45}$$

3. Regulator computes the price forecast w.r.t. the confidence parameter $\chi[t]$:

$$p^{+(j)}[t] = \frac{\zeta^{(j)}}{\chi[t]} \left(-z^{(j)}[t]\right)_+, \quad j = 1, \ldots, n, \tag{46}$$

where $\zeta^{(j)}$ are positive scaling coefficients.
4. Regulator updates

$$p[t+1] = \frac{t+1}{t+2} p[t] + \frac{1}{t+2} p^+[t] \tag{47}$$

by combining the previous price with the forecast. □

First, we give an interpretation for the price forecast (46). Recall that $z^{(j)}[t]$ represents the aggregated excess supply for good j accumulated up to time t. If $z^{(j)}[t] \geq 0$, i.e. supply exceeds demand, then $p^{+(j)}[t] = 0$ for good j. In case of $z^{(j)}[t] < 0$, the price forecast $p^{+(j)}[t]$ is proportional to the accumulated and aggregated excess demand with positive scaling coefficients $\zeta^{(j)}$. Here, $\chi[t]$ plays the role of a confidence parameter. Namely, $\chi[t]$'s express to which extent the regulator takes into account the excess demands while forecasting prices.

Secondly, let us interpret the price update (47). Due to the latter, the next price is a convex combination of the previous price and the price forecast. With time advancing, the proportion of the previous price becomes nearly one, but the fraction of the forecast vanishes. Hence, we conclude that our price update corresponds to a behavior of an experienced regulator. Such regulator credits the experience much more than the current forecast. Further, from (47) we have

$$p[t+1] = \frac{1}{t+2} \left(p[0] + \sum_{r=0}^{t} p^+[r] \right). \tag{48}$$

The latter means that the prices generated by (REG) can be viewed as historical averages of preceding forecasts. This averaging pattern is also quite natural to assume for regulator's behavior while adjusting prices. Moreover, as it will be shown later, this price adjustment based on averaging successively leads to equilibrium prices.

Along with the prices $\{(p_1[t], \ldots, p_K[t])\}_{t \geq 0}$ generated by method (TRA), we consider the corresponding historical averages of participation levels:

$$\alpha_k[t] \stackrel{\text{def}}{=} \frac{1}{t+1} \sum_{r=0}^{t} \alpha_k(p_k[r]), \quad \beta_i[t] \stackrel{\text{def}}{=} \frac{1}{t+1} \sum_{r=0}^{t} \beta_i(p[r]).$$

Note that $\alpha_k[t] \in [0,1]$ is the frequency of successful production attempts by kth producer up to time t. Analogously, $\beta_i[t] \in [0,1]$ is the frequency of successful consumption attempts by ith consumer up to time t. We denote by

$$\gamma[t] = (\alpha[t], \beta[t]) \stackrel{\text{def}}{=} \left(\{\alpha_k[t]\}_{k=1}^K, \{\beta_i[t]\}_{i=1}^I \right)$$

the system of average participation levels. The historical averages of production and consumption bundles are defined as follows:

$$\tilde{y}_k[t] \stackrel{\text{def}}{=} \frac{1}{t+1} \sum_{r=0}^{t} \alpha_k(p_k[r]) y_k(p_k[r]), \quad \tilde{x}_i[t] \stackrel{\text{def}}{=} \frac{1}{t+1} \sum_{r=0}^{t} \beta_i(p[r]) x_i(p[r]).$$

Due to convexity, $\tilde{y}_k[t] \in \alpha_k[t]\mathscr{Y}_k$ and $\tilde{x}_i[t] \in \beta_i[t]\mathscr{X}_i$. We denote by

$$\widetilde{F}[t] = (\tilde{y}[t], \tilde{x}[t]) \stackrel{\text{def}}{=} \left(\{\tilde{y}_k[t]\}_{k=1}^{K}, \{\tilde{x}_i[t]\}_{i=1}^{I} \right)$$

the average market flow. Overall, the sequence

$$(\alpha[t], \tilde{y}[t], \beta[t], \tilde{x}[t]) \in \mathscr{A}, \quad t \geq 0,$$

is feasible for the adjoint problem (**A**).

Now, we are ready to prove the main convergence result for (REG).

Theorem 9 (Convergence of Price Regulation). *At a productive market, let the regulator apply in (REG) nondecreasing confidence parameters satisfying*

$$\chi[t] - \chi[t-1] \to 0, \quad \chi[t] \to \infty.$$

Then, the sequence of prices, average participation levels, and the average market flow

$$\left(p[t], \gamma[t], \widetilde{F}[t]\right)$$

converges toward the set of market equilibria. The achievable rate of convergence is of the order $O\left(\frac{1}{\sqrt{t}}\right)$.

Proof. The iteration scheme (REG) is a variant of the quasi-monotone subgradient method (SM). Hence, we may obtain the convergence for (REG) by means of Theorem 8. For that, let us discuss the applicability of conditions (S1)–(S3).

On (S1): The price forecast (57) can be derived by means of the Euclidean prox-function for $\mathbb{R}^n_+$:

$$d(p) \stackrel{\text{def}}{=} \frac{1}{2} \sum_{j=1}^{n} \frac{1}{\zeta^{(j)}} \left(p^{(j)}\right)^2.$$

In fact, for $z[t] \in \mathbb{R}^n$, $\chi[t] > 0$ we consider the minimization problem as from step 3. in (SM):

$$\min_{p \in \mathbb{R}^n_+} \left\{ \langle z[t], p \rangle + \chi[t] d(p) \right\}.$$

Its unique solution is the price forecast (46) as from step 3. in (REG):

$$p^{+(j)}[t] = \frac{\zeta^{(j)}}{\chi[t]} \left(-z^{(j)}[t]\right)_{+}, \quad j = 1, \dots, n.$$

On (S2): It follows from Theorem 7 that the total excessive revenue is representable as a maximum of concave functions:

$$\mathscr{E}(p) = \max_{(\alpha,\tilde{y},\beta,\tilde{x}) \in \mathscr{A}} \quad \Phi\left(\alpha, \tilde{y}, \beta, \tilde{x}\right) + \varphi\left(p, \tilde{y}, \tilde{x}\right),$$

where

$$\varphi\left(p, \tilde{y}, \tilde{x}\right) = \left\langle p, \sum_{k=1}^{K} \tilde{y}_k - \sum_{i=1}^{I} \tilde{x}_i \right\rangle.$$

Note that φ is linear w.r.t. p. In particular, due to Theorem 7, the adjoint problem for the total revenue minimization (**P**) is the welfare maximization (**A**).
On (S3): The welfare maximization problem (**A**) satisfies the Slater condition in view of the market productivity (cf. Definition 2).

Overall, we apply Theorem 8 to deduce that $p[t]$ converges toward the solution set of (**P**), and $\left(\gamma[t], \widetilde{F}[t]\right)$ converges toward the solution set of (**A**) by order $O\left(\frac{1}{\sqrt{t}}\right)$. In view of the duality from Theorem 7, the assertion follows. □

4.2 *Trade*

Aiming to avoid the assumption of price regulation, we decentralize prices by introducing the trade design:

> kth producer settles and updates individual prices p_k, and consumers buy at the lowest purchase price $\min\limits_{k=1,\dots,K} p_k$.

Recall that for vectors $p_1, \dots, p_K \in \mathbb{R}^n$, we denote by $\min\limits_{k=1,\dots,K} p_k \in \mathbb{R}^n$ the vector with coordinates

$$\left(\min_{k=1,\dots,K} p_k\right)^{(j)} = \min_{k=1,\dots,K} p_k^{(j)}, \quad j = 1, \dots, n.$$

The trade design incorporates the feature of Bertrand competition, namely, that consumers search for lowest prices, e.g. [8]. Following the framework of Bertrand competition, we assume that consumers are able to undertake global price search across the producers.

For the trade design, the total excessive revenue depends on the producers' prices $(p_k)_{k=1}^K$ as follows:

$$\mathscr{E}(p_1,\ldots,p_K) \stackrel{\text{def}}{=} \sum_{k=1}^K \mathscr{E}_k(p_k) + \sum_{i=1}^I \mathscr{E}_i\left(\min_{k=1,\ldots,K} p_k\right) =$$

$$\sum_{k=1}^K \max_{y_k \in \mathscr{Y}_k} \left(\langle p_k, y_k\rangle - c_k(y_k)\right)_+ + \sum_{i=1}^I \max_{x_i \in \mathscr{X}_i} \left(w_i - \left\langle \min_{k=1,\ldots,K} p_k, x_i \right\rangle\right)_+ . \qquad (49)$$

The decentralization of prices makes the corresponding subdifferential information about excess demands available to producers. In fact, note that the total excessive revenue $\mathscr{E}$ from (49) is convex in the variables $(p_k)_{k=1}^K$. Let us obtain an expression for its convex subgradients $\nabla_{p_k}\mathscr{E}(p_1,\ldots,p_K)$ w.r.t. p_k:

$$\nabla_{p_k}\mathscr{E}(p_1,\ldots,p_K) = \tilde{y}_k - \sum_{i=1}^I \mu_{ik} \circ \tilde{x}_i, \quad k = 1,\ldots,K, \qquad (50)$$

where $\mu_{ik} \circ \tilde{x}_i = \left(\mu_{ik}^{(j)} \tilde{x}_i^{(j)}\right)_{j=1}^{(n)}$. Here, $\tilde{y}_k \in S_k(p_k)$ is the supply of kth producer w.r.t. the individual price p_k, and $\tilde{x}_i \in D_i\left(\min_{k=1,\ldots,K} p_k\right)$ is the demand of ith consumer w.r.t. the lowest purchase price $\min_{k=1,\ldots,K} p_k$. Moreover,

$$(\mu_{ik})_{k=1}^K \in M(p_1,\ldots,p_K),$$

where

$$M(p_1,\ldots,p_K) \stackrel{\text{def}}{=} \left\{ (\mu_k)_{k=1}^K \in [0,1]^{n\times K} \;\middle|\; \begin{array}{l} \sum_{k=1}^K \mu_k^{(j)} = 1, \\ \mu_k^{(j)} = 0 \text{ if } p_k^{(j)} \neq \min_{k=1,\ldots,K} p_k^{(j)} \\ j = 1,\ldots,n,\, k = 1,\ldots,K \end{array} \right\}.$$

Note that $\mu_{ik}^{(j)}$ can be interpreted as the share of ith consumer's demand from kth producer for good j. Indeed, the shares $\mu_{ik}^{(j)}$ for good j sum up to 1 over all producers $k = 1,\ldots,K$. Moreover, the share $\mu_{ik}^{(j)}$ vanishes if the kth producer's price $p_k^{(j)}$ exceeds the lowest purchase price $\min_{k=1,\ldots,K} p_k^{(j)}$ for good j.

We claim that the subdifferential information in (50) is known to kth producer. First, note that $\tilde{y}_k$ is kth producer's production. Despite of the fact that the shares μ_{ik} and the demands $\tilde{x}_i$ cannot be estimated by kth producer, their aggregated product $\sum_{i=1}^I \mu_{ik} \circ \tilde{x}_i$ is perfectly available to him. Indeed, $\sum_{i=1}^I \mu_{ik} \circ \tilde{x}_i$ forms the bundle of goods demanded by all consumers from kth producer. Altogether, the subgradients $\nabla_{p_k}\mathscr{E}(p_1,\ldots,p_K)$ represent the *individual excess* of kth producer's supply over all consumers' demands. Overall, we obtain:

Theorem 10 (Producers' Excess Supply and Total Excessive Revenue).

$$\partial_{p_k}\mathscr{E}(p_1,\ldots,p_K) = S_k(p_k) - \sum_{i=1}^{I} \mu_{ik} \circ D_i\left(\min_{k=1,\ldots,K} p_k\right), \quad k = 1,\ldots,K,$$

with demand shares $(\mu_{ik})_{k=1}^{K} \in M(p_1,\ldots,p_K)$.

Due to Theorem 10, the subdifferential of $\mathscr{E}(p_1,\ldots,p_K)$ is completely available to kth producer. This fact suggests to adjust prices by solving the minimization problem

$$\min_{p_1,\ldots,p_K\in\mathbb{R}^n_+} \mathscr{E}(p_1,\ldots,p_K). \qquad \textbf{(PD)}$$

Note that the minimization problem **(PD)** is stated w.r.t. the decentralized producers' prices $(p_k)_{k=1}^{K}$, while previously in **(P)** one minimizes over the common prices p.

We relate the minimization problem **(PD)** to **(P)**. For that, let us call function $f(x)$, $x \in \mathbb{R}^n$, nondecreasing (nonincreasing) in x if $f(x) \geq f(y)$ ($f(x) \leq f(y)$) for any $x \geq y$.

Lemma 5 (Decentralization I). *Let function of* $K+1$ *vector variables*

$$F(p_0,p_1,\ldots,p_K), \quad p_k \in \mathbb{R}^n, k = 0,\ldots,K,$$

be (a) nonincreasing in p_0, *and (b) nondecreasing in all other variables* p_k, $k = 1,\ldots,K$. *Then,*

$$\min_{p_1,\ldots,p_K\in\mathbb{R}^n_+} F\left(\min_{k=1,\ldots,K} p_k, p_1,\ldots,p_K\right) = \min_{p\in\mathbb{R}^n_+} F(p,\ldots,p).$$

Proof. Indeed,

$$\min_{p_1,\ldots,p_K\in\mathbb{R}^n_+} F\left(\min_{k=1,\ldots,K} p_k, p_1,\ldots,p_K\right) \overset{a)}{=}$$

$$\min_{p,p_1,\ldots,p_K\in\mathbb{R}^n_+} \left\{F(p,p_1,\ldots,p_K) \mid p_k \geq p, k = 1,\ldots,K\right\} \overset{b)}{=}$$

$$\min_{p\in\mathbb{R}^n_+} F(p,\ldots,p).$$

□

Next Theorem 11 states that the minimization of the total excessive revenue remains invariant under the trade design.

Theorem 11 (Total Revenue and Trade). *Problems* (**P**) *and* (**PD**) *are equivalent, i.e.*

$$\min_{p_1,\dots,p_K \in \mathbb{R}^n_+} \mathscr{E}(p_1,\dots,p_K) = \min_{p \in \mathbb{R}^n_+} \mathscr{E}(p). \tag{51}$$

Moreover,

(i) if $(p_k)_{k=1}^K$ *solves* (**PD**)*, then* $\min\limits_{k=1,\dots,K} p_k$ *solves* (**P**)*,*
(ii) if p *solves* (**P**)*, then* $(p,\dots,p)$ *solves* (**PD**)*.*

Proof. We set

$$F(p_0,p_1,\dots,p_K) \stackrel{\text{def}}{=} \sum_{k=1}^K \mathscr{E}_k(p_k) + \sum_{i=1}^I \mathscr{E}_i(p_0).$$

Note that F is nonincreasing in p_0, and nondecreasing in p_k, $k=1,\dots,K$. Applying Lemma 5 and in view of

$$F\left(\min_{k=1,\dots,K} p_k, p_1,\dots,p_K\right) = \mathscr{E}(p_1,\dots,p_K), \quad F(p,\dots,p) = \mathscr{E}(p),$$

(51) holds.

Let $(p_k)_{k=1}^K$ solve (**PD**). Then,

$$\min_{p \in \mathbb{R}^n_+} \mathscr{E}(p) \le \mathscr{E}\left(\min_{k=1,\dots,K} p_k\right) \stackrel{(49)}{\le} \mathscr{E}(p_1,\dots,p_K).$$

By using (51), $\min\limits_{k=1,\dots,K} p_k$ solves (**P**).

Now, let p solve (**P**). Then,

$$\min_{p_1,\dots,p_K \in \mathbb{R}^n_+} \mathscr{E}(p_1,\dots,p_K) \le \mathscr{E}(p,\dots,p) = \mathscr{E}(p),$$

By using (51), $(p,\dots,p)$ solves (**PD**). □

Further, we show that the welfare maximization problem (**A**) turns out to be adjoint not only for (**P**), but also for (**PD**). The proof of this fact uses the following Lemma 6.

Lemma 6. *For* $y_k, x \in \mathbb{R}^n_+$, $k=1,\dots,K$*, the inequality*

$$\sum_{k=1}^K y_k \ge x \tag{52}$$

is equivalent to

$$\sum_{k=1}^K \langle p_k, y_k\rangle \ge \left\langle \min_{k=1,\dots,K} p_k, x\right\rangle \text{ for all } p_k \in \mathbb{R}^n_+, k=1,\dots,K. \tag{53}$$

Proof. (i) Let (52) be satisfied. For $p_k \in \mathbb{R}^n_+$, $k = 1, \ldots, K$, we have

$$\sum_{k=1}^{K} \langle p_k, y_k \rangle \geq \left\langle \min_{k=1,\ldots,K} p_k, \sum_{k=1}^{K} y_k \right\rangle \geq \left\langle \min_{k=1,\ldots,K} p_k, x \right\rangle.$$

The first inequality is due to $y_k \in \mathbb{R}^n_+$, and $\min\limits_{k=1,\ldots,K} p_k \geq p_k$, $k = 1, \ldots, K$. The second inequality is due to (52) and $\min\limits_{k=1,\ldots,K} p_k \in \mathbb{R}^n_+$.

(ii) Let (53) be satisfied. Setting there $p_k = p \in \mathbb{R}^n_+$, we get

$$\left\langle p, \sum_{k=1}^{K} x_k \right\rangle \geq \langle p, y \rangle \quad \text{for all } p \in \mathbb{R}^n_+.$$

Hence, (52) is fulfilled. □

The welfare maximization (**A**) remains adjoint for the total revenue minimization (**PD**) under the trade design.

Theorem 12 (Adjoint for (PD)). *The welfare maximization* (**A**) *is adjoint for the total revenue minimization* (**PD**)*:*

$$\min_{p_1,\ldots,p_K \in \mathbb{R}^n_+} \mathscr{E}(p_1, \ldots, p_K) =$$

$$\max_{(\alpha, \tilde{y}, \beta, \tilde{x}) \in \mathscr{A}} \left\{ \Phi\left(\alpha, \tilde{y}, \beta, \tilde{x}\right) \;\middle|\; \sum_{k=1}^{K} \tilde{y}_k \geq \sum_{i=1}^{I} \tilde{x}_i \right\}. \tag{54}$$

Proof.

$$\min_{p_1,\ldots,p_K \in \mathbb{R}^n_+} \mathscr{E}(p_1, \ldots, p_K) =$$

$$= \min_{p_1,\ldots,p_K \in \mathbb{R}^n_+} \max_{(\alpha, \tilde{y}, \beta, \tilde{x}) \in \mathscr{A}} \Phi\left(\alpha, \tilde{y}, \beta, \tilde{x}\right) + \sum_{k=1}^{K} \langle p_k, \tilde{y}_k \rangle - \left\langle \min_{k=1,\ldots,K} p_k, \sum_{i=1}^{I} \tilde{x}_i \right\rangle,$$

$$= \max_{(\alpha, \tilde{y}, \beta, \tilde{x}) \in \mathscr{A}} \Phi\left(\alpha, \tilde{y}, \beta, \tilde{x}\right) + \min_{p_1,\ldots,p_K \in \mathbb{R}^n_+} \sum_{k=1}^{K} \langle p_k, \tilde{y}_k \rangle - \left\langle \min_{k=1,\ldots,K} p_k, \sum_{i=1}^{I} \tilde{x}_i \right\rangle,$$

$$= \max_{(\alpha,\tilde{y},\beta,\tilde{x}) \in \mathscr{A}} \left\{ \Phi\left(\alpha, \tilde{y}, \beta, \tilde{x}\right) \;\middle|\; \begin{array}{l} \sum\limits_{k=1}^{K} \langle p_k, \tilde{y}_k \rangle \geq \left\langle \min\limits_{k=1,\ldots,K} p_k, \sum\limits_{i=1}^{I} \tilde{x}_i \right\rangle \\ \text{for all } p_k \in \mathbb{R}^n_+, k = 1, \ldots, K \end{array} \right\}.$$

Applying Lemma 6, we get the assertion (54). □

We describe how producers may efficiently adjust their individual prices $(p_k)_{k=1}^K$ to arrive at an equilibrium price. This price adjustment corresponds to the quasi-monotone subgradient method from Sect. 3. It is applied to the minimization of the total excessive revenue (**PD**) under the trade design.

Let kth producer choose a sequence of positive confidence parameters $\{\chi_k[t]\}_{t\geq 0}$, $k = 1, \ldots, K$. We consider the following iteration:

Pricing via Trade (TRA)

1. Producers determine their current excess supplies $\nabla_{p_k}\mathscr{E}(p_1[t], \ldots, p_K[t])$:

(a) kth producer computes an optimal tentative production bundle

$$y_k(p_k[t]) \in \mathscr{Y}_k^*(p_k[t]),$$

and participation level

$$\alpha_k(p_k[t]) = \begin{cases} 1, \text{ if } \pi_k(p_k[t]) \geq \kappa_k, \\ 0, \text{ if } \pi_k(p_k[t]) < \kappa_k, \end{cases}$$

indicating whether $y_k(p_k[t])$ is implemented.
The production bundle is $\alpha_k(p_k[t])y_k(p_k[t])$, i.e. either $y_k(p_k[t])$ or zero.

(b) ith consumer identifies the lowest purchase prices

$$p[t] = \min_{k=1,\ldots,K} p_k[t],$$

computes an optimal tentative consumption bundle

$$x_i(p[t]) \in \mathscr{X}_i^*(p[t]),$$

and participation level

$$\beta_i(p[t]) = \begin{cases} 1, \text{ if } e_i(p[t]) \leq w_i, \\ 0, \text{ if } e_i(p[t]) < w_i, \end{cases}$$

indicating whether $x_i(p[t])$ is implemented.
The consumption bundle is $\beta_i(p[t])x_i(p[t])$, i.e. either $x_i(p[t])$ or zero.

(c) ith consumer decides on demand shares

$$(\mu_{ik}[t])_{k=1}^K \in M(p_1[t], \ldots, p_K[t]),$$

and demands from kth producer the bundle

$$\mu_{ik}[t] \circ \beta_i(p[t])x_i(p[t]), \quad k = 1, \ldots, K.$$

(d) kth producer computes the current excess supply

$$\nabla_{p_k}\mathcal{E}(p_1[t],\ldots,p_K[t]) = \alpha_k(p_k[t])y_k(p_k[t]) - \sum_{i=1}^{I}\mu_{ik}[t]\circ\beta_i(p[t])x_i(p[t]). \tag{55}$$

2. kth producer accumulates the excess supplies

$$z_k[t] = z_k[t-1] + \nabla_{p_k}\mathcal{E}(p_1[t],\ldots,p_K[t]), \quad z_k[-1] = 0. \tag{56}$$

3. kth producer computes the price forecast w.r.t. the confidence parameter $\chi_k[t]$:

$$p_k^{+(j)}[t] = \frac{\zeta_k^{(j)}}{\chi_k[t]}\left(-z_k^{(j)}[t]\right)_+, \quad j = 1,\ldots,n, \tag{57}$$

where $\zeta_k^{(j)}$ are positive scaling coefficients.

4. kth producer updates

$$p_k[t+1] = \frac{t+1}{t+2}p_k[t] + \frac{1}{t+2}p_k^+[t] \tag{58}$$

by combining the previous individual price with the forecast. □

Along with the prices $\{(p_1[t],\ldots,p_K[t])\}_{t\geq 0}$ generated by method (TRA), we consider the corresponding historical averages of participation levels:

$$\alpha_k[t] \stackrel{\text{def}}{=} \frac{1}{t+1}\sum_{r=0}^{t}\alpha_k(p_k[r]), \quad \beta_i[t] \stackrel{\text{def}}{=} \frac{1}{t+1}\sum_{r=0}^{t}\beta_i(p[r]).$$

Note that $\alpha_k[t] \in [0,1]$ is the frequency of successful production attempts by kth producer up to time t. Analogously, $\beta_i[t] \in [0,1]$ is the frequency of successful consumption attempts by ith consumer up to time t. We denote by

$$\gamma[t] = (\alpha[t],\beta[t]) \stackrel{\text{def}}{=} \left(\{\alpha_k[t]\}_{k=1}^K, \{\beta_i[t]\}_{i=1}^I\right)$$

the system of average participation levels. The historical averages of production and consumption bundles are defined as follows:

$$\tilde{y}_k[t] \stackrel{\text{def}}{=} \frac{1}{t+1}\sum_{r=0}^{t}\alpha_k(p_k[r])y_k(p_k[r]), \quad \tilde{x}_i[t] \stackrel{\text{def}}{=} \frac{1}{t+1}\sum_{r=0}^{t}\beta_i(p[r])x_i(p[r]).$$

Due to convexity, $\tilde{y}_k[t] \in \alpha_k[t]\mathcal{Y}_k$ and $\tilde{x}_i[t] \in \beta_i[t]\mathcal{X}_i$. We denote by

$$\widetilde{F}[t] = (\tilde{y}[t],\tilde{x}[t]) \stackrel{\text{def}}{=} \left(\{\tilde{y}_k[t]\}_{k=1}^K, \{\tilde{x}_i[t]\}_{i=1}^I\right)$$

the average market flow. Overall, the sequence

$$(\alpha[t], \tilde{y}[t], \beta[t], \tilde{x}[t]) \in \mathscr{A}, \quad t \geq 0,$$

is feasible for the adjoint problem (**A**).

Now, we are ready to prove the main convergence result for (TRA).

Theorem 13 (Convergence of Pricing via Trade). *At a productive market, let producers apply in (TRA) nondecreasing confidence parameters satisfying*

$$\chi_k[t] - \chi_k[t-1] \to 0, \quad \chi_k[t] \to \infty, \quad k = 1, \ldots, K.$$

Then, the sequence of lowest purchase prices, average participation levels, and the average market flow

$$\left(\min_{k=1,\ldots,K} p_k[t], \gamma[t], \widetilde{F}[t] \right)$$

converges toward the set of market equilibria. The achievable rate of convergence is of the order $O\left(\frac{1}{\sqrt{t}}\right)$.

Proof. The iteration scheme (TRA) is a variant of the quasi-monotone subgradient method (SM). Hence, we may obtain the convergence for (TRA) by means of Theorem 8. For that, let us discuss the applicability of conditions (S1)–(S3).

On (S1): The price forecast (57) can be derived by means of the Euclidean prox-functions for $\mathbb{R}^n_+$:

$$d_k(p) \stackrel{\text{def}}{=} \frac{1}{2} \sum_{j=1}^{n} \frac{1}{\zeta_k^{(j)}} \left(p^{(j)}\right)^2, \quad k = 1, \ldots, K.$$

In fact, for $z_k[t] \in \mathbb{R}^n$, $\chi_k[t] > 0$ we consider the minimization problem as from step 3. in (SM):

$$\min_{p_1,\ldots,p_K \in \mathbb{R}^n_+} \left\{ \sum_{k=1}^{K} \langle z_k[t], p_k \rangle + \chi_k[t] d_k(p_k) \right\}.$$

Its unique solution is the price forecast (57) as from step 3. in (TRA):

$$p_k^{+(j)}[t] = \frac{\zeta_k^{(j)}}{\chi_k[t]} \left(-z_k^{(j)}[t] \right)_+, \quad j = 1, \ldots, n, k = 1, \ldots, K.$$

On (S2): It follows from Theorem 12 that the total excessive revenue is representable as a maximum of concave functions:

$$\mathscr{E}(p_1, \ldots, p_K) = \max_{(\alpha,\tilde{y},\beta,\tilde{x}) \in \mathscr{A}} \Phi(\alpha, \tilde{y}, \beta, \tilde{x}) + \varphi(p_1, \ldots, p_K, \tilde{y}, \tilde{x}),$$

where

$$\varphi(p_1,\ldots,p_K,\tilde{y},\tilde{x}) = \sum_{k=1}^{K} \langle p_k, \tilde{y}_k \rangle - \left\langle \min_{k=1,\ldots,K} p_k, \sum_{i=1}^{I} \tilde{x}_i \right\rangle.$$

Although φ is not linear w.r.t. $(p_1,\ldots,p_K)$, but it is *partially linear*, i.e.

$$\varphi(p,\ldots,p,\tilde{y},\tilde{x}) = \left\langle p, \sum_{k=1}^{K} \tilde{y}_k - \sum_{i=1}^{I} \tilde{x}_i \right\rangle.$$

The partial linearity of φ suffices for the analogous convergence analysis as in Sect. 3 (see [11] for details). In particular, due to Theorem 12, the adjoint problem for the total revenue minimization (**PD**) remains unchanged under the trade design, i.e. it is the welfare maximization (**A**).
On (S3): The welfare maximization problem (**A**) satisfies the Slater condition in view of the market productivity (cf. Definition 2).

Overall, we apply Theorem 8 to deduce that the sequence $(p_k[t])_{k=1}^{K}$ converges toward the solution set of (**PD**), and $\left(\gamma[t], \widetilde{F}[t]\right)$ converges toward the solution set of (**A**) by order $O\left(\frac{1}{\sqrt{t}}\right)$. Due to Theorem 11, $\min\limits_{k=1,\ldots,K} p_k[t]$ converges toward the solution set of (**P**). In view of the duality from Theorem 12, the assertion follows.

□

4.3 Auction

Analogously, we proceed with the auction design:

*i*th consumer settles and updates his individual prices p_i, and producers sell at the highest offer price $\max\limits_{i=1,\ldots,I} p_i$.

Recall that for vectors $p_1,\ldots,p_I \in \mathbb{R}^n$, we denote by $\max\limits_{i=1,\ldots,I} p_i \in \mathbb{R}^n$ the vector with coordinates

$$\left(\max_{i=1,\ldots,I} p_i\right)^{(j)} = \max_{i=1,\ldots,I} p_i^{(j)}, \quad j = 1,\ldots,n.$$

The auction design incorporates the dominant aspect in auction theory that highest bidders are first served [6]. Following the auction framework, we assume that producers are able to undertake global price search across the consumers.

Here, the total excessive revenue depends on the consumers' prices $(p_i)_{i=1}^I$ as follows:

$$\mathcal{E}(p_1,\ldots,p_I) \stackrel{\text{def}}{=} \sum_{k=1}^K \mathcal{E}_k\left(\max_{i=1,\ldots,I} p_i\right) + \sum_{i=1}^I \mathcal{E}_i(p_i) =$$

$$\sum_{k=1}^K \max_{y_k \in \mathcal{Y}_k} \left(\left\langle \max_{i=1,\ldots,I} p_i, y_k \right\rangle - c_k(y_k)\right)_+ + \sum_{i=1}^I \max_{x_i \in \mathcal{X}_i} (w_i - \langle p_i, x_i \rangle)_+ . \tag{59}$$

The decentralization of prices makes the corresponding subdifferential information about excess demands available to consumers. In fact, note that the total revenue $\mathcal{E}$ from (59) is convex in the variables $(p_i)_{i=1}^I$. Let us obtain an expression for its convex subgradients $\nabla_{p_i}\mathcal{E}(p_1,\ldots,p_I)$ w.r.t. p_i:

$$\nabla_{p_i}\mathcal{E}(p_1,\ldots,p_I) = \sum_{k=1}^K \lambda_{ik} \circ \tilde{y}_k - \tilde{x}_i, \quad k = 1,\ldots,K. \tag{60}$$

where $\lambda_{ik} \circ \tilde{y}_k = \left(\lambda_{ik}^{(j)}\tilde{y}_k^{(j)}\right)_{j=1}^{(n)}$. Here, $\tilde{x}_i \in D_i(p_i)$ is the demand of ith consumer w.r.t. his individual price p_i, and $\tilde{y}_k \in S_k\left(\max_{i=1,\ldots,I} p_i\right)$ is the supply of kth producer w.r.t. the highest offer price $\max_{i=1,\ldots,I} p_i$. Moreover,

$$(\lambda_{ik})_{i=1}^I \in L(p_1,\ldots,p_I),$$

where

$$L(p_1,\ldots,p_I) \stackrel{\text{def}}{=} \left\{ (\lambda_i)_{i=1}^I \in [0,1]^{n\times I} \;\middle|\; \begin{array}{c} \sum_{i=1}^I \lambda_i^{(j)} = 1, \\ \lambda_i^{(j)} = 0 \text{ if } p_i^{(j)} \neq \max_{i=1,\ldots,I} p_i^{(j)} \\ j = 1,\ldots,n, i = 1,\ldots,I \end{array} \right\}.$$

Note that $\lambda_{ik}^{(j)}$ can be interpreted as the share of kth producer's supply to ith consumer for good j. Indeed, the shares $\lambda_{ik}^{(j)}$ for good j sum up to 1 over all consumers $i = 1,\ldots,I$. Moreover, the share $\lambda_{ik}^{(j)}$ vanishes if the ith consumer's price $p_i^{(j)}$ is less than the highest offer price $\max_{i=1,\ldots,I} p_i^{(j)}$ for good j.

We claim that the subdifferential information in (50) is known to ith consumer. First, note that $\tilde{x}_i$ is his consumption bundle. Despite of the fact that the shares λ_{ik} and the supplies $\tilde{y}_k$ cannot be estimated by ith consumer, their aggregated product $\sum_{k=1}^K \lambda_{ik} \circ \tilde{y}_k$ is perfectly available to him. Indeed, $\sum_{k=1}^K \lambda_{ik} \circ \tilde{y}_k$ forms the bundle of goods supplied by all producers to ith consumer. Altogether, the subgradients $\nabla_{p_i}\mathscr{E}(p_1,\ldots,p_I)$ represent the *individual excess* of ith consumer's supply over his demands.

Theorem 14 (Consumers' Excess Supply and Total Excessive Revenue).

$$\partial_{p_i}\mathscr{E}(p_1,\ldots,p_I) = \sum_{k=1}^{K} \lambda_{ik} \circ S_k\left(\max_{i=1,\ldots,I} p_I\right) - D_i\left(p_i\right), \quad i = 1,\ldots,I,$$

with supply shares $(\lambda_{ik})_{i=1}^{I} \in L\left(p_1,\ldots,p_I\right)$.

Due to Theorem 14, the subdifferential of $\mathscr{E}(p_1,\ldots,p_I)$ is completely available to ith consumer. This fact suggests to adjust prices by solving the minimization problem

$$\min_{p_1,\ldots,p_I\in\mathbb{R}^n_+} \mathscr{E}(p_1,\ldots,p_I). \qquad \textbf{(PA)}$$

Note that the minimization problem (**PA**) is stated w.r.t. the decentralized consumers' prices $(p_i)_{i=1}^{I}$, while previously in (**P**) one minimizes over the common prices p.

We relate the minimization problem (**PA**) to (**P**). For that, let us call function $f(x)$, $x \in \mathbb{R}^n$, nondecreasing (nonincreasing) in x if $f(x) \geq f(y)$ ($f(x) \leq f(y)$) for any $x \geq y$.

Lemma 7 (Decentralization II). *Let function of* $I+1$ *vector variables*

$$G(p_0, p_1,\ldots,p_I), \quad p_i \in \mathbb{R}^n, i = 0,\ldots,I,$$

be (a) nondecreasing in p_0*, and (b) nonincreasing in all other variables* p_i*,* $i = 1,\ldots,I$*. Then,*

$$\min_{p_1,\ldots,p_I\in\mathbb{R}^n_+} G\left(\max_{i=1,\ldots,I} p_i, p_1,\ldots,p_I\right) = \min_{p\in\mathbb{R}^n_+} G(p,\ldots,p).$$

Proof. Indeed,

$$\min_{p_1,\ldots,p_I\in\mathbb{R}^n_+} G\left(\max_{i=1,\ldots,I} p_i, p_1,\ldots,p_I\right) \overset{\text{a)}}{=}$$

$$\min_{p,p_1,\ldots,p_I\in\mathbb{R}^n_+} \left\{G\left(p, p_1,\ldots,p_I\right) \mid p_i \leq p, i = 1,\ldots,I\right\} \overset{\text{b)}}{=}$$

$$\min_{p\in\mathbb{R}^n_+} G(p,\ldots,p).$$

□

Next Theorem 15 states that the minimization of the total excessive revenue remains invariant under the auction design.

Theorem 15 (Total Revenue and Auction). *Problems* (**P**) *and* (**PA**) *are equivalent, i.e.*

$$\min_{p_1,\ldots,p_I \in \mathbb{R}^n_+} \mathscr{E}(p_1,\ldots,p_I) = \min_{p \in \mathbb{R}^n_+} \mathscr{E}(p). \tag{61}$$

Moreover,

(i) if $(p_i)_{i=1}^I$ *solves* (**PA**)*, then* $\max\limits_{i=1,\ldots,I} p_i$ *solves* (**P**)*,*
(ii) if p *solves* (**P**)*, then* $(p,\ldots,p)$ *solves* (**PA**)*.*

Proof. We set

$$G(p_0,p_1,\ldots,p_I) \stackrel{\mathrm{def}}{=} \sum_{k=1}^K \mathscr{E}_k(p_0) + \sum_{i=1}^I \mathscr{E}_i(p_i).$$

Note that G is nondecreasing in p_0, and nonincreasing in p_i, $i = 1,\ldots,I$. Applying Lemma 7 and in view of

$$G\left(\max_{i=1,\ldots,I} p_i, p_1,\ldots,p_I\right) = \mathscr{E}(p_1,\ldots,p_I), \quad G(p,\ldots,p) = \mathscr{E}(p),$$

(61) holds.

Let $(p_i)_{i=1}^I$ solve (**PA**). Then,

$$\min_{p \in \mathbb{R}^n_+} \mathscr{E}(p) \le \mathscr{E}\left(\max_{i=1,\ldots,I} p_i\right) \stackrel{(59)}{\le} \mathscr{E}(p_1,\ldots,p_I).$$

By using (61), $\max\limits_{i=1,\ldots,I} p_k$ solves (**P**).

Now, let p solve (**P**). Then,

$$\min_{p_1,\ldots,p_I \in \mathbb{R}^n_+} \mathscr{E}(p_1,\ldots,p_I) \le \mathscr{E}(p,\ldots,p) = \mathscr{E}(p),$$

By using (61), $(p,\ldots,p)$ solves (**PA**). □

Further, we show that the welfare maximization problem (**A**) turns out to be adjoint not only for (**P**), but also for (**PA**). The proof of this fact uses the following Lemma 8.

Lemma 8. *For* $x_i, y \in \mathbb{R}^n_+$, $i = 1,\ldots,I$, *the inequality*

$$\sum_{i=1}^I x_i \le y \tag{62}$$

is equivalent to

$$\sum_{i=1}^{I} \langle p_i, x_i \rangle \leq \left\langle \max_{i=1,\ldots,I} p_i, y \right\rangle \text{ for all } p_i \in \mathbb{R}^n_+, i = 1, \ldots, I. \tag{63}$$

Proof. (i) Let (62) be satisfied. For $p_i \in \mathbb{R}^n_+$, $i = 1, \ldots, I$, we have

$$\sum_{i=1}^{I} \langle p_i, x_i \rangle \leq \left\langle \max_{i=1,\ldots,I} p_i, \sum_{i=1}^{I} x_i \right\rangle \leq \left\langle \max_{i=1,\ldots,I} p_i, y \right\rangle.$$

The first inequality is due to $x_i \in \mathbb{R}^n_+$, and $p_i \leq \max_{i=1,\ldots,I} p_i$, $i = 1, \ldots, I$. The second inequality is due to (62) and $\max_{i=1,\ldots,I} \in \mathbb{R}^n_+$.

(ii) Let (63) be satisfied. Setting there $p_i = p \in \mathbb{R}^n_+$, we get

$$\left\langle p, \sum_{i=1}^{I} x_i \right\rangle \leq \langle p, y \rangle \quad \text{for all } p \in \mathbb{R}^n_+.$$

Hence, (62) is fulfilled. □

Theorem 16 (Adjoint for (PA)). *The welfare maximization (**A**) is adjoint for the total revenue minimization (**PA**):*

$$\min_{p_1,\ldots,p_I \in \mathbb{R}^n_+} \mathscr{E}(p_1, \ldots, p_I) = \max_{(\alpha, \tilde{y}, \beta, \tilde{x}) \in \mathscr{A}} \left\{ \Phi\left(\alpha, \tilde{y}, \beta, \tilde{x}\right) \;\middle|\; \sum_{k=1}^{K} \tilde{y}_k \geq \sum_{i=1}^{I} \tilde{x}_i \right\}. \tag{64}$$

Proof. We obtain:

$$\min_{p_1,\ldots,p_I \in \mathbb{R}^n_+} \mathscr{E}(p_1, \ldots, p_I) =$$

$$\begin{aligned}
&= \min_{p_1,\ldots,p_I \in \mathbb{R}^n_+} \max_{(\alpha, \tilde{y}, \beta, \tilde{x}) \in \mathscr{A}} \Phi\left(\alpha, \tilde{y}, \beta, \tilde{x}\right) + \left\langle \max_{i=1,\ldots,I} p_i, \sum_{k=1}^{K} \tilde{y}_k \right\rangle - \sum_{i=1}^{I} \langle p_i, \tilde{x}_i \rangle, \\
&= \max_{(\alpha, \tilde{y}, \beta, \tilde{x}) \in \mathscr{A}} \Phi\left(\alpha, \tilde{y}, \beta, \tilde{x}\right) + \min_{p_1,\ldots,p_I \in \mathbb{R}^n_+} \left\langle \max_{i=1,\ldots,I} p_i, \sum_{k=1}^{K} \tilde{y}_k \right\rangle - \sum_{i=1}^{I} \langle p_i, \tilde{x}_i \rangle, \\
&= \max_{(\alpha, \tilde{y}, \beta, \tilde{x}) \in \mathscr{A}} \left\{ \Phi\left(\alpha, \tilde{y}, \beta, \tilde{x}\right) \;\middle|\; \begin{array}{l} \sum_{i=1}^{I} \langle p_i, \tilde{x}_i \rangle \leq \left\langle \max_{i=1,\ldots,I} p_i, \sum_{k=1}^{K} \tilde{y}_k \right\rangle \\ \text{for all } p_i \in \mathbb{R}^n_+, i = 1, \ldots, I \end{array} \right\}.
\end{aligned}$$

Applying Lemma 8, we get the assertion (64). □

Analogously, we describe how consumers may efficiently adjust their individual prices $(p_i)_{i=1}^I$ to arrive at an equilibrium price. This price adjustment also corresponds to the quasi-monotone subgradient method from Sect. 3. It is applied to the minimization of the total excessive revenue (**PA**) under the auction design.

Let ith producer choose a sequence of positive confidence parameters $\{\chi_i[t]\}_{t\geq 0}$, $i = 1, \ldots, I$. We consider the following iteration:

Pricing via Auction (AUC)

1. Consumers determine their current excess supplies $\nabla_{p_i}\mathscr{E}(p_1[t], \ldots, p_i[t])$:

(a) ith consumer computes an optimal tentative consumption bundle

$$x_i(p_i[t]) \in \mathscr{X}_i^*(p_i[t]),$$

and participation level

$$\beta_i(p_i[t]) = \begin{cases} 1, \text{ if } e_i(p_i[t]) \leq w_i, \\ 0, \text{ if } e_i(p_i[t]) < w_i, \end{cases}$$

indicating whether $x_i(p_i[t])$ is implemented.

The consumption bundle is $\beta_i(p_i[t])x_i(p_i[t])$, i.e. either $x_i(p_i[t])$ or zero.

(b) kth producer identifies the highest offer prices

$$p[t] = \max_{i=1,\ldots,I} p_i[t],$$

and computes an optimal tentative production bundle

$$y_k(p[t]) \in \mathscr{Y}_k^*(p[t]),$$

and participation level

$$\alpha_k(p[t]) = \begin{cases} 1, \text{ if } \pi_k(p[t]) \geq \kappa_k, \\ 0, \text{ if } \pi_k(p[t]) < \kappa_k, \end{cases}$$

indicating whether $y_k(p[t])$ is implemented.

The production bundle is $\alpha_k(p[t])y_k(p[t])$, i.e. either $y_k(p[t])$ or zero.

(c) kth producer decides on supply shares

$$(\lambda_{ik}[t])_{i=1}^I \in L(p_1[t], \ldots, p_I[t]),$$

and supplies to ith consumer the bundle

$$\lambda_{ik}[t] \circ \alpha_k(p[t])y_k(p[t]).$$

(d) ith consumer computes the current excess supply

$$\nabla_{p_i} \mathscr{E}(p_1[t], \dots, p_I[t]) = \sum_{k=1}^{K} \lambda_{ik}[t] \circ \alpha_k(p[t]) y_k(p[t]) - \beta_i(p_i[t]) x_i(p_i[t]). \quad (65)$$

2. ith consumer accumulates the excess supplies

$$z_i[t] = z_i[t-1] + \nabla_{p_i} \mathscr{E}(p_1[t], \dots, p_I[t]), \quad z_i[-1] = 0. \quad (66)$$

3. ith consumer computes the price forecast w.r.t. the confidence parameter $\chi_i[t]$:

$$p_i^{+(j)}[t] = \frac{\zeta_i^{(j)}}{\chi_i[t]} \left(-z_i^{(j)}[t] \right)_+, \quad j = 1, \dots, n, \quad (67)$$

where $\zeta_i^{(j)}$ are positive scaling coefficients.

4. ith consumer updates

$$p_i[t+1] = \frac{t+1}{t+2} p_i[t] + \frac{1}{t+2} p_i^+[t] \quad (68)$$

by combining the previous individual price with the forecast. □

Along with the prices $\{(p_i[t])\}_{t \geq 0}$, $i = 1, \dots, I$, generated by method (AUC), we consider the corresponding historical averages of participation levels:

$$\alpha_k[t] \stackrel{\text{def}}{=} \frac{1}{t+1} \sum_{r=0}^{t} \alpha_k(p[r]), \quad \beta_i[t] \stackrel{\text{def}}{=} \frac{1}{t+1} \sum_{r=0}^{t} \beta_i(p_i[r]).$$

Note that $\alpha_k[t] \in [0,1]$ is the frequency of successful production attempts by kth producer up to time t. Analogously, $\beta_i[t] \in [0,1]$ is the frequency of successful consumption attempts by ith consumer up to time t. We denote by

$$\gamma[t] = (\alpha[t], \beta[t]) \stackrel{\text{def}}{=} \left(\{\alpha_k[t]\}_{k=1}^{K}, \{\beta_i[t]\}_{i=1}^{I} \right)$$

the system of average participation levels. The historical averages of production and consumption bundles are defined as follows:

$$\tilde{y}_k[t] \stackrel{\text{def}}{=} \frac{1}{t+1} \sum_{r=0}^{t} \alpha_k(p[r]) y_k(p[r]), \quad \tilde{x}_i[t] \stackrel{\text{def}}{=} \frac{1}{t+1} \sum_{r=0}^{t} \beta_i(p_i[r]) x_i(p_i[r]).$$

Due to convexity, $\tilde{y}_k[t] \in \alpha_k[t] \mathscr{Y}_k$ and $\tilde{x}_i[t] \in \beta_i[t] \mathscr{X}_i$. We denote by

$$\widetilde{F}[t] = (\tilde{y}[t], \tilde{x}[t]) \stackrel{\text{def}}{=} \left(\{\tilde{y}_k[t]\}_{k=1}^{K}, \{\tilde{x}_i[t]\}_{i=1}^{I} \right)$$

the average market flow. Overall, the sequence

$$(\alpha[t], \tilde{y}[t], \beta[t], \tilde{x}[t]) \in \mathscr{A}, \quad t \geq 0,$$

is feasible for the adjoint problem (**A**).

Now, we are ready to prove the main convergence result for (AUC).

Theorem 17 (Convergence of Pricing via Auction). *At a productive market, let consumers apply in (AUC) nondecreasing confidence parameters satisfying*

$$\chi_i[t] - \chi_i[t-1] \to 0, \quad \chi_i[t] \to \infty, \quad i = 1, \ldots, I.$$

Then, the sequence of highest offer prices, average participation levels, and the average market flow

$$\left(\max_{i=1,\ldots,I} p_i[t], \gamma[t], \widetilde{F}[t] \right)$$

converges toward the set of market equilibria. The achievable rate of convergence is of the order $O\left(\frac{1}{\sqrt{t}}\right)$.

Proof. The iteration scheme (AUC) is a variant of the quasi-monotone subgradient method (SM). Hence, we may obtain the convergence for (AUC) by means of Theorem 8. For that, let us discuss the applicability of conditions (S1)–(S3).

On (S1): The price forecast (67) can be derived by means of the Euclidean prox-functions for $\mathbb{R}^n_+$:

$$d_i(p) \overset{\text{def}}{=} \frac{1}{2} \sum_{j=1}^{n} \frac{1}{\zeta_i^{(j)}} \left(p^{(j)}\right)^2, \quad i = 1, \ldots, I.$$

In fact, for $z_i[t] \in \mathbb{R}^n$, $\chi_i[t] > 0$ we consider the minimization problem as from step 3. in (SM):

$$\min_{p_1,\ldots,p_I \in \mathbb{R}^n_+} \left\{ \sum_{i=1}^{I} \langle z_i[t], p_i \rangle + \chi_i[t] d_i(p_i) \right\}.$$

Its unique solution is the price forecast (67) as from step 3. in (AUC):

$$p_i^{+(j)}[t] = \frac{\zeta_i^{(j)}}{\chi_i[t]} \left(-z_i^{(j)}[t] \right)_+, \quad j = 1, \ldots, n, i = 1, \ldots, I.$$

On (S2): It follows from Theorem 16 that the total excessive revenue is representable as a maximum of concave functions:

$$\mathscr{E}(p_1, \ldots, p_I) = \max_{(\alpha,\tilde{y},\beta,\tilde{x}) \in \mathscr{A}} \Phi\left(\alpha, \tilde{y}, \beta, \tilde{x}\right) + \varphi\left(p_1, \ldots, p_I, \tilde{y}, \tilde{x}\right),$$

where

$$\varphi\left(p_1, \ldots, p_I, \tilde{y}, \tilde{x}\right) = \left\langle \max_{i=1,\ldots,I} p_i, \sum_{k=1}^{K} \tilde{y}_k \right\rangle - \langle p_i, \tilde{x}_i \rangle .$$

Although φ is not linear w.r.t. $(p_1, \ldots, p_I)$, but it is *partially linear*, i.e.

$$\varphi\left(p, \ldots, p, \tilde{y}, \tilde{x}\right) = \left\langle p, \sum_{k=1}^{K} \tilde{y}_k - \sum_{i=1}^{I} \tilde{x}_i \right\rangle .$$

The partial linearity of φ suffices for the analogous convergence analysis as in Sect. 3 (see [12] for details). In particular, due to Theorem 16, the adjoint problem for the total revenue minimization (**PA**) remains unchanged under the auction design, i.e. it is the welfare maximization (**A**).

On (S3): The welfare maximization problem (**A**) satisfies the Slater condition in view of the market productivity (cf. Definition 2).

Overall, we apply Theorem 8 to deduce that the sequence $(p_i[t])_{i=1}^{I}$ converges toward the solution set of (**PA**), and $\left(\gamma[t], \widetilde{F}[t]\right)$ converges toward the solution set of (**A**) by order $O\left(\frac{1}{\sqrt{t}}\right)$. Due to Theorem 15, $\max_{i=1,\ldots,I} p_i[t]$ converges toward the solution set of (**P**). In view of the duality from Theorem 16, the assertion follows.

□

5 Conclusions

We presented the excessive revenue model of a competitive market. Its crucial advantage is that it can be written in potential form. The convex potential is the total excessive revenue of market's participants. Equilibrium prices, which balance supply and demand, arise as the minimizers of the total excessive revenue. The latter constitutes the least revenue principle in analogy to extremal principles in physics. The least revenue principle allowed us to efficiently adjust prices by application of Convex Analysis. For that, we used quasi-monotone methods for nonsmooth convex minimization of the total excessive revenue. They represent implementable behavioral schemes for the real-life activities of producers and consumers due to the trade or auction. Thus, the main features of our price adjustment are as follows:

- *Reliability* refers to the fact that the price adjustment leads to equilibrium prices, and corresponding supply equals demand on average.
- *Computability* of price adjustment means that we can guarantee the convergence of the proposed price adjustment mechanisms at an explicitly stated (nonasymptotic) rate, which in fact is the best convergence rate achievable in large-scale nonsmooth convex minimization.

- *Decentralization* explains how market participants can successively update prices by themselves via trade or auction rather than by relying on a central authority.

Acknowledgements The authors would like to thank the referees for their precise and constructive remarks.

Appendix

Proof of Lemma 2:. We define the average linearization terms ℓ_t and ψ_t for f:

$$\begin{aligned} \ell_t(x) &\stackrel{\text{def}}{=} \sum_{r=0}^{t} f(x[r]) + \langle \nabla f(x[r]), x - x[r] \rangle , \\ \psi_t &\stackrel{\text{def}}{=} \min_{x \in X} \{ \ell_t(x) + \chi[t] d(x) \} . \end{aligned}$$

First, we show by induction that for all $t \geq 0$ it holds:

$$f(x[t]) - \frac{\psi_t}{t+1} \leq \rho_t. \tag{69}$$

Let us assume that condition (69) is valid for some $t \geq 0$. Then,

$$\begin{aligned} \psi_{t+1} &= \min_{x \in X} \{ \ell_t(x) + f(x_{t+1}) + \langle \nabla f(x[t+1]), x - x[t+1] \rangle + \chi[t+1] d(x) \} \\ &\stackrel{(37)}{\geq} \min_{x \in X} \{ \ell_t(x) + \chi[t] d(x) + f(x[t+1]) + \langle \nabla f(x[t+1]), x - x[t+1] \rangle \} \\ &\stackrel{(33)}{\geq} \min_{x \in X} \left\{ \psi_t + \frac{1}{2} \chi[t] \left\| x - x^+[t] \right\|^2 + f(x[t+1]) + \langle \nabla f(x[t+1]), x - x[t+1] \rangle \right\} \\ &\stackrel{(69)}{\geq} \min_{x \in X} \left\{ \begin{array}{l} (t+1) f(x[t]) - (t+1)\rho_t \\ + \frac{1}{2} \chi[t] \left\| x - x^+[t] \right\|^2 + f(x[t+1]) + \langle \nabla f(x[t+1]), x - x[t+1] \rangle \end{array} \right\} \\ &\stackrel{(31)}{\geq} \min_{x \in X} \left\{ \begin{array}{l} (t+1) \left[f(x[t+1]) + \langle \nabla f(x[t+1]), x[t] - x[t+1] \rangle \right] - (t+1)\rho_t \\ + \frac{1}{2} \chi[t] \left\| x - x^+[t] \right\|^2 + f(x[t+1]) + \langle \nabla f(x[t+1]), x - x[t+1] \rangle \end{array} \right\} . \end{aligned}$$

Since $(t+2)x[t+1] = (t+1)x[t] + x^+[t]$, we obtain

$$\psi_{t+1} \geq (t+2) f(x[t+1]) - (t+1)\rho_t$$

$$+ \min_{x \in X} \left\{ \langle \nabla f(x[t+1]), x - x^+[t] \rangle + \frac{1}{2} \chi[t] \left\| x - x^+[t] \right\|^2 \right\}$$

$$\geq (t+2) f(x[t+1]) - (t+1)\rho_t - \frac{1}{2\chi[t]} \| \nabla f(x[t+1]) \|_*^2 .$$

$$= (t+2) f(x[t+1]) - (t+2)\rho_{t+1}.$$

It remains to note that

$$\psi_0 = \min_{x \in X} \{ f(x[0]) + \langle \nabla f(x[0]), x - x[0] \rangle + \chi[0] d(x) \} \overset{(35)}{\geq} f(x[0]) - \rho_0.$$

Now, we relate the term $\frac{\psi_t}{t+1}$ from (69) to the adjoint problem (32). It holds due to convexity of $\varphi(\cdot, a)$, $a \in A$:

$$f(x[r]) + \langle \nabla f(x[r]), x - x[r] \rangle =$$

$$\overset{(29), (30)}{=} \Phi\left(a(x[r])\right) + \varphi\left(x[r], a(x[r])\right) + \left\langle \nabla_x \varphi\left(x[r], a(x[r])\right), x - x[r] \right\rangle$$

$$\leq \Phi\left(a(x[r]) + \varphi\left(x, a(x[r])\right)\right).$$

Hence, we obtain due to concavity of Φ and $\varphi(x, \cdot)$, $x \in X$:

$$\ell_t(x) \leq \sum_{r=0}^{t} \Phi\left(a(x[r]) + \varphi\left(x, a(x[r])\right)\right) \leq (t+1) \left[\Phi\left(a[t]\right) + \varphi\left(x, a[t]\right) \right].$$

Finally, we get

$$\frac{\psi_t}{t+1} \leq \Phi\left(a[t]\right) + \min_{x \in X} \left\{ \varphi\left(x, a[t]\right) + \frac{\chi[t]}{t+1} d(x) \right\} = \Phi\left(a[t]\right) - \delta_t(a[t]). \tag{70}$$

Altogether, (69) and (70) provide the formula (38). □

The following result on the quadratic penalty for the maximization problem (39) will be needed.

Lemma 9. *Under (S1)–(S3) it holds for* $\kappa > 0$*:*

$$\max_{a \in A} \left[\Phi(a) - \frac{\kappa}{2} \sum_{j=1}^{n} \left(h_j(a) \right)_+^2 \right] \leq f^* + \frac{1}{2\kappa} \sum_{j=1}^{n} x^{*(j)},$$

where x^* *solves the minimization problem (28).*

Proof. Let a^* be an optimal solution of (39). Due to the Slater condition, there exist some Lagrange multipliers $x^{*(j)}, j = 1, \ldots, n$ such that

$$\left\langle \nabla \Phi(a^*) - \sum_{j=1}^{n} x^{*(j)} \nabla h_j(a^*), a^* - a \right\rangle \geq 0, \quad \text{for all } a \in A, \tag{71}$$

$$x^{*(j)} \geq 0, \quad h_j(a^*) \leq 0, \quad \sum_{j=1}^{n} x^{*(j)} h_j(a^*) = 0. \tag{72}$$

Note that the vector of Lagrange multipliers $x^* = \left(x^{*(j)}, j = 1, \ldots, n\right)$ solves the minimization problem (28). Due to the concavity of Φ and the convexity of h_j, $j = 1, \ldots, n$, it holds for all $a \in A$:

$$\Phi(a) \leq \Phi(a^*) + \langle \nabla \Phi(a^*), a - a^* \rangle, \tag{73}$$

$$h_j(a) \geq h_j(a^*) + \left\langle \nabla h_j(a^*), a - a^* \right\rangle. \tag{74}$$

We estimate

$$\begin{aligned} \Phi(a) &\overset{(73)}{\leq} \Phi(a^*) + \langle \nabla \Phi(a^*), a - a^* \rangle \overset{(71)}{\leq} f^* + \sum_{j=1}^{n} x^{*(j)} \left\langle \nabla h_j(a^*), a - a^* \right\rangle \\ &\overset{(74)}{\leq} f^* + \sum_{j=1}^{n} x^{*(j)} \left(h_j(a) - h_j(a^*)\right) \overset{(72)}{=} f^* + \sum_{j=1}^{n} x^{*(j)} h_j(a), \quad a \in A. \end{aligned}$$

Hence,

$$\begin{aligned} \max_{a \in A} \left[\Phi(a) - \frac{\kappa}{2} \sum_{j=1}^{n} \left(h_j(a)\right)_+^2 \right] &\leq f^* + \max_{a \in A} \sum_{j=1}^{n} \left[x^{*(j)} h_j(a) - \frac{\kappa}{2} \left(h_j(a)\right)_+^2 \right] \\ &\leq f^* + \sum_{j=1}^{n} \max_{b_j \in \mathbb{R}} \sum_{j=1}^{n} \left[x^{*(j)} b_j - \frac{\kappa}{2} \left(b_j\right)_+^2 \right] \\ &= f^* + \sum_{j=1}^{n} \frac{1}{2\kappa} x^{*(j)}. \end{aligned}$$

□

Proof of Lemma 3:. First, we estimate the penalty term δ_t and the remainder term ρ_t, $t \geq 0$ under (S1)–(S3). It holds:

$$\begin{aligned}
\delta_t(a[t]) &= -\min_{x \in X}\left\{\varphi(x, a[t]) + \frac{\chi[t]}{t+1} d(x)\right\} \\
&= -\min_{x \in \mathbb{R}^n_+}\left\{-\sum_{j=1}^{n} x^{(j)} h_j(a[t]) + \frac{\chi[t]}{t+1} \cdot \frac{1}{2}\sum_{j=1}^{n}\left(x^{(j)}\right)^2\right\} \\
&= \frac{t+1}{\chi[t]}\sum_{j=1}^{n}\left(h_j(a[t])\right)_+^2, \\
\rho_t &= \frac{1}{t+1}\sum_{r=0}^{t}\frac{1}{2\chi[r-1]}\|\nabla f(x[r])\|_*^2 \\
&= \frac{1}{t+1}\sum_{r=0}^{t}\frac{1}{2\chi[r-1]}\sum_{j=1}^{n}\left(h_j(a[r])\right)^2 \\
&\le C_2\frac{1}{t+1}\sum_{r=0}^{t}\frac{1}{\chi[r-1]}.
\end{aligned}$$

The latter inequality follows due to the compactness of the adjoint set A and the convexity of $h_j, j = 1, \ldots, n$ with

$$C_2 = \frac{1}{2}\max_{a \in A}\sum_{j=1}^{n}\left(h_j(a)\right)^2. \tag{75}$$

Substituting into (38), we get the right-hand side of (40):

$$f(x[t]) - \Phi(a[t]) + \frac{t+1}{\chi[t]}\sum_{j=1}^{n}\left(h_j(a[t])\right)_+^2 \le C_2\frac{1}{t+1}\sum_{r=0}^{t}\frac{1}{\chi[r-1]}.$$

Now, we estimate this dual gap from below by using

$$\begin{aligned}
\Phi\,(a[t]) - \frac{t+1}{\chi[t]}\sum_{j=1}^{n}\left(h_j(a[t])\right)_+^2 &\le \max_{a \in A}\left[\Phi(a) - \frac{t+1}{\chi[t]}\sum_{j=1}^{n}\left(h_j(a)\right)_+^2\right] \\
&\overset{\text{Lemma 9}}{\le} f^* + C_1\frac{\chi[t]}{t+1},
\end{aligned}$$

where

$$C_1 = \frac{1}{4}\sum_{j=1}^{n} x^{*(j)} \tag{76}$$

and x^* is a solution of the minimization problem (28).

Finally, we get the left-hand side of (40)

$$f(x[t]) - \Phi(a[t]) + \frac{t+1}{\chi[t]}\sum_{j=1}^{n}\left(h_j(a[t])\right)_+^2 \ge f(x[t]) - f^* - C_1\frac{\chi[t]}{t+1}.$$

□

Proof of Lemma 4:. Since $\chi[t] - \chi[t-1] \to 0$, it holds by averaging that $\frac{1}{t+1}\sum_{r=0}^{t} \chi[r] - \chi[r-1] \to 0$. Thus,

$$\frac{1}{t+1}\chi[t] = \frac{1}{t+1}\sum_{r=0}^{t} \chi[r] - \chi[r-1] + \frac{1}{t+1}\chi[-1] \to 0.$$

From $\chi[t] \to \infty$ we have $\frac{1}{\chi[t]} \to 0$, and also by averaging, $\frac{1}{t+1}\sum_{r=0}^{t} \frac{1}{\chi[r-1]} \to 0$.

The convergence of the order $O\left(\frac{1}{\sqrt{t}}\right)$ can be achieved in (42) by choosing $\chi[t] = O(\sqrt{t})$. In fact, we obtain:

$$\frac{1}{t+1}\sum_{r=0}^{t} \frac{1}{\chi[r-1]} = \frac{1}{t+1}\left(\frac{1}{\chi[-1]} + \frac{1}{\chi[0]}\right) + \frac{1}{t+1}\sum_{r=1}^{t} \frac{1}{\sqrt{r}}.$$

Immediately, we see that $\frac{1}{t+1}\left(\frac{1}{\chi[-1]} + \frac{1}{\chi[0]}\right) \to 0$ as of the order $O\left(\frac{1}{t}\right)$. Note that for a convex univariate function $\xi(r)$, $r \in \mathbb{R}$, and integer bounds a, b, we have

$$\sum_{r=a}^{b} \xi(r) \leq \int_{a-1/2}^{b+1/2} \xi(s) ds. \tag{77}$$

Hence, we get

$$\frac{1}{t+1}\sum_{r=1}^{t} \frac{1}{\sqrt{r}} \overset{(77)}{\leq} \frac{1}{t+1}\int_{1-1/2}^{t+1/2} \frac{1}{\sqrt{s}} ds$$

$$= \frac{2}{t+1}\sqrt{s}\Big|_{1/2}^{t+1/2} = \frac{2}{t+1}\left(\sqrt{t+1/2} - \sqrt{1/2}\right) \to 0.$$

Here, the order of convergence is $O\left(\frac{1}{\sqrt{t}}\right)$. By assuming $\chi[t] = O(\sqrt{t})$, the convergence $\frac{\chi[t]}{t+1} = \frac{\sqrt{t}}{t+1} \to 0$ is also of the order $O\left(\frac{1}{\sqrt{t}}\right)$. □

References

1. Feynman, R.P., Leighton, R., Sands, M.: The Feynman Lectures on Physics. The Definitive and Extended Edition. Addison Wesley, Reading (2005)
2. Friedman, M., Friedman, R.: Free to Choose: A Personal Statement. Harcourt Brace Jovanovich, New York (1980)
3. Gale, D.: The Theory of Linear Economic Models. McGraw Hill, New York (1960)
4. Hiriart-Urruty, J.-B., Lemarchal, C.: Fundamentals of Convex Analysis. Springer, Berlin (2001)
5. Kreps, D.M.: Microeconomic Foundations I: Choice and Competitive Markets. Princeton University Press, Princeton (2012)
6. Krishna, V.: Auction Theory. Academic Press, Burlington (2010)
7. Krishna, V., Sonnenschein, H.: Duality in consumer theory. In: Chipman, J., McFadden, D., Richter, M. (eds.) Preferences, Uncertainty and Optimality, pp. 44–55. Westview Press, Boulder, CO (1990)
8. Mas-Colell, A., Whinston, M.D., Green, J.R.: Microeconomic Theory. Oxford University Press, New York (1995)
9. Mullainathan, S., Thaler, R.H.: Behavioral economics. In: Smelser, N.J., Baltes, P.B. (eds.) International Encyclopedia of Social and Behavioral Sciences. Elsevier, Amsterdam (2001)
10. Nesterov, Yu.: Introductory Lectures on Convex Optimization. Kluwer, Boston (2004)
11. Nesterov, Yu., Shikhman, V.: Algorithm of price adjustment for market equilibrium. CORE Discussion Paper 2015/1, 25 p. (2015)
12. Nesterov, Yu., Shikhman, V.: Computation of Fisher-Gale equilibrium by auction. CORE Discussion Paper 2015/35, 32 p. (2015)
13. Nesterov, Yu., Shikhman, V.: Quasi-monotone subgradient methods for nonsmooth convex minimization. J. Optim. Theory Appl. **165**, 917–940 (2015)
14. Nesterov, Yu., Shikhman, V.: Excessive revenue model of competitive markets. In: Mordukhovich, B.S., Reich, S., Zaslavski, A.J. (eds.) Nonlinear Analysis and Optimization, Contemporary Mathematics, vol. 659, pp. 189–219. AMS, Providence, RI (2016)
15. Rockafellar, R.T.: Convex Analysis. Princeton University Press, Princeton (1970)
16. Rosen, J.B.: Existence and uniqueness of equilibrium points for concave n-Person games. Econometrica **31**, 520–534 (1965)
17. Rubinstein, A.: Equilibrium in the jungle. Econ. J. **117**, 883–896 (2007)
18. Rubinstein, A.: Lecture Notes in Microeconomic Theory: The Economic Agent. Princeton University Press, Princeton (2012)
19. Samuelson, P.A.: The stability of equilibrium: comparative statics and dynamics. Econometrica **9**, 97–120 (1941)
20. Seel, N.M.: Incremental learning. In: Seel, N.M. (ed.) Encyclopedia of the Sciences of Learning, p. 1523. Springer, New York (2012)
21. Spearman, M.L.: Of physics and factory physics. Prod. Oper. Manag. **23**, 1875–1885 (2014)
22. Zălinescu, C.: Convex Analysis in General Vector Spaces. World Scientific, Singapore (2002)

The Legendre Transformation in Modern Optimization

Roman A. Polyak

Abstract The Legendre transform (LET) is a product of a general duality principle: any smooth curve is, on the one hand, a locus of pairs, which satisfy the given equation and, on the other hand, an envelope of a family of its tangent lines.

An application of the LET to a strictly convex and smooth function leads to the Legendre identity (LEID). For strictly convex and three times differentiable function the LET leads to the Legendre invariant (LEINV).

Although the LET has been known for more then 200 years both the LEID and the LEINV are critical in modern optimization theory and methods.

The purpose of the paper is to show the role of the LEID and the LEINV play in both constrained and unconstrained optimization.

Keywords Legendre transform • Duality • Lagrangian • Self-Concordant function • Nonlinear rescaling • Lagrangian transform

1 Introduction

Application of the duality principle to a strictly convex $f : \mathbb{R} \to \mathbb{R}$, leads to the Legendre transform

$$f^*(s) = \sup_{x \in \mathbb{R}} \{sx - f(x)\},$$

which is often called the Legendre-Fenchel transform (see [21, 29, 30]).

The LET, in turn, leads to two important notions: the Legendre identity

$$f^{*'}(s) \equiv f'^{-1}(s)$$

and the Legendre invariant

R.A. Polyak (✉)
Department of Mathematics, The Technion – Israel Institute of Technology, 32000 Haifa, Israel
e-mail: rpolyak@techunix.technion.ac.il; rpolyak@gmu.edu

B. Goldengorin (ed.), *Optimization and Its Applications in Control and Data Sciences*, Springer Optimization and Its Applications 115,
DOI 10.1007/978-3-319-42056-1_15

$$\text{LEINV}(f) = \left| \frac{d^3f}{dx^3}\left(\frac{d^2f}{dx^2}\right)^{-\frac{3}{2}} \right| = \left| -\frac{d^3f^*}{ds^3}\left(\frac{d^2f^*}{ds^2}\right)^{-\frac{3}{2}} \right|.$$

Our first goal is to show a number of duality results for optimization problems with equality and inequality constraints obtained in a unified manner by using LEID.

A number of methods for constrained optimization, which have been introduced in the past several decades and for a long time seemed to be unconnected, turned out to be equivalent. We start with two classical methods for equality constrained optimization.

First, the primal penalty method by Courant [16] and its dual equivalent—the regularization method by Tichonov [60].

Second, the primal multipliers method by Hestenes [28] and Powell [52], and its dual equivalent—the quadratic proximal point method by Moreau [39], Martinet [35, 36] Rockafellar [56, 57] (see also [2, 7, 24, 27, 44, 45, 58] and references therein).

Classes of primal SUMT and dual interior regularization, primal nonlinear rescaling (NR) and dual proximal points with φ-divergence distance functions, primal Lagrangian transformation (LT) and dual interior ellipsoids methods turned out to be equivalent.

We show that LEID is a universal tool for establishing the equivalence results, which are critical, for both understanding the nature of the methods and establishing their convergence properties.

Our second goal is to show how the equivalence results can be used for convergence analysis of both primal and dual methods.

In particular, the primal NR method with modified barrier (MBF) transformation leads to the dual proximal point method with Kullback-Leibler entropy divergence distance (see [50]). The corresponding dual multiplicative algorithm, which is closely related to the EM method for maximum likelihood reconstruction in position emission tomography as well as to image space reconstruction algorithm (see [17, 20, 62]), is the key instrument for establishing convergence of the MBF method (see [31, 46, 50, 53]).

In the framework of LT the MBF transformation leads to the dual interior proximal point method with Bregman distance (see [37, 49]).

The kernel $\varphi(s) = -\ln s + s - 1$ of the Bregman distance is a self-concordant (SC) function. Therefore the corresponding interior ellipsoids are Dikin's ellipsoids.

Application LT for linear programming (LP) calculations leads to Dikin's type method for the dual LP (see [18]).

The SC functions have been introduced by Yuri Nesterov and Arkadi Nemirovski in the late 1980s (See [42, 43]).

Their remarkable SC theory is the centerpiece of the interior point methods (IPMs), which for a long time was the main stream in modern optimization. The SC theory establishes the IPMs complexity for large classes of convex optimization problem from a general and unique point of view.

It turns out that a strictly convex $f \in C^3$ is self-concordant if LEINV(f) is bounded. The boundedness of LEINV(f) leads to the basic differential inequality, four sequential integrations of which produced the main SC properties.

The properties, in particular, lead to the upper and lower bounds for f at each step of a special damped Newton method for unconstrained minimization SC functions. The bounds allow establishing global convergence and show the efficiency of the damped Newton method for minimization SC function.

The critical ingredients in these developments are two special SC function: $w(t) = t - \ln(t+1)$ and its *LET* $w^*(s) = -s - \ln(1-s)$.

Usually two stages of the damped Newton method is considered (see [43]). At the first stage at each step the error bound $\Delta f(x) = f(x) - f(x^*)$ is reduced by $w(\lambda)$, where $0 < \lambda < 1$ is the Newton decrement. At the second stage $\Delta f(x)$ converges to zero with quadratic rate. We consider a middle stage where $\Delta f(x)$ converges to zero with superlinear rate, which is explicitly characterized by $w(\lambda)$ and $w^*(\lambda)$.

To show the role of LET and LEINV(f) in unconstrained optimization of SC functions was our third goal.

The paper is organized as follows.

In the next section along with LET we consider LEID and LEINV.

In Sect. 3 penalty and multipliers methods and their dual equivalents applied for optimization problems with equality constraints.

In Sect. 4 the classical SUMT methods and their dual equivalents—the interior regularization methods—are applied to convex optimization problem.

In Sect. 5 we consider the Nonlinear Rescaling theory and methods, in particular, the MBF and its dual equivalent—the prox with Kullback-Leibler entropy divergence distance.

In Sect. 6 the Lagrangian transform (LT) and its dual equivalent—the interior ellipsoids method—are considered. In particular, the LT with MBF transformation, which leads to the dual prox with Bregman distance.

In Sect. 7 we consider LEINV, which leads to the basic differential inequality, the main properties of the SC functions and eventually to the damped Newton method.

We conclude the paper (survey) with some remarks, which emphasize the role of LET, LEID and LEINV in modern optimization.

2 Legendre Transformation

We consider LET for a smooth and strictly convex scalar function of a scalar argument $f : \mathbb{R} \to \mathbb{R}$.

For a given $s = \tan\varphi$ let us consider line $l = \{(x, y) \in \mathbb{R}^2 : y = sx\}$. The corresponding tangent to the curve L_f with the same slope is defined as follows:

$$T(x, y) = \{(X, Y) \in \mathbb{R}^2 : Y - f(x) = f'(x)(X - x) = s(X - x)\}.$$

In other words $T(x, y)$ is a tangent to the curve $L_f = \{(x, y) : y = f(x)\}$ at the point (x, y): $f'(x) = s$. For $X = 0$, we have $Y = f(x) - sx$. The conjugate function

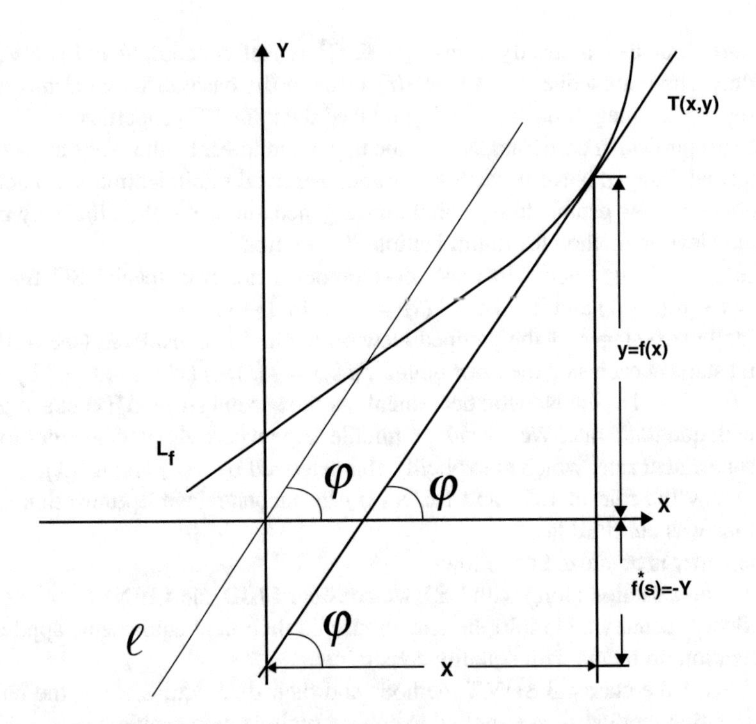

Fig. 1 Legendre transformation

$f^* : (a,b) \to \mathbb{R}, -\infty < a < b < \infty$ at the point s is defined as $f^*(s) = -Y = -f(x) + sx$. Therefore (see Fig. 1)

$$f^*(s) + f(x) = sx. \tag{1}$$

More often f^* is defined as follows

$$f^*(s) = \max_{x \in \mathbb{R}} \{sx - f(x)\}. \tag{2}$$

Keeping in mind that $T(x,y)$ is the supporting hyperplane to the epi$f = \{(y,x) : y \geq f(x)\}$ the maximum in (2) is reached at x: $f^{'}(x) = s$, therefore the primal representation of (1) is

$$f^*(f^{'}(x)) + f(x) \equiv f^{'}(x)x. \tag{3}$$

For a strictly convex f we have $f^{''}(x) > 0$, therefore due to the Inverse Function Theorem the equation $f^{'}(x) = s$ can be solved for x, that is

$$x(s) = f^{'-1}(s). \tag{4}$$

Using (4) from (3) we obtain the dual representation of (1)

$$f^*(s) + f(x(s)) \equiv sx(s). \tag{5}$$

Also, it follows from $f''(x) > 0$ that $x(s)$ in (2) is unique, so f^* is as smooth as f. The variables x and s are not independent, they are linked through equation $s = f'(x)$.

By differentiating (5) we obtain

$$f^{*'}(s) + f'(x(s))x'(s) \equiv x(s) + sx'(s). \tag{6}$$

In view of $f'(x(s)) = s$, from (4) and (6) we obtain the following identity,

$$f^{*'}(s) \equiv f'^{-1}(s), \tag{7}$$

which is called the Legendre identity (LEID).

From (4) and (7) we obtain

$$\frac{df^*(s)}{ds} = x. \tag{8}$$

On the other hand, we have

$$\frac{df(x)}{dx} = s. \tag{9}$$

From (8) and (9) it follows

$$a)\ \frac{d^2f^*(s)}{ds^2} = \frac{dx}{ds} \text{ and } b)\ \frac{d^2f(x)}{dx^2} = \frac{ds}{dx}. \tag{10}$$

From

$$\frac{dx}{ds} \cdot \frac{ds}{dx} = 1$$

and (10) we get

$$\frac{d^2f^*}{ds^2} \cdot \frac{d^2f}{dx^2} = 1, \tag{11}$$

so the local curvatures of f and f^* are inverses to each other.

The following Theorem established the relations of the third derivatives of f and f^*, which leads to the notion of Legendre invariant.

Theorem 1. *If $f \in C^3$ is strictly convex then*

$$\frac{d^3f^*}{ds^3} \cdot \left(\frac{d^2f^*}{ds^2}\right)^{-3/2} + \frac{d^3f}{dx^3} \cdot \left(\frac{d^2f}{dx^2}\right)^{-3/2} = 0. \tag{12}$$

Proof. By differentiating (11) in x we obtain

$$\frac{d^3f^*}{ds^3} \cdot \frac{ds}{dx} \cdot \frac{d^2f}{dx^2} + \frac{d^2f^*}{ds^2} \cdot \frac{d^3f}{dx^3} = 0.$$

In view of (10b) we have

$$\frac{d^3f^*}{ds^3} \cdot \left(\frac{d^2f}{dx^2}\right)^2 + \frac{d^2f^*}{ds^2} \cdot \frac{d^3f}{dx^3} = 0. \tag{13}$$

By differentiating (11) in s and keeping in mind (10a) we obtain

$$\frac{d^3f^*}{ds^3}\frac{d^2f}{dx^2} + \left(\frac{d^2f^*}{ds^2}\right)^2 \frac{d^3f}{dx^3} = 0. \tag{14}$$

Using (11), from (13) and (14) we have

$$\frac{d^3f^*}{ds^3} \cdot \frac{d^2f}{dx^2} + \frac{1}{\left(\frac{d^2f}{dx^2}\right)^2}\frac{d^3f}{dx^3} = 0$$

or

$$\frac{d^3f^*}{ds^3}\left(\frac{d^2f}{dx^2}\right)^3 + \frac{d^3f}{dx^3} = 0.$$

Keeping in mind $\frac{d^2f}{dx} > 0$ from the last equation follows

$$\frac{d^3f^*}{ds^3}\left(\frac{d^2f}{dx^2}\right)^{\frac{3}{2}} + \frac{d^3f}{dx^3}\left(\frac{d^2f}{dx^2}\right)^{-\frac{3}{2}} = 0.$$

Using (11) again we obtain (12).

Corollary 1. *From (12) we have*

$$-\frac{d^3f^*}{ds^3}\left(\frac{d^2f^*}{ds^2}\right)^{-3/2} = \frac{d^3f}{dx^3}\left(\frac{d^2f}{dx^2}\right)^{-3/2}.$$

The Legendre Invariant is defined as follows

$$\text{LEINV}(f) = \left|-\frac{d^3f^*}{ds^3}\left(\frac{d^2f^*}{ds^2}\right)^{-3/2}\right| = \left|\frac{d^3f}{dx^3}\left(\frac{d^2f}{dx^2}\right)^{-3/2}\right|. \tag{15}$$

For a strictly convex $f \in C^3$ boundedness of LEINV(f) defines the class of self-concordant (SC) functions introduced by Yuri Nesterov and A. Nemirovski in the late 1980s .

3 Equality Constrained Optimization

Let f and all c_i: $\mathbb{R}^n \to \mathbb{R}$, $i = 1, \ldots, m$ be continuously differentiable. We consider the following optimization problem with equality constrains

$$\begin{aligned} &\min f(x) \\ &\text{s. t. } c_i(x) = 0, i = 1, \ldots, m. \end{aligned} \tag{16}$$

We assume that (16) has a regular solution x^* that is

$$\text{rank } \nabla c(x^*) = m < n,$$

where $\nabla c(x)$ is the Jacobian of the vector-function $c(x) = (c_1(x), \ldots, c_m(x))^T$. Then (see, for example [45]) there exists $\lambda^* \in \mathbb{R}^m$:

$$\nabla_x L(x^*, \lambda^*) = 0, \; \nabla_\lambda L(x^*, \lambda^*) = c(x^*) = 0,$$

where

$$L(x, \lambda) = f(x) + \sum_{i=1}^{m} \lambda_i c_i(x)$$

is the classical Lagrangian, which corresponds to (16).

It is well known that the dual function

$$d(\lambda) = \inf\{L(x, \lambda) | x \in \mathbb{R}^n\} \tag{17}$$

is closed and concave. Its subdifferential

$$\partial d(\lambda) = \{g : d(u) - d(\lambda) \leq (g, u - \lambda), \forall u \in \mathbb{R}^m\} \tag{18}$$

at each $\lambda \in \mathbb{R}^n$ is a non-empty, bounded and convex set. If for a given $\lambda \in \mathbb{R}^m$ the minimizer

$$x(\lambda) = \arg\min\{L(x, \lambda) | x \in \mathbb{R}^n\}$$

exists then

$$\nabla_x L(x(\lambda), \lambda) = 0. \tag{19}$$

If the minimizer $x(\lambda)$ is unique, then the dual function

$$d(\lambda) = L(x(\lambda), \lambda)$$

is differentiable and the dual gradient

$$\nabla d(\lambda) = \nabla_x L(x(\lambda), \lambda)\nabla_\lambda x(\lambda) + \nabla_\lambda L(x(\lambda), \lambda),$$

where $\nabla_\lambda x(\lambda)$ is the Jacobian of vector-function $x(\lambda) = (x_1(\lambda), \dots, x_n(\lambda))^T$. In view of (19) we have

$$\nabla d(\lambda) = \nabla_\lambda L(x(\lambda), \lambda) = c(x(\lambda)). \tag{20}$$

In other words, the gradient of the dual function coincides with the residual vector computed at the primal minimizer $x(\lambda)$.

If $x(\lambda)$ is not unique, then for any $\hat{x} = x(\lambda) \in \text{Argmin}\{L(x, \lambda)|x \in \mathbb{R}^n\}$ we have

$$c(\hat{x}) \in \partial d(\lambda).$$

In fact, let

$$u : d(u) = L(x(u), u) = \min_{x \in \mathbb{R}^n} L(x, u), \tag{21}$$

then for any $\lambda \in \mathbb{R}^m$ we have

$$d(u) = \min\{f(x) + \sum_{i=1}^{m} u_i c_i(x)|x \in \mathbb{R}^n\} \leq f(\hat{x}) + \sum_{i=1}^{m} u_i c_i(\hat{x}) = f(\hat{x}) + \sum \lambda_i c_i(\hat{x})$$
$$+(c(\hat{x}), u - \lambda) = d(\lambda) + (c(\hat{x}), u - \lambda)$$

or

$$d(u) - d(\lambda) \leq (c(\hat{x}), u - \lambda), \forall u \in \mathbb{R}^m,$$

so (18) holds for $g = c(\hat{x})$, therefore

$$c(\hat{x}) \in \partial d(\hat{\lambda}). \tag{22}$$

The dual to (16) problem is

$$\begin{aligned} &\max d(\lambda) \\ &\text{s. t. } \lambda \in \mathbb{R}^m, \end{aligned} \tag{23}$$

which is a convex optimization problem independent from convexity properties of f and c_i, $i = 1, \ldots, m$ in (16).

The following inclusion

$$0 \in \partial d(\lambda^*) \tag{24}$$

is the optimality condition for the dual maximizer λ^* in (23).

3.1 Penalty Method and its Dual Equivalent

In this section we consider two methods for solving optimization problems with equality constraints and their dual equivalents.

In 1943 Courant introduced the following penalty function and correspondent method for solving (16) (see [16]).

Let $\pi(t) = \frac{1}{2}t^2$ and $k > 0$ be the penalty (scaling) parameter, then Courant's penalty function $P : \mathbb{R}^n \times \mathbb{R}_{++} \to \mathbb{R}$ is defined by the following formula

$$P(x,k) = f(x) + k^{-1}\sum_{i=1}^{m} \pi(kc_i(x)) = f(x) + \frac{k}{2}\|c(x)\|^2, \tag{25}$$

where $\|\cdot\|$ is Euclidean norm. At each step the penalty method finds unconstrained minimizer

$$x(k) : P(x(k),k) = \min_{x \in \mathbb{R}^n} P(x,k). \tag{26}$$

We assume that for a given $k > 0$ minimizer $x(k)$ exists and can be found from the system $\nabla_x P(x,k) = 0$. Then

$$\nabla_x P(x(k),k) =$$

$$\nabla f(x(k)) + \sum_{i=1}^{m} \pi'(kc_i(x(k)))\nabla c_i(x(k)) = 0. \tag{27}$$

Let

$$\lambda_i(k) = \pi'(kc_i(x(k)), \ i = 1,..,m. \tag{28}$$

From (27) and (28) follows

$$\nabla_x P(x(k),k) = \nabla f(x(k)) + \sum_{i=1}^{m} \lambda_i(k)\nabla c_i(x(k)) = \nabla_x L(x(k),\lambda(k)) = 0, \tag{29}$$

which means that $x(k)$ satisfies the necessary condition to be a minimizer of $L(x, \lambda(k))$. If $L(x(k), \lambda(k)) = \min_{x \in \mathbb{R}^n} L(x, \lambda(k))$, then $d(\lambda(k)) = L(x(k), \lambda(k))$ and

$$c(x(k)) \in \partial d(\lambda(k)). \tag{30}$$

Due to $\pi''(t) = 1$ the inverse function π'^{-1} exists. From (28) follows

$$c_i(x(k)) = k^{-1}\pi'^{-1}(\lambda_i(k)),\ i = 1, \ldots, m. \tag{31}$$

From (30), (31) and the LEID $\pi'^{-1} = \pi^{*'}$ we obtain

$$0 \in \partial d(\lambda(k)) - k^{-1}\sum_{i=1}^{m} \pi^{*'}(\lambda_i(k))e_i, \tag{32}$$

where $e_i = (0, \ldots, 1, .., 0)$.

The inclusion (32) is the optimality condition for $\lambda(k)$ to be the unconstrained maximizer of the following unconstrained maximization problem

$$d(\lambda(k)) - k^{-1}\sum_{i=1}^{m} \pi^*(\lambda_i(k)) = \max\{d(u) - k^{-1}\sum_{i=1}^{m} \pi^*(u_i) : u \in \mathbb{R}^m\}. \tag{33}$$

Due to $\pi^*(s) = \max_t\{st - \frac{1}{2}t^2\} = \frac{1}{2}s^2$ the problem (33) one can rewrite as follows

$$d(\lambda(k)) - \frac{1}{2k}\sum_{i=1}^{m} \lambda_i^2(k) = \max\{d(u) - \frac{1}{2k}\|u\|^2 : u \in \mathbb{R}^m\}. \tag{34}$$

Thus, Courant's penalty method (26) is equivalent to Tikhonov's (see [60]) regularization method (34) for the dual problem (23).

The convergence analysis of (34) is simple because the dual $d(u)$ is concave and $D(u, k) = d(u) - \frac{1}{2k}\|u\|^2$ is strongly concave.

Let $\{k_s\}_{s=0}^{\infty}$ be a positive monotone increasing sequence and $\lim_{s\to\infty} k_s = \infty$. We call it a regularization sequence. The correspondent sequence $\{\lambda_s\}_{s=0}^{\infty}$:

$$\lambda_s = \arg\max\{d(u) - \frac{1}{2k_s}\|u\|^2 : u \in \mathbb{R}^m\} \tag{35}$$

is unique due to the strong concavity of $D(u, k)$ in u.

Theorem 2. *If $L^* = \text{Argmax}\{d(\lambda)|\lambda \in \mathbb{R}^m\}$ is bounded and $f, c_i \in C^1$, $i = 1, \ldots, m$, then for any regularization sequence $\{k_s\}_{s=0}^{\infty}$ the following statements hold*

(1) $\|\lambda_{s+1}\| > \|\lambda_s\|$;
(2) $d(\lambda_{s+1}) > d(\lambda_s)$;

(3) $\lim_{s\to\infty} \lambda_s = \lambda^* = \arg\min_{\lambda\in L^*} \|\lambda\|$.

Proof. It follows from (35) and strong concavity of $D(u,k)$ in $u \in \mathbb{R}^m$ that

$$d(\lambda_s) - (2k_s)^{-1}\|\lambda_s\|^2 > d(\lambda_{s+1}) - (2k_s)^{-1}\|\lambda_{s+1}\|^2$$

and

$$d(\lambda_{s+1}) - (2k_{s+1})^{-1}\|\lambda_{s+1}\|^2 > d(\lambda_s) - (2k_{s+1})^{-1}\|\lambda_s\|^2. \tag{36}$$

By adding the inequalities we obtain

$$0.5(k_s^{-1} - k_{s+1}^{-1})[\|\lambda_{s+1}\|^2 - \|\lambda_s\|^2] > 0. \tag{37}$$

Keeping in mind $k_{s+1} > k_s$ from (37) we obtain (1).

From (36) we have

$$d(\lambda_{s+1}) - d(\lambda_s) > (2k_{s+1})^{-1}[\|\lambda_{s+1}\|^2 - \|\lambda_s\|^2] > 0, \tag{38}$$

therefore from (1) follows (2).

Due to concavity d from boundedness of L^* follows boundedness of any level set $\Lambda(\lambda_0) = \{\lambda \in \mathbb{R}^m : d(\lambda) \geq d(\lambda_0)\}$ (see Theorem 24 [22]). From (2) follows $\{\lambda_s\}_{s=0}^{\infty} \subset \Lambda(\lambda_0)$, therefore for any converging subsequence $\{\lambda_{s_i}\} \subset \{\lambda_s\}_{s=0}^{\infty}$: $\lim_{s_i\to\infty} \lambda_{s_i} = \hat{\lambda}$ we have

$$d(\lambda_{s_i}) - (2k_{s_i})^{-1}\|\lambda_{s_i}\|^2 > d(\lambda^*) - (2k_{s_i})^{-1}\|\lambda^*\|^2. \tag{39}$$

Taking the limit in (39) when $k_{s_i} \to \infty$ we obtain $d(\hat{\lambda}) \geq d(\lambda^*)$, therefore $\hat{\lambda} = \lambda^* \in L$. In view of (2) we have $\lim_{s\to\infty} d(\lambda_s) = d(\lambda^*)$.

It follows from (1) that $\lim_{s\to\infty} \|\lambda_s\| = \|\lambda^*\|$. Also from

$$d(\lambda_s) - (2k_s)^{-1}\|\lambda_s\|^2 > d(\lambda^*) - (2k_s)^{-1}\|\lambda^*\|^2$$

follows

$$\|\lambda^*\|^2 - \|\lambda_s\|^2 > 2k_s(d(\lambda^*) - d(\lambda_s)) \geq 0, \quad \forall \lambda^* \in L^*,$$

therefore $\lim_{s\to\infty} \|\lambda_s\| = \min_{\lambda\in L^*} \|\lambda\|$.

Convergence of the regularization method (34) is due to unbounded increase of the penalty parameter $k > 0$, therefore one can hardly expect solving the problem (23) with high accuracy.

3.2 Augmented Lagrangian and Quadratic Proximal Point Method

In this section we consider Augmented Lagrangian method (see [28, 52]), which allows eliminate difficulties associated with unbounded increase of the penalty parameter.

The problem (16) is equivalent to the following problem

$$f(x) + k^{-1}\sum_{i=1}^{m}\pi(kc_i(x)) \to \min \tag{40}$$

$$\text{s.t. } c_i(x) = 0, \quad i = 1, \ldots, m. \tag{41}$$

The correspondent classical Lagrangian $\mathscr{L} : \mathbb{R}^n \times \mathbb{R}^m \times \mathbb{R}_{++} \to \mathbb{R}$ for the equivalent problem (40)–(41) is given by

$$\mathscr{L}(x, \lambda, k) = f(x) - \sum_{i=1}^{m}\lambda_i c_i(x) + k^{-1}\sum_{i=1}^{m}\pi(kc_i(x)) =$$

$$f(x) - \sum_{i=1}^{m}\lambda_i c_i(x) + \frac{k}{2}\sum_{i=1}^{m}c_i^2(x).$$

$\mathscr{L}$ is called Augmented Lagrangian (AL) for the original problem (16).

We assume that for a given $(\lambda, k) \in \mathbb{R}^m \times \mathbb{R}^1_{++}$ the unconstrained minimizer $\hat{x}$ exists, that is

$$\hat{x} = \hat{x}(\lambda, k) : \nabla_x \mathscr{L}(\hat{x}, \lambda, k) = \nabla f(\hat{x}) - \sum_{i=1}^{m}(\lambda_i - \pi'(kc_i(\hat{x})))\nabla c_i(\hat{x}) = 0. \tag{42}$$

Let

$$\hat{\lambda}_i = \hat{\lambda}_i(\lambda, k) = \lambda_i - \pi'(kc_i(\hat{x})), i = 1, \ldots, m. \tag{43}$$

Then from (42) follows $\nabla_x L(\hat{x}, \hat{\lambda}) = 0$, which means that $\hat{x}$ satisfies the necessary condition for $\hat{x}$ to be a minimizer of $L(x, \hat{\lambda})$. If $L(\hat{x}, \hat{\lambda}) = \min_{x \in \mathbb{R}^n} L(x, \hat{\lambda})$ then $d(\hat{\lambda}) = L(\hat{x}, \hat{\lambda})$ and

$$c(\hat{x}) \in \partial d(\hat{\lambda}). \tag{44}$$

From (43) follows

$$c(\hat{x}) = \frac{1}{k}\pi'^{-1}(\hat{\lambda} - \lambda). \tag{45}$$

Using LEID and (45) we obtain

$$0 \in \partial d(\hat{\lambda}) - k^{-1} \sum_{i=1}^{m} \pi^{*'}(\hat{\lambda}_i - \lambda) e_i,$$

which is the optimality condition for $\hat{\lambda}$ to be the maximizer in the following unconstrained maximization problem

$$d(\hat{\lambda}) - k^{-1} \sum_{i=1}^{m} \pi^{*}(\hat{\lambda}_i - \lambda_i) = \max\{d(u) - k^{-1} \sum_{i=1}^{m} \pi^{*}(u_i - \lambda_i) : u \in \mathbb{R}^n\}. \quad (46)$$

In view of $\pi^*(s) = \frac{1}{2}s^2$ we can rewrite (46) as follows

$$\hat{\lambda} = \arg\max\{d(u) - \frac{1}{2k}\|u - \lambda\|^2 : u \in \mathbb{R}^n\} \quad (47)$$

Thus the multipliers method (42)–(43) is equivalent to the quadratic proximal point (prox) method (47) for the dual problem (23) (see [4, 27, 35, 39, 55–58] and references therein)

If $\hat{x}$ is a unique solution to the system $\nabla_x L(x, \hat{\lambda}) = 0$, then $\nabla d(\hat{\lambda}) = c(\hat{x})$ and from (45) follows

$$\hat{\lambda} = \lambda + k\nabla d(\hat{\lambda}),$$

which is an implicit Euler method for solving the following system of ordinary differential equations

$$\frac{d\lambda}{dt} = k\nabla d(\lambda), \ \lambda(0) = \lambda_0. \quad (48)$$

Let us consider the prox-function $p : \mathbb{R}^m \to \mathbb{R}$ defined as follows

$$p(\lambda) = d(u(\lambda)) - \frac{1}{2k}\|u(\lambda) - \lambda\|^2 = D(u(\lambda), \lambda) =$$

$$\max\{d(u) - \frac{1}{2k}\|u - \lambda\|^2 : u \in \mathbb{R}^n\}.$$

The function $D(u, \lambda)$ is strongly concave in $u \in \mathbb{R}^m$, therefore $u(\lambda) = \arg\max\{D(u, \lambda) : u \in \mathbb{R}^n\}$ is unique. The prox-function p is concave and differentiable. For its gradient we have

$$\nabla p(\lambda) = \nabla_u D(u(\lambda), \lambda) \cdot \nabla_\lambda u(\lambda) + \nabla_\lambda D(u, \lambda),$$

where $\nabla_\lambda u(\lambda)$ is the Jacobian of $u(\lambda) = (u_1(\lambda), \ldots, u_m(\lambda))^T$. Keeping in mind $\nabla_u D(u(\lambda), \lambda) = 0$ we obtain

$$\nabla p(\lambda) = \nabla_\lambda D(u, \lambda) = \frac{1}{k}(u(\lambda) - \lambda) = \frac{1}{k}(\hat{\lambda} - \lambda)$$

or

$$\hat{\lambda} = \lambda + k\nabla p(\lambda). \tag{49}$$

In other words, the prox-method (47) is an explicit Euler method for the following system

$$\frac{d\lambda}{dt} = k\nabla p(\lambda),\ \lambda(0) = \lambda_0.$$

By reiterating (49) we obtain the dual sequence $\{\lambda_s\}_{s=0}^{\infty}$:

$$\lambda_{s+1} = \lambda_s + k\nabla p(\lambda_s), \tag{50}$$

generated by the gradient method for maximization the prox function p. The gradient ∇p satisfies Lipschitz condition with constant $L = k^{-1}$. Therefore we have the following bound $\Delta p(\lambda_s) = p(\lambda^*) - p(\lambda_s) \leq O(sk)^{-1}$ (see, for example, [45]).

We saw that the dual aspects of the penalty and the multipliers methods are critical for understanding their convergence properties and LEID is the main instrument for obtaining the duality results.

It is even more so for constrained optimization problems with inequality constraints.

4 SUMT as Interior Regularization Methods for the Dual Problem

The sequential unconstrained minimization technique (SUMT) (see [22]) goes back to the 1950s, when R. Frisch introduced log-barrier function to replace a convex optimization with inequality constraints by a sequence of unconstrained convex minimization problems.

Let f and all-c_i, $i = 1, \ldots, m$ be convex and smooth. We consider the following convex optimization problem

$$\begin{aligned} &\min f(x) \\ &\text{s. t. } x \in \Omega, \end{aligned} \tag{51}$$

where $\Omega = \{x \in \mathbb{R}^n : c_i(x) \geq 0,\ i = 1, \ldots, m\}$.

From this point on we assume

A. The solution set $X^* = \text{Argmin}\{f(x) : x \in \Omega\}$ is not empty and bounded.
B. Slater condition holds, i.e. there exists $x_0 \in \Omega$: $c_i(x_0) > 0$, $i = 1, \ldots, m$.

By adding one constraint $c_0(x) = M - f(x) \geq 0$ with M large enough to the original set of constraints $c_i(x) \geq 0$, $i = 1, \ldots, m$ we obtain a new feasible set, which due to the assumption A convexity f and concavity c_i, $i = 1, \ldots, m$ is bounded (see Theorem 24 [22]) and the extra constraint $c_0(x) \geq 0$ for large M does not effect X^*.

So we assume from now on that Ω is bounded. It follows from KKT's Theorem that under Slater condition the existence of the primal solution

$$f(x^*) = \min\{f(x)|x \in \Omega\}$$

leads to the existence of $\lambda^* \in \mathbb{R}^m_+$ that for $\forall x \in \mathbb{R}^n$ and $\lambda \in \mathbb{R}^m_+$ we have

$$L(x^*, \lambda) \leq L(x^*, \lambda^*) \leq L(x, \lambda^*) \tag{52}$$

and λ^* is the solution of the dual problem

$$d(\lambda^*) = \max\{d(\lambda)|\lambda \in \mathbb{R}^m_+\}. \tag{53}$$

Also from B follows boundedness of the dual optimal set

$$L^* = \text{Argmax}\{d(\lambda) : \lambda \in \mathbb{R}^m_+\}.$$

From concavity d and boundedness L^* follows boundedness of the dual level set $\Lambda(\bar{\lambda}) = \{\lambda \in \mathbb{R}^m_+ : d(\lambda) \geq d(\bar{\lambda})\}$ for any given $\bar{\lambda} \in \mathbb{R}^m_+$: $d(\bar{\lambda}) < d(\lambda^*)$.

4.1 Logarithmic Barrier

To replace the constrained optimization problem (51) by a sequence of unconstrained minimization problems R. Frisch in 1955 introduced (see [23]) the log-barrier penalty function $P : \mathbb{R}^n \times \mathbb{R}_{++} \to \mathbb{R}$ defined as follows

$$P(x, k) = f(x) - k^{-1} \sum_{i=1}^{m} \pi(kc_i(x)),$$

where $\pi(t) = \ln t$, ($\pi(t) = -\infty$ for $t \leq 0$) and $k > 0$. Due to convexity f and concavity c_i $i = 1, \ldots, m$ the function P is convex in x. Due to Slater condition, convexity f, concavity c_i and boundedness Ω the recession cone of Ω is empty that is for any $x \in \Omega$, $k > 0$ and $0 \neq d \in \mathbb{R}^n$ we have

$$\lim_{t\to\infty} P(x + td, k) = \infty. \tag{54}$$

Therefore for any $k > 0$ there exists

$$x(k) : \nabla_x P(x(k), k) = 0. \tag{55}$$

Theorem 3. *If A and B hold and* $f, c_i \in C^1$, $i = 1, \dots, m$, *then interior log-barrier method* (55) *is equivalent to the interior regularization method*

$$\lambda(k) = \arg\max\{d(u) + k^{-1} \sum_{i=1}^{m} \ln u_i : u \in \mathbb{R}^m_+\} \tag{56}$$

and the following error bound holds

$$\max\{\Delta f(x(k)) = f(x(k)) - f(x^*), \Delta d(\lambda(k)) = d(\lambda^*) - d(\lambda(k))\} = mk^{-1}. \tag{57}$$

Proof. From (54) follows existence $x(k) : P(x(k), k) = \min\{P(x, k) : x \in \mathbb{R}^n\}$ for any $k > 0$.

Therefore

$$\nabla_x P(x(k), k) = \nabla f(x(k)) - \sum_{i=1}^{m} \pi'(k_i(x(k)) \nabla c_i(x(k)) = 0. \tag{58}$$

Let

$$\lambda_i(k) = \pi'(kc_i(x(k)) = (kc_i(x(k)))^{-1}, \ i = 1, .., m. \tag{59}$$

Then from (58) and (59) follows $\nabla_x P(x(k), k) = \nabla_x L(x(k), \lambda(k)) = 0$, therefore $d(\lambda(k)) = L(x(k), \lambda(k))$. From $\pi''(t) = -t^2 < 0$ follows existence of π'^{-1} and from (59) we have $kc(x(k)) = \pi'^{-1}(\lambda_i(k))$. Using LEID we obtain

$$c_i(x(k)) = k^{-1} \pi^{*'}(\lambda_i(k)), \tag{60}$$

where $\pi^*(s) = \inf_{t>0}\{st - \ln t\} = 1 + \ln s$. The subdifferential $\partial d(\lambda(k))$ contains $-c(x(k))$, that is

$$0 \in \partial d(\lambda(k)) + c(x(k)). \tag{61}$$

From (60) and (61) follows

$$0 \in \partial d(\lambda(k)) + k^{-1} \sum_{i=1}^{m} \pi^{*'}(\lambda_i(k)) e_i. \tag{62}$$

The last inclusion is the optimality criteria for $\lambda(k)$ to be the maximizer in (56).

The maximizer $\lambda(k)$ is unique due to the strict concavity of the objective function in (56).

Thus, SUMT with log-barrier function $P(x,k)$ is equivalent to the interior regularization method (56).

For primal interior trajectory $\{x(k)\}_{k=k_0>0}^{\infty}$ and dual interior trajectory $\{\lambda(k)\}_{k=k_0>0}^{\infty}$ we have

$$f(x(k)) \geq f(x^*) = d(\lambda^*) \geq d(\lambda(k)) = L(x(k),\lambda(k)) = f(x(k)) - (c(x(k)),\lambda(k)).$$

From (59) follows $\lambda_i(k)c_i(x(k)) = k^{-1}, i = 1,\ldots,m$, hence for the primal–dual gap we obtain

$$f(x(k)) - d(\lambda(k)) = (c(x(k)),\lambda(k)) = mk^{-1}.$$

Therefore for the primal and the dual error bounds we obtain (57). □

The main idea of the interior point methods (IPMs) is to stay "close" to the primal $\{x(k)\}_{k=0}^{\infty}$ or to the primal–dual $\{x(k),\lambda(k)\}_{k=0}^{\infty}$ trajectory and increase $k > 0$ at each step by a factor $(1-\frac{\alpha}{\sqrt{n}})^{-1}$, where $\alpha > 0$ is independent of n. In case of LP at each step we solve a system of linear equations, which requires $O(n^{2.5})$ operations. Therefore accuracy $\varepsilon > 0$ IPM are able to achieve in $O(n^3 \ln \varepsilon^{-1})$ operations.

In case of log-barrier transformation the situation is symmetric, that is both the primal interior penalty method (55) and the dual interior regularization method (56) are using the same log-barrier function.

It is not the case for other constraints transformations used in SUMT.

4.2 *Hyperbolic Barrier*

The hyperbolic barrier

$$\pi(t) = \begin{cases} -t^{-1}, t > 0 \\ -\infty, t \leq 0, \end{cases}$$

has been introduced by Carroll in the 1960s, (see [12]). It leads to the following hyperbolic penalty function

$$P(x,k) = f(x) - k^{-1}\sum_{i=1}^{m} \pi(kc_i(x)) = f(x) + k^{-1}\sum_{i=1}^{m}(kc_i(x))^{-1},$$

which is convex in $x \in \mathbb{R}^n$ for any $k > 0$. For the primal minimizer we obtain

$$x(k) : \nabla_x P(x(k),k) = \nabla f(x(k)) - \sum_{i=1}^{m} \pi'(kc_i(x(k)))\nabla c_i(x(k)) = 0. \tag{63}$$

For the vector of Lagrange multipliers we have

$$\lambda(k) = (\lambda_i(k) = \pi'(kc_i(x(k)) = (kc_i(x(k)))^{-2},\ i = 1, \ldots, m). \tag{64}$$

We will show later that vectors $\lambda(k)$, $k \geq 1$ are bounded. Let $L = \max_{i,k} \lambda_i(k)$.

Theorem 4. *If A and B hold and f, $c_i \in C^1$, $i = 1, .., m$, then hyperbolic barrier method* (63) *is equivalent to the parabolic regularization method*

$$d(\lambda(k)) + 2k^{-1} \sum_{i=1}^{m} \sqrt{\lambda_i(k)} = \max\{d(u) + 2k^{-1} \sum_{i=1}^{m} \sqrt{u_i} : u \in \mathbb{R}^m_+\} \tag{65}$$

and the following bounds holds

$$\max\{\Delta f(x(k)) = f(x(k)) - f(x^*),$$
$$\Delta d(\lambda(k)) = d(\lambda^*) - d(\lambda(k))\} \leq m\sqrt{L}k^{-1}. \tag{66}$$

Proof. From (63) and (64) follows

$$\nabla_x P(x(k), k) = \nabla_x L(x(k), \lambda(k)) = 0,$$

therefore $d(\lambda(k)) = L(x(k), \lambda(k))$.

From $\pi''(t) = -2t^{-3} < 0$, $\forall t > 0$ follows existence of π'^{-1}.

Using LEID from (64) we obtain

$$c_i(x(k)) = k^{-1}\pi'^{-1}(\lambda_i(k)) = k^{-1}\pi^{*'}(\lambda_i(k)),\ i = 1, \ldots, m,$$

where $\pi^*(s) = \inf_t\{st - \pi(t)\} = 2\sqrt{s}$.

The subgradient $-c(x(k)) \in \partial d(\lambda(k))$ that is

$$0 \in \partial d(\lambda(k)) + c(x(k)) = \partial d(\lambda(k)) + k^{-1} \sum_{i=1}^{m} \pi^{*'}(\lambda_i(k)) e_i. \tag{67}$$

The last inclusion is the optimality condition for the interior regularization method (65) for the dual problem.

Thus, the hyperbolic barrier method (63) is equivalent to the parabolic regularization method (65) and $D(u, k) = d(u) + 2k^{-1} \sum_{i=1}^{m} \sqrt{u_i}$ is strictly concave.

Using considerations similar to those in Theorem 2 and keeping in mind strict concavity of $D(u, k)$ in u from (65) we obtain

$$\sum_{i=1}^{m} \sqrt{\lambda_i(1)} > \ldots \sum_{i=1}^{m} \sqrt{\lambda_i(k)} > \sum_{k=1}^{m} \sqrt{\lambda_i(k+1)} > \ldots$$

Therefore the sequence $\{\lambda(k)\}_{k=1}^{\infty}$ is bounded, so there exists $L = \max_{i,k} \lambda_i(k) > 0$. From (64) for any $k \geq 1$ and $i = 1, \ldots, m$ we have

$$\lambda_i(k)c_i^2(x(k)) = k^{-2}$$

or

$$(\lambda_i(k)c_i(x(k)))^2 = k^{-2}\lambda_i(k) \leq k^{-2}L.$$

Therefore

$$(\lambda(k), c(x(k))) \leq m\sqrt{L}k^{-1}.$$

For the primal interior sequence $\{x(k)\}_{k=1}^{\infty}$ and dual interior sequence $\{\lambda(k)\}_{k=1}^{\infty}$ we have

$$f(x(k)) \geq f(x^*) = d(\lambda^*) \geq L(x(k), \lambda(k)) = d(\lambda(k)),$$

therefore

$$f(x(k)) - d(\lambda(k)) = (c(x(k)), \lambda(k))) \leq m\sqrt{L}k^{-1},$$

which leads to (66). □

In spite of similarity bounds (57) and (65) are fundamentally different because L can be very large for problems where Slater condition is "barely" satisfied, that is the primal feasible set is not "well defined".

This is one of the reasons why log-barrier function is so important.

4.3 Exponential Penalty

Exponential penalty $\pi(t) = -e^{-t}$ has been used by Motzkin in 1952 (see [40]) to transform a systems of linear inequalities into an unconstrained convex optimization problem in order to use unconstrained minimization technique for solving linear inequalities.

The exponential transformation $\pi(t) = -e^{-t}$ leads to the exponential penalty function

$$P(x, k) = f(x) - k^{-1} \sum_{i=1}^{m} \pi(kc_i(x)) = f(x) + k^{-1} \sum_{i=1}^{m} e^{-kc_i(x)},$$

which is for any $k > 0$ convex in $x \in \mathbb{R}^n$.

For the primal minimizer we have

$$x(k) : \nabla_x P(x(k), k) = \nabla f(x(k)) - \sum_{i=1}^{m} e^{-kc_i(x(k))} \nabla c_i(x(k)) = 0. \tag{68}$$

Let us introduce the Lagrange multipliers vector

$$\lambda(k) = (\lambda_i(k) = \pi^{'}(c_i(x(k)) = e^{-kc_i(x(k))},\ i = 1, \dots, m) \tag{69}$$

From (68) and (69) we have

$$\nabla_x P(x(k), k) = \nabla_x L(x(k), \lambda(k)) = 0.$$

Therefore from convexity $L(x, \lambda(k))$ in $x \in \mathbb{R}^n$ follows $d(\lambda(k)) = \min\{L(x, \lambda(k)) |x \in \mathbb{R}^n\} = L(x(k), \lambda(k))$ and $-c(x(k)) \in \partial d(\lambda(k))$, therefore

$$0 \in c(x(k)) + \partial d(\lambda(k)). \tag{70}$$

From $\pi^{''}(t) = -e^{-t} \neq 0$ follows the existence $\pi^{'-1}$, therefore using LEID from (69) we obtain

$$c_i(x(k)) = k^{-1} \pi^{'-1}(\lambda_i(k)) = k^{-1} \pi^{*'}(\lambda_i(k)),\ i = 1, \dots, m.$$

Inclusion (70) we can rewrite as follows

$$\partial d(\lambda(k)) + k^{-1} \sum \pi^{*'}(\lambda(k)) e_i = 0.$$

Keeping in mind $\pi^*(s) = \inf_t\{st - \pi(t)\} = \inf\{st + e^{-t}\} = -s \ln s + s$ from the last inclusion we obtain

$$d(\lambda(k)) - k^{-1} \sum_{i=1}^{m} \lambda_i(k)(\ln(\lambda_i(k) - 1)) =$$

$$\max\{d(u) - k^{-1} \sum_{i=1}^{m} u_i(\ln u_i - 1) : u \in \mathbb{R}^m_+\}. \tag{71}$$

It means that the exponential penalty method (68) is equivalent to the interior regularization method (71) with Shannon entropy regularization function $r(u) = -\sum_{i=1}^m u_i(\ln u_i - 1)$.

The convergence of the dual sequence $\{\lambda(k)\}_{k=0}^{\infty}$ can be proven using arguments similar to those used in Theorem 2.

We conclude the section by considering smoothing technique for convex optimization.

4.4 Log-Sigmoid (LS) Method

It follows from Karush-Kuhn-Tucker's Theorem that under Slater condition for x^* to be a solution of (51) it is necessary and sufficient existence $\lambda^* \in \mathbb{R}^m$, that the pair $(x^*; \lambda^*)$ is the saddle point of the Lagrangian, that is (52) hold.

From the right inequality of (52) and complementarity condition we obtain

$$f(x^*) \leq f(x) - \sum_{i=1}^{m} \lambda_i^* \min\{c_i(x), 0\} \leq$$

$$f(x) - \max_{1 \leq i \leq m} \lambda_i^* \sum_{i=1}^{m} \min\{c_i(x), 0\}$$

for any $x \in \mathbb{R}^n$. Therefore for any $r > \max_{1 \leq i \leq m} \lambda_i^*$ we have

$$f(x^*) \leq f(x) - r \sum_{i=1}^{m} \min\{c_i(x), 0\}, \forall x \in \mathbb{R}^n. \tag{72}$$

The function

$$Q(x, r) = f(x) - r \sum_{i=1}^{m} \min\{c_i(x), 0\}$$

is called exact penalty function.

Due to concavity c_i, $i = 1, \ldots, m$ functions $q_i(x) = \min\{c_i(x), 0\}$ are concave. From convexity f and concavity q_i, $i = 1, \ldots, m$ follows convexity $Q(x, r)$ in $x \in \mathbb{R}^n$. From (72) follows that solving (51) is equivalent to solving the following unconstrained minimization problem

$$f(x^*) = Q(x^*, r) = \min\{Q(x, r) : x \in \mathbb{R}^n\}. \tag{73}$$

The function $Q(x, r)$ is non-smooth at x^*. The smoothing techniques replace Q by a sequence of smooth functions, which approximate $Q(x, r)$. (see [3, 14, 47, 48] and references therein)

Log-sigmoid (LS) function $\pi : \mathbb{R} \to \mathbb{R}$ is defined by

$$\pi(t) = \ln S(t, 1) = \ln(1 + e^{-t})^{-1},$$

is one of such functions. We collect the log-sigmoid properties in the following assertion

Assertion 1. *The following statements are holds*

1. $\pi(t) = t - \ln(1 + e^t) < 0,\ \pi(0) = -\ln 2$
2. $\pi'(t) = (1 + e^t)^{-1} > 0,\ \pi'(0) = 2^{-1}$
3. $\pi''(t) = -e^t(1 + e^t)^{-2} < 0,\ \pi''(0) = -2^{-2}$.

The smooth penalty method employs the scaled LS function

$$k^{-1}\pi(kt) = t - k^{-1}\ln(1 + e^{kt}), \tag{74}$$

which is a smooth approximation of $q(t) = \min\{t, 0\}$.

In particular, from (74) follows

$$0 < q(t) - k^{-1}\pi(kt) < k^{-1}\ln 2. \tag{75}$$

It means that by increasing $k > 0$ the approximation can be made as accurate as one wants.

The smooth penalty function $P : \mathbb{R}^n \times \mathbb{R}_{++} \to \mathbb{R}$ defined by

$$P(x, k) = f(x) - k^{-1}\sum_{i=1}^{m}\pi(kc_i(x)) \tag{76}$$

is the main instrument in the smoothing technique.

From Assertion 1 follows that P is as smooth as f and c_i, $i = 1, .., m$.

The LS method at each step finds

$$x(k) : P(x(k), k) = \min\{P(x, k) : x \in \mathbb{R}^n\} \tag{77}$$

and increases $k > 0$ if the accuracy obtained is not satisfactory.

Without loss of generality we assume that f is bounded from below. Such assumption does not restrict the generality, because the original objective function f can be replaced by an equivalent $f(x) := \ln(1 + e^{f(x)}) \geq 0$.

Boundedness of Ω together with Slater condition, convexity f and concavity c_i, $i = 1, \ldots, m$ make the recession cone of Ω empty, that is (54) holds for $P(x, k)$ given by (76), any $k > 0$, $d \in \mathbb{R}^n$ and any $x \in \Omega$.

Therefore minimizer $x(k)$ in (77) exists for any $k > 0$ that is

$$\nabla_x P(x(k), k) = \nabla f(x(k)) - \sum_{i=1}^{m}\pi'(kc_i(x(k)))\nabla c_i(x(k)) =$$

$$= \nabla f(x(k)) - \sum_{i=1}^{m}(1 + e^{kc_i(x(k))})^{-1}\nabla c_i(x(k)) = 0.$$

Let

$$\lambda_i(k) = (1 + e^{kc_i(x(k))})^{-1},\ i = 1, \ldots, m, \tag{78}$$

then

$$\nabla_x P(x(k); k) = \nabla f(x(k)) - \sum_{i=1}^{m} \lambda_i(k) \nabla c_i(x(k)) = 0.$$

From (78) follows $\lambda_i(k) \leq 1$ for any $k > 0$. Therefore, generally speaking, one can't expect finding a good approximation for optimal Lagrange multipliers, no matter how large the penalty parameter $k > 0$ is.

If the dual sequence $\{\lambda(k)\}_{k=k_0}^{\infty}$ does not converges to $\lambda^* \in L^*$, then in view of the last equation one can't expect convergence of the primal sequence $\{x(k)\}_{k=k_0}^{\infty}$ to $x^* \in X^*$.

To guarantee convergence of the LS method we have to modify $P(x, k)$. Let $0 < \alpha < 0.5$ and

$$P(x, k) := P_\alpha(x, k) = f(x) - k^{-1+\alpha} \sum_{i=1}^{m} \pi(kc_i(x)). \tag{79}$$

It is easy to see that the modification does not effect the existence of $x(k)$. Therefore for any $k > 0$ there exists

$$x(k) : \nabla_x P(x(k), k) = \nabla f(x(k)) - k^\alpha \sum \pi^{'}(kc(x(k))) \nabla c_i(x(k)) = 0. \tag{80}$$

Theorem 5. *If A and B hold and f, $c_i \in C^1$, $i = 1, \ldots, m$, then the LS method (80) is equivalent to an interior regularization method*

$$d(\lambda(k)) + k^{-1} \sum_{i=1}^{m} \pi^*(k^{-\alpha} \lambda_i(k)) =$$

$$\max\{d(u) + k^{-1} \sum_{i=1}^{m} \pi^*(k^{-\alpha} u_i) : 0 \leq u_i \leq k^\alpha,\ i = 1, \ldots, m\}.$$

Proof. Let

$$\lambda_i(k) = k^\alpha \pi^{'}(kc_i(x(k))) = k^\alpha (1 + e^{kc_i(x(k))})^{-1},\ i = 1, \ldots, m. \tag{81}$$

From (80) and (81) follows

$$\begin{aligned} \nabla_x P(x(k), k) =& \nabla f(x(k)) - \sum_{i=1}^{m} \lambda_i(k) \nabla c_i(x(k)) = \\ & \nabla_x L(x(k), \lambda(k)) = 0. \end{aligned} \tag{82}$$

From (81) we have

$$\pi'(kc_i(x(k)) = k^{-\alpha}\lambda_i(k). \tag{83}$$

Due to $\pi''(t) < 0$ there exists π'^{-1}, therefore

$$c_i(x(k)) = k^{-1}\pi'^{-1}(k^{-\alpha}\lambda_i(k)).$$

Using LEID we obtain

$$c_i(x(k)) = k^{-1}\pi^{*'}(k^{-\alpha}\lambda_i(k)), \tag{84}$$

where

$$\pi^*(s) = \inf_t\{st - \pi(t)\} = -[(1-s)\ln(1-s) + s\ln s]$$

is Fermi-Dirac (FD) entropy function (see, for example, [54]).

From (82) follows $d(\lambda(k)) = L(x(k), \lambda(k))$, also the subdifferential $\partial d(\lambda(k))$ contains $-c(x(k))$, that is

$$0 \in c(x(k)) + \partial d(\lambda(k)). \tag{85}$$

Combining (84) and (85) we obtain

$$0 \in \partial d(\lambda(k)) + k^{-1}\sum_{i=1}^{m}\pi^{*'}(k^{-\alpha}\lambda_i(k))e_i. \tag{86}$$

The inclusion (86) is the optimality criteria for the following problem

$$d(\lambda(k)) + k^{-1}\sum_{i=1}^{m}\pi^*(k^{-\alpha}\lambda_i(k)) =$$

$$\max\{d(u) + k^{-1}r(u) : 0 \le u_i \le k^\alpha,\ i = 1,..,m\}, \tag{87}$$

where $r(u) = \sum_{i=1}^{m}\pi^*(k^{-\alpha}u_i)$.

In other words the LS method (80)–(81) is equivalent to the interior regularization method (87) with FD entropy function used for dual regularization. The FD function is strongly concave inside the cube $\{u \in \mathbb{R}^m : 0 \le u_i \le k^\alpha, i = 1, \dots, m\}$.

It follows from (87) that for any regularization sequence $\{k_s\}_{s=0}^{\infty}$ the Lagrange multipliers $0 < \lambda_i(k_s) < k_s^\alpha$, $i = 1, \dots, m$ can be any positive number, which underlines the importance of modification (79).

Theorem 6. *Under conditions of Theorem 5 for any regularization sequence* $\{k_s\}_{s=0}^{\infty}$*, the primal sequence*

$$\{x_s\}_{s=0}^{\infty} : \nabla_x P(x_s, k_s) = \nabla f(x_s) - \sum_{i=1}^{m} \lambda_{i,s} \nabla c_i(x_s) = 0 \tag{88}$$

and the dual sequence

$$\{\lambda_s\}_{s=0}^{\infty} : d(\lambda_s) + k_s^{-1} r(\lambda_s) =$$
$$\max\{d(u) + k_s^{-1} r(u) : 0 \leq u_i \leq k^{\alpha},\ i = 1, \ldots, m\} \tag{89}$$

the following statements hold

(1) a) $d(\lambda_{s+1}) > d(\lambda_s)$*; b)* $r(\lambda_{s+1}) < r(\lambda_s)$*;*
(2) $\lim_{s\to\infty} d(\lambda_s) = d(\lambda^*)$ *and* $\lambda^* = \arg\min\{r(\lambda) : \lambda \in L^*\}$*;*
(3) the primal–dual sequence $\{x_s, \lambda_s\}_{s=0}^{\infty}$ *is bounded and any limit point is the primal–dual solution.*

Proof. (1) From (89) and strong concavity $r(u)$ follows

$$d(\lambda_{s+1}) + k_{s+1}^{-1} r(\lambda_{s+1}) > d(\lambda_s) + k_{s+1}^{-1} r(\lambda_s) \tag{90}$$

and

$$d(\lambda_s) + k_s^{-1} r(\lambda_s) > d(\lambda_{s+1}) + k_s^{-1} r(\lambda_{s+1}). \tag{91}$$

Therefore

$$(k_{s+1}^{-1} - k_s^{-1})(r(\lambda_{s+1}) - r(\lambda_s)) > 0.$$

From $k_{s+1} > k_s$ and last inequality follows $r(\lambda_{s+1}) < r(\lambda_s)$, therefore from (90) follows

$$d(\lambda_{s+1}) > d(\lambda_s) + k_{s+1}^{-1}(r(\lambda_s) - r(\lambda_{s+1})) > d(\lambda_s). \tag{92}$$

(2) The monotone increasing sequence $\{d(\lambda_s)\}_{s=0}^{\infty}$ is bounded from above by $f(x^*)$. Therefore there is $\lim_{s\to\infty} d(\lambda_s) = \bar{d} \leq f(x^*) = d(\lambda^*)$.
From (89) follows

$$d(\lambda_s) + k_s^{-1} r(\lambda_s) \geq d(\lambda^*) + k_s^{-1} r(\lambda^*). \tag{93}$$

From (92) follows $\{\lambda_s\}_{s=0}^{\infty} \subset \Lambda(\lambda_0) = \{\lambda \in \mathbb{R}_+^m : d(\lambda) \geq d(\lambda_0)\}$. The set $\Lambda(\lambda_0)$ is bounded due to the boundedness of L^* and concavity d. Therefore there exists $\{\lambda_{s_i}\}_{i=1}^{\infty} \subset \{\lambda_s\}_{s=0}^{\infty}$ that $\lim_{s_i\to 0} \lambda_{s_i} = \bar{\lambda}$. By taking the limit in the

correspondent subsequence in (93) we obtain $d(\bar{\lambda}) \geq d(\lambda^*)$, that is $d(\bar{\lambda}) = d(\lambda^*)$.
From $\lim_{s\to\infty} d(\lambda_{s_i}) = d(\lambda^*)$ and 1a) follows $\lim_{s\to\infty} d(\lambda_s) = d(\lambda^*)$.
From (93) follows

$$d(\lambda^*) - d(\lambda_s) \leq k_s^{-1}(r(\lambda^*) - r(\lambda_s)),\ \forall \lambda^* \in L^*, \tag{94}$$

therefore (94) is true for $\lambda^* = \arg\min\{r(\lambda)|\lambda \in L^*\}$.

(3) We saw already the dual sequence $\{\lambda_s\}_{s=0}^{\infty}$ is bounded. Let us show that the primal is bounded too. For a given approximation x_s let consider two sets of indices $I_+(x_s) = \{i : c_i(x_s) \geq 0\}$ and $I_-(x_s) = \{i : c_i(x_s) < 0\}$.
Then keeping in mind $f(x_s) \geq 0$ we obtain

$$\begin{aligned} P(x_s, k_s) &= f(x_s) + k_s^{-1+\alpha} \sum_{i\in I_-(x_s)} \ln(1 + e^{-k_s c_i(x_s)}) \\ &+ k_s^{-1+\alpha} \sum_{i\in I_+(x_s)} \ln(1 + e^{-k_s c_i(x_s)}) \\ &\geq f(x_s) - k_s^{\alpha} \sum_{i\in I_-(x_s)} c_i(x_s) + k_s^{-1+\alpha} \sum_{i\in I_-(x_s)} \ln(1 + e^{k_s c_i(x_s)}) \\ &\geq f(x_s) - k_s^{\alpha} \sum_{i\in I_-(x_s)} c_i(x_s) \geq -k_s^{\alpha} \sum_{i\in I_-(x_s)} c_i(x_s). \end{aligned} \tag{95}$$

On the other hand,

$$P(x_s, k_s) \leq P(x^*, k_s) = f(x^*) - k_s^{-1+\alpha} \sum_{i=1}^{m} \pi(k_s c_i(x^*))$$

$$= f(x^*) + k_s^{-1+\alpha} \sum_{i=1}^{m} \ln(1 + e^{-k_s c_i(x^*)}) \leq f(x^*) + k_s^{-1+\alpha} m \ln 2. \tag{96}$$

From (95) and (96) follows

$$k_s^{\alpha} \sum_{i\in I_-(x_s)} |c_i(x_s)| \leq f(x^*) + k_s^{-1+\alpha} m \ln 2. \tag{97}$$

Therefore for any $s \geq 1$ we have

$$\max_{i\in I_-(x_s)} |c_i(x_s)| \leq k_s^{-\alpha} f(x^*) + k_s^{-1} m \ln 2. \tag{98}$$

It means that the primal sequence $\{x_s\}_{s=0}^{\infty}$ is bounded due to the boundedness of Ω. In other words, the primal–dual sequence $\{x_s, \lambda_s\}_{s=0}^{\infty}$ is bounded.

Let consider a converging subsequence $\{x_{s_i}, \lambda_{s_i}\}_{i=0}^{\infty}$: $\bar{x} = \lim_{i\to\infty} x_{s_i}$; $\bar{\lambda} = \lim_{i\to\infty} \lambda_{s_i}$. From (81) follows $\bar{\lambda}_i = 0$ for $i : c_i(\bar{x}) > 0$ and $\bar{\lambda}_i \geq 0$ for $i : c_i(\bar{x}) = 0$. From (82) follows $\nabla_x L(\bar{x}, \bar{\lambda}) = 0$, therefore $(\bar{x}, \bar{\lambda})$ is KKT's pair, that is $\bar{x} = x^*$, $\bar{\lambda} = \lambda^*$. □

The equivalence primal SUMT and dual interior regularization methods not only allows to prove convergence in a unified and simple manner, but also provide important information about dual feasible solution, which can be used to improve numerical performance. One can't, however, expect finding solution with high accuracy because finding the primal minimizer for large $k > 0$ is a difficult task for the well known reasons.

The difficulties, to a large extend, one can overcome by using the Nonlinear Rescaling theory and methods (see [31, 46, 47, 50, 53, 59] and references). One can view NR as an alternative to SUMT.

5 Nonlinear Rescaling and Interior Prox with Entropy like Distance

The NR scheme employs smooth, strictly concave and monotone increasing functions $\psi \in \Psi$ to transform the original set of constraints into an equivalent set. The transformation is scaled by a positive scaling (penalty) parameter. The Lagrangian for the equivalent problem is our main instrument.

At each step NR finds the primal minimizer of the Lagrangian for the equivalent problem and uses the minimizer to update the Lagrange multipliers (LM). The positive scaling parameter can be fixed or updated from step to step. The fundamental difference between NR and SUMT lies in the role of the LM vector.

In case of SUMT the LM vector is just a by product of the primal minimization. It provides valuable information about the dual vector but it does not effect the computational process. Therefore without unbound increase of the scaling parameter, which is the only tool to control the process, one can not guarantee convergence.

In the NR scheme on the top of the scaling parameter the LM vector is a critical extra tool, which controls computations.

The NR methods converges under any fixed scaling parameter, just due to the LM update (see [31, 46, 50, 53]). If one increases the scaling parameter from step to step, as SUMT does, then instead of sublinear the superlinear convergence rate can be achieved.

The interplay between Lagrangians for the original and the equivalent problems allows to show the equivalence of the primal NR method and dual proximal point method with φ-divergence entropy type distance. The kernel of the distance $\varphi = -\psi^*$, where ψ^* is the LET of ψ. The equivalence is the key ingredient of the convergence analysis.

We consider a class Ψ of smooth functions $\psi : (a, \infty) \to \mathbb{R}$, $-\infty < a < 0$ with the following properties

(1) $\psi(0) = 0$; (2) $\psi'(t) > 0$, $\psi(0) = 1$; (3) $\psi''(t) < 0$; (4) $\lim_{t\to\infty} \psi'(t) = 0$; (5) $\lim_{t\to a+} \psi'(t) = \infty$.

From (1)–(3) follows

$$\Omega = \{x \in \mathbb{R}^n : c_i(x) \geq 0,\ i = 1, \ldots, m\} = \{x \in \mathbb{R}^n : k^{-1}\psi(kc_i(x)) \geq 0,\ i = 1, \ldots, m\}$$

for any $k > 0$.

Therefore (51) is equivalent to

$$\begin{aligned} &\min f(x) \\ &\text{s.t. } k^{-1}\psi(kc_i(x)) \geq 0,\ i = 1, \ldots, m. \end{aligned} \tag{99}$$

The Lagrangian $\mathcal{L} : \mathbb{R}^n \times \mathbb{R}^m_+ \times \mathbb{R}_{++} \to \mathbb{R}$ for (99) is defined as follows

$$\mathcal{L}(x, \lambda, k) = f(x) - k^{-1} \sum_{i=1}^{m} \lambda_i \psi(kc_i(x)).$$

The properties of $\mathcal{L}(x, \lambda, k)$ at the KKT pair (x^*, λ^*) we collect in the following Assertion.

Assertion 2. *For any $k > 0$ and any KKT pair (x^*, λ^*) the following holds*

1° $\mathcal{L}(x^*, \lambda^*, k) = f(x^*)$
2° $\nabla_x \mathcal{L}(x^*, \lambda^*, k) = \nabla f(x^*) - \sum_{i=1}^m \lambda_i^* \nabla c_i(x^*) = \nabla_x L(x^*, \lambda^*) = 0$
3° $\nabla^2_{xx} \mathcal{L}(x^*, \lambda^*, k) = \nabla^2_{xx} L(x^*, \lambda^*) + k\nabla c^T(x^*)\Lambda^*\nabla c(x^*)$,

where $\nabla c(x^*) = J(c(x^*))$ is the Jacobian of $c(x) = (c_1(x), \ldots, c_m(x))^T$ and $\Lambda^* = I \cdot \lambda^*$.

Remark 1. The properties 1^0–3^0 show the fundamental difference between NR and SUMT. In particular, for log-barrier penalty

$$P(x, k) = f(x) - k^{-1} \sum_{i=1}^{m} \ln c_i(x)$$

neither P nor its gradient or Hessian exist at the solution x^*. Moreover, for any given $k > 0$ we have

$$\lim_{x \to x^*} P(x, k) = \infty.$$

On the other hand, $\mathcal{L}(x, \lambda^*, k)$ is an exact smooth approximation for the non-smooth

$$F(x, x^*) = \max\{f(x) - f(x^*), -c_i(x),\ i = 1, .., m\},$$

that is, for any given $k > 0$ we have

$$\min_{x\in\mathbb{R}^n} F(x, x^*) = F(x^*, x^*) = \min_{x\in\mathbb{R}^n}(\mathscr{L}(x, \lambda^*, k) - f(x^*)) = 0.$$

5.1 NR and Dual Prox with φ-Divergence Distance

In this section we consider the NR method and its dual equivalent—the prox method with φ-divergence distance for the dual problem.

Let $\psi \in \Psi$, $\lambda_0 = e = (1, \ldots, 1) \in \mathbb{R}^m_{++}$ and $k > 0$ are given. The NR step consists of finding the primal minimizer

$$\hat{x} :\equiv \hat{x}(\lambda, k) : \nabla_x \mathscr{L}(\hat{x}, \lambda, k) = 0 \tag{100}$$

following by the Lagrange multipliers update

$$\hat{\lambda} \equiv \hat{\lambda}(\lambda, k) = (\hat{\lambda}_1, \ldots, \hat{\lambda}_m) : \hat{\lambda}_i = \lambda_i \psi'(kc_i(\hat{x})),\ i = 1, \ldots, m. \tag{101}$$

Theorem 7. *If condition A and B hold and f, $c_i \in C^1$, $i = 1, \ldots, m$, then the NR method (100)–(101) is:*

(1) well defined;
(2) equivalent to the following prox method

$$d(\hat{\lambda}) - k^{-1}D(\hat{\lambda}, \lambda) = \max\{d(u) - k^{-1}D(u, \lambda) | u \in \mathbb{R}^m_{++}\}, \tag{102}$$

where $D(u, \lambda) = \sum_{i=1}^m \lambda_i \varphi(u_i/\lambda_i)$ is φ-divergence distance function based on kernel $\varphi = -\psi^*$.

Proof. (1) Due to the properties (1)–(3) of ψ, convexity f and concavity of all c_i, the Lagrangian $\mathscr{L}$ is convex in x. From boundedness of Ω, Slater condition and properties (3) and (5) of ψ follows emptiness of the Ω recession cone. It means that for any nontrivial direction $d \in \mathbb{R}^n$ and any $(\lambda, k) \in \mathbb{R}^{m+1}_{++}$ we have

$$\lim_{t\to\infty} \mathscr{L}(x + td, \lambda, k) = \infty$$

for any $x \in \Omega$. Hence for a given $(\lambda, k) \in \mathbb{R}^{m+1}_{++}$ there exists $\hat{x} \equiv \hat{x}(\lambda, k)$ defined by (100) and $\hat{\lambda} \equiv \hat{\lambda}(\lambda, k)$ defined by (101). Due to 2) of ψ we have $\lambda \in \mathbb{R}^m_{++} \Rightarrow \hat{\lambda} \in \mathbb{R}^m_{++}$, therefore NR method (100)–(101) is well defined.

(2) From (100) and (101) follows

$$\nabla_x \mathscr{L}(\hat{x}, \hat{\lambda}, k) = \nabla f(\hat{x}) - \sum_{i=1}^m \lambda_i \psi'(kc_i(\hat{x}))\nabla c_i(\hat{x}) = \nabla_x L(\hat{x}, \hat{\lambda}) = 0,$$

therefore

$$\min_{x\in\mathbb{R}^n} L(x,\hat{\lambda}) = L(\hat{x},\hat{\lambda}) = d(\hat{\lambda}).$$

The subdifferential $\partial d(\hat{\lambda})$ contains $-c(\hat{x})$, that is

$$0 \in c(\hat{x}) + \partial d(\hat{\lambda}). \tag{103}$$

From (101) follows $\psi'(kc_i(\hat{x})) = \hat{\lambda}_i/\lambda_i,\ i = 1,\dots,m$.

Due to (3) of ψ there exists an inverse ψ'^{-1}. Using LEID we obtain

$$c_i(\hat{x}) = k^{-1}\psi'^{-1}(\hat{\lambda}_i/\lambda_i) = k^{-1}\psi^{*'}(\hat{\lambda}_i/\lambda_i) \tag{104}$$

combining (103) and (104) we have

$$0 \in \partial d(\hat{\lambda}) + k^{-1}\sum_{i=1}^{m}\psi^{*'}\left(\hat{\lambda}_i/\lambda_i\right)e_i. \tag{105}$$

The inclusion (105) is the optimality criteria for $\hat{\lambda}$ to be a solution of problem (102). □

Remark 2. It follows from 1° and 2° of Assertion 2, that for any $k > 0$ we have $x^* = x(\lambda^*, k)$ and $\lambda^* = \lambda(\lambda^*, k)$, that is $\lambda^* \in \mathbb{R}^m_+$ is a fixed point of the mapping $\lambda \to \hat{\lambda}(\lambda, k)$.

Along with the class Ψ of transformations ψ we consider a class Φ of kernels $\varphi = -\psi^*$, with properties induced by properties of ψ. We collect them in the following Assertion.

Assertion 3. *The kernel $\varphi \in \Phi$ are strictly convex on $\mathbb{R}_+$ and possess the following properties on* $]0,\infty[$.

(1) $\varphi(s) \geq 0$, $\min_{s\geq 0}\varphi(s) = \varphi(1) = 0$,
(2) $\varphi'(1) = 0$;
(3) $\varphi''(s) > 0$.

Assertion 3 follows from properties (1)–(3) of ψ and (11).

The general NR scheme and corresponding methods were introduced in the early 1980s (see [46] and references therein). Independently the prox methods with φ-divergence distance has been studied by Teboulle (see [59]). The equivalence of NR and prox methods with φ-divergence distance was established in [50].

In the following section we consider an important particular case of NR—the MBF method.

5.2 Convergence of the MBF Method and its Dual Equivalent

For reasons, which will be clear later, we would like to concentrate on the NR method with transformation $\psi(t) = \ln(t+1)$, which leads to the MBF theory and methods developed in [46] (see also [10, 25, 26, 31, 34, 38, 41, 53] and references therein). The correspondent Lagrangian for the equivalent problem $\mathcal{L}$: $\mathbb{R}^n \times \mathbb{R}^m_+ \times \mathbb{R}_{++} \to \mathbb{R}$ is defined by formula

$$\mathcal{L}(x,\lambda,k) = f(x) - k^{-1}\sum_{i=1}^{m}\lambda_i \ln(kc_i(x)+1).$$

For a given $k>0$ and $\lambda_0 = e = (1,\dots,1) \in \mathbb{R}^m_{++}$ the MBF method generates the following primal–dual sequence $\{x_s,\lambda_s\}_{s=0}^{\infty}$:

$$x_{s+1} : \nabla_x \mathcal{L}(x_{s+1},\lambda_s,k) =$$

$$\nabla f(x_{s+1}) - \sum_{i=1}^{m}\lambda_{i,s}(kc_i(x_{s+1})+1)^{-1}\nabla c_i(x_{s+1}) = 0 \tag{106}$$

$$\lambda_{s+1} : \lambda_{i,s+1} = \lambda_{i,s}(kc(x_{s+1})+1)^{-1},\ i=1,\dots,m. \tag{107}$$

The Hausdorff distance between two compact sets in $\mathbb{R}^m_+$ will be used later.

Let X and Y be two bounded and closed sets in $\mathbb{R}^n$ and $d(x,y) = \|x-y\|$ is the Euclidean distance between $x \in X, y \in Y$. Then the Hausdorff distance between X and Y is defined as follows

$$d_H(X,Y) := \max\{\max_{x\in X}\min_{y\in Y} d(x,y), \max_{y\in Y}\min_{x\in X} d(x,y)\} =$$

$$\max\{\max_{x\in X} d(x,Y), \max_{y\in Y} d(y,X)\}.$$

For any pair of compact sets X and $Y \subset \mathbb{R}^n$

$$d_H(X,Y) = 0 \Leftrightarrow X = Y.$$

Let $Q \subset \mathbb{R}^m_{++}$ be a compact set, $\hat{Q} = \mathbb{R}^m_{++} \setminus Q$, $S(u,\epsilon) = \{v \in \mathbb{R}^m_+ : \|u-v\| \leq \epsilon\}$ and

$$\partial Q = \{u \in Q | \exists v \in Q : v \in S(u,\epsilon), \exists \hat{v} \in \hat{Q} : \hat{v} \in S(u,\epsilon)\}, \forall \epsilon > 0$$

be the boundary of Q.

Let $A \subset B \subset C$ be convex and compact sets in $\mathbb{R}^m_+$. The following inequality follows from the definition of Hausdorff distance.

$$d_H(A,\partial B) < d_H(A,\partial C) \tag{108}$$

Along with the dual sequence $\{\lambda_s\}_{s=0}^{\infty}$ we consider the corresponding convex and bounded level sets $\Lambda_s = \{\lambda \in \mathbb{R}^m_+ : d(\lambda) \geq d(\lambda_s)\}$ and their boundaries $\partial\Lambda_s = \{\lambda \in \Lambda_s : d(\lambda) = d(\lambda_s)\}$.

Theorem 8. *Under condition of Theorem 7 for any given $k > 0$ and any $\lambda_0 \in \mathbb{R}^m_{++}$ the MBF method (106)–(107) generates such primal–dual sequence $\{x_s, \lambda_s\}_{s=0}^{\infty}$ that:*

(1) $d(\lambda_{s+1}) > d(\lambda_s)$, $s \geq 0$
(2) $\lim_{s\to\infty} d(\lambda_s) = d(\lambda^*)$, $\lim_{s\to\infty} f(x_s) = f(x^*)$
(3) $\lim_{s\to\infty} d_H(\partial\Lambda_s, L^*) = 0$
(4) there exists a subsequence $\{s_l\}_{l=1}^{\infty}$ such that for $\bar{x}_l = \sum_{s=s_l}^{s_{l+1}} (s_{l+1} - s_l)^{-1} x_s$ we have $\lim_{l\to\infty} \bar{x}_l = \bar{x} \in X^*$, i.e. the primal sequence converges to the primal solution in the ergodic sense.

Proof. (1) It follows from Theorem 7 that method (106)–(107) is well defined and it is equivalent to following proximal point method

$$d(\lambda_{s+1}) - k^{-1} \sum_{i=1}^{m} \lambda_{i,s} \varphi(\lambda_{i,s+1}/\lambda_{i,s}) =$$

$$\max\{d(u) - k^{-1} \sum_{i=1}^{m} \lambda_{i,s} \varphi(u_i/\lambda_{i,s}) : u \in \mathbb{R}^m_{++}\}, \tag{109}$$

where $\varphi = -\psi^* = -\inf_{t>-1}\{st - \ln(t+1)\} = -\ln s + s - 1$ is the MBF kernel.

The φ-divergence distance function

$$D(\lambda, u) = \sum_{i=1}^{m} \lambda_i \varphi(u_i/\lambda_i) = \sum_{i=1}^{m} [-\lambda_i \ln u_i/\lambda_i + u_i - \lambda_i],$$

which measures the divergence between two vectors λ and u from $\mathbb{R}^m_{++}$ is, in fact, the Kullback-Leibler (KL) distance (see [20, 50, 59]). The MBF kernel $\varphi(s) = -\ln s + s - 1$ is strictly convex on $\mathbb{R}_{++}$ and $\varphi'(1) = 0$, therefore $\min_{s>0} \varphi(s) = \varphi(1) = 0$, also

a) $D(\lambda, u) > 0$, $\forall \lambda \neq u \in \mathbb{R}^m_{++}$
b) $D(\lambda, u) = 0 \Leftrightarrow \lambda = u$.

From (109) for $u = \lambda_s$ follows

$$d(\lambda_{s+1}) \geq d(\lambda_s) + k^{-1} \sum_{i=1}^{m} \lambda_{i,s} \varphi(\lambda_{i,s+1}/\lambda_{i,s}). \tag{110}$$

Therefore the sequence $\{d(\lambda_s)\}_{s=0}^{\infty}$ is monotone increasing, unless $\varphi(\lambda_{i,s+1}/\lambda_{i,s}) = 0$ for all $i = 1, \ldots, m$, but in such case $\lambda_{s+1} = \lambda_s = \lambda^*$. The monotone increasing sequence $\{d(\lambda_s)\}_{s=0}^{\infty}$ is bounded from above by $f(x^*)$, therefore there exists $\lim_{s\to\infty} d(\lambda_s) = \bar{d} \le f(x^*)$.

(2) Our next step is to show that $\bar{d} = f(x^*)$.
From $-c(x_{s+1}) \in \partial d(\lambda_{s+1})$ and concavity of the dual function d follows

$$d(\lambda) - d(\lambda_{s+1}) \le (-c(x_{s+1}), \lambda - \lambda_{s+1}), \ \forall \lambda \in \mathbb{R}^m_{++}.$$

So for $\lambda = \lambda_s$ we have

$$d(\lambda_{s+1}) - d(\lambda_s) \ge (c(x_{s+1}), \lambda_s - \lambda_{s+1}). \tag{111}$$

From the update formula (107) follows

$$(\lambda_{i,s} - \lambda_{i,s+1}) = kc_i(x_{s+1})\lambda_{i,s+1}, \ i = 1, \ldots, m, \tag{112}$$

therefore from (111) and (112) we have

$$d(\lambda_{s+1}) - d(\lambda_s) \ge k \sum_{i=1}^{m} c_i^2(x_{s+1})\lambda_{i,s+1}. \tag{113}$$

From Slater condition follows boundedness of L^*. Therefore from concavity d follows boundedness of the dual level set

$$\Lambda(\lambda_0) = \{\lambda \in \mathbb{R}^m_+ : d(\lambda) \ge d(\lambda_0)\}.$$

It follows from the dual monotonicity (110) that the dual sequence $\{\lambda_s\}_{s=0}^{\infty} \in \Lambda(\lambda_0)$ is bounded.
Therefore there exists $L > 0 : \max\limits_{i,s} \lambda_{i,s} = L$. From (113) follows

$$d(\lambda_{s+1}) - d(\lambda_s) \ge kL^{-1}(c(x_{s+1}), \lambda_{s+1})^2. \tag{114}$$

By summing up (114) from $s = 1$ to $s = N$ we obtain

$$d(\lambda^*) - d(\lambda_0) \ge d(\lambda_{N+1}) - d(\lambda_0) > kL^{-1} \sum_{s=1}^{N} (\lambda_s, c(x_s))^2,$$

which leads to asymptotic complementarity condition

$$\lim_{s\to\infty} (\lambda_s, c(x_s)) = 0. \tag{115}$$

On the other hand, from (110) follows

$$d(\lambda^*) - d(\lambda_0) \geq d(\lambda_N) - d(\lambda_0) \geq k^{-1} \sum_{s=1}^{N} D(\lambda_s, \lambda_{s+1}). \tag{116}$$

Therefore $\lim_{s\to\infty} D(\lambda_s, \lambda_{s+1}) = 0$, which means that divergence (entropy) between two sequential LM vectors asymptotically disappears, that is the dual sequence converges to the fixed point of the map $\lambda \to \hat{\lambda}(\lambda, k)$, which duc to Remark 2, is λ^*.

We need few more steps to prove it. Let us show first that

$$D(\lambda^*, \lambda_s) > D(\lambda^*, \lambda_{s+1}), \ \forall s \geq 0 \tag{117}$$

unless $\lambda_s = \lambda_{s+1} = \lambda^*$.

We assume $x \ln x = 0$ for $x = 0$, then

$$D(\lambda^*, \lambda_s) - D(\lambda^*, \lambda_{s+1}) = \sum_{i=1}^{m} \left(\lambda_i^* \ln \frac{\lambda_{i,s+1}}{\lambda_{i,s}} + \lambda_{i,s} - \lambda_{i,s+1} \right).$$

Invoking the update formula (107) we obtain

$$D(\lambda^*, \lambda_s) - D(\lambda^*, \lambda_{s+1}) = \sum_{i=1}^{m} \lambda_i^* \ln(kc_i(x_{s+1}) + 1)^{-1} + k \sum_{i=1}^{m} \lambda_{i,s+1} c_i(x_{s+1}).$$

Keeping in mind $\ln(1+t)^{-1} = -\ln(1+t) \geq -t$ we have

$$D(\lambda^*, \lambda_s) - D(\lambda^*, \lambda_{s+1}) \geq k \sum_{i=1}^{m} (\lambda_{i,s+1} - \lambda_i^*) c_i(x_{s+1}) =$$

$$k(-c(x_{s+1}), \lambda^* - \lambda_{s+1}). \tag{118}$$

From concavity d and $-c(x_{s+1}) \in \partial d(\lambda_{s+1})$ follows

$$0 \leq d(\lambda^*) - d(\lambda_{s+1}) \leq (-c(x_{s+1}), \lambda^* - \lambda_{s+1}). \tag{119}$$

Combining (118) and (119) we obtain

$$D(\lambda^*, \lambda_s) - D(\lambda^*, \lambda_{s+1}) \geq k(d(\lambda^*) - d(\lambda_{s+1})) > 0. \tag{120}$$

Assuming that $d(\lambda^*) - \bar{d} = \rho > 0$ and summing up the last inequality from $s = 0$ to $s = N$ we obtain $D(\lambda^*, \lambda_0) \geq kN\rho$, which is impossible for $N > 0$ large enough.

Therefore $\lim_{s\to\infty} d(\lambda_s) = \bar{d} = d(\lambda^*)$, which together with asymptotic complementarity (115) leads to

$$d(\lambda^*) = \lim_{s\to\infty} d(\lambda_s) = \lim_{s\to\infty} [f(x_s) - (\lambda_s, c(x_s))] =$$

$$\lim_{s\to\infty} f(x_s) = f(x^*). \tag{121}$$

(3) The dual sequence $\{\lambda_s\}_{s=0}^{\infty}$ is bounded, so it has a converging subsequence $\{\lambda_{s_i}\}_{i=0}^{\infty}$: $\lim_{i\to\infty} \lambda_{s_i} = \bar{\lambda}$. It follows from the dual convergence in value that $\bar{\lambda} = \lambda^* \in L^*$, therefore $\{\lambda \in \mathbb{R}^m_+ : d(\lambda) = d(\bar{\lambda})\} = L^*$.
From (110) follows $L^* \subset \ldots \subset \Lambda_{s+1} \subset \Lambda_s \subset \ldots \subset \Lambda_0$, therefore from (108) we obtain a monotone decreasing sequence $\{d_H(\partial\Lambda_s, L^*)\}_{s=0}^{\infty}$, which has a limit, that is

$$\lim_{s\to\infty} d_H(\partial\Lambda_s, L^*) = \rho \geq 0,$$

but $\rho > 0$ is impossible due to the continuity of the dual function and the convergence of the dual sequence in value.

(4) Let us consider the indices subset $I_+ = \{i : \bar{\lambda}_i > 0\}$, then from (115) we have $\lim_{s\to\infty} c_i(x_s) = c_i(\bar{x}) = 0, \quad i \in I_+$. Now we consider the indices subset $I_0 = \{i : \bar{\lambda}_i = 0\}$.
There exists a subsequence $\{\lambda_{s_l}\}_{l=1}^{\infty}$ that $\lambda_{i,s_{l+1}} \leq 0.5\lambda_{i,s_l}$, $i \in I_0$.
Using again the update formula (107) we obtain

$$\lambda_{s_{l+1}} \prod_{s=s_l}^{s_{l+1}} (kc_i(x_s) + 1) = \lambda_{i,s_l} \geq 2\lambda_{s_{l+1}}, \; i \in I_0.$$

Invoking the arithmetic-geometric means inequality we have

$$\frac{1}{s_{l+1} - s_l} \sum_{s=s_l}^{s_{l+1}} (kc_i(x_s) + 1) \geq \left(\prod_{s=s_l+1}^{s_{l+1}} (kc_i(x_s) + 1) \right)^{1/(s_{l+1}-s_l)} \geq 2^{(1/s_{l+1}-s_l)} > 1.$$

Therefore

$$\frac{k}{(s_{l+1} - s_l)} \sum_{s=s_l}^{s_{l+1}} c_i(x_s) > 0 \; i \in I_0.$$

From concavity c_i we obtain

$$c_i(\bar{x}_{l+1}) = c_i \left(\sum_{s=s_l+1}^{s_{l+1}} \frac{1}{s_{l+1} - s_l} x_s \right) \geq \frac{1}{s_{l+1} - s_l} \sum_{s=s_l+1}^{s_{l+1}} c_i(x_s) > 0, \; i \in I_0. \tag{122}$$

On the other hand, from convexity of f we have

$$f(\bar{x}_{l+1}) \leq \frac{1}{s_{l+1} - s_l} \sum_{s=s_l+1}^{s_{l+1}} f(x_s). \tag{123}$$

Without loosing generality we can assume that $\lim_{l \to \infty} \bar{x}_l = \bar{x} \in \Omega$. It follows from (121) that

$$f(\bar{x}) = \lim_{l \to \infty} f(\bar{x}_l) \leq \lim_{s \to \infty} f(x_s) = \lim_{s \to \infty} d(\lambda_s) = d(\lambda^*) = f(x^*).$$

Thus $f(\bar{x}) = f(x^*) = d(\lambda^*) = d(\bar{\lambda})$ and $\bar{x} = x^*$, $\bar{\lambda} = \lambda^*$. The proof of Theorem 8 is completed. □

We conclude the section with few remarks.

Remark 3. Each $\psi \in \Psi$ leads to a particular NR method for solving (51) as well as to an interior prox method for solving the dual problem (53). In this regard NR approach is source of methods for solving (53), which arises in a number of application such as non-negative least square, statistical learning theory, image space reconstruction, maximum likelihood estimation in emission tomography (see [17, 20, 62] and references therein).

Remark 4. The MBF method leads to the multiplicative method (107) for the dual problem. If the dual function d has a gradient, then $\nabla d(\lambda_{s+1}) = -c(x_{s+1})$. Formulas (107) can be rewritten as follows

$$\lambda_{i,s+1} - \lambda_{i,s} = k\lambda_{i,s+1}[\nabla d(\lambda_{s+1})],\ i = 1, \ldots, m, \tag{124}$$

which is, in fact, implicit Euler method for the following system of ordinary differential equations

$$\frac{d\lambda}{dt} = k\lambda \nabla d(\lambda),\ \lambda(0) = \lambda_0. \tag{125}$$

Therefore the dual MBF method (124) is called (see (1.7) in [20]) implicit multiplicative algorithm.

The explicit multiplicative algorithm (see (1.8) in [20]) is given by the following formula

$$\lambda_{i,s+1} = \lambda_{i,s}(1 - k[\nabla d(\lambda_s)]_i)^{-1},\ i = 1, \ldots, m. \tag{126}$$

It has been used by Eggermond [20] for solving non-negative least square, by Daube-Witherspoon and Muchlehner [17] for image space reconstruction (ISRA) and by Shepp and Vardi in their EM method for finding maximum likelihood estimation in emission tomography [62].

Remark 5. Under the standard second order sufficient optimality condition there exists $k_0 > 0$ that for $k \geq k_0$ the MBF method (106)–(107) converges with linear rate

$$\|x_{s+1} - x^*\| \leq \frac{c}{k}\|\lambda_s - \lambda^*\|; \; \|\lambda_{s+1} - \lambda^*\| \leq \frac{c}{k}\|\lambda_s - \lambda^*\|$$

and $c > 0$ is independent on $k \geq k_0$. By increasing k from step to step one obtains superlinear convergence rate (see [46]).

6 Lagrangian Transformation and Interior Ellipsoid Methods

The Lagrangian transformation (LT) scheme employs a class $\psi \in \Psi$ of smooth strictly concave, monotone increasing functions to transform terms of the Classical Lagrangian associated with constraints. The transformation is scaled by a positive scaling parameter.

Finding a primal minimizer of the transformed Lagrangian following by the Lagrange multipliers update leads to a new class of multipliers methods.

The LT methods are equivalent to proximal point methods with Bregman or Bregman type distance function for the dual problem. The kernel of the correspondent distance is $\varphi = -\psi^*$.

Each dual prox, in turn, is equivalent to an interior ellipsoid methods. In case of the MBF transformation $\psi(t) = \ln(t + 1)$ the dual prox is based on Bregman distance $B(u, v) = \sum_{i=1}^{m}(-\ln(u_i/v_i) + u_i/v_i - 1)$ with MBF kernel $\varphi = -\psi^* = -\ln s + s - 1$, which is SC function. Therefore the interior ellipsoids are Dikin's ellipsoids (see [18, 37, 42, 43, 49]).

Application of LT with MBF transformation for LP leads to Dikin's affine scaling type method for the dual LP.

6.1 Lagrangian Transformation

We consider a class Ψ of twice continuous differentiable functions $\psi : R \to R$ with the following properties

(1) $\psi(0) = 0$
(2) a) $\psi'(t) > 0$, b) $\psi'(0) = 1$, $\psi'(t) \leq at^{-1}, a > 0, t > 0$
(3) $-m_0^{-1} \leq \psi''(t) < 0$, $\forall t \in]-\infty, \infty[$
(4) $\psi''(t) \leq -M^{-1}$, $\forall t \in]-\infty, 0[$, $0 < m_0 < M < \infty$.

For a given $\psi \in \Psi$ and $k > 0$, the LT $\mathscr{L} : \mathbb{R}^n \times \mathbb{R}^m_+ \times \mathbb{R}_{++} \to \mathbb{R}$ is defined by the following formula

$$\mathscr{L}(x, \lambda, k) := f(x) - k^{-1} \sum_{i=1}^{m} \psi(k\lambda_i c_i(x)). \tag{127}$$

It follows from (2a) and (3), convexity f, concavity c_i, $i = 1, \ldots, m$ that for any given $\lambda \in \mathbb{R}^m_{++}$ and any $k > 0$ the LT is convex in x.

6.2 Primal Transformations and Dual Kernels

The well known transformations

- exponential [7, 40, 61] $\hat{\psi}_1(t) = 1 - e^{-t}$;
- logarithmic MBF [46] $\hat{\psi}_2(t) = \ln(t + 1)$;
- hyperbolic MBF [46] $\hat{\psi}_3(t) = t/(t + 1)$;
- log-sigmoid [48] $\hat{\psi}_4(t) = 2(\ln 2 + t - \ln(1 + e^t))$;
- Chen-Harker-Kanzow-Smale [48] (CHKS) $\hat{\psi}_5(t) = t - \sqrt{t^2 + 4\eta} + 2\sqrt{\eta}$, $\eta > 0$, unfortunately, do not belong to Ψ.

The transformations $\hat{\psi}_1, \hat{\psi}_2, \hat{\psi}_3$ do not satisfy (3) ($m_0 = 0$), while transformations $\hat{\psi}_4$ and $\hat{\psi}_5$ do not satisfy (4) ($M = \infty$). A slight modification of $\hat{\psi}_i, i = 1, \ldots, 5$, however, leads to $\psi_i \in \Psi$ (see [6]).

Let $-1 < \tau < 0$, we will use later the following truncated transformations $\psi_i : \mathbb{R} \to \mathbb{R}$ are defined as follows

$$\psi_i(t) := \begin{cases} \hat{\psi}_i(t), \infty > t \geq \tau \\ q_i(t), -\infty < t \leq \tau, \end{cases} \tag{128}$$

where $q_i(t) = a_i t^2 + b_i t + c_i$ and $a_i = 0.5\hat{\psi}_i''(\tau)$, $b_i = \hat{\psi}_i'(\tau) - \tau\hat{\psi}''(\tau)$, $c_i = \hat{\psi}_i'(\tau) - \tau\hat{\psi}_i'(\tau) + 0.5\tau^2\hat{\psi}_i''(\tau)$.

It is easy to check that for truncated transformations ψ_i, $i = 1, \ldots, 5$ the properties (1)–(4) hold, that is $\psi_i \in \Psi$.

In the future along with transformations $\psi_i \in \Psi$ their conjugate

$$\psi_i^*(s) := \begin{cases} \hat{\psi}_i^*(s), & s \leq \hat{\psi}_i'(\tau) \\ q_i^*(s) = (4a_i)^{-1}(s - b_i)^2 - c_i, & s \geq \hat{\psi}_i'(\tau), i = 1, \ldots, 5, \end{cases} \tag{129}$$

will play an important role, where $\hat{\psi}^*{}_i(s) = \inf_t\{st - \hat{\psi}_i(t)\}$ is the LET of $\hat{\psi}_i$.

With the class of primal transformations Ψ we associate the class of dual kernels

$$\varphi \in \Phi = \{\varphi = -\psi^* : \psi \in \Psi\}.$$

Using properties (2) and (4) one can find $0 < \theta_i < 1$ that

$$\hat{\psi}_i'(\tau) - \hat{\psi}_i'(0) = -\hat{\psi}_i''(\tau\theta_i)(-\tau) \geq -\tau M^{-1}, i = 1, \ldots, 5$$

or

$$\hat{\psi}_i'(\tau) \geq 1 - \tau M^{-1} = 1 + |\tau| M^{-1}.$$

Therefore from (129) for any $0 < s \leq 1 + |\tau|M^{-1}$ we have

$$\varphi_i(s) = \hat{\varphi}_i(s) = -\hat{\psi}_i^*(s) = \inf_t \{st - \hat{\psi}_i(t)\}, \tag{130}$$

where kernels

- exponential $\hat{\varphi}_1(s) = s \ln s - s + 1, \hat{\varphi}_1(0) = 1$;
- logarithmic MBF $\hat{\varphi}_2(s) = -\ln s + s - 1$;
- hyperbolic MBF $\hat{\varphi}_3(s) = -2\sqrt{s} + s + 1, \hat{\varphi}_3(0) = 1$;
- Fermi-Dirac $\hat{\varphi}_4(s) = (2 - s)\ln(2 - s) + s \ln s, \hat{\varphi}_4(0) = 2 \ln 2$;
- CMKS $\hat{\varphi}_5(s) = -2\sqrt{\eta}(\sqrt{(2 - s)s} - 1), \hat{\varphi}_5(0) = 2\sqrt{\eta}$
are infinitely differentiable on $]0, 1 + |\tau|M^{-1}[$.

To simplify the notations we omit indices of ψ and φ.

The properties of kernels $\varphi \in \Phi$ induced by (1)–(4) can be established by using (11).

We collect them in the following Assertion

Assertion 4. *The kernels $\varphi \in \Phi$ are strictly convex on $\mathbb{R}^m_+$, twice continuously differentiable and possess the following properties*

(1) $\varphi(s) \geq 0, \ \forall s \in]0, \infty[$ *and* $\min_{s \geq 0} \varphi(s) = \varphi(1) = 0$;

(2) a) $\lim_{s \to 0+} \varphi'(s) = -\infty$, *b)* $\varphi'(s)$ *is monotone increasing and*
c) $\varphi'(1) = 0$;

(3) a) $\varphi''(s) \geq m_0 > 0, \quad \forall s \in]0, \infty[$, *b)* $\varphi''(s) \leq M < \infty, \quad \forall s \in [1, \infty[$.

Let $Q \subset \mathbb{R}^m$ be an open convex set, $\hat{Q}$ is the closure of Q and $\varphi : \hat{Q} \to \mathbb{R}$ be a strictly convex closed function on $\hat{Q}$ and continuously differentiable on Q, then the Bregman distance $\mathbb{B}_\varphi : \hat{Q} \times Q \to R_+$ induced by φ is defined as follows (see [8]),

$$\mathbb{B}_\varphi(x, y) = \varphi(x) - \varphi(y) - (\nabla\varphi(y), x - y). \tag{131}$$

Let $\varphi \in \Phi$, then $B_\varphi : \mathbb{R}^m_+ \times \mathbb{R}^m_{++} \to \mathbb{R}_+$, defined by

$$B_\varphi(u, v) := \sum_{i=1}^m \varphi(u_i/v_i),$$

we call Bregman type distance induced by kernel φ. Due to $\varphi(1) = \varphi'(1) = 0$ for any $\varphi \in \Phi$, we have

$$\varphi(t) = \varphi(t) - \varphi(1) - \varphi'(1)(t-1), \tag{132}$$

which means that $\varphi(t) : \mathbb{R}_{++} \to \mathbb{R}_{++}$ is Bregman distance between $t > 0$ and 1.

By taking $t_i = \frac{u_i}{v_i}$ from (132) we obtain

$$B_\varphi(u, v) = B_\varphi(u, v) - B_\varphi(v, v) - (\nabla_u B_\varphi(v, v), u - v), \tag{133}$$

which justifies the definition of the Bregman type distance.

For the MBF kernel $\varphi_2(s) = -\ln s + s - 1$ we obtain the Bregman distance,

$$\mathbb{B}_2(u, v) = \sum_{i=1}^m \varphi_2(u_i/v_i) = \sum_{i=1}^m (-\ln u_i/v_i + u_i/v_i - 1) = \sum_{i=1}^m [-\ln u_i + \ln v_i + (u_i - v_i)/v_i], \tag{134}$$

which is induced by the standard log-barrier function $F(t) = -\sum_{i=1}^m \ln t_i$.

After Bregman's introduction his function in the 1960s (see [8]) the prox method with Bregman distance has been widely studied (see [9, 11, 13, 15, 19, 37, 48–50] and reference therein).

From the definition of $\mathbb{B}_2(u, v)$ follows

$$\nabla_u \mathbb{B}_2(u, v) = \nabla F(u) - \nabla F(v).$$

For $u \in \hat{Q}$, $v \in Q$ and $w \in Q$ the following three point identity established by Chen and Teboulle in [15] is an important element in the analysis of prox methods with Bregman distance

$$\mathbb{B}_2(u, v) - \mathbb{B}_2(u, w) - \mathbb{B}_2(w, v) = (\nabla F(v) - \nabla F(w), w - u). \tag{135}$$

The properties of Bregman's type distance functions we collect in the following Assertion.

Assertion 5. *The Bregman type distance satisfies the following properties:*

(1) $B_\varphi(u, v) \geq 0, \forall u \in \mathbb{R}^m_+, v \in \mathbb{R}^m_{++}, B_\varphi(u, v) = 0 \Leftrightarrow u = v, \ \forall v, u \in \mathbb{R}^m_{++}$; $B_\varphi(u, v) > 0$ *for any* $u \neq v$
(2) $B_\varphi(u, v) \geq \frac{1}{2} m_0 \sum_{i=1}^m (\frac{u_i}{v_i} - 1)^2, \forall u_i \in \mathbb{R}^m_+, v_i \in \mathbb{R}^m_{++}$;
(3) $B_\varphi(u, v) \leq \frac{1}{2} M \sum_{i=1}^m (\frac{u_i}{v_i} - 1)^2, \forall u \in \mathbb{R}^m_+, u \geq v > 0$;
(4) for any fixed $v \in \mathbb{R}^m_{++}$ *the gradient* $\nabla_u B_\varphi(u, v)$ *is a barrier function of* $u \in \mathbb{R}^m_{++}$, *i.e.*

$$\lim_{u_i \to 0_+} \frac{\partial}{\partial u_i} B_\varphi(u, v) = -\infty, i = 1, \ldots, m.$$

The properties (1)–(4) directly following from the properties of kernels $\varphi \in \Phi$ given in Assertion 4.

6.3 Primal LT and Dual Prox Methods

Let $\psi \in \Psi$, $\lambda_0 \in \mathbb{R}^m_{++}$ and $k > 0$ are given. The LT method generates a primal–dual sequence $\{x_s, \lambda_s\}_{s=1}^{\infty}$ by formulas

$$x_{s+1} : \nabla_x \mathscr{L}(x_{s+1}, \lambda_s, k) = 0 \tag{136}$$

$$\lambda_{i,s+1} = \lambda_{i,s}\psi'(k\lambda_{i,s}c_i(x_{s+1})), i = 1, \dots, m. \tag{137}$$

Theorem 9. *If conditions A and B hold and f, c_i, $i = 1, \dots, m$ continuously differentiable then:*

(1) *the LT method (136)–(137) is well defined and it is equivalent to the following interior proximal point method*

$$\lambda_{s+1} = \arg\max\{d(\lambda) - k^{-1}B_{\varphi}(\lambda, \lambda_s) | \lambda \in \mathbb{R}^m_{++}\}, \tag{138}$$

where

$$B_{\varphi}(u, v) := \sum_{i=1}^{m} \varphi(u_i/v_i)$$

and $\varphi = -\psi^*$.

(2) *for all $i = 1, \dots, m$ we have*

$$\lim_{s\to\infty} (\lambda_{i,s+1}/\lambda_{i,s}) = 1. \tag{139}$$

Proof. (1) From assumptions A, convexity of f, concavity of c_i, $i = 1, \dots, m$ and property (4) of $\psi \in \Psi$ for any $\lambda_s \in \mathbb{R}^m_{++}$ and $k > 0$ follows boundedness of the level set $\{x : \mathscr{L}(x, \lambda_s, k) \le \mathscr{L}(x_s, \lambda_s, k)\}$. Therefore, the minimizer x_s exists for any $s \ge 1$. It follows from property (2a) of $\psi \in \Psi$ and (137) that $\lambda_s \in \mathbb{R}^m_{++} \Rightarrow \lambda_{s+1} \in \mathbb{R}^m_{++}$. Therefore the LT method (136)–(137) is well defined. From (136) follows

$$\nabla_x \mathscr{L}(x_{s+1}, \lambda_s, k) =$$

$$\nabla f(x_{s+1}) - \sum_{i=1}^{m} \lambda_{i,s}\psi'(k\lambda_{i,s}c_i(x_{s+1}))\nabla c_i(x_{s+1})) = 0. \tag{140}$$

From (136) and (137) we obtain

$$\nabla_x \mathscr{L}(x_{s+1}, \lambda_s, k) = \nabla f(x_{s+1}) - \sum_{i=1}^{m} \lambda_{i,s+1}\nabla c_i(x_{s+1}) = \nabla_x L(x_{s+1}, \lambda_{s+1}) = 0,$$

therefore

$$d(\lambda_{s+1}) = L(x_{s+1}, \lambda_{s+1}) = \min\{L(x, \lambda_{s+1})|x \in \mathbb{R}^n\}.$$

From (137) we get

$$\psi'(k\lambda_{i,s}c_i(x_{s+1})) = \lambda_{i,s+1}/\lambda_{i,s}, i = 1, \dots, m.$$

In view of property (3) for any $\psi \in \Psi$ there exists an inverse ψ'^{-1}, therefore

$$c_i(x_{s+1}) = k^{-1}(\lambda_{i,s})^{-1}\psi'^{-1}(\lambda_{i,s+1}/\lambda_{i,s}), i = 1, \dots, m. \tag{141}$$

Using LEID $\psi'^{-1} = \psi^{*\prime}$ we obtain

$$c_i(x_{s+1}) = k^{-1}(\lambda_{i,s})^{-1}\psi^{*\prime}(\lambda_{i,s+1}/\lambda_{i,s}), \ \ i = 1, \dots, m. \tag{142}$$

Keeping in mind

$$-c(\lambda_{s+1}) \in \partial d(\lambda_{s+1})$$

and $\varphi = -\psi^*$ we have

$$0 \in \partial d(\lambda_{s+1}) - k^{-1}\sum_{i=1}^{m}(\lambda_{i,s})^{-1}\varphi'(\lambda_{i,s+1}/\lambda_{i,s})e_i.$$

The last inclusion is the optimality criteria for $\lambda_{s+1} \in \mathbb{R}^m_{++}$ to be the solution of the problem (138). Thus, the LT method (136)–(137) is equivalent to the interior proximal point method (138).

(2) From (1) of Assertion 5 and (138) follows

$$d(\lambda_{s+1}) \geq k^{-1}B_\varphi(\lambda_{s+1}, \lambda_s) + d(\lambda_s) > d(\lambda_s), \ \ \forall s > 0. \tag{143}$$

Summing up last inequality from $s = 0$ to $s = N$, we obtain

$$d(\lambda^*) - d(\lambda_0) \geq d(\lambda_{N+1}) - d(\lambda_0) > k^{-1}\sum_{s=0}^{N} B_\varphi(\lambda_{s+1}, \lambda_s),$$

therefore

$$\lim_{s\to\infty} B(\lambda_{s+1}, \lambda_s) = \lim_{s\to\infty}\sum_{i=1}^{m}\varphi(\lambda_{i,s+1}/\lambda_{i,s}) = 0. \tag{144}$$

From (144) and 2) of Assertion 5 follows

$$\lim_{s\to\infty} \lambda_{i,s+1}/\lambda_{i,s} = 1, \ i = 1, \dots, m. \tag{145}$$

□

Remark 6. From (130) and (145) follows that for $s \geq s_0 > 0$ the Bregman type distance functions B_{φ_i} used in (138) are based on kernels φ_i, which correspond to the original transformations $\hat{\psi}_i$.

The following Theorem establishes the equivalence of LT multipliers method and interior ellipsoid methods (IEMs) for the dual problem.

Theorem 10. *It conditions of Theorem 9 are satisfied then:*

(1) for a given $\varphi \in \Phi$ there exists a diagonal matrix $H_\varphi = diag(h^i_\varphi)_{i=1}^m$ with $h^i_\varphi > 0, i = 1, \dots, m$ that $B_\varphi(u, v) = \frac{1}{2}\|u - v\|^2_{H_\varphi}$, where $\|w\|^2_{H_\varphi} = w^T H_\varphi w$;
(2) The Interior Prox method (138) is equivalent to an interior quadratic prox (IQP) in the rescaled from step to step dual space, i.e.

$$\lambda_{s+1} = \arg\max\{d(\lambda) - \frac{1}{2k}\|\lambda - \lambda_s\|^2_{H^s_\varphi} | \lambda \in \mathbb{R}^m_+\}, \tag{146}$$

where $H^s_\varphi = diag(h^{i,s}_\varphi) = diag(2\varphi''(1 + \theta^s_i(\lambda_{i,s+1}/\lambda_{i,s} - 1))(\lambda_{i,s})^{-2})$ and $0 < \theta^s_i < 1$;
(3) The IQP is equivalent to an interior ellipsoid method (IEM) for the dual problem;
(4) There exists a converging to zero sequence $\{r_s > 0\}_{s=0}^\infty$ and step $s_0 > 0$ such that, for $\forall s \geq s_0$, the LT method (136)–(137) with truncated MBF transformation $\psi_2(t)$ is equivalent to the following IEM for the dual problem

$$\lambda_{s+1} = \arg\max\{d(\lambda) | \lambda \in E(\lambda_s, r_s)\}, \tag{147}$$

where $H_s = diag(\lambda_{i,s})_{i=1}^m$ and

$$E(\lambda_s, r_s) = \{\lambda : (\lambda - \lambda_s)^T H_s^{-2}(\lambda - \lambda_s) \leq r_s^2\}$$

is Dikin's ellipsoid associated with the standard log-barrier function $F(\lambda) = -\sum_{i=1}^m \ln \lambda_i$ for the dual feasible set $\mathbb{R}^m_+$.

Proof. (1) It follows from $\varphi(1) = \varphi'(1) = 0$ that

$$B_\varphi(u, v) = \frac{1}{2}\sum_{i=1}^m \varphi''(1 + \theta_i(\frac{u_i}{v_i} - 1))(\frac{u_i}{v_i} - 1)^2, \tag{148}$$

where $0 < \theta_i < 1$, $i = 1, \dots, m$.
Due to (3a) from Assertion 4, we have $\varphi''(1 + \theta_i(\frac{u_i}{v_i} - 1)) \geq m_0 > 0$, and due to property (2a) of $\psi \in \Psi$, we have $v \in \mathbb{R}^m_{++}$, therefore

$$h^i_\varphi = 2\varphi''(1 + \theta_i(\frac{u_i}{v_i} - 1))v_i^{-2} > 0, i = 1, \dots, m.$$

We consider the diagonal matrix $H_\varphi = diag(h^i_\varphi)^m_{i=1}$, then from (148) we have

$$B_\varphi(u,v) = \frac{1}{2}\|u-v\|^2_{H_\varphi}. \quad (149)$$

(2) By taking $u=\lambda$, $v=\lambda_s$ and $H_\varphi = H^s_\varphi$ from (138) and (149) we obtain (146)
(3) Let's consider the optimality criteria for the problem (146). Keeping in mind $\lambda_{s+1} \in \mathbb{R}^m_{++}$ we conclude that λ_{s+1} is an unconstrained maximizer in (146). Therefore one can find $g_{s+1} \in \partial d(\lambda_{s+1})$ that

$$g_{s+1} - k^{-1}H^s_\varphi(\lambda_{s+1}-\lambda_s) = 0. \quad (150)$$

Let $r_s = \|\lambda_{s+1}-\lambda_s\|_{H^s_\varphi}$, we consider an ellipsoid

$$E_\varphi(\lambda_s, r_s) = \{\lambda : (\lambda-\lambda_s)^T H^s_\varphi(\lambda-\lambda_s) \le r^2_s\}$$

with center $\lambda_s \in \mathbb{R}^m_{++}$ and radius r_s. It follows from 4) of Assertion 5 that $E(\lambda_s, r_s)$ is an interior ellipsoid in $\mathbb{R}^m_{++}$, i.e. $E_\varphi(\lambda_s, r_s) \subset \mathbb{R}^m_{++}$.
Moreover $\lambda_{s+1} \in \partial E_\varphi(\lambda_s, r_s) = \{\lambda : (\lambda-\lambda_s)^T H^s_\varphi(\lambda-\lambda_s) = r^2_s\}$, therefore (150) is the optimality condition for the following optimization problem

$$d(\lambda_{s+1}) = \max\{d(\lambda) | \lambda \in E_\varphi(\lambda_s, r_s)\} \quad (151)$$

and $(2k)^{-1}$ is the optimal Lagrange multiplier for the only constraint in (151). Thus, the Interior Prox method (138) is equivalent to the IEM (151).
(4) Let us consider the LT method (136)–(137) with truncated MBF transformation.
From (139) follows that for $s \ge s_0$ only Bregman distance

$$\mathbb{B}_2(\lambda,\lambda_s) = \sum_{i=1}^m \left(-ln\frac{\lambda_i}{\lambda^s_i} + \frac{\lambda_i}{\lambda^s_i} - 1\right)$$

is used in the LT method (136)–(137). Then

$$\nabla^2_{\lambda\lambda}\mathbb{B}_2(\lambda,\lambda_s)|_{\lambda=\lambda_s} = H^{-2}_s = (I\cdot\lambda_s)^{-2}.$$

In view of $\mathbb{B}_2(\lambda_s,\lambda_s) = 0$ and $\nabla_\lambda \mathbb{B}_2(\lambda_s,\lambda_s) = 0^m$, we obtain

$$\mathbb{B}_2(\lambda,\lambda_s) = \frac{1}{2}(\lambda-\lambda_s)^T H^{-2}_s(\lambda-\lambda_s) + o(\|\lambda-\lambda_s\|^2) =$$

$$= Q(\lambda,\lambda_s) + o(\|\lambda-\lambda_s\|^2).$$

It follows from (139) that for a any $s \geq s_0$ the term $o(\|\lambda_{s+1} - \lambda_s\|^2)$ can be ignored. Then the optimality criteria (150) can be rewritten as follows

$$g_{s+1} - k^{-1}H_s^{-2}(\lambda_{s+1} - \lambda_s) = 0.$$

Therefore

$$d(\lambda_{s+1}) = \max\{d(\lambda)|\lambda \in E(\lambda_s, r_s)\},$$

where $r_s^2 = Q(\lambda_{s+1}, \lambda_s)$ and

$$E(\lambda_s, r_s) = \{\lambda : (\lambda - \lambda_s)H_s^{-2}(\lambda - \lambda_s) = r_s^2\}$$

is Dikin's ellipsoid. The proof is completed □

The results of Theorem 10 were used in [49] for proving convergence LT method (136)–(137) and its dual equivalent (138) for Bregman type distance function.

Now we consider the LT method with truncated MBF transformation ψ_2.

It follows from (130) and (139) that for $s \geq s_0$ only original transformation $\psi_2(t) = \ln(t+1)$ is used in LT method (136)–(137), therefore only Bregman distance $\mathbb{B}_2(u, v) = \sum_{i=1}^m(-\ln(u_i/v_i) + u_i/v_i - 1)$ is used in the prox method (138).

In other words, for a given $k > 0$ the primal–dual sequence $\{x_s, \lambda_s\}_{s=s_0}^\infty$ is generated by the following formulas

$$x_{s+1} : \nabla_k \mathscr{L}(x_{s+1}, \lambda_s, k) =$$
$$\nabla f(x_{s+1}) - \sum_{i=1}^m \lambda_{i,s}(1 + k\lambda_{i,s}c_i(x_{s+1}))^{-1}\nabla c_i(x_{s+1}) = 0 \tag{152}$$

$$\lambda_{s+1} : \lambda_{i,s+1} = \lambda_{i,s}(1 + k\lambda_{i,s}c_i(x_{s+1}))^{-1},\ i = 1, \ldots, m. \tag{153}$$

The method (152)–(153) Matioti and Gonzaga called M^2BF (see [37]).

Theorem 11. *Under condition of Theorem 9 the M^2BF method generates such primal–dual sequence that:*

(1) $d(\lambda_{s+1}) > d(\lambda_s),\ s \geq s_0$
(2) *a)* $\lim_{s\to\infty} d(\lambda_s) = d(\lambda^*)$; *b)* $\lim_{s\to\infty} f(x_s) = f(x^*)$ *and*

$$c) \quad \lim_{s\to\infty} d_H(\partial \Lambda_s, L^*) = 0$$

(3) there is a subsequence $\{s_l\}_{l=1}^\infty$ *that for* $\bar{\lambda}_{i,s} = \lambda_{i,s}\left(\sum_{s=s_l}^{s_{l+1}} \lambda_{i,s}\right)^{-1}$ *the sequence* $\{\bar{x}_{l+1} = \sum_{s=s_l}^{s_{l+1}} \bar{\lambda}_{i,s}x_s\}_{l=0}^\infty$ *converges and* $\lim_{l\to\infty} \bar{x}_l = \bar{x} \in X^*$.

Proof. (1) From Theorem 10 follows that LT (152)–(153) is equivalent to the prox method (138) with Bregman distance. From (138) with $\lambda = \lambda_s$ we obtain

$$d(\lambda_{s+1}) \geq d(\lambda_s) + k^{-1} \sum_{i=1}^{m} (-\ln(\lambda_{i,s+1}/\lambda_{i,s}) + \lambda_{i,s+1}/\lambda_{i,s} - 1). \tag{154}$$

The Bregman distance is strictly convex in u, therefore from (154) follows $d(\lambda_{s+1}) > d(\lambda_s)$ unless $\lambda_{s+1} = \lambda_s \in \mathbb{R}^m_{++}$, then $c_i(x_{s+1}) = 0$, $i = 1,..,m$ and $(x_{s+1}, \lambda_{s+1}) = (x^*, \lambda^*)$ is a KKT pair.

(2) The monotone increasing sequence $\{d(\lambda_s)\}_{s=s_0}^{\infty}$ is bounded from above by $f(x^*)$, therefore there exists $\bar{d} = \lim_{s\to\infty} d(\lambda_s) \leq d(\lambda^*) = f(x^*)$.
The first step is to show that $\bar{d} = d(\lambda^*)$.
Using $\nabla_u \mathbb{B}_2(v, w) = \nabla F(v) - \nabla F(w)$ for $v = \lambda_s$ and $w = \lambda_{s+1}$ we obtain

$$\nabla_\lambda \mathbb{B}_2(\lambda, \lambda_{s+1})_{/\lambda=\lambda_s} = \nabla\varphi(\lambda_s) - \nabla\varphi(\lambda_{s+1}) = \left(-\sum_{i=1}^{m} \lambda_{i,s}^{-1} e_i + \sum_{i=1}^{m} \lambda_{i,s+1}^{-1} e_i \right).$$

From the three point identity (135) with $u = \lambda^*$, $v = \lambda_s$, $w = \lambda_{s+1}$ follows

$$\begin{aligned} \mathbb{B}_2(\lambda^*, \lambda_s) - \mathbb{B}_2(\lambda^*, \lambda_{s+1}) - \mathbb{B}_2(\lambda_{s+1}, \lambda_s) = \\ (\nabla\varphi(\lambda_s) - \nabla\varphi(\lambda_{s+1}), \lambda_{s+1} - \lambda^*) = \\ \sum_{i=1}^{m} (-\lambda_{i,s}^{-1} + \lambda_{i,s+1}^{-1})(\lambda_{i,s+1} - \lambda_i^*). \end{aligned} \tag{155}$$

From the update formula (153) follows

$$kc_i(x_{s+1}) = -\lambda_{i,s}^{-1} + \lambda_{i,s+1}^{-1}, \; i = 1, \ldots, m.$$

Therefore, keeping in mind, $\mathbb{B}_2(\lambda_s, \lambda_{s+1}) \geq 0$ we can rewrite (155) as follows

$$\mathbb{B}_2(\lambda^*, \lambda_s) - \mathbb{B}_2(\lambda^*, \lambda_{s+1}) \geq k(c(x_{s+1}), \lambda_{s+1} - \lambda^*).$$

From $-c(x_{s+1}) \in \partial d(\lambda_{s+1})$ we obtain

$$d(\lambda) - d(\lambda_{s+1}) \leq (-c(x_{s+1}), \lambda - \lambda_{s+1}), \forall \lambda \in \mathbb{R}^m_+. \tag{156}$$

For $\lambda = \lambda^*$ from (156) we get

$$(c(x_{s+1}), \lambda_{s+1} - \lambda^*) \geq d(\lambda^*) - d(\lambda_{s+1}).$$

Hence,

$$\mathbb{B}_2(\lambda^*, \lambda_s) - \mathbb{B}_2(\lambda^*, \lambda_{s+1}) \geq k(d(\lambda^*) - d(\lambda_{s+1})). \tag{157}$$

Assuming $\lim_{s\to\infty} d(\lambda_s) = \bar{d} < d(\lambda^*)$ we have $d(\lambda^*) - d(\lambda_s) \geq \rho > 0, \forall s \geq s_0$. Summing up (157) from $s = s_0$ to $s = N$ we obtain

$$\mathbb{B}_2(\lambda^*, \lambda_{s_0}) - k(N - s_0)\rho \geq \mathbb{B}_2(\lambda^*, \lambda_{N+1}),$$

which is impossible for large N. Therefore

$$\lim_{s\to\infty} d(\lambda_s) = d(\lambda^*). \tag{158}$$

From (156) with $\lambda = \lambda_s$ we obtain

$$d(\lambda_s) - d(\lambda_{s+1}) \leq (-c(x_{s+1}), \lambda_s - \lambda_{s+1}).$$

Using the update formula (153) from last inequality we obtain

$$d(\lambda_{s+1}) - d(\lambda_s)) \geq (c(x_{s+1}), \lambda_s - \lambda_{s+1}) =$$

$$k\sum_{i=1}^{m} \lambda_{i,s}\lambda_{i,s+1}c_i(x_{s+1}) = k\sum_{i=1}^{m} \lambda_{i,s}/\lambda_{i,s+1}(\lambda_{i,s+1}c_i(x_{s+1}))^2. \tag{159}$$

Summing up (159) from $s = s_0$ to $s = N$ we have

$$d(\lambda^*) - d(\lambda_{s_0}) > d(\lambda_{N+1}) - d(\lambda_{s_0}) \geq k\sum_{s=s_0}^{N}\sum_{i=1}^{m} \lambda_{i,s}/\lambda_{i,s+1}(\lambda_{i,s+1}c_i(x_{s+1}))^2.$$

Keeping in mind (139) we obtain asymptotic complementarity condition

$$\lim_{s\to\infty} (\lambda_s, c(x_s)) = 0. \tag{160}$$

Therefore

$$d(\lambda^*) = \lim_{s\to\infty} d(\lambda_s) = \lim_{s\to\infty} [f(x_s) - (\lambda_s, c(x_s))] = \lim_{s\to\infty} f(x_s),$$

that is

$$\lim_{s\to\infty} f(x_s) = d(\lambda^*) = f(x^*). \tag{161}$$

From Slater condition follows boundedness of L^*. Therefore from concavity of d follows boundedness $\Lambda(\lambda_0) = \{\lambda \in \mathbb{R}^m_+ : d(\lambda) \geq d(\lambda_0)\}$. For any monotone increasing sequence $\{d(\lambda_s)\}_{s=s_0}^{\infty}$ follows boundedness $\Lambda_s = \{\lambda \in \mathbb{R}^m_+ : d(\lambda) \geq d(\lambda_s)\}$ and $\Lambda_0 \supset \ldots \supset \Lambda_s \supset \Lambda_{s+1} \supset \ldots \supset L^*$. Therefore from (108) we have

$$d_H(L^*, \partial\Lambda_s) > d_H(L^*, \partial\Lambda_{s+1}),\ s \geq s_0. \tag{162}$$

From (161) and (162) and continuity of d follows

$$\lim_{s\to\infty} d_H(L^*, \partial\Lambda_s) = 0.$$

(3) The dual sequence $\{\lambda_s\}_{s=0}^{\infty} \subset \Lambda(\lambda_0)$ is bounded, therefore there is a converging subsequence $\{\lambda_{s_l}\}_{l=1}^{\infty}$: $\lim_{l\to\infty} \lambda_{s_l} = \bar{\lambda}$.
Consider two subsets of indices $I_+ = \{i : \bar{\lambda}_i > 0\}$ and $I_0 = \{i : \bar{\lambda}_i = 0\}$. From the asymptotic complementarity (160) follows $\lim_{s\to\infty} c_i(x_s) = 0,\ i \in I_+$.
There exist such subsequence $\{s_l\}_{l=1}^{\infty}$ that for any $i \in I_0$ we have $\lim_{l\to\infty} \lambda_{i,s_l} = 0$, therefore without loosing the generality we can assume that

$$\lambda_{i,s_{l+1}} \le 0.5\lambda_{i,s_l},\ i \in I_0.$$

Using the update formula (153) we obtain

$$\lambda_{s_{l+1}} \prod_{s=s_l}^{s_{l+1}} (k\lambda_{i,s} c_i(x_s) + 1) = \lambda_{i,s_l} \ge 2\lambda_{i,s_{l+1}},\ i \in I_0.$$

Invoking the arithmetic-geometric means inequality for $i \in I_0$ we obtain

$$\frac{1}{s_{l+1} - s_l} \sum_{s=s_l}^{s_{l+1}} (k\lambda_{i,s} c_i(x_s) + 1) \ge \left(\prod_{s=s_l}^{s_{l+1}} (k\lambda_{i,s} c_i(x_s) + 1)\right)^{\frac{1}{s_{l+1}-s_l}} \ge 2^{\frac{1}{s_{l+1}-s_l}}$$

or

$$\sum_{s=s_l}^{s_{l+1}} \lambda_{i,s} c_i(x_s) > 0,\ i \in I_0.$$

Using Jensen inequality and concavity c_i we obtain

$$c_i(\bar{x}_{l+1}) = c_i\left(\sum_{s=s_l}^{s_{l+1}} \bar{\lambda}_{i,s} x_s\right) \ge \sum_{s=s_l}^{s_{l+1}} \bar{\lambda}_{i,s} c_i(x_s) > 0,$$

where $\bar{\lambda}_{i,s} = \lambda_{i,s}\left(\sum_{s=s_l}^{s_{l+1}} \lambda_{i,s}\right)^{-1} \ge 0$, $\sum_{s=s_l}^{s_{l+1}} \bar{\lambda}_{i,s} = 1,\ i \in I_0$. Keeping in mind $\lim_{s\to\infty} c_i(x_s) = 0,\ i \in I_+$ we conclude that the sequence $\{\bar{x}_{l+1}\}_{l=0}^{\infty}$ is asymptotically feasible, therefore it is bounded. Without loosing generality we can assume that $\lim_{l\to\infty} \bar{x}_l = \bar{x} \in \Omega$.
From convexity f follows

$$f(\bar{x}_{l+1}) \le \sum_{s=s_l}^{s_{l+1}} \bar{\lambda}_{i,s} f(x_s).$$

Therefore from (161) follows

$$f(\bar{x}) = \lim_{l\to\infty} f(\bar{x}_{l+1}) \leq \lim_{s\to\infty} f(x_s) = \lim_{s\to\infty} d(\lambda_s) = d(\lambda^*) = f(x^*).$$

Thus, $f(\bar{x}) = f(x^*)$, hence $d(\lambda^*) = d(\bar{\lambda})$ and $\bar{x} = x^*, \bar{\lambda} = \lambda^*$. □

The items (1) and (2a) of Theorem 11 were proven by Matioli and Gonzaga (see Theorem 3.2 in [37]).

6.4 Lagrangian Transformation and Affine Scaling Method for LP

Let $a \in \mathbb{R}^n, b \in \mathbb{R}^m$ and $A : \mathbb{R}^n \to \mathbb{R}^m$ are given. We consider the following LP problem

$$x^* \in X^* = Argmin\{(a,x)|c(x) = Ax - b \geq 0\} \tag{163}$$

and the dual LP

$$\lambda^* \in L^* = Argmin\{(b,\lambda)|r(\lambda) = A^T\lambda - a = 0, \lambda \in \mathbb{R}^m_+\}. \tag{164}$$

The LT $\mathscr{L} : \mathbb{R}^n \times \mathbb{R}^m \times \mathbb{R}_{++} \to \mathbb{R}$ for LP is defined as follows

$$\mathscr{L}(x,\lambda,k) := (a,x) - k^{-1}\sum_{s=1}^{m} \psi(k\lambda_i c_i(x)), \tag{165}$$

where $c_i(x) = (Ax - b)_i = (a_i, x) - b_i, \quad i = 1,\ldots,m$.

We assume that $X^* \neq \phi$ is bounded and so is the dual optimal set L^*.

The LT method generate primal–dual sequence $\{x_{s+1}, \lambda_{s+1}\}_{s=0}^{\infty}$ by the following formulas

$$x_{s+1} : \nabla_x \mathscr{L}(x_{s+1}, \lambda_s, k_s) = 0 \tag{166}$$

$$\lambda_{s+1} : \lambda_{i,s+1} = \lambda_{i,s}\psi'(k_s\lambda_{i,s}c_i(x_{s+1})), i = 1,\ldots,m. \tag{167}$$

Theorem 12. *If the primal optimal X^* is bounded, then the LT method* (166)–(167) *is well defined for any transformation $\psi \in \Psi$. For the dual sequence $\{\lambda_s\}_{s=0}^{\infty}$ generated by* (167) *the following statements hold true:*

(1) the LT method (166)–(167) *is equivalent to the following Interior Prox*

$$k(b,\lambda_{s+1}) - B_\varphi(\lambda_{s+1},\lambda_s) = \max\{k(b,\lambda) - B_\varphi(\lambda,\lambda_s)|A^T\lambda = 0\},$$

where $B_\varphi(u,v) = \sum_{i=1}^m \varphi(\frac{u_i}{v_i})$ is the Bregman type distance;

(2) *there exists* $s_0 > 0$ *that for any* $s \geq s_0$ *the LT method with truncated MBF transformation* $\psi_2(t)$ *is equivalent to the affine scaling type method for the dual LP.*

Proof. (1) We use the vector form for formula (167) assuming that the multiplication and division are componentwise, i.e. for vectors $a, b \in \mathbb{R}^n$, the vector $c = ab = (c_i = a_i b_i, \quad i = 1, \ldots, n)$ and the vector $d = a/b = (d_i = a_i/b_i, i = 1, \ldots, n)$. From (167) follows

$$\frac{\lambda_{s+1}}{\lambda_s} = \psi'(k\lambda_s c(x_{s+1})). \tag{168}$$

Using again the inverse function formula we obtain

$$k\lambda_s c(x_{s+1}) = \psi'^{-1}(\lambda_{s+1}/\lambda_s). \tag{169}$$

It follows from (166) and (167) that

$$\nabla_x \mathscr{L}(x_{s+1}, \lambda_s, k) = a - A^T \psi'(k\lambda_s c(x_{s+1}))\lambda_s = a - A^T \lambda_{s+1}$$
$$= \nabla_x L(x_{s+1}, \lambda_{s+1}) = 0,$$

therefore

$$d(\lambda_{s+1}) = L(x_{s+1}, \lambda_{s+1}) = (a, x_{s+1}) - (\lambda_{s+1}, Ax_{s+1} - b) =$$
$$(a - A^T\lambda_{s+1}, x_{s+1}) + (b, \lambda_{s+1}) = (b, \lambda_{s+1}).$$

Using LEID $\psi'^{-1} = \psi^{*\prime}$ and $\varphi = -\psi^*$ we can rewrite (169) as follows

$$-kc(x_{s+1}) - (\lambda_s)^{-1}\varphi'(\lambda_{s+1}/\lambda_s) = 0. \tag{170}$$

Keeping in mind $A^T\lambda_{s+1} = a, -c(x_{s+1}) \in \partial d(\lambda_{s+1})$ and $\lambda_{s+1} \in \mathbb{R}^m_{++}$ we can view (170) as the optimality criteria for the following problem

$$k(b, \lambda_{s+1}) - B_\varphi(\lambda_{s+1}, \lambda_s) = \max\{kd(\lambda) - B_\varphi(\lambda, \lambda_s) | A^T\lambda = a\}, \tag{171}$$

where $B_\varphi(\lambda, \lambda_s) = \sum_{i=1}^{q} \varphi(\lambda_i/\lambda_{i,s})$ is Bregman type distance.

(2) Let's consider the LT method with truncated MBF transformation $\psi_2(t)$. It follows from (139) that there exists s_0 that for any $s \geq s_0$ only MBF kernel $\varphi_2 = -\ln s + s - 1$ and correspondent Bregman distance

$$\mathbb{B}_2(\lambda, \lambda_s) = \sum_{i=1}^{q}(-ln\frac{\lambda_i}{\lambda_{i,s}} + \frac{\lambda_i}{\lambda_{i,s}} - 1)$$

will be used in (171). Using considerations similar to those in item 4) Theorem 10 we can rewrite (171) as follows

$$k(b, \lambda_{s+1}) = \arg\max\{k(b, \lambda) | \lambda \in E(\lambda_s, r_s), A^T \lambda = a\}, \tag{172}$$

where $r_s^2 = Q(\lambda_{s+1}, \lambda_s)$ and $E(\lambda_s, r_s) = \left\{\lambda : (\lambda - \lambda_s)^T H_s^{-2} (\lambda - \lambda_s) \leq r_s\right\}$ is Dikin's ellipsoid and (172) is affine scaling type method for the dual LP (see [18]).

In the final part of the paper we will show the role of LET and LEINV in unconstrained minimization of SC functions. For the basic SC properties and damped Newton method see [42] and [43].

7 Legendre Invariant and Self-Concordant Functions

We consider a closed convex function $F \in C^3$ defined on an open convex set $\mathrm{dom}\, F \subset \mathbb{R}^n$. For a given $x \in \mathrm{dom}\, F$ and direction $u \in \mathbb{R}^n \setminus \{0\}$ we consider the restriction

$$f(t) = F(x + tu)$$

of F, which is defined on $\mathrm{dom} f = \{t : x + tu \in \mathrm{dom}\, F\}$. Along with f, let us consider its derivatives

$$\begin{aligned} f'(t) &= (\nabla F(x + tu), u)\,, \\ f''(t) &= (\nabla^2 F(x + tu)u, u)\,, \\ f'''(t) &= (\nabla^3 F(x + tu)[u]u, u)\,, \end{aligned}$$

where ∇F is the gradient of F, $\nabla^2 F$ is the Hessian of F and

$$\nabla^3 F(x)[u] = \lim_{\tau \to +0} \tau^{-1} \left[\nabla^2 F(x + \tau u) - \nabla^2 F(x)\right].$$

Then,

$$\begin{aligned} DF(x)[u] &:= (\nabla F(x), u) = f'(0)\,, \\ D^2 F(x)[u, u] &:= (\nabla^2 F(x)u, u) = f''(0)\,, \\ D^3 F(x)[u, u, u] &:= (\nabla^3 F(x)[u]u, u) = f'''(0)\,. \end{aligned}$$

Function F is self-concordant if there is $M > 0$ such that the inequality

$$D^3 F(x)[u, u, u] \leq M(\nabla^2 F(x)u, u)^{\frac{3}{2}}$$

holds for any $x \in \mathrm{dom}\, F$ and any $u \in \mathbb{R}^n$.

If for a SC function F the dom F does not contain a straight line, then the Hessian $\nabla^2 F(x)$ is positive definite at any $x \in \operatorname{dom} F$. We assume that such condition holds, so for any $x \in \operatorname{dom} F$ and any $u \in \mathbb{R}^n \setminus \{0\}$ we have

$$(\nabla^2 F(x)u, u) = f''(0) > 0, \tag{173}$$

that is F is strictly convex on dom F.

A strictly convex function F is self-concordant (SC) if the Legendre invariant of the restriction $f(t) = F(x + tu)$ is bounded, i.e. for any $x \in \operatorname{dom} F$ and any direction $u = y - x \in \mathbb{R}^n \setminus \{0\}$ there exist $M > 0$ that

$$\mathrm{LEINV}(f) = |f'''(t)|(f''(t))^{-\frac{3}{2}} \le M, \ \forall t : x + tu \in \operatorname{dom} F. \tag{174}$$

Let us consider the log-barrier function $F(x) = -\ln x$, then for any $x \in \operatorname{dom} F = \{x : x > 0\}$ we have $F'(x) = -x^{-1}$, $F''(x) = x^{-2}$, $F'''(x) = -2x^{-3}$ and

$$\mathrm{LEINV}(F) = \left|F'''(x)\right| \left(F''(x)\right)^{-3/2} \le 2. \tag{175}$$

Therefore, $F(x) = -\ln x$ is self-concordant with $M = 2$.

The following function

$$g(t) = (\nabla^2 F(x + tu)u, u)^{-1/2} = \left(f''(t)\right)^{-1/2},$$

is critical for the self-concordance (SC) theory.

For any $t \in \operatorname{dom} f$, we have

$$g'(t) = \frac{d\left[(f''(t))^{-1/2}\right]}{dt} = -\frac{1}{2} f'''(t)(f''(t))^{-3/2}.$$

It follows from (175) that

$$0.5\,\mathrm{LEINV}(f) = |g'(t)| \le 1, \quad \forall\, t \in \operatorname{dom} f. \tag{176}$$

The differential inequality (176) is the key element for establishing basic bounds for SC functions.

The other important component of the SC theory is two local scaled norms of a vector $u \in \mathbb{R}^n$. The first local scaled norm is defined at each point $x \in \operatorname{dom} F$ as follows

$$\|u\|_x = \left(\nabla^2 F(x)u, u\right)^{1/2}.$$

The second scaled norm is defined by formula

$$\|v\|_x^* = \left(\left(\nabla^2 F(x)\right)^{-1} v, v\right)^{1/2}.$$

From (173) follows that the second scaled norm is well defined at each $x \in \operatorname{dom} F$. The following Cauchy-Schwarz (CS) inequality for scaled norms will be often used later.

Let matrix $A \in \mathbb{R}^{n \times n}$ be symmetric and positive definite, then $A^{1/2}$ exists and

$$\begin{aligned}|(u, v)| &= \left|(A^{1/2}u,\, A^{-1/2}v)\right| \leq \left\|A^{1/2}u\right\| \left\|A^{-1/2}v\right\| \\ &= \left(A^{1/2}u,\, A^{1/2}u\right)^{1/2} \left(A^{-1/2}v,\, A^{-1/2}v\right)^{1/2} \\ &= (Au, u)^{1/2} \left(A^{-1}v, v\right)^{1/2} = \|u\|_A \; \|v\|_{A^{-1}} \,.\end{aligned}$$

By taking $A = \nabla^2 F(x)$, for any $u, v \in \mathbb{R}^n$ one obtains the following CS inequality:

$$|(u, v)| \leq \|u\|_x \; \|v\|_x^* \,.$$

The following Proposition will be used later.

Proposition 1. *A function F is self-concordant if and only if for any $x \in \operatorname{dom} F$ and any $u_1, u_2, u_3 \in \mathbb{R}^n \setminus \{0\}$ we have*

$$\left|D^3 F(x)\,[u_1, u_2, u_3]\right| \leq 2 \prod_{i=1}^{3} \|u_i\|_x \,, \tag{177}$$

where $D^3 F(x)[u_1, u_2, u_3] = (\nabla^3 F(x)[u_1]u_2, u_3)$.

The following theorem establishes one of the most important facts about SC functions: any SC function is a barrier function on $\operatorname{dom} F$. The opposite statement is, generally speaking, not true, that is not every barrier function is self-concordant. For example, the hyperbolic barrier $F(x) = x^{-1}$ defined on $\operatorname{dom} F = \{x : x > 0\}$ is not a SC function.

Theorem 13. *Let F be a closed convex function on an open $\operatorname{dom} F$. Then, for any $\bar{x} \in \partial(\operatorname{dom} F)$ and any sequence $\{x_s\} \subset \operatorname{dom} F$ such that $x_s \to \bar{x}$, we have*

$$\lim_{s \to \infty} F(x_s) = \infty \,. \tag{178}$$

Proof. From convexity of F follows

$$F(x_s) \geq F(x_0) + (\nabla F(x_0),\, x_s - x_0)$$

for any given $x_0 \in \operatorname{dom} F$.

So, the sequence $\{F(x_s)\}$ is bounded from below. If (177) is not true, then the sequence $\{F(x_s)\}$ is bounded from above. Therefore, it has a limit point $\bar{F}$. Without loss of generality, we can assume that $z_s = (x_s, F(x_s)) \to \bar{z} = (\bar{x}, \bar{F})$. Since the function F is closed, we have $\bar{z} \in \operatorname{epi} F$, but it is impossible because $\bar{x} \notin \operatorname{dom} F$. Therefore for any sequence

$$\{x_s\} \subset \operatorname{dom} F : \lim_{s \to \infty} x_s = \bar{x} \in \partial(\operatorname{dom} F)$$

we have (177). It means that F is a barrier function on the $cl(\operatorname{dom} F)$. $\square$

For any $x \in \operatorname{dom} F$, and any $u \in \mathbb{R}^n \setminus \{0\}$ from (173) follows

$$\left(\nabla^2 F(x)u, u\right) = \|u\|_x^2 > 0 \tag{179}$$

and for $\forall t \in \operatorname{dom} f$ we have

$$g(t) = \left(\nabla^2 F(x + tu)u, u\right)^{-1/2} = \|u\|_{x+tu}^{-1} > 0. \tag{180}$$

7.1 *Basic Bounds for SC Functions*

In this section the basic bounds for SC functions will be obtained by integration of inequalities (176) and (177).

First Integration Keeping in mind $f''(t) > 0$ from (176), for any $s > 0$, we obtain

$$-\int_0^s dt \le \int_0^s d\left(f''(t)^{-1/2}\right) \le \int_0^s dt\,.$$

Therefore

$$f''(0)^{-1/2} - s \le f''(s)^{-1/2} \le f''(0)^{-1/2} + s \tag{181}$$

or

$$\left(f''(0)^{-1/2} + s\right)^{-2} \le f''(s) \le \left(f''(0)^{-1/2} - s\right)^{-2}. \tag{182}$$

The left inequality in (182) holds for all $s \ge 0$, while the right inequality holds only for $0 \le s < f''(0)^{-1/2}$.

Let $x, y \in \operatorname{dom} F$, $y \ne x$, $u = y - x$ and $y(s) = x + s(y - x)$, $0 \le s \le 1$, so $y(0) = x$ and $y(1) = y$. Therefore,

$$f''(0) = \left(\nabla^2 F(x)(y - x), y - x\right) = \|y - x\|_x^2$$

and

$$f''(0)^{1/2} = \|y - x\|_x\,.$$

Also,

$$f''(1) = \left(\nabla^2 F(y)(y - x), y - x\right) = \|y - x\|_y^2$$

and

$$f''(1)^{1/2} = \|y - x\|_y\,.$$

From (181), for $s = 1$ follows

$$f''(0)^{-1/2} - 1 \leq f''(1)^{-1/2} \leq f''(0)^{-1/2} + 1,$$

or

$$\frac{1}{\|y-x\|_x} - 1 \leq \frac{1}{\|y-x\|_y} \leq \frac{1}{\|y-x\|_x} + 1\,.$$

From the right inequality follows

$$\|y-x\|_y \geq \frac{\|y-x\|_x}{1+\|y-x\|_x}\,. \tag{183}$$

If $\|y-x\|_x < 1$, then from the left inequality follows

$$\|y-x\|_y \leq \frac{\|y-x\|_x}{1-\|y-x\|_x}\,. \tag{184}$$

By integrating (176) we get

$$g(t) \geqslant g(0) - |t|\,, \quad t \in \operatorname{dom} f\,. \tag{185}$$

For $x + tu \in \operatorname{dom} F$ from (180) follows $g(t) > 0$. From Theorem 13 follows $F(x+tu) \to \infty$ when $x + tu \to \partial(\operatorname{dom} F)$. Therefore, $(\nabla^2 F(x+tu)u, u)$ cannot be bounded when $x + tu \to \partial(\operatorname{dom} F)$. Therefore from (180) follows $g(t) \to 0$ when $x + tu \to \partial(\operatorname{dom} F)$. It follows from (185) that any $t : |t| < g(0)$ belongs to $\operatorname{dom} f$, i.e.

$$(-g(0), g(0)) = \left(-\|u\|_x^{-1},\ \|u\|_x^{-1}\right) \subset \operatorname{dom} f\,.$$

Therefore, the set

$$E^0(x,1) = \left\{y = x + tu : t^2\,\|u\|_x^2 < 1\right\}$$

is contained in $\operatorname{dom} F$. In other words, the *Dikin's ellipsoid*

$$E(x,r) = \left\{y \in \mathbb{R}^n : \|y-x\|_x^2 \leq r\right\},$$

is contained in $\operatorname{dom} F$ for any $x \in \operatorname{dom} F$ and any $r < 1$.

One can expect that, for any $x \in \operatorname{dom} F$ and any $y \in E(x,r)$, the Hessians $\nabla^2 F(x)$ and $\nabla^2 F(y)$ are "close" enough if $0 < r < 1$ is small enough. The second integration allows to establish the corresponding bounds.

Second Integration Let us fix $x \in \operatorname{dom} F$, for a given $y \in \operatorname{dom} F$ $(y \neq x)$ consider direction $u = y - x \in \mathbb{R}^n \setminus \{0\}$. Let $y(t) = x + tu = x + t(y - x)$, then for $t \geq 0$ and $y(t) \in \operatorname{dom} F$ we have

$$\psi(t) = \|u\|^2_{y(t)} = (F''(y(t))u, u).$$

It follows from Proposition 1 that

$$\left|\psi'(t)\right| = D^3F(y(t))[y - x, u, u] \leq 2\|y - x\|_{y(t)} \, \|u\|^2_{y(t)} = 2\|y - x\|_{y(t)} \, \psi(t).$$

First of all, $\|y(t) - x\|_x \leq \|y - x\|_x$ for any $t \in [0, 1]$. Keeping in mind that $y - x = t^{-1}(y(t) - x)$ and assuming $\|y - x\|_x < 1$ from (184) follows

$$\begin{aligned}\left|\psi'(t)\right| &\leq \frac{2}{t}\|y(t) - x\|_{y(t)} \, \psi(t) \leq \frac{2}{t}\frac{\|y(t) - x\|_x}{1 - \|y(t) - x\|_x}\psi(t) \\ &\leq 2\frac{\|y - x\|_x}{1 - t\|y - x\|_x}\psi(t).\end{aligned}$$

Therefore for $0 < t < \|y - x\|_x^{-1}$ follows

$$\frac{|\psi'(t)|}{\psi(t)} \leq \frac{2\|y - x\|_x}{1 - t\|y - x\|_x}.$$

By integrating the above inequality we get

$$-2\int_0^s \frac{\|y - x\|_x}{1 - t\|y - x\|_x}dt \leq \int_0^s \frac{\psi'(t)}{\psi(t)}dt \leq 2\int_0^s \frac{\|y - x\|_x}{1 - t\|y - x\|_x}dt,$$

for $0 < s < \|y - x\|_x^{-1}$, hence

$$2\ln(1 - s\|y - x\|_x) \leq \ln\psi(s) - \ln\psi(0) \leq -2\ln(1 - s\|y - x\|_x).$$

For $s = 1$, we have

$$\psi(0)(1 - \|y - x\|_x)^2 \leq \psi(1) \leq \psi(0)(1 - \|y - x\|_x)^{-2}. \tag{186}$$

In view of $\psi(0) = (\nabla^2F(x)u, u)$ and $\psi(1) = (\nabla^2F(y)u, u)$ for any $u \in \mathbb{R}^n \setminus \{0\}$ from (186) follows

$$(1 - \|y - x\|_x)^2 (\nabla^2F(x)u, u) \leq (\nabla^2F(y)u, u) \leq (1 - \|y - x\|_x)^{-2} (\nabla^2F(x)u, u).$$

Therefore the following matrix inequality holds

$$(1 - \|y - x\|_x)^2 \nabla^2F(x) \preccurlyeq \nabla^2F(y) \preccurlyeq \nabla^2F(x)(1 - \|y - x\|_x)^{-2}, \tag{187}$$

where $A \succcurlyeq B$ means that $A - B$ is nonnegative definite. Note that (187) takes place for any $x, y \in \operatorname{dom} F$.

In order to find the upper and the lower bounds for the matrix

$$G = \int_0^1 \nabla^2 F(x + \tau(y - x))d\tau \tag{188}$$

let us consider (187) for $y := x + \tau(y - x)$.

From the left inequality (187) follows

$$G = \int_0^1 \nabla^2 F(x + \tau(y - x))d\tau \succcurlyeq \nabla^2 F(x) \int_0^1 (1 - \tau \|y - x\|_x)^2 \, d\tau \, .$$

Therefore, for $r = \|y - x\|_x < 1$, we have

$$G \succcurlyeq \nabla^2 F(x) \int_0^1 (1 - \tau r)^2 d\tau = \nabla^2 F(x) \left(1 - r + \frac{r^2}{3}\right) . \tag{189}$$

From the right inequality (187) follows

$$G \preccurlyeq \nabla^2 F(x) \int_0^1 (1 - \tau r)^{-2} d\tau = \nabla^2 F(x) \frac{1}{1 - r} , \tag{190}$$

i.e. for any $x \in \operatorname{dom} F$, the following inequalities hold:

$$\left(1 - r + \frac{r^2}{3}\right) \nabla^2 F(x) \preccurlyeq G \preccurlyeq \frac{1}{1 - r} \nabla^2 F(x) \, . \tag{191}$$

The first two integrations produced two very important facts.

1. For any $x \in \operatorname{dom} F$, Dikin's ellipsoid

$$E(x, r) = \{y \in \mathbb{R}^n : \|y - x\|_x^2 \leq r\}$$

is contained in $\operatorname{dom} F$, for any $0 \leq r < 1$.

2. For any $x \in \operatorname{dom} F$ and any $y \in E(x, r)$ from (187) follows

$$(1 - r)^2 \nabla^2 F(x) \preccurlyeq \nabla^2 F(y) \preccurlyeq \frac{1}{(1 - r)^2} \nabla^2 F(x) \, , \tag{192}$$

i.e. the function F is almost quadratic inside the ellipsoid $E(x, r)$ for small $0 \leq r < 1$.

The bounds for the gradient $\nabla F(x)$, which is a monotone operator in $\mathbb{R}^n$, we establish by integrating (182).

Third Integration From (182), for $0 \le t < f(0)^{-1/2} = \|y-x\|_x^{-1}$ and $0 \le s \le 1$ we obtain

$$\int_0^s \left(f''(0)^{-1/2} + t\right)^{-2} dt \le \int_0^s f''(t)dt \le \int_0^s \left(f''(0)^{-1/2} - t\right)^{-2} dt,$$

or

$$f'(0) + f''(0)^{1/2}\left(1 - \left(1 + sf''(0)^{1/2}\right)^{-1}\right)$$
$$\le f'(s) \le f'(0) - f''(0)^{1/2}\left(1 - \left(1 - sf''(0)^{1/2}\right)^{-1}\right).$$

The obtained inequalities we can rewrite as follows

$$f'(0) + w'(f''(0)^{\frac{1}{2}}s) \le f'(s) \le f'(0) + w^{*'}(f''(0)^{\frac{1}{2}}s), \tag{193}$$

where $\omega(t) = t - \ln(1+t)$ and $\omega^*(s) = \sup_{t>-1}\{st - t + \ln(1+t)\} = -s - \ln(1-s) = \omega(-s)$ is the LET of $\omega(t)$.

From the right inequality (193), for $s = 1$ follows

$$f'(1) - f'(0) \le -f''(0)^{1/2}\left(1 - \frac{1}{1 - f''(0)^{1/2}}\right) = \frac{f''(0)}{1 - f''(0)^{1/2}}.$$

Recalling formulas for $f'(0), f'(1), f''(0)$, and $f''(1)$ we get

$$(\nabla F(y) - \nabla F(x), y - x) \le \frac{\|y-x\|_x^2}{1 - \|y-x\|_x} \tag{194}$$

for any x and $y \in \text{dom}\,F$.

From the left inequality in (193), for $s = 1$ follows

$$f'(1) - f'(0) \ge f''(0)^{1/2}\left(1 - \frac{1}{1 + f''(0)^{1/2}}\right) = \frac{f''(0)}{1 + f''(0)^{1/2}}$$

or

$$(\nabla F(y) - \nabla F(x), y - x) \ge \frac{\|y-x\|_x^2}{1 + \|y-x\|_x}. \tag{195}$$

Fourth Integration In order to establish bounds for $F(y) - F(x)$ it is enough to integrate the inequalities (193).

Taking the integral of the right inequality (193), we obtain

$$f(s) \le f(0) + f'(0)s + \omega^*\left(f''(0)^{1/2}s\right)$$

$$= f(0) + f'(0)s - f''(0)^{1/2}s - \ln\left(1 - f''(0)^{1/2}s\right)$$
$$= U(s)\,. \tag{196}$$

In other words, $U(s)$ is an upper bound for $f(s)$ on the interval $[0, f''(0)^{-1/2})$. Recall that $f''(0)^{-1/2} = \|y - x\|_x^{-1} > 1$. For $s = 1$ from (196) follows

$$f(1) - f(0) \leq f'(0) + \omega^*\left(f''(0)^{1/2}\right) = f'(0) + \omega^*\left(\|y - x\|_x\right)\,. \tag{197}$$

Keeping in mind $f(0) = F(x), f(1) = F(y)$, from (197), we get

$$F(y) - F(x) \leq (\nabla F(x), y - x) + \omega^*\left(\|y - x\|_x\right)\,. \tag{198}$$

Integration of the left inequality (193) leads to the lower bound $L(s)$ for $f(s)$

$$f(s) \geq f(0) + f'(0)s + \omega\left(f''(0)^{1/2}s\right)$$
$$= f(0) + f'(0)s + f''(0)^{1/2}s - \ln\left(1 + f''(0)^{1/2}s\right)$$
$$= L(s)\,,\ \forall\, s \geq 0\,. \tag{199}$$

For $s = 1$, we have

$$f(1) - f(0) \geq f'(0) + \omega\left(f''(0)^{1/2}\right)$$

or

$$F(y) - F(x) \geq (\nabla F(x), y - x) + \omega\left(\|y - x\|_x\right)\,. \tag{200}$$

We conclude the section by considering the existence of the minimizer

$$x^* = \arg\min\{F(x) : x \in \operatorname{dom} F\} \tag{201}$$

for a self-concordant function F.

It follows from (173) that the Hessian $\nabla^2 F(x)$ is positive definite for any $x \in \operatorname{dom} F$, but the existence of $x^* : \nabla F(x^*) = 0$, does not follow from strict convexity of F.

However, it guarantees the existence of the local norm $\|v\|_x^* = \left(\left(\nabla^2 F(x)\right)^{-1} v, v\right)^{1/2} > 0$ at any $x \in \operatorname{dom} F$.

For $v = \nabla F(x)$, one obtains the following scaled norm of the gradient $\nabla F(x)$,

$$\lambda(x) = \left(\nabla^2 F(x)^{-1} \nabla F(x), \nabla F(x)\right)^{1/2} = \|\nabla F(x)\|_x^* > 0\,,$$

which plays an important role in SC theory. It is called Newton decrement of F at the point $x \in \operatorname{dom} F$.

Theorem 14. *If* $\lambda(x) < 1$ *for some* $x \in \operatorname{dom} F$ *then the minimizer* x^* *in (201) exists.*

Proof. For $u = y - x \neq 0$ and $v = \nabla F(x)$, where x and $y \in \operatorname{dom} F$ from CS inequality $|(u, v)| \leq \|v\|_x^* \|u\|_x$ follows

$$|(\nabla F(x), y - x)| \leq \|\nabla F(x)\|_x^* \|y - x\|_x . \tag{202}$$

From (200) and (202) and the formula for $\lambda(x)$ follows

$$F(y) - F(x) \geq -\lambda(x) \|y - x\|_x + \omega(\|y - x\|_x) .$$

Therefore, for any $y \in \mathscr{L}(x) = \{y \in \mathbb{R}^n : F(y) \leq F(x)\}$ we have

$$\omega(\|y - x\|_x) \leq \lambda(x) \|y - x\|_x ,$$

i.e.

$$\|y - x\|_x^{-1} \omega(\|y - x\|_x) \leq \lambda(x) < 1 .$$

From the definition of $\omega(t)$ follows

$$1 - \frac{1}{\|y - x\|_x} \ln(1 + \|y - x\|_x) \leq \lambda(x) < 1 .$$

The function $1 - \tau^{-1} \ln(1 + \tau)$ is monotone increasing for $\tau > 0$. Therefore, for a given $0 < \lambda(x) < 1$, the equation

$$1 - \lambda(x) = \tau^{-1} \ln(1 + \tau)$$

has a unique root $\bar{\tau} > 0$. Thus, for any $y \in \mathscr{L}(x)$, we have

$$\|y - x\|_x \leq \bar{\tau} ,$$

i.e. the level set $\mathscr{L}(x)$ at $x \in \operatorname{dom} F$ is bounded and closed due to the continuity of F. Therefore, x^* exists due to the Weierstrass theorem. The minimizer x^* is unique due to the strict convexity of $F(x)$ for $x \in \operatorname{dom} F$. □

The theorem presents an interesting result: a local condition $\lambda(x) < 1$ at some $x \in \operatorname{dom} F$ guarantees the existence of x^*, which is a global property of F on the $\operatorname{dom} F$. The condition $0 < \lambda(x) < 1$ will plays an important role later.

Let us briefly summarize the basic properties of the SC functions established so far.

1. The SC function F is a barrier function on $\operatorname{dom} F$.
2. For any $x \in \operatorname{dom} F$ and any $0 < r < 1$, there is a Dikin's ellipsoid inside $\operatorname{dom} F$, i.e.

$$E(x, r) = \left\{y : \|y - x\|_x^2 \leq r\right\} \subset \operatorname{dom} F .$$

3. For any $x \in \operatorname{dom} F$ and small enough $0 < r < 1$, the function F is almost quadratic inside of the Dikin's ellipsoid $E(x, r)$ due to the bounds (192).
4. The gradient ∇F is a strictly monotone operator on $\operatorname{dom} F$ with upper and lower monotonicity bounds given by (194) and (195).
5. For any $x \in \operatorname{dom} F$ and any direction $u = y - x$, the restriction $f(s) = F(x + s(y - x))$ is bounded by $U(s)$ and $L(s)$ (see (196) and (199)).
6. Condition $0 < \lambda(x) < 1$ at any $x \in \operatorname{dom} F$ guarantees the existence of a unique minimizer x^* on $\operatorname{dom} F$.

It is quite remarkable that practically all important properties of SC functions follow from a single differential inequality (176), which is, a direct consequence of the boundedness of LEINV(f).

We conclude the section by showing that Newton method can be very efficient for global minimization of SC functions, in spite of the fact that F is not strongly convex.

7.2 *Damped Newton Method for Minimization of SC Function*

The SC functions are strictly convex on $\operatorname{dom} F$. Such a property, generally speaking, does not guarantee global convergence of the Newton method. For example, $f(t) = \sqrt{1 + t^2}$ is strictly convex, but Newton method for finding $\min_t f(t)$ diverges from any starting point $t_0 \notin]-1, 1[$.

Turns out that SC properties guarantee convergence of the special damped Newton method from any starting point. Moreover, such method goes through three phases. In the first phase each step reduces the error bound $\Delta f(x) = f(x) - f(x^*)$ by a constant, which is independent on $x \in \operatorname{dom} F$. In the second phase the error bound converges to zero with at least superlinear rate. The superlinear rate is characterized explicitly through $w(\lambda)$ and its LET $w^*(\lambda)$, where $0 < \lambda < 1$ is the Newton decrement. At the final phase the damped Newton method practically turns into standard Newton method and the error bound converges to zero with quadratic rate.

The following bounds for the restriction $f(s) = F(x + su)$ at $x \in \operatorname{dom} F$ in the direction $u = y - x \in \mathbb{R}^n \setminus \{0\}$ is our main tool

$$\underset{s \geq 0}{\mathscr{L}(s)} \leq f(s) \leq \underset{0 \leq s \leq f''(0)^{-(1/2)}}{U(s)} . \tag{203}$$

Let $x \in \operatorname{dom} F, f(0) = F(x)$ and $x \neq x^*$, then there exists $y \in \operatorname{dom} F$ such that for $u = y - x \neq 0$ we have

$$\begin{aligned} &a) \;\; f'(0) = (\nabla F(x), u) < 0\,, \text{ and} \\ &b) \;\; f''(0) = \left(\nabla^2 F(x)u, u\right) = \|u\|_x^2 = d^2 > 0. \end{aligned} \tag{204}$$

We would like to estimate the reduction of F, as a result of one Newton step with $x \in \operatorname{dom} F$ as a starting point.

Let us consider the upper bound

$$U(s) = f(0) + f'(0)s - ds - \ln(1 - ds),$$

for $f(s)$. The function $U(s)$ is strongly convex in s on $[0, d)^{-1}$. Also, $U'(0) = f'(0) < 0$ and $U'(s) \to \infty$ for $s \to d^{-1}$. Therefore, the equation

$$U'(s) = f'(0) - d + d(1 - ds)^{-1} = 0 \tag{205}$$

has a unique solution $\bar{s} \in [0, d^{-1})$, which is the unconstrained minimizer for $U(s)$. From (205) we have

$$\bar{s} = -f'(0)d^{-2}\left(1 - f'(0)d^{-1}\right)^{-1} = \Delta(1 + \lambda)^{-1}$$

where $\Delta = -f'(0)d^{-2}$ and $0 < \lambda = -f'(0)d^{-1} < 1$. On the other hand, the unconstrained minimizer $\bar{s}$ is a result of one step of the damped Newton method for finding $\min_{s \geq 0} U(s)$ with step length $t = (1 + \lambda)^{-1}$ from $s = 0$ as a starting point. It is easy to see that

$$U\left((1 + \lambda)^{-1}\Delta\right) = f(0) - \omega(\lambda).$$

From the right inequality in (203), we obtain

$$f\left((1 + \lambda)^{-1}\Delta\right) \leq f(0) - \omega(\lambda). \tag{206}$$

Keeping in mind (204) for the Newton direction $u = y - x = -(\nabla^2 F(x))^{-1}\nabla F(x)$ we obtain

$$\Delta = -\frac{f'(0)}{f''(0)} = -\frac{(\nabla F(x), u)}{(\nabla^2 F(x)u, u)} = 1.$$

In view of $f(0) = F(x)$, we can rewrite (206) as follows:

$$F\left(x - (1 + \lambda)^{-1}(\nabla^2 F(x))^{-1}\nabla F(x)\right) \leq F(x) - \omega(\lambda). \tag{207}$$

In other words, finding an unconstrained minimizer of the upper bound $U(s)$ is equivalent to one step of the damped Newton method

$$x_{k+1} = x_k - (1 + \lambda(x_k))^{-1}\left(\nabla^2 F(x_k)\right)^{-1}\nabla F(x_k) \tag{208}$$

for minimization of $F(x)$ on $\operatorname{dom} F$. Moreover, our considerations are independent from the starting point $x \in \operatorname{dom} F$. Therefore, for any starting point $x_0 \in \operatorname{dom} F$ and

$k \geq 1$, we have

$$F(x_{k+1}) \leq F(x_k) - \omega(\lambda). \tag{209}$$

The bound (209) is universal, i.e. it is true for any $x_k \in \operatorname{dom} F$.
Let us compute $\lambda = f'(0)f''(0)^{-1/2}$ for the Newton direction

$$u = -\nabla^2 F(x)^{-1} \nabla F(x).$$

We have

$$\begin{aligned}
\lambda \equiv \lambda(x) &= -f'(0)f''(0)^{-1/2} \\
&= -\frac{(\nabla F(x), u)}{(\nabla^2 F(x)u, u)^{1/2}} \\
&= \left(\nabla^2 F(x)^{-1} \nabla F(x), \nabla F(x)\right)^{1/2} \\
&= \|\nabla F(x)\|_x^* .
\end{aligned}$$

We have seen already that it is critical that $0 < \lambda(x_k) < 1, \ \forall\, k \geq 0$.

The function $\omega(t) = t - \ln(1+t)$ is a monotone increasing, therefore for a small $\beta > 0$ and $1 > \lambda(x) \geq \beta$, from (209) we obtain reduction of $F(x)$ by a constant $\omega(\beta)$ at each damped Newton step. Therefore, the number of damped Newton steps is bounded by

$$N \leq (\omega(\beta))^{-1}(F(x^0) - F(x^*)).$$

The bound (209), however, can be substantially improved for

$$x \in S(x^*, r) = \{x \in \operatorname{dom} F : F(x) - F(x^*) \leq r\}$$

and $0 < r < 1$.

Let us consider the lower bound

$$L(s) = f(0) + f'(0)s + ds - \ln(1 + ds) \leq f(s), \quad s \geq 0.$$

The function $L(s)$ is strictly convex on $s \geq 0$. If $0 < \lambda = -f'(0)d^{-1} < 1$, then

$$L'\left(\Delta(1-\lambda)^{-1}\right) = 0.$$

Therefore,

$$\bar{\bar{s}} = \Delta(1-\lambda)^{-1} = \arg\min\{L(s) \mid s \geq 0\}$$

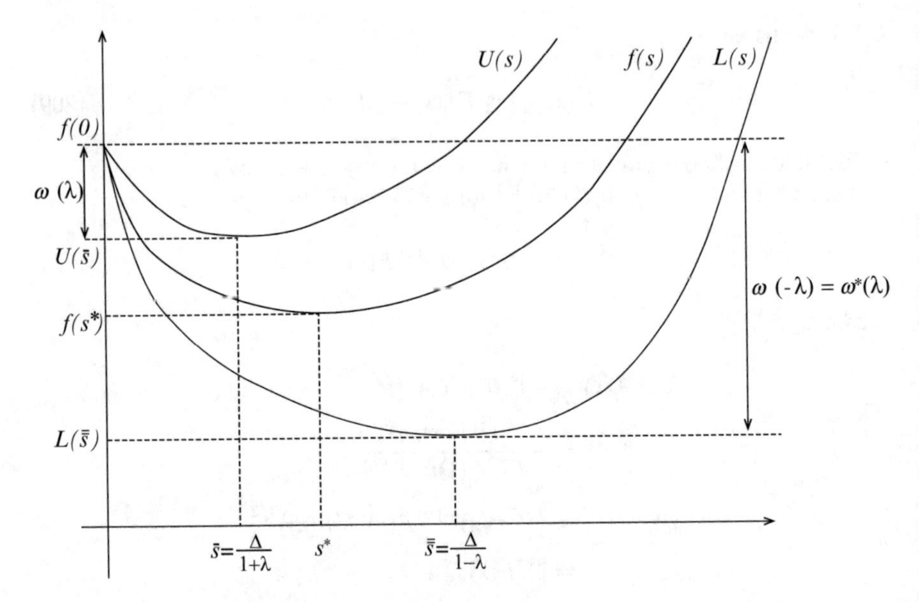

Fig. 2 Lower and upper bounds of a self-concordant function

and

$$L(\bar{\bar{s}}) = f(0) - \omega(-\lambda)\,.$$

Along with $\bar{s}$ and $\bar{\bar{s}}$ we consider (see Fig. 2)

$$s^* = \arg\min\{f(s) \mid s \geq 0\}\,.$$

For a small $0 < r < 1$ and $x \in S(x^*, r)$, we have $f(0) - f(s^*) < 1$, hence $f(0) - f(\bar{s}) < 1$. The relative progress per step is more convenient to measure on the logarithmic scale

$$\kappa = \frac{\ln(f(\bar{s}) - f(s^*))}{\ln(f(0) - f(s^*))}\,.$$

From $\omega(\lambda) < f(0) - f(s^*) < 1$ follows $-\ln\omega(\lambda) > -\ln(f(0) - f(s^*))$ or $\ln(f(0) - f(s^*)) > \ln\omega(\lambda)$. From $f(\bar{s}) \leq f(0) - \omega(\lambda)$ and $f(s^*) \geq f(0) - \omega(-\lambda)$ follows (see Fig 2)

$$f(\bar{s}) - f(s^*) \leq \omega(-\lambda) - \omega(\lambda)\,.$$

Hence,

$$\ln(f(\bar{s}) - f(s^*)) < \ln(\omega(-\lambda) - \omega(\lambda))$$

and

$$\kappa(\lambda) \leq \frac{\ln(\omega(-\lambda) - \omega(\lambda))}{\ln \omega(\lambda)}$$
$$= \frac{\ln\left(-2\lambda + \ln(1+\lambda)(1-\lambda)^{-1}\right)}{\ln(\lambda - \ln(1+\lambda))}.$$

For $0 < \lambda \leq 0.5$, we have

$$\kappa(\lambda) \leq \frac{\ln\left(\frac{2\lambda^3}{3} + \frac{2\lambda^5}{5}\right)}{\ln\left(\frac{\lambda^2}{2} - \frac{\lambda^3}{3} + \frac{\lambda^4}{4}\right)}.$$

In particular, $\kappa(0.40) \approx 1.09$. Thus, the sequence $\{x_k\}_{k=0}^{\infty}$ generated by the damped Newton method (208) with $\lambda(x_k) = 0.40$ converges in value at least with 1.09 Q-superlinear rate, that is for the error bound the $\Delta(x_k) = F(x_k) - F(x^*) < 1$, we have $\Delta(x_{k+1}) \leq (\Delta(x_k))^{1.09}$.

Due to $\lim_{k\to\infty} \lambda(x_k) = \lim_{k\to\infty} \|\nabla F(x_k)\|_x = 0$ from some point on, method (208) practically turns into the classical Newton method

$$x_{k+1} = x_k - \nabla^2 F(x_k)^{-1} \nabla F(x_k), \tag{210}$$

which converges with quadratic rate.

Instead of waiting for this to happen, there is a way of switching, at some point, from (208) to (210) and guarantee that from this point on, only Newton method (210) is used. Using such a strategy it is possible to achieve quadratic convergence earlier.

The following Theorem characterize the neighborhood at x^* when quadratic convergence accuracy.

Theorem 15. *Let* $x \in \operatorname{dom} F$ *and*

$$\lambda(x) = \left(\nabla^2 F(x)^{-1} \nabla F(x), \nabla F(x)\right)^{1/2} < 1,$$

then,

1. *the point*

$$\hat{x} = x - \nabla^2 F(x)^{-1} \nabla F(x) \tag{211}$$

   belongs to dom F*;*
2. *the following bound holds*

$$\lambda(\hat{x}) \leq \left(\frac{\lambda(x)}{1 - \lambda(x)}\right)^2. \tag{212}$$

Proof. 1. Let $p = \hat{x} - x = -\nabla^2 F(x)^{-1}\nabla F(x)$, $\lambda = \lambda(x)$, then

$$\|p\|_x = \left(\nabla^2 F(x)p, p\right)^{1/2} = \left(\nabla F(x),\ \nabla^2 F(x)^{-1}\nabla F(x)\right)^{1/2}$$
$$= \|\nabla F(x)\|_x^* = \lambda(x) = \lambda < 1\,;$$

therefore, $\hat{x} \in \operatorname{dom} F$.

2. First of all, note that if $A = A^T \succ 0$, $B = B^T \succ 0$ and $A \succcurlyeq B$, then

$$A^{-1} - B^{-1} = -A^{-1}(A - B)B^{-1} \preccurlyeq 0\,.$$

For $y = \hat{x}$ from the left inequality in (187), we obtain

$$\lambda(\hat{x}) = \|\nabla F(\hat{x})\|_{\hat{x}}^* \le (1 - \|p\|_x)^{-1}\left(\nabla^2 F(x)^{-1}\nabla F(\hat{x}),\ \nabla F(\hat{x})\right)^{1/2}$$
$$= (1 - \|p\|_x)^{-1}\,\|\nabla F(\hat{x})\|_x^*\,.$$

We can then rewrite (211) as follows

$$\nabla^2 F(x)\,(\hat{x} - x) + \nabla F(x) = 0\,.$$

Therefore,

$$\nabla F(\hat{x}) = \nabla F(\hat{x}) - \nabla F(x) - \nabla^2 F(x)(\hat{x} - x)\,.$$

Then, using (188) and formula (185) (see p. 6 [45]), we obtain

$$\nabla F(\hat{x}) - \nabla F(x) = \int_0^1 \left(\nabla^2 F(x + \tau(\hat{x} - x))\right)(\hat{x} - x)d\tau = G(\hat{x} - x)\,.$$

Hence,

$$\nabla F(\hat{x}) = \left(G - \nabla^2 F(x)\right)(\hat{x} - x) = \hat{G}(\hat{x} - x) = \hat{G}p$$

and $\hat{G}^T = \hat{G}$.

From CS inequality follows

$$\|\nabla F(\hat{x})\|_x^{*2} = \left(\nabla^2 F(x)^{-1}\hat{G}p, \hat{G}p\right) = \left(\hat{G}\nabla^2 F(x)^{-1}\hat{G}p, p\right)$$
$$\le \left\|\hat{G}\nabla^2 F(x)^{-1}\hat{G}p\right\|_x^* \|p\|_x\,. \tag{213}$$

Then

$$\left\|\hat{G}\nabla^2 F(x)^{-1}\hat{G}p\right\|_x^* = \left(\hat{G}\nabla^2 F(x)^{-1}\hat{G}p,\ \nabla^2 F(x)^{-1}\hat{G}\nabla^2 F(x)^{-1}\hat{G}p\right)^{1/2}$$
$$= \left(H(x)^2\nabla^2 F(x)^{-1/2}\hat{G}p,\ \nabla^2 F(x)^{-1/2}\hat{G}p\right)^{1/2}$$
$$\leq \|H(x)\| \left(\nabla^2 F(x)^{-1/2}\hat{G}p, \nabla^2 F(x)^{-1/2}\hat{G}p\right)^{1/2}$$
$$= \|H(x)\| \left(\nabla^2 F(x)^{-1}\hat{G}p, \hat{G}p\right)$$
$$= \|H(x)\| \left(\nabla^2 F(x)^{-1}\nabla F(\hat{x}), \nabla F(\hat{x})\right)^{1/2}$$
$$= \|H(x)\|\ \|\nabla F(\hat{x})\|_x^*\ ,$$

where $H(x) = \nabla^2 F(x)^{-1/2}\hat{G}\nabla^2 F(x)^{-1/2}$, therefore $\nabla^2 F(x)^{\frac{1}{2}} H(x) \nabla^2 F^{\frac{1}{2}}(x) = \hat{G}$.

From (213) and the last inequality we obtain

$$\|\nabla F(\hat{x})\|_x^* \leq \|H(x)\|\ \|p\|_x = \lambda\|H(x)\|\,.$$

It follows from (191)

$$\left(-\lambda + \frac{\lambda^2}{3}\right)\nabla^2 F(x) \preccurlyeq \hat{G} = G - \nabla^2 F(x) \preccurlyeq \frac{\lambda}{1-\lambda}\nabla^2 F(x)\,.$$

Then,

$$\|H(x)\| \leq \max\left\{\frac{\lambda}{1-\lambda}, -\lambda + \frac{\lambda^2}{3}\right\} = \frac{\lambda}{1-\lambda}\,.$$

Therefore,

$$\lambda^2(\hat{x}) \leq \frac{1}{(1-\lambda)^2}\ \|\nabla F(\hat{x})\|_x^{*2} \leq \frac{1}{(1-\lambda)^2}\lambda^2\|H(x)\|^2 \leq \frac{\lambda^4}{(1-\lambda)^4}$$

or

$$\lambda(\hat{x}) \leq \frac{\lambda^2}{(1-\lambda)^2}\,.$$

We saw already that $\lambda = \lambda(x) < 1$ is the main ingredient for the damped Newton method (208) to converge. To retain the same condition for $\lambda(\hat{x})$, it is sufficient to require $\lambda(\hat{x}) \leq \lambda \leq \lambda^2/(1-\lambda)^2$. The function $[\lambda/(1-\lambda)]^2$ is positive and monotone increasing on $(0, 1)$. Therefore, to find an upper bound for λ it is enough to solve the equation $\lambda/(1-\lambda)^2 = 1$. In other words, for any $\lambda = \lambda(x) < \bar{\lambda} = \frac{3-\sqrt{5}}{2}$, we have

$$\lambda(\hat{x}) \leq \left(\frac{\lambda}{1-\lambda}\right)^2.$$

Thus, the damped Newton method (208) follows three major stages in terms of the rate of convergence. First, it reduces the function value by a constant at each step. Then, it converges with superlinear rate and, finally, in the neighborhood of the solution it converges with quadratic rate.

The Newton area, where the Newton method converges with the quadratic rate is defined as follows:

$$N(x^*, \beta) = \left\{ x : \lambda(x) = \|\nabla F(x)\|_x^* \leq \beta < \bar{\lambda} = \frac{3-\sqrt{5}}{2} \right\}. \tag{214}$$

To speed up the damped Newton method (208) one can use the following switching strategy. For a given $0 < \beta < \bar{\lambda} = (3-\sqrt{5})/2$, one uses the damped Newton method (208) if $\lambda(x_k) > \beta$ and the "pure" Newton method (210) when $\lambda(x_k) \leq \beta$.

8 Concluding Remarks

The LEID is an universal instrument for establishing the duality results for SUMT, NR and LT methods. The duality result, in turn, are critical for both understanding the convergence mechanisms and the convergence analysis.

In particular, the update formula (107) and concavity of the dual function d leads to the following bound

$$d(\lambda_{s+1}) - d(\lambda_s) \geq (kL)^{-1}\|\lambda_{s+1} - \lambda_s\|^2,$$

which together with $d(\lambda_{s+1}) - d(\lambda_s) \to 0$ shows that the Lagrange multipliers do not change much from same point on. It means that if Newton method is used for primal minimization then, from some point on, usually after very few Lagrange multipliers update the approximation for the primal minimizer x_s is in the Newton area for the next minimizer x_{s+1}.

Therefore it takes few and, from some point on, only one Newton step to find the next primal approximation and update the Lagrange multipliers.

This phenomenon is called—the "hot" start (see [46]). The neighborhood of the solution where the "hot" start occurs has been characterized in [38] and observed in [5, 10, 25, 41].

It follows from Remark 5 that, under standard second order optimality condition, each Lagrange multipliers update shrinks the distance between the current and the optimal solution by a factor, which can be made as small as one wants by increasing $k > 0$.

In contrast to SUMT the NR methods requires much less computational effort per digit of accuracy at the end of the process then at the beginning.

Therefore NR methods is used when high accuracy needed (see, for example, [1]).

One of the most important features of NR methods is their numerical stability. It is due to the stability of the Newton's area, which does not shrink to a point in the final phase. Therefore one of the most reliable NLP solver PENNON is based on NR methods (see [32–34]).

The NR method with truncated MBF transformation has been widely used for both testing the NLP software and solving real life problems (see [1, 5, 10, 25, 32–34, 38, 41]). The numerical results obtained strongly support the theory, including the "hot" start phenomenon.

The NR as well as LT are primal exterior points methods. Their dual equivalence are interior points methods.

In particular, the LT with MBF transform $\psi(t) = \ln(t+1)$ leads to the interior prox with Bregman distance, which is based on the self-concordant MBF kernel $\varphi(s) = -\psi^*(s) = -\ln s + s - 1$. Application of this LT for LP calculations leads to Dikin's type interior point method for the dual LP. It establishes, eventually, the remarkable connection between exterior and interior point methods (see [37, 49]).

On the other hand, the LEINV is in the heart of the SC theory—one of the most beautiful chapters of the modern optimization.

Although the Legendre Transformation was introduced more than 200 years ago, we saw that LEID and LEINV are still critical in modern optimization both constrained and unconstrained.

References

1. Alber, M., Reemtsen, R.: Intensity modulated radiotherapy treatment planning by use of a barrier-penalty multiplier method. Optim. Methods Softw. **22**(3), 391–411 (2007)
2. Antipin, A.S.: Methods of Nonlinear Programming Based on the Direct and Dual Augmentation of the Lagrangian. VNIISI, Moscow (1979)
3. Auslender, R., Cominetti, R., Haddou, M.: Asymptotic analysis for penalty and barrier methods in convex and linear programming. Math. Oper. Res. **22**(1), 43–62 (1997)
4. Bauschke, H., Matouskova, E., Reich, S.: Projection and proximal point methods, convergence results and counterexamples. Nonlinear Anal. **56**(5), 715–738 (2004)
5. Ben-Tal, A., Nemirovski, A.: Optimal design of engineering structures. Optima **47**, 4–9 (1995)
6. Ben-Tal, A., Zibulevski, M.: Penalty-barrier methods for convex programming problems. SIAM J. Optim. **7**, 347–366 (1997)
7. Bertsekas, D.: Constrained Optimization and Lagrange Multiplier Methods. Academic Press, New York (1982)
8. Bregman, L.: The relaxation method for finding the common point of convex sets and its application to the solution of problems in convex programming. USSR Comput. Math. Math. Phys. **7**, 200–217 (1967)
9. Bregman, L., Censor, Y., Reich, S.: Dykstra algorithm as the nonlinear extension of Bregman's optimization method. J. Convex Anal. **6**(2), 319–333 (1999)

10. Breitfeld, M., Shanno, D.: Computational experience with modified log-barrier methods for nonlinear programming. Ann. Oper. Res. **62**, 439–464 (1996)
11. Byrne, C., Censor, Y.: Proximity function minimization using multiple Bregman projections with application to split feasibility and Kullback-Leibler distance minimization. Ann. Oper. Res. **105**, 77–98 (2001)
12. Carroll, C.: The created response surface technique for optimizing nonlinear-restrained systems. Oper. Res. **9**(2), 169–184 (1961)
13. Censor, Y., Zenios, S.: The proximal minimization algorithm with d-functions. J. Optim. Theory Appl. **73**, 451–464 (1992)
14. Chen, C., Mangasarian, O.L.: Smoothing methods for convex inequalities and linear complementarity problems. Math. Program. **71**, 51–69 (1995)
15. Chen, G., Teboulle, M.: Convergence analysis of a proximal-like minimization algorithm using Bregman functions. SIAM J. Optim. **3**(4), 538–543 (1993)
16. Courant, R.: Variational methods for the solution of problems of equilibrium and vibrations. Bull. Am. Math. Soc. **49**, 1–23 (1943)
17. Daube-Witherspoon, M., Muehllehner, G.: An iterative space reconstruction algorithm suitable for volume ECT. IEEE Trans. Med Imaging **5**, 61–66 (1986)
18. Dikin, I.: Iterative solutions of linear and quadratic programming problems. Sov. Math. Dokl. **8**, 674–675 (1967)
19. Eckstein, J.: Nonlinear proximal point algorithms using Bregman functions with applications to convex programming. Math. Oper. Res. **18**(1), 202–226 (1993)
20. Eggermont, P.: Multiplicative iterative algorithm for convex programming. Linear Algebra Appl. **130**, 25–32 (1990)
21. Ekeland, I.: Legendre duality in nonconvex optimization and calculus of variations. SIAM J. Control. Optim. **16**(6), 905–934 (1977)
22. Fiacco, A., Mc Cormick, G.: Nonlinear Programming, Sequential Unconstrained Minimization Techniques. SIAM, Philadelphia (1990)
23. Frisch, K.: The logarithmic potential method for convex programming. Memorandum of May 13 1955, University Institute of Economics, Oslo (1955)
24. Goldshtein, E., Tretiakov, N.: Modified Lagrangian Functions. Nauka, Moscow (1989)
25. Griva, I., Polyak, R.: Primal-dual nonlinear rescaling method with dynamic scaling parameter update. Math. Program. Ser. A **106**, 237–259 (2006)
26. Griva, I., Polyak, R.: Proximal point nonlinear rescaling method for convex optimization. Numer. Algebra Control Optim. **1**(3), 283–299 (2013)
27. Guler, O.: On the convergence of the proximal point algorithm for convex minimization. SIAM J. Control Optim. **29**, 403–419 (1991)
28. Hestenes, M.R.: Multipliers and gradient methods. J. Optim. Theory Appl. **4**, 303–320 (1969)
29. Hiriat-Urruty, J., Martinez-Legaz, J.: New formulas for the Legendre-Fenchel transform. J. Math. Anal. Appl. **288**, 544–555 (2003)
30. Ioffe, A., Tichomirov, V.: Duality of convex functions and extremum problems. Uspexi Mat. Nauk **23**(6)(144), 51–116 (1968)
31. Jensen, D., Polyak, R.: The convergence of a modify barrier method for convex programming. IBM J. Res. Dev. **38**(3), 307–321 (1999)
32. Kocvara, M., Stingl, M.: PENNON. A code for convex nonlinear and semidefinite programming. Optim. Methods Softw. **18**(3), 317–333 (2003)
33. Kocvara, M., Stingl, M.: Recent progress in the NLP-SDP code PENNON. In: Workshop Optimization and Applications, Oberwalfach (2005)
34. Kocvara, M., Stingl, M.: On the solution of large-scale SDP problems by the modified barrier method using iterative solver. Math. Program. Ser. B **109**, 413–444 (2007)
35. Martinet, B.: Regularization d'inequations variationelles par approximations successive. Rev. Fr. Inf. Res. Ofer **4**(R3), 154–159 (1970)
36. Martinet, B.: Determination approachee d'un point fixe d'une application pseudo-contractante. C.R. Acad. Sci. Paris **274**(2), 163–165 (1972)

37. Matioli, L., Gonzaga, C.: A new family of penalties for augmented Lagrangian methods. Numer. Linear Algebra Appl. **15**, 925–944 (2008)
38. Melman, A., Polyak, R.: The Newton modified barrier method for QP problems. Ann. Oper. Res. **62**, 465–519 (1996)
39. Moreau, J.: Proximite et dualite dans un espace Hilbertien. Bull. Soc. Math. France **93**, 273–299 (1965)
40. Motzkin, T.: New techniques for linear inequalities and optimization. In: project SCOOP, Symposium on Linear Inequalities and Programming, Planning Research Division, Director of Management Analysis Service, U.S. Air Force, Washington, D.C., no. 10, (1952)
41. Nash, S., Polyak, R., Sofer, A.: A numerical comparison of barrier and modified barrier method for large scale bound-constrained optimization. In: Hager, W., Hearn, D., Pardalos, P. (eds.) Large Scale Optimization, State of the Art, pp. 319–338. Kluwer Academic, Dordrecht (1994)
42. Nesterov, Yu., Nemirovsky, A.: Interior Point Polynomial Algorithms in Convex Programming. SIAM, Philadelphia (1994)
43. Nesterov, Yu.: Introductory Lectures on Convex Optimization. Kluwer Academic, Norwell, MA (2004)
44. Polyak, B.: Iterative methods using lagrange multipliers for solving extremal problems with constraints of the equation type. Comput. Math. Math. Phys. **10**(5), 42–52 (1970)
45. Polyak, B.: Introduction to Optimization. Optimization Software, New York (1987)
46. Polyak, R.: Modified barrier functions (theory and methods). Math Program. **54**, 177–222 (1992)
47. Polyak, R.: Log-sigmoid multipliers method in constrained optimization. Ann. Oper. Res. **101**, 427–460 (2001)
48. Polyak, R.: Nonlinear rescaling vs. smoothing technique in convex optimization. Math. Program. **92**, 197–235 (2002)
49. Polyak, R.: Lagrangian transformation and interior ellipsoid methods in convex optimization. J. Optim. Theory Appl. **163**(3), 966–992 (2015)
50. Polyak, R., Teboulle, M.: Nonlinear rescaling and proximal-like methods in convex optimization. Math. Program. **76**, 265–284 (1997)
51. Polyak, B., Tret'yakov, N.: The method of penalty estimates for conditional extremum problems. Comput. Math. Math. Phys. **13**(1), 42–58 (1973)
52. Powell, M.J.D.: A method for nonlinear constraints in minimization problems. In: Fletcher (ed.) Optimization, pp. 283–298. Academic Press, London (1969)
53. Powell, M.: Some convergence properties of the modified log barrier methods for linear programming. SIAM J. Optim. **50**(4), 695–739 (1995)
54. Ray, A., Majumder, S.: Derivation of some new distributions in statistical mechanics using maximum entropy approach. Yugosl. J. Oper. Res. **24**(1), 145–155 (2014)
55. Reich, S., Sabach, S.: Two strong convergence theorems for a proximal method in reflexive Banach spaces. Numer. Funct. Anal. Optim. **31**, 22–44 (2010)
56. Rockafellar, R.T.: A dual approach to solving nonlinear programming problems by unconstrained minimization. Math. Program. **5**, 354–373 (1973)
57. Rockafellar, R.T.: Augmented Lagrangians and applications of the proximal points algorithms in convex programming. Math. Oper. Res. **1**, 97–116 (1976)
58. Rockafellar, R.T.: Monotone operators and the proximal point algorithm. SIAM J. Control Optim. **14**, 877–898 (1976)
59. Teboulle, M.: Entropic proximal mappings with application to nonlinear programming. Math. Oper. Res. **17**, 670–690 (1992)
60. Tikhonov, A.N.: Solution of incorrectly formulated problems and the regularization method. Sov. Math. (Translated) **4**, 1035–1038 (1963)
61. Tseng, P., Bertsekas, D.: On the convergence of the exponential multipliers method for convex programming. Math. Program. **60**, 1–19 (1993)
62. Vardi, Y., Shepp, L., Kaufman, L.: A statistical model for positron emission tomography. J. Am. Stat. Assoc. **80**, 8–38 (1985)

Printed in the United States
By Bookmasters